Human Baroreflexes
in Health and Disease

Monographs of the Physiological Society

43. Dwain L. Eckberg and Peter Sleight, *Human Baroreflexes in Health Disease*, 1992

Human Baroreflexes in Health and Disease

Dwain L. Eckberg

Professor of Medicine and Physiology
Hunter Holmes McGuire Department of Veterans Affairs Medical Center
and Medical College of Virginia
Richmond, Virginia

and

Peter Sleight

Field Marshall Alexander Professor of Cardiovascular Medicine
Oxford University
Oxford

With a Foreword by
Björn Folkow

Professor of Physiology
University of Göteborg
Göteborg, Sweden

CLARENDON PRESS · OXFORD

1992

Oxford University Press, Walton Street, Oxford OX2 6DP

Oxford New York Toronto
Delhi Bombay Calcutta Madras Karachi
Petaling Jaya Singapore Hong Kong Tokyo
Nairobi Dar es Salaam Cape Town
Melbourne Auckland

and associated companies in
Berlin Ibadan

Oxford is a trade mark of Oxford University Press

Published in the United States
by Oxford University Press, New York

A catalogue record for this book is
available from the British Library

Library of Congress Cataloging in Publication Data

(Data available)

ISBN 0–19–857693–5

Set by H Charlesworth & Co Ltd, Huddersfield

Printed and bound in Great Britain by
Biddles Ltd,
Guildford & King's Lynn

D.L.E.

To my girls,
Rebecca,
Sarah, Marya, Leah,
and little Alexandra

P.S.

To my family, Gillian, Christopher, and James, who have all helped in this long endeavour

Foreword

by
BJÖRN FOLKOW
Professor of Physiology
Department of Physiology
University of Göteborg
Göteborg, Sweden

Virtually all human activities have their trends and fashions, and science is certainly no exception, where biomedical research provides abundant examples. The latest wave of enthusiasm — and, at that, a very justified one — emerges from the breakthroughs in molecular biology which have revolutionized genetics, microbiology, and cell biology. The present success in these areas has its risks, however, because it often seems to fascinate the scientific community to the extent that other important aspects of life sciences are disregarded. After all, above the cellular level, the infinitely more complex hierarchies of function and control preside which characterize higher organisms — with *homo sapiens* topping the list — and in many of these areas we still do not know more than 'the tip of an iceberg seen above the water's surface'.

This book, for the best of reasons, puts humans in the very centre, by dealing with one of our most important and complex control mechanisms for keeping cardiovascular function in proper balance during daily life. Moreover, the baroreflex mechanism is for humans of particular relevance: after some 5–10 million years in the erect position to free the forelimbs for more useful purposes, *homo sapiens* is still occasionally troubled by the fact that more than two-thirds of the blood volume and vascular circuits are thereby placed below the level of the heart. This is almost unique among mammals, and for hydrostatic reasons, it presents a serious challenge to our pressure-controlling devices, with the baroreflex in the very centre.

No doubt, we are in the end made up of cells, and thus ultimately depend on their subcellular and molecular functions. However, even an isolated blood vessel or heart is something far more than the mere sum of cells, as they form differentiated tissue complexes capable of remarkably sophisticated functions. Further, heart and vessels together form a still more complex organ system *in vivo*, governed by their own efficient autoregulatory systems. This forms a local level (I) of organ system control which, in a way, resembles community rule in human society.

This local control level is, however, steadily under the supervision of still higher control levels whereby the cardiovascular system is adjusted to the needs of other organ systems, and to those of the body as a whole. Thus, based on information from pressure-, volume-, and chemical-sensitive cardiovascular receptors, the bulbar level of the brainstem integrates pressure, resistance, and cardiac output and filling via efferent nervous and hormonal links. This second level (II) of control — the main topic of this book — allows for redistributions of flow, adjusts the system to imposed loads, and maintains perfusion pressures at fairly constant levels. In emergencies the system favours myocardial and cerebral over other vascular beds. In a way, level II functions like the governmental administrative supervision in human society, which also can in emergencies overrule local controls and impose other priorities.

The third main level (III) of cardiovascular control is organized at the limbic — hypothalamic levels of the brain. Besides getting all intrinsic information from levels I and II, level III also receives information from the individual's environment via telereceptors (vision, hearing, smell). If again compared with human society, level III resembles a nation's government and its foreign service: just as a nation, when reacting to international threats, can according to prearranged plans mobilize armed forces and adjust communications, distribution, and production to support them, the biological control level III can induce appropriate anticipatory adjustments to environmental stimuli by selecting the most fitting among a variety of genetically imprinted 'emotional' response patterns. Like the all-out mobilizations in human societies, these prearranged response patterns engage in parallel all three efferent links controlled by the brain, including (a) the somatomotor link, adjusting behaviour to the imposed situation; (b) the autonomic nervous link, adjusting not only the cardiovascular system, but also other organ systems to support (a); and (c) the hormonal link, adjusting metabolism, the fluid-electrolyte state, and nutritional depots, again to support (a).

Of these three main levels of control, which together determine cardiovascular performance in higher organisms, this book deals with level II, and particularly with how the all-important 'barostat' control functions in humans. However, as the three control levels continually communicate and influence each other, and in fact establish a greater 'functional unit', such interactions are appropriately outlined in the book with respect to their influence on, and interactions with the baroreflex.

Therefore, this book is also a splendid example of how important, and how exceedingly complex, the myriad servocontrols in higher organisms really are — being of an entirely different dimension from those controlling events inside cells. To draw a rough parallel: although a modern jumbo jet comprises manifold *per se* sophisticated components, it is nevertheless something much beyond the sum of such components. They remain a 'heap of junk' until assembled and trimmed to the perfect level of coordination that characterizes the flying wonder — particularly with a skilled pilot at the helm. If boards of directors in the airplane industry got so fascinated by new types of microcomponents and their production as to forget what flight engineers, experts on servosystems, aerodynamics, and test pilots mean for the industry, there would soon be no one around to produce an airplane — still less, persons to make it airborne. Now, biomedical science is responsible for a similar wide range — from molecules up to humans and their environment — and a balance between areas and levels of research must be maintained to avoid long-term shortcomings.

A final word about the authors. They have both devoted decades of their research to experimental studies in humans, with particular emphasis on the highly complex machinery that guards our blood pressure level — whatever happens concerning body position, increases of activity, accidental fluid losses, and the like. They are truly world authorities in this field and, thus, are ideally fitted to 'make the topic airborne'. So, readers, enjoy the flight!

Preface

We thank our academic institutions, the Hunter Holmes McGuire Department of Veterans Affairs Medical Center, the Medical College of Virginia, Virginia Commonwealth University, and Oxford University for allowing us time to pursue writing this monograph.

We thank granting agencies, whose generous support enabled us to do and report some of the research that we describe. These include the British Heart Foundation, the British Medical Research Council, the Litchfield Foundation, the National Aeronautics and Space Administration, the National Institutes of Health, and the Swedish Medical Research Council. We thank John P. Chalmers for providing a temporary academic home for us at Flinders University of South Australia in Adelaide.

This book was written with help from many people, whom we thank.

We thank Björn Folkow for writing the Foreword ('You asked for it and I suffered from it.')

We thank scientists who allowed us to include their unpublished data; these include Thomas J. Ebert, John Ernsting, Thomas R. Porter, Peter J. Schwartz, Urs Scherrer, William J. Welch, Mary A. Woo, and their coworkers. We also thank in our Figure captions other scientists and their publishers who allowed us to reproduce published data.

We thank our friends who expertly refereed chapters for us. These include Joachim O. Arndt, the late Chandler McC. Brooks, Vernon S. Bishop, James Conway, Mark A. Creager, Thomas J. Ebert, Kenneth A. Ellenbogen, David S. Goldstein, John M. Karemaker, John Ludbrook, Christopher A. Morley, Eric N. Prystowsky, David A. Ruggiero, J. Philip Saul, Peter J. Schwartz, K. Michael Spyer, Ilan J. van Lieshout, Menashe B. Waxman, Wouter Weiling, and Robert Zelis. David S. Goldstein deserves special thanks; he reviewed not one chapter, but two, and not once, but (due to the vagaries of the US Postal Service) twice.

We thank the reference workers in libraries located on several continents who helped us with our literature searches. Their libraries include the University of Virginia Medical Library, Charlottesville; Zentralbibliothek der Medizin, Cologne; Flinders University of South Australia Medical Library, Adelaide; The Bodleian and Radcliffe Science Libraries of Oxford University, Oxford; the College of Physicians Library, Philadelphia; Hunter Holmes McGuire Department of Veterans Affairs Medical Center Library and Tompkins-McCaw Library of the Medical College of Virginia, Richmond; and The National Library of Medicine, Bethesda. We also thank Frederick K. Lautzenheiser, Associate Archivist, The Cleveland Clinic

Foundation. Finally, we thank Gillian Sleight for literature searches and editorial help.

We thank the photographers who helped us, including Juan-Carlos Chazo, James A. Brownie, and Donald B. Sprague, the artists who drew some of the figures, Robert L. Shepherd, and Carolyn R. Bledsoe. We especially thank Larry A. Beightol, who painstakingly redrew many of the graphs we reproduced.

We thank Lloyd Wolfinbarger, Jr and the LifeNet Virginia Tissue Bank for providing a peroneal nerve specimen, and Nitya R. Ghatak for making the electron micrographs of it. We thank James J. Gibbons and Mark D. Snyder for photographs of their peroneal nerve dissections.

We thank Iona S. Bong, and the secretaries who have helped us, including Cheryl G. Belcher, Chandra D. Crawley, Tiffany A. Rudd, and E. Jean Ryland.

We thank the Chairman of the Editorial Board for Monographs of The Physiological Society, Allen G. Brown, and Society referees for their constructive suggestions regarding our book. Finally, we thank the staff of Oxford University Press, who guided us through this endeavour (but who at the beginning, did not warn us about what we were getting into).

Richmond/Oxford D.L.E.
October 1991 P.S.

Contents

Previous Volumes in this Series

Published by Edward Arnold

1. H. Barcroft and H.J.C. Swan, *Sympathetic Control of Human Blood Vessels*, 1953*
2. A.C. Burton and O.G. Edholm, *Man in a Cold Environment*, 1955*
3. G.W. Harris, *Neural Control of the Pituitary Gland*, 1955
4. A.H. James, *Physiology of Gastric Digestion*, 1957*
5. B. Delisle Burns, *The Mammalian Cerebral Cortex*, 1958*
6. G.S. Brindley, *Physiology of the Retina and Visual Pathway*, 1960 (2nd edition 1970)
7. D.A. McDonald, *Blood flow in arteries*, 1960*
8. A.S.V. Burgen and N.G. Emmelin, *Physiology of the Salivary Glands*, 1961
9. Audrey U. Smith, *Biological effects of freezing and supercooling*, 1961*
10. W.J. O'Connor, *Renal Function*, 1962*
11. R.A. Gregory, *Secretory Mechanisms of the Gastro-Intestinal Tract*, 1962*
12. C.A. Keele and Desiree Armstrong, *Substances producing Pain and Itch*, 1964*
13. R. Whittam, *Transport and Diffusion in Red Blood Cells*, 1964*
14. J. Grayson and D. Mendel, *Physiology of the Splanchnic Circulation*, 1965*
15. B.T. Donovan and J.J. van der Werff ten Bosch, *Physiology of Puberty*, 1965*
16. I. de Burgh Daly and Catherine Hebb, *Pulmonary and Bronchial Vascular Systems*, 1966*
17. I.C. Whitfield, *The Auditory Pathway*, 1967*
18. L.E. Mount, *The Climatic Physiology of the Pig*, 1968*
19. J.I. Hubbard, R. Llinàs, and D. Quastel, *Electrophysiological Analysis of Synaptic Transmission*, 1969*
20. S.E. Dicker, *Mechanisms of Urine Concentration and Dilution in Mammals*, 1970*
21. G. Kahlson and Elsa Rosengren, *Biogenesis and Physiology of Histamine*, 1971*
22. A.T. Cowie and J.S. Tindal, *The Physiology of Lactation*, 1971*
23. Peter B.C. Matthews, *Mammalian Muscle Receptors and their Central Actions*, 1972*
24. C.R. House, *Water Transport in Cells and Tissues*, 1974*
25. P.P. Newman, *Visceral Afferent Functions of the Nervous System*, 1974*
28. M.J. Purves, *The Physiology of the Cerebral Circulation*, 1972*

Published by Cambridge University Press

29. D. McK. Kerslake, *The Stress of Hot Environments*, 1972*
30. M.R. Bennett, *Autonomic Neuromuscular Transmission*, 1972*
31. A.G. Macdonald, *Physiological Aspects of Deep Sea Biology*, 1975*
32. M. Peaker and J.L. Linzell, *Salt Glands in Birds and Reptiles*, 1975*
33. J.A. Barrowman, *Physiology of the Gastro-intestinal Lymphatic System*, 1978*
35. J.T. Fitzsimmons, *The Physiology of Thirst and Sodium Appetite*, 1979
39. R.J. Linden and C.T. Kappagoda, *Atrial Receptors*, 1982
40. R. Elsner and B. Gooden, *Diving and Asphyxia*, 1983

Published by Academic Press

34. C.G. Phillips and R. Porter, *Corticospinal Neurones, Their Role in Movements*, 1977*
36. O.H. Petersen, *The Electrophysiology of Gland Cells*, 1980*
37. W.R. Keatinge and M. Clare Harman, *Local Mechanisms Controlling Blood Vessels*, 1980*
38. H. Hensel, *Thermoreception and Temperature Regulation*, 1981
41. R.C. Woledge, N. Curtin, and E. Homsher, *Energetic Aspects of Muscle Contraction*, 1984
42. J. R. Henderson, *"Mémoir sur le Pancreas", by Claude Bernard*, (Translation), 1985

Volumes marked with an asterisk are now out of print.

PART A

General Considerations

1

Introduction

Man is the measure of all things
Protagoras, as quoted by Plato
in *Theaetetus*

Nerve endings in the walls of the carotid sinuses and the aortic arch transduce arterial pressure changes and provide the central nervous system with a steady stream of encoded information. On the basis of this information, efferent autonomic neural activity is modulated finely, and the neurohumoral milieu of the heart and blood vessels is adjusted on a second-to-second basis. The arterial baroreflex may be the most important of the cardiovascular control mechanisms, because the baroreflex, above all other reflex mechanisms is the one whose speed is most adequate to respond rapidly to the abrupt transients of arterial pressure that occur in daily life.

Although the earliest research into baroreflex mechanisms was done with human subjects, this field of physiology is dominated by research conducted in animals. Responses from healthy animals may not accurately reflect responses from humans with cardiovascular diseases, in whom baroreflex autonomic regulatory mechanisms may be distorted almost beyond recognition. Nevertheless, the role of human research is ambiguous. In the not too distant past, a rhetorical question posed in the New Testament (John 1:46) might have been paraphrased as 'Can any good thing come out of human baroreflex research?' Alternatively, it might have been legitimate to paraphrase Samuel Johnson and compare efforts of human baroreflex research workers with those of dogs that walk on their hind legs: it is not done well; but one is surprised to find it done at all.

PROBLEMS WITH HUMAN BAROREFLEX RESEARCH

Difficulties encountered by scientists who study human baroreflex physiology and pathophysiology are daunting; indeed, it is likely that many studies have been conducted without a full appreciation of the complexities underlying human responses to the outwardly simple interventions that are used. First, many stimuli applied to human volunteers cannot be quantified adequately. For example, there is no standard quantification of the intensity of carotid massage. Second, stimuli applied to human subjects tend to be complex. For example, the simple act of standing alters central command to neuromuscular effectors, inputs from cardiopulmonary receptors, arterial baroreceptors, otolith and labyrinth receptors, and changes of respiration. Third, human volunteers are 'closed loop' preparations; that is, responses to reflex forcings applied to one receptor area alter sensory traffic from other areas, and such alterations of afferent input to the central nervous system modify the net responses that are being studied. Fourth, human research is usually 'black box' research. An input is changed, an output is measured, and inferences are drawn regarding how the change of input was translated into the change of output. Although this method is a time-honoured approach to the study of complex systems, it usually does not yield precise definition of mechanisms. Fifth, conclusions drawn from human studies rely importantly upon statistical inference. It usually is not possible to establish definitively cause-and-effect relations. Moreover, human research usually does not benefit from longitudinal follow-up of experimental subjects. For example, although baroreflex abnormalities have been documented in diverse patient groups, including those with hypertension, acute myocardial infarction, and diabetes mellitus, such data are scientifically incomplete, because they do not include data obtained *prior to*, as well as *following*, the onset of the clinical conditions.

ADVANTAGES OF HUMAN RESEARCH

Notwithstanding difficulties attending human studies, there are compelling reasons to conduct research with human subjects. First, the ultimate goal of scientists involved in *medical* research may be to understand humans themselves; it may not be sufficient to understand physiological phenomena in animals only and then extrapolate to humans. Human studies may be necessary, even if they merely confirm what is known from other research. Second, arguments abound regarding the applicability of animal models to human disease. Such arguments are moot when patients are studied. Third, in some respects, interpretation of results obtained from human subjects is easier than interpretation of results obtained from animals. For example, most human studies are conducted without use of general anaesthesia, muscle paralysis, artificial ventilation, and surgical dissection of tissues. Thus there is no need to assume that these particular experimental provocations did not distort the results obtained. Fourth, research committees sharply circumscribe the scope of interventions that can be used in humans. This oversight may have the unanticipated advantage that scientists studying human subjects must limit their inquiries to interventions that perturb physiology in minor ways. Thus, although human researchers may not be able to study the effects of toxic concentrations of drugs on their subjects, they are protected from the pitfall of assuming that responses to toxic drug doses have therapeutic relevance. Finally, human subjects have a remarkable advantage over most animals: they can *cooperate* with the experimenter. On a simple level, they can hold still when the investigator hurts them (as when a needle is inserted). On a more complex level, they can hold their breath on command, breathe on cue to any depth specified, perform complex mental arithmetic, or grip a dynamometer to produce whatever intensity of exertion is required.

THE HUMAN BAROREFLEX IN PERSPECTIVE

Fortunately, there is no compulsion to choose *either* human *or* animal studies. It is within this context that this book takes its place among earlier single- or few-authored monographs† devoted to baroreflex physiology (Hering 1927; Heymans 1929; Koch 1931; Heymans *et al.* 1933; Heymans and Neil 1958). This is the *first* volume dedicated to *human* baroreflexes. Therefore, by far the most quotations are from human research and the number of citations to important earlier work done in animals is necessarily limited. In some instances, however, data from humans are meagre. This is particularly true for central nervous system autonomic pathways and afferent baroreceptor activity. In such cases, data from animals are quoted more completely than otherwise, and attempts are made to relate them to such data as are available from humans. In some instances, we take licence from deficiencies in the human database; we extrapolate further than we should and assume that humans are not exempt from physiological laws laid down for other species.

We attempt to bridge the gap that exists between the scientific literature derived from animals and that derived from humans. Some studies quoted were published in clinical literature that may be unfamiliar to basic scientists, and some were published in basic science journals that may be unfamiliar to clinicians. We hope that this book pulls this literature together in a single source, and thereby catalogues much that is important to human baroreflex physiology.

Many experimental methods available for use in humans have been developed recently. Some are ingenious and yield results that earlier might have been thought impossible to obtain from human volunteers. Development of these new methods has increased the respectability and scientific credibility of human baroreflex research. We hope that our book will help to fix a place for human studies in the understanding of baroreflex physiology

† In addition, two multiauthored books were published earlier (Kezdi 1967*a*; Sleight 1980), and a third has just been published (Persson, P.B. and Kirchheim, H.R. (1991), *Baroreceptor reflexes, integrative functions and clinical aspects*, Springer-Verlag).

and pathophysiology. We intend it as a celebration of the advances that have been made. At the same time, we hope that it describes clearly the existing deficiencies in the understanding of baroreflex mechanisms, and points a way for future research in an exciting and important area of medical science.

2

History

The arterial baroreflex has captured the interest of scientists and naturalists since antiquity, and the historical development of understanding of baroreflex mechanisms yields a penetrating insight into the nature of scientific discovery. An important thread of continuity in this development is the deep and abiding interest of generations of scientists in the *human* response to carotid massage.

ORIGINS OF CAROTID MASSAGE

The Greek word καροτιδεσ (*karotides*) derives from the root verb καρο (*karo*), to plunge into deep, heavy sleep. The word was used by numerous Greek scientists and physicians, including Aristotle, Dioscordies, Erotianius, Clemens of Rome, Rufus of Ephesus, Plutarch, Aretaeus of Cappadocia, and Galen. Probably the clearest ancient exposition of the effects of carotid sinus stimulation was by Rufus of Ephesus, who practised medicine during the reign of Trajan (AD 53–117). This physician wrote in his treatise *Names of the Parts of the Human Body*, 'The ancients named the vessels traversing the neck, carotids, because when you press upon them [people] become sleepy and speechless' (Daremberg 1854). Thus the earliest, albeit unstructured, baroreflex research may have been conducted upon human subjects. Rufus's description of carotid massage was followed by this statement, which, from a modern perspective, strikes close to the physiological mechanism underlying responses to carotid massage: 'But in our age, it is found

found that this is an affection of the sensitive nerves which lie near to the arteries, and not to the arteries themselves . . .' (As recently as 1947, Schlager and Meier reported on the use of *bilateral* carotid sinus massage as a treatment for insomnia in Balinese islanders. Figure 2.1 shows such therapeutic application of carotid massage.)

The first modern description of carotid massage is found in the treatise *An Inquiry into the Symptoms and Causes of the Syncope Angiosa Commonly Called Angina Pectoris*, published by the English physician Caleb Hillier Parry in 1799. Neck massage was applied during one patient's paroxysm of angina with the result that '. . . his heart, in general, soon begins to beat more slowly; and at length, he feels threatenings of actual syncope'. Carotid massage does not figure prominently in this medical text, and there is no evidence that physicians who reported subsequently upon carotid massage were aware of the precedent established by Parry for this manoeuvre.

The technique of carotid massage was resurrected by Augustus Volney Waller (1862), who in his discussion of potential mechanisms erroneously excluded involvement of the carotid sinus in the bradycardic response, on the basis that cardiac slowing occurs when the carotid artery is only partially obliterated, and does not occur when blood flow in the carotid artery is interrupted completely.

In 1836, the English surgeon Sir Astley Cooper tied the carotid and

FIG. 2.1. Carotid massage as performed by Balinese islanders as therapy for insomnia. Reproduced with permission (Schlager and Meier 1947).

vertebral arteries of a dog and reported that '. . . its heart beat quickly and forcibly'. Cooper correctly ascribed these changes to diminished cerebral perfusion; however, based upon this precedent, subsequent authors erroneously ascribed the tachycardia which occurs with bilateral carotid artery occlusion (but without concomitant ligation of the *vertebral* arteries) to reduced cerebral blood flow.

Although the first devices for measuring and recording arterial pressure were not developed until the mid-nineteenth century, by Carl Ludwig (1847) and Vierordt (1854), François Magendie (1854) inferred from other data that bilateral carotid artery occlusion leads to hypertension. Thus these very early studies established the relationship between carotid hypotension, and tachycardia and systemic hypertension.

ELECTRICAL VAGUS NERVE STIMULATION

As early as 1669, an English surgeon, Richard Lower, who was a pupil of William Harvey, determined that vagal ligature speeds the heart rate. (In this research, he followed up on the observation of another of his mentors, Thomas Willis (1664), that vagal section in the dog is followed by loss of consciousness, and 'great trembling' of the heart.) In 1817, Philip observed that electrical stimuli applied to structures within the central nervous system provoke acceleration or slowing of the heart rate. Although Philip did not discover nervous pathways from the brain to the heart, he correctly concluded that the brain exerts control over cardiac function.

Electric currents were not applied to autonomic nerves until after 1833, when Faraday (1839) invented his electromagnetic induction device. In 1837, Volkmann (1853) showed that electric vagal stimuli slow the heart; however, he dismissed this finding because at that time nerves were regarded as exclusively stimulatory. (Indeed, Whytt (1775) applied mechanical vagus stimuli in animals whose hearts had stopped, in an effort to stimulate resumption of cardiac activity.) In 1845, the brothers Eduard Fredrich and Ernst Heinrich Weber, and in 1846, Julius L. Budge described experiments in which they applied shocks to the medulla or vagal roots (the Webers) or to the vagus nerves (Budge) and provoked bradycardia. The Webers enjoyed a slight priority in the discovery, but also introduced the revolutionary concept that nerves could be *inhibitory* as well as excitatory. Walter H. Gaskell subsequently (1883) showed that vagus nerve stimulation also reduces the strength of cardiac contractions. These studies identified the vagus nerves as efferent inhibitory pathways in the baroreceptor – cardiac reflex arc.

In 1854, Vierordt invented the first device for registering human arterial pressure. Marey's modification of this equipment (shown in Fig. 2.2) and its use in experimental animals led to his seminal observation (Marey 1859)

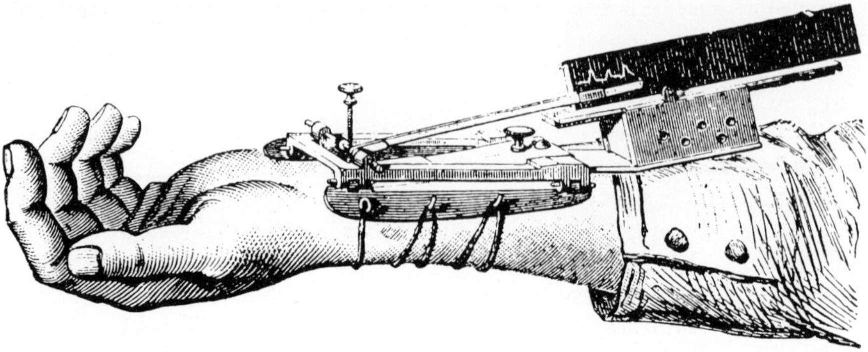

FIG. 2.2. Marey's modification of Vierordt's sphygmograph (Vierordt 1854).

that heart rate slows when blood pressure rises. In 1870 (as reported in 1874), Nawrocki confirmed and extended this observation by showing that heart rate increases when blood pressure falls (Trzebski 1987). This was also confirmed by one of Marey's assistants, François-Franck (1876), who also showed that since both heart rate speeding and slowing could be prevented by atropine, they were mediated by changes of vagal traffic.

CONTROVERSY REGARDING THE MECHANISM OF CAROTID MASSAGE

Czermak in 1866 used Marey's device to record his own arterial pressure during massage of a small round pulsating swelling on the right side of his neck. Czermak's remarkable record, reproduced in Fig. 2.3, which may be the first published recording (albeit indirect) of human arterial pressure and baroreflex responses, shows clearly that in humans, carotid massage provokes hypotension as well as bradycardia. Czermak's article, entitled 'Ueber mechanische Vagus-Reizung beim Menschen' (On the mechanism of the vagus test in man), firmly established carotid massage in the armamentarium of practising physiologists. (He also described symptoms suggestive of transient ischaemic attacks, including flashes of light, followed by a loss of visual field in his right eye and a feeling of faintness, which he attributed to cerebral ischaemia.) Czermak was an eminent and highly respected scientist, and his opinions did not die easily. His emphatically stated view, that bradycardia during neck massage was caused by direct stimulation of vagus nerve fibres, was probably shaped by familiarity with the earlier electrical vagus nerve stimulation studies of the Weber brothers and Budge.

However, experimental results which conflicted with this view began to appear immediately. In the same year that Czermak's study was published, the Russian physiologist Elie Cyon and his mentor Carl Ludwig (1866)

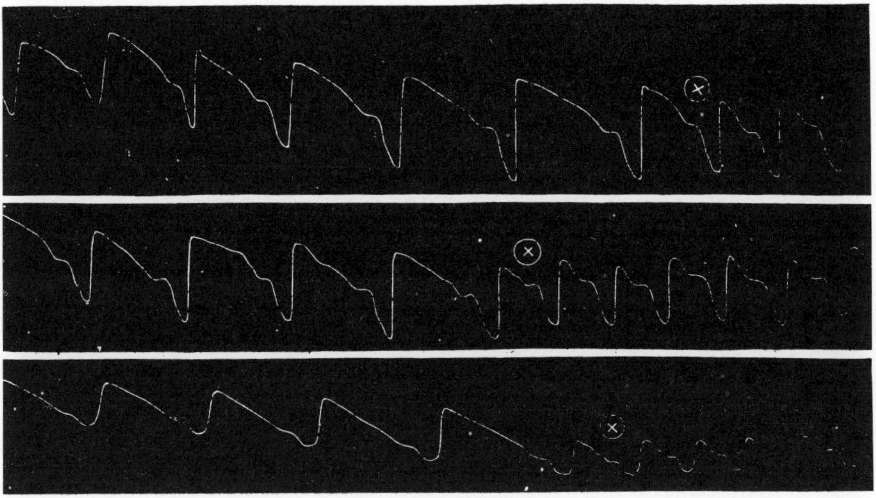

FIG. 2.3. Czermak's (1866) indirect recording of his own blood pressure, measured during carotid massage (onset denoted by x).

reported that electrical stimulation of the central end of the transected aortic depressor nerve provokes bradycardia and hypotension. Bradycardia did not occur after atropine pretreatment, and therefore was assumed to be vagal. Hypotension persisted after atropine, and therefore was assumed to be mediated by reduction of trophic vasoconstrictor activity, whose existence had been inferred earlier by Pourfois du Petit (1727), Bernard (1851), and Schiff (1853). Cyon and Ludwig (1866) erroneously believed that afferent fibres in the depressor nerve originated in the heart (rather than in the aortic wall). Nevertheless, their study pointed clearly to the existence of a reflex whose activation produced haemodynamic changes similar to those produced by carotid massage, and whose afferent and efferent limbs were in part non-vagal.

In 1870, Concato reported that bradycardia could be evoked more easily by pressure applied to the carotid artery than to the vagus nerve. (He also demonstrated in humans the reflex dilatation of facial vessels during neck massage.) Nevertheless, Quincke five years later (1875) reaffirmed, with reservations, Czermak's view that neck massage directly stimulates vagal cardiac fibres. Quincke's reservations arose from his concern that pressure applied to the carotid artery should not directly involve the vagus nerves.

However, further support for the *reflex* nature of haemodynamic responses to neck massage came from studies that focused on central nervous system components of the reflex arc. In a remarkable experiment, Claude Bernard (1858) showed that blood pressure falls when the cervical spinal cord is cut. He inferred that blood pressure is supported by background discharges of sympathetic nerves which result from activity generated in the central nervous system above the level of transection.

Goltz in 1864 confirmed that reflex arterial pressure adjustments require the integrity of the spinal cord, and Owsjannikow (1871) and Dittmar (1873) reported that neural structures involved with cardiovascular control are located in sharply circumscribed areas of the brainstem. Cyon and Ludwig (1866) transected successive levels of the brain and concluded that reflex cardiovascular control centres are located in the medulla. Bayliss (1893) cast doubt on the notion that responses to neck massage are mediated by cerebral ischaemia with his finding that bilateral carotid occlusion does not reduce medullary blood flow importantly. The idea that changes of cerebral blood flow contribute to baroreflex responses was finally laid to rest in 1924 by C. Heymans and A. Ladon, who showed that reflex responses triggered by changes of carotid sinus pressure may occur in a dog whose cerebral perfusion is maintained by a donor dog (Heymans and Ladon 1925).

Moreover, Pagano and Siciliano found, in independent studies published in 1900 in the same issue of the same journal (*Archives Italiennes de Biologie*), that ligation of all carotid branches which supply the medulla leads to neither tachycardia nor bradycardia. In addition, both authors drew attention to the role of the carotid sinus in cardiovascular responses. Pagano found that reflex bradycardia could be produced by topical application of irritant chemicals (including silver nitrate, nicotine, and sodium carbonate) to the carotid sinus. Siciliano concluded from his studies that pressure upon the common carotid arteries provokes tonic activity in cardiac inhibitory nerves.

In 1879, Pavlov (who also had trained with Carl Ludwig) published further evidence for the existence of a pressor reflex. He found that if both vagus nerves are cut during hypotension caused by haemorrhage, a further reduction of blood pressure occurs (McDowall 1924).

TWENTIETH CENTURY RESEARCH

Considerable progress was made in defining the nature of the baroreceptors in the early twentieth century. Koster and Tschermak (1902) used nerve degeneration techniques to show that afferent aortic depressor nerve fibres originate mainly in the wall of the aorta, rather than in the heart, as Cyon and Ludwig had suggested (1866). This view was also supported by Tello (1924). Koster and Tschermak (1902) also recorded action potentials from the depressor nerve during experimental distension of the isolated aortic arch. Einthoven *et al.* (1908) subsequently recorded depressor nerve activity in animals with intact aortic arches and demonstrated that this activity occurs in bursts which have cardiac periodicity.

In a remarkable study published in 1912, Sollman and Brown made a large contribution toward the understanding of carotid baroreflex mechanisms. They followed up on a chance observation that traction on a ligated carotid artery produces hypotension and bradycardia. They applied stretch to different areas of the carotid arterial tree and localized responses to the earliest portion of the internal carotid artery. They then showed that responses occur during stretch of a segment of the internal carotid artery which is isolated from all other parts of the carotid system, but which retains its connections to the carotid 'nerve plexus'.

Sollman and Brown did not name the carotid sinus nerve; this nerve had been identified as a depressor branch of the glossopharyngeal nerve earlier by Knoll (1885) and was named later the 'carotid sinus nerve' (*Karotissinus-nerven*) by Knoll's pupil, Heinrich Ewald Hering (1924). However, Sollman and Brown (1912) described the plexus of nerves in the region of the carotid sinus and showed that ligation of these nerves abolishes reflex responses provoked by manipulation of the carotid sinus.

HEINRICH HERING AND KOCH

The important experiments of Heinrich Hering and of his assistant, E. Koch, confirmed studies of earlier workers, led to a synthesis of extant knowledge, and thereby inaugurated the modern era of baroreflex research. In 1905, Hering became interested in Czermak's vagus pressure test (which in his earliest article he renamed *der Karotisdruckversuch* — the carotid pressure test) when he provoked profound bradycardia by massaging the neck of an elderly woman patient (Hering did not publish this work until 1923). In 1923, Hering followed up on this clinical observation and showed that in anaesthetized dogs and rabbits, vigorous manipulation of the exposed vagus nerve does *not* lead to cardiac slowing, but that very gentle massage of the carotid sinus does. (A similar conclusion was drawn much later by Carlsten *et al.* (1957), who reported that 'fairly rough handling' of the cervical vagus nerve in patients undergoing neck surgery does not alter heart rate.) Hering localized the origin of reflex responses to nerve endings at the carotid bifurcation and showed that transection of the carotid sinus nerve abolishes these reflex responses.

In 1923, Koch demonstrated in a simple but elegant experiment conducted with 250 (*sic*) human volunteers, that reflex responses can be evoked only when the neck is massaged directly over the area of the bifurcation of the carotid sinus (Fig. 2.4). Hering and Koch recognized the functional relationship between carotid sinus and aortic arch baroreceptors and repeated experiments conducted earlier by others on aortic baroreflex mechanisms. They stressed particularly that blood pressure and heart rate changes are integral components of the response to baroreceptor stimulation. They suggested that aortic and carotid baroreceptors work in concert, and that,

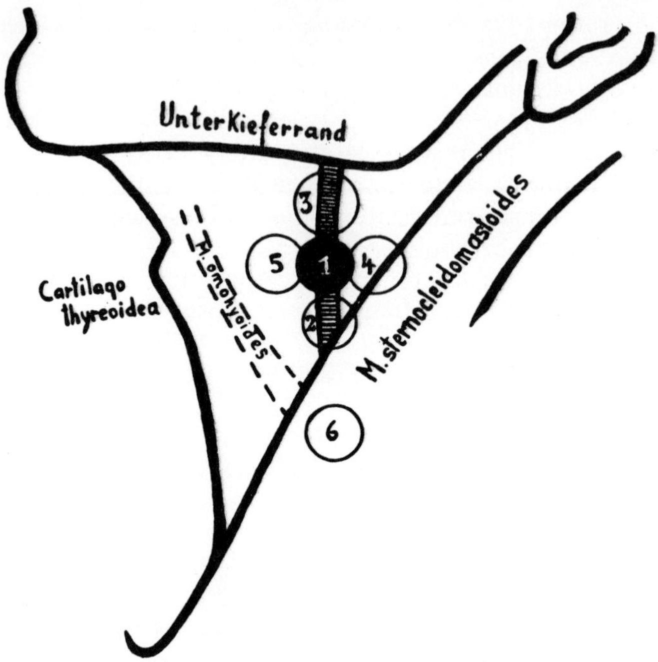

FIG. 2.4. Drawing from Koch (1923) showing that responses are maximal when neck massage is applied in the region of the carotid bifurcation.

of the two expressions of the reflex, blood pressure regulation is more important than heart rate regulation.

Koch subsequently (1931) used the isolated blind carotid sinus preparation developed by Moissejeff (1926) to open the carotid sinus baroreflex loop (with bilateral vagotomy and section of the contralateral carotid sinus nerve) to define the reflex consequences of variations in carotid sinus pressure. He showed that as carotid pressure is raised in steps at 20 s intervals from low to high levels, R – R intervals inscribe a sigmoid function (and arterial pressures inscribe a reverse sigmoid function). This *Blutdruckcharacteristik* (blood pressure characteristic), shown in Fig. 2.5, was to become the standard approach for quantification of baroreflex responses. Although their contribution is rarely recognized, Lim and Hsu also published reverse sigmoid carotid sinus – arterial pressure relations in 1931.

The ability to make recordings of nerve traffic was tied closely to the ability to amplify electronic signals and register them. The earliest recordings of baroreceptor nerve traffic were made by Lewandowsky (1898), Koster and Tschermak (1902), Einthoven *et al.* 1908; and Einthoven (1911). In general, these authors documented the occurrence of pulse synchronous changes of aortic nerve voltage. Although Einthoven (1911) devised an instrument, a string galvanometer suspended in a vacuum, which could

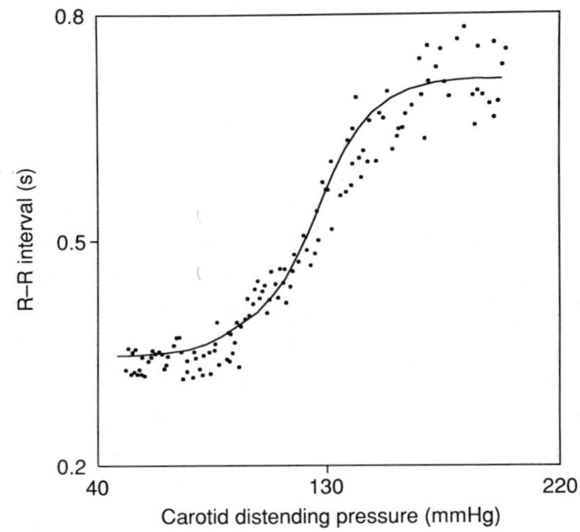

FIG. 2.5. Koch's *Blutdruckcharacteristik* (1931).

follow high frequency events, high fidelity measurements of nerve traffic did not become available routinely until Adrian (1926) developed his capillary electrometer and amplifier. Although Adrian used his equipment to document an increase of aortic nerve activity after an injection of adrenaline, he did not have equipment to enable him to record electronically the arterial pressure stimuli that increased baroreceptor nerve activity.

CHEMICAL MEDIATORS

Although the studies of the Weber brothers and of Budge showed as early as 1845 that vagus nerves may be involved in baroreflex responses (E. Weber and Weber 1845; Budge 1846), and Bezold and Hirt (1867) showed that faradic stimulation of the thoracic spinal cord led to cardioacceleration and hypertension, delineation of chemical mediators of neural responses did not occur until much later. Elliott (1905) proposed that adrenaline, a principal component in extracts of the suprarenal medulla, may be the active substance released from sympathetic nerve endings. (It was not until 1945 that Gernandt and colleagues proved that arterial pressure elevations reduce levels of efferent sympathetic nervous traffic.) Later, Cannon and Rosenblueth (1933) and von Euler and Liljestrand (1943) showed that the sympathetic neural transmitter is noradrenaline (or 'sympathin') rather than adrenaline. Dale (1914) first suggested that acetylcholine may be a neurotransmitter and Loewi (1921) proved that this transmitter (or *vagusstoff*) specifically mediates vagal cardiac inhibition.

HUMAN BARORECEPTOR STIMULATION

As discussed above, *human* baroreflex research figured prominently in development of the earliest understanding of baroreflex mechanisms. However, by the middle of the twentieth century, methods available for study of baroreflex mechanisms in humans lacked the precision and reproducibility of methods available for animal research. Two developments changed this. First, in 1957, two flight lieutenants in the British Royal Air Force, J. Ernsting and D. J. Parry, developed a neck chamber to counteract neck distension during the positive pressure breathing used to treat pilots (Ernsting and Parry 1957; Ernsting 1966). Pressure and suction delivered to this chamber could be quantitated with great accuracy with an electronic pressure gauge, and R – R interval and arterial pressure responses could also be measured. The original neck chamber, worn by the late D.J. Parry, is shown in Fig. 2.6.

Second, in 1969, Smyth *et al.* showed that when arterial pressure is raised with bolus injections of a pressor drug, the R – R interval prolongation that occurs can be related, with high correlation coefficients, to preceding systolic pressures with linear regression analysis. An original record from Smyth's article is depicted in Fig. 2.7. The neck chamber and bolus pressor injection techniques, and variations upon them (see Chapter 5), for the first time

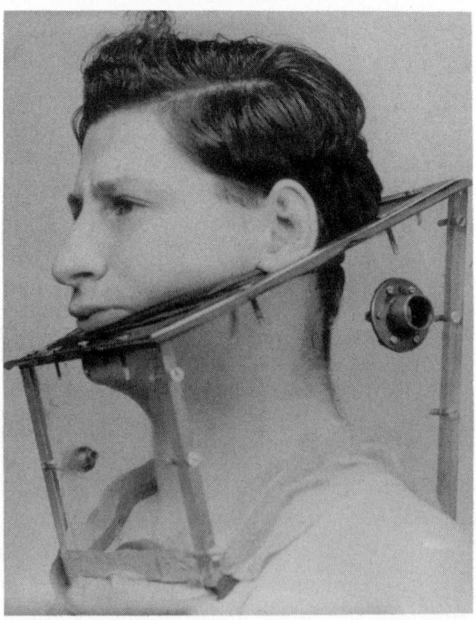

FIG. 2.6. The original neck chamber, developed by Ernsting and Parry (photograph kindly provided by J. Ernsting).

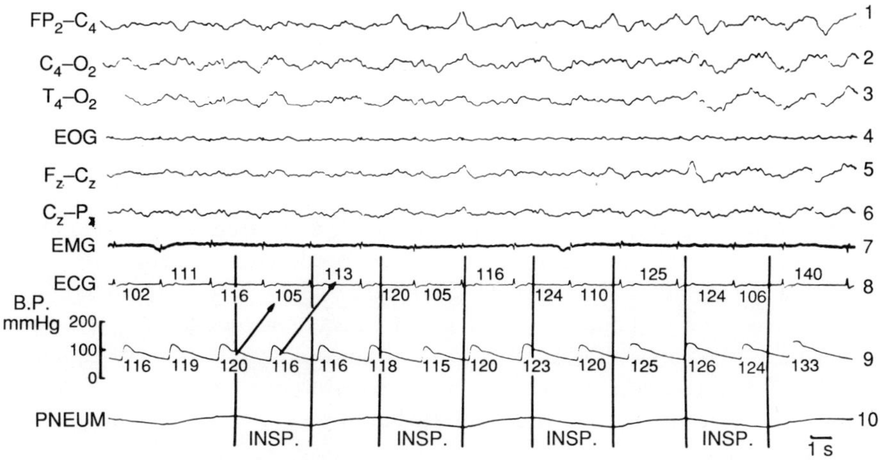

FIG. 2.7. Original record depicting reflex responses to a bolus phenylephrine injection. The electroencephalogram was recorded to identify stages of sleep. Reproduced with permission (Smyth *et al.* 1969).

permitted researchers to perturb human baroreflexes in precise and measurable ways, measure responses accurately, and publish scientifically credible data.

SUMMARY

The intellectual history of the baroreflex provides a fascinating insight into the nature of scientific progress. Much of the early interest in baroreflex physiology arose from curiosity about *human* responses to carotid massage, transmitted from teacher to pupil. Thus in 1861 Waller revived the technique of neck massage, which had lain dormant since Parry described it in 1799. His pupil E. Hering probably passed curiosity about this response to his assistant, J.N. Czermak, who in turn interested his assistant H.I. Quincke in the response. H.E. Hering was taught by his father and Quincke, and he trained his assistant E. Koch.

Advances in understanding of baroreflex mechanisms were tied closely to the development of research tools, including a device for delivery of an electromagnetic induction current (Faraday 1839), a kymograph for recording physiological data (C. Ludwig 1847), a blood pressure sensor (Vierordt 1854; Marey 1859), and a means to record electrical activity in nerve fibres (Einthoven *et al.* 1908). The availability of these research tools determined in part the types of experiments that could be undertaken.

The path towards understanding of baroreflex mechanisms did not proceed relentlessly upward. The name Czermak gave to neck massage, the 'vagus pressure test' (or *Vagusdruckversuch*) implied its mechanism. The work of the Webers and Budge with electrical vagus nerve stimulation (which entered into the mainstream of baroreflex thinking without having been quoted by either Waller or Czermak) made the causative association between neck massage and vagus stimulation eminently reasonable. This association was enunciated with high authority (by Czermak (1866)), and early contradictory studies, which showed that pressure on the carotid sinus is more likely to cause cardiac slowing than pressure on the vagus nerve (Concato 1870; Quincke 1875) were forgotten or ignored, and subsequently were repeated. Eventually, scientists (H.E. Hering and E. Koch) appeared who with incisiveness and lucidity synthesized extant knowledge and derived a cohesive, unified appreciation of baroreflex physiology. Hering subsequently (1924*a,b*) injected new heterodoxy (that baroreceptor malfunction *causes* human hypertension) and thus initiated a new cycle of experiment and debate.

3

Baroreflex anatomy

I will praise thee; for I am fearfully and wonderfully made: marvellous are thy works; and that my soul knoweth right well.

Psalm 139:14 (Authorized Version)

Antonio Scarpa's magnificent 1794 anatomical drawing showing arterial baroreceptor areas is reproduced in Fig. 3.1.

BARORECEPTOR ANATOMY

During embryonic development, aortic and carotid baroreceptor areas arise from the third and fourth branchial arches (Koch 1931). Neural structures presumed to be baroreceptors are dense in the arch of the aorta between the brachiocephalic artery and the ligamentum arteriosum (Roever 1869; Tello 1924), the carotid sinus (De Castro 1926; Sunder-Plassmann 1930; Koch 1931), and other areas, including the common carotid, subclavian, and innominate arteries (Kahn 1929; Nonidez 1937). Although these latter arteries have innervation, the density of innervation is not as great as that of the carotid sinus area (Shin *et al.* 1987), and almost all baroreflex control is thought to be exerted by receptors in the carotid sinus and aortic receptor areas (Heymans *et al.* 1944). The anatomy of arterial baroreceptors has been reviewed in many articles (e.g. De Castro 1940; Ábrahám 1953; S.M. Hilton 1963; Rees 1967*b*; B. Winter 1972) and books (de Boissezon 1942; W.E. Adams 1958; Grigor'eva 1962; Ábrahám 1969).

Macroscopic anatomy

The familiar fusiform dilatation of the carotid sinus at the origin of the internal carotid artery was recognized during the first and middle thirds of

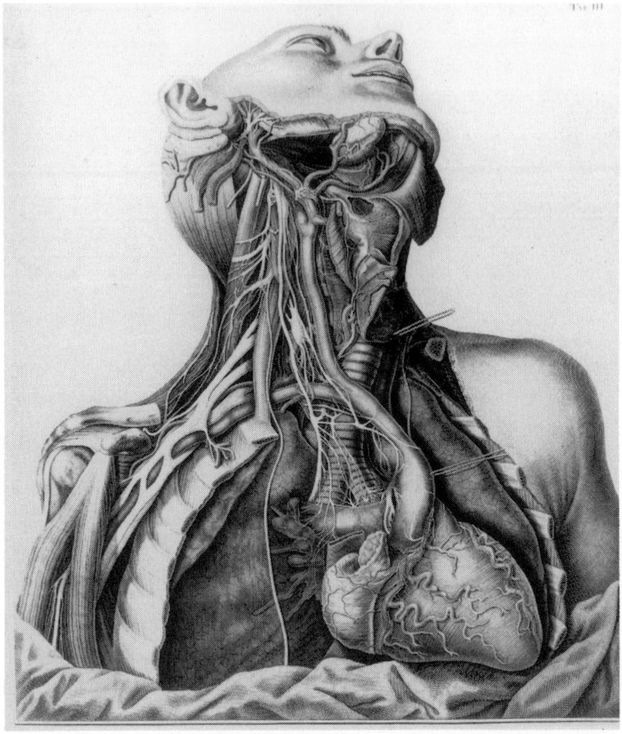

FIG. 3.1. Anatomy of arterial baroreceptor areas published by Antonio Scarpa in 1794.

the nineteenth century (Burns 1811; Luschka 1862). In 1866, Patrick Manson called attention to a 'Peculiar affection of the internal carotid artery in connexion with disease of the brain', and in 1876, Meyer published his paper, 'Ueber aneurysmatische Veränderungen der Carotis interna Geisteskranker' (On aneurysmic changes of the internal carotid artery of the mentally ill). Meyer, but not Manson (or Luschka), regarded the swelling of the internal carotid artery as an affliction related to mental disease. However, one year after Meyer's report, Schäfer (1877) dispelled this notion permanently when he showed that dilatation of the carotid sinus is also present in healthy people.

Meyer, whose observations (1876) were published fully 65 years after Burns's monograph (1811), is generally credited with discovery of the carotid sinus. His contribution may be that he recognized, perhaps for the first time, that internal carotid thinning and dilatation are more than simply physiological changes that occur after arterial branches. Presciently, he wrote, '. . . eine Art manometrischen Maßstabs für die während des Lebens erlittenen Circulationsstörungen gelten' ((the aneurysmic internal carotid artery changes) can be regarded as a kind of manometric measure for the (intracranial) disturbances of circulation suffered through a lifetime). W.E. Adams's book, *The comparative morphology of the carotid body and carotid sinus* (1958), includes a meticulously crafted review of the early history of the carotid sinus.

Binswanger (1879) carried out careful anatomical examinations of carotid sinuses in 200 patients. He found that the site of carotid dilatation is variable in the extent to which it is confined to the internal carotid artery (this was found in about 40 per cent of the cases). Figure 3.2 shows several variations of human carotid sinus anatomy published by Boyd (1937).

Both Binswanger (1879) and later De Castro (1928; 1940) compared the thinning and the increase in fibro-elastic fibres of the muscular layer with those of the subjacent common carotid artery. A subsequent biochemical study by Fischer and Llaurado (1966) and an electron microscope study by Shin *et al.* (1987) showed that the carotid sinus has the highest connective tissue content of any segment of the arterial tree. These studies suggest that the carotid sinus is a transitional vessel whose mix of components lies between those of muscular and elastic arteries.

Although Burns (1811) clearly documented the predilection of the human carotid sinus to develop *atherosclerosis*, virtually all studies of carotid thinning and attendant augmented compliance do not factor in this confounding variable. (Indeed, with one exception that we know of (Angell-James 1974), the important influence of carotid atherosclerosis on baroreceptor transduction has not been studied in any species.) The studies of Heath *et al.* (1973) and Winson *et al.* (1974) show that although in humans the carotid sinus wall thickness is less than that of the common carotid artery, total wall thickness may be augmented in a major way when the presence of athero-

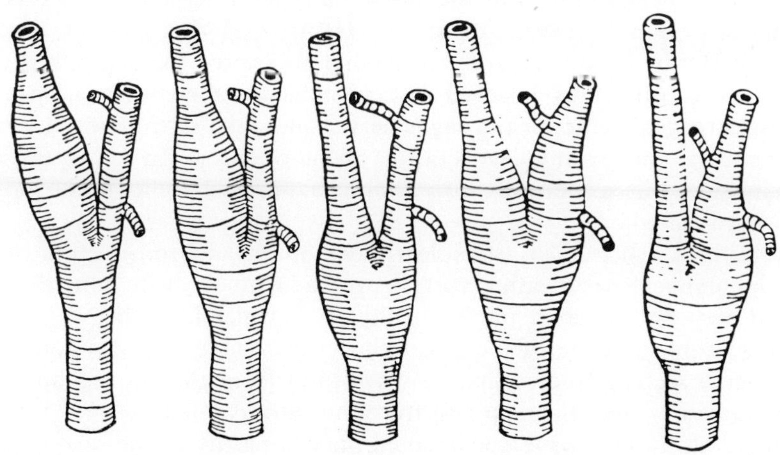

FIG. 3.2. Variations of human carotid sinus anatomy published by Boyd in 1937. Reproduced with permission.

sclerotic disease is included in the measurement. Although neural structures may be preserved in atherosclerotic arteries (Sunder-Plassmann 1930), atherosclerotic thickening would be expected to reduce distensibility (to a level *less* (not more) than that of the common carotid artery) and thereby impair baroreceptor transduction in the carotid sinus. This notion was suggested first by H.E. Hering (1927). Muratori (1967) found that in all except the youngest subjects, the prevalence of atherosclerotic change markedly altered these relations and rendered the sinus less distensible than the nearby common carotid artery (Winson *et al.* 1974). Thus it is not clear that the supranormal carotid compliance found in healthy animals is present also in adult humans, particularly older ones whose compliance may be reduced severely by atherosclerosis.

Baroreceptor nerves

The aortic depressor nerve was described before the carotid sinus nerve. Although discovery of the depressor nerve is usually attributed to Cyon and Ludwig (1866), priority probably belongs to Theile (1852), whose findings were cited by Cyon and Ludwig. Nevertheless, the importance of the seminal article by Cyon and Ludwig in the development of understanding of baroreflex physiology cannot be overestimated. Although they thought that depressor nerve afferent fibres arose from the heart rather than from the aorta as was shown subsequently (Tello 1924), they anticipated the true function of the depressor nerve, and called it a 'cardiac safety valve' that responded to and buffered rises in pressure by reflexly slowing the heart rate and dilating the blood vessels.

Before the studies of Heinrich E. Hering (1923; 1924*a,b*), there had been little interest in the branch of the glossopharyngeal nerve described as the *Sinusnerv* by Knoll (1885), a pupil of Heinrich Hering's father, Ewald. Heinrich Hering, who is regarded as having discovered the baroreflex function of the carotid sinus, believed that carotid sinus afferent fibres leave the glossopharyngeal nerve for the vagus nerve, since high section of the former nerve causes scarcely any degeneration of the receptors. In retrospect, it is probable that this section was *rostral to* the axons' cell bodies in the glossopharyngeal ganglion.

In his 1928 article, DeCastro clearly demonstrated (from section of the glossopharyngeal nerve either rostral or caudal to the ganglion) that the carotid body and sinus are innervated by axons which have their cell bodies in the sensory ganglion of the glossopharyngeal nerve. He also concluded from nerve section studies that axons to and from the carotid body are more numerous than those to and from the carotid sinus.

Numerous authors have documented the complexity of the carotid sinus nerve and its division into branches (Gerard and Billingsley 1923; Boyd 1937; Meijling 1938; D.M. McDonald 1983*a*). (Those of us who have recorded from the sinus nerve are only too well aware of the profusion of small nerves in the region of the carotid bifurcation, and can well sympathize with the early anatomists who failed to recognize the special nature of the twig from the glossopharyngeal nerve.)

However, the available evidence suggests that in *humans* there may be a functional division of the carotid sinus nerve into two branches (medial and lateral), which supply the carotid body and the carotid sinus. B. Winter (1972) promoted bilateral section of the medial branches supplying the carotid bodies as a surgical therapy for asthma. Although he did not provide the relevant data, he indicated that after such operations, baroreflex function, as judged from responses to the Valsalva manoeuvre, the cold pressor test, and carotid massage, is intact.

Most information on the composition of baroreceptor nerves comes from specimens taken from cats (Gerard and Billingsley 1923; De Castro 1951) and rats (D.M. McDonald 1983*a,b*). McDonald's superb, careful studies of the carotid sinus nerve in rats suggest that the carotid sinus nerve divides into multiple branches which travel to the carotid sinus in different ways. Some traverse the carotid body before terminating in the carotid sinus wall, and others join sympathetic (ganglio-glomerular) nerves first. The complex relation between the carotid sinus nerve and the wall of the carotid sinus is illustrated in Fig. 3.3 (D.M. McDonald 1983*a*). DeCastro (1951), Rees (1967*b*), and D.M. McDonald (1983*a*) all found that most (about 90 per cent) of afferent carotid sinus fibres travel in the carotid sinus nerve; a small proportion of fibres travel to and from the sinus via minute vagal branches (De Castro 1926). Similarly, Nonidez (1941) found that, although most aortic nerve fibres are located in the depressor nerve, some follow other 'devious' paths.

The adventitia of the carotid sinus wall is supplied by a rich network of blood vessels (J.T. Hansen 1987), which D.M. McDonald and Larue (1983) characterize as 'vasa vasora'. Nerve endings in the carotid sinus lie close to these capillaries and appear to be supplied by them. A single arteriole usually runs within the carotid sinus nerve itself, and a venule usually accompanies the carotid sinus nerve outside the perineurium (D.M. McDonald 1983*a*).

Figure 3.4 is an electron micrograph that shows a cross-section of a rat carotid sinus nerve (D.M. McDonald 1983*a*). This section illustrates several typical features of baroreceptor nerves: unmyelinated fibres are dispersed throughout the nerve and greatly outnumber myelinated fibres (De Castro 1926), and the nerve contains a single blood vessel and is enveloped by perineurium (Nonidez 1941). In rats (D.M. McDonald 1983*a*) and cats (Eyzaguirre and Uchizono 1961), at the mid-point of the nerve, the total number of axons is about 500 and 2000. These axons are either afferent or efferent, and include sympathetic fibres (Eyzaguirre and Uchizono 1961; Reis and Fuxe 1968; D.M. McDonald and Mitchell 1975). A great many baroreceptor axons are unmyelinated (Rees *et al.* 1978), and most of these

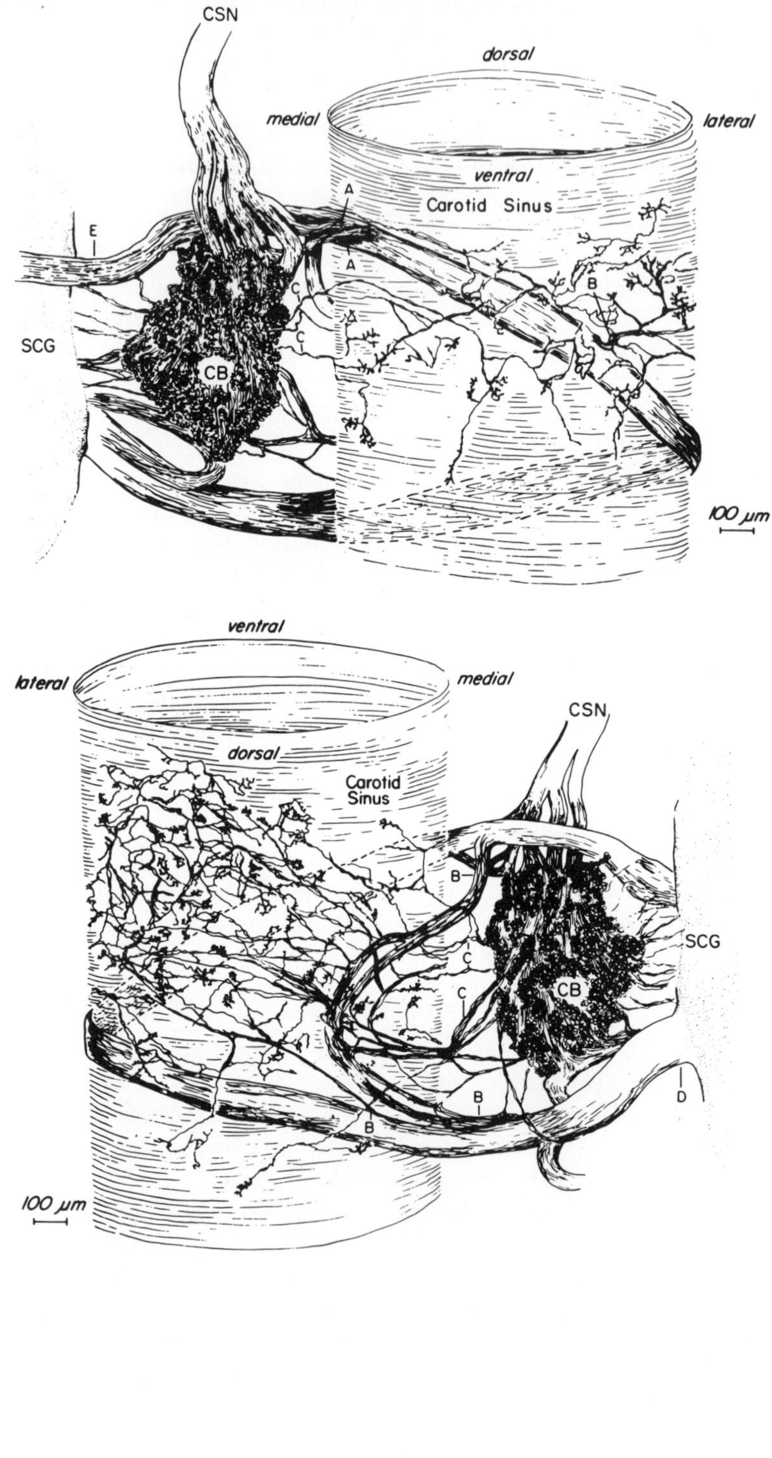
CSN
dorsal
medial
lateral
ventral
Carotid Sinus
A
A
E
C
C
B
SCG
CB
100 μm
ventral
lateral
medial
CSN
dorsal
Carotid
Sinus
B
SCG
C
C
CB
B
B
D
100 μm

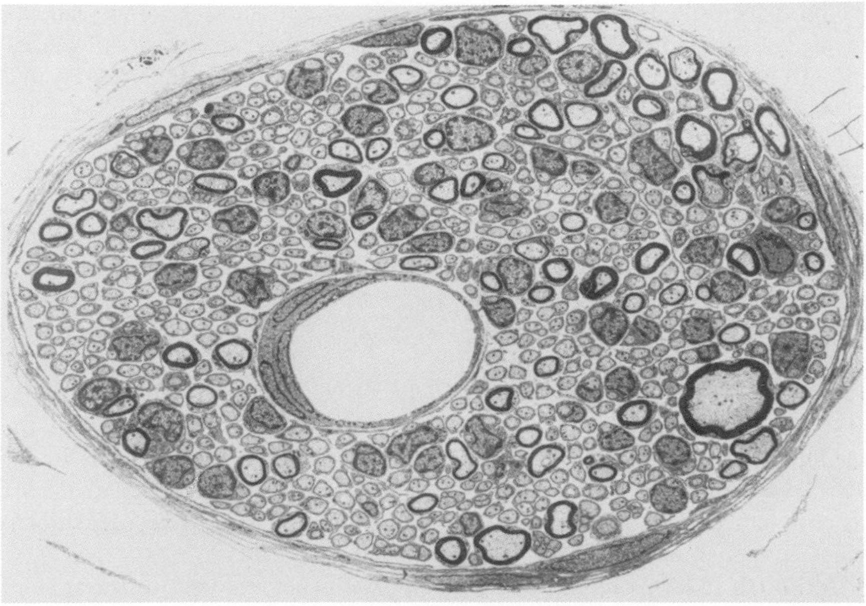

FIG. 3.4. Electron micrograph of carotid sinus nerve of rat. Reproduced with permission (D.M. McDonald 1983*a*).

persist after sympathectomy (Rees 1967*b*). To our knowledge, there are no published studies of human carotid sinus nerve ultrastructure.

Heinrich Hering conducted his earliest studies of carotid sinus baroreflex function without knowledge of the nervous connections of the carotid sinus. It remained for DeCastro (1926), who began his work soon after Hering published his first article on carotid baroreflex function, and subsequently Sunder-Plassmann (1930) and Hering's assistant Koch (1931) to provide neuroanatomical explanations for Hering's physiological observations.

Terminal structure of baroreceptors

The carotid sinus nerve enters the wall of the carotid sinus at right angles (Ábrahám 1969). Myelinated nerves tend to traverse the adventitia to the

---

FIG. 3.3. Entrance of the carotid sinus nerve into the wall of the carotid artery in the rat. CB = carotid body; CNS = carotid sinus nerve; SCG = superior cervical sympathetic ganglion; D = point of emergence of sympathetic nerve from superior cervical sympathetic ganglion; A = point at which two branches of the carotid sinus nerve join a sympathetic nerve; B = point at which carotid sinus nerve branches leave the sympathetic nerve; C = point at which other axons enter the carotid sinus nerve from the carotid body. Reproduced with permission (D.M. McDonald 1983*a*).

media–adventitia border, lose their myelin sheaths and become 'naked', divide into numerous branches, and arborise profusely. In several species, noradrenergic fibres in the carotid sinus nerve terminate in multi-layered adventitial plexuses near smooth muscle (Rees 1967*a*; Reis and Fuxe 1968; Stanton and Hinrichsen 1980).

Virtually all histological studies of baroreceptor anatomy, since the earliest ones by DeCastro (1926, 1928), Sunder-Plassmann (1930), and Nonidez (1937), agree that most of the neural structures are located in the arterial *adventitia*. However, there also is abundant evidence that some axons enter the media (Koster and Tschermak 1902; Ábrahám 1969) (or 'invaginate' from the adventitia into the media (Ábrahám 1969; Krauhs 1979)). (Some authors suggest that these axons extend as far as the intima (Eyzaguirre and Lewin 1961; D.M. McDonald and Mitchell 1975; Knoche *et al.* 1980).) Rees (1968) suggests that it may be unnecessary for nerve fibres to penetrate the media to be influenced by smooth muscle. There are fenestrations in the elastic layers that divide media and adventitia, and in these areas, axons are in contact with medial smooth muscle cells. Moreover, Rees also documented apposition of axons and smooth muscle cells (1967*b*).

Although it seems highly likely that nerve structures within the walls of the carotid sinus and aorta subserve baroreceptor function, to our knowledge, this has never been proven; there have been no intraneural recordings made from these nerves during arterial wall stretch (Guharay and Sachs 1984). Accordingly, some authors (Rees 1967*b*; Krauhs 1979) are careful to label the neurones they describe as 'presumptive' baroreceptor neurones.

DeCastro (1928) acknowledged the masterly studies of Heinrich Hering in elucidating the physiology of the carotid sinus region, but disagreed with the speculation of Drüner (1925) that the receptors lay in the carotid body. DeCastro carried out the first serious studies of carotid sinus innervation, in Cajal's laboratory in Madrid (De Castro 1926). He studied the carotid body and carotid sinus innervation in *humans* with both methylene blue and silver nitrate reduction; in addition, he used degeneration experiments after nerve section in the rat. The first studies in man were on a fetus and in men of 20, 24, 30, and 45 years, who died accidentally. The tissue was removed within four hours of death (De Castro 1928).

DeCastro described many features typical of presumptive baroreceptor nerves. He noted a rich innervation of the sinus adventitia and indicated that thick myelinated fibres are fewer in number than receptors, because axons divide into multiple terminal branches. DeCastro proposed that nerves end in one of two ways. First, some terminals, which appear to arise from medium or fine myelinated or from unmyelinated axons, branch diffusely and have extensive arborizations. Second, some terminals arise from thicker myelinated axons, are more circumscribed, with a field of about one third or one fourth that of the first type, and comprise glomus- or grape-like structures. Subsequent authors have argued for (Böck and Gorgas 1976) and against (Ábrahám 1967; Knoche *et al.* 1980) rigid classifications of putative baroreceptors into discrete types.

Light- and electron-microscopic studies document similarities of baroreceptors located in the carotid and aortic areas to stretch receptors located elsewhere. Figure 3.5 is an electron micrograph that shows many features of neural structures in the carotid sinus wall (Shin *et al.* 1987). Baroreceptors are characterized by densely packed mitochondria (m) (Rees 1967*b*), which are thought to reflect the high metabolic rate of stretch-sensitive organs that fire rapidly with each cardiac cycle. Other features include dense cored vesicles, plates, coils, rings, lamellated bodies (lb), filaments, vacuoles, and glycogen granules (g) (Rees 1967*b*; Shin *et al.* 1987).

Figure 3.6 shows light-microscopic features of an extensively arborized axon terminal in the wall of a carotid sinus obtained from a patient who died of a pulmonary embolism, published by Ábrahám (1969). Figure 3.7 depicts what may be the only published three-dimensional reconstruction of a baroreceptor neurite, from serial ultrathin electronmicroscopic sections of a rat aorta (Krauhs 1979). Although these neurites may not all be stretch-sensitive, they do appear to be representative of a common type of nerve ending (DeCastro's Type I). There is remarkable similarity between such nerves in humans and rats.

The mode of termination of putative baroreceptors may be of especial interest. First, it is clear that baroreceptors are not strictly comparable to other types of sensory receptors such as Pacinian corpuscles and muscle spindles, that have specialized cells associated with them. Baroreceptors may terminate as free ('naked' (Rees 1967*b*)) axolemmas which are exposed to vascular wall elements over a relatively long course. Terminations of baroreceptor axons are enveloped by an elastic layer which is closely apposed to deformable elements of the vascular wall including elastic fibres, smooth muscle cells, and collagen bundles (Knoche and Schmitt 1964; Rees 1967*b*). These layers may play a critical role in the transfer of force from vascular wall elements to the sensory nerve terminals.

The anatomical relation of baroreceptors to vascular wall elements has been studied by numerous authors. DeCastro found that neurites wrap around collagen bundles, and are present in several planes in the adventitia. The endings appear to be arranged along collagen bundles which run parallel to the axis of the vessel. More recently, Nonidez (1941) and Böck and Gorgas (1976) described branching neurites that lie between and penetrate through gaps in the elastic layer.

Sensory nerves in the wall of the carotid sinus also are associated with efferent sympathetic nerves. Classical histofluorescence studies have documented the presence of catecholamines in the walls of the carotid sinuses of animals (Rees 1967*a*; Reis and Fuxe 1968; Böck and Gorgas 1976). Catecholaminergic fibres appear to be sympathetic because histofluorescence is lost after superior ganglionectomy. In addition, it is apparent that the carotid

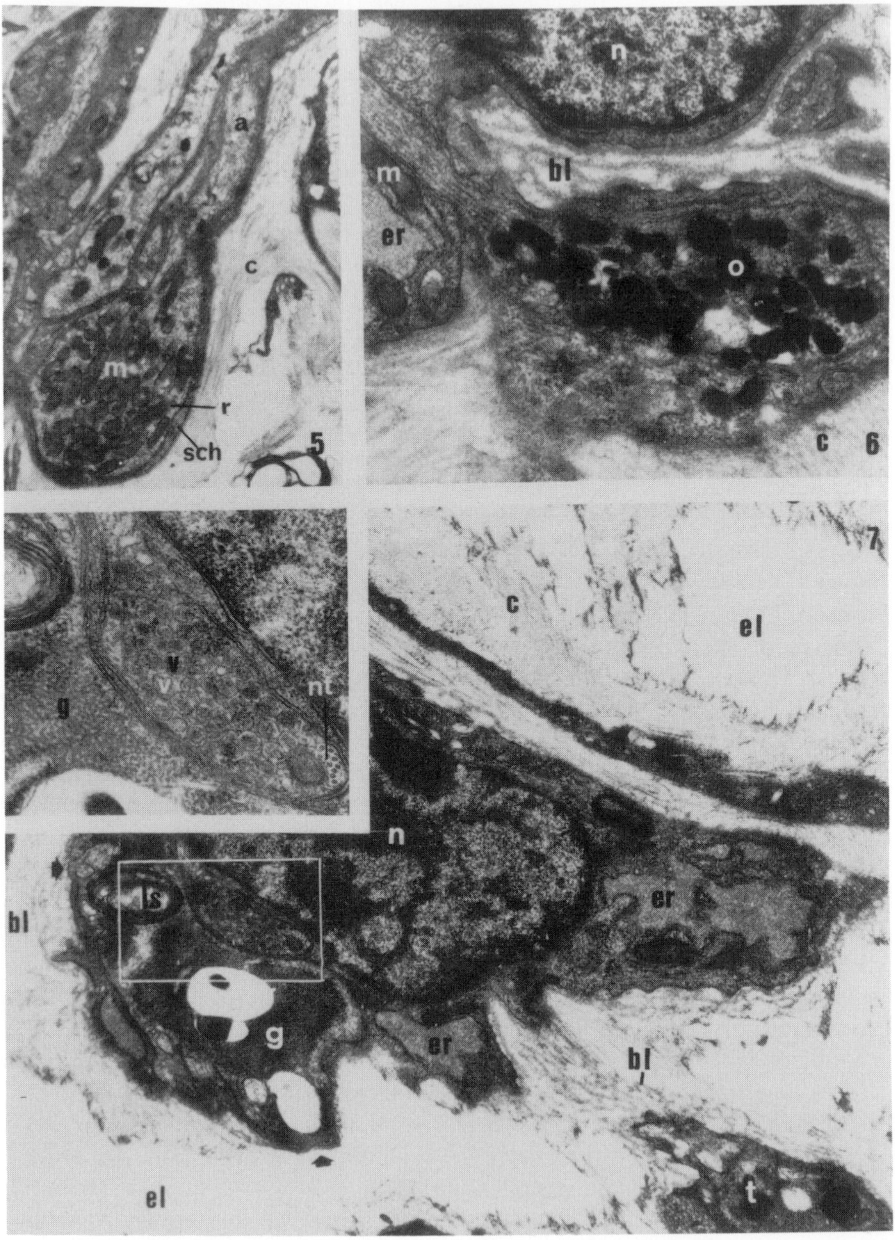

FIG. 3.5. Electron micrograph of carotid sinus wall of rat. r = putative baroreceptor nerve terminal; a = axon; o = osmophilic bodies; m = mitochondria; er = endoplasmic reticulum; bl = basal laminae; c = dense collagen fibrils; n = nucleus of Schwann cell; el = medial elastic laminae; g = glycogen granules; ls = lamellar system; t = terminal; c = collagen; n = Schwann cell nucleus; nt = neurotubuli; v = vesicles. Reproduced with permission (Shin *et al.* 1987).

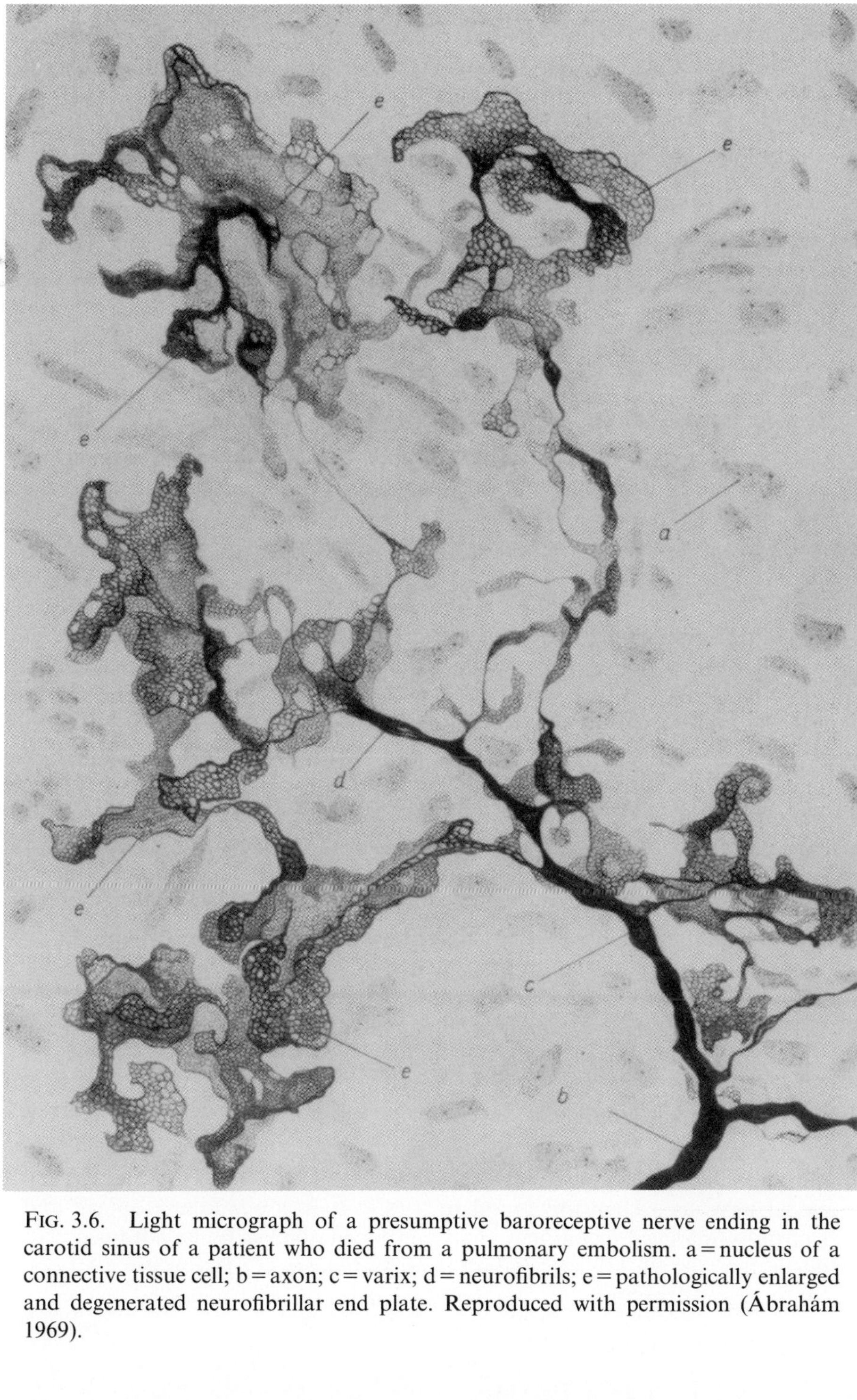

FIG. 3.6. Light micrograph of a presumptive baroreceptive nerve ending in the carotid sinus of a patient who died from a pulmonary embolism. a = nucleus of a connective tissue cell; b = axon; c = varix; d = neurofibrils; e = pathologically enlarged and degenerated neurofibrillar end plate. Reproduced with permission (Ábrahám 1969).

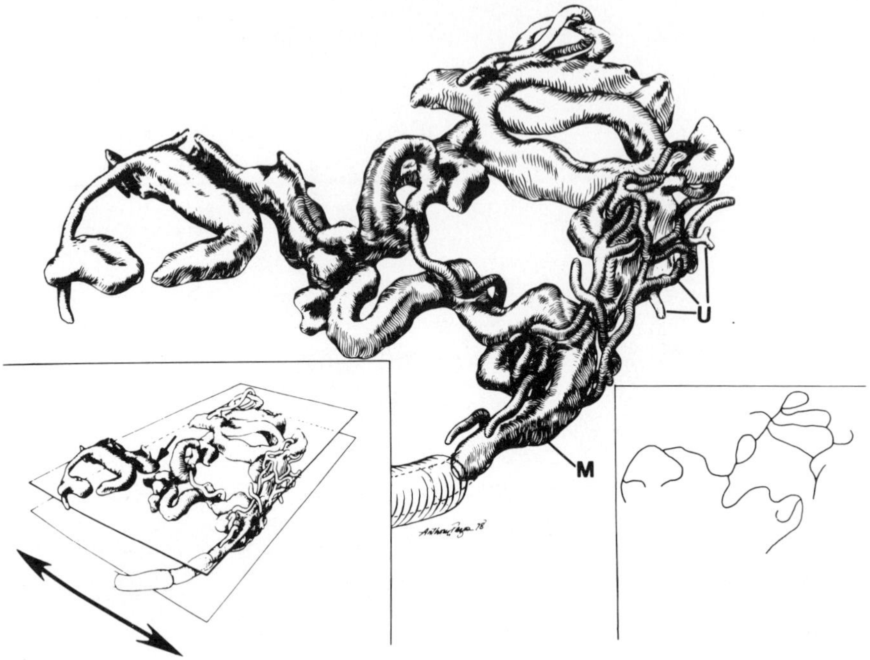

FIG. 3.7. Drawing of a computer-aided reconstruction of a rat aortic nerve terminal made from an electron micrograph. M = myelinated fibre; U = unmyelinated fibres. Inset at left shows that terminal was located between two sheets of elastic lamina. Small arrow indicates where terminal goes through a fenestration into another layer of adventitia. Double-headed arrow indicates presumed direction of stimulation. Reproduced with permission (Krauhs 1979).

sinus has non-noradrenergic efferent innervation; Kummer (1990) documented the presence of neuropeptide Y and vasoactive intestinal polypeptide-like immunoreactivity in association with catecholaminergic fibres in the media of guinea pig carotid sinuses. The role of these peptides in baroreceptor transduction is not known with certainty. Minson *et al.* (1990) speculate that neuropeptide Y sensitizes baroreceptor fibres, since intravenous injections increase the gain of integrated baroreflex responses.

CENTRAL NERVOUS SYSTEM

The most penetrating analyses of baroreceptor reflex pathways within the central nervous system are found in studies that combine anatomical and electrophysiological techniques. This combined approach has not been possible in humans, and therefore data from humans tend to focus on anatomy, without regard to function. Knowledge of central baroreflex pathways in humans derives from several types of measurements. There are some published data on responses of patients to electrical stimuli delivered via plunge electrodes into the brain (Kelly 1972). Some articles describe the effects of pathological lesions upon cardiovascular function (Yamour *et al.* 1980; Talman 1985). The most explicit anatomical information in humans comes from immunocytochemical localizations of neuroactive substances in post-mortem brain specimens.

The literature describing human responses to electrical brain stimulation is devoted primarily to psychiatric aspects of responses. In general, such stimuli have been delivered to the frontal cortex, rather than to medullary sites that are more directly involved with baroreflex control, and autonomic responses have not been studied rigorously. In an exception to this, Heath and Mickle (1960) remark that electrical stimulation of the rostral hypothalamus (*sic*) provokes tachycardia. Pool and Ransohoff (1949) indicate that stimulation of rostral cingulate gyri generally leads to blood pressure elevation and cardiodeceleration. With the possible exceptions noted below, pathological lesions in patients are not confined to discrete neural tracts, and autonomic changes resulting from such lesions are described in only the most general terms.

Moreover, published data appear to point towards important species differences in central anatomy; therefore, conclusions derived from animals may not be strictly applicable to humans. For this brief review, we rely heavily on animal studies, to which the reader is referred for more comprehensive discussions of central structures participating in baroreflex responses. We focus on functional connections rather than structure, and attempt to provide a broad outline rather than explicit detail. Studies across species identify a variety of central structures that are critically important to modulation of baroreflex function (Calaresu *et al.* 1975; Spyer 1981; Dampney 1981*b*; Loewy and Spyer 1990). These are discussed next.

Baroreceptor projections: the solitary tract nucleus

Several critical features of central baroreflex anatomy are depicted in Fig. 3.8. As mentioned, carotid baroreceptor cell bodies are located in the inferior, or petrossal, ganglion of the glossopharyngeal nerve, and aortic baroreceptor cell bodies are located in the inferior, or nodose, ganglion of the vagus nerve. The carotid sinus nerve comprises about two-thirds unmyelinated, and one-third myelinated fibres; the majority of myelinated fibres subserve chemoreception (Fidone and Sato 1969). Afferent baroreceptor neurones terminate in the solitary tract nucleus, which extends over almost the entire length of the medulla. Neurones in the solitary tract nucleus subserving cardiovascular regulation occupy primarily the densely packed portion near the obex.

Conclusions regarding the primary termination of baroreceptor afferent

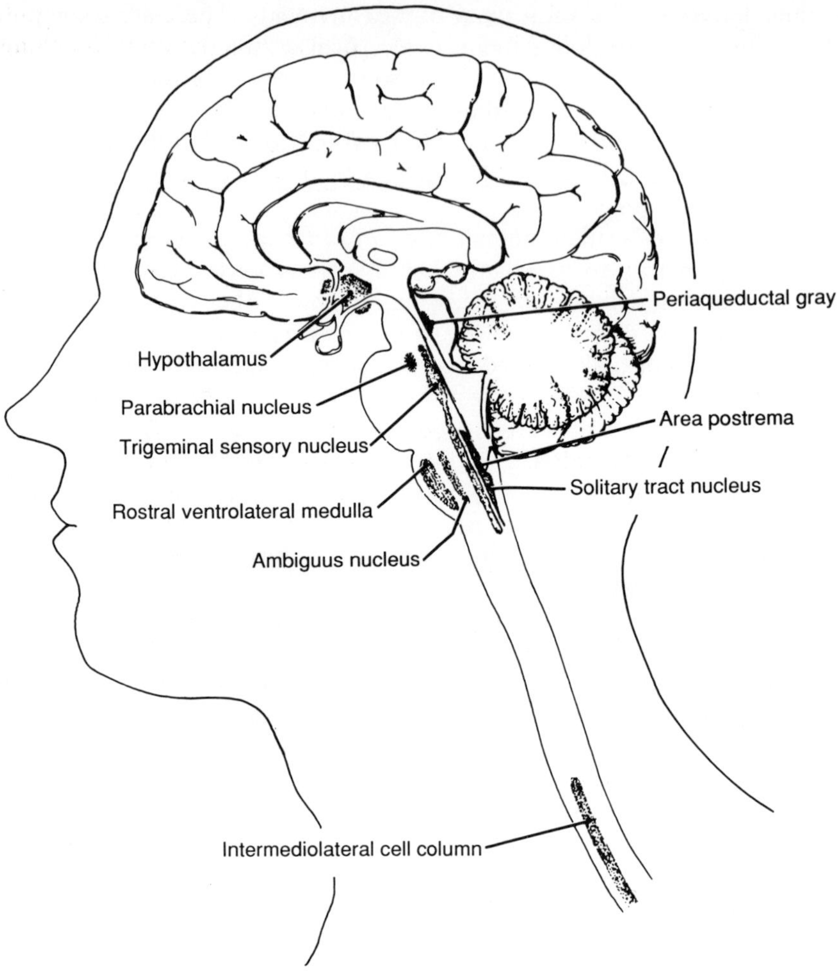

FIG. 3.8. Major areas involved with baroreflexes.

fibres are based upon data from several types of experiments. First, transection of carotid sinus or aortic depressor nerves leads to degeneration of axons in the solitary tract nucleus (Oberholzer 1960; Cottle 1964). Since in most species, carotid sinus and aortic depressor nerves contain chemoreceptor as well as baroreceptor fibres, such degeneration studies establish *limits* rather than precise locations of terminations of baroreceptor afferent fibres. After aortic (Ciriello 1983) or carotid sinus (A.J. Berger 1979; Panneton and Loewy 1980) nerve section, fibre degeneration (or transport of horseradish peroxidase applied to the proximal cut nerve end (Ciriello 1983)) end just rostral or caudal to the obex, primarily (but not exclusively) ipsilaterally.

Second, lesions placed in the solitary tract nucleus (which destroy fibres *passing through* as well as those terminating in this area — see below) abolish baroreflex responses (Oberholzer 1960; Miura and Reis 1972). In fact, bilateral destruction of this area causes fulminant, often lethal hypertension in rats (Doba and Reis 1973) and striking chronic lability of blood pressure in cats (Reis *et al.* 1977). Conversely, electrical stimulation of parts of the solitary tract nucleus mimics the effects of baroreceptor stimulation and provokes hypotension and bradycardia (Oberholzer 1960). Limited data from humans tend to validate the notion that the solitary tract nucleus is important for baroreflex control. Surgical damage in the region of the solitary tract nucleus may be followed by resting tachycardia, supine hypertension, orthostatic hypotension, and elevated plasma noradrenaline levels (Langford *et al.* 1987). Although human diseases do not usually create precisely delimited anatomical defects, some patients with medullary lesions have been described who had severe hypertension (Pamphlett and Harper 1985), or what was thought to be unusually large blood pressure lability (Kemp 1957; Magnus *et al.* 1977). Fein and Frishman (1980) reported two patients whose long-standing hypertension was cured by surgical mobilization of loops of ectatic posterior cerebellar arteries that were thought to have compressed solitary tract nuclei. Pamphlett and Harper (1985) described pathological findings in a boy who developed severe hypertension secondary to focal necrotizing encephalomyelopathy. This patient's bilateral medullary lesions involved his ambiguus nuclei, reticular formations, and trigeminal and solitary nuclei and tracts.

Third, carotid sinus or aortic depressor neurones can be activated antidromically *only* by stimuli applied in the region of the solitary tract nucleus. Such research shows that myelinated and unmyelinated baroreceptors project primarily to the ipsilateral medial and dorsolateral (Lipski *et al.* 1975; Jordan and Spyer 1977; Jordan and Spyer 1980; Donoghue *et al.* 1982; Donoghue *et al.* 1984; Spyer *et al.* 1984), and commissural (A.J. Berger 1979; Housley *et al.* 1987) portions of the solitary tract nucleus.

Fourth, electrical stimulation of the central ends of the carotid sinus or aortic nerves evokes postsynaptic potentials or single unit firing primarily in the vicinity of the solitary tract nucleus (F.D. Anderson and Berry 1956; D.R. Humphrey 1967; Miura and Reis 1969). With electrical stimuli, the assumption is made that short latencies between stimuli and responses indicate simple (possibly monosynaptic) pathways. This assumption is strengthened by the finding that electrolytic destruction of a very small area of the solitary tract nucleus where orthodromic firing is most likely to be found, abolishes systemic baroreflex responses (D.R. Humphrey 1967; Miura and Reis 1972). However, the same methodological limitation of nerve degeneration studies applies to afferent nerve stimulation; because in most species carotid sinus and aortic nerves contain chemoreceptor as well as baroreceptor fibres, baroreceptor nerve stimulation defines limits for baroreceptor terminations rather than distinct boundaries. However, anatomical localizations can be improved by the testing of responses of neurones identified by electrical stimuli to changes of systemic arterial (D.R. Humphrey 1967; Miura and Reis 1972) or carotid sinus (Biscoe and Sampson 1970) pressure.

Finally, latencies between the onset of sudden carotid sinus pressure increases and medullary neurone firings are shortest when recordings are made in the solitary tract nucleus (Lipski *et al.* 1975). However, surprisingly few solitary tract nucleus neurones have typical spontaneous pulse-synchronous firing patterns (Mifflin *et al.* 1988*a*). This theme iterates in discussions of firing of other central neurones thought to participate in baroreflex responses. One explanation for absence of clear pulse rhythmicity in solitary tract neurones is that incoming baroreceptor afferent traffic arrives at different times, according to the distances of receptive fields from the medulla, the conduction velocities of afferent fibres, and the numbers of interposed branching points (Donoghue *et al.* 1984). However, some neurones have been shown to have pulse-synchronous firing patterns similar to those of baroreceptor afferent neurones (Czachurski *et al.* 1988). Also, activity of some solitary tract nucleus neurones during step increases of carotid sinus pressure increases abruptly and then declines steadily (Lipski *et al.* 1975; Lipski and Trzebski 1975); as mentioned in Chapter 6, similar patterns of firing are observed in carotid sinus nerve recordings during step carotid sinus pressure changes (Bronk and Kaltreider 1931).

Taken together, evidence obtained with the diverse methods described above identifies the solitary tract nucleus as the critically important, *exclusive* first relay station (and, as will be seen, integrating area) for baroreceptor information. Afferent fibres from cardiopulmonary baroreceptors as well as those from carotid and aortic chemoreceptors also terminate in the solitary tract nucleus (Miura and Reis 1972; Baertschi *et al.* 1975; Lipski *et al.* 1975; Kalia and Mesulam 1980; Donoghue *et al.* 1984). Presumably, there is greater opportunity for reflex interactions to occur in areas where afferent terminations overlap than in areas where there is no overlap (Gabriel and Seller 1970). Trigeminal receptors, whose stimulation also increases vagal-cardiac nerve activity, do *not* terminate in the solitary tract nucleus (Kumada *et al.* 1977).

Interneuronal pathways and mechanisms

Mono- and polysynaptic interneuronal connections between primary baroreceptor terminations in the solitary tract nucleus and efferent sympathetic and vagal nuclei are of vital importance in modulation of cardiovascular responses to changing levels of arterial pressure. Intermediate neurones in baroreflex responses cannot be identified as easily as primary afferent or efferent neurones, since they do not degenerate after transection of afferent baroreceptor or efferent sympathetic or vagal nerves. Therefore, anatomical information regarding interneurones carrying baroreceptor information has been derived from other research approaches.

One of these is to place electrolytic lesions in the solitary tract nucleus and investigate the pattern of degeneration (Torvik 1957; Morest 1967; Cottle and Calaresu 1975). Such research documents unequivocally the presence of monosynaptic paths between the solitary tract nucleus and numerous other areas of the brain. Another method is to inject tritiated amino acids (Cowan *et al.* 1972; Ricardo and Koh 1978) or horseradish peroxidase (C.A. Ross *et al.* 1985) into the solitary tract nucleus and identify sites of anterograde or retrograde axonal conduction of the materials.

Anatomical studies (summarized by Palkovits and Záborszky (1977)) identify monosynaptic interconnections within the solitary tract nucleus (Loewy and Burton 1978), as well as between this nucleus and the dorsal vagal and ambiguus nuclei (Morest 1967; Cottle and Calaresu 1975; Loewy and Burton 1978; Norgren 1978; Beckstead *et al.* 1980), dorsal and lateral reticular formations (Morest 1967), the hypothalamus (Spyer 1972; M.R. Thomas and Calaresu 1972; Ricardo and Koh 1978; ter Horst *et al.* 1984), the rostral ventrolateral medulla (Sawchenko and Swanson 1982; C.A. Ross *et al.* 1985), and the intermediolateral region of the spinal cord (Cottle and Calaresu 1975; Loewy and Burton 1978; C.A. Ross *et al.* 1983; Dampney *et al.* 1987). These extensive monosynaptic connections tend to be reciprocal; areas that receive inputs from the solitary tract nucleus also tend to send projections back (Loewy 1990). This arrangement establishes an anatomical basis for cells that receive information from baroreceptors to influence the flow of such sensory input. The existence of widespread connections within the solitary tract nucleus itself points to the probability that this primary relay station itself is a major centre for integration of baroreceptor reflexes (Mifflin *et al.* 1988*b*).

The quality of anatomical and physiological information derived from studies involving lesions is limited, however. Electrolytic lesions destroy secondary as well as primary fibres, and also neurones merely passing through. Moreover, degeneration extends only to the first synapse; thus the technique is not helpful in defining multisynaptic pathways, which no doubt contribute importantly to baroreflex control. Also, this strictly anatomical information provides no clues regarding *function*; there is no guarantee that neurones damaged by lesions in the solitary tract nucleus are involved in baroreflex responses. However, destruction of portions of the solitary tract nucleus may document the *functional* importance of discrete neural pathways to other areas. For example, critically placed lesions in the solitary tract may abolish vasodepressor responses to carotid sinus nerve stimulation without affecting vagally mediated bradycardia (Pórszász *et al.* 1962; D.R. Humphrey 1967). This points towards early functional division of interneuronal pathways from baroreceptor terminations to medullary vagal and spinal sympathetic motoneurones. Alternative methods to electrolytic lesions include injections of radioactive tracers and horseradish peroxidase. Although radioactive tracers do not seem to be taken up by neurones passing through (Cowan *et al.* 1972), horseradish peroxidase may be taken up by such neurones if they are damaged.

Because the techniques described above identify only monosynaptic pathways, other approaches have been necessary to delineate polysynaptic pathways. Attempts to identify functional baroreflex links are based upon a variety of electrophysiological techniques, including measurements of baseline pulse synchronous firing patterns, post-event histograms (with recordings triggered by electrocardiographic R-waves or arterial pressure pulses), interneuronal responses to changes of afferent baroreceptor input (D.R. Humphrey 1967; Biscoe and Sampson 1970), frequency histograms or autocorrelograms (Mannard and Polosa 1973), neuronal responses to local injections of excitant amino acids (Goodchild *et al.* 1982), and measurements of metabolic activity with 2-deoxyglucose during baroreflex interventions (Ciriello *et al.* 1982). A variety of important insights have been drawn from such experimental approaches.

First, although cardiac rhythms can be detected in widely scattered areas in the brainstem (but are very difficult to find (Hellner and Baumgarten 1961; Salmoiraghi 1962; D.R. Humphrey 1967; Seller and Illert 1969; Miura and Reis 1972)), they are most likely to be found in the solitary tract nucleus (Salmoiraghi 1962; Miura and Reis 1972). However, the presence of cardiac rhythmicity does not prove that an interneurone is influenced by arterial baroreceptors, and the absence of cardiac rhythmicity does not exclude the possibility that a neurone may be influenced by baroreceptors. Indeed, the paucity of such rhythms counsels against assigning a label such as 'cardiovascular' to neurones with cardiac rhythms.

Neurones identified by signal-averaging may increase or decrease their rates of firing when arterial pressure rises. Since these rhythms may disappear after section of afferent baroreceptor nerves (Koepchen *et al.* 1967), such neurones are truly 'baroreceptive'. However, other neurones with cardiac periodicities may subserve different functions at other instants in time. Stroh-Werz and coworkers (1977) recorded systolic firing of 28 fibres in the solitary tract nucleus. Most of these neurones discharged with multiple activity peaks, and therefore presumably were influenced by multiple afferent inputs with different latencies.

Some neurones fired with respiratory periodicities; this was taken as evidence that some modulation of baroreceptor influences by respiratory activity occurs within the solitary tract nucleus. This conclusion is disputed by data published by Mifflin *et al.* (1988*a*), who found no influence of central respiratory activity or lung inflation on post-synaptic potentials of solitary tract neurones stimulated by baroreceptor inputs. (Earlier studies seem to have ruled out the possibility that respiratory influences are exerted on baroreceptor neurones *pre*synaptically, before they terminate in the solitary tract nucleus (Rudomin 1967; Barillot 1970; Jordan and Spyer 1979; Richter *et al.* 1986)).

One neurone with cardiac periodicity studied by Stroh-Werz *et al.* (1977) continued firing after bilateral vagal cooling, and its activity was thought to result from carotid sinus baroreceptor stimulation. Two other neurones stopped firing after vagal cooling, and their activity was thought to result from aortic baroreceptor stimulation. In another study (Stroh-Werz *et al.* 1976), this group showed that medullary units which fire in synchrony with arterial pulses may fire intermittently, even at identical arterial pressures. An additional complexity is that some interneurones fire with cardiac periodicities after all afferent baroreceptor nerves have been cut (Pórszász and Pórszász-Gibiszer 1968).

As discussed by Koepchen and associates (1975), alternating cardiovascular and respiratory rhythms may be found in recordings from single interneurones, whose activity may shift from one pattern to another with changes of arterial pressure, and be coordinated in a variable, or 'sliding', manner. Thus functional specificity of interneurones may be quantitative, rather than qualitative (Koepchen *et al.* 1981). Koepchen and coworkers (1975) recorded from a single brainstem neurone which was excited by tactile stimulation and inspiration, and inhibited by blood pressure elevation.

One particularly fruitful electrophysiological approach used to characterize function of interneurones involved in baroreflex control is to measure changes of their intracellular potentials provoked by electrical carotid sinus or aortic nerve stimulation. Such responses are usually categorized as excitatory or inhibitory post-synaptic potentials (abbreviated EPSP's and IPSP's) or combinations thereof (Mifflin and Felder 1990). (In all likelihood, interneuronal modulation is much more varied than this useful but simple construct implies (Kitahama *et al.* 1985)). Moreover, there is no guarantee that electrical stimulation of a baroreceptor afferent nerve excites or inhibits a given interneurone directly, without interposition of one or more other interneurones.)

Nonetheless, this genre of research has yielded a variety of important insights into complex central interneuronal mechanisms. First, it is apparent that multiple afferent autonomic sensory inputs converge on to the same interneurones (Biscoe and Sampson 1970; J.R. Hildebrandt 1974). Thus myelinated and unmyelinated carotid baroreceptor (Donoghue *et al.* 1984), vagal (Jordan *et al.* 1988), and cardiopulmonary (Keith *et al.* 1975; Donoghue *et al.* 1981) fibres may converge on to the same ipsilateral and contralateral (Mifflin and Felder 1988) interneurones. Felder (1986) documented convergence of renal, carotid sinus, and aortic afferent neurones on to the same interneurones in the solitary tract nucleus. Although it seems likely that in other areas of the medulla, both chemoreceptor and baroreceptor inputs converge onto the same neurones (Lipski *et al.* 1976), such convergence has been sought, but has not been found, for interneurones located within the solitary tract nucleus (Miura and Reis 1972).

Second, as discussed persuasively by Mifflin and Felder (1990), the convergence of multiple inputs on to the same interneurone establishes a solid anatomical basis for reflex 'interactions', which may have far-reaching implications. For example, since lengths of afferent baroreceptor nerves and their conduction velocities and degrees of branching differ substantially, there is considerable temporal dispersion of inputs arriving from arterial baroreceptors (Donoghue *et al.* 1984). A wealth of published evidence (D.R. Humphrey 1967; Lipski and Trzebski 1975; Gabriel and Seller 1970; McCall *et al.* 1977; Donoghue *et al.* 1985; Mifflin and Felder 1988) indicates that early-arriving baroreceptor inputs may alter responsiveness of interneurones to later inputs. Therefore, through this mechanism, an input from one receptive field may alter responses to an input from another, and provide an anatomical substrate for some of the baroreflex interactions discussed in Chapter 7. A variation on this theme is that inputs from multiple receptive fields summate to provoke integrated autonomic neural responses (Donoghue *et al.* 1985; Felder and Heesch 1987). Presumably because of this, when an input from one field suddenly disappears, as after carotid sinus (O'Leary and Scher 1990) or cardiopulmonary (Persson *et al.* 1987) baroreceptor denervation, remaining inputs may assume progressively greater importance and, with passage of time, re-establish normal baroreflex responses and arterial pressure levels.

Another electrophysiological method used to define baroreflex pathways is to measure latencies between electrical stimuli applied to afferent baroreceptor nerves and evoked responses (J.R. Hildebrandt 1974). The great majority of fibres in the solitary tract nucleus excited by electrical stimulation of carotid sinus or aortic nerves have latencies that point to polysynaptic pathways (Lipski *et al.* 1975). Long latencies for central transmission also point toward probably very complex, multisynaptic interneuronal pathways between primary baroreceptor terminations and efferent motoneurones. However, there is no guarantee that even pathways with the shortest latencies are monosynaptic. Nor, for that matter is it clear that monosynaptic pathways have greater physiological importance than polysynaptic pathways.

Latencies between electrical stimulation of carotid sinus nerves or arterial pressure changes and efferent sympathetic (J.H. Green and Heffron 1968*b*; Coote and Downman 1969) or vagal (Iriuchijima and Kumada 1964; Kunze 1972; Gebber *et al.* 1973) activity also are long and variable. McAllen and Spyer (1978*a*) observed a bimodal distribution of latencies between carotid sinus nerve stimuli and efferent *vagal* responses, with short and long peaks between 20 and 60, and between 70 and 110 ms.

Gebber *et al.* (1973) studied latencies between electrical stimuli applied at various locations in the spinal cord and hypothalamus and evoked responses of postganglionic *sympathetic* vasoconstrictor neurones, and deduced the existence of two functionally different pathways that modulate sympathetic outflow (to the carotid sinus nerve). One set of fibres was, and the other set was not, inhibited by pharmacological arterial pressure elevation. Fibres inhibited by baroreceptor stimulation had long (greater than 50 ms) latencies, presumably because they were at the end of long polysynaptic pathways, and long-lasting evoked potentials. Fibres not inhibited by baroreceptors had short latencies (less than 50 ms), short evoked responses, and required trains, rather than single stimuli, to elicit firing.

Inferential evidence from humans supports the existence of baroreflex-sensitive and -insensitive pathways. Very small elevations of arterial pressure reduce plasma noradrenaline concentrations to a minimum level (Eckberg *et al.* 1986) and abolish muscle sympathetic nerve activity (Eckberg *et al.* 1988). Since larger elevations of pressure do not lead to further reductions of plasma noradrenaline levels, some neuronal pools that release noradrenaline must be uninfluenced by baroreceptor input.

Specific central areas important to baroreflex control

Ciriello and coworkers (1982, 1983) and Kostreva (1983) identified central structures involved with baroreflex control by measuring metabolic activity with 2-deoxyglucose during baroreceptor stimulation or electrical stimulation of afferent baroreceptor nerves. These studies document changes of central neuronal activity in the solitary tract nucleus, dorsal motor nucleus of the vagus, ambiguus nucleus, ventrolateral medulla, parabrachial and paraventricular nuclei, hypothalamus, amygdala, and the dorsal horns of the thoracolumbar spinal cord. The increases of metabolic activity that were found in the spinal cord suggest that arterial pressure elevations trigger inhibitory interneuronal firing; since metabolic activity did not increase after sinoaortic denervation, these effects are mediated by arterial baroreceptors (Ciriello *et al.* 1982).

The precise location of *interneurones that drive sympathetic motonuclei* is not known with certainty in humans. However, there is much evidence to suggest that such cells are located in a restricted area of the medulla in animals (Feldberg and Guertzenstein 1976; Reis *et al.* 1984*a*). When the spinal cord is separated from the brainstem, blood pressure falls (Bernard 1858; Dittmar 1870; C.A. Ross *et al.* 1984*b*; Granata *et al.* 1985*b*) and spontaneous sympathetic firing declines to very low levels (Mannard and Polosa 1973). This suggests that sympathetic firing is in some way *stimulated* by inputs from the brainstem. In this connection, much attention has focused on a group of cells located in the *rostral ventrolateral medulla*, as a final, critically important neuronal pathway for modulation of sympathetic firing (Reis and Cuenod 1965; Ciriello *et al.* 1986). The exact location of this area (which is probably the classical 'tonic vasomotor centre' described by Owsjannikow (1871) and Dittmar (1873)), and its connections may vary according to species; the area (or its subdivisions) is known variously as the 'subretrofacial nucleus' (Dampney and McAllen 1988; Dampney 1990), the rostroventrolateral reticular nucleus (C.A. Ross *et al.* 1984*a,b*), the lateral paragigantocellularis nucleus (Olszewski and Baxter 1954; Andrezik *et al.* 1981; Lovick 1987), and the 'retrotrapezoid nucleus' (J.L. Feldman and Ellenberger 1988). By all accounts, this area comprises a very small number of cells, perhaps as few as 200 in the rat (the 'marvellous 200' — D.J. Reis, unpublished).

Attention was first called to this area by the observations of Dählstrom and Fuxe (1964) that it contains dense catecholamine histofluorescence. Subsequent studies documented the presence of the enzymes dopamine-*beta*-hydroxylase (Swanson and Hartman 1975), which converts dopamine to noradrenaline, and phenylethanolamine-*N*-methyltransferase (Hökfelt al. 1974; Saavedra *et al.* 1974; Ruggiero *et al.* 1985*b*), which converts noradrenaline to adrenaline. Neurones containing these enzymes are concentrated in contiguous areas of the ventrolateral medulla, designated C1 (Hökfelt *et al.* 1974) and A1 (Dählstrom and Fuxe 1964, 1965), and make direct connections with other areas involved with baroreflex responses, including the solitary tract nucleus (Dampney *et al.* 1982; C.A. Ross *et al.* 1985), parabrachial nuclei (Dampney *et al.* 1982), hypothalamus (Sawchenko and Swanson 1982; Day and Renaud 1984), and importantly, the thoracic intermediolateral column containing spinal sympathetic motonuclei (Amendt *et al.* 1979; Blessing *et al.* 1981; C.A. Ross *et al.* 1981; C.A. Ross *et al.* 1984*a*). Moreover, central areas with monosynaptic outputs to the intermediolateral cell columns also seem to receive inputs from this area (Calaresu *et al.* 1984).

Barman and Gebber (1983) and Guyenet *et al.* (D.L. Brown and Guyenet 1985; Haselton and Guyenet 1989*b*) showed that firing of spontaneously active rostral ventrolateral medullary neurones (in the area designated C1) may be time-locked with (and therefore entrained by) cardiac activity, and may occur prior to that of spinal preganglionic sympathetic neurones. Neurones in this area may have cardiac rhythms, and fire in inverse relation to arterial pressure with threshold, linear, and saturation ranges (D.L. Brown and Guyenet 1985). After baroreceptor denervation, firing of these neurones may decline or disappear (D.L. Brown and Guyenet 1985). Firing that persists after baroreceptor denervation is no longer time-locked to cardiac activity, but is free-running; however, even after baroreceptor denervation, firing in these neurones predicts firing in postganglionic splanchnic neurones (Morrison *et al.* 1988). It is not known whether the medullary circuits that drive spinal sympathetic motoneurones are simple pacemakers, or are more complex ensembles of neurones that interact to generate rhythmic activity (Gebber 1990).

Electrical stimulation of the rostral ventrolateral medulla or injections of excitant amino acids (which stimulate cell bodies but not axons passing through [Fries and Zieglgänsberger 1974]) provokes large increases of arterial pressure and heart rate (Dampney and Moon 1980; Dampney *et al.* 1982; C.A. Ross *et al.* 1983), and destruction of this area or injection of inhibitory amino acids leads to profound hypotension (Guertzenstein and Silver 1974; Dampney and Moon 1980; C.A. Ross *et al.* 1984*b*) and abolishes vasodepressor responses to baroreceptor stimulation (Dampney 1981*a*; McAllen *et al.* 1982; Granata *et al.* 1983; Cechetto and Calaresu 1985; Granata *et al.* 1985*b*). Cardioacceleration provoked by chemical stimulation of the ventrolateral medulla is thought to be mediated by neuronally released noradrenaline or adreno-medullary secretion of adrenaline, because it is prevented by pretreatment with the beta-blocker propranolol (Lovick 1987). Cells in this area do not seem to be dependent upon stimulation from higher centres; vasodepressor reflexes persist after mid-collicular (rostral to the medulla) decerebration and bilateral destruction of parabrachial nuclei (Granata *et al.* 1985*b*). After destruction of the rostral ventrolateral medulla, subsequent cervical spinal cord transection does not lead to further blood pressure reduction (C.A. Ross *et al.* 1983). Focal injection of kainic acid, a chemical that sequentially excites and then destroys neurones (Olney 1978), increases arterial pressure and renal sympathetic nerve activity and then reduces arterial pressure to levels found after high spinal cord transection (McAllen *et al.* 1982). Many (but not all (D.L. Brown and Guyenet 1984; Agarwal and Calaresu 1990)) neurones in this area are inhibited (or hyperpolarized (Dembowsky and McAllen 1990)) by baroreceptor stimulation or electrical stimulation of the solitary tract nucleus (Caverson *et al.* 1984; Morrison *et al.* 1988).

There is some evidence that neurones in the rostral ventrolateral medulla are organized according to target organ rather than spinal level (Lovick 1987; Dampney and McAllen 1988; McAllen and Dampney 1990). Also, there is evidence that cells in this area are involved in pressor response to cerebral ischaemia (Dampney and Moon 1980), the somatopressor reflex (Sato and Schmidt 1973; Kozelka *et al.* 1981; Stornetta *et al.* 1989), baroreceptor-mediated release of arginine vasopressin (Day and Renaud 1984), fastigial cardiovascular responses (Dormer *et al.* 1989), the centrally mediated effects of clonidine (Ernsberger *et al.* 1988), and control of respiration (Caverson *et al.* 1984; Millhorn and Eldridge 1986; Haselton and Guyenet 1989*a*; Guyenet *et al.* 1990).

The *caudal portion of the ventrolateral medulla* contains noradrenergic (A1) (Dählstrom and Fuxe 1964) and non-catecholaminergic neurones (retroambigual area) (Dählstrom and Fuxe 1964; Ruggiero *et al.* 1989), but appears to be involved with sympathoinhibition, rather than sympathoexcitation. Injection of the excitotoxic agent kainic acid provokes sequential initial reductions and later increases of arterial pressure and sympathetic nerve activity (Granata *et al.* 1985*a*). Destruction of this area provokes pressure elevations, and electrical or chemical stimulation provokes pressure reductions (Blessing *et al.* 1981; Blessing and Reis 1982; Blessing *et al.* 1982). Neurones in the A1 area of the caudal ventrolateral medulla area do not appear to project directly to the spinal cord (Blessing *et al.* 1981; C.A. Ross *et al.* 1981). Moreover, sympathetic responses to stimulation or inhibition of this area are abolished by destruction of neurones in the rostral ventrolateral medulla (Granata *et al.* 1986). Thus it seems likely that sympathoinhibition resulting from stimulation of neurones in the caudal portion of the ventrolateral medulla is mediated by neurones in the rostral portion. Day *et al.* (1983) found that the vasodepressor area correlates with the retroambigual area (a source of afferent neurones to the rostral ventrolateral medulla (C.A. Ross *et al.* 1985)), and not with the A1 area, which projects to the vasopressin cell group in the hypothalamus but not to the spinal cord.

Evidence from animals points toward important cerebellar influences upon baroreceptor reflex mechanisms (Bradley *et al.* 1987*a,b*). The fastigial nucleus of the *cerebellum* sends monosynaptic fibres to the solitary tract nucleus (Batton *et al.* 1977). Electrical stimulation of portions of the vermis of the cerebellum or the fastigial nucleus provokes cardioacceleration secondary to increased sympathetic and reduced vagal traffic to the heart (Lisander and Martner 1971) and arterial pressure elevation (Bradley *et al.* 1987*a*; Paton *et al.* 1990). (Cardiovascular responses to electrical cerebellar stimuli are preserved after kainic acid destruction of perikarya, however; this raises the possibility that fastigial pressor responses are mediated in part by axons passing through the cerebellum (R.T. Henry and Connor 1989).) Cerebellar stimulation reduces hypertensive responses to bilateral carotid artery occlusion (Moruzzi 1940; Miura and Reis 1970) and may abolish firing of neurones in the solitary tract nuclei caused by electrical stimulation of afferent carotid sinus or aortic nerves (Paton *et al.* 1990). We are aware of one attempt to document cerebellar influences on baroreceptor reflexes in humans; R. Winter and coworkers (1982) found that caloric stimulation of labyrinthine receptors does not alter R–R interval lengthening caused by phenylephrine injections.

There is considerable anatomical and electrophysiological evidence linking *hypothalamic centres* and their outflows through the central core of the midbrain, the 'periaqueductal gray' (Bandler and McCulloch 1984; S.M. Hilton and Redfern 1986; Carrive *et al.* 1988), with baroreceptor responses. There are discrete connections between the solitary tract nucleus and the hypothalamus (Ricardo and Koh 1978), and between the hypothalamus and preganglionic sympathetic and vagal motoneurones (Spyer 1972; Saper *et al.* 1976). One study (S.M. Hilton and Spyer 1971) shows that stimulation of pre-optic neurones and the anterior hypothalamus elicits vasodepressor and bradycardic responses similar to those triggered by baroreceptor stimulation. Another (Klevans and Gebber 1970) shows that electrical stimuli can be delivered to septal and pre-optic areas or the anterior hypothalamus, which alone do not influence baseline heart rate or blood pressure, but which enhance baroreflex-mediated bradycardia strikingly. Neurones within these areas also may be activated by electrical stimulation of afferent baroreceptor nerves (Barker *et al.* 1970; Spyer 1972; M.R. Thomas and Calaresu 1972), and lesions within these areas severely reduce the pressure range over which baroreceptors modulate arterial pressure (S.M. Hilton and Spyer 1971; B.B. Kent *et al.* 1971). Baroreceptor–hypothalamic interconnections also appear to modulate antidiuretic hormone release (Barker *et al.* 1970), and may be involved with putative baroreceptor modulation of cognitive activity (Lacey and Lacey 1962) and emotion (O.A. Smith *et al.* 1980).

Stimulation of the hypothalamic *'defence area'* (W.R. Hess and Brügger 1943) provokes long-lasting inhibition of cells in the solitary tract nucleus that are excited by baroreceptor inputs (Mifflin *et al.* 1988*b*). The hypothalamic defence area appears to function as a conduit for reactions originating in the *amygdala* (S.M. Hilton and Zbrożyna 1963; Timms 1981; Cechetto and Calaresu 1985). Hopkins and Holstege (1978) and Schwaber *et al.* (1980) documented the existence of direct pathways from the central nucleus of the amygdala through the hypothalamus, to the solitary tract nucleus and (in rabbits) the dorsal motor vagal nucleus. G.E. Cox *et al.* (1986) studied the physiological significance of these pathways; they found that many solitary tract neurones excited by electrical stimulation of the central nucleus of the amygdala also receive inputs from aortic afferent baroreceptor fibres. It is not known if isolated temporal lobe seizures in humans (Blumhardt *et al.* 1986), which may provoke tachycardia as early as 2 s after their onset (Marshall *et al.* 1983), do so by opposing the restraining effects of afferent baroreceptor activity.

Mifflin *et al.* (1988*b*) and Jordan *et al.* (1988) studied the effects of electrical stimuli delivered to the hypothalamic defence area on responses of neurones in the solitary tract nucleus to electrical carotid sinus nerve stimuli or carotid pressure elevations. They found that without exception, hypothalamic stimuli inhibited interneuronal fibre responses to baroreceptor stimuli. They postulated that such interactions occurring in the region of the initial afferent baroreceptor relay station may function importantly to modulate baroreflex function.

Neurones in the *parabrachial nuclei* appear to function as important relay stations between the amygdala (Hopkins and Holstege 1978), cerebellum (Somana and Walberg 1979; Paton and Spyer 1990), hypothalamus (Norgren and Leonard 1973), and solitary tract nuclei (Granata and Kitai 1989). Chemical or electrical stimulation of parabrachial nuclei provokes bradycardia, and increases of arterial pressure, cardiac output, efferent renal sympathetic nerve traffic, and peripheral vascular resistance (which are not distributed evenly among different vascular beds (R.B. Hamilton *et al.* 1981; Mraovitch *et al.* 1982; Hade *et al.* 1988; Paton and Spyer 1990)). Conditioning electrical stimuli delivered to parabrachial neurones may inhibit solitary tract nucleus neuronal responses to carotid sinus nerve stimuli (Felder and Mifflin 1988).

Other central areas may influence baroreflex control by responding to neuroactive substances released at distant sites. A.J. Dunn (1978) suggests that the cerebral ventricular system functions as more than a 'sewer' (or, for that matter, than a 'cistern'). Cerebrospinal fluid flows in a fixed direction, has high concentrations of peptides (Kee and Wood 1985) which might influence distant structures, and does not contain proteolytic enzymes. It is of interest that structures important for baroreflex function, including the solitary tract and dorsal motor vagal nuclei, lie subjacent to the fourth ventricle, and that injections of drugs into the fourth ventricle or into nearby structures provoke profound changes of autonomic cardiovascular nervous outflow (Loeschcke and Koepchen 1958; Feldberg and Guertzenstein 1976).

Moreover, structures in several circumventricular locations have fenestrated capillaries that lack tight endothelial junctions, which permit blood-borne neuroactive macromolecules to breach the blood brain barrier and influence central neurones (Broadwell and Brightman 1976; Miselis *et al.* 1987). Some of these areas have monosynaptic connections with other structures important in baroreflex modulation; the subfornical organ sends projections to the median pre-optic nucleus of the hypothalamus (Miselis *et al.* 1979; R.W. Lind 1987), and the area postrema, which protrudes into the fourth ventricle, sends projections to the solitary tract nucleus (Barnes and Ferrario 1981) and the rostral ventrolateral medulla (C.A. Ross *et al.* 1985).

It seems likely that the *subfornical organ* modulates baroreflex function by allowing macromolecules into the central nervous system. Lesions of the subfornical organ reduce pressor responses caused by intravenous injections of angiotensin II (Mangiapane and Simpson 1980*b*), and injection of angiotensin II in the region of the subfornical organ leads to arterial pressure elevations (Mangiapane and Simpson 1980*a*). Central actions of angiotensin II are also discussed in Chapters 5 and 8.

Several types of evidence point to the highly vascular (Brizzee and Neal 1954) area postrema as an important link in baroreflex control (Barnes and Ferrario 1981; Ferrario *et al.* 1987). It seems likely that some influence of the area postrema on baroreflexes is mediated by circulating peptides that gain access to the central nervous system through its capillary network. Microinjections of angiotensin II into the area postrema increase arterial pressure. Destruction of the area postrema eliminates vasopressin- but not phenylephrine-induced sympathoinhibition (Undesser *et al.* 1985; Applegate *et al.* 1987) and pressor responses to intravenous (Scroop *et al.* 1971) or intravertebral (Joy and Lowe 1970; Gildenberg *et al.* 1973) angiotensin II injections.

The area postrema receives inputs from carotid sinus (Panneton and Loewy 1980) and vagus nerves (Ciriello *et al.* 1981) and sends projections to the solitary tract nucleus (Clemente and Van Breemen 1955; Morest 1967;

Vigier and Rouviere 1979; Barnes and Ferrario 1981). (It should be noted that the anatomical location of the the area postrema may confound attempts to study its functional importance. Since the area postrema and solitary tract nuclei are supplied by the vertebral artery, lesions of the area postrema may influence baroreflex mechanisms by interfering with blood flow to the solitary tract nucleus. Similarly, since the commissural nucleus of the solitary tract nucleus directly underlies the area postrema, destruction or electrical stimulation of the area postrema may also destroy or stimulate neurones in the solitary tract nucleus (A.K. Johnson and Loewy 1990).)

Chronic ablation of the area postrema leads to transient arterial pressure elevation followed by sustained mild hypotension and heart rate speeding (Ferrario *et al.* 1979) in dogs, but mild hypotension, bradycardia, and baroreflex enhancement in awake rats (Skoog and Mangiapane 1988). Electrical or chemical stimulation of the area postrema (in anaesthetized rabbits) inhibits renal sympathetic nerve activity and augments baroreflex-mediated reductions of renal sympathetic nerve activity (Hasser *et al.* 1987).

Chemical mediators

Incoming traffic carried over afferent baroreceptor neurones releases chemicals that alter properties of other neurones and eventually provoke changes of efferent vagal and sympathetic outflow. An ever-lengthening list (Iversen 1983; Schmitt 1984; Brody 1986; Diz *et al.* 1987; J.L. Reid and Rubin 1987) of chemicals (referred to generically as 'informational' or 'neuroactive' substances, and in many instances located with molecular genetic probes) includes such classical neurotransmitters as: acetylcholine, noradrenaline, and glutamate (Zhang *et al.* 1990); neuropeptides, such as somatostatin (Chigr *et al.* 1989), angiotensin II, beta-endorphin, arginine vasopressin; histamine (Akins and Bealer 1991); and substance P (Chan *et al.* 1990); and neurohormones, such as thyroid- (Mattila and Buñag 1986; S.-i. Tsunoda *et al.* 1988) and corticotropin-releasing (S.-i. Tsunoda *et al.* 1988) factors.

Understanding of central autonomic chemical transactions is increasing incrementally. The classical construct, that each afferent nerve impulse releases a single neurotransmitter which traverses a synapse and excites or inhibits firing of a contiguous neurone by altering its ionic permeability in an all-or-none fashion, may have been supplanted by more recent research. Refinements of this early view include the possibilities that cells release multiple transmitters (Burn and Rand 1965), non-synaptic release of chemicals occurs, responses of secondary neurones are graded rather than all-or-none, released chemicals modulate responsiveness of neurones to other chemicals, and chemicals are released for diffusion to distant neurones (Dismukes 1979) such as those in the circumventricular organs discussed above.

There is a substantial number of studies describing anatomical localization of chemical mediator substances in *human* brains. Such investigations have been facilitated by the development of highly specific immunohistochemical techniques, which can be used with post-mortem specimens, especially if the time from death until staining is short. In such studies, specific areas in the central nervous system are identified according to their location and cytoarchitecture, and correlated with similar areas in animals known to be involved in baroreflex mechanisms. The major deficiency of studies involving human specimens is that they provide no clues regarding the functional significance of the chemicals localized. In contrast, studies conducted in animals provide the opportunity for analysis of function as well as anatomy.

Even in animals, however, conclusions regarding chemical mediation of baroreflex responses are highly inferential. There is a bewildering array of chemicals thought to be involved in autonomic and baroreflex regulation, and understanding of functional interrelations among various chemicals (which may coexist in the same neurones) is highly conjectural. In what follows, we focus on several prime candidates for roles as chemical modulators of baroreflex function. We discuss in detail anatomical studies involving human specimens. For more complete treatments of chemical mediation of baroreflex function, the reader is referred to studies resulting from animal investigations, referred to below.

Catecholamines may be of particular importance. In humans, catecholamine areas have been identified with antibodies to tyrosine hydroxylase and phenylethanolamine-*N*-methyltransferase (Pearson *et al.* 1983; Arango *et al.* 1988), enzymes that are involved importantly in the synthesis of dopamine, noradrenaline and adrenaline; [3H]*para*-aminoclonidine (Unnerstall *et al.* 1984; Probst *et al.* 1985), an agent which binds to *alpha*2-adrenergic receptors; and antibodies to phenylethanolamine-*N*-methyltransferase (PNMT) (Lew *et al.* 1977; Kitahama *et al.* 1985; Arango *et al.* 1988), the final enzyme involved in the synthesis of adrenaline. These studies document the presence of substantial concentrations of catecholamines in areas of human brainstems that are probably important to baroreflex mechanisms, including the solitary tract nucleus, dorsal motor vagal nucleus, area postrema, locus coeruleus, hypothalamus, medullary raphe, ventrolateral medulla, the medullary ventral surface, and spinal intermediolateral cell columns.

Figure 3.9 is a montage of darkfield photomicrographs of a human brainstem stained with antibodies to tyrosine hydroxylase, published by Arango *et al.* (1988). The montage shows a tract of dense immunoreactivity extending from the region of the solitary tract nucleus (upper left) to the rostral ventrolateral medulla (lower right). This careful human study documents important localization of catecholamine neurones in medullary regions shown in animals to be involved in baroreflex regulation and sympathetic control.

Muscarinic *cholinergic* receptors are found in several areas of the human brain, including the solitary tract nucleus, the dorsal motor vagal and ambiguus nuclei, the parabrachial nuclei, and the hypothalamus (Cortes *et*

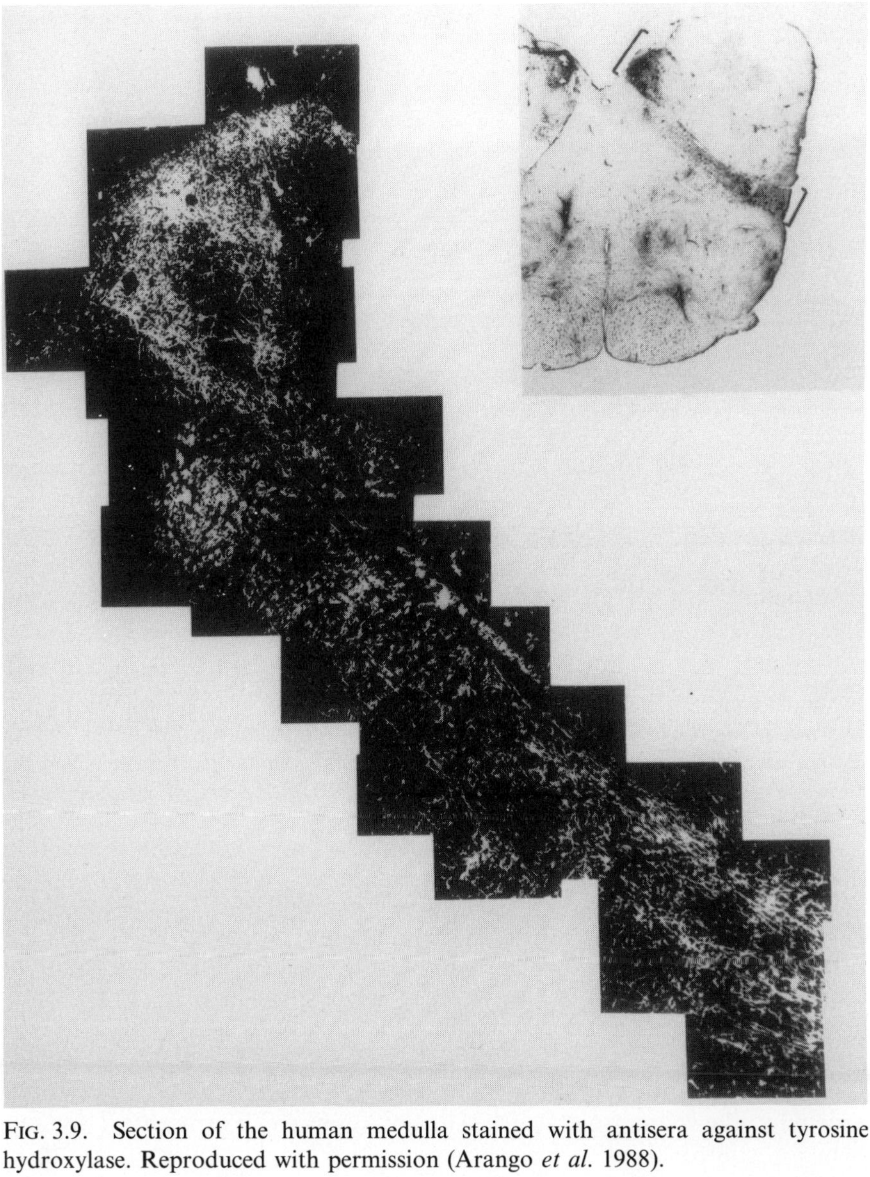

FIG. 3.9. Section of the human medulla stained with antisera against tyrosine hydroxylase. Reproduced with permission (Arango *et al.* 1988).

al. 1984, 1986; Hyde *et al.* 1988). An autoradiograph from a post-mortem human brain treated with the muscarinic receptor antagonist, *N*-[3H]methyl-scopolamine, published by Cortes *et al.* (1986), is shown in Fig. 3.10. Conceivably, the presence of muscarinic cholinergic receptors in this area of the human brainstem provides a substrate for central influences of cholinergic blocking drugs such as atropine sulphate, that readily cross the blood–brain

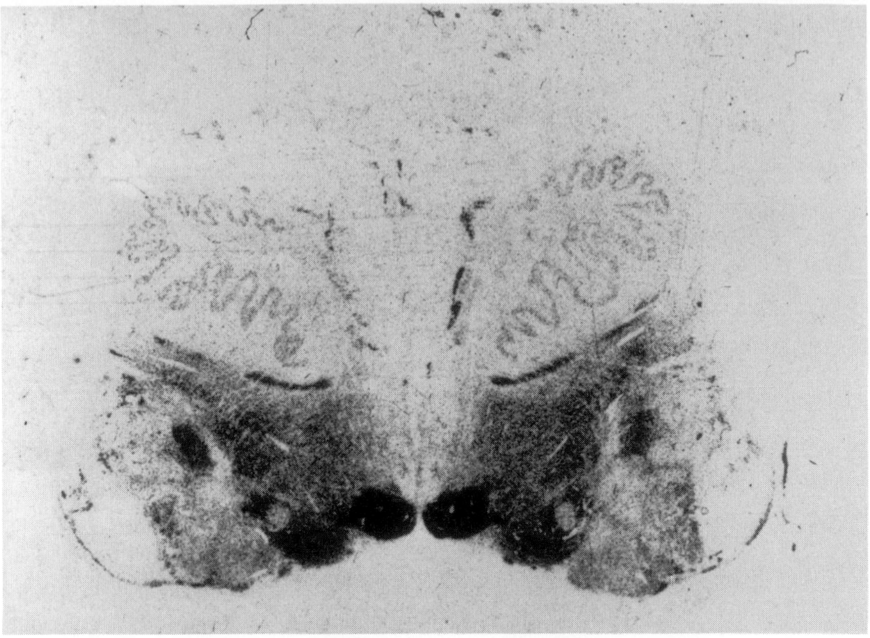

FIG. 3.10. Autoradiograph of a human medulla labelled with a muscarinic cholinergic blocker. The level of this coronal section is approximately the same as that depicted in Fig. 3.11. The dense staining in the central dorsal portion of the section (darker area) is located primarily in the regions of the solitary tract nucleus and the dorsal vagal motor and ambiguus nuclei. Reproduced with permission (Cortes *et al.* 1986).

barrier (see Chapter 8). There is substantial evidence from animals that acetylcholine participates in central modulation of sympathetic activity. Although acetylcholinesterase is concentrated most heavily in the region of the ambiguus nucleus, it is also found in other areas involved importantly in blood pressure regulation, including the locus coeruleus (Palkovits and Jacobowitz 1974).

The identity of the *neurotransmitter released by afferent baroreceptor traffic* arriving in the solitary tract nucleus is not known with certainty. This issue is highly controversial, in part because of the large number (over 20) of putative neurotransmitters identified in the solitary tract nucleus (Palkovits 1981; Howe 1985). Among these, one candidate has received particular attention: the excitatory amino acid *glutamate*. Several lines of evidence support a role for glutamate. The neurochemical and electrophysiological studies of Talman *et al.* (1980, 1984) and others (Siemers *et al.* 1982) show that glutamate is present in the solitary tract nucleus, and the studies of Monaghan and Cotman (1985) document the presence of [^{3}H]glutamate binding sites. Electrical vagus nerve stimulation appears to release glutamate in the solitary tract nucleus (Granata and Reis 1983; Granata *et al.* 1984). Microinjections of glutamate into the solitary tract nucleus mimic baroreceptor stimulation and provoke hypotension and bradycardia. (Baroreflex mimicry does not prove that glutamate is the primary transmitter, however, since injections of a variety of substances, including catecholamines (Zandberg *et al.* 1979), serotonin (W.A. Wolf *et al.* 1981), and acetylcholine (Criscione *et al.* 1983) into the solitary tract nucleus also provoke hypotension and bradycardia.) Injections of several putative glutamate antagonists abolish both sympathetically and vagally mediated baroreflex responses and provoke hypertension and cardioacceleration (Talman *et al.* 1981; S.J. Humphrey and McCall 1984; Guyenet *et al.* 1987). Finally, uptake of glutamate in the solitary tract nucleus is abolished by ganglionectomy and is enhanced by electrical stimulation of afferent vagal nerves (Talman *et al.* 1980, 1984).

Some studies dispute a role for glutamate as the neuroactive substance released by baroreceptor afferent activity. These studies (Leone and Gordon 1989; Talman 1989; Andresen and Yang 1990) support the notion that excitatory amino acids are involved in transmission of baroreceptor information in the solitary tract nucleus, but suggest that the specific transmitter is not glutamate.

Several other substances are present in the human solitary tract nucleus, and may be involved directly or indirectly in modulation of baroreceptor responses. These include: angiotensin II (Allen *et al.* 1988); serotonin (Palacios *et al.* 1983); substance P (Nomura *et al.* 1987; Rikard-Bell *et al.* 1990; Yew *et al.* 1990); oxytocin, vasopressin, and neurophysin (Sofroniew *et al.* 1981; Jenkins *et al.* 1984; Loup *et al.* 1989); enkephalin (Yew *et al.* 1990); atrial natriuretic peptide (Ermirio *et al.* 1989); and *gamma*-aminobutyric acid (GABA).

The case for *gamma-aminobutyric acid* as a neurotransmitter that inhibits the activity of baroreceptor neurones in the solitary tract nucleus seems to be particularly strong. Neurones in this region contain *gamma*-aminobutyric acid (Gale *et al.* 1980; Siemers *et al.* 1982; Maley and Newton 1985; Meeley *et al.* 1985; Q. Wang and Li 1988) and *gamma*-aminobutyric acid binding sites (Gale *et al.* 1980). The *gamma*-aminobutyric acid antagonist bicuculine potentiates baroreflex effects on the heart (Barman and Gebber 1979). Levels of *gamma*-aminobutyric acid appear to be increased by baroreceptor stimulation (Kubo and Kihara 1987). Microinjection or iontophoresis of *gamma*-aminobutyric acid into the solitary tract nuclei reduces arterial pressure (Kubo and Kihara 1987) and opposes the inhibitory effects of defence area stimulation on solitary tract neurones, which presumably are involved in baroreceptor reflex control (Jordan *et al.* 1988).

There also is evidence for a role for *gamma*-aminobutyric acid in modulating baroreflex-mediated changes of efferent vagal and sympathetic motoneurone activity. Glutamic acid decarboxylase, the enzyme that synthesizes *gamma*-aminobutyric acid (Ruggiero *et al.* 1985*a*) is present in neurones in the rostral ventrolateral medulla area that is critically involved in modulation of sympathetic activity. *Gamma*-aminobutyric acid binding sites are also present in the dorsal motor vagal and ambiguus nuclei, and the medullary reticular nuclei (Gale *et al.* 1980). The *gamma*-aminobutyric acid blocking agent bicuculine enhances baroreflex-mediated bradycardia (Gillis *et al.* 1980). Intravenous injections of *gamma*-aminobutyric acid antagonists increase blood pressure and cardiac sympathetic nerve activity (McCall 1986), and local applications or injections of *gamma*-aminobutyric acid antagonists to the ventral surface of the rostral ventrolateral medulla increase blood pressure (Reis *et al.* 1984*b*; C.A. Ross *et al.* 1984*b*; Sun and Guyenet 1987), and block inhibition of sympathetically related cells in the rostral ventrolateral medulla (Sun and Guyenet 1985) and renal sympathetic nerve activity (K.A. Yamada *et al.* 1984) caused by baroreceptor stimulation. Some evidence points to a site of action of *gamma*-aminobutyric acid rostral to the ventrolateral medulla. The effects of intravenous *gamma*-aminobutyric acid antagonists are abolished by midcollicular transections (McCall 1986). This suggests that *gamma*-aminobutyric acid inhibits forebrain structures that excite sympathostimulatory neurones projecting to the brainstem.

The presence of the *opiate* enkephalin in the solitary tract nucleus of humans (Yew *et al.* 1990), and the presence of opioid-containing nerve terminals (Kalia *et al.* 1984; Hökfelt *et al.* 1977) and receptors (Atweh and Kuhar 1977) in the solitary tract nucleus and hypothalamus of animals support a role for endogenous opiates in modulation of baroreflex function. Microinjections of opioid agonists into the solitary tract nucleus oppose sympathetically mediated pressor responses to bilateral carotid artery occlusion (Hassen and Feuerstein 1987) and sympathoinhibitory responses to electrical aortic nerve stimulation (Gordon 1990). Some evidence suggests that opioids are important in human autonomic regulation. The stereospecific *mu*-opioid receptor blocker, naloxone hydrochloride, opposes cardiovascular responses to clonidine administration (Farsang *et al.* 1984). Importantly, naloxone has also been proposed as a remedy for haemorrhagic hypotension to prevent vasovagal reactions (Holaday 1983). Hypotheses regarding possible central opiate participation in vasovagal reactions are discussed in Chapter 10.

Vagal nuclei

Most research into the anatomical location of vagal cardiac nuclei has been conducted in *cats*. It is clear that in this species, cardiac vagal nuclei are located primarily in the *ambiguus nucleus* (so-named because its borders are indistinct (Kōsaka 1909)). Horseradish peroxidase applied to the heart or transected cervical vagus nerves accumulates bilaterally in the dorsal motor and ambiguus nuclei (Kalia and Mesulam 1980; Hopkins 1987). During electrical stimulation of the aortic nerve or medial reticular formation, this area shows (by accumulation of 2-deoxyglucose) increased metabolic activity (Ciriello *et al.* 1986; Gonzalez-Lima 1988). Destruction of this nucleus causes degeneration of cardiac vagal fibres (Szentagothai 1952) and abolishes baroreceptor–cardiac reflex responses (T.M. Lee *et al.* 1972). Electrical stimulation of the dorsal motor vagal nucleus usually does not provoke bradycardia (Calaresu and Pearce 1965), and destruction of this nucleus does not abolish baroreflex bradycardia (Kerr 1967). Moreover, electrical stimuli applied to the ambiguus nucleus to provoke bradycardia also trigger antidromic activity in efferent cardiac vagal motoneurones (McAllen and Spyer 1976). There appears to be some laterality of cardiomotor nuclei in ambiguus nuclei; electrical stimulation of the right ambiguus nucleus inhibits primarily the sinoatrial node, and stimulation of the left inhibits primarily the atrioventricular node (M.E. Thompson *et al.* 1987).

In cats, neurones in the ambiguus nucleus with cardiac periodicities and projections to cardiac vagal branches have conduction velocities of about 3 to 16 m/s and probably comprise small myelinated fibres. It is likely that small numbers of these motoneurones exert great effects upon heart rate; they are located in a very small circumscribed region of the ventrolateral ambiguus nucleus (Hopkins 1987; Spyer and Jordan 1987) whose major input appears to be from the solitary tract nucleus (Stuesse and Fish 1984), and cardiac slowing predictably follows their spontaneous firing (McAllen and Spyer 1978*b*).

Controversy exists regarding the possibility that efferent cardiac vagal nuclei are located also in the *dorsal motor nucleus* (Kalia 1981). Published evidence suggests that in dogs and monkeys, vagal–cardiac motonuclei arise from both ambiguus and dorsal motor nuclei (Gunn *et al.* 1968; G.A.G. Mitchell and Warwick 1955). The location of vagal–cardiac motoneurones in humans is unknown. Medullary locations involved with autonomic cardiovascular control are shown in Fig. 3.11.

There is some evidence that the level of 'tonic' activity of vagal–cardiac motoneurones depends on the level of stimulatory autonomic sensory input (Polosa *et al.* 1979). Kunze (1972) showed in chloralose-anaesthetized cats that vagal–cardiac motoneurones that are silent at usual blood pressures fire when arterial pressure is raised to extremely high levels (mean pressures above 150 mmHg). However, it should be noted that vagal–cardiac motoneurones are exquisitely sensitive to general anaesthetics (S.F. Vatner *et al.* 1971), and their silence may well be artifactual, secondary to anaesthesia. Iriuchijima and Kumada documented tonic firing in the great majority (54 of 58) of vagal–cardiac neurones when morphine and chloralose were used for anaesthesia (Iriuchijima and Kumada 1963).

There is also evidence that of autonomic sensory inputs that stimulate vagal–cardiac motoneurones, those from baroreceptors are very important, if not prepotent. Kollai and Koizumi recorded vagal–cardiac nerve traffic in

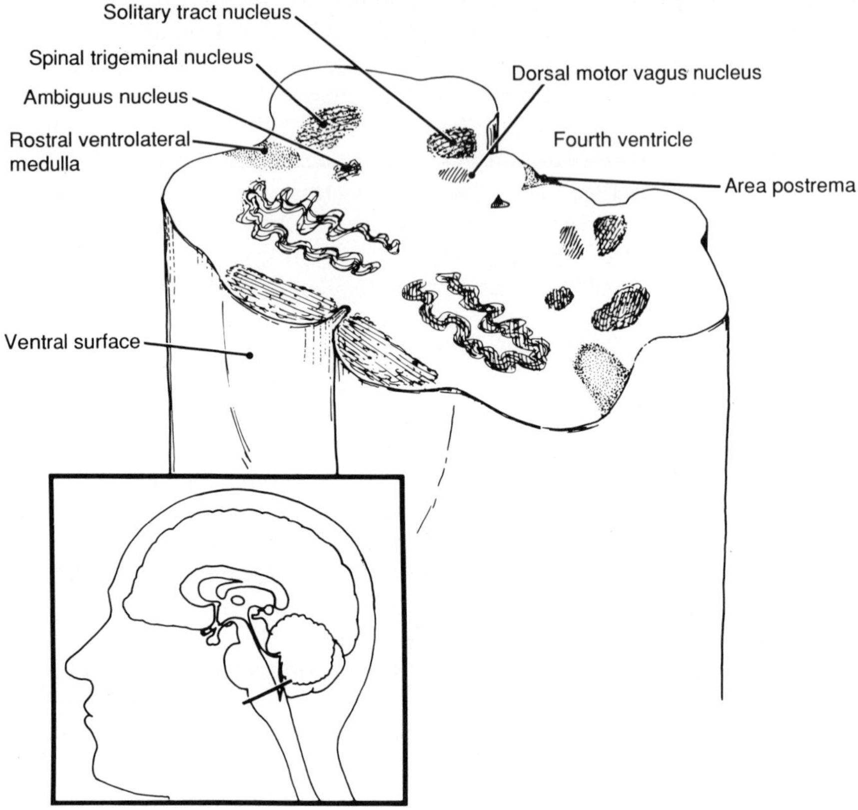

FIG. 3.11. Anatomical location of medullary structures involved in baroreflex control. These anatomical landmarks are shown in a human, but are derived from experiments conducted in animals, after D.L. Brown and Guyenet (1984).

chloralose-anaesthetized dogs and showed progressive diminution of firing rates as individual baroreceptor (and chemoreceptor) afferent tracts were cut (Kollai and Koizumi 1979). From this study, it did not appear that vagal–cardiac nuclei are totally dependent for their activity upon stimulation by baroreceptors, because some activity persisted after transection of all baroreceptor afferent nerves. Eckberg *et al.* (1988) showed in unanaesthetized human volunteers that both R–R interval and R–R interval standard deviations (non-invasive estimates of vagal–cardiac traffic — see Chapter 8) decline to low levels in parallel, during progressive reductions of arterial pressure.

Sympathetic nuclei

All preganglionic sympathetic motonuclei are located in the spinal cord. The majority of cardiovascular sympathetic nuclei are located in or near the intermediolateral column of the thoracic and upper lumbar segments (Calaresu *et al.* 1975). This is shown in Fig. 3.12. As with neurones in the rostral ventrolateral medulla (Lovick 1987; Dampney and McAllen 1988), sympathetic preganglionic neurones in the spinal cord appear to be organized according to target organ (Barman and Wurster 1975). Sympathetic impulses modulating sinus node function, ventricular contractility, and arteriolar vasoconstriction arise from nuclei in multiple spinal segments, but are concentrated primarily between the fifth cervical and the fourth lumbar segments (Calaresu *et al.* 1975). In cats, neurones involved with cardioacceleration are located between the first and eighth spinal thoracic segments (J.L. Henry and Calaresu 1972). Almost all of these latter neurones are located on the right-hand side (J.L. Henry and Calaresu 1974).

The principal neurotransmitter for sympathetic *pre*ganglionic neurones is

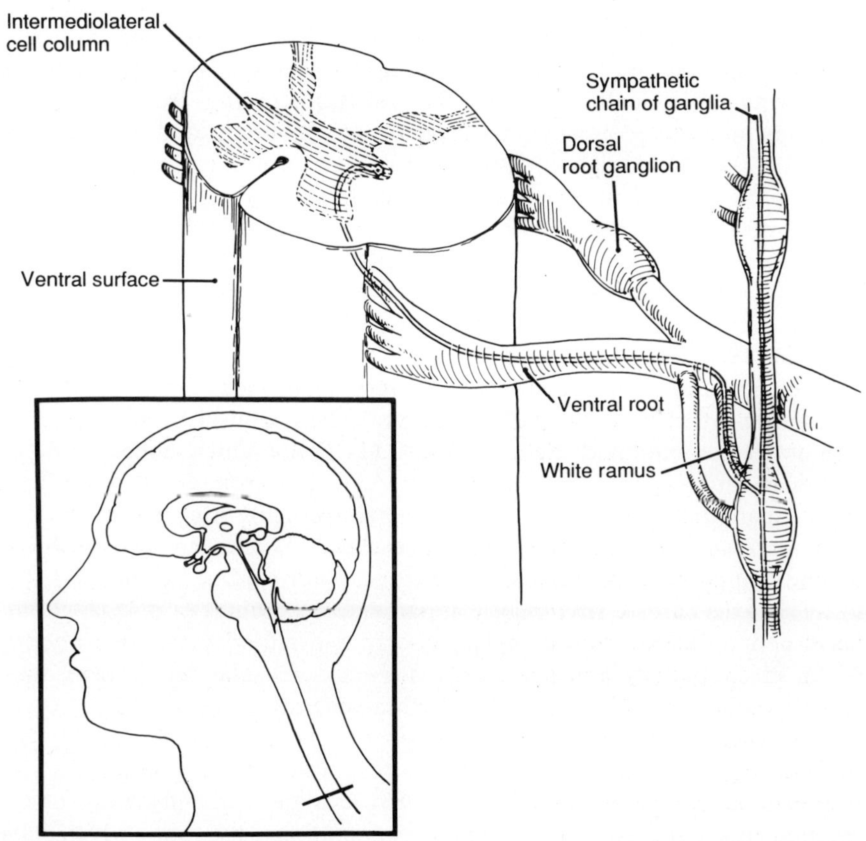

FIG. 3.12. Location of sympathetic motoneurones in spinal cord. These anatomical landmarks are shown in a human, but are derived from experiments conducted in animals, after Caverson *et al.* (1984).

acetylcholine (Feldberg and Gaddum 1934). Although the principal neurotransmitter for sympathetic *post*ganglionic neurones is noradrenaline, neurones supplying skin (Dale and Feldberg 1934) and muscle (Bolme *et al.* 1970) release acetylcholine. Moreover, immunohistochemical analyses have shown that pre- and postganglionic sympathetic neurones containing classical neurotransmitters also contain a host of peptides (Hökfelt *et al.* 1980; Burnstock 1986; Schultzberg and Lindh 1988) which may function as co-transmitters or neuromodulators. Inferential evidence from humans suggests that acetylcholine (Zelis *et al.* 1976; Sanders *et al.* 1989*b*) and neuro peptide Y (Eckberg *et al.* 1988) are released by postganglionic muscle sympathetic neurones.

Most studies of preganglionic sympathetic traffic indicate that the majority (perhaps 70 per cent) of these fibres are silent (Koizumi and Sato 1972; Spyer 1981; Jänig and Szulczyk 1980), despite the presence of subthreshold stimulatory inputs (McLachlan and Hirst 1980). Of fibres with tonic activity, many appear to fire irregularly, and at wide intervals (Coote and Westbury 1979; Polosa *et al.* 1979), and fibres that discharge in synchrony with cardiac activity may be difficult to locate (Seller 1973). Recordings from such active fibres document cardiac rhythms in about 15 (Mannard and Polosa 1973; Seller 1973) to 60 (Jänig and Szulczyk 1980) per cent. Respiratory modulation is present in about 50 per cent of such fibres (Polosa *et al.* 1979). This stands in marked contrast with the spontaneous firing (about 90 per cent (Koizumi and Sato 1972)) and obvious synchronization found in postganglionic sympathetic nerves in animals (Irisawa and Ninomiya 1967; Koizumi and Sato 1972; Barman and Gebber 1976; Jänig *et al.* 1983) and humans (Eckberg 1976; Porter *et al.* 1990).

Preganglionic spinal sympathetic neurones can be located with antidromic stimulation of cervical sympathetic nerves. These neurones can be excited (see above) or inhibited (Seller 1973; D.G. Taylor and Gebber 1975) by electrical stimuli applied in the brainstem. As with neurones in the brainstem, cardiac periodicity can be found with signal-averaging triggered by electrocardiographic R waves or arterial pulses (Seller 1973). It is unusual, however, to find sympathetic preganglionic neurones with cardiac periodicity; it is likely that the cardiac rhythmicity apparent in recordings of postganglionic nerve activity arises from complex, poorly understood ganglionic filtering of encoded, spatially and temporally dispersed irregular inputs from individual components of sympathetic rhythm generating networks.

Spontaneous, infrequent, irregular firing of postganglionic muscle sympathetic nerves recorded in patients with chronic, high, complete spinal transections (Stjernberg and Wallin 1983) may be a manifestation of the spontaneous firing patterns of spinal preganglionic neurones which have been freed from central, primarily baroreceptor-controlled influences. Recordings of muscle and skin sympathetic nerve activities from one spinal injury patient are shown in Fig. 3.13 (Stjernberg *et al.* 1986).

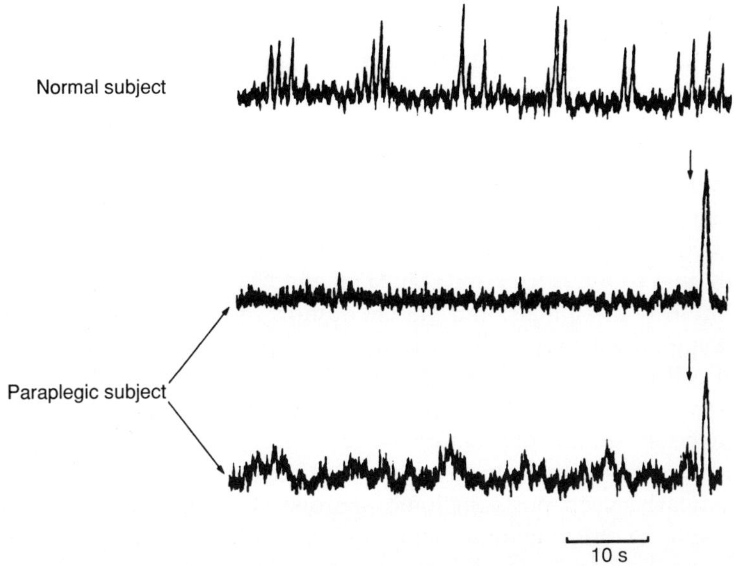

FIG. 3.13. Muscle sympathetic nerve activity in healthy subject (top tracing) and patient with a complete high spinal transection (lower two tracings). The arrow indicates the onset of urinary bladder compression. Such data provide evidence that in healthy subjects, higher centres stimulate and entrain spinal sympathetic motoneurones. Reproduced with permission (Stjernberg *et al.* 1986).

Connections between the brainstem and higher structures and spinal sympathetic motonuclei are important to sympathetic regulation. Anatomical and electrophysiological evidence (summarized by Loewy (1982)) points towards the existence of monosynaptic connections between sympathetic preganglionic neurones in the intermediolateral cell column and solitary tract nuclei (Loewy and Burton 1978), other regions of the medulla (Dählstrom and Fuxe 1965; J.L. Henry and Calaresu 1974), and hypothalamus (Saper *et al.* 1976). Histochemical stains show that these pathways contain a variety of neurotransmitters, including noradrenaline, adrenaline, and serotonin (Dählstrom and Fuxe 1965; Loewy and McKellar 1980).

It has long been recognized that electrical stimulation of certain portions of the medulla increases blood pressure (Ranson and Billlingsley 1916; R.S. Alexander 1946; Gebber *et al.* 1973; J.L. Henry and Calaresu 1974). Foreman and Wurster (1973) showed that destruction of descending spinal sympathetic pathways, which interrupts fibres from the medulla, reduces blood pressure severely. A clinical counterpart of this physiology may be found in patients with permanent complete high cervical spinal cord transections, who have severely reduced levels of baseline (muscle) sympathetic nerve traffic (Stjernberg and Wallin 1983), and low circulating plasma catecholamine levels and orthostatic hypotension (Frankel *et al.* 1972; Mathias *et al.* 1975). As discussed above, these experimental and clinical observations suggest that higher central nervous system structures function in part to *stimulate* preganglionic sympathetic motoneurone firing.

The anatomy of sympathetic fibres as they leave the spinal cord is illustrated in Fig. 3.12. Preganglionic sympathetic neurones may be myelinated or unmyelinated and may have widely varying conduction velocities (Fernandez de Molina *et al.* 1965; Polosa 1967; Wurster 1977; Coote 1988). Conduction velocities of preganglionic spinal sympathetic neurones have been found to be about 3 to 5 m/s for renal (McAllen and May 1990) and muscle (Jänig 1985) outflows in animals, and muscle outflow in humans (Wallin and Rea 1988). Sympathetic preganglionic fibres derive primarily from the intermediolateral cell column and leave the spinal cord in the ventral roots. They then form white rami and terminate in one or more sympathetic ganglia in the paravertebral chain of sympathetic ganglia. Each preganglionic sympathetic fibre is thought to connect with about 300 to 400 postganglionic neurones; similarly, each postganglionic neurone is thought to receive inputs from about 25 preganglionic neurones, each of which may derive from several spinal segments (Faden and Petras 1978; Polosa *et al.* 1979). Most sympathetic postganglionic fibres re-enter spinal nerves and travel with other efferent and afferent fibres to the periphery.

Postganglionic sympathetic fibres comprise unmyelinated fibres which are believed to travel in mixed peripheral nerves in bundles. In the peroneal nerve, sympathetic fibres destined for muscle beds appear to travel with other afferent and efferent nerves in fascicles subserving muscle. Similarly, sympathetic fibres destined for skin tend to travel with other skin nerves in discrete fascicles. An electronmicrograph of a cross-section of a human peroneal nerve is shown in Fig. 8.1.

SUMMARY

Nerves presumed to function as baroreceptors in humans and other species are most abundant in the adventitia of the carotid sinus and aorta, which areas (in absence of atherosclerosis) are especially distensible and elastic. Putative baroreceptor terminations are complex structures, which are packed with mitochondria and other organelles, and enveloped by elastic layers interwoven between and among planes of collagen and smooth muscle.

Afferent information regarding stretch of baroreceptors arrives in the central nervous system primarily in the ipsilateral solitary tract nucleus. Traffic carried over individual afferent baroreceptor neurones arrives at different times, and the activity profiles of interneurones receiving baroreceptor information tend to lack pulse-synchrony. Absence of pulse synchrony is a feature of most central neurones involved in baroreflex function; mechanisms responsible for reconstitution of pulse-synchronous firing in efferent preganglionic vagal and postganglionic sympathetic neurones are understood poorly. There are surprisingly numerous, accurate maps showing locations of chemical mediators in human brains; however, the precise roles of these many substances in baroreflex control is understood poorly (a qualification that may apply to data from animals as well as those from humans). Several central structures are involved importantly in baroreflex modulation in animals, including the hypothalamus and amygdala, the parabrachial nuclei, the subfornical organ, the cerebellum, and the rostral ventrolateral medulla. Many interconnections among these structures are monosynaptic; although the presence of monosynaptic linkages may point to the importance of specific regions in baroreflex control, we know of no attempts to test this notion scientifically. The location of vagal–cardiac motoneurones in humans is unknown. In cats these neurones are found exclusively in the ambiguus nucleus; in other species, they are found also in the dorsal motor nucleus of the vagus. Sympathetic preganglionic motoneurones are located in the intermediolateral column of the spinal cord.

PART B

Methods for Human Research

4

Valsalva's manoeuvre

And men must walk at least before they dance.
Alexander Pope, Letter to Lord Bolingbroke

A diverse and powerful array of quantitative techniques is available for research into human arterial baroreflex control mechanisms: afferent sensory input can be modified by numerous physical and pharmacological manoeuvres; central autonomic modulation can be altered; changes of efferent autonomic activity can be measured directly or inferred from changes of effector organ function; and efferent mediation of responses can be inferred from changes of response patterns produced by autonomic blocking drugs or from measurements of neurotransmitters.

This is the first of two chapters devoted to methods used to modify arterial baroreceptor afferent traffic in humans. The research method described in this chapter, the Valsalva manoeuvre, is relatively non-selective; its use modifies the activity of several major receptor groups, including arterial baroreceptors. Techniques described in the next chapter modify the activity of arterial baroreceptors more selectively than the one described in this chapter. However, differences among these tests are probably more quantitative than qualitative, since it is likely that *all* experimental interventions available currently alter the activity of multiple receptor groups.

The effects of straining against a closed glottis were known to the ancient Romans. For example, Caesar's disconsolate prisoner, Cato of Utica, discoursed about methods available to him for committing suicide (he ultimately used his dirk):

When they besought him to do no violence to himself but to go to sleep without his dirk, he replied still more plausibly, 'Could I not strangle myself with my clothing if I wished to, or knock my brains out against the wall, or throw myself headlong to the ground, or *destroy myself by holding my breath*?' (Appian of Alexandria 1958).

Dawson (1943) published an excellent history (which, however, has a spectacularly inaccurate bibliography) of Valsalva's manoeuvre. In it, he quotes the anatomist Giovanni Battista Morgagni's reference to the writings of his mentor, Antonio Maria Valsalva (1666-1723). 'If the glottis be closed after a deep inspiration, and a strenuous and prolonged expiratory effort be then made, such pressure can be exerted upon the heart and intrathoracic vessels that the movement and flow of the blood are temporarily arrested.'

It is likely that Valsalva's manoeuvre is the most widely used test of human arterial baroreflex function; responses to this test have been studied extensively in normal subjects and patients with cardiovascular diseases. The Valsalva manoeuvre holds great attraction because it is safe (Sharpey-Schafer 1955; Levin 1966), can be performed without sophisticated equipment, and yields reproducible, quantitative results (Levin 1966; Baldwa and Ewing 1977). The uniquely human Valsalva manoeuvre triggers haemodynamic responses by means of abrupt transient voluntary elevations of intrathoracic and intra-abdominal pressures provoked by straining. A representative Valsalva manoeuvre (M.L. Smith, J.M. Fritsch, L.A. Beightol, K.A. Ellenbogen, and D.L. Eckberg, unpublished) is illustrated in Fig. 4.1.

This figure shows typical arterial pressure changes (in this case measured

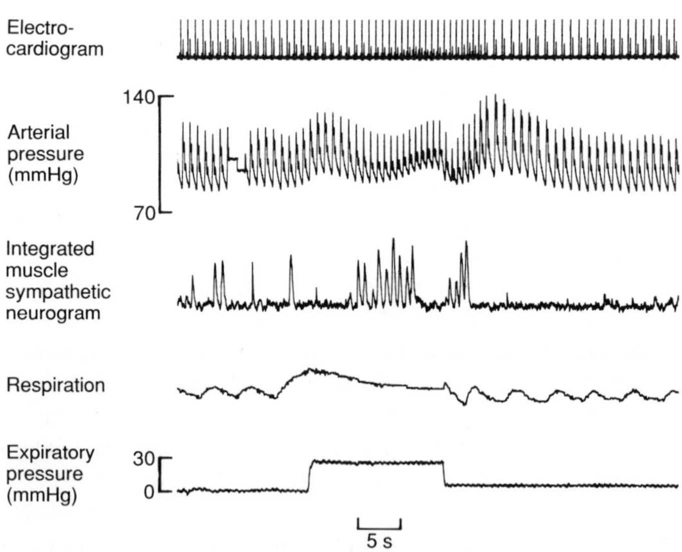

FIG. 4.1. Valsalva manoeuvre performed by one healthy young subject. Reproduced with permission (M.L. Smith *et al.* 1989*b*).

with a non-invasive photoplethysmographic device (Imholz *et al.* 1988)). In general, muscle sympathetic nerve activity changes in reciprocal relation to changes of arterial pressure. Release of Valsalva straining brings a steep pressure elevation and sympathetic neural silence, which continues beyond the time when arterial pressure returns to baseline levels (M.L. Smith *et al.* unpublished).

THE VALSALVA MANOEUVRE IS HETEROGENEOUS

This widely used test, as described in published reports, is surprisingly heterogeneous; the methods used, and therefore the results obtained (Leon *et al.* 1970; Korner *et al.* 1976) are widely variable. Straining may be initiated after a maximal inspiration (Sharpey-Schafer 1955; D.S. Goldstein *et al.* 1982), after a 'full' inspiration (Burger 1926; Judson *et al.* 1955), or at the end of a normal inspiration (Korner *et al.* 1976; M.L. Smith *et al.*, unpublished). (Tidal volume has not been measured in most published studies and therefore the actual depth of the breath preceding the test is usually unknown.) Mouth pressure during the period of straining (which provides an excellent estimate of intrapleural pressure (Elisberg *et al.* 1951)) is commonly raised to 40 mmHg (Ruskin *et al.* 1968; Brooker *et al.* 1974; Parisi *et al.* 1976; Pratt *et al.* 1983), but may be raised by a 'vigorous expiratory effort' (Delius *et al.* 1972*b*), or to 'a constant level' (D.J. Stone *et al.* 1965), to 40-50 mmHg (McIntosh *et al.* 1954), to 40-60 mmHg (Judson *et al.* 1955), or to a level in excess of 35 mmHg (Leon *et al.* 1970). The duration of straining is commonly 10 s (Judson *et al.* 1955; Stucki *et al.* 1955; Duke *et al.* 1976), but it may be 12-15 s (Brooker *et al.* 1974), 15 s (D.G. Greene and Bunnell 1953; Elisberg 1963; M.L. Smith *et al.*, unpublished), 20 s (Burger 1926; Parisi *et al.* 1976), 15-30 s (Stucki *et al.* 1955), 30 s (Stegemann *et al.* 1988), 40 s (Fernandez *et al.* 1988), 22-73 s (McIntosh *et al.* 1954), or for 'as long as the patient is able' (H. Goldberg *et al.* 1952). In some studies, the duration (Fernandez *et al.* 1987) or the intensity (Korner *et al.* 1979; M.L. Smith *et al.*, unpublished) is varied to elaborate stimulus–response relations.

Other aspects of Valsalva testing, including the position of the glottis during the period of straining, may not be uniform among studies. Some protocols employ a small vent in the mouthpiece that allows air to escape during the Valsalva manoeuvre through an open glottis, and thus prevents volunteers from increasing mouth rather than chest pressure (Mackenzie *et al.* 1931; S.A. Smith *et al.* 1987). Also, although the goal of this modification is laudatory, it may introduce an additional (possibly only theoretical) complexity, because as straining continues, the level of exertion and fatigue increase and afferent traffic from cardiopulmonary and thoracic stretch receptors diminishes. It is possible that the breathing pattern after release of Valsalva straining influences important late reflex responses (Eckberg and

Orshan 1977; Eckberg *et al.* 1980). Some studies (S.A. Smith *et al.* 1987, M.L. Smith *et al.*, unpublished) have been done in which subjects attempted to control breathing in the post-straining period; these attempts met with only partial success, because straining accompanied by asphyxia provides subjects with a powerful stimulus to breathe deeply and rapidly. Although it is clear that body position during Valsalva straining exerts an important influence on haemodynamic responses (Mäntysaari *et al.* 1984; Stegemann *et al.* 1988; Ten Harkel *et al.* 1990), most reports do not indicate subjects' positions.

THE FOUR PHASES OF THE VALSALVA MANOEUVRE

Notwithstanding methodological differences that exist among investigators who use the Valsalva manoeuvre, there appears to be universal acceptance of the proposal of W.F. Hamilton *et al.* (1936) that responses of normal subjects to the manoeuvre be divided into four phases. These sequential phases (shown clearly in Fig. 4.1) are as follows:
(1) a brief rise of arterial pressure and reduction of heart rate immediately after the onset of straining;
(2) a fall, and later partial or complete recovery, of arterial pressure to baseline levels, and a cardioacceleration during the period of straining;
(3) a sudden, very brief further reduction of arterial pressure and elevation of heart rate immediately following the release of straining;
(4) a terminal, sustained elevation of arterial pressure above control levels and concomitant slowing of heart rate.

Three aspects of Valsalva manoeuvre pressure changes deserve emphasis. First, with lesser degrees of straining, systolic and diastolic pressures *may remain above baseline levels* throughout Valsalva straining (Paulev *et al.* 1988; M.L. Smith *et al.*, unpublished). Second, pressure changes described above apply to pressures measured in systemic arteries; absolute distending pressures in the aorta, pulmonary vessels, and heart are probably *below* baseline levels throughout straining. Third, excepting the pressure changes that occur at the onset of straining (when carotid distending pressures rise and intrathoracic aorta distending pressures fall), pressures in systemic arteries and intrathoracic vessels *change* in the same direction. Figure 4.2 depicts transoesophageal echographic changes of thoracic aorta circumference in one subject provoked by four identical Valsalva manoeuvres.

Phase 1. It is likely that several mechanisms contribute to abrupt elevation of arterial pressure during the few seconds of Phase 1. Increased intrathoracic and intra-abdominal pressures compress the aorta and probably propel relatively incompressible blood into the peripheral vascular bed (Little *et al.* 1985). Left ventricular stroke volume may (Ruskin *et al.* 1968; D. Robertson *et al.* 1977) or may not (Greenfield *et al.* 1967) increase. Increased left ventricular ejection does not appear to result importantly from squeezing

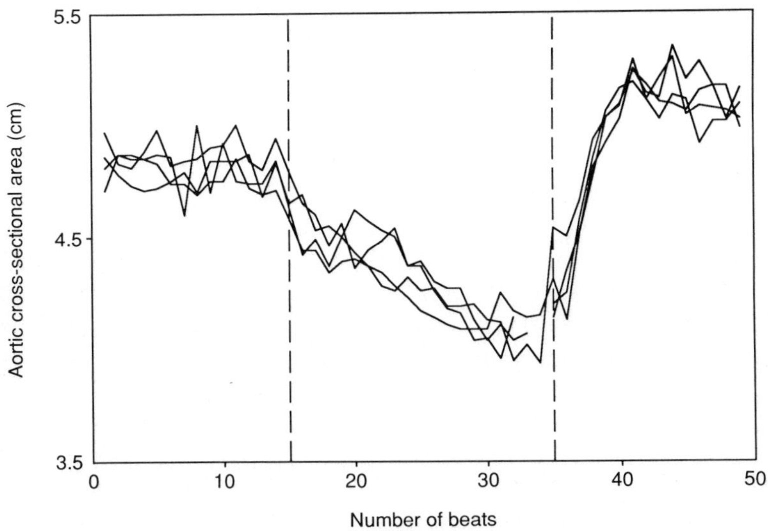

FIG. 4.2. Oesophageal echographic thoracic aorta cross-sectional area during four 15 s periods of Valsalva straining (dashed lines) at −30 mmHg in one healthy subject.

of blood from the pulmonary vascular bed into the left ventricle (Sarnoff *et al.* 1948), since increases of left ventricular end-diastolic volume during Phase 1 are small and variable (D. Robertson *et al.* 1977). An alternative explanation is that reduced afterload transiently improves left ventricular function (Buda *et al.* 1979; Karam *et al.* 1984).

Peripheral arterial pressure increases with the onset of straining (Murgo *et al.* 1981). Arterial pressure elevation during Phase 1 of the Valsalva manoeuvre does not appear to be mediated sympathetically, because it is not preceded by elevations of (muscle) sympathetic nerve activity (Delius *et al.* 1972*b*; M.L. Smith *et al.*, unpublished), and occurs in normal subjects after administration of alpha-adrenergic blocking drugs (Korner *et al.* 1976), and in patients with high cervical spinal cord transections (W.E. Watson 1962). Elevations of arterial pressure, which may be as large as increases of intrathoracic pressure (W.F. Hamilton *et al.* 1936), are associated with brief prolongations of R–R intervals. A minority of subjects experience cardioacceleration during Phase 1, presumably secondary to the exertion involved (J.M. Karemaker, unpublished). R–R interval prolongation is neurally mediated, because it is virtually absent in transplanted human hearts (Bernardi *et al.* 1989), and it results from increased vagal–cardiac nervous activity, since it is prevented by pretreatment with large doses of atropine sulphate (Leon *et al.* 1970). Although reductions of efferent sympathetic nervous outflow occur (Fig. 4.1 (Delius *et al.* 1972*b*; M.L. Smith *et al.*, unpublished)), they probably do not contribute to R–R interval prolongation, because of the relatively long latency between changes of sympathetic–cardiac traffic and sinus node responses (Spear and Moore 1973).

Phase 2. The arterial pressure elevation of Phase 1 is succeeded by pressure reduction in Phase 2, as venous return to the heart is impeded, venous valves close, and venous blood accumulates in the limbs (Candel and Ehrlich 1953; Stucki *et al.* 1955; Sharpey-Schafer 1965) and neck (Burger 1926), and peripheral venous (Korner *et al.* 1976; Korner *et al.* 1979) and cerebrospinal (Burge 1921; W.F. Hamilton *et al.* 1944) pressures rise. Vein-to-artery circulation time is increased (Stucki *et al.* 1955). Cardiac (Dietlen 1906), left atrial (D. Robertson *et al.* 1977; M.W. Cooper *et al.* 1980) and left ventricular (Brooker *et al.* 1974; Parisi *et al.* 1976; D. Robertson *et al.* 1977; M.W. Cooper *et al.* 1980) dimensions, left ventricular stroke volume (Ruskin *et al.* 1968; Brooker *et al.* 1974; Parisi *et al.* 1976; Little *et al.* 1985) and contractility (Hoshino *et al.* 1988), aortic cross-sectional area (Greenfield *et al.* 1967), and arterial pressure decline. Reductions of arterial pressure may be by as much as 80 per cent of pre-Valsalva levels (Sharpey-Schafer 1965). During Phase 2, pressures in all baroreceptive areas fall.

Cardiac output falls (Korner *et al.* 1976) despite cardioacceleration (caused primarily by withdrawal of cholinergic restraint (Leon *et al.* 1970)). Efferent sympathetic nervous outflow to limb muscles (Delius *et al.* 1972*b*; M.L. Smith *et al.*, unpublished) (but probably not skin (Roddie *et al.* 1957)), and total peripheral resistance (Mellander and Johansson 1968; Delius *et al.* 1972*b*; Korner *et al.* 1976; D.S. Goldstein *et al.* 1986; S.A. Smith *et al.* 1987) increase. Increases of sympathetic activity are proportional to the duration (Fernandez *et al.* 1987) and intensity (M.L. Smith *et al.*, unpublished) of straining. Within about 4 s after the augmentation of efferent sympathetic activity, the decline of arterial pressure is arrested (Delius *et al.* 1972*b*), peripheral vascular resistance increases (Korner *et al.* 1976), arterial pressure begins to return towards control levels (despite the continued decline of left ventricular stroke volume (Parisi *et al.* 1976)), and muscle sympathetic nerve activity declines (M.L. Smith *et al.*, unpublished). For unexplained reasons, arterial pressure *and* muscle sympathetic nerve activity increase during the final seconds of Valsalva straining (M.L. Smith *et al.*, unpublished).

Although Valsalva manoeuvres last only seconds, there is considerable evidence that ongoing reflex activity modifies the baroreflex stimulus. Arterial pressure falls more if cardioacceleration is prevented by cardiac autonomic blockade with propranolol and atropine (Korner *et al.* 1976) or atropine alone (D.S. Goldstein and Keiser 1984), or if reflex vasoconstriction is absent because of prior sympathectomy (Wilkins and Culbertson 1947) or disease (Sharpey-Schafer 1956), or prevented by pretreatment with tetraethylammonium chloride (D.G. Greene and Bunnell 1953) or phentolamine and guanethedine (Korner *et al.* 1976).

Phase 3. Immediately after expiratory pressure is released, peripheral arterial pressure falls transiently; this brief period forms Phase 3. Pressure falls after alpha-adrenergic blockade (Korner *et al.* 1976) and, therefore, as with the initial pressure elevation, results from mechanical rather than neural factors, including sudden augmentation of left ventricular afterload (Buda *et al.* 1979). During this phase, there is an additional, sometimes large burst of efferent muscle sympathetic nerve activity (Wallin *et al.* 1974*a*; M.L. Smith *et al.*, unpublished) and a further minor increase of heart rate.

Phase 4. During Phase 4, venous return to the heart (Wexler *et al.* 1968), left ventricular stroke volume (Brooker *et al.* 1974; Parisi *et al.* 1976), and cardiac output (Greenfield *et al.* 1967; Ruskin *et al.* 1968; Brooker *et al.* 1974) return towards normal levels. Systemic and pulmonary (G.deJ. Lee *et al.* 1954) arterial pressures return to and then rise above control levels, because now normal left and right ventricular stroke volumes are ejected into constricted systemic and pulmonary arterial beds. In the systemic circulation, this rise of pressure is prevented or reduced substantially by pharmacological adrenergic blockade (Korner *et al.* 1979) or surgical sympathectomy (Wilkins and Culbertson 1947). In the pulmonary circulation, the rise is unaffected by ganglionic blockade (G.deJ. Lee *et al.* 1954). Arterial pressure elevation leads to protracted reduction of efferent muscle sympathetic nerve activity (Delius *et al.* 1972*b*; M.L. Smith *et al.*, unpublished) and to bradycardia (mediated by increased cholinergic restraint (Elisberg *et al.* 1953; Leon *et al.* 1970)).

Although the Valsalva manoeuvre has been used almost exclusively as a means to study vagal baroreflex mechanisms, there is growing support for use of Valsalva straining as a quantitative tool for gauging baroreflex control of *sympathetic* outflow. M.L. Smith *et al.*, unpublished) showed that increases of muscle sympathetic nerve activity are proportional to the intensity of straining, and, importantly, that Phase 4 pressure elevations are proportional to preceding increases of sympathetic traffic. The latter observation suggests that it may be appropriate to use Phase 4 pressure elevations as indices of the magnitude of preceding sympathetic responses to straining. Fernandez *et al.* (1988) showed that after prolonged Valsalva straining (40 s), elevations of plasma noradrenaline and adrenaline levels are proportional to the intensity of straining.

AUTONOMIC BASES FOR RESPONSES TO THE VALSALVA MANOEUVRE

The Valsalva manoeuvre is a very complex physiological perturbation that alters intracardiac and intravascular distending pressures (sometimes in opposite directions), afferent barosensory traffic, central modulation of autonomic traffic, arterial pressure, and heart rate. Therefore this test has the potential to yield a wealth of information regarding autonomic function.

Figure 4.3 shows average peripheral arterial pressure and peroneal nerve muscle sympathetic activity from one subject during five periods of 30 mmHg Valsalva straining for 15 s. In this subject, diastolic pressure remained *above* baseline levels during straining, and sympathetic activity varied in relation to changes, rather than absolute levels of pressure. Sympathetic bursts tended to be absent in the post-strain period, even after pressure had returned to baseline levels.

Arterial pressure elevation during Phase 1 appears to be due to mechanical rather than neural factors. It is likely that during this phase, afferent traffic from carotid baroreceptors increases and afferent traffic from aortic baroreceptors decreases. Therefore cardiac deceleration probably results exclusively from increased afferent traffic from *carotid* baroreceptors. There are theoretical reasons for preferring Phase 1 to subsequent phases of the Valsalva manoeuvre for analysis of human baroreceptor–cardiac reflex responses. The arterial pressure stimulus arises from mechanical rather than neural factors; therefore, it occurs in patients with cardiovascular diseases, and in patients who may not have intact efferent sympathetic neural pathways. This phase is brief, however, and its quantitative analysis requires beat-by-

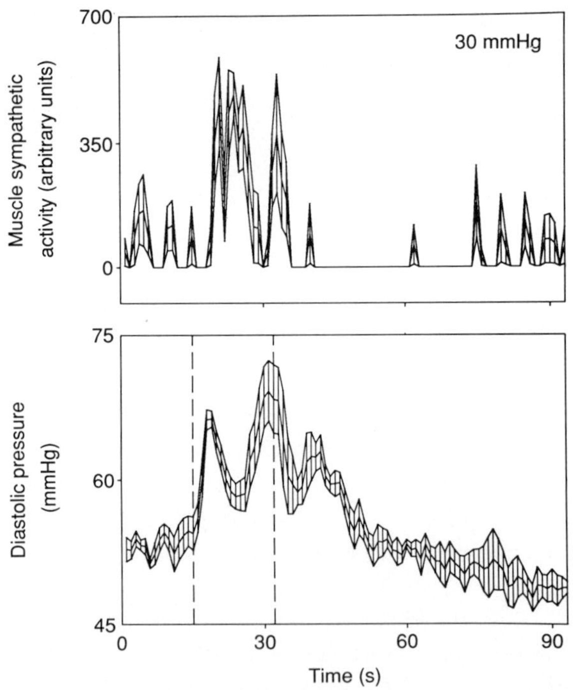

FIG. 4.3. Average peroneal nerve muscle sympathetic activity and diastolic pressure in one subject during five Valsalva manoeuvres of 30 mmHg for 15 s (dashed lines). Reproduced with permission (M.L. Smith *et al.* 1989*b*).

beat measurement of arterial pressure. As will be seen, some registration of arterial pressure is necessary for any interpretation of Valsalva responses; the intensity of straining probably does not provide an adequate index of the intensity of the stimulus for Valsalva responses.

Until recently, beat-by-beat measurements of arterial pressure could be obtained only with arterial cannulation; the development of the Peňáz volume clamp method (Peňáz 1973), the commercial availability of volume clamp devices (Finapres, Ohmeda, Englewood, CO, USA; USM-801, Ueda, Tokyo, Japan), and their validation in studies involving the Valsalva manoeuvre (Imholz *et al.* 1988; Parati *et al.* 1989), render arterial cannulation for Valsalva testing unnecessary. Arterial pressures measured directly and indirectly during Valsalva manoeuvres with the Peňáz device are shown for two subjects in Fig. 4.4.

As straining continues, reductions of distending pressures occur in all baroreceptor areas; however, because of increases of intrathoracic pressures, absolute reductions of distending pressures are greater in aortic and cardiopulmonary than carotid barosensory areas. Cardiac acceleration during continued straining may result from inhomogeneous changes of barosensory input profiles (which include reductions of afferent carotid baroreceptor traffic). Responses of conscious rabbits to 'Valsalva-like' manoeuvres suggest that cardioacceleration during straining is mediated primarily by reduced afferent traffic from aortic baroreceptors (Blombery and Korner 1982).

Cardioacceleration does not appear to be modulated importantly by mechanical factors such as altered sinus node (Brooks *et al.* 1966) or sinus node artery (James and Nadeau 1963) stretch secondary to changing right atrial volumes; cardioacceleration is negligible in patients with transplanted hearts, and it can be prevented in subjects with normally–innervated hearts by atropine (Leon *et al.* 1970). However, this does not exclude the possibility that responses may be mediated in part by altered stretch of cardiopulmonary receptors. Cardiac acceleration may result also from inhibition of efferent vagal activity associated with physical exertion (Freyschuss 1970) and deep inspiration before the manoeuvre (Eckberg and Orshan 1977).

There is some evidence that increased sympathetic efferent activity and vasoconstriction, upon which subsequent pressure elevation in Phase 4 depends (Korner *et al.* 1976), is mediated partly by reduced activity of cardiopulmonary receptors (Roddie and Shepherd 1958; Sundlöf and Wallin 1978*a*; Zoller *et al.* 1972). Some patients may develop normal elevations of arterial pressure and subnormal cardiac slowing after the Valsalva manoeuvre because, although their cardiopulmonary receptors function normally to restore pressure towards normal, their arterial baroreflex responses to pressure increases are impaired. Alternatively, even if vagal baroreceptor–cardiac reflexes are normal, bradycardia does not occur without prior adrenergically mediated vasoconstriction and pressure elevation.

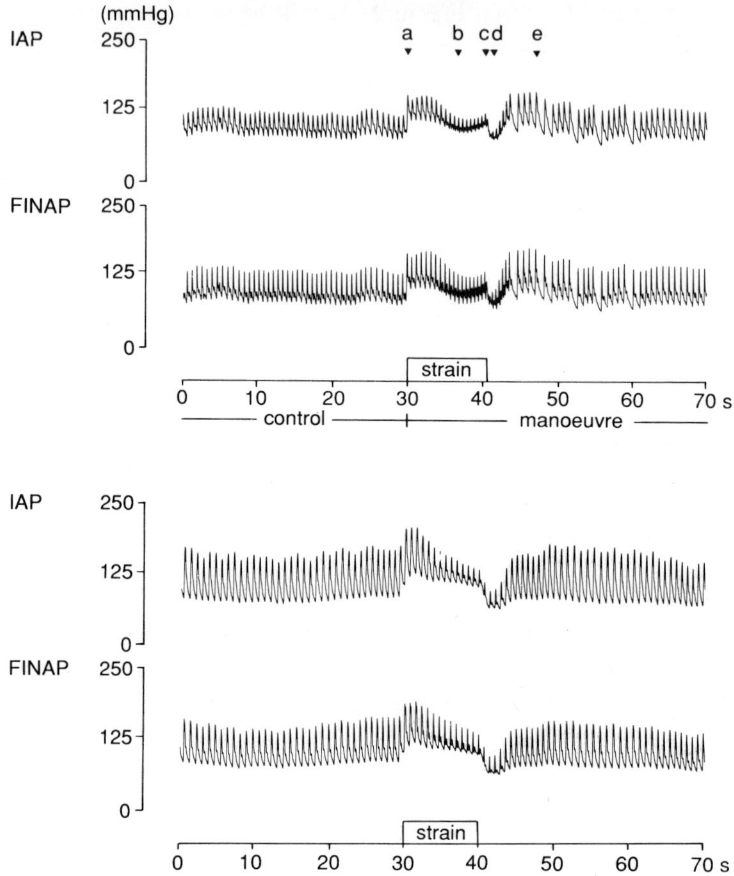

FIG. 4.4. Arterial pressures measured simultaneously with an intra-arterial catheter and a non-invasive photoplethysmographic device. IAP = intra-arterial pressure; FINAP = photoplethysmographic pressure. Reproduced with permission (Imholz *et al.* 1988).

HAEMODYNAMIC RESPONSES OF PATIENTS WITH CARDIOVASCULAR DISEASES

Abnormal patterns of responses to the Valsalva manoeuvre are shown in Figs. 4.5 and 4.6. Figure 4.5 shows an original arterial pressure 'square wave' of a heart failure patient published by Sharpey-Schafer (1955). Figure 4.6 shows the response of a patient with autonomic failure who lacked an appropriate sympathetic response during Valsalva straining and consequently lacked a Phase 4 pressure overshoot. (This patient illustrates an important principle of Valsalva testing: Phase 4 analyses of R–R interval changes are meaningless if earlier sympathetic vasoconstriction does not

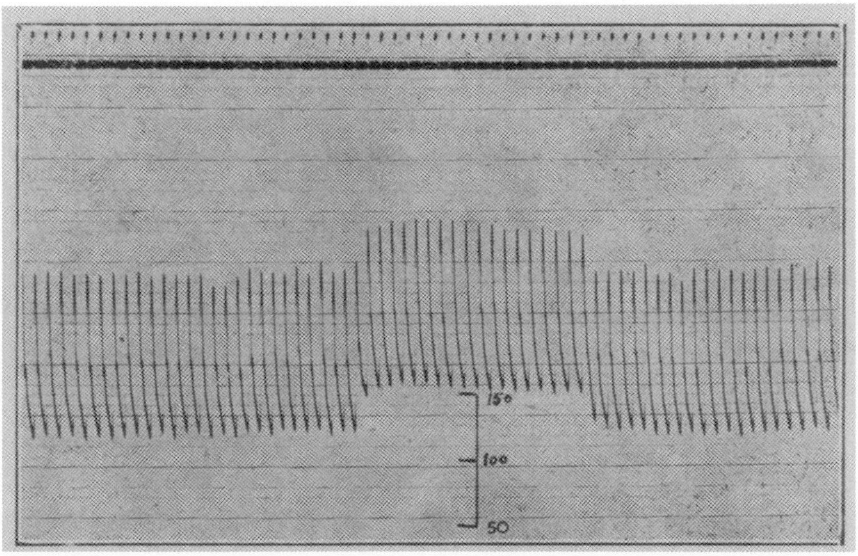

FIG. 4.5. Arterial pressure 'square wave' measured during Valsalva straining in a patient with congestive heart failure. Reproduced with permission (Sharpey-Schafer 1955).

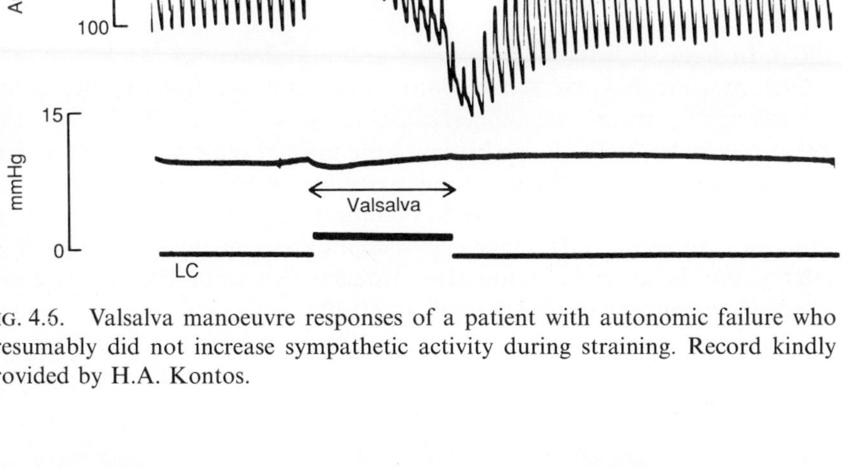

FIG. 4.6. Valsalva manoeuvre responses of a patient with autonomic failure who presumably did not increase sympathetic activity during straining. Record kindly provided by H.A. Kontos.

occur.) Responses of most patients tend to differ from those of healthy subjects in quantitative rather than qualitative ways. For example, a minority of healthy subjects have 'square wave' pressure responses in the supine position (Ten Harkel *et al.* 1990).

Since early pressure elevations result from mechanical rather than neural factors, in contrast to later phases of the Valsalva manoeuvre, Phase 1 may be used to compare responses of patients with those of healthy subjects. Patients with advanced cardiovascular diseases (the most striking abnormalities are found in patients with frank congestive heart failure (Morand 1972)) experience normal elevations of arterial pressure at the onset of straining (H. Goldberg *et al.* 1952; Judson *et al.* 1955; Elisberg 1963), but subnormal reductions of heart rate (Elisberg 1963). Hypertensive patients have greater and more long-lasting Phase 1 R–R interval prolongation than normotensive subjects (Korner *et al.* 1979).

In Phase 2, as straining continues, arterial pressure may remain elevated in cardiac patients with advanced heart failure (the 'square-wave' response shown in Fig. 4.5 (W.F. Hamilton *et al.* 1944; Sharpey-Schafer 1955; Zema *et al.* 1980)), or it may fall slightly (H. Goldberg *et al.* 1952; Sharpey-Schafer 1955). Heart rate may remain constant (Sharpey-Schafer 1955), or it may accelerate slightly (H. Goldberg *et al.* 1952; Judson *et al.* 1955; Elisberg 1963; Parisi *et al.* 1976). Left atrial (D. Robertson *et al.* 1977) and left ventricular (Parisi *et al.* 1976; Little *et al.* 1985) dimensions decline, but insignificantly (in contrast to the significant reductions of left heart volumes that occur in healthy people). Heart failure patients may not experience a reduction of left ventricular volume during straining, because their excessive pulmonary vascular volumes serve as reservoirs to supply blood to the left heart during reductions of venous return to the right heart. Right ventricular stroke volume declines (Little *et al.* 1985), but left ventricular stroke volume remains nearly fixed (Parisi *et al.* 1976), and circulation time may be prolonged only slightly (Stucki *et al.* 1955). Arterial pressure may not rise during the latter portion of Phase 2 (D.J. Stone *et al.* 1965).

During Phase 4, after release of straining, arterial pressure may return to, but not rise above, control levels (H. Goldberg *et al.* 1952; D.J. Stone *et al.* 1965). In some patients (at least those with mitral stenosis (McIntosh *et al.* 1954)), pressure may rise above control levels; however, the rate of rise and the extent of pressure elevation (H. Goldberg *et al.* 1952; McIntosh *et al.* 1954) may be less than those which occur in normal subjects. Phase 4 arterial pressure elevation may occur without associated cardiac slowing (McIntosh *et al.* 1954). This response pattern has not been sought and studied systematically in cardiac patients; it may be a counterpart of the subnormal R–R interval prolongation that is found in cardiac patients after bolus intravenous injections of pressor drugs (Eckberg *et al.* 1971).

CRITICISMS OF THE VALSALVA MANOEUVRE

The Valsalva manoeuvre has been used clinically since the eighteenth century (Dawson 1943) and has yielded important insights into human autonomic physiology and pathophysiology. It has great advantages as a tool for human research. It is simple and does not require much equipment. It can be used to trigger and measure cardiac and sympathetic responses to elevations and reductions of arterial pressure mediated by both sympathetic and parasympathetic systems. The range of pressures that can be used is vastly greater than can be used with all other techniques (world-class weight lifters may increase their systemic arterial pressure to more than 400 mmHg (MacDougall *et al.* 1985)!). During Phase 4 pressure elevations, changes of R–R intervals can be related to changes of systolic pressures to derive slopes that correlate extremely closely with those derived after bolus phenylephrine injections (S.A. Smith *et al.* 1986). Phase 4 pressure elevations are proportional to preceding changes of muscle sympathetic nerve activity, and therefore can be used to estimate changes of sympathetic traffic, in absence of direct recordings (M.L. Smith *et al.*, unpublished). Nevertheless, the Valsalva manoeuvre has its detractors.

Four phases of responses can be identified regardless of how the Valsalva manoeuvre is performed. However, with this, as with other tests of human autonomic function, responses must be related to stimulus intensity. For some measurements derived from the Valsalva manoeuvre, it may be impossible to quantitate the intensity of stimulation (or, reduction of stimulation) in the multiple receptor areas involved. From the preceding discussion, it can be seen that simple measurements of mouth pressure do not convey adequate information regarding the intensity of Valsalva stimuli. Haemodynamic responses of patients may be very different from those of healthy subjects, even though mouth pressures are identical. Moreover, even if stimulus intensity is gauged by measurements of intraarterial and transmural cardiopulmonary pressures during the Valsalva manoeuvre, it still may be impossible to quantitate intensities of stimulation. Reflex adjustments occur during the Valsalva manoeuvre, and these minimize measured pressure changes.

In healthy people, reduction of arterial pressure during Phase 2 is much greater after administration of autonomic blocking drugs than before (Korner *et al.* 1976), and the pressure rise during Phase 4 is much greater if parasympathetic reflex adjustments are prevented by atropine (Elisberg 1963; D.S. Goldstein and Keiser 1984). Therefore ongoing reflex adjustments modify measured haemodynamic responses to the Valsalva manoeuvre, and arterial pressure changes are *functions* of the driving stimuli rather than the stimuli themselves. (The problem that concurrent reflex adjustments modify ongoing autonomic stimuli is not unique to the Valsalva manoeuvre, however; it exists whenever autonomic stimuli last long enough for reflex adjustments to occur.) Reflex adjustments that may occur during the Valsalva manoeuvre are illustrated schematically in Fig. 4.7.

Probably the least controversial use of the Valsalva manoeuvre is for measurement of vagal baroreceptor–cardiac reflex responses during Phase 4. During straining, noradrenaline is released endogenously, and this, in conjunction with increased rates of venous return to the heart, provokes ramps of arterial pressure that are comparable to those provoked by intravenous bolus injections of pressors. Baroreflex slopes derived from Phase 4 of the Valsalva manoeuvre correlate closely with slopes derived with bolus pressor injections (T.G. Pickering and Sleight 1969; Palmero *et al.* 1981; Trimarco *et al.* 1983; S.A. Smith *et al.* 1987), and their use does not depend upon explicit understanding of the haemodynamic forces that provoked them. Moreover, pressure elevations during Phase 4 affect all barosensory areas, but pressure elevations during straining differentially affect carotid and aortic and cardiopulmonary areas). Therefore measurements obtained after Valsalva straining reflect greater vagal stimulation than interventions such as carotid massage or neck suction, that stimulate only one afferent baroreceptor area (Waxman *et al.* 1980*c*).

The high correlations between baroreflex slopes derived from the bolus phenylephrine method and those derived from Phase 4 of the Valsalva manoeuvre found by T.G. Pickering and Sleight (1969) and Palmero *et al.* (1981) suggest that both methods provide quantitative assessments of baroreflex gain. S.A. Smith *et al.* (1987) also found a highly significant correlation

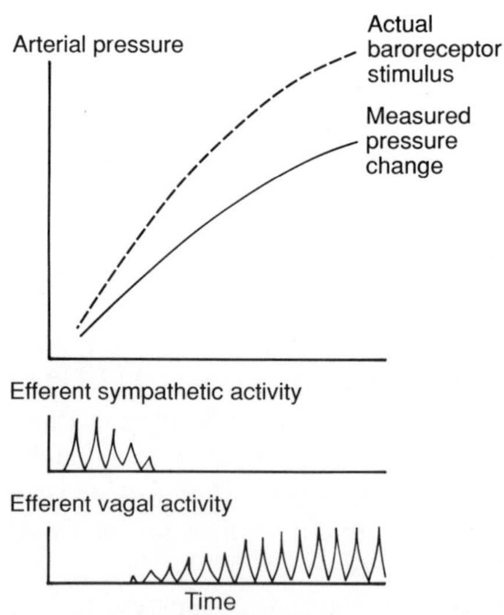

FIG. 4.7. Schematic illustration of ongoing reflex adjustments during and following Valsalva straining. Reproduced with permission (Eckberg 1980*b*).

between results obtained with the two methods, and contributed an important discussion of methodology. We support the method used by Palmero *et al.* (1981), who obtained a linear regression correlation coefficient of 0.91 when data collection was begun during the Phase 4 pressure elevation *when R–R intervals began to lengthen* (not when arterial pressure began to rise). This makes excellent sense, because arterial pressures may fall below the threshold level for baroreceptor activation (Eckberg 1980*a*) during straining, and the earliest arterial pressure–R–R interval pairs may fall in the threshold rather than in the 'linear' portion of the arterial pressure–sinus node response relation.

If measurements are confined to baroreflex slopes derived from post-Valsalva pressure overshoots, it should be possible to use the test to compare responses of healthy subjects and cardiovascular patients. However, as indicated, this requires measurement or estimation of arterial pressure. Arterial pressure has not been measured in many studies of Valsalva responses. Probably, comparisons of the Valsalva responses of cardiovascular patients and those of healthy subjects which were made without measurement of arterial pressure are unintelligible. First, if the Valsalva manoeuvre is to be used to differentiate between abnormal and normal subjects, stimuli and the groups studied must be comparable. This minimum condition was not met in most published studies. (Indeed, this requirement may not be met even in comparisons between groups of healthy subjects; very few studies consider the influence of *age*, which may influence both the post-Valsalva pressure overshoot and R–R interval response (Shimada *et al.* 1986).) Many reports do not describe explicitly the conditions of the Valsalva manoeuvre used. It may be hazardous to guess what type of test was used, because, as described earlier, there is no agreement in the medical literature on what a standard Valsalva manoeuvre comprises. The magnitude of responses to the Valsalva manoeuvre depends importantly on the particulars of the test used, as reflected by such factors as the depth of inspiration before straining, the magnitude of intrathoracic pressure elevations, and the duration of straining. Valid quantitative comparisons between studies may be inappropriate if the particulars of Valsalva manoeuvres used were different.

Second, even if conditions of testing are described explicitly, it may be impossible to interpret the results; in many studies (H. Goldberg *et al.* 1952; McIntosh *et al.* 1954; Judson *et al.* 1955; Stucki *et al.* 1955; D.J. Stone *et al.* 1965; Leon *et al.* 1970), conditions of the Valsalva manoeuvre were not controlled rigidly, and the intensities and durations of straining were determined by the subjects themselves. In at least one study (McIntosh *et al.* 1954), patients with heart disease were not able to maintain straining as long as normal subjects. If the Valsalva technique was not the same for patients and normal subjects, then quantitative comparisons of results are of questionable value.

Moreover, even if specific parameters of the Valsalva manoeuvre are controlled rigidly, the intensity of stimulation of receptor areas in normal subjects and patients with cardiovascular diseases may be dissimilar. In heart failure patients, post-straining pressure elevation varies inversely with the degree of left ventricular impairment, as gauged by end-diastolic pressure and ejection fraction (Zema *et al.* 1980). In patients with atrial septal defects (G.deJ. Lee and Gimlette 1957; Kronik *et al.* 1979), or in the 18 per cent of healthy persons who have patent foramina ovale (Lynch *et al.* 1984), abnormal arterial pressure responses to the Valsalva manoeuvre may be due in part to intracardiac shunting during the manoeuvre. In patients with congestive heart failure, arterial pressure (Sharpey-Schafer 1955) and left ventricular dimensions (Parisi *et al.* 1976; Little *et al.* 1985) may not decline during Phase 2. In such patients, the Valsalva manoeuvre probably does not provoke the same alterations of afferent autonomic input as it does in normal subjects. If arterial pressures and cardiac chamber pressures and volumes do not change to the same extent in both groups, receptors may not be altered to the same extent; therefore, failure of patients with cardiovascular diseases to respond appropriately to the Valsalva manoeuvre cannot be attributed necessarily to reduced autonomic responsiveness.

SUMMARY

This chapter critically evaluates Valsalva's manoeuvre. Notwithstanding the major theoretical problems outlined, there is much to commend this test of human baroreflex function. We suggest that the Valsalva manoeuvre deserves reconsideration, and perhaps rehabilitation. This method is non-pharmacological and can be used to obtain quantitative information about both sympathetic and vagal baroreflex systems. The Phase 4 pressure elevation is particularly attractive as a means to gauge vagal baroreflex responses. During intense straining, arterial pressures may fall below threshold levels; therefore, although we not are aware of any studies of this, the Valsalva manoeuvre may be useful to define baroreceptor–cardiac thresholds. Also, since the Phase 4 pressure rise is proportional to straining, it may be very large, and it may encompass a large portion of the stimulus–response relation. Moreover, since during and after straining, pressures change from lower to higher, results are not confounded by hysteresis.

However, as much of this chapter indicates, the simplicity of the Valsalva manoeuvre may be illusory. It is an exceedingly complex global test of the mechanical status of the circulation, the integrity of several populations of autonomic receptors (including arterial baroreceptors, cardiopulmonary receptors, and probably chemoreceptors), complex central autonomic interactions, and efferent sympathetic and parasympathetic effectors. An unusual response may point towards abnormal cardiovascular hydraulics as well as towards autonomic dysfunction. In part, because of the complexity of the Valsalva manoeuvre, other experimental approaches have been used that perturb the human baroreceptors more selectively. These are discussed in the next chapter.

5

Selective methods

The Valsalva manoeuvre, upright tilt, and other interventions that reduce venous return to the heart provoke complex changes of afferent autonomic sensory input that are mediated by multiple receptor groups. This chapter deals with more selective methods used to change human afferent arterial baroreceptor activity.

CAROTID MASSAGE

As Chapter 2 indicates, the technique of stimulating baroreceptors with carotid massage dates from antiquity; in modern times also, this research approach antedates both neck chamber and vasoactive drug methods. Carotid massage is simple to perform but difficult to quantify. Most workers who use carotid massage have not attempted to validate the technique with comparisons of responses to those elicited by more widely accepted methods. However, in one study, a significant correlation between R–R interval responses to carotid sinus massage and neck suction was found in patients with carotid sinus syncope (Dehn *et al.* 1984). We are not aware of any study that has reproduced and validated this observation. Also, we are not aware of any evidence that responses to carotid massage are reproducible.

Carotid massage is used widely in clinical medicine, and some reports attest to its safety (Lown and Levine 1961). Precise data on the safety of carotid massage are not available; however, scattered reports suggest that the technique is not entirely safe. One complication is stroke secondary to embolization of atheromatous material (Beal *et al.* 1981; Bastulli and Orlowski 1985). Another is ventricular fibrillation (which tends to occur in digitalis-intoxicated patients (Porus and Marcus 1963)). The widest clinical uses for carotid massage are to diagnose carotid sinus syncope (Lown and Levine 1961) and angina pectoris (S.A. Levine 1962), and to treat for supraventricular dysrhythmias (Waxman *et al.* 1980*c*) and pulmonary edema (Alzamora-Castro *et al.* 1955). The technique also has been used to identify autonomic abnormalities in patients with cardiovascular diseases. Patients with coronary heart disease may have supranormal (not subnormal) cardiac slowing during carotid massage (Sigler 1942; K.A. Brown *et al.* 1980).

CAROTID OCCLUSION

Hering (1927) introduced the method of manual carotid artery occlusion in humans, and G.W. Pickering *et al.* applied this method to study of hypertensive patients in 1936. In 1957, Roddie and Shepherd reported the haemodynamic responses of seven healthy subjects and one patient to unilateral or bilateral common carotid artery occlusion. In these remarkable experiments, carotid compression was usually performed by the subject, and pressure was measured with a catheter introduced into the common carotid artery and advanced to the carotid sinus. Mean carotid pressure fell from about 80 to about 40 mmHg (or to about 20 mmHg if both carotid arteries were compressed simultaneously). Mean arterial pressure, pulse pressure, heart rate, respiration, and hand (primarily skin) and calf (primarily muscle) blood flow increased, and central venous pressure did not change. In another remarkable study, Tydén *et al.* (1980*b*) manually occluded surgically exposed carotid arteries before and after carotid endarterectomy and showed that average mean internal carotid pressure fell by 45 mmHg before and 34 mmHg after surgery. Figure 5.1 depicts responses of one subject studied by Roddie and Shepherd (1957) to unilateral carotid artery compression.

Although the results of the study of Roddie and Shepherd (1957) are quoted widely, we are not aware of any groups that use this technique currently. Carotid occlusion may have been supplanted by methods employing neck chambers. Therefore the method is primarily of historical interest. Several problems limit its usefulness. Measurements of human carotid artery pressure are usually not made for research purposes; the carotid bifurcation is prone to atherosclerosis, and catheterization of the carotid sinus carries some risk. Without intracarotid pressure measurements, it may be impossible to know how carotid occlusion affects intracarotid pressure. Carotid pressure changes are variable and inconstant, even with carotid artery ligation. The carotid pressure that is achieved is determined by flow from extracarotid cerebral vessels, as well as by the degree of reflex arterial pressure elevation that results from carotid occlusion (Chungcharoen

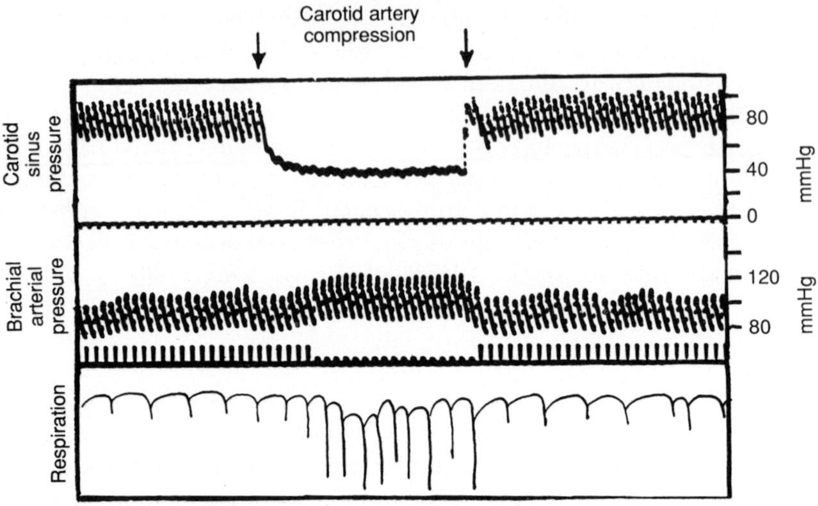

FIG. 5.1. Haemodynamic responses of one patient to unilateral carotid occlusion.Reproduced with permission (Roddie and Shepherd 1957).

et al. 1952). The arterial pressure elevation that occurs during carotid occlusion does not result simply from increased sympathetic activity; the study of Isaacson and Reid (1990) suggests that it is modulated importantly by angiotensin II. Ongoing reflex adjustments tend to increase carotid pressure during occlusion. Also, carotid occlusion is not a specific test for baroreflex function, since if carotid pressure falls to very low levels, increased chemoreceptor activity, as well as reduced baroreceptor activity may contribute to reflex responses (von Euler and Liljestrand 1943).

ELECTRICAL CAROTID SINUS NERVE STIMULATION

In the mid-1960s, a surgically implantable carotid sinus nerve stimulator was developed for use in treatment of hypertension (S.I. Schwartz *et al.* 1967; Brest *et al.* 1972), angina pectoris (Braunwald *et al.* 1967; Dunning 1971), and paroxysmal supraventricular tachycardia (Braunwald *et al.* 1969). In patients who had these devices (and in some studied during neck surgery (Carlsten *et al.* 1958)), bilateral electrical carotid sinus nerve stimuli were delivered as a means to discover baroreflex mechanisms (S.E. Epstein *et al.* 1969; Eckberg *et al.* 1972; Korsukewitz *et al.* 1974; C. Borst and Karemaker 1983). Since the carotid sinus nerve contains chemoreceptor as well as baroreceptor nerve fibres, carotid sinus nerve stimulation is not a selective test for baroreflex function. This approach to study of baroreflex physiology is also mentioned as an historical footnote. Therapeutic carotid sinus nerve stimulation was supplanted by drugs, electronic devices, and myocardial revascularization procedures. The equipment is no longer manufactured, and there are few surviving patients with functioning implanted stimulators. Insights into physiological mechanisms derived from carotid sinus nerve stimulation are discussed in other chapters.

Three quantitative approaches (and many variations thereupon) are available for study of human arterial baroreflex function: pharmacological elevations or reductions of arterial pressure (Smyth *et al.* 1969); neck suction or pressure (Ernsting and Parry 1957); and methods that use computers to discover possible baroreflex engagement during spontaneous fluctuations of arterial pressure and R–R intervals. The technique for stimulating carotid baroreceptors with neck suction was described in a preliminary communication by Ernsting and Parry in 1957. The technique for stimulating all barosensory areas pharmacologically (the 'Oxford method') was described first by Smyth *et al.* in 1969. Although the Oxford technique was described second, it gained earlier widespread acceptance, and is discussed first. Techniques for quantifying baroreflex function with computer analyses of spontaneous arterial pressure and R–R interval changes are described last.

BOLUS INJECTIONS OF VASOACTIVE DRUGS

With the Oxford method, an alpha-adrenergic agonist (usually phenylephrine, less commonly angiotensin-II, rarely methoxamine or noradrenaline) is injected intravenously as a bolus, and prolongations of successive R–R intervals after the injection are plotted as functions of preceding systolic pressures. This method is illustrated in Fig. 2.7. The slope of the linear regression of R–R interval on systolic arterial pressure is considered to provide an index of arterial baroreflex responsiveness or 'sensitivity' (more properly, 'gain' (Sagawa 1978)). R–R (or P–P) interval prolongation after bolus injections of vasoactive drugs is mediated by increased efferent vagal–cardiac nerve activity. Bradycardia persists after sympathetic denervation in patients after high spinal anaesthesia (N.M. Greene and Bachand 1971) and rats after chemical sympathectomy (Julien *et al.* 1990), is unchanged or even augmented after beta-adrenergic blockade (and therefore it probably does not result importantly from sympathetic withdrawal), and is nearly abolished after muscarinic blockade (Robinson *et al.* 1966; Eckberg *et al.* 1971; T.G. Pickering *et al.* 1972*b*; Eckberg *et al.* 1976; A.C. Simon *et al.* 1977).

The time course of one patient's systolic pressure and R–R interval responses to a bolus injection of phenylephrine is shown in Fig. 5.2 (LaRovere *et al.* 1988). Data used for linear regression analysis during the rapid increase of pressure and R–R intervals are shown on the right. In this

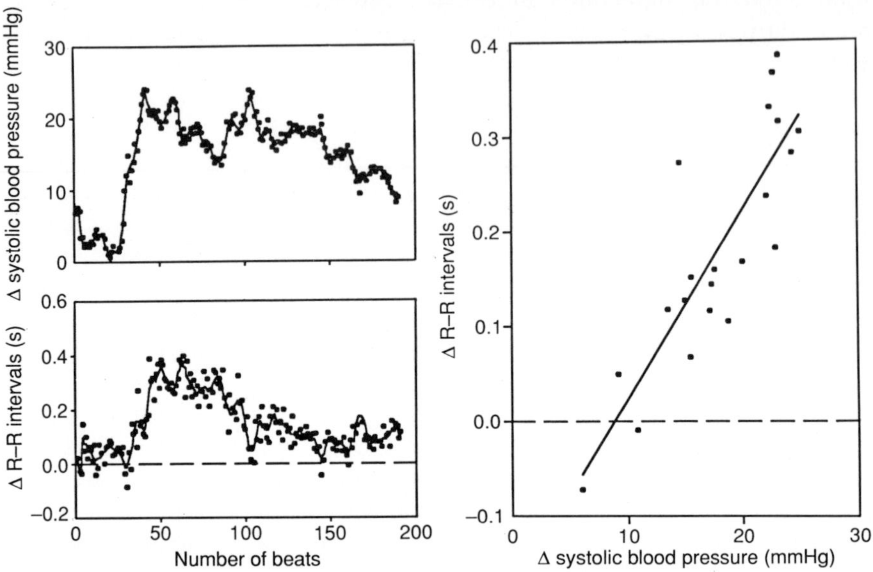

FIG. 5.2. Time course of systolic pressure and R–R interval changes after a bolus injection of phenylephrine. Adapted with permission (LaRovere *et al.* 1988).

patient, the slope of the R–R interval–systolic pressure relation was 19 ms/mmHg, and the correlation coefficient was 0.85.

Advantages of the bolus pressor technique

The Oxford technique holds several important advantages over other research approaches. Chief among these is that the ultimate stimulus, the arterial pulse, is physiological. This method requires no unusual equipment other than a pressure transducer and an amplifier. It involves no voluntary physical exertion or patient co-operation, as does the Valsalva manoeuvre. In contrast with both Valsalva and neck chamber methods, it is unobtrusive; subjects need not be aware that arterial pressure is being perturbed and that baroreflex measurements are being made. Until recently, this method required arterial cannulation. Now, as the study of Parati *et al.* (1989) shows, phenylephrine baroreflex tests can be performed accurately with non-invasive finger phlethysmography.

Participation of reflex mechanisms in addition to the arterial baroreflex is probably small. Respiration during the pressure change can be controlled or accounted for. (However, respiration is usually not measured, and if measured, is ignored in calculations of regression slopes.) Even though intracardiac pressures change (Petrov *et al.* 1979), cardiac receptors do not appear to play an important role in mediating responses (but see below). It is clear that cardiopulmonary (primarily left ventricular) receptors *can* mediate bradycardic responses to rising arterial pressures (Aviado and Schmidt 1955). One explanation for lack of their participation after bolus pressor injections, advanced by Ludbrook (1984), is that arterial baroreceptor thresholds are lower than cardiopulmonary baroreceptor thresholds, and that initial pressure elevations trigger arterial baroreflex responses that prevent pressures from rising above threshold levels for cardiopulmonary baroreceptor firing. This explanation may not be totally satisfactory, however, since bradycardia does not occur after phenylephrine injections given to animals with denervated sinoaortic but intact cardiac baroreceptors (N. Alexander and DeCuir 1966; Korner *et al.* 1972).

Heart rate does not change, or accelerates slightly, if arterial pressure elevation is prevented and ganglionic blockade given after phenylephrine injections (Varma *et al.* 1960); this militates against important direct influences of the pressor drugs used on sinus node function or baroreceptors. Nevertheless, there are many theoretical objections to the bolus pressor technique.

Objections to the bolus pressor technique

The Oxford method is pharmacological. The method perturbs the reflex it is measuring, and may be subject to the vagaries of drug actions. The bolus vasoactive drug injection technique is useful only to study baroreflex influences on sinus node function; since the drugs injected alter arterial pressure, they cannot be used to evaluate baroreflex control of arterial pressure.

There is substantial evidence that some vasoactive drugs alter the R–R interval–systolic pressure relation by means other than pressure elevation or reduction. In this regard, angiotensin-II is a major offender. In the mid- to late 1960s, angiotensin-II had just been released for human use, and it was the earliest drug used by the Oxford group (Smyth *et al.* 1969). Angiotensin-II seemed a logical choice, because it is a powerful pressor agent and is found normally in human plasma. However, it was not long until problems with the use of angiotensin-II became apparent, and the Oxford group abandoned angiotensin-II in favour of phenylephrine hydrochloride. The group's earliest study (Smyth *et al.* 1969) indicated that although the initial baroreflex slope was linear, the terminal R–R interval change departed from linearity in a bizarre fashion. A study by a different group (Ismay *et al.* 1979) suggested that even the initial baroreflex response may be unpredictable when angiotensin is used.

A variety of mechanisms may be responsible for R–R interval responses to angiotensin-II injections. Fujii and Vatner (1985) showed in conscious dogs that a major part of the pressor response to intravenous angiotensin is mediated by alpha-adrenergic agonists. Injections of angiotensin-II into the vertebral arteries of patients provokes arterial pressure elevation that are greater than those which follow injections of the same doses intravenously; also, vertebral artery injections of angiotensin-II may provoke tachycardia, not bradycardia (Ueda *et al.* 1969). These data point to important central effects, which seem to occur because angiotensin gains access to the central nervous system via the area postrema (see Chapter 3 (S. Matsukawa and Reid 1990)). Angiotensin-II seems particularly to affect vagal responses to rising pressure. Microinjections of angiotensin-II into the solitary tract nucleus augment R–R interval prolongation after intravenous phenylephrine (Campagnole-Santos *et al.* 1988). After intravenous bolus injections of phenylephrine, vagal–cardiac nerve activity almost always increases; however, after intravenous bolus injections of angiotensin-II, vagal firing may decrease, remain the same, or increase to lesser extent than after phenylephrine (Lumbers *et al.* 1979; Lumbers and Potter 1982). Angiotensin-II is bound to and transported by vagus nerve receptors (Diz and Ferrario 1988). Because of these practical and theoretical problems, use of angiotensin-II for human baroreflex testing has largely been abandoned.

Second, drugs used to elicit baroreflex responses may directly modify baroreceptor transduction. Although phenylephrine has become the mainstay for pharmacological human baroreflex testing (Smyth *et al.* 1969; Eckberg *et al.* 1971; A.C. Simon *et al.* 1977), this agent is not without its detractors. A study conducted by the Oxford group (to its credit) indicates that this does occur (Peveler *et al.* 1983). They found that bolus injections of phenylephrine similar to those used in humans lead to reductions (not increases) of carotid sinus diameter at times when arterial pressure is rising. These increases of wall tension are associated with increases of subsequent R–R intervals. Direct effects of phenylephrine on baroreceptor transduction may not be of great practical importance, however, because Lumbers *et al.* (1979) found that increases of baroreceptor afferent activity provoked by intravenous phenylephrine and inflation of a balloon in the descending aorta are comparable. Also, Fritsch *et al.* (1986) found that R–R interval–systolic pressure slopes after spontaneous bursts of muscle sympathetic nerve activity (conceived as microinjections of noradrenaline) are similar to those occurring after bolus phenylephrine injections. Finally, infusions of phenylephrine do not seem to alter R–R interval responses to neck suction (Ferguson *et al.* 1985; Fritsch *et al.* 1989).

A third theoretical objection to the Oxford method is that injections of vasoactive drugs may alter cardiopulmonary afferent activity (Beck *et al.* 1961; McKay *et al.* 1984). There is some evidence, drawn from studies conducted in anaesthetized cats, that cardiopulmonary receptors contribute to reflex responses to intravenous boluses of phenylephrine (Petrov *et al.* 1979). This concern is probably relevant also to neck chamber methods of altering afferent baroreceptor activity; any intervention that alters heart rate, preload, afterload, or myocardial contractility would be expected to modify cardiopulmonary receptor activity. The practical consequences of cardiopulmonary receptor participation in human baroreflex testing, if any, are unknown.

A fourth concern with the Oxford method is that vasoactive drugs alter baroreflex responses through effects on central or efferent limbs of the reflex arc. As mentioned, angiotensin-II indisputably alters the central nervous system portion of the reflex arc. It is less well appreciated that phenylephrine also may exert central effects (Imaizumi *et al.* 1984*a*). Vasoactive drugs also may exert direct effects upon sinus node function. Phenylephrine achieved wide acceptance because it was believed to be a pure alpha-adrenergic agonist (Ahlquist 1948). However, phenylephrine and methoxamine alter heart rate when they are infused directly (usually in high concentrations) into the sinus node artery of anaesthetized dogs (White *et al.* 1973*a*). Phenylephrine and noradrenaline cause cardioacceleration, which can be prevented by beta-adrenergic blockade, and methoxamine causes cardiac deceleration, which can be prevented by atropine.

White and Eckberg studied the direct effects of phenylephrine in a two-dog preparation (arterial blood from one dog perfused the sinus node artery of another) and published their results as a preliminary communication (White *et al.* 1973*a*). A bolus of phenylephrine was given intravenously to one dog, and it provoked reflex bradycardia as expected. Later, arterial blood from that dog (containing diluted phenylephrine) arrived at the sinus node artery of the other dog and caused cardioacceleration.

There is inferential evidence that phenylephrine influences sinus node function directly in humans. When phenylephrine is given intravenously to patients with severe heart disease, who may have abnormal parasympathetic reflex responses (Eckberg *et al.* 1971), cardioacceleration rather than cardiodeceleration also may occur. Aksamit *et al.* (1987) showed that in a patient with sinoaortic baroreceptor denervation secondary to radiation therapy, phenylephrine injections cause cardioacceleration. Phenylephrine also reduces sinus node responses to electrical vagal stimulation (McGrattan *et al.* 1987) and causes cardioacceleration after muscarinic blockade (Keys and Violante 1942). The occurrence of cardioacceleration after phenylephrine injections supports the view that this prototypic alpha-adrenergic agonist (Ahlquist 1948) has some measurable beta-adrenergic agonist properties (White *et al.* 1973*a*).

A final concern regarding the influence of drugs used to test baroreflex function is that if repeated injections of vasoactive drugs exceed the subject's metabolic and excretory capacities, the agents will accumulate, and alter baseline autonomic activity. Although Hjemdahl developed a method for measurement of phenylephrine in plasma (Hjemdahl *et al.* 1979), we are not aware that any serial measurements of phenylephrine concentrations during baroreflex testing have been made. D.L. Eckberg (unpublished) had one experience which suggests that phenylephrine may accumulate during serial baroreflex testing. He gave a series of phenylephrine injections to one healthy young adult, and followed these injections with an intravenous blocking dose of atropine sulphate. Atropine led to a major elevation of arterial pressure (diastolic pressures in this subject rose from normotensive levels to above 150 mmHg). Since atropine usually does not provoke major elevations of arterial pressure (Gilman *et al.* 1980), this response was attributed to accumulation of phenylephrine. Presumably, high plasma levels of phenylephrine did not lead to pressure elevation before atropine because they were buffered by baroreflex mechanisms.

Theoretically, accumulation of vasoactive drugs could be hazardous for human volunteers. In practice, as far as the authors are aware, no permanent complication has occurred because of untoward actions of vasoactive drugs. We are aware of isolated instances in which vasoactive drugs have seemed to provoke premature ventricular and atrial beats.

A strong argument can be made that the proper way to study baroreflex responses is to work out entire sigmoid R–R interval–systolic pressure relations (Koch 1931) rather than portions thereof. The Oxford method involves only moderate pressure changes, and treats the baroreceptor stimulus–sinus node response relation as a linear function, rather than as the non-linear sigmoid function that it is. The appropriateness of this usage is supported indirectly by the usually high correlation coefficients that are obtained with linear regression analyses. Moreover, some data (Eckberg 1980*a*; Rea and Eckberg 1987) suggest that in many healthy young adults, systolic pressure lies on the linear portion of the sigmoid function. Therefore, in such subjects, pressure elevations will traverse the linear portion of this relation, appropriately. In addition, in most subjects, the linear range extends well beyond the 20-30 mmHg pressure elevations that usually follow bolus phenylephrine injections (Rea and Eckberg 1987). (It should be noted that the 'linear' portion of this relation is an abstraction; logistic analysis of human sigmoid baroreceptor–cardiac relations (Rea and Eckberg 1987; Pawelczyk and Raven 1989) suggests that the centre of the 'linear' portion is merely the position where concavity becomes convexity.)

This theoretical problem is not dealt with entirely by high correlation coefficients, however. During ramps of arterial pressure changes, the correlation coefficient may be high, even though the baroreflex slope is not measured in the 'linear' portion of the sigmoid relation. This possibility is illustrated in Fig. 5.3. The *left panel* shows the original *Blutdruckcharacteristik* (blood pressure characteristic), as published by Koch in 1931. Koch obtained these data by applying stepwise increases of pressure in a vascularly isolated carotid sinus preparation in an anaesthetized dog. He raised pressure from about 51 to 193 mmHg and prolonged R–R intervals from about 0.33 to 0.78 s. The *right panel* shows linear regressions computed for each successive 30 pairs of data; the average pressure range for the four data sets was 35 (range: 27 to 49) mmHg. The correlation coefficient for the first relation was only 0.16. Correlation coefficients for the last three relations were 0.86, 0.91,

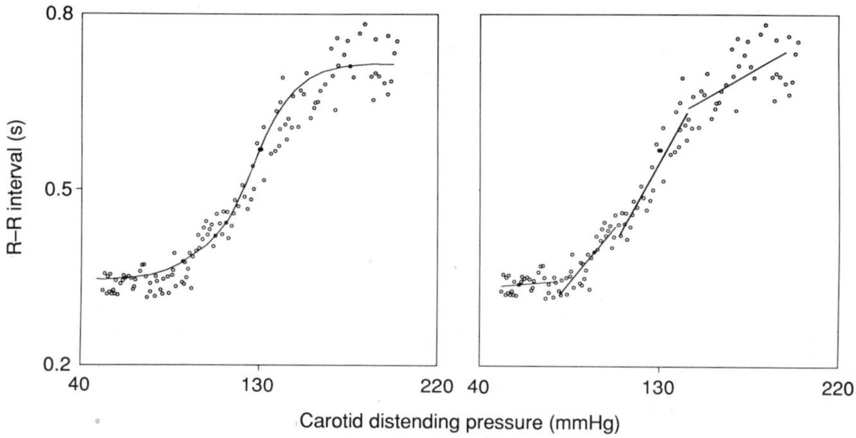

FIG. 5.3. Koch's blood pressure characteristic. Lines in right panel were derived by application of linear regression analyses to each set of 30 pairs of data published originally by Koch (1931). Adapted with permission.

and 0.60 (all $P < 0.01$), and slopes were 4.2, 6.2, and 2.0 ms/mmHg. Thus significant correlations were obtained over most of Koch's relation, but with a threefold difference among slopes. Data from humans (Karemaker and Borst 1980) indicate that *large* bolus phenylephrine doses may provoke curvilinear R–R interval responses, which nonetheless may have high correlation coefficients.

Since bolus pressor drug injections do not define the entire sigmoid R–R interval–systolic pressure relation, it may be difficult to interpret changes of slopes that result from experimental interventions. For example, baroreflex slopes obtained with the Oxford method decline during isometric exercise (J.D. Bristow *et al.* 1971*b*; T.G. Pickering *et al.* 1971; Cunningham *et al.* 1972*a*; Sleight *et al.* 1979; Somers *et al.* 1985). However, from published data, it is not possible to determine if the reduction of baroreflex slopes represents true reduction of reflex gain, or a shift of the subject on the stimulus–response relation into non-linear threshold or saturation regions. Some evidence suggests that the operational point (defined with a neck chamber) does not shift in parallel with acute shifts of baseline arterial pressure. Pressure elevations during phenylephrine infusions shift the operational point towards the saturation range, and pressure reductions during nitroprusside infusions shift the operational point toward the threshold range (Fritsch *et al.* 1989).

Also, since the bolus pressor technique does not define the entire sigmoid R–R interval–arterial pressure relation, it does not indicate where cardiovascular patients (who may have subnormal slopes (Eckberg *et al.* 1971)) lie. Data from heart failure patients obtained with neck pressure changes

(Sopher *et al.* 1990) suggest that their subnormal baroreflex slopes may result in part because they operate in the threshold rather than the linear portion of their baroreceptor stimulus–cardiac response relations. In such patients, baroreflex slopes may yield little information about maximum baroreflex gain because their arterial pressures remain below the linear portion of the stimulus–response relation.

These objections apply to baroreflex testing with *rapid* bolus injections of pressor or depressor drugs. As will be seen, methods used to study the effects of vasoactive drugs are heterogeneous; with more prolonged infusions of vasoactive drugs (Korner *et al.* 1974) or sequential infusions of depressors followed by pressors, entire sigmoid baroreceptor stimulus–sinus node response relations may be defined. These variations are discussed below.

The technique of bolus injections of vasoactive drugs may not be equally suitable for evaluation of R–R interval responses to both falling and rising pressures. First, such usage treats resting pressure as a 'watershed' and requires that subjects' resting pressures lie in the middle of the linear portion of the sigmoid baroreceptor–cardiac reflex relation. Although some subjects appear to lie in such equipoise (Rea and Eckberg 1987), not all do. Healthy young adults (Eckberg 1980*a*) may lie at, and patients with heart failure (Sopher *et al.* 1990) may lie below, the threshold region (as defined by responses to graded neck suction), and endurance athletes may lie near the saturation region (B.D. Levine *et al.* 1991). Champion athletes may have lower than normal baroreflex slopes after pressor drug injections, simply on the basis of their resting positions on the baroreceptor stimulus–sinus node response relation. In the case of athletes, relatively shallow slopes after phenylephrine injections might say nothing about slopes in the linear range of the R–R interval–systolic pressure relation.

Second, use of both pressor and depressor drugs ignores the well-known hysteresis of baroreceptors (see Chapter 6). It may be inappropriate to join together plots of responses to rising and falling pressures at their starting points. An example of one subject's responses to phenylephrine and nitroglycerine is shown in Fig. 5.4 (T.G. Pickering *et al.* 1972*c*). Based upon recordings made from baroreceptor nerves during elevations and subsequent reductions of pressures (Arndt *et al.* 1975; Sleight 1981; Mukharji *et al.* 1987), baroreceptor hysteresis would be expected to yield lower slopes of baroreflex responses when pressure is falling than when it is rising, independent of reflex 'gain' (Masuda *et al.* 1984).

Third, sequential injections of pressors and depressors may not be ideal for study of *sympathetic* physiology. At identical pressures, *sympathetic* activity is greater when pressure is falling than when it is rising (Sundlöf and Wallin 1978*b*). Therefore, there may be systematic differences between sympathetic responses to rising and falling pressures that are independent of absolute levels of pressure. Also, changes of sympathetic activity should be related more to diastolic, than systolic pressures (Smyth *et al.* 1969).

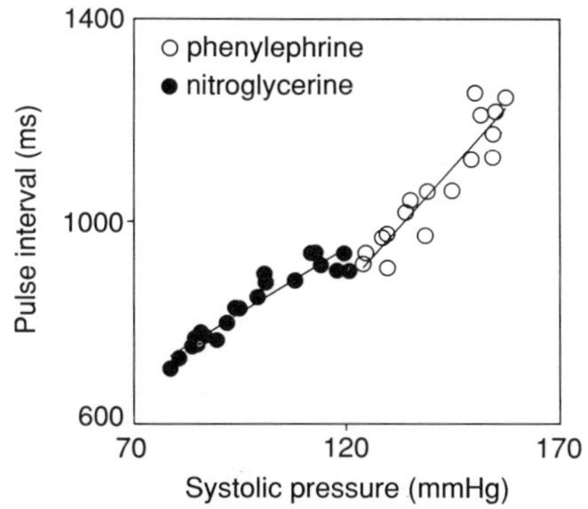

FIG. 5.4. Comparison of R–R interval responses of one subject to intravenous bolus injections of phenylephrine and nitroglycerine. Adapted with permission (T.G. Pickering *et al.* 1972*c*).

Usually, studies employing vasoactive drug injections focus on systolic rather than diastolic pressure.

In addition to problems with vasoactive drug methods mentioned above, there are also problems related to the use of any *ramp* method for studying baroreflex function. Vagal–cardiac nervous traffic flows as pulse-synchronous bursts from the central nervous system to the heart (Iriuchijima and Kumada 1964; Jewett 1964; Katona *et al.* 1970; Kunze 1972). It is well worked out in animals (G.L. Brown and Eccles 1934; J.V.O. Reid 1969; Dong and Reitz 1970; Iano *et al.* 1973; Jalife and Moe 1979) and humans (Eckberg 1976) that the influence of acetylcholine on the sinus node depends critically on the timing of its release at the sinus node and on the rate of its degradation by acetylcholinesterase (Trimarco *et al.* 1981; see also Chapter 9). This physiology influences how sinus node responses to ramped baroreceptor stimuli are analysed. Stated simply, the question is: which heart beat does a given arterial pulse influence most?

The question was a practical one in the early days of bolus pressor injections in Oxford; it was not known which R–R interval an arterial pressure should be paired with — the *same* interval in which the pulse occurred? the next one? or a later one? The first part of this question is: what is the *latency* between an arterial baroreceptor stimulus and the earliest sinus node inhibition that occurs? As Chapter 9 indicates, there is disagreement in published studies about the latency of human baroreceptor–cardiac reflex responses. The question of latency is complicated also by central differences in latencies with rising and falling pressures; latencies after abrupt reductions of pressure are about 100 ms longer than latencies after abrupt elevations of pressure in animals (Katona and Barnett 1969) and humans (Eckberg 1980*a*). Moreover, latencies between arterial pulses and their muscle sympathetic nerve responses seem not to be fixed, but to vary (slightly) with ongoing haemodynamic changes (Fagius *et al.* 1987; Eckberg *et al.* 1988). The shortest baroreceptor–cardiac reflex latencies (which are in line with data from experimental animals (G.L. Brown and Eccles 1934)), about 250 to 300 ms, were found by Eckberg (Eckberg 1976, 1980*a*) and Sopher (1990) and their co-workers. (Others have placed human baroreceptor–cardiac reflex latency at about 475 ms (T.G. Pickering and Davies 1973; C. Borst and Karemaker 1983).) Whether the latency is longer or shorter, however, all data from humans establish the probability that an arterial pulse can influence the *same* sinus node cycle in which it occurs.

Another important question is: how long does sinus node inhibition persist after an arterial pulse? This question is complicated by the fact that a single arterial pulse does not stand out from its peers as an isolated event. Rather, human arterial pressure tends to rise and fall as ramps (Sundlöf and Wallin 1977; Fritsch *et al.* 1986). Therefore, studies of the duration of sinus node inhibition relate both to the duration of inhibition arising from a single pulse and to the duration of the ramp of which that pulse is a part. Several studies speak to this question (T.G. Pickering and Davies 1973; Eckberg and Eckberg 1982; S.A. Smith *et al.* 1986; Ferrari *et al.* 1987).

The first was by T.G. Pickering and Davies (1973). They calculated regression coefficients and slopes with different pairs of systolic pressures and R–R intervals, with the scheme illustrated in Fig. 5.5. For the first analysis, successive systolic pressures were plotted as functions of R–R intervals that preceded them by two beats. For the second analysis, systolic pressures were plotted with the R–R interval that immediately preceded the pulse. This iteration was continued to the fifth arterial pulse. Correlation coefficients suggested that the best fit of data was found when systolic pulses were related to the same R–R interval in which the arterial pulses occurred, or the next one. Data published by Eckberg and Eckberg (1982), Fritsch *et al.* (1986), and Ferrari *et al.* (1987) support these findings, and confirm the intuitively obvious suggestion made by T.G. Pickering and Davies (1973)

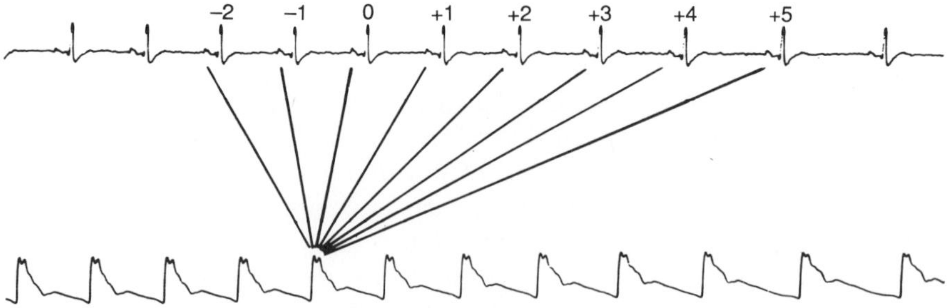

FIG. 5.5. Scheme used by T.G. Pickering and Davies (1973) to determine which R–R interval offset should be related to each arterial pulse. Adapted with permission.

that the choice of offset depends upon the baseline R–R interval. If it is long (greater than 800 ms, or the heart rate is slower than 75 beats/min), there is enough time for acetylcholine released in response to a systolic pulse to influence the same cardiac cycle in which it occurs. If it is short (less than 800 ms), systolic pulses will exert more influence on the next than on the same R–R interval (that is, with an offset of one). At resting human heart rates, offsets of more than two usually are not necessary. Since there is a direct relation between baseline R–R intervals and baroreflex gain (T. Bennett *et al.* 1978; Raczkowska *et al.* 1983), subjects with very short R–R intervals, in whom greater offsets might be necessary, tend to have grossly subnormal baroreflex responses (T.G. Pickering and Davies 1973).

However, there is another issue that these studies do not address. Prolongation of R–R intervals by arterial pressure ramps is progressive; therefore, an offset chosen on the basis of the control R–R interval measured before the ramp began may not be the best choice for later R–R intervals. This issue was joined by S.A. Smith *et al.* (1986). They devised a method for iterative computer calculations of regressions that took into account the progressive R–R interval prolongations that occurred during ramps. Their method allowed them to calculate 60 separate correlation coefficients and slopes from each bolus injection of phenylephrine. Results of one such analysis are shown in Fig. 5.6. Slopes calculated with the new method were compared with slopes calculated with the traditional Oxford method (offset not indicated). The correlation coefficient was an astonishing 0.97. Slopes calculated with the new method were slightly less (by 7 per cent) than those calculated with the Oxford method. S.A. Smith *et al.* (1986) concluded (and

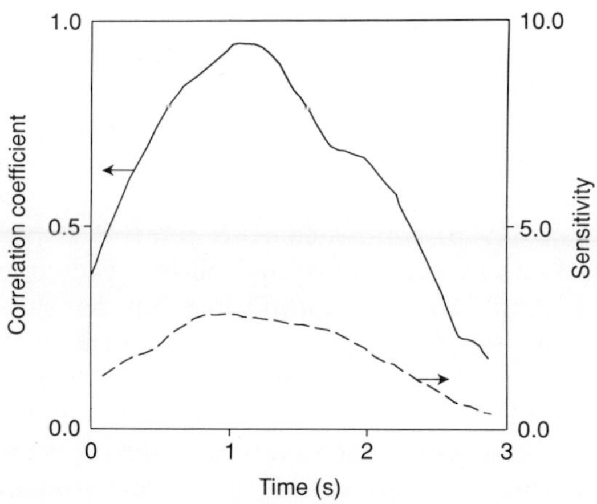

FIG. 5.6. Variation on the Oxford bolus phenylephrine technique developed by S.A. Smith *et al.* (1986). Baroreflex sensitivity is read at point at which correlation coefficient is maximal. Adapted with permission.

we concur) that the interesting theoretical issues which their study addresses may have little practical significance for the testing of human baroreflexes. (This study looked only at responses to *rising* pressure, and the results do not apply necessarily to responses to falling pressure. Conceivably, timing of arterial pulses in relation to sinus node depolarization may be a more important factor when pressures are falling, and may contribute to the uniformly shallower slopes calculated with this intervention.)

As with methods described in the previous chapter, the technique of stimulating arterial baroreceptors with injections of pressor drugs is not uniform from study to study. First, as mentioned, different drugs may be used. Second, methods for delivering vasoactive drugs vary. The original approach used by Smyth *et al.* (1969) was to inject angiotensin-II as boluses (over 5 to 10 s). This is the approach (but with phenylephrine) used by most workers. However, variations on this approach have been developed. One is to inject vasoactive drugs as *infusions*. Korner *et al.* (1974) gave graded infusions over about 15-20 s and measured 'steady-state' R–R interval responses. Dohi *et al.* (1983) gave infusions of phenylephrine over 2 min at rates sufficient to raise systolic pressure by 60 mmHg. The infusion method has the advantage that it may define an entire sigmoid R–R interval–systolic pressure relation. The method also has potential disadvantages. The rate of arterial pressure change is probably less than that which follows bolus injections of vasoactive drugs. Conceivably, reduced baroreflex gain during infusions (see below) reflects ongoing baroreflex resetting (Fritsch *et al.* 1989). Also, because of ongoing reflex adjustments, the pressure range measured during vasoactive drug infusions may be small (Greenway and Innes 1981). Prolonged infusions of vasoactive drugs cause heart rate changes mediated by both sympathetic and parasympathetic branches of the baroreflex arc (Dohi *et al.* 1983), and the relative contribution of each may be unknown. Finally, infusions of vasoactive drugs take longer than bolus injections; this factor may limit the usefulness of infusions to study short-lived effects of acute interventions. Average stimulus–response relations from one normotensive and two hypertensive groups published by Korner are depicted in Fig. 11.2.

Sullebarger *et al.* (1990) contrasted R–R interval responses of healthy subjects to boluses and infusions of phenylephrine. They found that baroreflex slopes after both methods are reproducible, but that slopes after boluses are more than twice slopes after phenylephrine infusions. They postulated that baroreflex resetting during infusions may reduce vagal responses.

An additional permutation on the vasoactive drug technique was developed by T.J. Ebert (1990). He gave *sequential* intravenous bolus injections of the vasodilator, nitroprusside and the vasoconstrictor phenylephrine. The durations of injections were intermediate between those used by Smyth *et al.* (1969) and Korner *et al.* (1974) and their co-workers. The rationale for Ebert's method is that the initial dose of nitroprusside lowers pressure below the threshold range, and the subsequent dose of phenylephrine progressively increases pressure and carries it through the threshold and linear ranges, and possibly into the saturation range. Thus this variation has the potential to define the entire human baroreceptor–cardiac stimulus–response relation with a duration of drug effect that may be intermediate between that of bolus and infusion methods. It also has the theoretical advantage that since R–R intervals are measured only as pressures are rising, influences of baroreceptor hysteresis are avoided (see Chapter 6). There has not yet been a systematic comparison between slopes derived from this method and slopes derived from the conventional pressor injection technique.

A fourth variation on the vasoactive drug method is to administer drugs as constant (lasting minutes) infusions (S.H. Grossman *et al.* 1982). This method has the advantage that changes of concentrations of neurotransmitters such as noradrenaline, adrenaline, and neuropeptide Y can be measured. This method in part sets aside one of the major objections to vasoactive drug methods, that they cannot be used evaluate blood pressure control. Data from the study of S.H. Grossman *et al.* (1982) are depicted in Fig. 5.7.

Although most workers evaluate baroreflex responses to pressor injections with least-squares linear regression, there may be subtle differences in the choice of data to be analysed. The regression analysis of arterial pressure and pulse interval may begin before the injection is made (Trimarco *et al.* 1981), as soon as the injection is completed (Smyth *et al.* 1969), or with the onset of the arterial pressure elevation (Eckberg *et al.* 1971). We are aware of no study in which analyses were begun when the regression of pulse

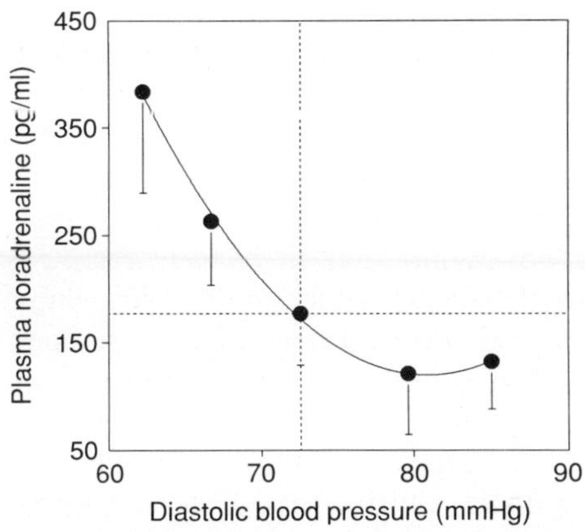

FIG. 5.7. Changes of plasma noradrenaline levels in a group of normotensive volunteers before (horizontal dotted line) and after two rates of infusions each of phenylephrine and nitroprusside. Adapted with permission (S.H. Grossman *et al.* 1982).

interval on arterial pressure became linear (in subjects whose systolic arterial pressures lay below their thresholds for baroreceptor–cardiac reflex activation). Exclusion of systolic pressures that fall below the threshold for baroreflex activation should improve the correlation coefficients of these relations. Also, we are aware of no group that routinely excludes from analyses R–R intervals that fall during inspiration. Exclusion of R–R intervals occurring during inspiration increases the slope slightly (Smyth *et al.* 1969).

It is a common practice to correlate each systolic pressure with the R–R interval of the beat that follows the systolic pulse (Smyth *et al.* 1969; Eckberg *et al.* 1971). As discussed, this is appropriate for subjects with relatively rapid (above 80 beats/min) heart rates, but it may be inappropriate for subjects with slower baseline heart rates in whom the maximum vagal sinus node inhibition may occur in the same R–R interval in which the systolic pulse occurs (T.G. Pickering and Davies 1973; Eckberg and Eckberg 1982). Inter-individual comparisons of baroreflex slopes after phenylephrine injections (D.S. Goldstein *et al.* 1982) may be difficult to interpret if a single offset is used for subjects with widely divergent baseline R–R intervals (Ferrari *et al.* 1987).

A possibly decisive argument in support of ramped methods of baroreceptor stimulation is that human pressure changes occur as ramps; therefore, pressure changes provoked by drugs elicit responses that are similar to those that occur in everyday life. Moreover, parallel changes of pressure and R–R intervals reflect baroreflex involvement; indeed, in the absence of functioning baroreflexes, R–R intervals *shorten* when arterial pressure rises (Persson *et al.* 1988). As mentioned in the previous chapter, several studies have compared Oxford slopes with those that occur physiologically. T.G. Pickering and Sleight (1969), T.G. Pickering (1970), Palmero *et al.* (1981), and S.A. Smith *et al.* (1987) found strong correlations between R–R interval–systolic pressure relations after phenylephrine and after Valsalva straining. S.A. Smith *et al.* (1987) discussed reasons why slopes derived from the two methods are not identical, and concluded that post-Valsalva slopes are highly dependent on the choice of pressure and R–R interval pairs for analysis.

Fritsch *et al.* (1986) found that in healthy resting subjects, ramps of arterial pressure triggered by bursts of muscle sympathetic activity are followed by proportional prolongations of R–R intervals. Slopes and correlation coefficients derived from linear regressions of R–R intervals on systolic pressure during these ramps were comparable to those measured after bolus intravenous pressor injections. Unfortunately, phenylephrine slopes were not obtained in the same subjects. Robbe *et al.* (1987) used spectral analyses of R–R interval changes during spontaneous pressure elevations measured with chronic recordings (see below) and found that the gain determined with their method correlated closely ($r = 0.94$) with slopes of baroreflex responses triggered by phenylephrine injections. Thus several types of proof establish strong correlations between baroreflex slopes measured after vasoactive drug injections and those measured after physiological pressure elevations. However, they fall short of proving that pressors, whether physiological or pharmacological, do not influence baroreceptor transduction.

NECK SUCTION OR PRESSURE

The technique of neck suction to stimulate carotid baroreceptors was described first in a preliminary publication by Ernsting and Parry in 1957 (their original chamber is depicted in Fig. 2.6). This approach is based upon the firmly established principle that afferent baroreceptor traffic can be increased by either positive pressure *inside* or negative pressure *outside* a baroreceptive artery (Fig. 6.3 (Angell James 1971*b*)). Positive or negative pressure changes applied to a chamber fixed over or around the neck are transmitted to the carotid arteries in human subjects and provoke nearly linear reductions or increases of carotid artery diameter (Kober and Arndt 1970). Responses to baroreflex forcings are assessed with electrocardiographic R–R or P–P intervals, heart rate changes, muscle sympathetic nerve activity, ventricular stroke volumes, and plasma neurotransmitter concentrations.

Advantages of the neck chamber technique

The technique of stretching or compressing carotid baroreceptors with neck suction or pressure holds a variety of major advantages over the pharmacological technique for altering baroreceptor activity. Perhaps the most important of these is that baroreflex-mediated changes of *arterial pressure* as well as heart rate can be measured. Neck suction is a non-invasive technique that does not require arterial cannulation. (However, the development of the non-invasive finger pressure technique by Peňáz (1962) may relegate this advantage to one that is merely of historical interest.)

Neck pressure changes are *mechanical* stimuli; since no drugs are involved, there is no concern about possible modifications of physiological baroreflex effects by the drugs used to elicit them (see above). Importantly, baroreceptors can be studied over a wider range of pressures with the neck chamber than with the pharmacological technique. For example, one algorithm used to study baroreflex responses (Sprenkle *et al.* 1986) employs neck pressures ranging from 40 to −65 mmHg, a difference of 105 mmHg. Because such a wide range of pressures can be used, the entire sigmoid baroreceptor stimulus–response relation is more likely to be defined by the neck chamber than by the bolus pharmacological technique. The neck chamber technique permits finer control of stimulus parameters than the pharmacological technique; with electronically controlled valves and closely fitting chambers, it is possible to control the rate of pressure change, intensity, timing, and duration of stimuli with extreme accuracy (Eckberg 1976; Eckberg 1977*b*).

Although there has been no multicentre prospective study of the risks of experimental neck suction and pressure, we are aware of no complications arising from use of this technique. Indeed, two groups (Angell-James and Lumley 1974; Tydén *et al.* 1981) have used neck suction to study patients before and after carotid endarterectomy for transient ischaemic attacks.

Disadvantages of the neck chamber technique

As with the pressor injection technique, there are pitfalls with the neck pressure technique. Certain theoretical objections have been dealt with in published studies. Altered chemoreceptor activity does not appear to contribute to sinus node responses to neck suction; responses are similar before and after inhalation of 100 per cent oxygen (Eckberg *et al.* 1975; Ludbrook *et al.* 1977). The sinus node response does not appear to represent a nonspecific reaction to a noxious stimulus; sinus node (Ludbrook *et al.* 1977; Eckberg 1980*a*) and muscle sympathetic nerve (Wallin and Eckberg 1982) responses to neck suction are directionally opposite (vagal–cardiac traffic is increased and muscle sympathetic nerve traffic is decreased). Responses to neck suction do not appear to be mediated by altered cerebral blood flow (Ludbrook *et al.* 1977). However, there are substantive disadvantages with the neck chamber technique for study of baroreflex mechanisms.

Neck chambers are not comfortable to wear; moreover, if complex positive and negative stimuli are delivered, the chamber must be affixed to the volunteer tightly. Subjects are aware of pressure changes that exceed about 5 mmHg. Therefore it is impossible to deliver neck pressure changes without subjects' knowledge. Dworkin (1988) devised an ingenious method to use neck suction for psychophysiological research that circumvents this problem: he delivers brief neck suction repetitively, during either systole or diastole. R–R interval changes are significantly larger for systolic stimuli. Since the subject is not aware of what phase of the cardiac cycle stimuli fall in, this approach may be a psychophysiological equivalent of a 'placebo' condition. The Dworkin approach is illustrated in Fig. 5.8.

By almost any reckoning, experimental neck pressure changes are unphysiological. Neck pressure changes produce a sudden shift of baseline carotid baroreceptor traffic, *upon which naturally occurring arterial pulses ride*. This is illustrated schematically in Fig. 5.9. Although the principal alternative to the neck chamber technique, the Oxford method, involves use of drugs, the ultimate baroreceptor stimulus, the arterial pulse, is physiological. Once the investigator has released the vasoactive drug into the circulation, arterial pressure changes and reflex responses occur apace, and there is little he or she can do to modify them. Thereafter, baroreceptors are

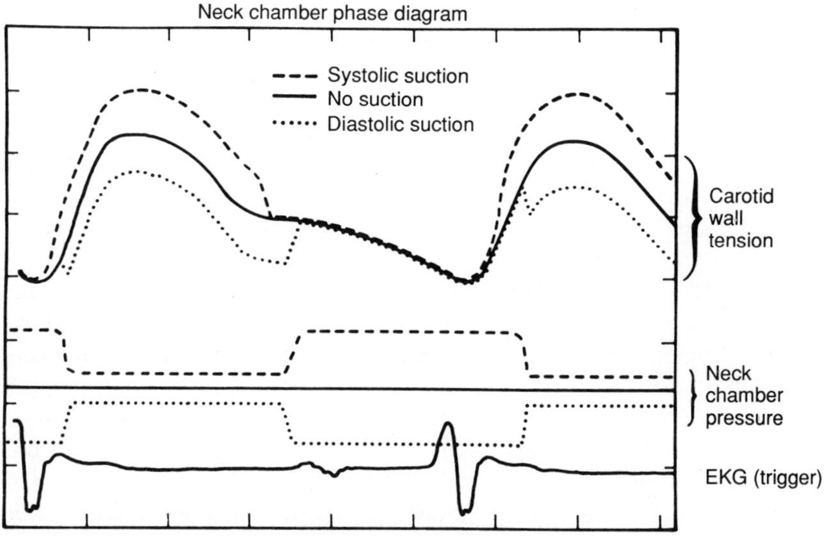

FIG. 5.8. Method developed by Dworkin (1988) to apply neck suction as a 'placebo'. Stimuli delivered during systole provoke greater R–R interval prolongation than those delivered during diastole (placebo). The subject is not aware at which time in the cardiac cycle neck suction is applied. Adapted with permission.

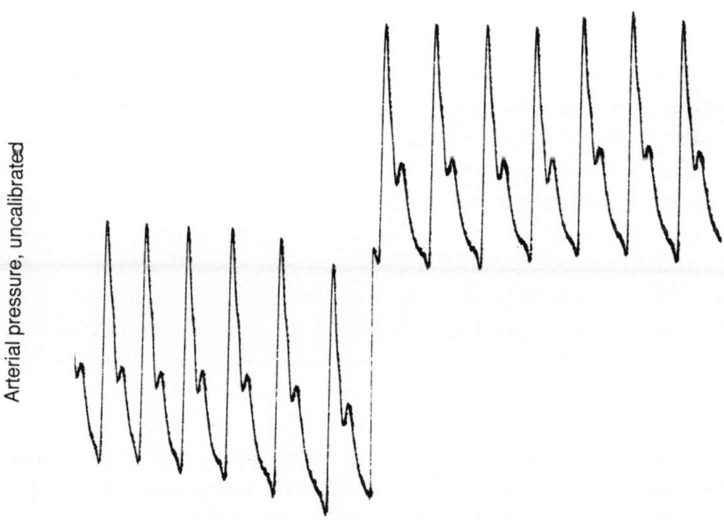

FIG. 5.9. Schematic representation of effect of abrupt neck suction on baroreceptor input signal.

stimulated with naturally occurring arterial pulses, with naturally occurring rates of pressure change, over a naturally occurring range of pressures.

Neck chamber techniques are not bound by such constraints. Even the most closely controlled changes of afferent baroreceptor profile provoked by neck pressure changes are unphysiological. *Sustained neck suction* increases afferent traffic from carotid baroreceptors, at a time when afferent traffic from aortic baroreceptors is decreasing (due to reflex reductions of arterial pressure). There is no physiological counterpart for this intervention (unless a subject stands on his or her head). It is clear that aortic baroreceptors modify responses to carotid forcings — an interaction that was noted as early as 1935 (Schneyer 1935*a,b*). There *is* a counterpart for *neck pressure*, however. When a person stands, carotid pressure and afferent carotid baroreceptor traffic decline more than aortic arch pressure and afferent aortic baroreceptor traffic. This same combination occurs with brief neck pressure. However, with standing, the disparities between pressures in the two barosensory areas are limited by the hydrostatic gradient that is produced, which is a function of the height of the subject. Hydrostatic gradients with standing cannot be as great as those produced by intense neck pressure.

With the neck chamber technique, the investigator is limited by the equipment at hand. If the equipment can be controlled finely, the attributes of stimuli can also be controlled finely. If, on the other hand, the available equipment does not permit fine control, the investigator and consequently the investigator's science are constrained. There is no reason to believe that humans are exempt from the rules of baroreceptor transduction described in Chapter 6; therefore, responses of subjects to neck pressure can be predicted on the basis of the attributes of the neck pressure changes that the equipment can deliver.

It seems almost certain that arterial baroreceptors are *primarily* responsible for autonomic responses to neck pressure changes. However, the possibility that other receptors *contribute* to these responses has not been excluded. For example, since it is well recognized that tracheal stretch (Brodie and Russell 1900; Marlot *et al.* 1982) (or superior laryngeal nerve stimulation (François-Franck 1876; Brodie and Russell 1900)) provokes bradycardia, tracheal receptors could contribute to the R–R interval prolongation that occurs with neck suction (or to the R–R interval shortening that occurs with neck pressure). It is not known if neck pressure changes alter tracheal dimensions. Jain and co-workers (1973) used aerosolized local anaesthetic to block reflexes arising from airway stimulation; this method could be, but has not been, used to prevent possible participation of tracheal receptors in reflex responses to neck pressure changes. It seems highly unlikely that tracheal receptors are primarily responsible for responses to neck pressure changes. For example, there would be no reason to expect that tracheal as well as carotid reflexes would be abnormal in hypertensive patients.

Although published studies (Eckberg and Orshan 1977; Eckberg *et al.* 1980; Trzebski *et al.* 1980; Eckberg *et al.* 1985) indicate that *respiration* influences vagal and sympathetic responses to neck pressure changes, they do not indicate clearly what should be done with respiration during human baroreflex research. One approach is to apply neck pressure changes without regard to respiration. An example of this is depicted in Fig. 3 in the article published by D.S. Goldstein *et al.* (1982). Although the authors did not remark it, this illustration shows clearly that inspiration during intense neck suction is associated with transient R–R interval shortening.

Eckberg usually delivers neck pressure changes about 5 s after the onset of held expiration. This approach was chosen because R–R intervals tend to be constant during held expiration (Eckberg 1976); therefore, R–R interval changes occurring during neck pressure changes are thought to result from changes of baroreceptor traffic. However, as discussed in Chapter 7, constant R–R intervals during apnoea do not necessarily mean that vagal traffic is constant. Fritsch and co-workers (1991) showed that muscle sympathetic nerve activity and arterial pressure rise steadily during held expiration. This is shown in Fig. 7.3. Constant R–R intervals during held expiration probably reflect increasing levels of sympathetic activity balanced nearly exactly by increasing levels of vagal activity. Not all subjects have constant R–R intervals during held breaths. Dornhorst (1952) and Trzebski (1980) and their co-workers reported that in some subjects, sinus arrhythmia persists during prolonged breath holding (at end of *in*spiration).

Another question that has not been dealt with satisfactorily in humans concerns the *effect* of baroreflex stimuli on breathing. Studies in animals (Heymans and Bouckaert 1930; Galdston *et al.* 1943) and humans (Melcher 1976; Trzebski *et al.* 1980) indicate clearly that mechanical baroreceptor stimuli alter respiratory drive. It is not known how baroreceptor-induced alterations of respiratory drive modify ongoing baroreflex responses. This problem is not unique to the neck chamber technique; changes of respiration occur also with pharmacological stimulation of baroreceptors (Cori and Buchwald 1930; Whelan and Young 1953; Joels 1975). Interactions between baroreflexes and respiration are discussed in Chapter 7.

The probability that neck pressure changes alter afferent traffic from cardiopulmonary and aortic receptors has been mentioned.

One of the prime rationales for studying baroreflexes with neck pressure changes is that it is possible to control precisely the actual pressures seen by baroreceptors. However, studies of Ludbrook and co-workers (1977) raise the possibility that pressures transmitted to the carotid sheath may not be the same as those registered in the neck chamber. These authors measured pressures with a 20 cm catheter whose tip was advanced to the vicinity of the carotid sinus in human volunteers; they found that neck pressure is reduced by 14 per cent and neck suction by 36 per cent in transmission through neck tissues. These data conflict with those of Thron *et al.* (1967), who measured pressure changes in the *oesophagus* in human volunteers, and with those of Shubrooks (1972), who measured neck tissue pressure with a surgically introduced balloon catheter in anaesthetized dogs with plaster neck chambers. Both Thron and Shubrooks reported nearly perfect transmission of neck chamber pressure changes through neck tissues.

Ludbrook's data are, however, open to criticism. First, although the needle over which the catheter was advanced was positioned in an area where pulsations were transmitted, the precise location of its tip was not determined. Second, the method used for recording neck tissue pressure is open to criticism. Ludbrook *et al.* (1977) flushed the catheter intermittently with saline. Pitfalls associated with techniques for measurement of tissue pressure were discussed exhaustively by Guyton *et al.* (1971). The tip of the catheter or needle may become occluded by tissue or gel, and spontaneous or induced pressure changes (such as those caused by neck pressure or suction) may cause fluid to be sucked out of the needle or catheter, and may change the relation between the tip and surrounding tissues.

Several different approaches have been developed to surmount these difficulties with tissue pressure measurements. These include a catheter with minute plastic wicks protruding from its tip (Scholander *et al.* 1968), a capsule connected to a catheter (Jarecki *et al.* 1978), and a catheter connected to a balloon (this was used by Shubrooks (1972), who found nearly perfect transmission of neck chamber pressures in dogs). Another method involves injection of a minute amount (about one microlitre) of fluid from a catheter to form a fluid interface between the catheter tip and the tissue; however, this interface persists only for about 30 to 60 s, after which time a new injection must be made. The catheter used by Ludbrook *et al.* (1977) was flushed intermittently with 0.1 to 0.2 ml saline. After each flush, measurements were deferred for 30 to 60 s; the authors do not indicate what the interval between flushes was.

A study by Hargens and co-workers (1987) provides support for the conclusions of Ludbrook *et al.* (1977), that pressure transmission through neck tissue is imperfect. Hargens advanced balloon catheters to various arm and leg depths of human cadavers and measured pressures during inflation of encircling blood pressure cuffs to levels between 100 and 500 mmHg. He found that pressure is uniformly reduced in transmission through the tissues of the extremities, and that the reductions are greatest deep in the limbs.

The results reported by Ludbrook *et al.* (1977) on this question are easily the most complete in the literature, and it seems unlikely that they will be superseded soon by other measurements. This study found that there was surprising uniformity of pressure transmission within and among subjects. Repeated applications of neck pressure changes in the same subject and in different subjects (including some who were 'asthenic' and others who were 'thickset') were likely to be attenuated to the same degree. These observations have several important implications. First, since pressure transmission is consistent within individuals, neck chamber techniques can be used to evaluate the effects of interventions on subjects. Second, it should be possible to use the neck chamber technique in groups of subjects who have different body habituses. Because of this finding, transmission of pressure should be the same in normotensive subjects and hypertensive patients, who tend to be heavier (Chiang *et al.* 1969). Third, since transmission does not appear to be influenced by neck size, it should be possible to use the neck chamber technique before and after interventions that may alter neck size, such as head-down tilt and exposure to microgravity in space.

The prime indication for use of pressure correction factors may be to compare responses to neck pressure with those to neck suction. Figure 5.10 shows average R–R interval responses of one subject to 21 applications of a complex neck pressure change (Sprenkle *et al.* 1986). Responses are plotted against measured neck chamber changes (solid line) and against neck pressure changes corrected for defects in transmission through neck tissue according to Ludbrook *et al.* (1977) (dashed line). (It is problematic whether correction factors were applied correctly to these data, because Ludbrook's corrections did not anticipate that negative pressure changes might begin during positive pressure changes.) The primary influence of application of correction factors is to increase the slope in the linear portion, and to reduce the pressure range over which baroreflexes are active.

It may not be necessary to correct for imperfect transmission of pressure in most neck chamber experiments. In studies in which only neck pressure

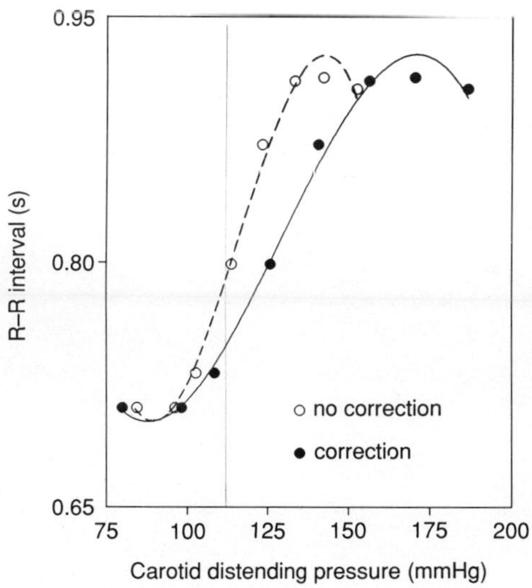

FIG. 5.10. Effect of applying factors to correct for imperfect transmission of neck chamber to the carotid sheath.

or suction is applied, it is probably irrelevant that there may be dissimilar transmission of pressure of the opposite sign. If the effects of *interventions* on responses to neck chamber pressure changes are being measured, it may be unnecessary to apply pressure corrections; presumably, defects in transmission should be equal before and after interventions, and therefore should not modify the significance of differences between responses before and after the interventions.

Techniques for study of baroreflex mechanisms with neck chambers may be more heterogeneous than other methods described earlier. For example, although Valsalva manoeuvres differ widely (see Chapter 4), *after Valsalva straining is released*, the reaction takes its course, without interference from the investigator. Similarly, after a bolus of a vasoactive drug is released into the circulation, pressure changes occur independent of the investigator. However, *all aspects of neck chamber stimuli are determined by the investigator and his or her equipment.*

First, neck chamber designs are numerous. Figure 2.6 shows the original neck chamber invented by Ernsting and Parry (1957). Subsequent developments in neck chamber design have tended to occur independent of each other in Europe and the United States. Three different types of neck chamber are depicted in Fig. 5.11. These include the one used by Thron *et al.* (1967) and Bjurstedt *et al.* (1975), which encloses the head and neck in a diving helmet (Fig. 5.11, left); the one used by Mancia (Ludbrook *et al.* 1977), which is made of plaster of paris and encircles the neck but does not enclose the head (Fig. 5.11, centre); and one developed by Eckberg *et al.* (1975), which is made of malleable sheet lead and encircles only the anterior and lateral aspects of the neck (Fig. 5.11, right). Bauernfeind and co-workers

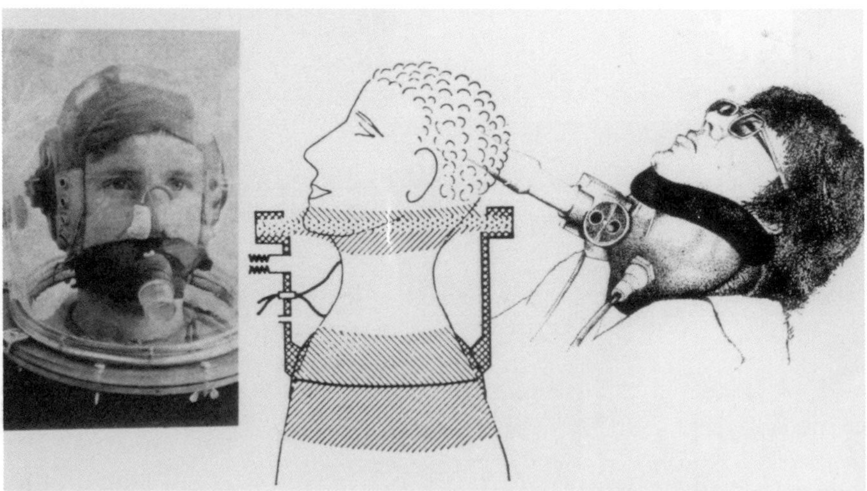

FIG. 5.11. Three types of neck chamber (see text for explanation). Reproduced with permission (Bjurstedt *et al.* 1975; Ludbrook *et al.* 1977; Eckberg *et al.* 1975).

(1985), in a preliminary communication, and Tafil-Klawe *et al.* (1990) and their co-workers devised a chamber that can be used to stimulate selectively right or left carotid baroreceptors (Tafil-Klawe reported that neck suction applied to the right side elicits more cardiac slowing than neck suction applied to the left).

To our knowledge, no one has compared responses of the same subjects to pressure changes applied to different neck chambers. Therefore there is an element of conjecture regarding differences among them. Clearly the first two deliver pressure all around the neck, and the third delivers pressure only to the anterior and lateral aspects of the neck. The practical importance of this difference is unknown. All three chambers share one common attribute: they all leak air. Therefore the vacuum or pressure source must withdraw or replenish air continuously. Usually, commercial vacuum cleaners are used for this purpose. According to published records, the third chamber (Fig. 5.11, right) provides a better vacuum and pressure seal than the first two; with this device, rates of pressure change can exceed 4000 mmHg/s (Eckberg 1977*b*). Also, the third unit tends to fit all subjects; the second may require modification for individual subjects (G. Mancia, personal communication).

A fourth neck chamber (Sprenkle *et al.* 1986) is shown in Fig. 5.12. This chamber is a descendant of the one shown in Fig. 5.11 (right). It differs from that one and from the others depicted in Fig. 5.11 in that, because of its contour and construction materials (Silastic, Dow-Corning, Midland, MI, USA), it forms nearly perfect vacuum and pressure seals. Because the

FIG. 5.12. Neck chamber developed for use on American Space Shuttle. Adapted with permission (Eckberg 1991).

seal is so good, pressure changes can be produced by a bellows (with a finite capacity) driven by a stepping motor. Use of the bellows–stepping motor combination allows delivery of complex wave forms, which can be controlled by computer software. Responses to a complex pressure sequence delivered by this equipment are shown in Fig. 5.13 (Fritsch *et al.* 1989). This unit was developed for research on board the American Space Shuttle and is available commercially (Engineering Development Lab, Newport News, VA, USA). Instructions for construction of this unit, and the one depicted on the right in Fig. 5.11 have been published (Sprenkle *et al.* 1986; Eckberg *et al.* 1975). Al-Timman and co-workers (1988) described a further modification of the neck chamber shown in Fig. 5.12, which may make it easier to construct.

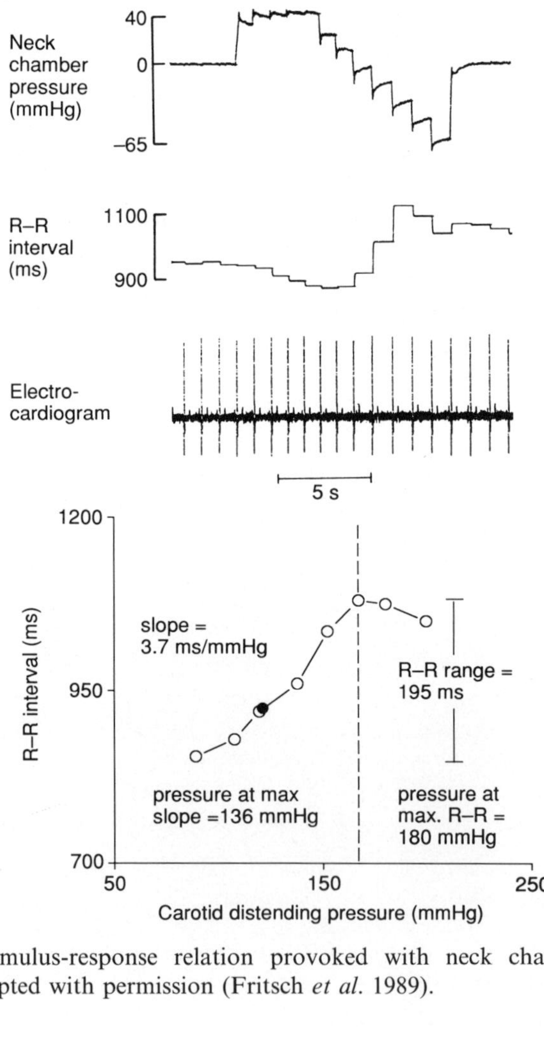

FIG. 5.13. Stimulus-response relation provoked with neck chamber shown in Fig. 5.12. Adapted with permission (Fritsch *et al.* 1989).

Figure 5.13 also indicates ways of reducing these data for statistical analyses. D.A. Ludwig and Convertino (1991) systematically evaluated mathematical means for characterizing the function depicted in this figure. They concluded that some information is redundant, and that over 90 per cent of the variance across the nine parameters that they evaluated can be explained by four measurements: baseline R–R interval and arterial pressure, baroreflex gain (maximum slope of the R–R interval/carotid distending pressure relation), and the ratio of baseline R–R interval to maximum R–R interval (operational point).

A cursory reading of Chapter 6 will indicate that the form of pressure changes delivered to baroreceptive areas importantly determines afferent baroreceptor responses. There is no reason to expect that human baroreceptors are different from those in other species; therefore, in humans as in other species, qualities of efferent autonomic responses to baroreflex forcings will be determined importantly by the qualities of pressure inputs.

It may be a mistake to assume that if the same pressure is applied in different studies, for example −30 mmHg, baroreceptor stimuli are the same. A 30 mmHg neck suction pulse begun at a time in the cardiac cycle when the sinus node is maximally responsive, with a maximum rate of pressure change of 2000 mmHg/s, lasting 0.6 s may be related only distantly to a 30 mmHg pulse of neck suction begun at a random time within the cardiac cycle, with a maximum rate of pressure change of 10 mmHg/s, lasting several min. Although the ultimate intensity of stimuli used across published studies may be comparable, other attributes of stimuli may be vastly different.

We suggest that all attributes of baroreflex stimuli applied experimentally must be described if responses to the intervention are to be interpreted properly. The mere indication of the neck pressure change achieved ultimately provides very little information about the actual baroreceptor stimulus. It is not surprising that different investigators publish widely divergent results from experiments employing ostensibly similar neck chamber pressures. For example, the average R–R interval prolongation obtained with −60 mmHg was about 6 ms in one study (Thron *et al.* 1967), about 400 ms in another (Eckberg 1980*a*), and 180 ms in a third (Sopher *et al.* 1990).

Many disparities among measurements obtained from healthy human subjects to neck suction can be explained on the basis of knowledge of factors that influence afferent and efferent baroreflex responses. When pressure changes occur rapidly, baroreceptor firing rates reach maximum levels early, and then adapt to levels much lower than those achieved at the beginning of stimulation. When pressure changes occur slowly, firing rates increase gradually and eventually reach the low, adapted firing rate that is achieved when pressure changes are rapid (Landgren 1952*a*). Low adapted firing frequencies translate into low rates of firing of efferent vagal–cardiac nerve activity (Kunze 1980), and these translate into small increases of R–

R intervals (Katona *et al.* 1970). Figure 5.14, published by Kunze (1980) from research conducted on anaesthetized cats, provides some indication of how peak and adapted baroreceptor firing frequencies are translated into efferent vagal–cardiac activity. The maximum adapted firing frequency was nearly an order of magnitude less than the peak response.

The *rate of neck chamber pressure change* varies considerably from study to study. For example, the rate of onset of neck suction may be extremely slow (from 0 to −30 mmHg may require 10-15 s (J. Wagner *et al.* 1968)) or extremely rapid (the transition from 0 to −30 may be accomplished within about 25 ms (Eckberg 1977*b*)). Because the magnitude of sinus node (Eckberg 1977*b*) and muscle sympathetic nerve (Båth *et al.* 1981) responses are determined importantly by the rate of onset of neck chamber stimuli, it is probably inappropriate to compare results from studies in which the rates of pressure change are dissimilar.

Similarly, the *duration* of baroreflex stimuli applied by different groups is highly variable. Neck suction is applied commonly for about 2 min (J. Wagner *et al.* 1968; Mancia *et al.* 1978*b*), but it may be applied for periods as brief as 0.1 s (Eckberg 1977*b*). As mentioned, one major difficulty in the study of *human* baroreflexes is that subjects' reflexes are intact, and available to modify integrated responses to stimuli. *We believe that all responses of humans to baroreflex forcings are compromised by ongoing reflex adjustments.* A corollary of this is that in studies of human baroreflex function, *it may not be possible to quantify accurately the driving stimulus.* Neck suction of only 0.1 s duration increases vagal–cardiac nerve traffic and delays the appearance of the next sinus beat (Eckberg 1977). Prolongation of that R–R interval prolongs diastolic runoff, lowers the next diastolic pressure (Rand-

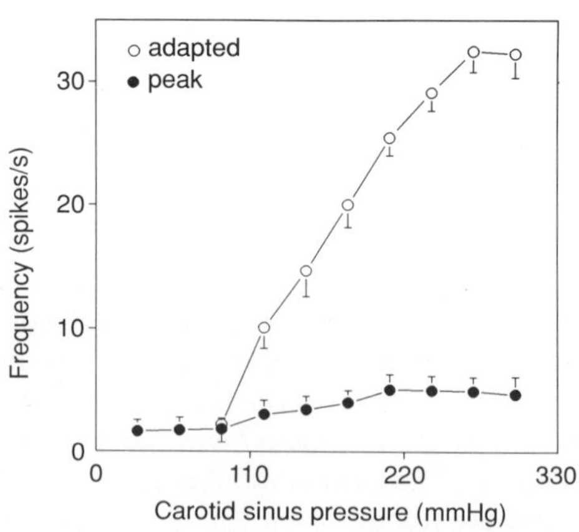

FIG. 5.14. Peak and adapted vagus nerve baroreceptor firing frequencies during a step arterial pressure increase measured in anaesthetized cats. Adapted with permission (Kunze 1980).

all *et al.* 1984), and increases the likelihood of occurrence of a muscle sympathetic burst (Sundlöf and Wallin 1977). By this means, an extremely brief baroreceptor stimulus can increase sympathetic response in the same R–R interval in which it is delivered.

Although contamination of reflex responses cannot be avoided entirely in humans, effects are minimized if short stimuli are used. In the example given, although sympathetic traffic might be increased in the same heart period in which the baroreceptor stimulus is given, its effects on arterial pressure will not be seen immediately. Before a sympathetic burst can alter arterial pressure, it must travel from sympathetic ganglia over unmyelinated fibres to effectors (the latency to the peroneal nerve in the leg is about 1.3 s (Fagius and Wallin 1980*a*)). Moreover, further delays are introduced by release of noradrenaline and by the time required for noradrenaline to act on vascular receptors and initiate smooth muscle contraction. Because of these delays, arterial pressure does not begin to rise until about the second heart beat after the sympathetic burst, and does not reach a peak until about the sixth beat (Båth *et al.* 1981; Wallin and Nerhed 1982). Therefore it seems likely that 2 or perhaps 3 s may elapse after a neck pressure change before sympathetically mediated changes of arterial pressure occur. As mentioned, R–R interval prolongation may alter diastolic pressure of the cardiac interval in which neck pressure is delivered.

In published literature on human baroreflex responses, the same words are used by different groups to convey vastly different meanings. In Eckberg's studies, the shortest neck pressure change applied has been 0.1 s, and in all but a few studies, the longest neck pressure change applied has been 5 s (Eckberg 1977*b*; Eckberg 1980*a*). In his nomenclature, 'late' usually refers to the last second of a 5 s pressure change. In contrast, in Mancia's studies, 'early' responses to neck pressure changes are taken between 5 and 15 s, and 'late' responses are those occurring between 90 and 120 s (Mancia *et al.* 1978*b*, 1985).

The conclusion of this argument is that, simply because humans are 'closed loop' preparations, baroreflex forcings should be as brief as possible. A corollary of this is that the longer the baroreflex forcing is, the more complex the mediation of responses will be, and the more difficult the response will be to interpret.

There is another aspect of baroreflex physiology that militates in favour of brief stimuli: baroreceptor *adaptation* (and its first cousin, *resetting*). Major (about 50 per cent) adaptation of firing frequency occurs in experimental animals within 0.1 s of the onset of a pressure step (Landgren 1952*a*), and baroreceptor-vagal and -sympathetic *resetting* in humans begins within seconds (M.L. Smith *et al.* 1989*b*). These concerns may apply more to neck *suction* than *pressure*. Mancia has shown convincingly that haemodynamic responses to neck pressure are sustained for at least 2 min (Mancia *et al.* 1985). This is not to say that viscoelastic creep does not occur when dimensions of baroreceptive areas are reduced; Franz showed that return of firing frequency towards baseline levels occurs both with increases and decreases of distending pressure (Franz 1969). This phenomenon is shown in Fig. 6.17.

The duration of step pressure changes used in human research is highly variable, but it must influence the magnitude of measured responses importantly. Many investigators continue neck suction (or pressure) for 2 min, and a few continue neck suction for as long as 20 min (Norsk *et al.* 1987). Figure 5.15 shows a record obtained from one subject of superimposed muscle sympathetic nerve responses to ten applications of 5 s neck suction. Muscle sympathetic nerve activity was inhibited totally but only briefly during neck suction; by the final second of suction, sympathetic traffic was returning towards baseline levels. After more prolonged neck suction, muscle sympathetic nerve activity may not be different from control, and cardioacceleration instead of cardiac deceleration may occur (Båth *et al.* 1981). Similarly, after 4 min of neck pressure or suction, plasma noradrenaline concentrations may be similar to what they were before the neck pressure change was applied (Mancia *et al.* 1979*c*).

Other neck pressure wave forms have been used. Sinusoidal neck suction has been used extensively, especially by Bevegård and Lindblad and their Swedish co-workers (Bevegård *et al.* 1977*a,c,d*; Lindblad 1980; Lindblad *et al.* 1981). Sinusoidal stimuli (with short periods) have several advantages. First, the effects of baroreceptor adaptation are less than they are with prolonged stimuli. Second, with use of sinusoidal stimuli with different

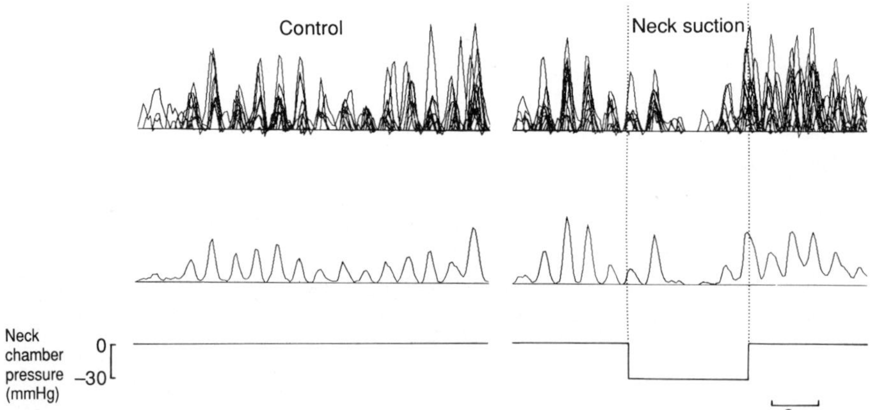

FIG. 5.15. Muscle sympathetic nerve activity of one healthy subject during quiet rest (left) and neck suction at − 30 mmHg for 5 s (right). Upper tracing was obtained by superimposing responses to ten applications of neck suction. Adapted with permission (Wallin and Eckberg 1982).

periods, it is possible to study the influence of rate of pressure change on the variables being measured. (However, it is also likely that in studies employing sinusoidal neck suction in humans, the maximum rate of change, even with sinusoids with short periods, does not reach the maximum rate of pressure change of the naturally occurring arterial pulse.) Third, use of sinusoidal neck suction permits signal averaging of responses to be done. A fourth (potential) advantage of sinusoids (or repetitive suction) is that subjects are not aware of when suction occurs in relation to cardiac activity. Muscle sympathetic nerve activity, arterial pressure, and R–R interval responses to sinusoids with different periods are shown in Fig. 5.16 (Båth *et al.* 1981).

Finally, neck pressure changes can be delivered as repetitive discrete stimuli. In such trains of stimuli, the intensity and the polarity (that is, neck suction or pressure) may be constant (Baskerville *et al.* 1979); the polarity may be constant, but the intensity of stimuli may be steadily increasing (that is, neck pressure may change as a repetitive ramp function (Eckberg and Eckberg 1982; T.J. Ebert *et al.* 1984*b*)); or polarity as well as intensity of stimuli may change. In an example of the last type, Eckberg (1980*a*) increased neck chamber pressure to about 40 mmHg, and about 5 s later, *reduced* neck chamber pressure briefly, by graded amounts. The rationale for this approach is that the initial positive pressure may reduce carotid distending pressure to below threshold levels, and subsequent neck chamber pressure reductions may define the entire stimulus–response relation. A response to one such forcing is shown in Fig. 5.17.

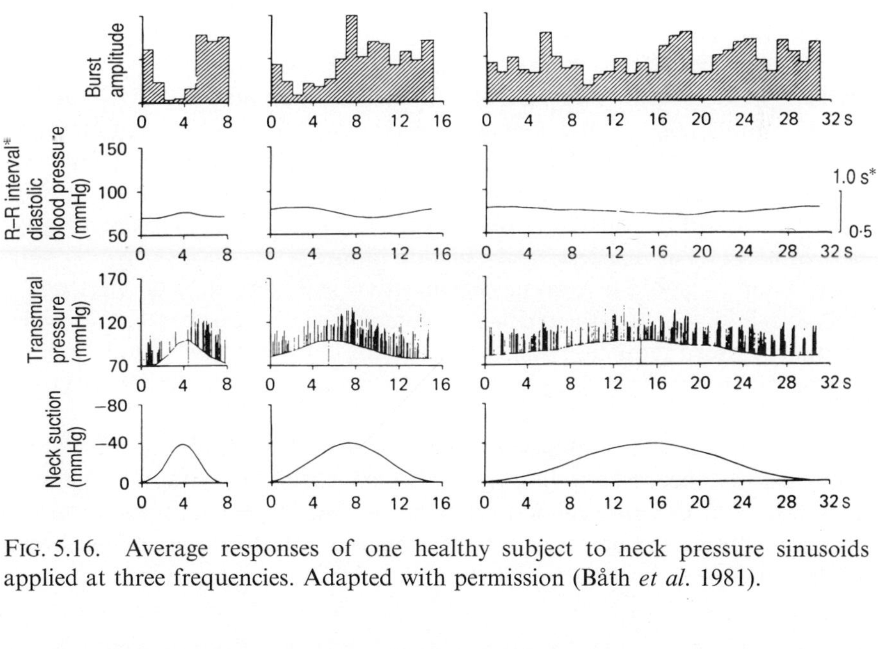

FIG. 5.16. Average responses of one healthy subject to neck pressure sinusoids applied at three frequencies. Adapted with permission (Båth *et al.* 1981).

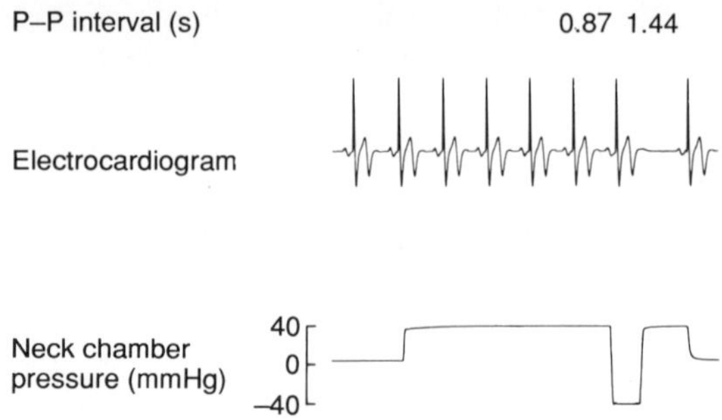

FIG. 5.17. Response of one healthy young subject to 0.6 s of neck suction of −40 mmHg, applied after a step increase of pressure to 40 mmHg. Adapted with permission (Eckberg 1980*a*).

Another variant of complex neck pressure changes is shown in Fig. 5.13. For such stimuli, pressure is increased by about 40 mmHg for about 5 s. Then, neck pressure is reduced with each R-wave, by fixed decrements of about 15 mmHg each, until a neck chamber pressure of about −65 mmHg is achieved. This method has the advantage that in many subjects, an entire stimulus–response relation can be defined, during a period lasting only about 15 s. (In some subjects, however, the 40 mmHg pressure applied at the beginning of the sequence does not lower their carotid distending pressures to below threshold levels. This is a limitation of the method that is dictated by subject comfort; neck pressures above 40 mmHg are noxious and tolerated poorly.)

This complex forcing perturbs sympathetic activity in a way that may minimize closed loop baroreflex adjustments. Fritsch *et al.* (1991) measured arterial pressure indirectly and muscle sympathetic nerve activity directly during complex neck chamber changes, as shown in Fig. 5.18. Their measurements help to define ongoing reflex adjustments during these complex stimuli. With the initial 5 s of neck pressure, muscle sympathetic nerve activity increases, R–R intervals decrease, and arterial pressure rises (by an average of 3 mmHg). The earliest neck suction appears to abolish sympathetic activity, and therefore, during the negative stairstep, arterial pressure is constant initially and then declines somewhat. It may be that absolute arterial pressure reductions occurring during baroreceptor stimulation are minimized by the preceding increases of muscle sympathetic nerve activity that occur.

There are limited published data on the reproducibility of human responses to baroreceptor stimuli. It is extremely important to know if responses to a forcing function are reproducible, especially if the influence of interventions is to be studied. Reproducibility can be discussed on several levels.

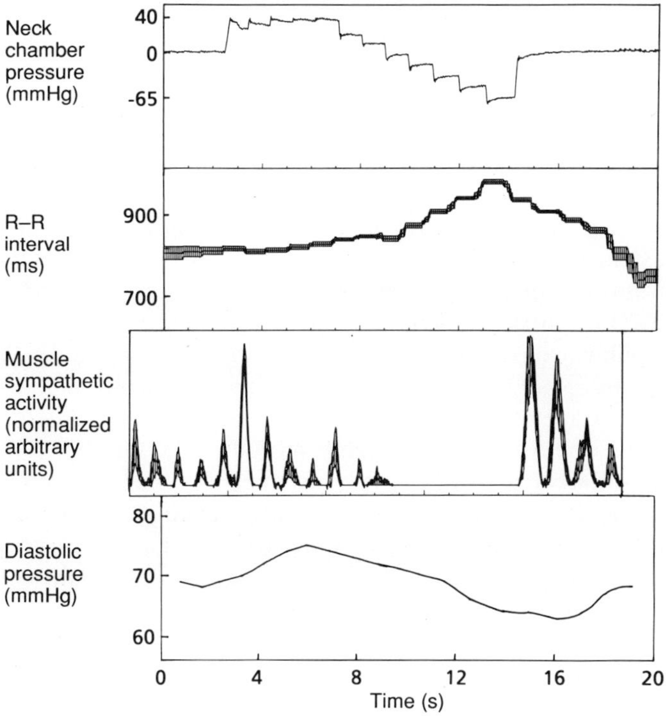

FIG. 5.18. Signal-averaged responses of one healthy subject to seven applications of the complex neck pressure profile generated by equipment depicted in Fig. 5.12. Reproduced with permission (Fritsch *et al.* 1991).

First, with closely fitting chambers whose pressure changes are controlled by an electronically controlled valve, the baroreflex *stimuli* are highly reproducible. Eckberg (1976) planimetered the area of neck stimuli of −60 mmHg for 0.58 s delivered ten times to each of eight subjects, and obtained a standard deviation of an observation of 1.6 per cent.

Two studies found poor short-term reproducibility of human baroreflex responses (D.S. Goldstein *et al.* 1982; Parati *et al.* 1983). In these studies, it was not clear that the rate of neck pressure change, the timing of the onset of stimuli within the cardiac cycle, and respiratory activity were controlled. Therefore it is likely that some variability of baroreflex responses results from influences known to affect baroreflex responses that were not controlled.

However, even if *all* of these factors are controlled, there remains substantial variability of human baroreflex responses. Sinus node responses to identical neck chamber pressure changes delivered at pressures that fall within the linear range of the stimulus–response relation are highly variable (Eckberg 1980*a*). One hypothesis advanced to explain this variability is that it is due to uncontrolled (uncontrollable?) central nervous system modulation of vagal responses to incoming baroreceptor information. Within this context, the variability of baroreceptor–cardiac reflex responses in the linear range may be an index of the ability of central factors to override incoming baroreflex information. Variability is small with intense stimuli, which presumably present such strong stimulation of vagal nuclei that their firing cannot be interdicted by central influences. Variability of baroreflex responses of one subject is shown in Fig. 5.19. These data, and those of D.S. Goldstein *et al.* (1982) and Parati *et al.* (1983), dictate that accurate characterization of human baroreceptor–cardiac reflexes requires that multiple, identical baroreceptor stimuli be delivered and responses be averaged.

Limited published data suggest that, as with plasma noradrenaline levels (de Champlain *et al.* 1976) and muscle sympathetic nerve activity (Wallin *et al.* 1981), average vagal baroreflex responses of groups of subjects studied on more than one occasion are highly reproducible. Gribbin *et al.* (1971) studied five subjects with phenylephrine injections on two occasions and found that R–R interval–systolic pressure relations were comparable. Eckberg (1976) drew similar conclusions from responses to graded neck suction of five subjects studied serially. In a more comprehensive study,

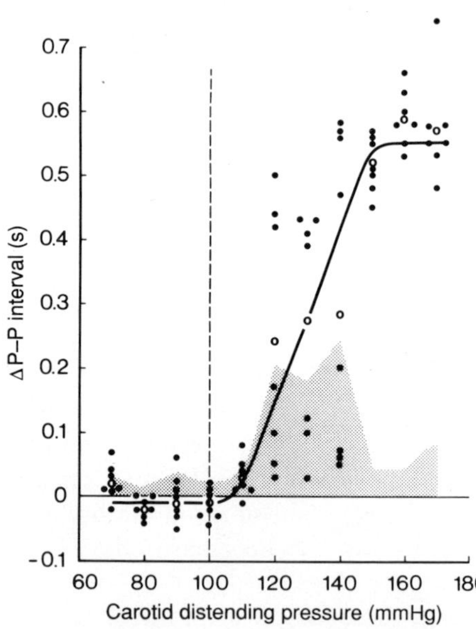

FIG. 5.19. R–R interval responses of one subject to multiple applications of different intensities of neck pressure applied according to the scheme depicted in Fig. 5.17. The stippled area indicates standard deviations of responses. Adapted with permission (Eckberg 1980*a*).

D.L. Eckberg *et al.* (1992) found that responses of 34 subjects studied twice were highly reproducible. These data are depicted in Fig. 10.1.

Although pressor drug infusions and neck suction seem to stimulate primarily arterial baroreceptors rather than other types of autonomic receptors, these methods share an intrinsic flaw with other techniques used to study autonomic reflexes in humans. Abrupt baroreceptor stimulation leads to reduction of efferent sympathetic nervous activity (Wallin and Eckberg 1982) and arterial pressure (Baskerville *et al.* 1979) within 2 s, and to cardioacceleration as early as 0.5 s after the onset of an abrupt reduction of afferent arterial baroreceptor activity (Eckberg 1980*a*). These nearly instantaneous reflex changes may alter afferent arterial baroreceptor traffic during the period of baroreceptor stimulation with pressor drugs or neck suction. The findings of Evans *et al.* (1979) and Greenway and Innes (1981) and their co-workers support this possibility. Evans *et al.* found that responses to comparable doses of pressor drugs are augmented by prevention of heart rate changes by cardiac pacing, and reflex adrenergic and cholinergic effects by administration of beta-adrenergic and muscarinic cholinergic blocking drugs. Their findings are summarized in Fig. 5.20.

Indirect evidence that ongoing baroreflex adjustments buffer arterial pressure responses to pressor injections can be found also in studies with human subjects: patients with subnormal baroreflex responses, including those with idiopathic orthostatic hypotension (H.N. Wagner and Braunwald 1956; Möhring *et al.* 1980) or diabetes (Eckberg *et al.* 1986), experience much greater pressure elevations during pressor infusions than people with intact reflex responses. Patients with autonomic dysfunction may experience major, prolonged elevations of arterial pressure after using dilute phenylephrine eye drops (Hines *et al.* 1981). Healthy subjects experience much greater pressure elevation with bolus phenylephrine injections after atropine than before (T.G. Pickering *et al.* 1972*b*); this suggests that cholinergic baroreflex mechanisms are at work *during the pressure ramp*, to limit the extent of pressure change. Therefore, as with the Valsalva manoeuvre, concurrent reflex adjustments may modify the intensity of the baroreflex stimuli delivered. Thus measured arterial (or neck chamber) pressure is a *function* of the stimulus rather than the stimulus itself.

COMPUTER ANALYSES THAT DISCOVER BAROREFLEX ENGAGEMENT

Two methods have been developed to discover and quantify baroreflex engagement in records of spontaneous fluctuations of R–R intervals and arterial pressures. Such approaches hold great promise because they can be used for continuous rather than intermittent analysis of baroreflex function, are unobtrusive, and do not require that assumptions be made regarding

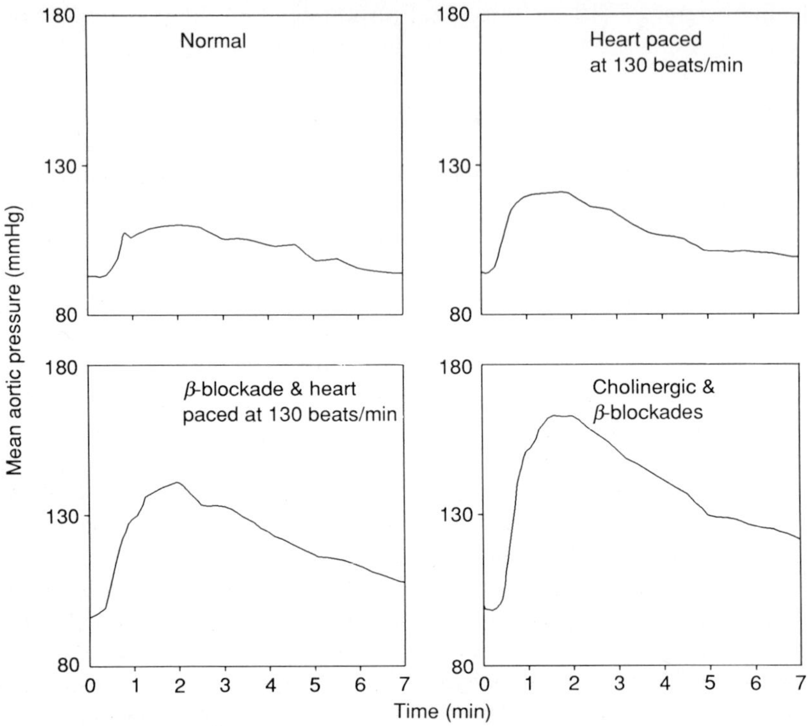

FIG. 5.20. Average arterial pressure responses of seven conscious dogs to bolus injections of phenylephrine. Adapted with permission (Evans *et al.* 1979).

the effects of drugs or mechanical devices used to trigger baroreflex responses.

The simpler of the two methods involves calculation of regressions between spontaneously occurring ramps of arterial pressure (either ascending or descending) and subsequent R–R intervals. Fritsch *et al.* (1986) and Parati *et al.* (1988) and their co-workers used such a method in healthy human subjects. In the study of Fritsch, continuous recordings of arterial pressure, electrocardiograms, and muscle sympathetic nerve activity were examined visually to find ramps of three or more increasing arterial pulses that were preceded, and presumably caused, by bursts of sympathetic activity. (Fritsch and co-workers arbitrarily limited their study to ramps preceded by sympathetic activity so that these ramps would be analogous to those occurring after bolus injections of pressor drugs. Their data do not exclude the possibility that similar analyses might have been made independent of the presence or absence of sympathetic bursts.) One such ramp is shown in Fig. 5.21. R–R interval changes with various offsets (see above) were related to preceding systolic pressures and linear regression analysis was performed.

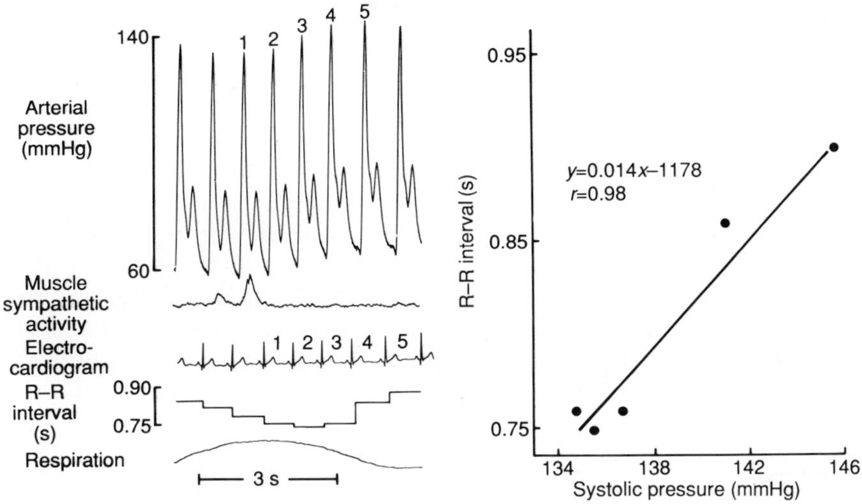

FIG. 5.21. Spontaneous five-beat ramp of arterial pressure of one subject. Adapted with permission (Fritsch *et al.* 1986).

The principal findings of this study are that the average slopes of these relations are similar to those measured after bolus injections of phenylephrine, are related inversely to baseline arterial pressure, and are related directly to the rate of arterial pressure change.

The study of Parati *et al.* (1988) was based on research conducted earlier by Bertineri *et al.* (1988) in cats. Bertineri and co-workers restricted their analyses to positive or negative pressure ramps comprising three or more pulses, and also required that there be concordant R–R interval changes (that is, they required that pressures and R–R intervals rise or fall together). About 15 per cent of heart beats satisfied these stringent criteria. The advantage of Bertineri's approach is that it does not require information regarding muscle sympathetic nerve activity, which is usually not available. More importantly, it permits analyses to be made automatically by computer. Bertineri *et al.* found that slopes of the relations they calculated were reduced substantially by sinoaortic denervation. (As mentioned, Persson *et al.* (1988) found that spontaneous arterial pressure increases are associated with R–R interval *decreases* after sinoaortic denervation.) These data provide strong evidence that calculations of baroreflex slopes during spontaneously occurring pressure changes yield valid quantitative information regarding baroreflex gain. The physiological basis for spontaneous pressure changes, discussed in Chapter 8, is far from clear.

More complicated computer methods were developed by Peňáz (1978), Sayers (1982), deBoer (1987), Robbe (1987), and Baselli (1988) and their co-workers. Their approaches are based upon the observation that high-frequency (above about 0.2 Hz) R–R interval or heart rate changes, which are known to be mediated vagally, are associated with (and probably caused by) arterial pressure changes occurring also at high (that is, respiratory) frequencies. Several treatments of data are used to quantify the relation between high-frequency changes in R–R interval and arterial pressure recordings. (These techniques are described in outline here; for more detailed descriptions, the reader is referred to the articles referenced above.) This process is illustrated by Fig. 5.22, published by de Boer *et al.* (1986). First, discrete measurements of sequential R–R intervals and arterial pressures are subjected to a fast Fourier transformation to derive power spectra. These power spectra express variability of the signals as functions of their frequency content. Figure 5.22a shows power spectra of respiration (dashed line) and

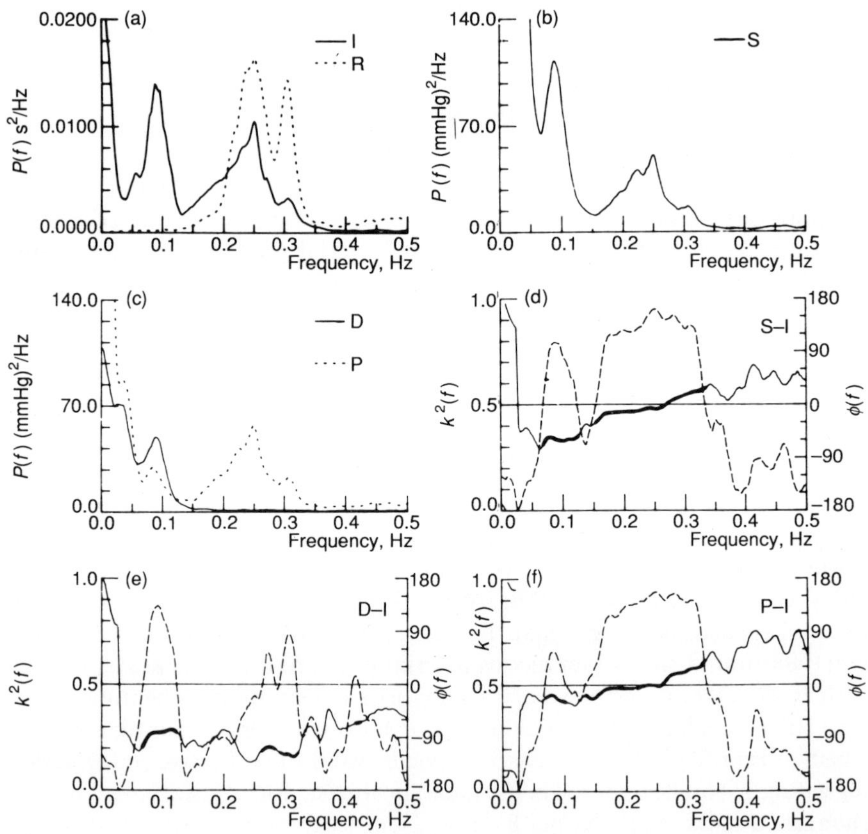

FIG. 5.22. Method developed by de Boer *et al.* (1986) to obtain baroreflex slopes from resting arterial pressure and electrocardiograms. See text for explanation. Adapted with permission.

R–R intervals (solid line). Figures 5.22b and c show power spectra of systolic and diastolic pressures. (Figure 5.22c also documents the intriguing observation of de Boer *et al.* (1987) that *diastolic* pressure does not fluctuate appreciably at respiratory frequencies.)

Second, cross-spectra between the two signals are calculated. Cross-spectra have two components. The (squared) coherence spectrum defines frequencies over which variability between two signals is related linearly (a value of 0 indicates no correlation, and a value of 1 indicates perfect correlation). The second component is the phase spectrum, which indicates how one signal leads or lags the other. In Fig. 5.22(d)-(f), coherence spectra (squared) are shown as dashed lines, and phase spectra are shown as solid lines. In this figure, a negative phase spectrum indicates that pressure variability leads R–R interval variability. (This is an oversimplification; as pointed out by de Boer *et al.* (1986), at a phase of zero, pressure may lead R–R interval by 360 degrees (3.33 s), or R–R interval may lead pressure by 360 or 720 degrees. The interpretation of these data depends upon clinical correlations.) The heavy solid lines indicate the phase spectra over which the coherence spectra have high values.

The final steps taken to derive baroreflex slopes differ among different groups. De Boer and co-workers (1986) use cross-spectral analyses to determine the frequency range over which high correlations exist between R–R intervals and arterial pressure. They then filter R–R intervals and arterial pressures to exclude changes occurring outside this frequency range and calculate the slope of the resulting regression. A plot of all pairs of data from one subject that were not filtered is shown in Fig. 5.23. Robbe *et al.* (1987) obtain a baroreflex slope (which they call the modulus) by measuring the ratio between changes of R–R interval time and changes in systolic pressure. Results obtained with this method correlate extremely well with those obtained with the phenylephrine method.

Advantages and disadvantages with studies of spontaneous pressure variations

The great advantage of these approaches is that they do not involve interventions; they do not require use of foreign or naturally occurring drugs or neurotransmitters, or unnatural neck pressure changes. Unlike the vasoactive drug and neck chamber methods, these methods do not perturb the reflex that they measure. A corollary of this is that, in conjunction with ambulatory intraarterial pressure monitoring, these methods permit study of baroreflex mechanisms as they are perturbed by physiological events, such as changes of posture, exertion, and mental arousal (Steptoe and Sawada 1989).

As with other approaches, there are disadvantages to studies of responses to spontaneous arterial pressure variability. As with the vasoactive drug

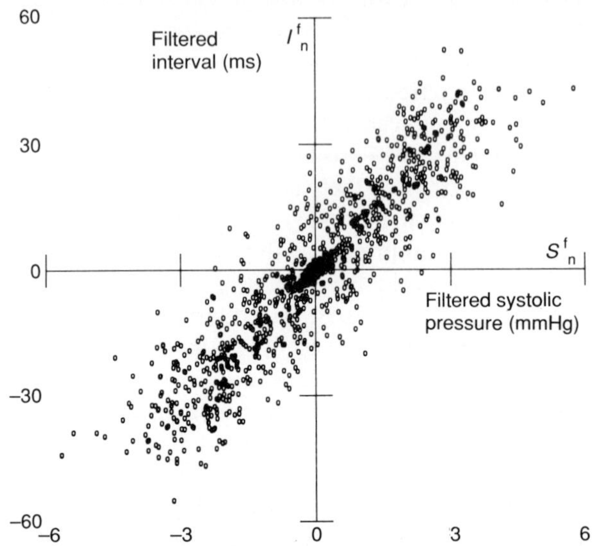

FIG. 5.23. Scattergram obtained by de Boer *et al.* (1986) to obtain baroreflex slopes from resting arterial pressure and electrocardiograms. Adapted with permission.

technique, this method can be used only to study baroreflex modulation of sinus node function; it cannot be used to study baroreflex control of arterial pressure or sympathetic activity. These methods require relatively long (usually between 1 and 5 min) periods of recording; thus their temporal resolution does not compete with that of other approaches which can measure responses occurring over periods lasting seconds. Also, the computer measurements described require that the time series be stationary. Thus these methods may not be applicable to studies involving physiological transients. Finally, as with vasoactive drug techniques, this method delineates only a portion of the sigmoid baroreceptor–cardiac reflex relation.

METHODS THAT HARNESS THE FORCE OF GRAVITY

Methods described above alter baroreceptor firing through changes of pressure inside or outside baroreceptive areas. Firing can also be altered by arterial pressure transients provoked by abrupt changes of body position. Several methods have been described. Lamberti *et al.* (1968) elegantly quantified arterial pressure responses of conscious dogs to static and dynamic head raising and lowering. A. Bhattacharya *et al.* (1979) perturbed heart rate in chronically instrumented tranquillized dogs by applying rhythmical whole-body oscillations. Lindqvist (1990) used an ingenious rocking tilt-table in his studies of healthy volunteers. Arguably the most technically demanding method to stimulate human baroreceptors was used by D. Linnarsson (unpublished), who measured R–R interval responses of healthy subjects to increased or decreased gravitational forces provoked by parabolic flights of a jet plane.

SUMMARY

The second half of the twentieth century has seen the development of a wealth of methods that make it possible for researchers to mount scientifically credible studies of human baroreflex function. We have discussed these new methods critically, and have underscored the failings of all of them. We believe that most of the problems we have discussed are of greater theoretical than practical importance. In the study of Robbe *et al.* (1987) the Oxford method and the computer analysis method both yielded slopes that reflected baroreflex gain. These methods and most neck suction protocols reduce complex sigmoid baroreceptor stimulus–sinus node response relations to linear functions. Other methods, including those employing 'steady-state' vasoactive drug infusions and sequential neck pressure–suction sequences may generate entire sigmoid baroreceptor–cardiac relations. Although the practical advantages of defining entire sigmoid baroreflex functions remain to be determined, published data are sufficient to indicate that not all populations operate on the linear portion of the relation, and that healthy subjects may operate on different portions of the relation during acute blood pressure changes. Unlike pharmacological methods, the neck chamber technique can be used used to study baroreflex control of arterial pressure. Drug infusion methods can be used to study baroreflex modulation of neurotransmitters, and all techniques can be used to study baroreflex control of sympathetic activity. Probably none of the available techniques deals adequately with the possibility that ongoing baroreflex adjustments modify the baroreflex responses being measured.

PART C
Physiology

6

Afferent baroreceptor activity

Adrian ran in and out, controlling the recording apparatus in the dark-room and developing the photographic plates. We were excited, both of us quite aware that what we now saw had never been observed before and that we were discovering a great secret of life, how the sensory nerves transmit their information to the brain.

Yngve Zotterman, *Touch, Tickle and Pain*

HUMAN BARORECEPTOR RECORDINGS

Hauss and Schmitt (1961, 1962*a,b*), Langrehr (1964), and Angell-James and Lumley (1974) have published the only direct recordings of human baroreceptor traffic in the literature. Their recordings, which probably include activity of chemoreceptor as well as baroreceptor nerve fibres, were made in patients undergoing surgery. The remarkable aortic nerve recording published by Langrehr (1964) is reproduced in Fig. 6.1. The left-hand tracing shows convincing pulse-synchronous firing in a multifibre preparation, and the right-hand tracing shows firing in a single fibre! (Presumably the occurrence of baroreceptor firing slightly before the beginning of the arterial pulse resulted from delays in registration of the pressure pulse.)

Angell-James and Lumley correlated *integrated* carotid sinus nerve activity with arterial pressure measured with a pressure gauge, and carotid sinus dimensions measured with a mechanical cantilever transducer applied directly to the artery. Their findings, depicted in Fig. 6.2, show that increases of pressure are associated with increases of carotid artery diameter and

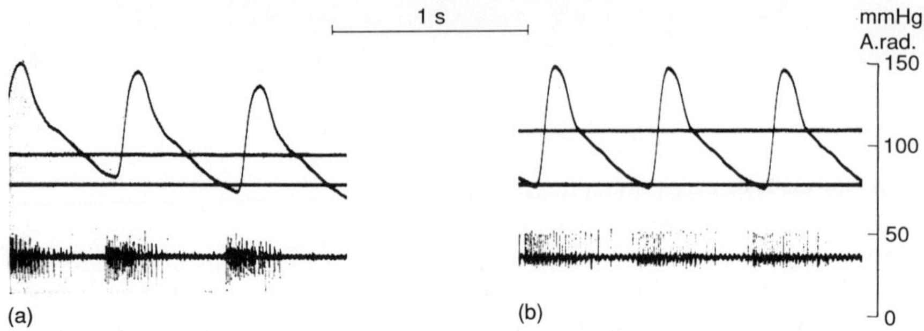

FIG. 6.1. The only convincing recordings of human multiple and single fibre baroreceptor nerve traffic extant. Reproduced with permission (Langrehr 1964).

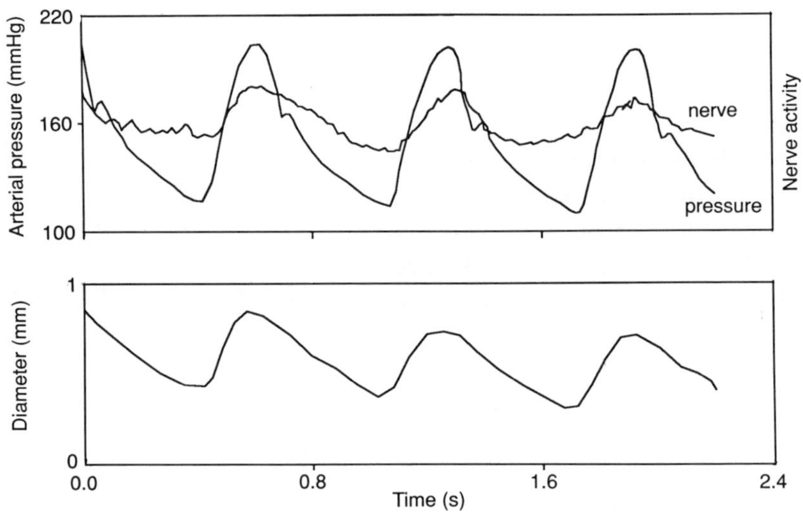

FIG. 6.2. Recording of integrated human carotid sinus nerve activity and diameter. Adapted with permission (Angell-James and Lumley 1974).

baroreceptor firing. Putative baroreceptor nerve recordings published by Hauss and Schmitt (1961, 1962*a,b*) are difficult to interpret; they show tonic rather than pulse-synchronous carotid sinus nerve firing.

Angell-James and Lumley (1974) also obtained inferential data on possible short-term functional consequences of changes of carotid sinus nerve activity. In six patients, post-operative correlations were made between changes of carotid sinus nerve traffic resulting from carotid endarterectomy and changes of blood pressure. In one patient, neither carotid sinus nerve traffic nor blood pressure changed. In two patients, reductions of nerve traffic after endarterectomy were associated with elevations of post-operative blood pressure. In three patients, increases of nerve traffic were associated with reductions of blood pressure. It seems likely that in the patients studied, changes of carotid sinus nerve traffic *caused* blood pressure changes. This notion is supported by the report of Tydén and co-workers (1980*a*) that post-endarterectomy hypotension and bradycardia (which they also thought were due to suddenly increased levels of afferent baroreceptor activity) are abolished by injections of lignocaine into the region of the carotid sinus.

Towne and Bernhard (1980) dispute the conclusion that post-operative hypertension is due to impaired baroreceptor transduction. Most of the carotid endarterectomy surgery in their series of 232 patients was done by two surgeons who used very different techniques. One routinely divided all visible baroreceptor nerves; the other attempted to protect and preserve them. The incidence of post-operative hypertension in patients of both surgeons was 24 per cent.

With these exceptions, almost all data on baroreceptor afferent activity have been obtained from animals. Paintal (1972) and Kirchheim (1976) dealt with this aspect of baroreflex function exhaustively. In what follows, we review the physiology of the afferent limb of the baroreflex arc, and emphasize aspects that may have particular relevance to human baroreflex testing. Since the number of direct recordings of human baroreceptor afferent traffic is vanishingly small, some data on integrated efferent neural or effector responses from human subjects are included in this chapter; this approach may be justified because of the great similarity between patterns of afferent baroreceptor nerve recordings from animals and patterns of efferent nerve or effector responses from human subjects.

THE TRANSDUCTION PROCESS

Function of baroreceptors is linked intimately to the mechanical properties of the tissues to which they are bound. The structure of baroreceptors is complex (see Fig. 3.7), and their manner of coupling to other tissues is understood poorly. Methods used for studying anatomical changes during baroreceptor transduction are primitive, and have been limited to measurements of arterial diameter changes and calculations of arterial wall stress based upon measurements of intra-arterial pressure and arterial diameter and wall thickness.

It is assumed that transduction occurs at the level of the baroreceptor nerve endings; however, as pointed out by A.M. Brown (1980), receptor potentials of baroreceptor nerves have never been recorded (published recordings are of axon potentials), and as pointed out by Sachs (1987), there is no definitive proof that baroreceptor nerve endings separated from cells of surrounding tissues produce mechanosensitive generator currents. Thus the critical first element in the baroreflex arc still must be regarded as a 'black box'; understanding of how baroreceptors convert pressure changes into electrical impulses is based almost exclusively upon inferences from axon recordings.

Two studies suggest strongly that *baroreceptors transduce stretch*. Hauss *et al.* (1949) measured afferent carotid sinus nerve activity in anaesthetized dogs before and after application of plaster casts encircling the carotid sinuses. They found that limitation of expansion of carotid sinuses by casts abolishes baroreceptor nerve traffic.

In a second report that is remarkable for the complexity of the experimental preparation, Angell-James (1971*b*) studied the effects of reductions of pressure outside the aorta on aortic nerve activity. She found that intra-aortic pressure elevations and extra-aortic pressure reductions produce comparable increases of aortic nerve traffic. This is illustrated in Fig. 6.3. Angell-James also found that when pressure is changed simultaneously and equally inside and outside the aorta, aortic nerve traffic does not change. These data (and those of Hauss *et al.* (1949)) speak against a role for *compression* of receptors in pressure transduction, and suggest that the lamellated, Pacinian-like structures described by Dropmann (1967) in baroreceptor areas may not have great functional importance. The data do not exclude the likelihood that the longitudinal stretch that occurs during pulsatile arterial expansion (Koushanpour and Kelso 1972; Arndt *et al.* 1975) also stimulates baroreceptors (J.H. Green and Heffron 1968*b*).

These results are relevant to the experimental study of *human* baroreflexes. As discussed in Chapter 5, research into human baroreflex function is based upon measurements of responses to changes of transmural pressure provoked either by changes of pressure within or without baroreceptor areas. The study of Angell-James (1971*b*) suggests that neck pressure changes (that is, pressure changes applied *outside* baroreceptive areas) should be as effective as intracarotid (or intra-aortic) pressure changes in altering baroreceptor nerve traffic.

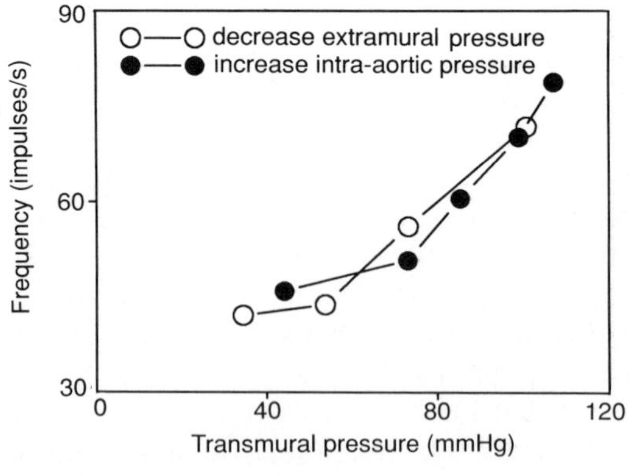

FIG. 6.3. Aortic nerve activity during pressure increases within the aorta and pressure reductions outside the aorta. Adapted with permission (Angell James 1971*b*).

Landgren (1952*a*) was the first to show (in anaesthetized cats) a relation between carotid sinus dimensions and single, probably myelinated baroreceptor nerve activity. His data show that although baroreceptor firing is proportional to vessel diameter, firing becomes maximal before vessel diameter is maximal. Koushanpour and Kelso (1972) showed that during progressive increases of pressure in an isolated carotid sinus preparation, calculated receptor strain continues to increase after baroreceptor firing reaches maximum levels. As discussed below, the simple direct relation between arterial diameter and baroreceptor firing probably masks great complexities introduced by arterial and baroreceptor geometry and passive viscoelastic and active myogenic changes occurring within the arterial wall.

The ultrasonic measurements of human carotid sinus diameters of Kober and Arndt (1970), Fig. 6.4, indicate that positive and negative pressure changes applied to a neck chamber yield nearly linear changes of human carotid sinus diameter. Thus, although few actual measurements of human baroreceptor nerve traffic have been reported, the data of Landgren (1952*a*) and others document a relation between carotid stretch and nerve firing frequency, and the data of Kober and Arndt (1970) document a relation between pressures applied to a neck chamber and actual changes of carotid artery diameter. It follows that the nearly linear increases of R–R intervals that occur in response to low to moderate intensities of graded neck suction (Eckberg 1980*a*) are probably related to linear increases of carotid baroreceptor firing.

Kober and Arndt (1970) used a *non-invasive* echo-ranging technique and found that carotid artery dimensions increase by about 14 per cent with each systolic pulse. Angell-James and Lumley (1974) and Greenfield *et al.* (1964) used intraoperative mechanical transducers and found that carotid or aortic dimensions or calculated cross-sectional areas increase by only about 2 to 3 per cent. Kober and Arndt speculate that the disparity between

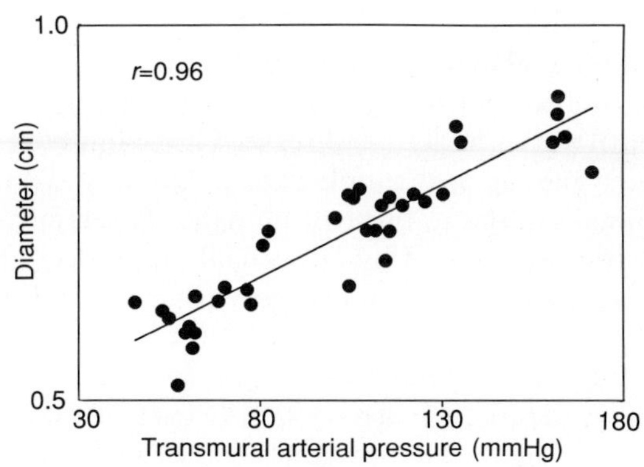

FIG. 6.4. Human carotid artery diameter during neck pressure and suction. Adapted with permission (Kober and Arndt 1970).

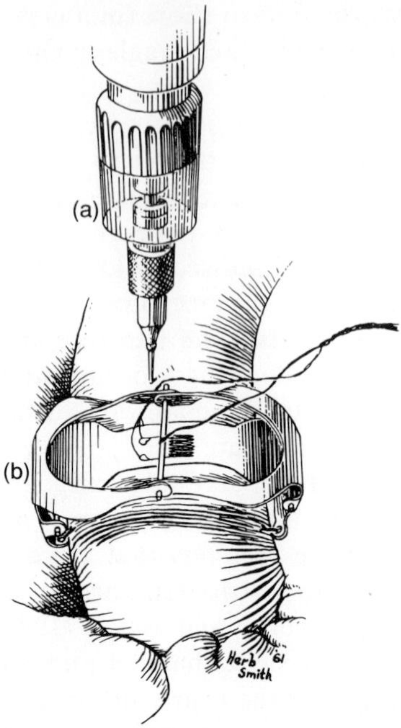

FIG. 6.5. Apparatus used to measure aortic diameter and pressure in patients undergoing thoracic surgery. Adapted with permission (Greenfield and Patel 1962).

their measurements and those obtained invasively by others exists because of reductions of compliance secondary to surgical dissection and direct application of measuring devices. (A similar conclusion was drawn by Shubrooks (1972), who applied external neck pressure changes to dogs and found that baroreflex gain is much larger when carotid distending pressure is increased with non-invasive methods than when it is increased with invasive methods.)

A complementary explanation is that in the patients of Angell-James and Lumley (1974) studied during carotid endarterectomy, compliance was reduced by extensive atherosclerotic disease. The ubiquity of atherosclerosis in adult humans argues against simple extrapolation of data on mechanical properties of animal arteries to those of humans. This issue is also discussed in Chapter 3. A transducer used by Greenfield and Patel (1962) to measure aortic compliance in patients during surgery is shown in Fig. 6.5.

RELATION BETWEEN PRESSURE AND BARORECEPTOR ACTIVITY

The earliest recordings of baroreceptor nerve activity were published in a remarkable article by Bronk and Stella (1932). Tracings of multifibre carotid sinus nerve activity (upper panel) and single fibre (probably myelinated) baroreceptor nerve activity (lower panel) from this article are reproduced in Fig. 6.6. The upper record shows that during each systole, baroreceptor firing begins with a barrage, which is succeeded by reduced activity, and then neural silence during diastole. The single fibre recording (lower panel) shows that during this experimental session, volleys of firing comprising (with one exception) three action potentials occurred regularly, at a rate of one volley per heart beat (the time scale is different from that of the upper panel). Other figures in this article show that peak baroreceptor nerve firing coincides with the upstroke of the systolic arterial pulse.

Many essential features of baroreceptor function are delineated in the article by Bronk and Stella (1932). Baroreceptor firing begins when arterial pressure exceeds some threshold level. Firing does not increase gradually from zero as pressure rises; rather, it begins at some frequency substantially greater than zero. The number of impulses increases linearly over a range of mean arterial pressures between about 60 and 120 mmHg (the linear regression correlation coefficient of the pressure–frequency relation in their Fig. 8 is 0.99). As pressure rises, total nerve activity increases through increased firing rates of individual fibres, recruitment of additional fibres whose thresholds for activation are higher, and prolongation of firing duration. When pressure is high (mean aortic pressure of 100-120 mmHg), firing

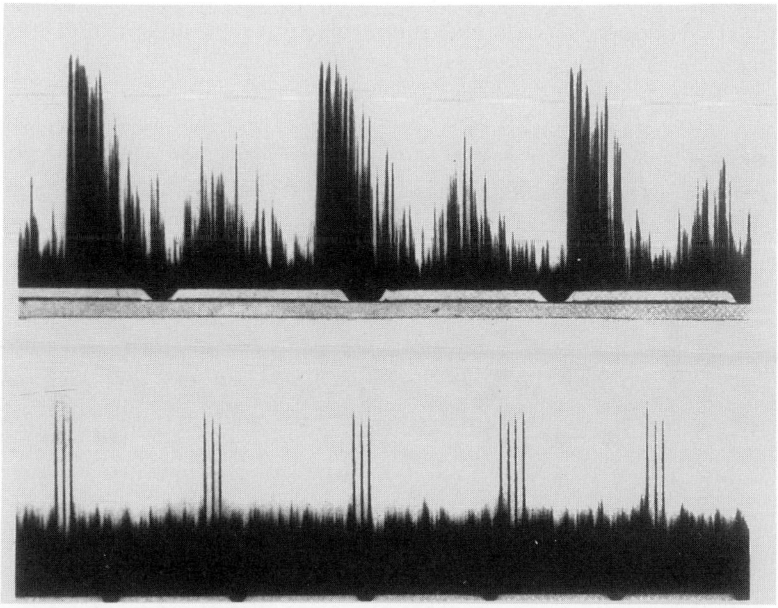

FIG. 6.6. Earliest published recordings of carotid sinus nerve action potentials. Reproduced with permission (Bronk and Stella 1932).

persists throughout the cardiac cycle. Rate sensitivity is evident in the preponderance of firing at the onset of systole, when the rate of pressure change and arterial expansion is greatest. Hysteresis is suggested by the observation that for identical pressures, firing rates are greater when pressure is rising than when it is falling.

Baroreceptors are extremely sensitive to intra-arterial pressure changes. One of the interventions used by Landgren in his landmark study (1952*a*), was to deliver a pressure step of *1 mmHg* to an isolated carotid sinus preparation. The response of one fibre to such a small stimulus is shown in Fig. 6.7. Such extreme sensitivity may be related to the observations of Hubbard (1958) that Pacinian corpuscle activation occurs with unobservably small deformations, and of Bishop *et al.* (1987*a*) that carotid baroreceptors exert powerful haemodynamic effects in response to changes of arterial pressure that are too small to measure. Guharay and Sachs (1984) studied stretch-activated skeletal muscle ion-channels and found that expansion of cultured membranes by as little as 2 per cent is likely to open membrane channels.

Sensitivity of baroreceptors to small mechanical perturbations may (or may not) also explain the data of Charlton and Baertschi (1982) and Hajduczok and his co-workers (1988) which suggest that baroreceptors detect *flow* as well as pressure. Hajduczok *et al.* (1988) found that increases of afferent baroreceptor firing with changes of flow at constant pressure of 25 mmHg are equal to those with increases of static distending pressure from 25 to 75 mmHg. Hajduczok's data may not resolve this issue definitively, however, because

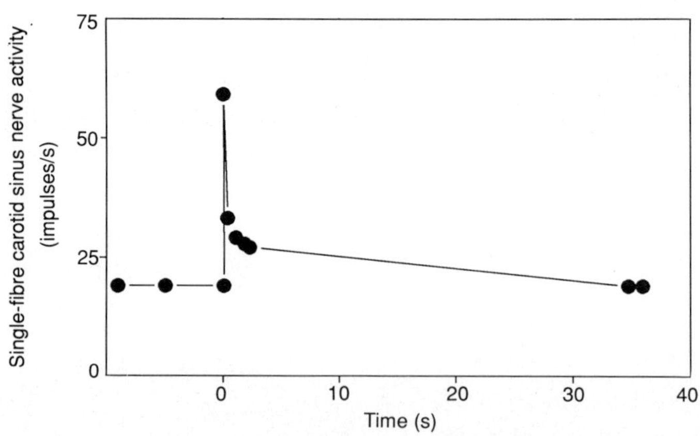

FIG. 6.7. Activity of a single, probably myelinated afferent carotid sinus nerve fibre during a step pressure increase of 1 mmHg. Adapted with permission (Landgren 1952*a*).

(1) most of the 50 mmHg carotid sinus pressure range he tested lay below or in the threshold region of his preparation;

(2) his measurements of nerve traffic 30 to 60 s after step pressure changes undoubtedly missed major early firing transients, as shown in Fig. 6.7.

Firing frequency as a function of artificial *step pressure changes* applied to isolated baroreceptive arteries has been measured by numerous authors (Bauer 1939; Ead *et al.* 1952; Landgren 1952*a*; Katona *et al.* 1968; Angell James 1971*a*). A coherent picture has emerged about how this simplest of pressure changes is encoded into afferent baroreceptor nerve traffic. Each receptor has a threshold for activation (which, however, fluctuates with the prevailing arterial pressure (Dorward *et al.* 1982)). When pressure exceeds this threshold (if the rate of pressure change is sufficiently large — see below), receptor firing begins. Thus, firing does not increase gradually in direct proportion to pressure; rather, firing begins abruptly at some fixed, initial frequency, which tends to be constant with similar pressure steps during acute experiments.

Baroreceptor discharge rates increase as pressure is raised above threshold levels. Firing frequency increases more steeply with earlier than with later increases of pressure (Ead *et al.* 1952; Sleight 1964; Angell James 1971*a*; Kalkoff *et al.* 1971; Arndt *et al.* 1975), and then reaches some maximum plateau frequency (termed 'saturation' frequency (Korner *et al.* 1974) or 'point of inflexion' (Angell James 1971*a*; Samodelov *et al.* 1979)). Uncommonly, some fibres continue to increase their firing rates in response to pressures considerably above normal. Angell-James (1971*a*) recorded from ten fibres that continued to increase their firing rates to 250 mmHg. She found a variety of responses in individual fibres as the saturation pressure (in multifibre preparations) is exceeded; some individual fibres continue to increase their firing rates, and others remain constant, decline (sometimes to zero activity), or begin firing intermittently. Responses of two fibres whose firing declined at high pressures are depicted in Fig. 6.8

Myelinated and unmyelinated baroreceptors

As mentioned, there is great variability among responses of individual baroreceptors to step pressure changes (Ead *et al.* 1952; Angell James 1971*a*; Buda *et al.* 1979; Samodelov *et al.* 1979). Angell-James (1971*a*) and A.M. Brown (1980) speculate that some of the differences among individual baroreceptors are due to differences in their linkages to surrounding tissues. Another likely possibility is that firing varies importantly on the basis of the physical characteristics of the receptors themselves. Baroreceptors are classified according to whether their axons are myelinated ('A' fibres) or unmyelinated ('C' fibres). The suggestion that there are baroreceptors with *unmyelinated* fibres was made first by Landgren (1952*a*), who showed that

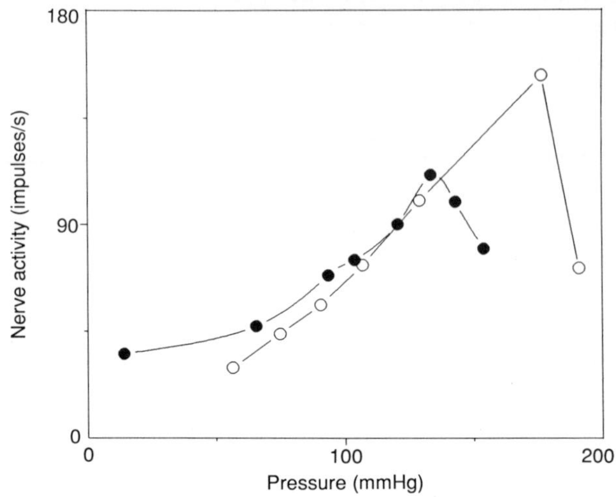

FIG. 6.8. Aortic nerve firing during changes of intra-aortic pressure. Adapted with permission (Angell James 1971*a*).

some fibres in multifibre preparations have low signal-to-noise ratios and fire at lower frequencies than others.

Additional inferential evidence that some baroreceptors are unmyelinated came from the studies of Douglas *et al.* (1956*a,b*) and Douglas and Schaumann (1956). They measured arterial pressure reductions provoked by electrical stimulation of aortic nerves and found that vasodepressor responses are highly non-linear. Arterial pressures fall gradually to an asymptotic floor as intensities of stimuli are increased. After this limit is reached, arterial pressure does not decline further, even though stimulus voltages are increased substantially. However, if voltage is raised high enough, further substantial reductions of pressure occur. From this response pattern, the authors deduced the existence of two groups of baroreceptors; both groups reflexly lower arterial pressure, and one comprises baroreceptor fibres smaller than those recognized before. Definitive histological and electrophysiological proof for the existence of unmyelinated baroreceptor fibres came from the study published by Fidone and Sato (1969).

The firing patterns of baroreceptors with myelinated and unmyelinated fibres are distinctly different. Typically, myelinated fibres begin to fire at lower pressures than unmyelinated fibres (Fidone and Sato 1969; Thoren and Jones 1977; Coleridge *et al.* 1987). Thorén *et al.* (1977) found with *in vivo* measurements that at a mean aortic pressure of 83 mmHg, 95 per cent of myelinated fibres but only 23 per cent of unmyelinated fibres fire. Also, increases of firing rates and maximum firing rates during pressure elevations are higher for myelinated than unmyelinated fibres (Arndt *et al.* 1977), and firing of myelinated fibres is more regular. In the study of Thorén *et al.* (1977) and one by Coleridge *et al.* (1987), there was little or no overlap of thresholds, firing rates at threshold, or firing rates above threshold pressures. Some of these differences are illustrated by Thorén's single fibre recordings of the myelinated (left) and unmyelinated (right) baroreceptor fibres shown in Fig. 6.9.

One implication of the different firing thresholds of myelinated and unmyelinated fibres is that autonomic responses to modest-to-moderate increases of arterial pressure are likely to be mediated primarily by myelinated fibres, whereas responses to large increases of pressure are likely to be mediated by both myelinated and unmyelinated fibres. Another is that if pressure elevations occur gradually, reflex changes initiated by lower-threshold myelinated fibres may prevent pressure from rising high enough to activate unmyelinated fibres. A third is that since minor arterial pressure elevations effectively silence human muscle sympathetic nerve activity (Eckberg *et al.* 1988), baroreceptor modulation of vasoconstrictor traffic to the muscle vascular bed may be mediated exclusively by myelinated baroreceptors.

Although baroreceptors with unmyelinated fibres tend to fire only at high pressures, when they do fire they can provoke powerful systemic arterial pressure reductions (Douglas and Schaumann 1956; Douglas *et al.* 1956*a,b*) and bradycardia (Angell James 1971*c*), at low firing frequencies. Regarding the latter, Kardon *et al.* (1975) studied R–R interval responses to electrical aortic nerve stimulation with voltages chosen to activate both myelinated and unmyelinated fibres, or only myelinated fibres, before and after interruption of efferent vagal and sympathetic nerves. Their results suggest that myelinated fibres modulate both sympathetic and parasympathetic autonomic outflows to the sinus node, but that unmyelinated fibres modulate primarily vagal outflow.

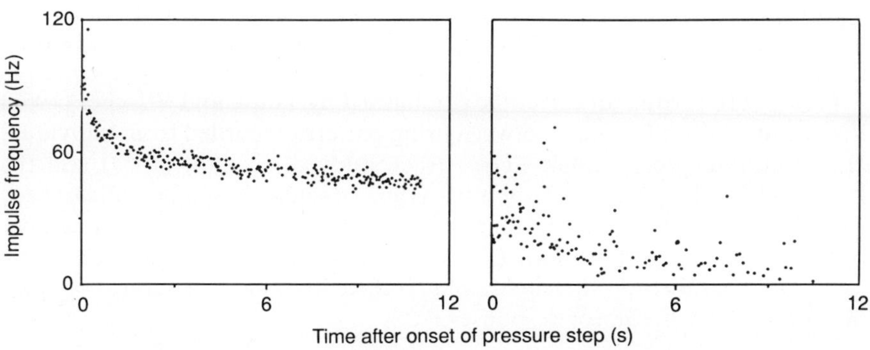

FIG. 6.9. Instantaneous firing frequencies of myelinated (left) and unmyelinated (right) single aortic nerves during 220 mmHg steps. Adapted with permission (Thorén *et al.* 1977).

Single and multifibre activity

A.M. Brown (1980) argues against use of multifibre recordings to evaluate baroreceptor function, on the bases that

(1) registrations of nerve traffic are not accurate when large numbers of fibres are activated synchronously;

(2) multifibre recordings are extremely susceptible to small changes of recording conditions;

(3) such recordings are insensitive to small voltage firings (such as those from unmyelinated fibres).

Others argue that it is difficult to study enough individual nerve fibres to gain statistical power sufficient to interpret results, and that baroreceptor responses of whole nerves differ from those of individual receptors in ways that are important determinants of integrated autonomic responses. Therefore, patterns of individual baroreceptor responses to pressure changes may not reflect whole nerve activity faithfully, unless large numbers of single fibres are studied, with proportional representation from unmyelinated as well as myelinated fibres.

Support for use of multifibre preparations comes from two studies which indicate that whole nerve responses comprise arithmetic sums of contributions from individual fibres (Kalkoff 1957; Sleight *et al.* 1977). Since threshold and saturation pressures differ widely among baroreceptors, the range of responses recorded from whole nerves is greater than that of component fibres, the activation threshold is lower, and the saturation pressure is higher (Trank and Visscher 1962; Arndt *et al.* 1975). Also, the *form* of multifibre nerve activity is different from that of individual fibres. The classic 'sigmoid' baroreceptor stimulus–response relation is found with multifibre or whole nerve preparations, not single fibres. Whereas individual fibres begin firing abruptly at some frequency and with increasing pressure increase their rates of firing nearly linearly, whole nerve responses have smoothly rounded threshold and saturation regions. Thus, notwithstanding their great theoretical disadvantages, whole nerve or multifibre recordings may portray a more accurate picture of global baroreceptor responses to pressure changes than single fibres.

Figure 6.10 is from an early study published by Trank and Visscher (1962) that pointed out differences between firing patterns recorded from individual and multifibre preparations. (The earlier study of Kalkoff (1957) and the later study of Irisawa and Ninomiya (1967) document similar differences.) This figure shows activity of a single fibre (the dotted line, a), and activity of two multifibre preparations (the solid and dashed lines, b and c). These latter two lines identify threshold, and all three lines identify saturation nonlinearities of baroreceptor responses.

Baroreceptor firing at very low pressures

Several workers have shown that during progressive reductions of arterial pressure, sympathetic activity in animals (Dorward *et al.* 1985; Morita and

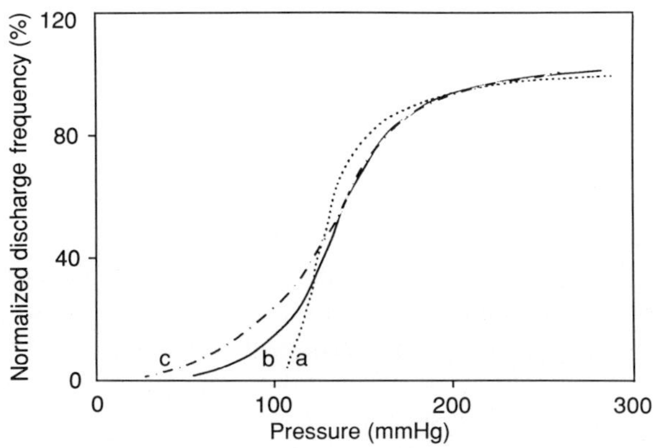

FIG. 6.10. Afferent activity recorded from single (a) and multifibre (b and c) carotid sinus nerve preparations during carotid pressure changes. Adapted with permission (Trank and Visscher 1962).

Vatner 1985), or plasma noradrenaline level in humans (Eckberg *et al.* 1986) first increases, and then *decreases*. (Changes of plasma noradrenaline are shown in Fig. 10.5.) Reductions of sympathetic outflow during progressive reductions of pressure, which may be related causally to vasovagal reactions, are generally ascribed to paradoxically increased firing rates of left ventricular receptors during hypotension. According to this construct, firing increases when receptors are 'crimped' by sympathetically driven supramaximal left ventricular contractions begun from small end-diastolic volumes (Gauer and Henry 1964). This possibility is supported by the important studies of Öberg and White (1970), Öberg and Thorén (1972), and Thorén (1976), who documented increased firing rates of ventricular receptors during contractions mounted at low pressures.

One case report from humans, however, suggests that increased ventricular receptor firing is not *necessary* for the occurrence of vasovagal reactions. Scherrer *et al.* (1990*b*) triggered a typical vasovagal reaction with lower body suction in a patient with an orthotopic heart transplant. In this patient, abrupt reduction of muscle sympathetic nerve activity during central blood volume depletion cannot be ascribed to paradoxical firing of ventricular receptors at small volumes, because the transplantation procedure denervates the donor heart ventricles.

This report reopens the question of what receptors trigger vasovagal reactions. Some studies (Landgren 1952*a*; McCubbin *et al.* 1956; Angell James 1971*c*; Thorén *et al.* 1977) document increased firing rates of *arterial baroreceptors* when pressure is reduced to very low levels. Such paradoxical firing has been observed from the earliest years of baroreflex research. This phenomenon may represent experimental artifact secondary to unphysiological emptying of arteries, and it may have no practical significance. Firing of arterial baroreceptors at very low pressures, however, establishes the possibility that as arterial pressure falls progressively, paradoxical reductions of sympathetic outflow may be mediated by arterial as well as ventricular receptors. J.H. Green (1967) showed that with low pulsatile pressures, baroreceptor firing may occur in late diastole. Also, when pressure is lowered to very low levels in isolated carotid sinus preparations, systemic arterial pressure *falls* (Heymans and Bouckaert 1929; Stella 1931).

There are several possible mechanisms to explain increased arterial baroreceptor firing at low arterial pressures. First, it may be due to 'crimping' of receptors at small diameters, the mechanism proposed for increases of left ventricular receptor activity during hypotension. Second, it may be due to a direct effect of sympathetic stimulation on baroreceptors; although modest increases of noradrenaline constrict vascular smooth muscle, unload baroreceptors, and *reduce* baroreceptor firing, large increases of sympathetic activity directly excite baroreceptors (Munch *et al.* 1987). Third, it may be due to rhythmic smooth muscle contractions provoked by increased sympathetic stimulation (Biamino and Kruckenberg 1969). It is not known if reductions of human arterial pressure during experimental hypotension are sufficient to increase arterial baroreceptor firing.

VISCOELASTIC INFLUENCES ON BARORECEPTORS

As indicated in Chapter 3, complex baroreceptor nerve endings are coupled to elastin, collagen, smooth muscle, and amorphous ground material. Therefore, responses of baroreceptors to deformation cannot be understood apart from an understanding of how their viscoelastic connections modify ongoing deformation. Because of viscoelastic factors, baroreceptors have memory (possibly only short-term (Dorward and Korner 1987)), and their instantaneous responses to pressures are conditioned by their immediate past histories of pressure changes. This possibility has already been hinted at above; sensitivity to rate of pressure change is in part a function of viscoelastic responses to deformation. (Seagard *et al.* (1990) suggest that some (primarily unmyelinated) baroreceptors provide long term information on absolute blood pressure levels.)

Rate of pressure change

Thresholds for baroreceptor firing are not fixed; they depend importantly on the prevailing static pressure (Dorward *et al.* 1982) and rate of dynamic pressure increases. The relation between baroreceptor threshold and the rate of pressure change was worked out well by Landgren (1952*a*). Figure 6.11

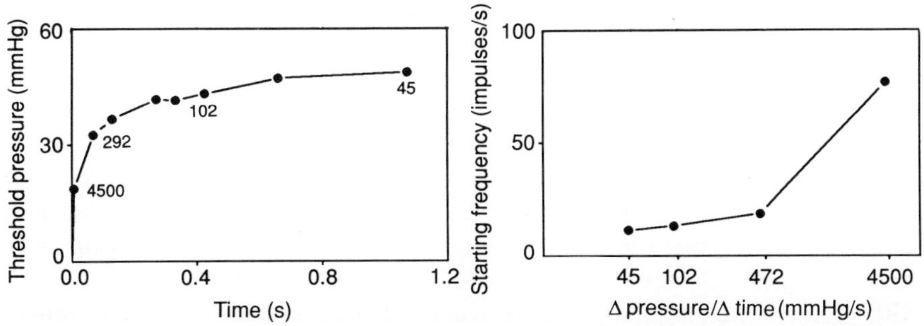

FIG. 6.11. Carotid sinus baroreceptor threshold pressure and initial firing frequency as a function of rate of pressure change. Numbers in the left panel refer to maximum d*P*/d*t*, in mmHg/s. Adapted with permission (Landgren 1952*a*).

summarizes Landgren's measurements. The left panel shows that firing begins earlier and at a lower threshold pressure when the rate of pressure change is rapid (4500 mmHg) than when it is low (45 mmHg/s). The right panel shows also that the starting frequency is higher when the rate of pressure change is greater.

Two studies with humans have employed different rates of neck suction to evaluate the influence of rate of pressure change on baroreflex responses. The results of one study are summarized in Fig. 6.12. In this study, baroreceptor stimuli, with a range of rates of pressure below and above those of the naturally occurring arterial pulse, were applied (Eckberg 1977*b*). This figure suggests that rate of pressure change is an important determinant of integrated human (vagal–cardiac) baroreflex responses, but only when rates are unphysiologically low, below about 300 mmHg/s. (D.T. Mason *et al.*

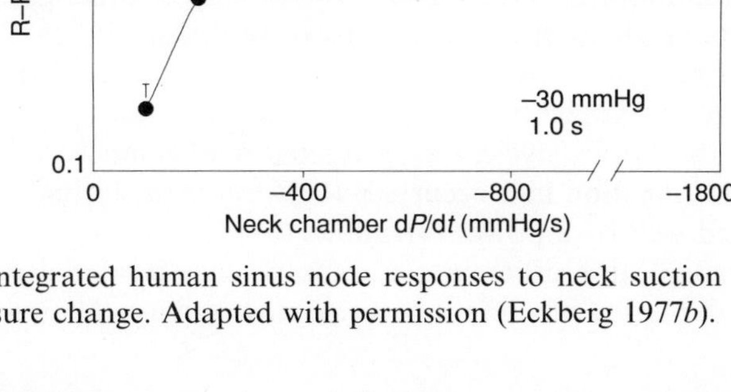

FIG. 6.12. Integrated human sinus node responses to neck suction with different rates of pressure change. Adapted with permission (Eckberg 1977*b*).

(1964) found that the average maximum d*P*/d*t* in human brachial arteries is about 800 mmHg/s.)

In a second study, Båth and co-workers (1981) measured changes of peroneal nerve muscle sympathetic activity during application of ramps of neck suction with different rates of pressure change, (*all* of which probably were below that of the naturally occurring pulse). In this study, maximal baroreflex inhibition of sympathetic activity occurred in response to stimuli with the most rapid onsets. Indeed, in some subjects, neck suction with extremely gradual onsets did not inhibit muscle sympathetic activity at all. Although it seems likely that in the studies of Eckberg and Båth, differences in vagal and sympathetic responses reflect differences in activation of rate-sensitive baroreceptors, this is an unproven assumption. A representative record from Båth's study is depicted in Fig. 5.16.

Adaptation

A closely related manifestation of the influence of viscoelastic factors on baroreflex function is the phenomenon of *adaptation.* The rapid firing that occurs during abrupt elevations of pressure (to above threshold levels) in baroreceptor areas does not persist undiminished. Rather, it decays progressively during maintained pressure elevations as a power function; that is, the relation between the log of the change of impulses/s and the log of time is linear. Landgren found that firing frequency declines 50 per cent by 0.1 s after an abrupt pressure rise, and 95 per cent by 30 to 60 s. Because in most studies, nerve activity does not decline to zero during sustained high pressures, baroreceptors are termed 'slow adapting' (Adrian 1926; Bauer 1939; Landgren 1952*a*). Adaptation is present also after sudden reductions of pressure (where it is called 'creep'), but may have a slower time course than that which occurs during sudden elevations of pressure (Bronk and Stella 1935); this constitutes yet another baroreceptor non-linearity. Figure 6.13 depicts adaptation as a power function in the top panel, and as nerve activity against the log of time in the lower panel (Landgren 1952*a*).

It has long been recognized that integrated human responses to sustained carotid baroreceptor stimulation with neck suction decay progressively (Thron *et al.* 1967; J. Wagner *et al.* 1968; Eckberg 1977*a*; Mancia *et al.* 1978*b*). (The stimulus in these human studies may be analogous to the step increases of pulsatile arterial pressure used by Potter (1982).) In one study conducted in humans (Eckberg 1977*a*), the time course of decay of sinus node inhibition provoked by 5 s pulses of neck suction, applied at different times during the cardiac cycle, was evaluated mathematically. In this study, after peak prolongation had occurred, R–R intervals declined at rates that were described well by a power function.

Although adaptation of integrated responses can be explained entirely on the basis of adaptation of afferent baroreceptor activity, the site(s) of adapta-

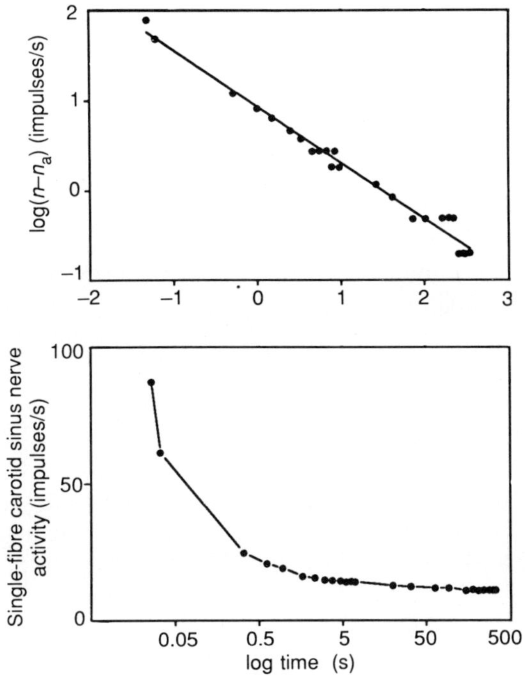

FIG. 6.13. Decay of carotid sinus firing during a sustained 120 mmHg pressure increase. Adaptation is described well by a power function (upper panel). Adapted with permission (Landgren 1952*a*).

tion in the intact humans studied is not known. Adaptation of integrated baroreflex responses occurs also within the central nervous system in animals (Richter *et al.* 1970; Potter 1982; Kunze 1986) and humans (Korsukewitz *et al.* 1974); the degree of inhibition of muscle sympathetic nerve activity declines during continuous electrical carotid sinus nerve stimulation (Wallin *et al.* 1975); and sinus node inhibition decays (or 'fades') during continuous electrical vagus nerve stimulation (S.C. Wang and Borison 1947; P. Martin *et al.* 1982; Salata and Jalife 1985). Moreover, neck suction, which is by any reckoning an unphysiological intervention, provokes reflex reduction of arterial pressure, which is sensed and opposed (C.V. Winder 1937) by aortic and cardiopulmonary receptors. Thus although adaptation of the baroreceptors themselves is sufficient to explain the decay of sinus node responses to neck suction, it is by no means the only possible explanation for this phenomenon.

Resetting

The phenomenon of adaptation is thought to result from viscoelastic relaxation occurring within the walls of arteries, which leads to progressive reduction of strain placed upon baroreceptors during pressure elevation. As indicated, adaptation begins instantaneously, and, according to mathematical models, continues as long as the new level of stretch is maintained. Landgren (1952*a*) followed the course of adaptation out to 10 min. When pressure changes are maintained for longer periods, changes of baroreceptor responses are termed *resetting*. Although it seems likely that adaptation and resetting are related phenomena (Coleridge *et al.* 1984), the precise time when adaptation becomes resetting has not been defined.

As discussed in Chapter 2, Hering (1924*a,b*) believed that hypertension may result from impaired function of baroreceptors. However, since the risky study of Lampen *et al.* (1949), it has been appreciated that baroreceptors continue to function, even in patients with malignant hypertension. The question remained, in what way does baroreceptor function change when pressure levels are changed chronically?

This issue was joined in the seminal study of McCubbin *et al.* (1956), who compared carotid sinus traffic in normotensive and chronic renal hypertensive dogs. Recordings from a normotensive and a renal hypertensive dog are depicted in Fig. 6.14. McCubbin found that baroreceptor firing is intermittent in both groups, but occurs at much higher pressures in hypertensive than in normotensive dogs. When pressure in normotensive dogs is raised acutely to hypertensive levels (third tracing from top), baroreceptor firing becomes continuous. McCubbin *et al.* believed that in their hypertensive dogs, resetting was due to hypertension-induced reductions of carotid arterial compliance. Subsequent research by others (Kezdi 1967*b*; Angell-James 1974; Andresen *et al.* 1978; Chapleau and Abboud 1987) shows that although changes of baroreceptor firing occur in chronically hypertensive animals secondary to reductions of compliance and arterial disease, changes of baroreceptor firing may occur also during acute pressure changes, in the absence of alterations of measured compliance (A.M. Brown *et al.* 1976; Sapru and Krieger 1979).

Subsequent studies by Aars (1968), Angell-James (1973, 1974), and Sleight and co-workers (1977) document similar abnormalities of firing patterns in single fibre aortic and carotid sinus nerves of rabbits with hypertension. Of interest, in the study of Sleight *et al.* (1977), is the finding that when baroreceptors of normotensive and hypertensive dogs were studied at their unanaesthetized arterial pressures, average impulse frequencies for the two groups were identical. Thus the central nervous systems of normotensive and hypertensive dogs receive the same information regarding arterial pressure; the afferent baroreceptor inputs of hypertensive dogs give no clue to the presence of hypertension.

A variety of studies points toward the existence of baroreceptor resetting of integrated baroreflex responses in *humans*. The earliest is the observation of the Oxford group that in hypertensive patients and normotensive subjects,

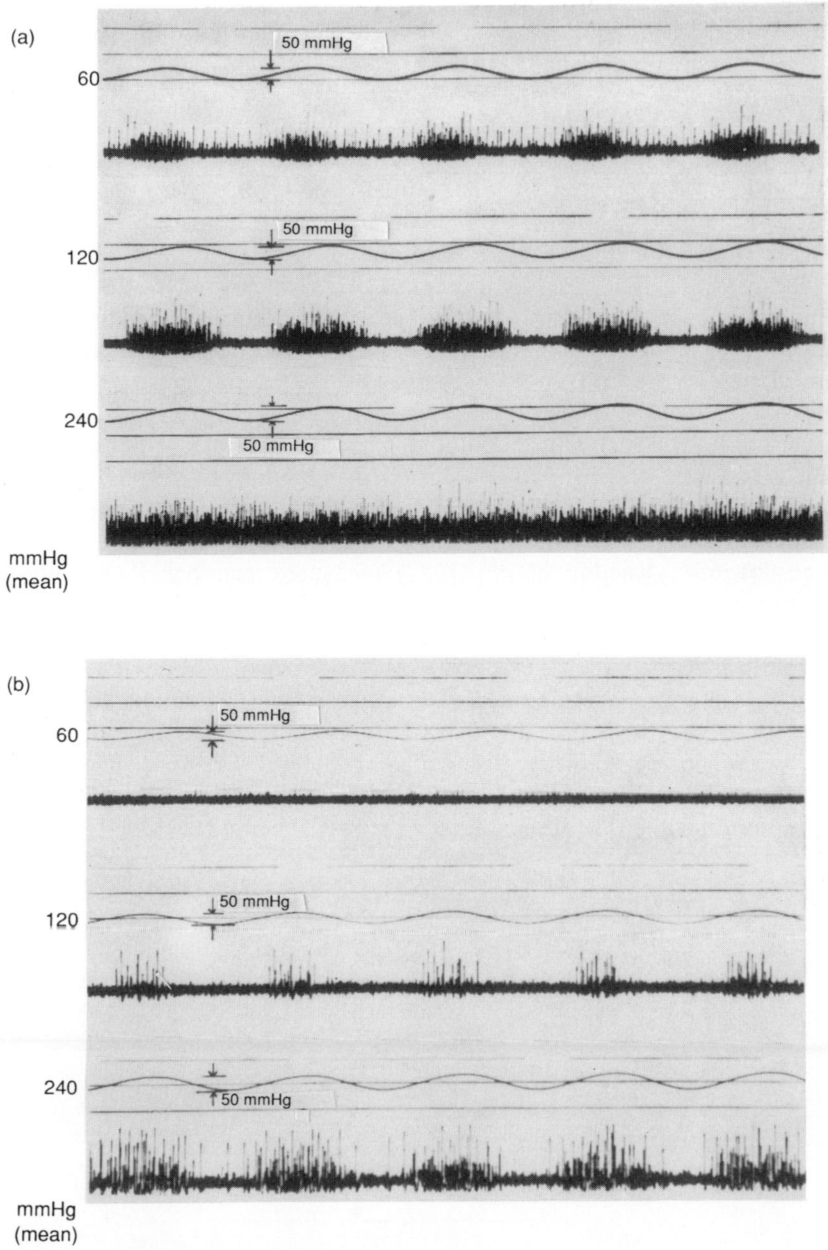

FIG. 6.14. Multifibre carotid sinus nerve recordings from (a) normotensive and (b) renal hypertensive dogs. Reproduced with permission (McCubbin *et al.* 1956).

heart rate is similar, but in hypertensive patients, post-phenylephrine R–R interval–systolic pressure relations are shifted to the right (J.D. Bristow *et al.* 1969*b*). In a subsequent study (J.D. Bristow *et al.* 1969*a*), these authors (and Korner *et al.* (1974)) drew a distinction between resetting — parallel shifting of the R–R interval–systolic pressure relation on the pressure axis — and reduction of the slope of this relation. Reduction of slopes was taken to indicate *pathophysiological* reflex impairment, rather than *physiological* reflex resetting.

Other studies focus on human baroreflex responses to *acute* changes of arterial pressure. T.G. Pickering *et al.* (1972*a*) measured slopes of post-phenylephrine R–R interval–systolic pressure relations in patients with chronic renal failure on haemodialysis. They found no significant change of these slopes before and after acute changes of arterial pressure resulting from haemodialysis. These results are depicted in Fig. 6.15. Since these workers did not define entire stimulus–response relations, it is not known if their patients' entire relations shifted during dialysis-induced pressure changes, or if their patients' relations were fixed, and measurements were obtained at different pressures on the 'linear' portion.

Subsequently, Kasting *et al.* (1987) measured sigmoid R–R interval responses to stereotyped neck pressure–suction sequences (Fig. 5.13 (Sprenkle *et al.* 1986)) in awake healthy young adults every three hours during a 24 h period. All stimulus–response relations for two subjects are depicted in Fig. 6.16. Kasting's results suggest that that in healthy humans, baroreceptor–cardiac reflex responses are involved in an ongoing, continuous process of resetting during spontaneous, circadian arterial pressure fluctuations. These data do not identify where in the reflex arc resetting occurs.

The anatomical basis for the ongoing process of acute resetting that occurs

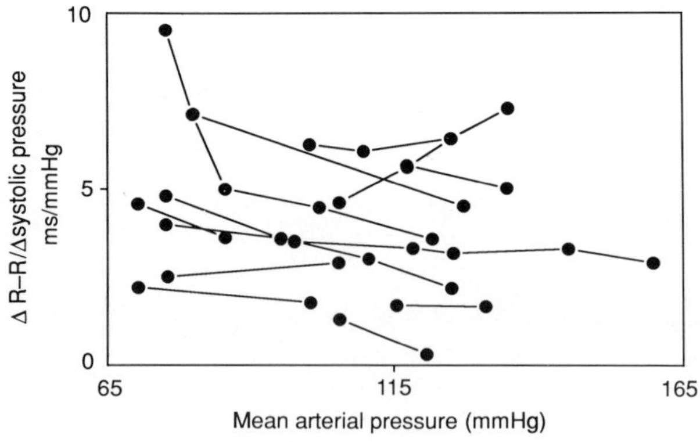

FIG. 6.15. Slopes of phenylephrine-induced R–R interval changes measured at different pressures in patients during haemodialysis. Adapted with permission (T.G. Pickering *et al.* 1972*a*).

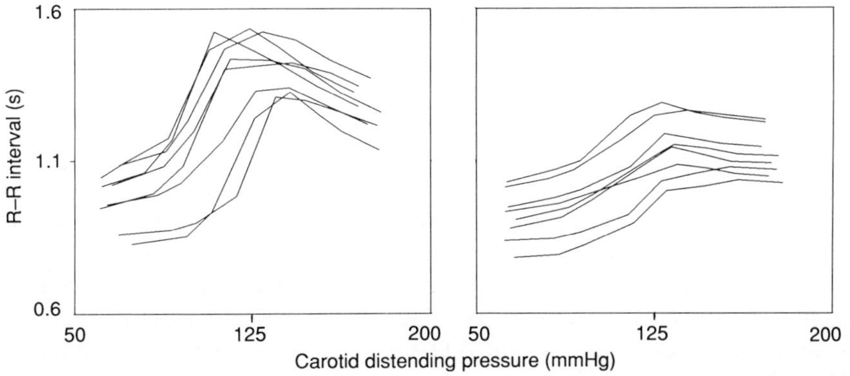

FIG. 6.16. Baroreceptor–cardiac reflex relation in two healthy subjects studied serially during circadian R–R interval and blood pressure changes. Adapted with permission (Kasting *et al.* 1987).

physiologically is unknown. It is caused by altered pressure and, presumably, mechanical changes within baroreceptive arteries rather than neurohumoral or electrolyte changes, because hypertension (Kezdi *et al.* 1973) or hypotension (Igler *et al.* 1981) limited to one carotid artery does not cause resetting of the contralateral normotensive artery. We are not aware of any studies which document parallel improvement of arterial compliance and arterial baroreflex responses after treatment of hypertension. A study by Mancia *et al.* (1988), however, suggests that regression of left ventricular hypertrophy resulting from chronic treatment of hypertension with beta-blocking drugs increases the gain of forearm vascular and plasma noradrenaline responses to lower body suction and passive leg raising.

Off response

The earliest recordings of Bronk and Stella (1932) indicate that at usual resting levels of arterial pressure, baroreceptors fire in systole and are silent in late diastole. Subsequently (1935), these authors investigated the intermittency of baroreceptor firing by measuring afferent discharges when pressures in isolated carotid sinuses are suddenly reduced from higher to lower levels. They found that after sudden reductions of pressure, impulse frequency does not fall to steady-state levels expected from the new pressure; rather, firing stops altogether, and then resumes at a low frequency which gradually increases to the expected level.

Landgren (1952*a*) also explored this phenomenon, which he termed 'post-excitatory depression'. He found that the duration of neural silence is a direct function of the extent of abrupt pressure reductions, and that firing does not resume at all, if pressure falls below the fibre's threshold level. An example of post-excitatory depression, recorded by A.M. Brown (1980) from an isolated rat aorta preparation, is depicted in Fig. 6.17, left (this figure also illustrates the phenomenon of adaptation occurring during a sustained pressure elevation). The off response is generally thought to be mediated by complex viscoelastic changes occurring in response to abrupt reductions of arterial wall dimensions. However, Brown's data suggest that it is mediated by ionic fluxes; the off response was abolished by perfusion of the aorta with a solution containing no potassium (Fig. 6.17, right). Brown's findings do not exclude an influence of potassium on smooth muscle; therefore, viscoelastic and ionic explanations for post-excitatory depression may not be mutually exclusive. Others speculate that post-excitatory depression is mediated by sodium–potassium ATPase mechanisms, since it is prevented by pretreatment with digoxin (Saum *et al.* 1976).

Pulsatile pressure

The preceding discussion focuses on neural responses to step increases or decreases of pressure. Baroreceptor response patterns are more complex when driving stimuli are *pulsatile*.

In 1952, Ead *et al.* published results which launched a genre of research into the physiological role of pulsatility in control of blood pressure. Their study shows that arterial pressure is lower when pulsatile pressure is substituted for steady pressure. However, these authors added an important caveat

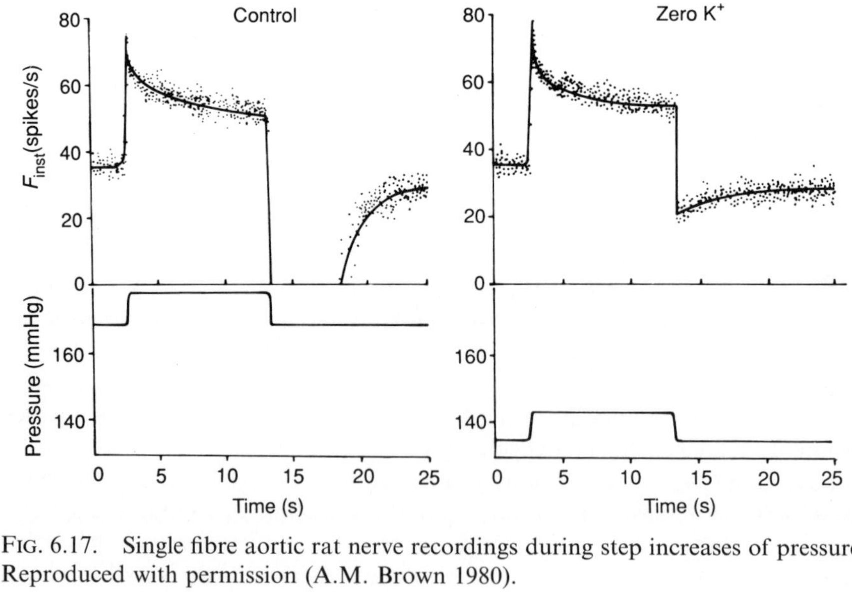

FIG. 6.17. Single fibre aortic rat nerve recordings during step increases of pressure. Reproduced with permission (A.M. Brown 1980).

which has been iterated by subsequent generations of researchers who have studied the effects of pulsatility on baroreceptor traffic (Angell James 1971*b*; Arndt *et al.* 1975, 1977; Chapleau and Abboud 1987), or systemic arterial pressure (Stegemann and Tibes 1969): The influence of pulsatility is dependent critically upon the level of pressure at which pulsatility is introduced. Addition of pulsations (with no change of mean carotid sinus pressure) leads to substantial reductions of systemic arterial pressure when pulsatility is introduced at pressures near the threshold region for baroreceptor activation. In this instance, pulsatility leads to net increases of afferent baroreceptor activity (Trank and Visscher 1962; Arndt *et al.* 1975; Angell James and Daly 1970). However, pulsatility may exert no influence on afferent baroreceptor activity (Koushanpour and McGee 1969) or systemic arterial pressure (Stegemann and Tibes 1969) when it is introduced at high mean carotid sinus pressures.

Indeed, Stegemann and Tibes (1969) observed that arterial pressures might even rise when pulsatile pressure changes are applied in the saturation range. Presumably, in their study, pulsatility at high pressures caused net reductions of afferent baroreceptor activity. Similarly, Dorward *et al.* (1985) observed increases of renal sympathetic nerve activity when arterial pressures are raised to high levels. Not infrequently in humans, carotid baroreceptor–cardiac reflex responses triggered by neck suction decline after peak R–R interval prolongation has occurred; it is not known if this is due to reductions of baroreceptor nerve firing rates at the highest transmural pressures, as described by Angell-James (1971*a*), vagal withdrawal secondary to reflex aortic hypotension, or other factors.

Subsequent to the study of Ead *et al.* (1952), Irisawa and Ninomiya (1967) integrated whole aortic and carotid sinus nerve activities electronically and expressed nerve activity as a function of instantaneous pressure. Their results illustrated graphically, as '*hysteresis* loops', what Bronk and Stella had described much earlier (1932), that nerve activity at identical pressures is lower when pressures are falling than when pressures are rising. Angell James (1971*b*) and Masuda *et al.* (1984) subsequently showed that hysteresis loops tend to be clockwise. Baroreceptor hysteresis has at least two components: changes determined by passive viscoelastic changes occurring within the walls of baroreceptive areas; and changes determined by the degree of smooth muscle contraction, which varies continuously with prevailing pressure (Bayliss 1902). Hysteresis cannot be explained simply on the basis of changes of calculated arterial wall stress, since at any given pressure, artery diameters (and hence, calculated stresses) are greater during diastole than systole in animals (R. Wagner and Kapal 1951; Remington 1955) and humans (Greenfield *et al.* 1964).

An elegant example of hysteresis of arterial diameter and baroreceptor firing published by Coleridge *et al.* (1984) is shown in Fig. 6.18. The clockwise loop (c) inscribed by aortic baroreceptor firing during the pressure

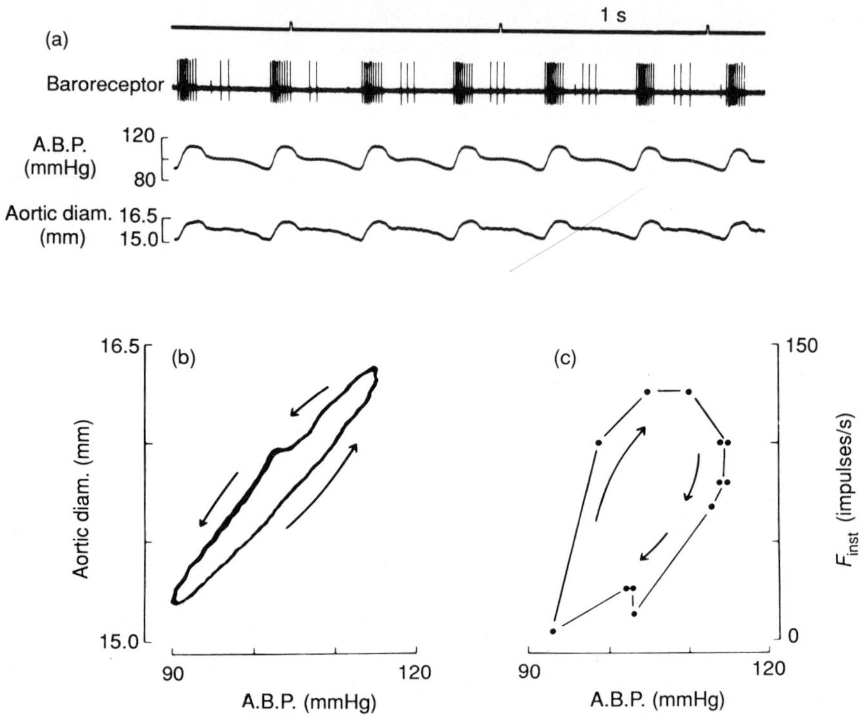

FIG. 6.18. Aortic diameter, pressure, and baroreceptor nerve activity during physiological pulsations. Reproduced with permission (Coleridge *et al.* 1984).

pulse contrasts with the counterclockwise loop (b) inscribed by aortic diameter. Increases of aortic diameter during diastole reflect creep of the aortic wall, which presumably translates into reductions of tension on baroreceptors. The degree of creep, however, was very small. Therefore simple dilatation of the aorta is unlikely to account for all of the large reduction of nerve traffic that occurred during diastole. This points toward the occurrence of ongoing, unmeasured changes of tension within the wall of the aorta within the cardiac cycle.

There is a substantial literature on the importance of individual components of pulsatile wave forms on baroreceptor activity. Angell-James (1971*a*) showed that as *pulse pressure* increases, systolic rates of baroreceptor firing increase, but with steadily diminishing increments. She explains this on several bases. First, for individual fibres, the slope of the pressure–firing frequency relation becomes less steep as pressures rise. Second, the rate of recruitment of previously silent fibres becomes progressively less as their thresholds for firing are exceeded during rising pressure. Third (and perhaps exceptionally), some fibres *reduce* their rates of firing as their saturation pressures are approached. This is shown in Fig. 6.8.

Angell-James (1971*b*) also observed that as pulse pressure increases, diastolic firing rates fall to zero, and the duration of diastolic silence increases. This observation may explain the findings of Gero and Gerova (1962) and Stegemann and Tibes (1969) that increases of pulse pressure (at constant mean pressures) lead to further reductions of systemic arterial pressure. Importantly, it also may explain the findings of Sundlöf and Wallin (1977), which indicate that whereas levels of muscle sympathetic outflow are related *inversely* to diastolic pressure, they are related *directly* to pulse pressure. Since the level of muscle sympathetic activity appears to be unrelated to systolic pressure (Sundlöf and Wallin 1977), it may not matter how much systolic firing is augmented by increases of pulse pressures. What does matter, however, is that increased pulse pressure prolongs periods of baroreceptor silence (Angell James 1971*b*), and this in turn augments sympathetic outflow.

The study of Chapleau and Abboud (1987) focuses on the off response that is a component of baroreceptor responses to pulsatility. The principal emphasis of their study is that neural silence occurs at higher pressures during pulsatile than during steady pressures. A similar observation was made also by Bronk and Stella in 1935. This introduces another non-linearity into baroreceptor transduction; average discharge rates increase less during systole than they decrease during diastole. This concept probably explains the relation between trends of arterial pressure (that is, the immediate past history) and levels of muscle sympathetic activity. Muscle sympathetic nerve activity is greater when pressures are falling than when they are rising (at identical pressures), probably because baroreceptors are more likely to be silent (Sundlöf and Wallin 1977).

Chapleau and Abboud show also that the likelihood that baroreceptors are silent is related inversely to diastolic pressure. When diastolic pressure is only about 10 mmHg above the static firing threshold, diastolic silence occurs 100 per cent of the time. However, when diastolic pressure is about 80 mmHg above the static firing frequency, diastolic silence does not occur, and baroreceptor firing rates are not different from those which occur during steady pressures. These findings contradict the observations of Arndt *et al.* (1975) that baroreceptor firing becomes continuous whenever diastolic pressure exceeds the threshold for baroreceptor activation.

Figure 6.19 from a study by Arndt *et al.* (1975) illustrates differences between responses of a single baroreceptor fibre to static and dynamic stretches applied to an isolated aorta preparation. Centre tracings in each of the insets show the degree of stretch; upper tracings show afferent firing; and lower tracings show afferent firing rates. When sinusoidal stretch is imposed above and below threshold (a), firing increases above static levels, and then ceases as tension is reduced. When sinusoidal stretch is begun in the 'linear' range (b), firing waxes and wanes, but is withal continuous. When sinusoidal stretch is imposed in the saturation range (c), the high rate

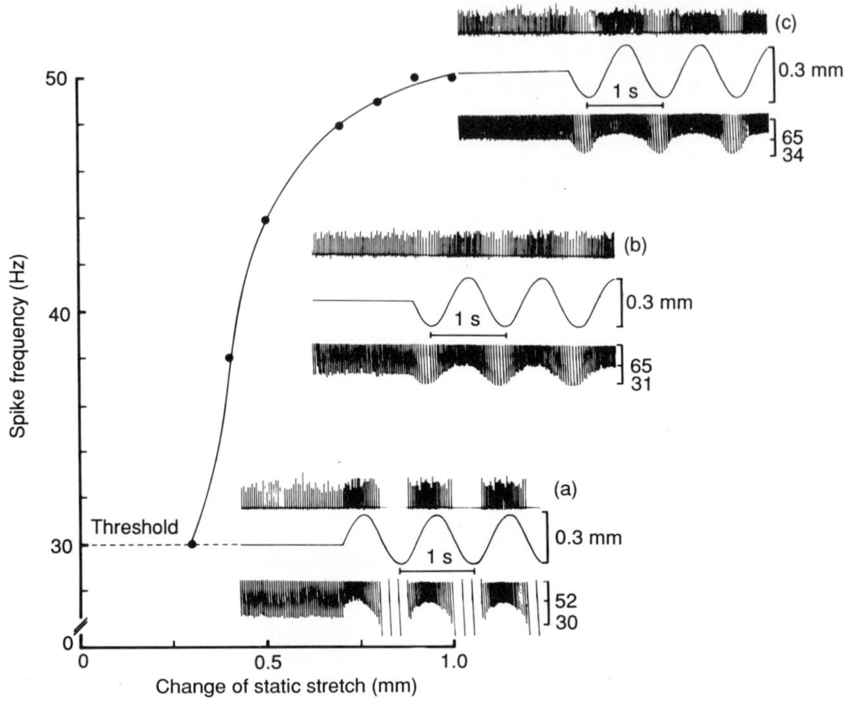

FIG. 6.19. Responses of a single baroreceptor fibre to steady and sinusoidal stretches applied to an isolated aorta preparation. See text for discussion. Reproduced with permission (Arndt *et al.* 1975).

of firing present with sustained stretch cannot increase much, but can decrease substantially during reductions of stretch. Thus introduction of pulsatility at high degrees of stretch may actually reduce baroreceptor firing rates. This observation probably explains the finding of Stegemann and Tibes (1969) that when pulsatility is introduced at high intracarotid pressures, systemic arterial pressure may actually rise.

Almost all of the studies quoted above from animals employed artificial sine waves. Additional complexities arise when *natural arterial pulses* are substituted for sine waves. Abboud *et al.* (1988) compared baroreceptor firing during sine waves and natural pulses. One obvious difference in responses to the two forcings is that a second burst of baroreceptor activity occurs during the naturally occurring dicrotic notch (this is shown in Fig. 6.6, and in Fig. 6.18(c), as the notch in the lowest portion of the baroreceptor firing loop). More importantly, heart rate slowing prolongs the duration of diastole more than systole. As a result, the diastolic period of neural silence during each heart beat is longer, and net afferent baroreceptor activity per heart beat is less. There are very limited data on the effect of heart rate changes on human muscle sympathetic nerve activity. Wallin *et al.* (1974*b*) found that during atrial fibrillation, sympathetic activity is greater at slower than at faster heart rates. Conceivably, in this study longer periods of afferent baroreceptor diastolic silence (as well as greater reductions of diastolic pressure) contributed to increased sympathetic outflow at slower heart rates.

VASOCONSTRICTORS AND BARORECEPTOR FUNCTION

It has long been recognized that vasoconstrictors, including adrenaline and noradrenaline, applied to the carotid sinus reduce systemic arterial pressure (Palme 1936; Heymans and van den Heuvel-Heymans 1950; Kalkoff *et al.* 1966; Peveler *et al.* 1983). This effect appears to be due to a reflex mediated by afferent traffic carried over the carotid sinus nerve, since it does not occur after carotid sinus nerve section (Palme 1936; Heymans and van den Heuvel-Heymans 1950). Hypotension occurs also with electrical (Landgren *et al.* 1951; Landgren 1952*b*; Koizumi and Sato 1969; Sampson and Mills 1970) and reflex (Felder *et al.* 1983) stimulation of efferent sympathetic nerves to the carotid sinus. The functional importance of this mechanism is unknown. Some studies employed 'industrial' doses of vasoconstrictors to elicit effects. However, it is likely that noradrenaline concentrations at synaptic clefts are much higher than those in plasma (Ljung 1969), and the studies of Kunze (1984) and Munch (1987) and their co-workers suggest that changes of noradrenaline levels exert effects on baroreceptor transduction at physiological pressures and noradrenaline concentrations.

The pressor used most commonly to stimulate human baroreceptors, phenylephrine, augments baroreceptor afferent responses to pressure changes when it is applied to the carotid sinus (Peveler *et al.* 1983). However, some evidence suggests that bolus phenylephrine used to elicit reflex bradycardia may *not* modify integrated baroreflex responses importantly. Ferguson *et al.* (1985) found in six subjects that steady infusions of phenylephrine do not alter R–R interval responses to graded neck pressure. Lumbers *et al.* (1979) found that arterial pressure elevations provoked by intravenous phenylephrine and aortic balloon inflation lead to similar augmentation of afferent carotid sinus nerve traffic. (The Australian literature on this point is conflicting, however; Faris (1980*a*) found that when pressure is raised with phenylephrine, slopes of R–R interval–systolic pressure relations are about 50 per cent less than slopes measured when arterial pressure is raised with inflation of a balloon in the descending aorta.) Moreover, slopes of R–R interval–systolic pressure relations after naturally occurring muscle sympathetic bursts (which might be regarded as micro-injections of noradrenaline) are similar to those after bolus injections of phenylephrine (Fritsch *et al.* 1986).

REPRODUCIBILITY OF BARORECEPTOR RESPONSES

Katona *et al.* (1968) showed that well-myelinated baroreceptor axons behave in a highly deterministic manner. They measured single fibre responses to precisely triggered pressure increases delivered to an isolated carotid sinus. They found that when reproducible stimuli are applied repeatedly during baroreceptor silence, baroreceptor firing begins at the same time, and recurs at predictable intervals. The standard deviation of the times of occurrence of bursts, used an index of the variability of firing, is very small initially, but increases steadily with later firings, as pressure stimuli are maintained. However, if step pressures are delivered at times when baroreceptors are firing, synchronization does not occur. Also, if the rate of pressure change is small, synchronization does not occur. Figure 6.20 shows signal-averaged carotid baroreceptor responses to naturally occurring arterial pulses (upper panel) and to three 'square wave' pressure steps, superimposed on an oscilloscope trace.

Kalkoff (1959) studied baroreceptor firing with whole nerve rather than single fibre preparations. Their conclusions are similar to those drawn later by Katona and co-workers (1968) and suggest that when mechanical deformations are imposed on baroreceptive areas, changes are encoded in baroreceptor nerve traffic with remarkable accuracy and fidelity. The implication of these two studies for *human* research is that when baroreceptor stimuli with rapid rates of pressure change are applied reproducibly, increments of afferent baroreceptor nerve traffic provoked by the stimuli should be constant and highly reproducible. A corollary is that widely divergent responses to identical baroreceptor stimuli (as shown in Fig. 5.19) are probably not due to variability of afferent baroreceptor nerve responses. Reproducibility of integrated human baroreflexes is discussed in Chapter 10.

SUMMARY

This chapter reviews how changes of baroreceptive artery dimensions, pressure, and calculated tension translate into changes of afferent nerve activity. There are striking similarities between response patterns recorded in afferent baroreceptor nerves in animals and directly or indirectly recorded response patterns in efferent sympathetic and vagal nerves in humans. Integrated human responses to changes of arterial distending pressure also are sigmoidal (vagal) or reverse-sigmoidal (sympathetic), with threshold, linear, and saturation ranges. Sustained baroreceptor stimulation in humans also

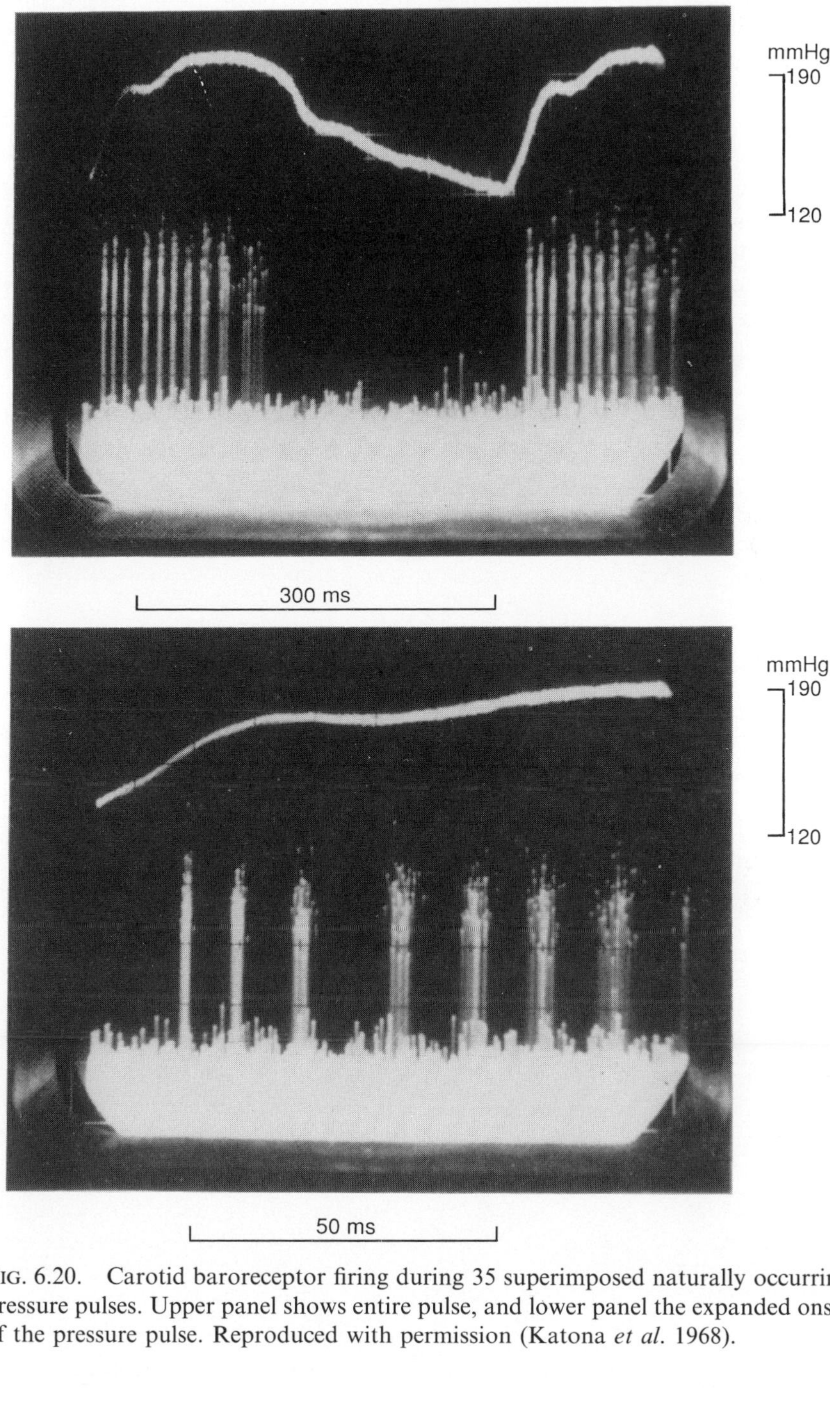

FIG. 6.20. Carotid baroreceptor firing during 35 superimposed naturally occurring pressure pulses. Upper panel shows entire pulse, and lower panel the expanded onset of the pressure pulse. Reproduced with permission (Katona *et al.* 1968).

trigger abrupt changes of efferent sympathetic and vagal traffic, which decay rapidly. Human stimulus–response relations also float with prevailing arterial pressures. They also reset during spontaneous or induced sustained increases or decreases of arterial pressure. Efferent sympathetic responses to pressure changes also show hysteresis, with sympathetic bursts more likely to appear when arterial pressure is falling (when afferent baroreceptor traffic is less) than when it is rising. Understanding of how arterial pressure modulates afferent baroreceptor activity in animals probably provides substantial insight into how changes of arterial distending pressure in humans are translated into integrated baroreflex responses.

7

Reflex interactions

RESPIRATION

Respiratory sinus arrhythmia

Respiratory sinus arrhythmia, vagally mediated R–R interval fluctuations synchronous with respiration, was first observed in animals. Stephen Hales (1733) first documented respiratory fluctuations of blood pressure, and the philosopher and physiologist Albrecht von Haller (1778) first observed respiratory fluctuations of heart rate. Ewald Hering (1871), Heinrich Hering's father, showed that with moderate lung inflations, cardioacceleration occurs during the inspiratory phase of respiration, and Fredericq (1882) showed that respiratory fluctuations of heart rate persist in animals with open chests, in the absence of respiratory movements and changes of pulmonary and thoracic stretch receptor activity. Heinrich Hering (1930) showed also that respiratory sinus arrhythmia is abolished by bilateral vagotomy and carotid artery occlusion. Schneyer (1935*a*), Schweitzer (1935), and Anrep *et al.* (1936) showed that respiratory sinus arrhythmia is dependent critically on the level of baroreceptor stimulation of vagal nuclei; both low and high arterial pressures tend to abolish sinus arrhythmia.

Respiratory–cardiovascular interactions can be studied particularly well in humans because *humans can cooperate* and vary their respiratory rates and tidal volumes on command. Average fluctuations of muscle sympathetic nerve activity, R–R intervals, and diastolic pressure during a single respiratory cycle, with frequency-controlled (rate = 12 breaths/min) breathing in supine, resting human subjects are shown in Fig. 7.1 (Eckberg *et al.* 1985). This figure indicates that during inspiration, sympathetic activity and R–R intervals decline in parallel in reciprocal relation to arterial pressure; during expiration, sympathetic activity and R–R intervals increase in parallel, as arterial pressure falls. The majority of respiratory R–R interval fluctuations are neurally mediated, because they are virtually absent in patients with heart transplants (Beck *et al.* 1969; Stinson *et al.* 1972; M.L. Smith *et al.* 1990). The unexpected reciprocal relation between vagal activity (R–R intervals) and arterial pressure, documented also by Melcher (1980), suggests that respiratory sinus arrhythmia is not due simply to baroreflex mediated

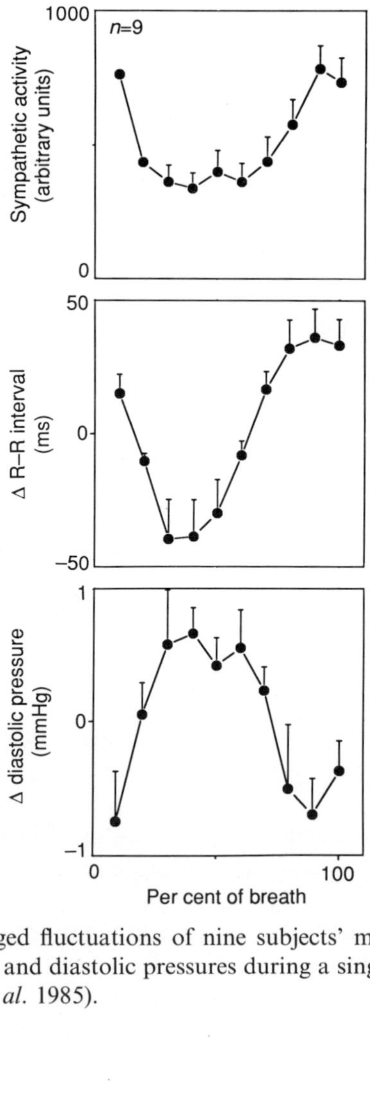

FIG. 7.1. Signal-averaged fluctuations of nine subjects' muscle sympathetic nerve activity, R–R intervals, and diastolic pressures during a single breath. Adapted with permission (Eckberg *et al.* 1985).

changes of vagal outflow (Matthes and Ebeling 1948; Mechelke 1953); a baroreflex mechanism would dictate that vagal activity changes *in parallel* with arterial pressure. Therefore other possible explanations for apparent inspiratory suppression and expiratory facilitation of autonomic outflow must be invoked.

Baroreflex control of sinus node function

Reciprocal changes of vagal activity and arterial pressure may result from respiratory interference with baroreceptor stimulation of vagal firing, as suggested first by studies conducted in anaesthetized dogs by Koepchen *et al.* (1961). The earliest indication of possible respiratory–baroreflex interactions in humans came from the study of Smyth *et al.* (1969) who found that baroreflex slopes after angiotensin-II injections (see Chapter 5) are slightly steeper when expiratory pulses are considered separately from inspiratory pulses. One interpretation of these data is that inspiration reduces baroreceptor stimulation of vagal motonuclei.

This possibility was studied prospectively by Eckberg and Orshan (1977), who delivered precise baroreceptor stimuli during different phases of respiration. Moderate, brief neck suction caused greater R–R interval prolongation in expiration than in inspiration, but intense neck suction caused comparable R–R interval prolongation in expiration and inspiration. Thus in humans as in animals (Anrep *et al.* 1936), respiratory gating of baroreceptor stimulation of vagal motoneurones is finite and can be overcome by strong baroreceptor stimulation. C. Borst and Karemaker (1983) electrically stimulated carotid sinus nerves in unsedated patients with carotid sinus nerve stimulators placed for treatment of angina. Their study failed to document differences in sinus node responses to stimuli delivered in inspiration and expiration. We speculate that respiratory variability of responses was not observed in this study because carotid sinus nerve stimuli were large and overwhelmed finite respiratory influences on vagal–cardiac motoneurones.

Subsequent studies (Eckberg *et al.* 1980, 1985) showed that central autonomic modulation varies continuously as a sinusoidal function during respiration; lengthened R–R intervals caused by neck suction, and augmented muscle sympathetic nerve activity caused by neck pressure, tend to be maximal during late inspiration and early expiration, and minimal during late expiration and early inspiration. The time course of respiratory gating of human vagal and sympathetic baroreflex responses is very similar to the time course of respiratory gating of the duration of sympathetic inhibition provoked by electrical carotid sinus nerve stimulation, as documented by Seller *et al.* (1968) in anaesthetized dogs.

Data from humans and a dog are depicted in Fig. 7.2. In the human study (left), baroreceptors were stimulated with brief neck suction; in the dog study (right), carotid baroreceptor (and chemoreceptor) nerves were stimu-

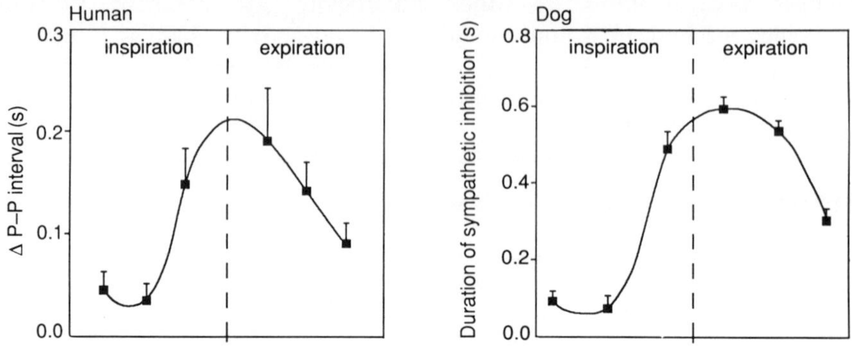

FIG. 7.2. Influence of respiration on human and canine autonomic responses to carotid stimulation. Left: changes of P–P interval plotted as a function of the timing of 0.6 s of 30 mmHg neck suction within the respiratory cycle in six healthy men and women. Right: duration of lumbar sympathetic inhibition after 2 s of electrical carotid sinus nerve stimulation applied at different times in the respiratory cycle, in one anaesthetized dog. Adapted with permission (Eckberg *et al.* 1980; Seller *et al.* 1968).

lated electrically. In both studies, stimuli were delivered during all phases of the respiratory cycle, and responses were plotted according the timing of stimuli in the respiratory cycle. These data show clearly that phasic respiratory suppression of baroreflex responses is not limited simply to the time in the respiratory cycle when inspiratory motoneurone activity occurs.

Another study (Eckberg *et al.* 1980) examined the influence of *respiratory rate* on vagal baroreflex control and found that differential inspiratory–expiratory baroreflex responsiveness is preserved at breathing rates of 3, 6, and 12 but abolished at 24 breaths/min. This study did not define the mechanisms responsible for loss of respiratory gating of vagal baroreflex responses during rapid breathing. One possibility is that baroreflex gating shifts within the respiratory cycle as a function of breathing rate.

Autonomic changes in absence of breathing

It is common practice to apply brief forcings during held inspiration or expiration to minimize respiration-related fluctuations of baroreflex responsiveness. Although Eckberg (1976) found that heart rate is constant during held *expiration*, Trzebski *et al.* (1980) and Hirsch and Bishop (1981) found that respiratory sinus arrhythmia may persist during (prolonged) held *inspiration*. One explanation for respiratory sinus arrhythmia during held inspiration is that it reflects persistence of arterial pressure fluctuations (Huikuri *et al.* 1989). Persistence of sinus arrhythmia in the absence of breathing suggests strongly that sinus arrhythmia results in part from central modulation of vagal activity (since with fixed chest expansion, afferent pulmonary and thoracic stretch receptor activity is probably

Subsequent research showed that the constant heart rate found in almost (but not quite) all subjects during held *expiration* masks complexities of human autonomic neural regulation. Fritsch *et al.* (1991) found that both muscle sympathetic nerve activity and arterial pressure increase progressively during held expiration. This is shown for one subject in Fig. 7.3. The degree of pressure elevation during held expiration probably was sufficient to increase vagal–cardiac inhibition. Therefore, since no R–R interval prolongation occurred, it is likely that increased vagal restraint was counterbalanced exactly by increased sympathetic stimulation of the sinoatrial node. The mechanisms for increases of sympathetic nerve activity during held expiration are unknown; decreases of arterial and central oxygen concentrations and increases of carbon dioxide content occur, but may be insufficient to explain these changes (Rowell *et al.* 1989).

Hirsch and Bishop (1981) and Eckberg (1983) studied the effects of breathing frequency and tidal volume on respiratory sinus arrhythmia. Both

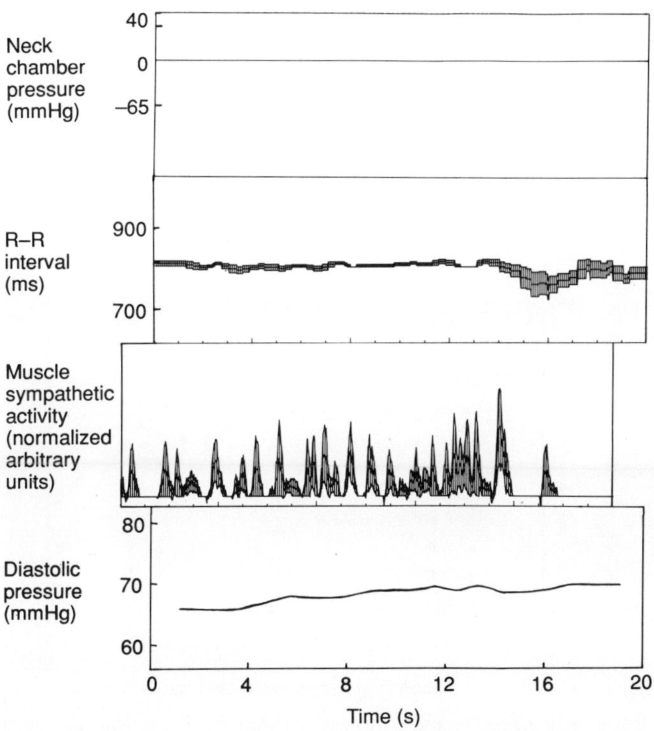

FIG. 7.3. Average autonomic responses of one healthy subject to ten periods of held expiration. Adapted with permission (Fritsch *et al.* 1991).

studies indicate that although tidal volume directly influences the magnitude of sinus arrhythmia, the influence of tidal volume is small. In Eckberg's study, a 50 per cent increase of tidal volume increased sinus arrhythmia (peak minus valley changes of R–R intervals) by only 15 per cent. In contrast, a reduction of respiratory rate from 20 to 6 breaths/min more than doubled the magnitude of sinus arrhythmia. Hirsch's Bode plots of sinus arrhythmia, breathing frequency, and tidal volume are shown in Fig. 7.4.

Dornhorst *et al.* (1952) published experimental records that underscore the complexity of sinus arrhythmia. They showed that *blood pressure* fluctuations

(1) increase in amplitude as respiratory rate slows (maximal increases occur at about six breaths/min);

(2) occur at different times in the respiratory cycle at different breathing rates;

(3) may occur in absence of heart rate fluctuations;

(4) may occur in absence of breathing (in confirmation of Fredericq's (1882) finding);

(5) may occur as submultiples of breathing rates.

Mechanisms underlying respiratory heart rate and blood pressure fluctuations and baroreflex contributions thereto are understood poorly (Koepchen 1984). De Boer *et al.* (1987) developed a model that predicted power and cross-spectra of respiratory sinus arrhythmia. Their model suggests that respiration directly affects cardiac output and blood pressure, and

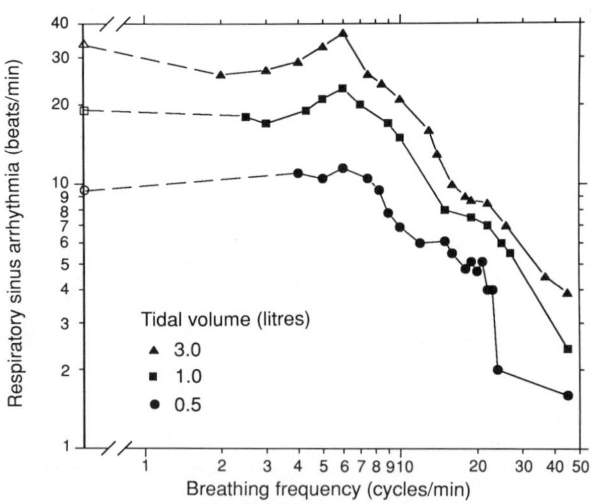

FIG. 7.4. Bode plots of average peak minus valley R–R interval changes of healthy subjects for different breathing rates and tidal volumes. Adapted with permission (Hirsch and Bishop 1981).

that sinus arrhythmia results from vagally mediated baroreflex responses to pressure changes. Some evidence suggests that fluctuating levels of afferent pulmonary activity does not importantly determine respiratory sinus arrhythmia; mechanical lung inflation does not produce sinus arrhythmia in patients during cardiopulmonary bypass (Ott *et al.* 1975) or in conscious volunteers (Freyschuss and Melcher 1975). Presumably the direct relation between respiratory sinus arrhythmia and tidal volume (Hirsch and Bishop 1981; Selman *et al.* 1982; Eckberg 1983) results from central respiratory drive.

Sympathetic outflow to muscle

Figure 7.1 illustrates respiratory fluctuations of human muscle sympathetic nerve activity (Eckberg *et al.* 1985). Power spectral analyses show that R–R interval and sympathetic nerve outflows have similar frequency contents, and that both may have substantial power at low (below 0.10 Hz) as well as high frequencies. A subsequent study (Eckberg *et al.* 1988) explored the effects of sustained changes of arterial pressure on respiratory modulation of muscle sympathetic outflow, during stepwise infusions of vasoactive drugs. Respiratory fluctuations of sympathetic activity for one subject are shown in Fig. 7.5. Although the absolute levels of sympathetic activity correlated inversely with arterial pressure, respiratory modulation of sympathetic outflow was present at all pressures studied. (However, very little sympathetic activity remained at the highest diastolic pressure studied, 85 mmHg, which was only 6 mmHg above baseline diastolic pressure.)

Influence of baroreceptor input on respiration

There is a substantial literature that documents interactions between arterial pressure and respiratory function. Arterial pressure elevation reduces tracheal tension and dilates the trachea (Nadel and Widdicombe 1962; Schultz *et al.* 1987). Data from animals show clearly that baroreceptor stimulation depresses ventilation (C.F. Schmidt 1932; C.V. Winder 1938; Brunner *et al.* 1982; Jung and Katona 1990); the relation between carotid sinus pressure and respiratory period and expiratory duration is sigmoid, and is very similar to the relation between arterial or carotid distending pressures and R–R interval responses (Koch 1931; Eckberg 1980*a*). Carotid massage leads to hyperpnea, which can be prevented by pretreatment with lignocaine injections in the carotid sinus area (Galdston *et al.* 1943). Baroreceptor stimulation with neck suction (a very obtrusive method) increases tidal volume and ventilation (Trzebski *et al.* 1980). The reason for increased rather than decreased ventilation during baroreceptor stimulation in humans is unclear; one possibility, suggested originally by C.V. Winder (1938), is

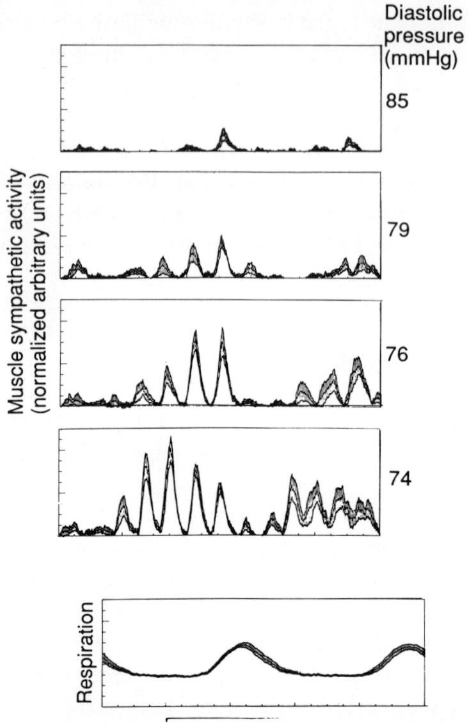

FIG. 7.5. Signal-averaged muscle sympathetic nerve activity measured in one subject during ten breaths at different pressure levels. For each sweep, data collection was triggered by an R-wave occurring in the first third of inspiration. Adapted with permission (Eckberg *et al.* 1988).

that respiration follows falling aortic rather than increasing carotid distending pressure during carotid baroreceptor stimulation.

CHEMORECEPTOR REFLEXES

Activation of chemoreceptors provokes a variety of changes of autonomic outflow, and presents several possibilities for interactions with arterial baroreceptors. For a comprehensive review, the reader should consult the chapter on interactions between respiration and the circulation by M.deB. Daly (1986) and the reviews by Heistad and Abboud (1980) and Rowell and Blackmon (1987). This aspect of physiological control is complex. First, experimental chemoreceptor reflex activation takes several forms; these include varying degrees of hypoxia with hypercapnia (asphyxia), hypoxia with hypocapnia (as might occur secondary to hypoxia-induced hyperventilation), and hypercapnia with normal or raised oxygen concentrations. Further, responses to chemoreceptor stimuli may be mediated by different chemoreceptor sensitive receptors, including chemoreceptors located in the regions of the carotid bifurcation and aortic arch, and chemically sensitive neurones located in the medulla (which are primarily responsive to changes of carbon dioxide levels). Finally, in addition to chemoreceptor reflex activation, changes of blood gas levels modulate ventilation, arterial pressure, heart rate, and vascular resistance. All of these factors may exert independent influences on baroreflex function.

Hypoxia

It is clear from studies in animals and humans that systemic hypoxia causes tachycardia (Asmussen and Chiodi 1941; Holton and Wood 1965; M.deB. Daly 1986). Bernthal *et al.* (1951) and M.deB. Daly and Scott (1958) found that in anaesthetized dogs, chemoreceptor reflex-induced tachycardia is secondary to hyperventilation; when central respiratory drive is abolished by artificial ventilation, hypoxia provokes bradycardia. This view is disputed in a provocative study published by Gupta and Singh (1987). They suggested that conclusions derived from animals depend importantly on the anaesthetic used. They speculated that hypoxia may *not* lead to vagal withdrawal and cardioacceleration in some animal preparations, because the anaesthetics used had already nearly abolished efferent vagal outflow. In support of this notion, they found that when anaesthetics are used which preserve vagal outflow, hypoxia leads to cardioacceleration.

Several groups have studied the effects of hypoxia on heart rate in humans. Slutsky and Rebuck (1978) found that isocapnic hypoxia produces tachycardia which is related linearly to the degree of hypoxia. Hedemark and Kronenberg (1982) found that these relations are unchanged by slow wave sleep, and that hypoxia does not cause arousal from sleep consistently, as does hypercapnia. This study tends to exclude anxiety secondary to hypoxia as the cause of cardioacceleration.

Eckberg *et al.* (1982) studied the effects of mild brief hypoxia on heart rate in healthy humans trained to control their respiratory frequencies and tidal volumes at constant levels. They found that extremely mild hypoxia (reductions of arterial oxygen saturation as low as about 0.1 per cent) consistently leads to vagal withdrawal, as reflected by reductions of respiratory sinus arrhythmia and R–R intervals, in the absence of changes of ventilation. This conclusion confirmed results published earlier by Asmussen and Chiodi (1941) and Dripps and Comroe (1947). Although respiratory rates and tidal volumes were not controlled in the earlier studies, some volunteers did not alter ventilation during exposure to hypoxia, and nevertheless experienced cardioacceleration. Moreover, in the studies of Dripps and Comroe (1947) and Eckberg *et al.* (1982), inhalation of 100 per cent oxygen led to cardiac slowing. Jennett used a different experimental paradigm and drew similar conclusions. She found that sudden relief of hypoxia in humans by a breath of oxygen provokes brief bradycardia (Jennett 1969). The view that chemoreceptor reflex activation causes cardioacceleration in humans independent of changes of ventilation is supported also by results of studies conducted in patients in whom the carotid bodies have been removed surgically as treatment for asthma; in such patients, hypoxia does not increase ventilation or alter blood pressure, but does increase heart rate (P.M. Gross *et al.* 1976).

Although these results challenge the notion that hypoxia provokes primary cardiac slowing, which is overridden by ventilation-induced cardiac speeding in humans, different data suggest otherwise. Drysdale and Strange-Petersen (1977) proposed that the bradycardia found by Jennett (1969) after a hyperoxic breath is due to transient hypoventilation; when subjects avoid hypoventilation, hyperoxia provokes minor but convincing cardioacceleration. Conscious patients with acute spinal cord injury may experience profound bradycardia when they are disconnected from the respirator (Frankel *et al.* 1975; Welply *et al.* 1975; Berk and Levy 1977). This response is attributed to a reflex initiated by chemoreceptors (M.deB. Daly 1986). These results suggest that in humans, *any* respiration during hypoxaemia increases heart rate. (Sleight (1976) proposed an alternative mechanism: when tetraplegic patients are disconnected from the respirator, the ensuing hypoxia increases sympathetic activity, which in turn increases ventricular receptor activity, which stimulates vagal–cardiac nerve activity.)

An alternative method to determine the effects of chemoreceptor stimulation on heart rate is to measure R–R intervals during breath holding (brief asphyxia). Raper *et al.* (1967) found no consistent heart rate change during breath holding. P.M. Gross *et al.* (1976) compared responses to breath holding in healthy subjects and patients with denervated carotid bodies; they found that after ventilation with 21 per cent oxygen, healthy people experience bradycardia, but carotid chemoreceptor denervated subjects experience tachycardia. This tachycardia (unlike the above bradycardia) increases linearly with the degree of hypoxia. Gross concluded that bradycardia in healthy subjects results from peripheral chemoreceptor stimulation. We note that this conclusion was not based on any statistical analysis of the data, however, and examination of individual results suggests that the 'normal' bradycardia is unimpressive and quite variable among different healthy subjects.

Several groups have sought and not found evidence that cardioacceleration during hypoxia results from impaired *baroreceptor–cardiac reflex* function. J.D. Bristow *et al.* (1971*a*) found no change of baroreflex slopes during hypoxia. J. Bhattacharya *et al.* (1973) held ventilation and arterial carbon dioxide levels constant and found no change of baroreflex slopes during progressive hypoxia. However, baroreflex slopes were reset in the direction of shorter R–R intervals at constant reference pressures. Similarly, Eckberg *et al.* (1982) found no change of R–R interval responses to brief, single intensity neck suction delivered during mild hypoxia.

A coherent picture of the *circulatory effects* of chemoreflex stimulation in humans has emerged. It is well established that severe hypoxaemia causes forearm vasodilatation (Abramson *et al.* 1943; Kontos *et al.* 1967; Lugliani *et al.* 1973). Black and Roddie (1958) showed that hypoxic vasodilatation persists after local anaesthesia of nerves to the forearm, and is therefore not neurally mediated. Richardson *et al.* (1967) showed that severe hypoxia increases cardiac output and forearm blood flow. Since forearm flow was not altered by adrenergic blockade, Richardson considered that direct effects, rather than reflex sympathetic withdrawal, caused vasodilatation. This view was further supported by his finding that plasma catecholamine levels (measured fluorometrically) did not increase during hypoxia. This finding was confirmed by Rowell *et al.* (1989).

The study of Rowell *et al.* (1989), which is discussed below, may be the only example of changes of human muscle sympathetic nerve activity that are not paralleled by changes of antecubital vein plasma noradrenaline concentrations. The mechanisms responsible for this finding were not defined. Conceivably, noradrenaline spillover increased, but plasma noradrenaline concentrations remained constant because of dilution secondary to increased forearm blood flow. Alternatively, noradrenaline spillover increased, but was balanced by increased neuronal re-uptake.

In the study of Richardson *et al.* (1967), hypoxaemia did not lower blood pressure, notwithstanding vasodilatation in the important muscle vascular bed. There appear to be at least two explanations for this. First, hypoxic vasodilatation is selective and does not involve all vascular beds (Robin *et al.* 1958; Rowell and Blackmon 1987). Second, hypoxic vasodilatation is opposed by reflex vasoconstriction triggered by chemoreceptors. Lugliani *et al.* (1973) found that during hypoxia, subjects with intact carotid chemoreceptors have minor increases, but patients with carotid body denervations have reductions of arterial pressure. Since patients with carotid body denervations have preserved arterial baroreflex function (as reflected by normal blood pressure overshoots and bradycardia after Valsalva straining), these data suggest that chemoreflexes oppose blood pressure reductions secondary to direct hypoxic vasodilatation. Since these patients had intact aortic bodies, arterial pressure elevations in intact subjects during hypoxia must have been mediated by *carotid* chemoreceptors.

It should be noted that the studies quoted above on possible interactions between chemoreflexes and baroreflexes in humans involve baroreflex control of *vagal* outflow. We are aware of no study of chemoreflex influences on baroreflex control of sympathetic outflow. This type of study is needed, because Heistad *et al.* (1974) found in anaesthetized dogs that chemoreflex-induced muscle vasoconstriction is enhanced by acute hypotension and inhibited by acute hypertension. This interaction could be studied elegantly in conscious humans with direct recordings of *muscle sympathetic nerve activity*, which are discussed next.

Until the technique of sympathetic microneurography was used, it was impossible to know whether efferent sympathetic activity decreases or increases in subjects exposed to hypoxia. The studies of Black and Roddie (1958) and Richardson *et al.* (1967) document forearm (primarily muscle) vasodilatation, unaffected by neural blockade during hypoxia. Contrariwise, studies conducted in anaesthetized animals document muscle vasoconstriction (Angell James and Daly 1969) and augmentation of efferent muscle sympathetic nerve activity (Gregor and Jänig 1977). At least four groups have published studies of chemoreflex influences on human muscle sympathetic nerve activity. The consensus of these studies is that hypoxia increases human muscle sympathetic nerve activity.

M. Saito *et al.* (1988) measured muscle sympathetic nerve activity in healthy volunteers studied in hypobaric chambers during simulation of altitudes of 4000, 5000, and 6000 metres for 30 min. At all levels, muscle sympathetic nerve activity increased. Somers and co-workers (1988*a*, 1989*a,b*) found that 14 per cent oxygen does not alter, but that 10 per cent oxygen increases muscle sympathetic nerve activity significantly. Rowell *et al.* (1989) exposed subjects to 12, 10, and 8 per cent oxygen for 20, 20, and 10 min. They found that 12 per cent oxygen raises levels of muscle sympathetic nerve activity, but that increases are gradual, and do not become significant until the twelfth minute. More severe hypoxia augments muscle sympathetic nerve activity earlier. Rowell's findings are depicted in Fig. 7.6. (In this study, subjects did not control their ventilation and developed respiratory alkalosis during ventilation with hypoxic gas.)

D.L. Eckberg, A. Trzebski, J.M. Fritsch, R.F. Rea, M.L. Smith, and L.A. Beightol (unpublished) used frequency- as well as time-domain techniques to evaluate muscle sympathetic nerve responses to brief hypoxia during controlled ventilation. Logistic regression analyses suggested that increased ventilation during hypoxia minimizes increases of sympathetic traffic. Power spectral analyses indicated that hypoxia shifts sympathetic outflow from higher to lower frequencies.

There are several unconnected observations which suggest that human sympathetic responses to hypoxia may have *pathophysiological* significance, a hypothesis advanced originally by Przybylski (1981). Chemoreceptor reflex stimulation of ventilation and sympathetic activity may be augmented in patients with borderline (Trzebski *et al.* 1982) and fixed (Somers *et al.* 1988*a*) hypertension. Patients with sleep apnea are exposed to frequent nocturnal chemoreflex activation and have episodic high levels of muscle sympathetic nerve traffic (Hedner *et al.* 1988). Such patients are more likely than healthy people to develop arterial hypertension (Burack *et al.* 1977). Conversely, patients with primary hypertension have a high incidence of sleep apnea.

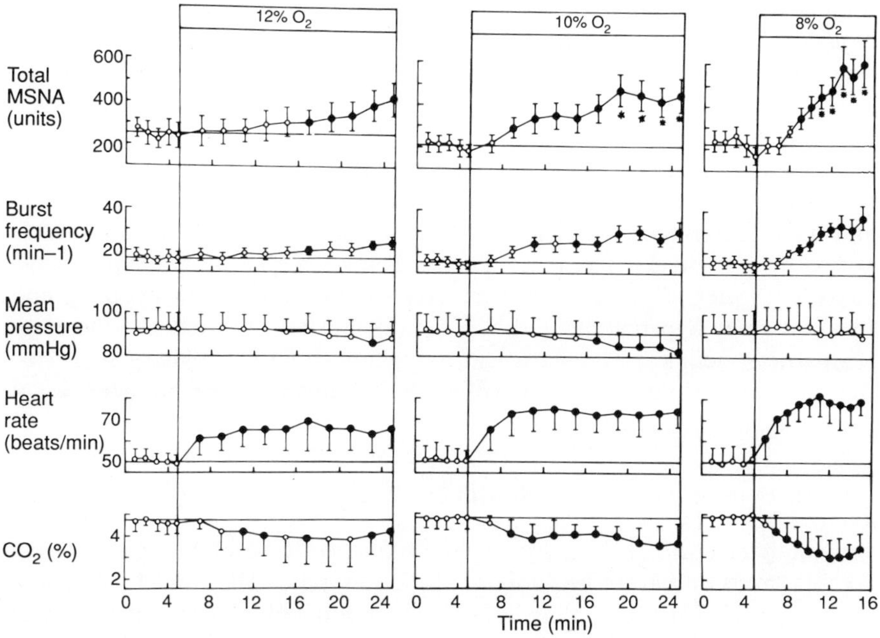

FIG. 7.6. Time course of changes of muscle sympathetic nerve activity (MSNA), mean arterial pressure, heart rate and end-tidal carbon dioxide during uncontrolled breathing at three levels of inspiratory oxygen. Closed circles = significantly different from control. Reproduced with permission (Rowell *et al.* 1989).

Thus sleep apnea and hypertension may be risk factors for each other (Fletcher *et al.* 1985; Przybylski 1981; Somers *et al.* 1988*b*).

There has been no study of the effects of chemoreflex activation on baroreflex control of human sympathetic activity. However, the influence of hypoxia on vasoconstriction has been studied. Henriksen and Rowell (1986) found that moderate hypoxia (10 to 11 per cent oxygen) does not alter postural reflex reduction of skeletal muscle blood flow. Heistad and Wheeler (1970) found that moderate hypoxia (10 per cent oxygen) reduces muscle vasoconstriction caused by intense lower body suction. In a subsequent study (Heistad *et al.* 1972), this group showed that some patients with chronic hypoxaemia have impaired vasoconstrictor responses to lower body suction, which can be improved by administration of oxygen.

Hypercapnia

Human responses to hypercapnia have not been studied as extensively as responses to hypoxia. Studies involving patients with carotid body denervations indicate that chemoreflex responses to hypoxia are mediated primarily or exclusively by carotid bodies (Lugliani *et al.* 1973). Such patients have preserved, but diminished ventilatory responses to hypercapnia (Lugliani *et al.* 1971). Reduction of responses of patients with carotid body denervations to hypercapnia provides inferential evidence that in humans, some responsibility for mediation of responses to hypercapnia lies with carotid chemoreceptors.

Human sympathetic nerve responses to hypercapnia have been measured by two groups. In both the series published by Somers *et al.* (1988*a*, 1989*a,b*) and the unpublished study by Eckberg *et al.*, hypercapnia increased muscle sympathetic nerve activity. The studies of Somers suggested that hypercapnia leads to greater increases of muscle sympathetic traffic than hypoxia, and that combined hypercapnia and hypoxia (asphyxia) synergistically increases sympathetic activity. The statistical modelling done by Eckberg suggested that increases of ventilation provoked by hypercapnia oppose increases of muscle sympathetic nerve activity.

Combinations of gaseous stimuli with exercise

Cunningham *et al.* (1972*b*) studied the influence of several respiratory gas combinations on baroreflex function during exercise. They and J. Bhattacharya *et al.* (1973) found that hypoxia does not reduce baroreflex gain at rest, notwithstanding the occurrence of substantial tachycardia (which during exercise is associated with *reduced* baroreflex gain). In contrast, hypercapnic hypoxia (asphyxia) reduces baroreflex sensitivity at rest. Reduction of baroreflex gain is much more pronounced with pure exercise stimuli than with pure gaseous stimuli, but hypoxia superimposed upon exercise results in significant further reductions of baroreflex gain.

The confounding effect of ventilation was studied by J.D. Bristow *et al.* (1971*a*) who examined the separate effects of increases of alveolar carbon dioxide levels and ventilation. Bristow's new finding was that the scatter of results is much smaller when ventilation is held constant. Voluntary control of breathing had little effect on steady-state pressures or pulse rate, but reduced baroreflex sensitivity. This finding may be related to the observation that baroreceptor–cardiac reflex slopes are related inversely to the level of arousal (Conway *et al.* 1983).

In summary, although a substantial body of evidence indicates that hypoxia causes cardioacceleration in humans, published data do not point towards impairment of baroreceptor–cardiac reflex control. Hypoxia combined with hypercapnia impairs baroreceptor–cardiac reflex control, and hypoxia combined with exercise leads to greater impairment of baroreflex function than exercise alone (discussed below). Although hypoxaemia dilates resistance vessels directly, it also increases muscle sympathetic nerve activity, and blood pressure does not change importantly. The possibility that hypoxia modifies baroreflex control of sympathetic activity has not been studied adequately in humans.

THE DIVING REFLEX

The diving reflex comprises a remarkable series of cardiovascular and metabolic changes that increases survival in the absence of breathing during immersion (Irving 1934). The evidence suggests that this orchestrated response pattern serves to prolong survival by reducing myocardial oxygen consumption (Gooden *et al.* 1974; Elsner *et al.* 1985) and shunting blood from less essential organs (Heistad and Wheeler 1970; Angell James and Daly 1972) to the heart and brain (Andersen 1963; Folkow *et al.* 1967; Andersen and Blix 1974). There are several excellent reviews of this physiology (Scholander 1940; Andersen 1966; Paulev 1969; Angell James and Daly 1972; Butler and Jones 1982).

Although the diving reflex can be studied during total body immersion, it may be studied more simply during immersion of the face, which avoids complexities resulting from translocation of intravascular volume from the peripheral to the central circulations (M. Epstein 1976, 1978; Greenleaf 1984). The elements of the diving reflex are apnoea, bradycardia, reductions of cardiac output and stroke volume, and vasoconstriction. Opposite changes occur following diving.

The diving reflex typically begins with facial cooling. In humans, cooling may result from immersion of the face in cold water (Heistad *et al.* 1968), contact with an ice bag (Kawakami *et al.* 1967; Moore *et al.* 1972), or convection with cold air (LeBlanc *et al.* 1976; Riggs *et al.* 1981). The extent of reflex changes is inversely proportional to the degree of cooling (Whayne and Killip 1967; Moore *et al.* 1972; Fagius and Sundlöf 1986). This discussion does not discriminate among diving reflex responses elicited by different methods.

The afferent limb of the diving reflex is the trigeminal nerve (Beau 1860; Richet 1898; Lombroso 1913; Andersen 1963), whose probable central terminations are shown in Fig. 3.8. Of potential importance for human research is the fact that, unlike most autonomic cardiovascular sensory inputs, trigeminal afferent pathways do *not* terminate in the solitary tract nucleus (Kumada *et al.* 1977). Therefore the diving reflex may offer a means to test the integrity of human vagal–cardiac reflexes that is independent of baroreceptors and their primary brainstem termination. (This is not to say that baroreceptors are not involved in human diving responses however, because as discussed below, arterial pressure elevations in humans during facial immersion may be substantial.)

Diving apnoea is reflex because it occurs in decerebrate ducks (F.M. Huxley 1913; Andersen 1963). Apnoea may not be necessary for diving reactions to occur. However, although cardiac slowing occurs when subjects breathe during face immersion (Whayne and Killip 1967; Heistad and Wheeler 1970; Eckberg *et al.* 1984), the magnitude of bradycardia is less (Feigl and Folkow 1963; Campbell *et al.* 1969). Breathing appears to oppose diving bradycardia through efferent respiratory motoneurone activity and afferent pulmonary and thoracic stretch receptor activity (Bamford and Jones 1976; Angell-James *et al.* 1981). In ducks, breathing during simulated diving abolishes bradycardia (Angell James and Daly 1972). In diving animals, chemoreceptor stimulation promotes bradycardia (Feigl and Folkow 1963; Jones and Purves 1970) but does not overcome trigeminal receptor inhibition of breathing (M.deB. Daly *et al.* 1977). In humans, bradycardia occurs before measurable changes of blood gases (Olsen *et al.* 1962).

Caputa and Cabanac (1979) made the curious observation that the heart rate response to face immersion in water at 15 degrees C depends on the subjects' core temperatures. When subjects are hypothermic, normal bradycardia is replaced by tachycardia. They suggested that the normothermic bradycardic response was partly mediated by a direct central nervous effect from cold blood reaching the brain via the ophthalmic veins.

The occurrence of *bradycardia* during diving has been recognized since the famous experiment of Paul Bert (1870) in which he placed a duck's head in water, palpated the duck's breast, and counted the heart rate. In diving animals, the heart rate may be reduced by as much as 90 per cent (Irving *et al.* 1942; Folkow *et al.* 1967), and in diving humans, the heart rate may be halved (Scholander *et al.* 1962). Extreme bradycardia (Finley *et al.* 1979), sinus arrest (Bergman *et al.* 1972), and circulatory collapse probably secondary to high grade atrioventricular block (Whayne and Killip 1967) occur in humans during face immersion. The onset of cardiac slowing may be nearly instantaneous (Scholander 1940). More commonly, heart rate slowing is progressive; after an initial abrupt heart rate slowing, additional slowing occurs (Folkow *et al.* 1967). Bradycardia is mediated by increased vagal–cardiac nerve traffic; it is abolished by vagal nerve transection (Richet 1894; Lombroso 1913) or atropine (Richet 1894; Van Citters *et al.* 1965; Folkow *et al.* 1967; Butler and Jones 1968) in diving animals, and by large doses of atropine in humans (Heistad *et al.* 1968; Finley *et al.* 1979). Face immersion bradycardia is not reduced by beta-blockade with propranolol (Finley *et al.* 1979).

Several factors influence the degree of bradycardia that occurs during face immersion. As with baroreflex bradycardia (Conway *et al.* 1983), mental arithmetic (A. Ross and Steptoe 1980; Fagius and Sundlöf 1986), or 'harassment' of the volunteer (S. Wolf *et al.* 1965) diminishes it. Curiously, bradycardia may persist despite vigorous exercise (Olsen *et al.* 1962; Scholander *et al.* 1962; Bergman *et al.* 1972; Finley *et al.* 1979). As with vagal baroreflex responses (Duke *et al.* 1976), respiratory sinus arrhythmia (Pfeifer *et al.* 1983; Shannon *et al.* 1987), and tachycardia caused by chemoreflex activation (Kronenberg and Drage 1973), diving bradycardia declines with advancing age (Hong *et al.* 1970; Rey 1971).

As discussed in Chapter 13, increased vagal–cardiac traffic alters cardiac rhythm. Not surprisingly, face immersion provokes a variety of vagally mediated changes of cardiac rhythm, including shifts of the atrial pacemaker away from the sinoatrial node (reflected by changes of P wave morphology), sinus arrest with emergence of subsidiary pacemakers, premature atrial beats, and even atrial fibrillation (Olsen *et al.* 1962; Scholander *et al.* 1962; Bergman *et al.* 1972). Simulated diving is standard therapy for paroxysmal supraventricular dysrhythmias (Bergman *et al.* 1972; *Lancet* 1981).

Two groups have studied heart rate responses to simulated diving in patients after myocardial infarction. Ryan *et al.* (1976) found normal bradycardia during face immersion in patients three months after infarction, and Gooden *et al.* (1978) found supranormal bradycardia in patients 3 to 13 days after infarctions. The significance of these observations, and the reasons for these divergent results, are unknown.

It is not entirely clear whether bradycardia is a primary adjustment secondary to trigeminal stimulation of vagal motoneurones, or if it is merely part of a larger reflex pattern that includes arterial pressure elevation, baroreceptor stimulation, and bradycardia. Finley *et al.* (1979) argue that since arterial pressure may remain unchanged during bradycardia caused by diving, increased baroreceptor activity is not essential for the response. Jones (1973) and F.M. Smith and Jones (1990) studied the effects of arterial baroreceptor denervation on diving responses of conscious ducks. In their studies, diving provoked hypotension in intact ducks. Baroreceptor denervated ducks experienced a more rapid development of bradycardia and more severe hypotension. Thus baroreceptors opposed haemodynamic changes provoked by trigeminal stimulation.

Eckberg *et al.* (1984) investigated interactions among trigeminal and carotid baroreceptor stimulation and breathing. Responses of one subject are illustrated in Fig. 7.7. Diving increased average R–R intervals, respiratory sinus arrhythmia, and R–R interval lengthening during neck suction. Eckberg proposed that the most 'economical' interpretation of these data is that trigeminal stimulation during facial immersion increases responsiveness of vagal–cardiac motoneurones to stimulation by autonomic sensory inputs, including particularly those from arterial baroreceptors. These data support the view that in humans as in animals, the effects of afferent sensory inputs on vagal motoneurones reflect the algebraic sum of all influences (Langhorst *et al.* 1980), as demonstrated in cats by McAllen and Spyer (1978*a*).

Unfortunately, this study employed only one intensity of baroreceptor stimulation; therefore it is not known how the diving reflex altered sigmoid baroreceptor–cardiac response relations. Indeed, we are not aware that such data have been obtained from any species; however, Angell-James *et al.* (1978) delivered single step pressure increases to isolated carotid sinuses (in two anaesthetized harbour seals during simulated diving) and also found augmented R–R interval lengthening. Therefore extant data suggest but do not prove that trigeminal stimulation increases vagal–cardiac respons-

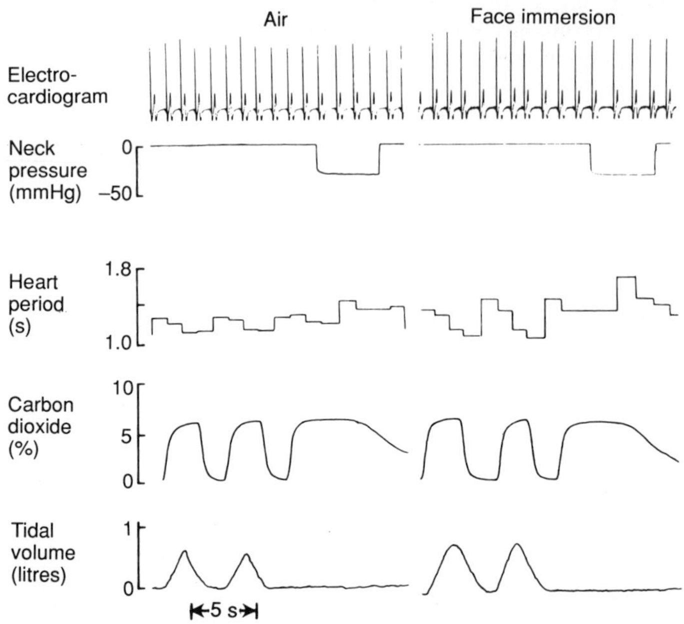

FIG. 7.7. Responses of one subject to controlled breathing (left of each panel) and neck suction before and during face immersion in ice water. Adapted with permission (Eckberg *et al.* 1984).

iveness, at higher levels of sympathetic activity and arterial pressure than usual.

Arterial pressure responses are highly variable in diving animals; arterial pressure usually increases in humans during simulated diving (Kawakami *et al.* 1967; Heistad *et al.* 1968). Since diving bradycardia may be profound (Folkow *et al.* 1967), either stroke volume or peripheral vascular resistance must increase considerably to prevent cardiovascular collapse. There is substantial indirect and direct evidence that sympathetic activity increases during diving and simulated diving. Prevention of vasoconstriction with adrenergic blockade, or catecholamine depletion with reserpine in diving animals, shortens survival (Andersen and Blix 1974).

Cardiac output and left ventricular stroke volume decrease moderately during simulated diving (Kawakami *et al.* 1967), despite venoconstriction (Djojosugito *et al.* 1969) and increased filling pressures (Johansen and Aakhus 1963). Folkow *et al.* (1967) argue that some reduction of left ventricular stroke volume in diving ducks is due to vagally mediated negative inotropism (Ferrante and Opdyke 1969). Forearm vascular resistance increases, from about 30 to 60 per cent (Heistad *et al.* 1968; Campbell *et al.* 1969). Vasoconstriction is is adrenergically mediated (Kobinger and Oda

1969; Andersen and Blix 1974), and is probably secondary to increases of sympathetic nerve activity (Fagius and Sundlöf 1986). In humans, increases of muscle sympathetic nerve activity occur in parallel with increases of arterial pressure (Fagius and Sundlöf 1986), a situation reminiscent of the simultaneous increases of sympathetic activity and pressure that occur during exercise (Mark *et al.* 1985) and late during Valsalva straining (M.L. Smith *et al.*, unpublished). Notwithstanding parallel increases of muscle sympathetic nerve activity and arterial pressure during face immersion, episodic reductions of pressure are answered by increases of sympathetic activity. This is shown well in the example from the study of Fagius and Sundlöf (1986), depicted in Fig. 7.8.

There is no correlation between vasoconstriction and bradycardia in diving seals (Murdaugh *et al.* 1968), or between increases of sympathetic activity and R–R intervals (Fagius and Sundlöf 1986) in humans. Scholander *et al.* (1962) showed that in professional divers, blood lactate levels do not increase during dives, but increase after dives. This suggests inferentially that lactate is sequestered in muscle vascular beds during dives and is washed out into the circulation when vasodilatation occurs after dives. In diving animals, splanchnic and skin blood flow decline markedly (Irving *et al.* 1942; Hollenberg and Uvnas 1963). Although Heistad *et al.* (1968) documented finger (primarily skin) vasoconstriction, Fagius and Sundlöf (1986) found that face immersion reduces skin sympathetic nerve activity.

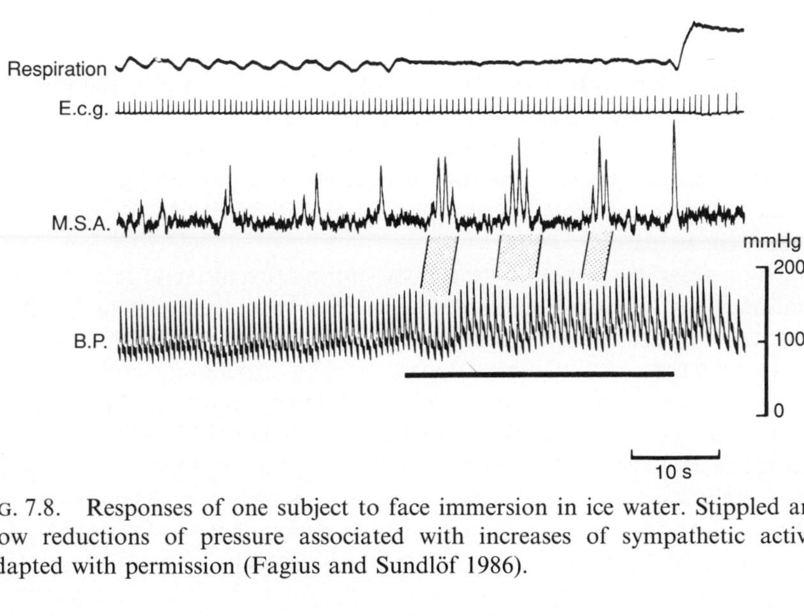

FIG. 7.8. Responses of one subject to face immersion in ice water. Stippled areas show reductions of pressure associated with increases of sympathetic activity. Adapted with permission (Fagius and Sundlöf 1986).

EXERCISE

Haemodynamic adjustments at the onset of exercise, including simultaneous increases of heart rate *and* blood pressure, support the strong likelihood that in some way, exercise modifies arterial baroreflex mechanisms. As will be seen, this conclusion is highly controversial. Several excellent reviews on this subject have been published (Christensen and Galbo 1983; Ludbrook 1983; H.L. Stone *et al.* 1985; Rowell 1986).

Autonomic responses to exercise

R–R intervals begin to shorten (heart rate begins to speed) within one second of the onset of isometric exercise (Gasser and Meek 1914; Freyschuss 1970; Petro *et al.* 1970). If muscle contraction begins early in a cardiac cycle, R–R interval shortening begins in the same cardiac cycle as exercise begins, with a latency estimated to be about 550 ms (Hollander and Bouman 1975). During sustained mild isometric exercise, the majority of heart rate speeding occurs during the first 15 s (Victor *et al.* 1987*b*; Eriksen *et al.* 1990).

Initial heart rate speeding is mediated by abrupt vagal withdrawal; it is reduced by large doses of atropine (Freyschuss 1970; Eckberg *et al.* 1972; Hollander and Bouman 1975; Fagraeus and Linnarsson 1976); it is present in patients with high spinal cord transections who have intact vagal–cardiac but interrupted sympathetic–cardiac control (Freyschuss 1970), and is absent in patients with Chagas' disease who are thought to have parasympathetic cardiac denervation but preserved sympathetic innervation (Gallo *et al.* 1987); and it is associated with loss of (vagal) R–R interval spectral power at respiratory frequencies (Arai *et al.* 1989). Heart rate speeding is proportional to the intensity of exercise (Victor *et al.* 1987*b*; Seals *et al.* 1988); however, after large doses of atropine, the proportionality to the intensity of mild to moderate exercise is lost, and the onset of heart rate speeding is delayed (Victor *et al.* 1987*b*). As the duration or intensity of exercise increases, sympathetic augmentation supplements vagal withdrawal to speed the heart (Robinson *et al.* 1966; Freyschuss 1970; C.E. Martin *et al.* 1974; Hume *et al.* 1979).

Cardioacceleration issues importantly (but not exclusively) from the volitional aspect of muscle exercise (*'central command'* (Krogh and Lindhard 1913)). This was studied elegantly by Goodwin *et al.* (1972) who increased or decreased the force of biceps contraction by stimulating muscle spindle afferents in the biceps or triceps with mechanical vibration. These stimuli decreased or increased central command, and changed blood pressure and heart rate responses in parallel with central command. Other studies point to the importance of central command in exercise-induced heart rate changes. Cardioacceleration is present (and even augmented (J.H. Mitchell *et al.* 1989)) when the intention to contract a muscle group is present but actual muscle contraction is reduced by neuromuscular blockade with succinylcholine (Freyschuss 1970) or curare (Victor *et al.* 1989). Heart rate speeding does not appear to result from afferent input from muscle nerves that are sensitive to chemical changes ('metaboreceptors' (Kaufman *et al.* 1983; Mense and Stahnke 1983; Kaufman *et al.* 1984)); when venous return from a muscle is prevented at the end of exercise, heart rate returns to normal rapidly, but arterial pressure remains elevated — see below (Alam and Smirk 1937; Rowell *et al.* 1976; Seals *et al.* 1988). (To our knowledge, the question 'why does heart rate return only to, but not below baseline levels in these studies, when arterial pressure remains elevated?' has neither been asked nor answered.)

Freyschuss (1970) showed that by the second heart beat after the onset of isometric exercise, systolic and diastolic arterial pressures are increased and pulse pressure is decreased. Arterial pressure elevation at the onset of exercise is succeeded by brief (about 10 second) reductions of left ventricular stroke volume, peripheral resistance, and blood pressure (L. Adams *et al.* 1987; Sprangers 1990); thereafter, pressure elevation is sustained. As with heart rate speeding, the degree of pressure elevation is proportional to the intensity of exercise (Victor *et al.* 1987*b*; Seals *et al.* 1988) and to the mass of exercising muscles (Rowell *et al.* 1976; J.H. Mitchell *et al.* 1980; Blomqvist *et al.* 1981; Rowell *et al.* 1981). The earliest elevations of pressure appear to be secondary to increases of cardiac output, which are in turn secondary to cardioacceleration and withdrawal of vagal restraint (Freyschuss 1970; Eriksen *et al.* 1990). Subsequent pressure elevations appear to be secondary to increases of sympathetic nerve activity.

At the onset of exercise, *blood flow* to non-exercising muscles increases (A.R. Lind *et al.* 1964; Bevegård and Shepherd 1966*b*; Eklund *et al.* 1974; Rusch *et al.* 1981), notwithstanding increases of muscle sympathetic nerve activity (which begin about 30 to 60 s after the onset of exercise (Sanders *et al.* 1989*b*). Sanders *et al.* (1989*b*) showed elegantly that increases of muscle blood flow early in exercise can be prevented by pretreatment with intraarterial atropine; this suggests strongly that flow is augmented by cholinergic vasodilatation. Subsequently (after about 1 min) vasoconstriction reduces flow in these muscles to below baseline levels (Bevegård and Shepherd 1966*b*; Eklund *et al.* 1974). Evidence that this vasoconstriction is neurally mediated comes from responses to blockade of local efferent nerves (D.A. Blair *et al.* 1961) or alpha-adrenergic receptors (Freyschuss 1970), which partially prevents arterial pressure elevations during exercise, and from the elevations of plasma catecholamine levels that occur (Vendsalu 1960; Christensen and Brandsborg 1973; Christensen and Galbo 1983).

Studies of plasma catecholamines during exercise show that although catecholamine concentrations rise in proportion to the level of exertion, the rise is exponential and does not begin immediately. Figure 7.9 shows that increases of plasma noradrenaline level are very small at low workloads.

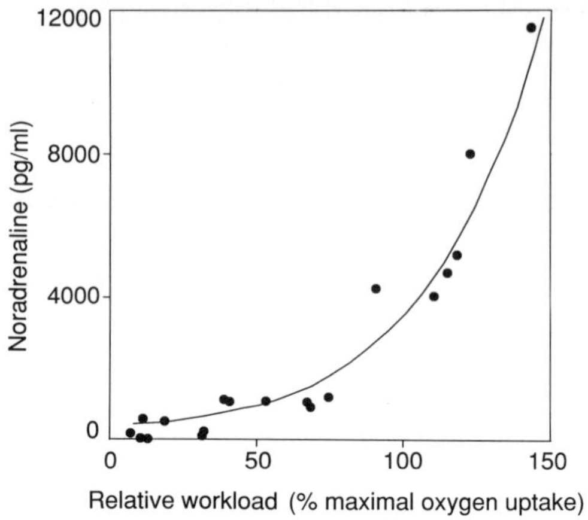

FIG. 7.9. Plasma noradrenaline levels as functions of the percentage of maximal aerobic workload. Adapted with permission (Häggendal *et al.* 1970).

During moderate exercise, catecholamine levels do not begin to rise until the heart rate has increased by about 30 beats/min (Christensen and Brandsborg 1973). To the extent that plasma catecholamine levels reflect sympathetic nerve activity (see Chapter 8), these data support the notion that the earliest heart rate increases during exercise are mediated by vagal withdrawal, and that sympathetic traffic (to the sinoatrial node) does not increase immediately at the onset of exercise. (However, it is also likely that there are delays between increases of neural catecholamine release and the onset of cardioacceleration and the elevation of plasma catecholamine levels in venous blood (Rea *et al.* 1990).)

The first *direct* evidence for increased human sympathetic nerve activity during exercise was published by Delius and co-workers (1972*b,c*). They showed that isometric contractions provoke parallel increases of muscle sympathetic nerve activity and arterial pressure, and reciprocal reductions of forearm (non-exercising) muscle blood flow.

There are several limitations to use of sympathetic microneurography to study human exercise physiology. Since the method of microneurography (see Chapter 8) dictates that the position of the recording electrode be stable, sympathetic nerve activity can be recorded only from nerves supplying *non-exercising* muscle groups. Therefore knowledge of sympathetic outflow to exercising muscles comes from less-direct methods. Also, sympathetic recordings during exercise have been limited primarily to nerves supplying skeletal muscle. (Very limited data (Delius *et al.* 1972*c*) indicate also that human skin sympathetic activity increases slightly during muscle exertion.) Human sympathetic recordings have not been made from nerves supplying other important organs, including the kidneys and viscera, whose vasoconstrictor traffic in all likelihood increases during exercise (Rowell 1973). Therefore the important new insights regarding exercise-related increases of muscle sympathetic nerve activity discussed below do not necessarily supply direct information on responses of other sympathetic outflows.

After about one minute of exercise, blood flow to non-exercising muscle groups diminishes, presumably secondary to vasoconstrictor sympathetic activity, and flow to exercising muscle groups increases, secondary to accumulation of vasodilatory metabolic products (A.R. Lind *et al.* 1964; Rowell 1974). With increasing levels of exertion, vasoconstriction secondary to increasing sympathetic traffic is opposed by metabolic vasodilatation. Vasoconstriction induced reflexly by lower body suction does not occur in exercising muscles if the degree of exercise is intense (Strandell and Shepherd 1967; Sinoway and Prophet 1990). Savard *et al.* (1989) showed that during graded exercise, noradrenaline spillover from exercising limbs increases, but without corresponding reductions of limb blood flow. Metabolic vasodilatation may not be prepotent, however; Rowell and Sheriff (1988) suggested, and Sinoway and Prophet (1990) proved, that metaboreceptor-initiated increases of sympathetic traffic can vasoconstrict maximally dilated muscle vascular beds. Metaboreceptor mediated vasoconstriction is seen as a crucial response to prevent extreme vasodilatation and hypotension during intense exercise.

Several factors influence the quality and quantity of sympathetic nerve responses to exercise; these include the duration, type (sustained or rhythmic), and intensity of exercise, and the sensation of fatigue. As mentioned, sympathetic activity does not begin to rise immediately after the onset of exercise, as does heart rate speeding. Increases of sympathetic activity begin after about 1 min of mild to moderate exercise, and then gradually increase, at a time when heart rate may have reached a plateau (Lipski and Trzebski 1975; Mark *et al.* 1985). The time course of changes of sympathetic outflow (to a non-working muscle), heart rate, and mean arterial pressure during sustained ischaemic isometric exercise measured in one subject is depicted in Fig. 7.10.

Delays in augmentation of sympathetic activity after the onset of exercise have at least three explanations. First, the level of exertion may be insufficient to increase sympathetic activity; initially inadequate levels of exertion may become adequate during sustained isometric contractions because of progressively increasing fatigue (R.S.J. Clarke *et al.* 1958; M. Saito *et al.* 1989). Second, baroreflexes may oppose increases of sympathetic activity. Third, accumulation of metabolites increases gradually during exertion. The second and third possibilities are discussed below. During sustained isometric exercise, arterial pressure increases further after sympathetic activity begins to rise, and the degree of correlation between pressure and sympathetic traffic is high (Seals *et al.* 1988).

Differences between sympathetic responses to sustained and rhythmic exercise appear to be more quantitative than qualitative. Afferent muscle

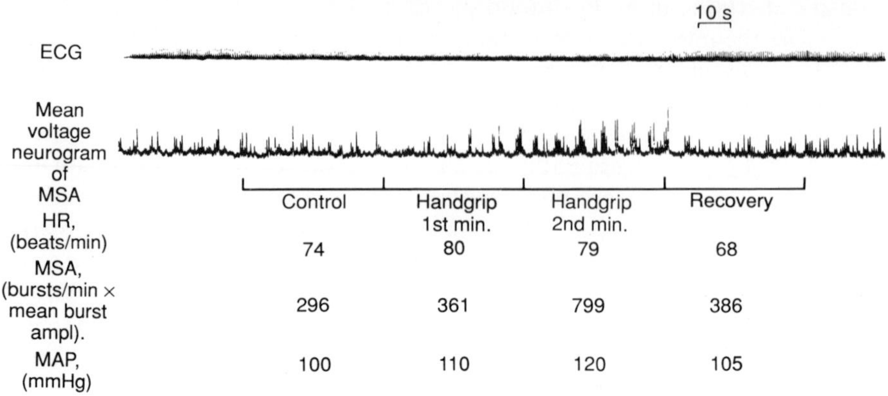

FIG. 7.10. Muscle sympathetic nerve activity and haemodynamic changes recorded from one subject during sustained handgrip at 30 per cent maximum voluntary capacity. Reproduced with permission (Mark *et al.* 1985).

nerve fibres thought to mediate reflex autonomic responses are activated by both sustained and rhythmic exercise (Kaufman *et al.* 1984). The article by Alam and Smirk (1937) on pressor responses to exercise shows that pressor responses are elicited when subjects perform rhythmic exercise at rapid rates without tourniquets, as well as sustained exercise with tourniquets. Although sympathetic activity increases with mild sustained but not mild rhythmic exercise, sympathetic activity does increase with intense rhythmic exercise (Victor *et al.* 1987*b*; Victor and Seals 1989).

Several lines of evidence support the notion that the degree of muscle *ischaemia* during exercise importantly determines sympathetic outflow. Passive muscle vibration does not increase sympathetic activity (Mark *et al.* 1985); therefore, exertion-related sympathoexcitation probably is not secondary to increases of afferent muscle mechanoreceptor traffic. Also (although this may be difficult to prove), sympathoexcitation does not appear to be caused by pain (Alam and Smirk 1937, 1938; Rowell *et al.* 1976). Isometric exercise at high levels of exertion is more likely to be associated with muscle ischaemia than rhythmic exercise, in part because mechanical compression of vessels impedes blood flow into and out of working muscles (Humphreys and Lind 1963; D.E. Donald *et al.* 1967). Within this context, intense rhythmic exercise may lead to muscle ischaemia because the rate of generation of metabolic products exceeds the rate of washout.

As mentioned, the possibility that increases of sympathetic activity during exercise result reflexly from accumulation of metabolites in working muscle was suggested first by Alam and Smirk (1937), who showed that when a tourniquet is placed on the thigh during isometric leg exercise, arterial pressure (but not heart rate) remains elevated. Freund *et al.* (1979) suggested that exercise-induced pressor responses are mediated by increased afferent nerve traffic from exercising muscle, because in three out of four patients, exercise-induced sympathoexcitation was abolished or reduced by epidural blockade of afferent nerve traffic from exercising muscle groups. Subsequently, several studies (Victor *et al.* 1987*b*; Victor *et al.* 1988; Scherrer *et al.* 1990*a*) showed that application of tourniquets after exercise leads to persistence of elevated levels of sympathetic traffic. This is illustrated by the example from one subject published by Scherrer *et al.* (1990*a*) depicted in Fig. 7.11. Muscle sympathetic nerve activity does not increase when tourniquets are applied without exertion (Victor *et al.* 1987*b*). Thus sympathetic stimulation requires the build-up of metabolites provoked by actual exertion.

The identity of the metabolic trigger(s) responsible for stimulation of sympathetic activity is not known with certainty. However, nuclear magnetic resonance studies by Victor *et al.* (1988) show that during exertion, muscle sympathetic nerve activity increases in parallel with reductions of intracellular pH, without obvious relation to levels of phosphocreatine, inorganic phosphate, or adenosine triphosphate. Some of the results of Victor's study are summarized in Fig. 7.12.

The precise role of *'central command'* in stimulation of muscle sympathetic motoneurones during exertion is controversial. Mark *et al.* (1985) showed that electrical stimulation of muscle contraction (without apparent volition) increases sympathetic activity; this suggests that central command is not

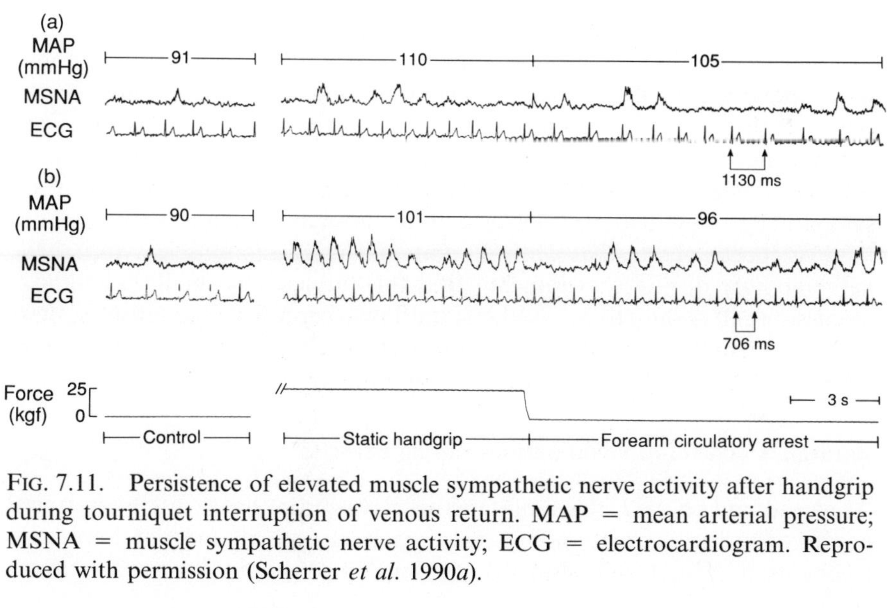

FIG. 7.11. Persistence of elevated muscle sympathetic nerve activity after handgrip during tourniquet interruption of venous return. MAP = mean arterial pressure; MSNA = muscle sympathetic nerve activity; ECG = electrocardiogram. Reproduced with permission (Scherrer *et al.* 1990*a*).

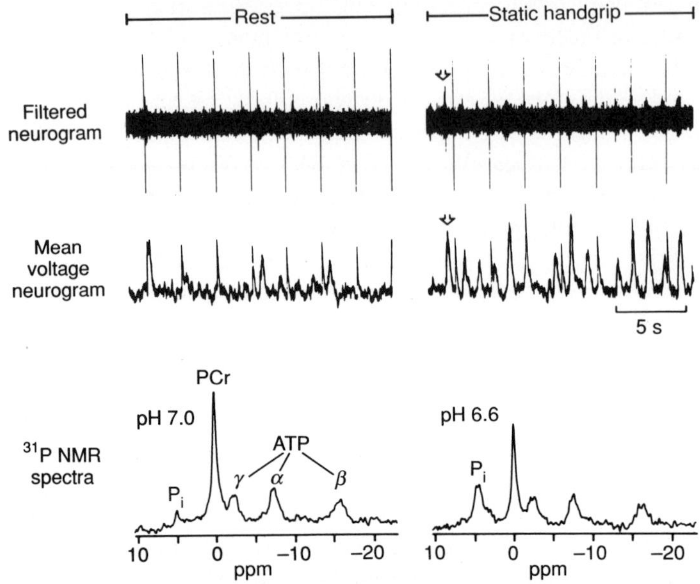

FIG. 7.12. Muscle sympathetic nerve activity before (left) and during (right) sustained handgrip, and associated ^{31}P nuclear magnetic resonance spectra. PCr, phosphocreatine; Pi, inorganic phosphate; ATP, adenosine triphosphate. The large spikes in the neurogram result from nuclear magnetic resonance imaging. Reproduced with permission (Victor *et al.* 1988).

necessary for exercise-related pressor changes. Mark suggested also that since during the first minute of isometric contraction, muscle sympathetic nerve activity may decline slightly, central command may actually *inhibit* sympathetic activity. This seems unlikely, however, since Victor *et al.* (1989) showed that when extreme efforts to produce muscle contraction are made with only negligible contraction (as after curare administration), muscle sympathetic nerve activity increases in a minor but significant way. Thus the evidence suggests that central command increases muscle sympathetic nerve activity during exertion, but that this influence is small. (The above discussion refers only to sympathetic outflow to non-working muscles; there are no measurements of sympathetic outflow to exercising muscles or to other important organs, including abdominal viscera, in humans.)

Baroreflex control of vagal outflow during exercise

As already mentioned, the simultaneous elevation of heart rate and arterial pressure during exercise points towards a change of arterial baroreflex mechanisms. Heart rate slowing, not speeding, should accompany rising pressure, and sympathetic activity reductions, not increases, should accompany elevations of pressure. However, it is intuitively obvious that *some* baroreflex control must be maintained during exercise, because increases of conductance in large working muscle masses could easily overcome the finite capacity of the heart to increase cardiac output (Andersen and Saltin 1985; Rowell *et al.* 1986; Savard *et al.* 1989). Without increases of vascular resistance, severe exercise, with diversion of blood flow to large muscle groups, could lead to circulatory collapse. The next paragraphs discuss the role of baroreflexes during muscular exertion. The major questions in this controversial area are:

(a) is baroreflex control of the circulation altered by exercise?

(b) if so, are there qualitative or quantitative differences between changes of heart rate and peripheral vascular resistance control?

(c) does acute exercise or long-term physical training affect the baroreflex control of the circulation?

Some controversy in this area arises not because of differences in experimental results but because of differences in interpretation of data. This is particularly true regarding measurements of heart rate, as opposed to pulse interval, to characterize reflex changes. Another area of controversy is over the range of baroreflex responses elicited; if measurements are made over only a portion of the sigmoid vagal baroreceptor–cardiac reflex relation (see Chapter 6), it may be impossible to know the position on the relation at which measurements were made, and in what way the entire relation changed. Other disagreements appear to be due to differences in the form of exercise (for example, isometric or dynamic), and particularly to the intensity of exercise.

The advantages of R–R intervals as measures of reflex vagal responses are dealt with in Chapter 8. Briefly, they are as follows:

1. The relation between pressure and R–R interval is linear, whereas that between heart rate and pressure is curvilinear, over the range of rates found in humans (at the high heart rates of rats this distinction may be less important). This is shown in Fig. 8.11.

2. Therefore, reflex increases of vagal activity or electrical vagal stimuli produce linear increases of R–R interval, but curvilinear reductions of heart rate.

It follows that measurements of heart rate cannot be used to estimate changes of vagus nerve traffic, as though they were simple linear functions. The slope of the linear relation between pulse interval and pressure is unaffected by pressure elevations with drugs (Smyth *et al.* 1969) or by increases of pulse interval produced by neck suction (Eckberg *et al.* 1975).

With the onset of *rhythmic muscular exercise*, blood pressure and heart rate rise together; therefore the normal relation between pressure and heart rate has been altered. It had been thought previously, however, that this represented a resetting of the reflex, which none the less was capable of operating normally at this higher 'set point'. Heymans and Neil (1958) stated 'It is a common mistake to claim that the baroreceptors are less active in exercise . . .' This opinion was reinforced by the work of Bevegård and Shepherd (1966*a*) and Robinson *et al.* (1966). Both groups found that reflex control of heart rate is effective during exercise. J.D. Bristow *et al.* (1971*b*) questioned the interpretation, rather than the results, of these two groups. Bristow measured phenylephrine baroreflex slopes in young healthy subjects during erect bicycle exercise, and documented progressive baroreflex impairment with increasing levels of exercise. At heart rates above 150 beats/min, there was *no* reflex cardiac slowing in response to a provoked pressure rise. Reflex sensitivity was reduced within 5 s of the onset of exercise; this pointed towards a central nervous mechanism, rather than some mechanism dependent on heart rate, blood pressure, oxygen consumption, or accumulation of metabolites in exercising muscle. When data of Bevegård and Shepherd (1966*a*) and of Robinson *et al.* (1966) were replotted with pulse interval as the dependent variable, the results of all three groups were in agreement. Importantly, during *intense* exercise, when heart rate approaches 150 beats/min (pulse interval of 400 ms), baroreflex gain approaches zero, whether the measure used is heart rate or pulse interval. Progressive reductions of R–R interval lengthening after bolus phenylephrine injections during increasing levels of exercise are shown in Fig. 7.13. R–R interval lengthening (but not arterial pressure reductions) secondary to bilateral electrical stimulation of the carotid sinus nerves is also greatly diminished during bicycle (S.E. Epstein *et al.* 1969) and treadmill exercise (Eckberg *et al.* 1972).

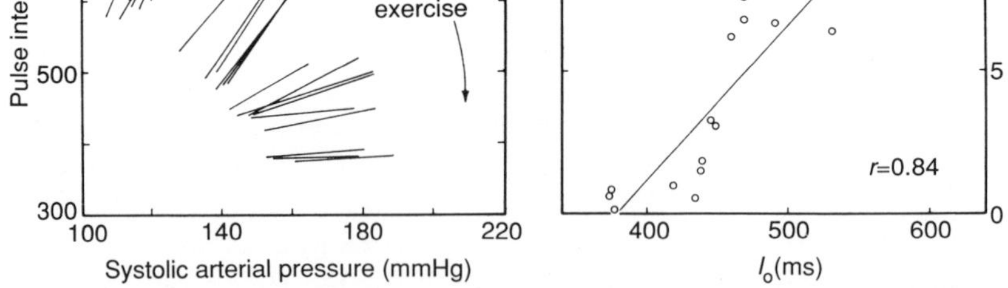

FIG. 7.13. Relation between R–R intervals and systolic pressures after bolus phenylephrine injections at steadily increasing workloads. I = R–R interval; I_o = baseline R–R interval; P = systolic pressure. Reproduced with permission (Cunningham *et al.* 1970).

The great disadvantage of R–R interval measurements in studies of reflex control is that changes of heart rate are of greater physiological relevance to haemodynamic responses to exercise than are changes of R–R intervals. Thus, with mild exertion, baroreflex mediated reductions of heart rate may be normal, even though baroreflex reductions of vagal outflow (as judged from changes of R–R intervals) are manifestly subnormal. However, as Fig. 7.13 indicates, this issue is moot at high levels of exercise, when neither R–R interval nor heart rate changes in response to baroreceptor stimulation.

Fiocchi and his colleagues in Leuven (1985) used a neck chamber (Eckberg *et al.* 1975) to examine the carotid baroreceptor reflex control of both heart rate and blood pressure during moderate and severe supine exercise. Their data help to resolve some of the previous controversy. They confirm that exercise reduces baroreflex control of R–R intervals during carotid baroreceptor stimulation (neck suction) and unloading (neck pressure). This however is much more striking at heavier levels of exercise (average heart rates of 155 beats/min) than at lower levels, particularly during neck suction. With neck pressure, increases of heart rate were abolished even during moderate exercise.

T.J. Ebert (1986) conducted a similar study, but concluded that baroreflex control of both R–R interval and arterial pressure is normal, and is merely reset to operate at higher than usual levels of pressure and heart rate. Here the disagreement is not due to methods of analysis, since Ebert used R–R intervals. Probably, differences resulted from Ebert's use of lower levels of exercise (20 per cent of maximal voluntary contraction) than those used in other studies (30 to 64 per cent maximum level of contraction (Cunningham *et al.* 1972*a*; Mancia *et al.* 1978*a*)), which showed reductions of baroreflex responses.

One great deficiency of all published studies employing injections of vasoactive drugs or graded neck pressure changes is that they define only *portions* of sigmoid baroreceptor–cardiac reflex relations (see Chapter 6). In these studies, neither threshold nor saturation pressure is defined with certainty, and it is not possible to know if flattening of baroreceptor–cardiac slopes during exercise is due to true reductions of baroreflex gain (as most authors claim), or to shifts of operating pressures into threshold or saturation ranges. Fiocchi's study provides a clue that exercise shifts baroreflex relations into the threshold region: in his study, R–R interval shortening with neck pressure was abolished, but R–R interval lengthening with neck suction was only reduced.

Baroreflex control of blood pressure and sympathetic activity during exercise

There have been few studies in humans on baroreflex control of blood pressure during *dynamic exercise*. Bevegård and Shepherd (1966*a*) used neck suction (50 mmHg) for 1 min during three grades of exercise (which produced heart rates of 97, 113, and 130 beats/min) for 5 min. Absolute reductions of arterial pressure during neck suction (15 to 16 mmHg) were identical at all three grades; the authors concluded that baroreflex control of blood pressure is unimpaired during moderate exercise. These authors also found that responses were sustained during 1 min of neck suction, and they concluded that there was little sign of buffering by other reflex areas. Reductions of blood pressure were due to small reductions of cardiac output and vasodilatation in non-exercising muscles.

Ewing *et al.* (1976) engaged baroreflexes with Valsalva manoeuvres before, during, and after 2 min of sustained handgrip at 30 per cent maximum voluntary capacity. Their results, summarized in Fig. 7.14, suggest that exercise does not impair baroreflex control of arterial pressure. (Although the authors did not comment on it, elevation of blood pressure above pre-Valsalva manoeuvre levels may have been less after 2 min of handgrip than before handgrip.) Ludbrook *et al.* (1978) used the Milan neck chamber shown in Fig. 5.11 (centre) to examine blood pressure responses of healthy

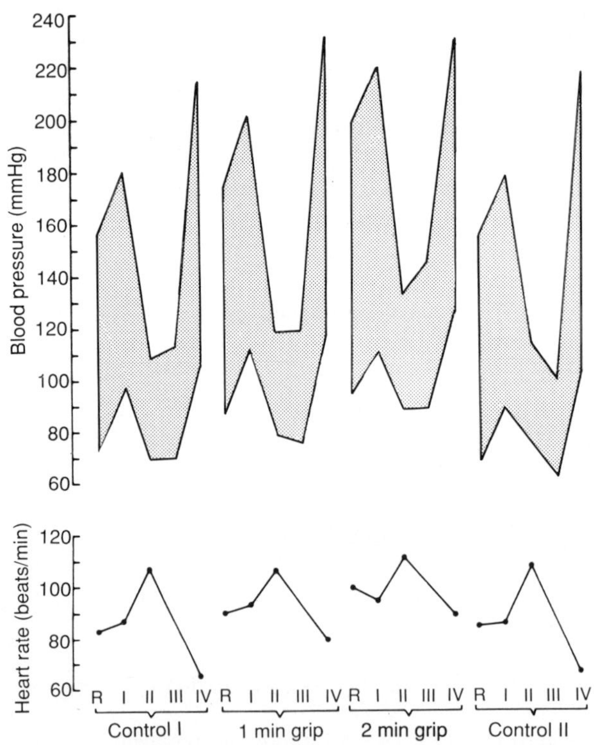

FIG. 7.14. Systolic and diastolic pressure and heart rates during Valsalva manoeuvres in one healthy subject. Reproduced with permission (Ewing *et al.* 1976).

men during 60 s of handgrip at 35 and 45 per cent, and 30 s at 65 per cent of maximum voluntary contraction. They found no evidence for a reduction of baroreflex control of arterial pressure at any level of isometric exercise.

Fiocchi and his colleagues (1985) re-examined this question, but with heavier levels of exertion, that produced heart rates as high as 155 beats/min. They found that blood pressure responses to neck suction, and to a lesser extent neck pressure, are diminished by heavy exercise. An alternative way to view these data is that *some* baroreflex control of arterial pressure (and presumably sympathetic activity) is preserved during exercise. As with data on the effects of exercise on heart rate, there is no published study in which the entire reverse sigmoid baroreceptor stimulus–arterial pressure or sympathetic response relation (Rea and Eckberg 1987) during exercise is worked out.

There is such a study in animals, however. An elegant experiment by Melcher and Donald (1981) conducted in conscious exercising dogs suggests that arterial pressure and heart rate control are not altered by exercise. These workers elaborated entire baroreceptor stimulus–arterial pressure response relations and concluded that although these relations are shifted to higher pressures during exercise (the baroreflex is 'reset'), reflex gain is not altered. One possible deficiency in this study is that although the authors obtained clear reflex responses from the pressurized isolated carotid sinuses in the conscious dogs they studied, they employed *static* rather than pulsatile pressures (Melcher and Donald 1981). Båth *et al.* (1981) showed that rapid, dynamic (sinusoidal) neck pressure changes provoke larger and more sustained changes of human muscle sympathetic nerve activity than slow or static pressure changes. Therefore it is possible that more physiological stimulation of baroreceptors than that used by Melcher and Donald might have shown differences of baroreflex control between rest and exercise. It is fair to conclude from their study, however, that substantial baroreflex buffering of arterial pressure is preserved during exercise.

O'Leary *et al.* (1991) measured haemodynamic responses to bilateral carotid artery occlusion in chronically instrumented conscious exercising dogs. They documented continuing baroreflex control of exercising muscle. They suggested that baroreflex mediated vasoconstriction maintains blood pressure during exercise whenever metabolically mediated vasodilatation outstrips the capacity of the heart to maintain adequate cardiac output.

Other studies indicate that there is no impairment of blood pressure control at moderate levels of dynamic exercise in dogs (McRitchie *et al.* 1976; Melcher and Donald 1981), and rather little impairment in rabbits (Faris *et al.* 1982). In the the last study, lack of baroreflex impairment may have resulted in part from milder degrees of exercise; it is difficult to induce rabbits to exercise on a treadmill (John Ludbrook, personal communication).

Direct measurements of human muscle sympathetic nerve traffic made by

Delius *et al.* (1972*b*) show that increases of sympathetic activity during exercise occur in parallel with increases of arterial pressure. Of great interest in their study, during exercise-induced elevations of arterial pressure, episodic reductions of pressure were answered by increases of sympathetic activity. (This is shown in their Fig. 4.) This finding supports that of O'Leary and co-workers (1991), that baroreflex mediated vasoconstriction occurs during active exercise. This observation raises at least two questions regarding baroreflex function during exercise: in what way do baroreceptors influence human sympathetic activity during exercise; and is the gain of baroreceptor–sympathetic reflex responses altered by exercise?

Two studies speak directly to the role of arterial baroreceptors in modulating human sympathetic activity during exercise, but address this question in different ways. Scherrer *et al.* (1990*a*) measured muscle sympathetic nerve activity during sustained 33 per cent maximal handgrip at three arterial pressures: during usual pressure elevations provoked by handgrip; during arterial pressures lower than those during handgrip, but above resting pressures (pressure was lowered during handgrip by nitroprusside infusion); and during pressures in excess of those produced by handgrip (pressure was raised during handgrip by phenylephrine infusion). Their findings are exemplified by responses of one subject shown in Fig. 7.15. When the blood pressure elevation that usually occurs during exercise was opposed by nitroprusside, both muscle sympathetic nerve activity and heart rate increased by more than 300 per cent. When the exercise-induced blood pressure rise was augmented by phenylephrine, both muscle sympathetic nerve activity and heart rate increases were reduced by more than 50 per cent. These data indicate elegantly that arterial baroreceptors continue to function during moderate static exercise and oppose both increases of sympathetic nerve activity and withdrawal of vagal activity. This conclusion is supported by a study by Sheriff *et al.* (1990) which showed in anaesthetized dogs that increases of arterial pressure during exercise-related muscle ischaemia are much greater after than before sinoaortic denervation.

Walker *et al.* (1980) found that low levels of lower body suction, which did not lower arterial pressure significantly, increased resting forearm resistance modestly, and greatly increased vasoconstrictor responses to moderate isometric handgrip. This result and the results of Scherrer *et al.* (1990*a*) suggest that reductions of sympathoinhibitory traffic during exercise augment sympathetic responses to exercise.

Eckberg and Wallin (1987) approached the question of baroreflex control of sympathetic activity during exercise differently. Since closed-loop adjustments confound interpretation of results from even the most sophisticated methods used to study human baroreflex control (see Chapter 5), they focused on muscle sympathetic nerve transients during only 5 s applications of neck pressure or suction. The principal findings of their study are depicted in the signal-averaged data (obtained during ten applications of neck pres-

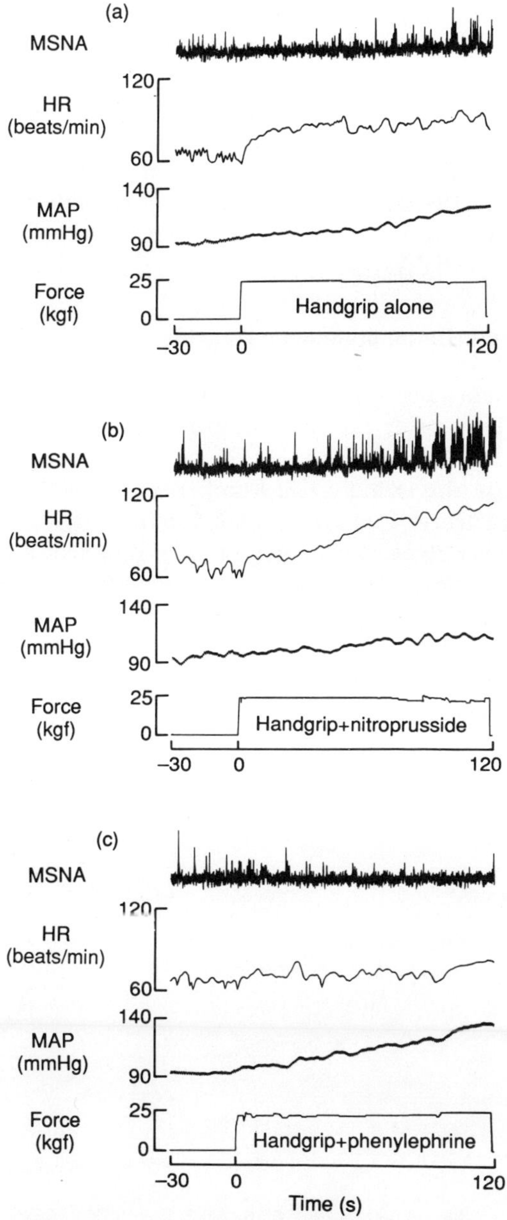

FIG. 7.15. Augmentation and inhibition of increased muscle sympathetic nerve activity produced during handgrip by (b) nitroprusside and (c) phenylephrine infusions. Adapted with permission (Scherrer *et al.* 1990*a*).

sure in each of ten subjects) shown in Fig. 7.16. Both the abrupt sympathetic burst (middle panels) and the vagally mediated R–R interval shortening (lowest panels) were reduced (but by small amounts) by sustained isometric handgrip at 30 per cent maximum voluntary capacity. Inhibition of muscle sympathetic nerve activity by neck suction was also slightly reduced by handgrip (in the healthy young volunteers, there was little sympathetic activity to inhibit). Taken together, the studies of Eckberg and Wallin (1987) and Scherrer *et al.* (1990*a*) suggest that human baroreflex mechanisms are impaired during mild to moderate exercise, but nevertheless buffer sympathoexcitation substantially.

Acute effects of exercise on autonomic control

Several types of data suggest that intense exercise alters autonomic function for at least one hour after exertion. Immediately after exercise (in supine subjects), there are episodic large (>30 beats/min), as yet unexplained, reductions of heart rate (Kauf 1923; Lamb 1963; Jennett *et al.* 1982). Such responses are depicted in Fig. 7.17 (Lamb 1963).

Blood pressure is reduced for longer periods. Bjurstedt *et al.* (1983) noted *severe* orthostatic hypotension and near-syncope with adequate central venous pressure levels in six subjects after exhausting exercise. Vasovagal reactions after extreme exertion (Öberg and Thorén 1972; 1973) probably should be distinguished from the minor normal post-exercise blood pressure reduction which is described next.

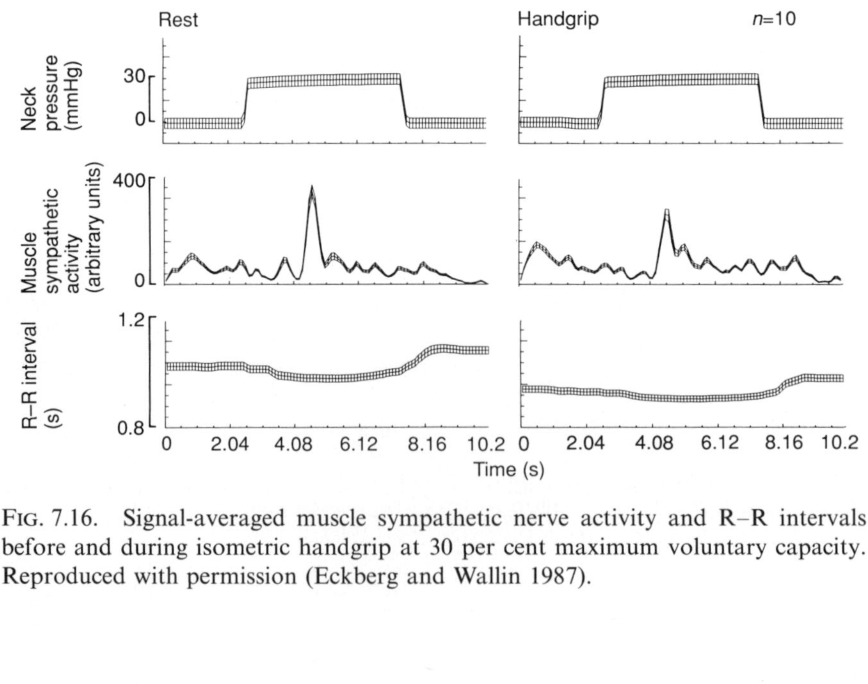

FIG. 7.16. Signal-averaged muscle sympathetic nerve activity and R–R intervals before and during isometric handgrip at 30 per cent maximum voluntary capacity. Reproduced with permission (Eckberg and Wallin 1987).

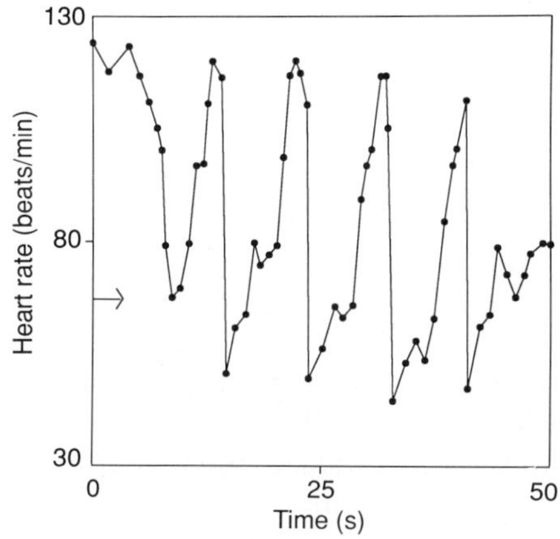

FIG. 7.17. Heart rate changes in one subject immediately after strenuous exercise (arrow). Adapted with permission (Lamb 1963).

Fitzgerald (1981) noted that his own blood pressure was reduced for 4 to 10 hours after 25 min of jogging. Wilcox *et al.* (1982) and T. Bennett (1984) confirmed this finding in prospective studies, and found that systolic pressure falls by about 25 per cent after 5 to 10 min of moderate exercise. Somers and Conway (1987*a*) and Somers *et al.* (1985) measured blood pressure non-invasively in subjects after intense exercise and documented modest systolic pressure reductions (8/8 in normotensive subjects and 18/14 mmHg in hypertensive subjects) until about 2 hours after exercise. Blood pressure reductions persisted after heart rates returned to normal, and in contrast to the results of Wilcox *et al.* (1982) were unrelated to pre-exercising blood pressure levels.

Somers *et al.* (1985) also examined the possible roles of post-exercise hypovolaemia, catecholamines, and atrial natriuretic factor in blood pressure reductions after exercise. They found that haematocrit, body temperature, catecholamines, and atrial natriuretic factor had returned to normal in most subjects by 1 hour. (The last result does not exclude a contribution from atrial natriuretic factor to post-exercise blood pressure reductions, however, since atrial natriuretic factor inhibits sympathetic outflow (Floras 1990), and vasodepression following infusions of atrial natriuretic factor in humans persists 2 to 3 hours after return of plasma levels to normal (Richards *et al.* 1985; E.A. Anderson *et al.* 1986).) Subsequently, Wilcox *et al.* (1987) found that post-exercise blood pressure reductions cannot be explained by changes of renin activity, vasopressin levels, or beta-adrenergic mechanisms.

Somers *et al.* (1985) also measured phenylephrine baroreflex slopes during post-exercise blood pressure reductions. Baroreflex slopes averaged 16 ms/mmHg pre-exercise, 11 ms/mmHg 10 min after exercise, and 23 ms/mmHg 1 hour after exercise ($P = < 0.01$, compared with the average pre-exercise levels). There was no significant relation between baroreflex changes and baseline blood pressures. Convertino and Adams (1991) measured carotid baroreceptor–cardiac reflex relations with ramped neck pressure–suction sequences (Sprenkle *et al.* 1986) and found that baroreflex responses remain significantly augmented for at least 24 hours after maximal exercise. T. Bennett (1984) found that heart rate and forearm vasoconstriction responses to lower body suction increase following exercise.

Floras *et al.* (1989) measured muscle sympathetic nerve activity directly before and after exercise. They found that an average systolic pressure reduction of 12 mmHg in eight borderline hypertensive subjects 1 hour after exercise was associated with a 30 to 40 per cent reduction of sympathetic nerve activity. In two subjects who had no reduction of blood pressure after exercise, sympathetic traffic was unchanged. Since in this study, reductions of pressure were not associated with reciprocal increases of heart rate, these results point towards post-exercise resetting of baroreflex relations.

In conclusion, acute, moderate dynamic exercise lowers arterial pressure for 1 to 2 hours. Arterial and probably cardiopulmonary baroreflex gain is increased. Pressure reductions may be due to reduced sympathetic nerve activity, which may be due in part to increases of arterial and cardiopulmonary baroreflex gain.

Chronic effects of exercise on autonomic control

Somers *et al.* (1991) and Somers and Conway (1987*a*) obtained home blood measurements in 15 mildly hypertensive subjects before and after they trained with the Canadian 5BX exercise program. Average work capacity increased 45 per cent, and weight, arm circumference, and 24 hour urinary sodium and potassium excretion did not change. Resting systolic and diastolic pressures fell by about 5 mmHg, and heart rates fell by about 7 beats/min. Ten of these 15 subjects stopped their exercise programs and were studied again, 4 to 5 months later. On detraining, subjects' weights increased significantly, and their work capacity decreased by 40 per cent. Systolic and diastolic pressures and heart rates returned to baseline levels with detraining.

An additional 16 hypertensive subjects were evaluated by Somers *et al.* (1991) with direct intra-arterial monitoring away from hospital. Each subject participated in two 24 hour studies. In some, the sequence was from unfit to fit; in others it was from fit to unfit. Fitness was associated with a 10/6 mmHg fall of arterial pressure and a waking heart rate reduction of 10 beats/min. There was a small but significant increase of phenylephrine baroreflex slopes, from 14 to 17.5 ms/mmHg, after training. Increased respiratory sinus arrhythmia (Schraufek *et al.* 1990), and baroreflex gain, calculated from power spectral analyses of R–R intervals and arterial pressure (Pagani *et al.* 1988) (see Chapter 8) have also been documented after physical training.

Somers *et al.* (1991) also examined the influence of physical training on blood pressure and heart rate during sleep in six subjects. In three, the sequence was from unfit to fit, and in the other three, the sequence was from fit to unfit. Somers found in trained subjects that although ambulatory blood pressures were lower, sleep blood pressures were higher, even though heart rates were lower. Thus there was a clear but unexplained difference between training effects on sleeping heart rates and on blood pressures.

Takeshita *et al.* (1986) found that athletes have greater reflex forearm vasoconstriction during lower body suction than non-athletes, and suggested that training augments cardiopulmonary reflex gain. Improvement of responses to lower body suction may be determined by the level of training, however; Giannattasio *et al.* (1990) found that responses to low-level lower body suction are subnormal in professional athletes. They speculated that reflex impairment results importantly from mechanical impairment of cardiac stretch receptors by cardiac hypertrophy. We are unaware of within-subject comparisons of the effect of training on cardiopulmonary reflexes. In another cross-sectional study, B.D. Levine *et al.* (1991) attempted to define the entire sigmoid carotid baroreceptor–cardiac relation (see Chapter 5) in subjects with different levels of physical fitness. He found that maximum baroreflex slopes are reduced in athletically fit people because they operate in the saturation portion of their baroreflex relations. Other groups, including healthy people who become physically inactive, patients with heart failure (G.W. Stone *et al.* 1986; Sopher *et al.* 1990), and patients with spinal cord injuries (Convertino *et al.* 1990), who may lie near the extreme end of the spectrum of training to detraining, have impaired baroreflex control. Heart failure patients are discussed in detail in Chapter 14.

In summary, exercise leads to withdrawal of vagal and augmentation of sympathetic activities. Moderate exercise impairs baroreflex control of R–R intervals; however, this may be of minor practical importance, because baroreflex control of heart rate is intact. In contrast, severe exercise abolishes baroreflex control of sinus node function, whether it be characterized by R–R intervals or heart rate. It is not known if flat baroreflex responses reflect true reductions of baroreflex gain, or merely shifts into the saturation or threshold portions of the arterial pressure–sinus node response relation. Baroreflex control of sympathetic activity is preserved during exercise and functions to limit increases of (muscle) sympathetic activity.

CARDIOPULMONARY REFLEXES

Nerve endings in the walls of human cardiac chambers and pulmonary arteries and veins (Johnston 1968) sense deformation and, with arterial baroreceptors, contribute to ongoing autonomic cardiovascular adjustments (I.deB. Daly 1927; Aviado and Schmidt 1955; Paintal 1963; Mancia and Donald 1975). These extra-arterial mechanoreceptors may be misnamed *'low pressure'* baroreceptors, because there is clear evidence that left ventricular mechanoreceptors sense systolic wall tension (Coleridge *et al.* 1964; Muers and Sleight 1972) and may indirectly mediate sympathetic responses to ventricular volume changes (Mohanty *et al.* 1987). Afferent cardiopulmonary neurones are both myelinated and unmyelinated, and travel to the central nervous system over diverse pathways, which include the aortic and vagus nerves, and nerves to and from the spinal cord (Malliani 1982; Longhurst 1984). Of nerves terminating in the medulla, the great majority terminate in the solitary tract nucleus. The physiological role of cardiopulmonary receptors has been covered in several excellent reviews (Korner 1971; Thorén 1979; Linden and Kappagoda 1982; Malliani 1982; Mark and Mancia 1983; Longhurst 1984; Hainsworth 1991).

Since most mechanoreceptors from cardiovascular structures synapse on the same cells in the solitary tract nucleus, they influence the same pool of central interneurones, and continuously modify the responsiveness of those neurones to incoming information. Studies in animals (Gabriel and Seller 1970; Kendrick *et al.* 1979) indicate clearly that input from aortic receptors may condition these secondary neurones and alter their responsiveness to input from carotid baroreceptors. Thus haemodynamic events are sensed by a multitude of widely disseminated receptors and are signalled by ongoing barrages of neural information that tend to arrive at the same place at different times. Not surprisingly, this anatomical and functional arrangement opens the possibility for reflex 'interactions' of great complexity and diversity.

The study of human cardiopulmonary receptors grew into a cottage industry following development of the technique of lower body negative pressure (LBNP) (Greenfield *et al.* 1963; E. Brown *et al.* 1966). Lower body suction holds great attraction because it is non-invasive and non-pharmacological, and it can be controlled finely and reversed rapidly. This genre of research is a lineal descendant of the Belfast Physiological School, which began systematic examination of cardiopulmonary reflexes with the elegantly simple studies of Roddie and Shepherd (1956) and Roddie *et al.* (1957). This group showed that central venous pressure elevations produced by raising subjects' legs lead to reflex forearm vasodilatation. Since this response occurs in the absence of arterial pressure changes, the authors postulated that the effects are mediated by low pressure receptors in the intrathoracic vascular bed. (Other methods used in humans to alter cardiopulmonary pressure, including leg raising (Roddie and Shepherd 1956), lower body positive pressure (Bevegård *et al.* 1977*b*), volume infusion (Stucki *et al.* 1955; C.A. Thompson *et al.* 1990), diuresis or haemorrhage (Barcroft *et al.* 1944; R.H. Murray *et al.* 1967; C.A. Thompson *et al.* 1990), occlusive thigh cuffs

(Kiowski and Julius 1978), and head-out body immersion (Gauer and Henry 1963; M. Epstein 1976, 1978; Parati *et al.* 1987), are not dealt with explicitly here.)

The complexity that these neural connections present to study of human baroreflexes is daunting. *We apply a stringent definition of a reflex interaction*: a change in the shape of the sigmoid arterial pressure–vagal (Eckberg 1980*a*) or reverse sigmoid arterial pressure–sympathetic (Rea and Eckberg 1987) relation provoked by activation of some other reflex mechanism. Such a change could involve gain (slope), range, or both. Since ethical considerations preclude use of many of the highly invasive approaches that are employed in animals, the task of delineating reflex interactions in humans has fostered the design of numerous ingenious research 'strategies'. We begin our discussion of arterial and cardiopulmonary baroreceptor interactions by examining critically assumptions underlying this aspect of human baroreflex research.

We ask several questions. First, since the human cardiovascular system is a continuous integrated whole, can any single portion be segregated from other portions and studied individually? Is it possible to stimulate one area without affecting other areas, and is it possible to ascribe any measured effect to one separate anatomical region? (These questions are integral to the issue of 'closed-loop' reflex adjustments; they are based on the possibility that events occurring in one part of the cardiovascular system are sensed everywhere and acted on in different ways by different receptors and effectors.) Second, is it possible to know if a reflex *'interaction'* has occurred?

The seminal study of Zoller *et al.* (1972) showed for the first time that low (5 and 10 mmHg) but not high (20 and 40 mmHg) levels of lower body suction reduce right heart (or thoracic vein) pressures without changing either heart rate or arterial pressure. (Subsequently, E.G. Nabel *et al.* (1987) showed that minor lower body suction also reduces left atrial dimensions.) Zoller's study suggested that progressive sequestration of blood in the lower body *sequentially* disengages cardiopulmonary and arterial baroreceptors, and, by implication, that there is a transition pressure level or *threshold* for blood sequestration below which both sets of receptors are disengaged. Zoller's observations have been confirmed by by numerous workers, including J.M. Johnson *et al.* (1974) and Pawelczyk and Raven (1989).

Assumptions underlying studies of human arterial and cardiopulmonary baroreceptor reflex interactions

Several assumptions undergird the notion that venous pressure can be reduced without alteration of arterial pressure, a concept related to that of 'non-hypotensive haemorrhage' (J.P. Henry *et al.* 1956). (Terence Bennett (1987) expressed his wonderment at this concept by saying that 'something unexplained must be happening'.) First, it is assumed that measurements of pressures in these studies are accurate. There is no doubt that all levels of lower body suction reduce central venous pressure.

Similarly, there is no doubt that low levels of lower body suction do not reduce arterial pressure. (However, we are aware of no attempt to measure aortic pressure during lower body suction or haemorrhage with a high-fidelity catheter-tip transducer.) J.M. Johnson *et al.* (1974) advanced saline-filled catheters into the aorta, displayed their data at high gain, and documented adequate frequency–response characteristics of their catheter–transducer system after each study. It should be noted that one of nine subjects was 'exceptional'; he increased mean arterial and pulse pressures at the beginning of ramped lower body suction at a rate of 1 mmHg/min. Johnson's widely quoted results are summarized in Fig. 7.18. Pawelczyk and Raven (1989) also found no significant change of arterial pressure; however, they noted a non-significant trend towards reduction of arterial pulse pressure during lower body suction of 15 mmHg in the eight subjects they studied. They argued that even if pulse pressure changes with low levels of lower body suction, small changes of *pulse* pressure would not be expected to alter carotid baroreflex function (Yamazaki *et al.* 1988).

The second assumption is that arterial baroreceptor disengagement cannot have occurred during lower body suction if the driving stimulus (arterial pressure) did not change. Here, the argument is not that arterial baroreceptors are insensitive; Fig. 6.7 documents a substantially increased firing rate of a single baroreceptor fibre after a 1 mmHg step increase of pressure (Landgren 1952*a*). Rather, the argument is that when arterial pressure changes, baroreceptors compensate only partly, and leave a residual 'error signal' that betrays the change in the driving stimulus that provoked the response (Korner 1971; Sagawa 1983).

This assumption is challenged by measurements of Peterson and Bishop (1974) of the autonomic responses of conscious dogs to acute circumflex coronary artery occlusion. They found that with arterial baroreceptors intact, the earliest detectable reduction of systolic pressure began 14 s after coronary occlusion; after sinoaortic denervation, the earliest reduction of systolic pressure occurred at 8 s. They concluded that in these dogs, baroreflex adjustments took place in the absence of measurable changes of arterial pressure.

The third assumption is that arterial baroreceptor engagement cannot have occurred if there is no evidence for changes of autonomic outflow. (Increases of muscle sympathetic nerve activity occur during minor lower body suction, but are ascribed to disengagement of cardiopulmonary receptors.) In most articles which show that arterial pressure does not fall with low levels of lower body suction, there also is no change of heart rate. (However, this finding is not universal; Walker *et al.* (1980) and Baily *et al.* (1990) documented small but significant increases of heart rate with lower body suction of 5 and 15 mmHg.) Since arterial pressures in resting humans

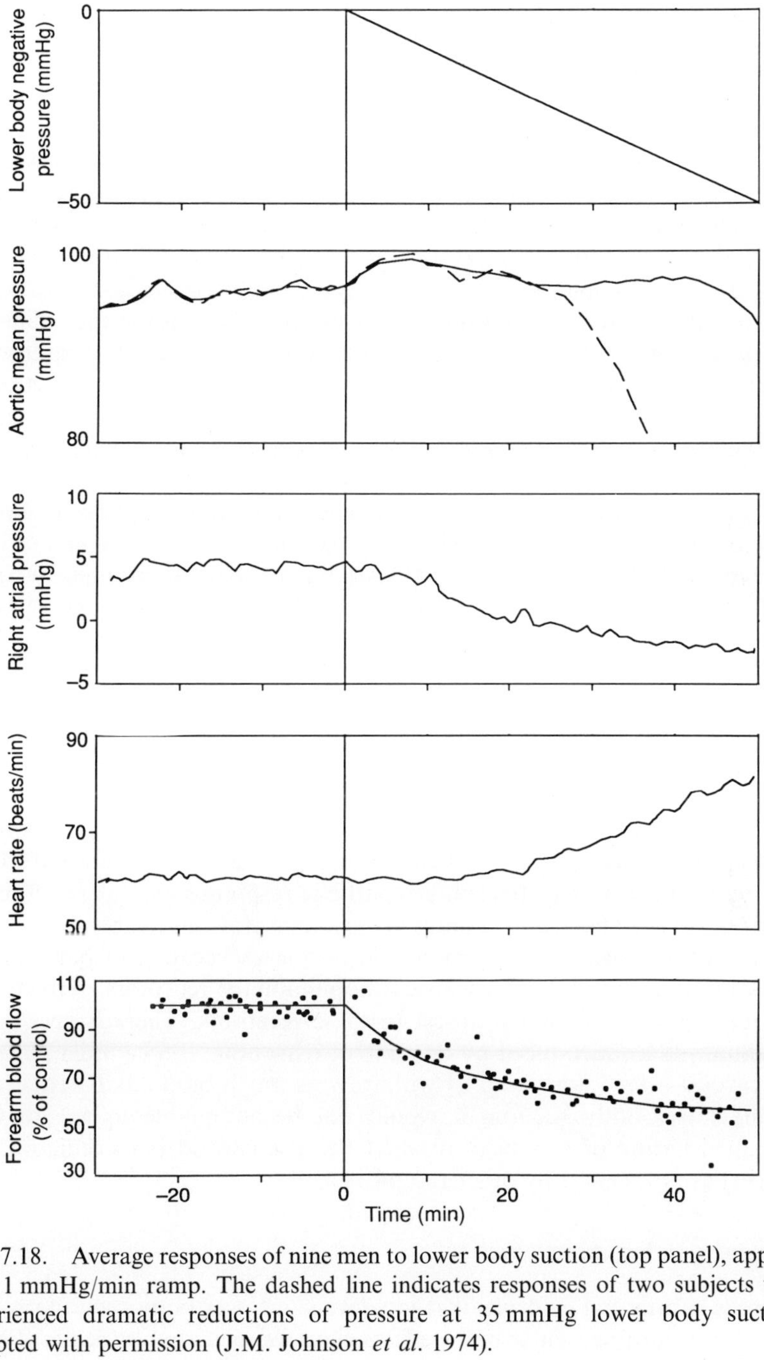

FIG. 7.18. Average responses of nine men to lower body suction (top panel), applied as a 1 mmHg/min ramp. The dashed line indicates responses of two subjects who experienced dramatic reductions of pressure at 35 mmHg lower body suction. Adapted with permission (J.M. Johnson *et al.* 1974).

lie above threshold levels for baroreflex-induced heart rate responses (Eckberg 1980*a*; Rea and Eckberg 1987), changes of afferent arterial baroreceptor input should be reflected by changes of heart rate.

The second and third assumptions are challenged by studies conducted in animals. Hakumäki *et al.* (1985) showed that in intact anaesthetized dogs, gradual haemorrhage does not change arterial pressure, but reduces aortic baroreceptor activity consistently. Thus this study challenges the assumption that reductions of venous volume can occur without concomitant reductions of arterial baroreceptor activity. As far as we know, there is no comparable challenge from human studies.

A fourth assumption, and one that depends entirely upon the validity of the preceding three, is that since muscle sympathetic nerve responses to lower body suction are sustained (Victor and Mark 1985), function of cardiopulmonary baroreceptors is qualitatively different from that of carotid arterial baroreceptors, whose responses (to neck pressure changes) are evanescent (Båth *et al.* 1981). (A variation on this argument, discussed below, is that since sympathetic responses to arterial pressure changes provoked by continuous pressor drug infusions are sustained (Ferguson *et al.* 1985; Sanders *et al.* 1988; Sanders *et al.* 1989*a*), but sympathetic responses to neck pressure changes are evanescent (Båth *et al.* 1981; Wallin and Eckberg 1982), function of aortic baroreceptors is qualitatively different from that of carotid baroreceptors.) There is some evidence that behaviour of atrial receptors may be different from that of arterial baroreceptors; two studies (Kappagoda *et al.* 1972; Gilmore and Zucker 1974) suggest that the rate of adaptation of firing frequency during prolonged stretch or pressure elevation is not as rapid in atrial as in arterial baroreceptors (see Chapter 6).

It may be that the issue raised by both prolonged lower body suction and pressor drug infusions is: how many receptor areas are affected? If *all* baroreceptor areas are affected, sympathetic responses are sustained, but if only one area is affected, responses are evanescent. Conceivably, lower body suction leads to sustained sympathetic responses because all baroreceptor areas are affected. The focus on cardiopulmonary receptors derives from the sequestration of blood closest to these receptors. There is an obvious potential pitfall introduced by this line of reasoning: if a cannula placed in the carotid artery were used to withdraw as much blood as is sequestered during lower body suction, it would not be acceptable to say that the sustained nature of responses resulted because carotid baroreceptors have properties not shared by other receptors.

Responses to lower body suction

There is substantial evidence that low levels of lower body suction provoke autonomic cardiovascular reflex effects that are mediated by reduced levels of cardiopulmonary mechanoreceptor afferent activity, without the participation of arterial baroreceptors. (The preliminary finding of Joyner and co-workers (1990*b*), that forearm vasoconstriction occurs during low-level lower body suction in patients with heart–lung transplants does not *necessarily* contradict this; conceivably, such reflex effects are mediated in innervated atrial remnants that are preserved.) Although this point is argued by some, there is no argument that *high* levels of lower body suction (greater than 20 mmHg) reduce arterial pressure and trigger reflex changes mediated by arterial as well as cardiopulmonary baroreceptors. Therefore we begin this section by describing responses to lower body suction, some of which are surely mediated by reductions of afferent *arterial* baroreceptor activity. There are two excellent reviews of human responses to lower body suction (E. Brown *et al.* 1966; Wolthuis *et al.* 1974).

Vacuum applied to the lower body (usually below the iliac crests) shifts blood caudally within the vascular compartment and from the intra-to the extravascular space (Foux *et al.* 1976), and increases volume in the legs (E. Brown *et al.* 1966; R.H. Murray *et al.* 1967; Musgrave *et al.* 1969) and buttocks and pelvis (E. Brown *et al.* 1966) in proportion to the intensity of suction. Proportional reductions of intrathoracic vein and right atrial pressures (Wolthuis *et al.* 1974) occur, associated with reductions of left atrial diameter (E.G. Nabel *et al.* 1987), left ventricular end-diastolic and -systolic dimensions, stroke volume (Nixon *et al.* 1982; Grassi *et al.* 1988), and cardiac output (Stevens and Lamb 1965). Heart rate (R.H. Murray and Shropshire 1970) increases and systolic arterial pressure reductions occur in proportion to levels of lower body suction levels greater than about 20 mmHg (Stevens and Lamb 1965; Musgrave *et al.* 1971), diastolic pressure changes are variable, and pulse pressure declines.

Lower body suction increases vascular resistance in human muscle (R.H. Murray *et al.* 1967; Baily *et al.* 1990), splanchnic (J.M. Johnson *et al.* 1974; Abboud *et al.* 1979), and coronary (Trimarco *et al.* 1988) vascular beds. Muscle sympathetic nerve traffic to both arm and leg increases in proportion to the level of vasoconstriction (Sundlöf and Wallin 1978; Victor and Leimbach 1987; Rea and Wallin 1989; Baily *et al.* 1990), and these changes are associated with proportional increases of venous plasma noradrenaline levels (Goldsmith *et al.* 1982; Grassi *et al.* 1985; Mohanty *et al.* 1988). Arterial noradrenaline levels also increase, primarily because of reductions of noradrenaline clearance (Baily *et al.* 1990). Maximal increases of forearm vascular resistance and muscle sympathetic nerve activity occur within the early minutes of lower body suction, and are sustained (Sundlöf and Wallin 1978a; Mohanty *et al.* 1988; Baily *et al.* 1990; Joyner *et al.* 1990*a*). Lower body suction also leads to transient venoconstriction (Samueloff *et al.* 1966; Tripathi *et al.* 1989).

There is good evidence that increased vascular resistance during lower body suction is mediated importantly by receptors in the left ventricle. Vasoconstriction during graded lower body suction is nearly absent in patients with orthotopic cardiac transplants who have denervated ventricles but innervated atria and pulmonary veins (Mohanty *et al.* 1987). There is also inferential evidence that sympathetic stimulation of left ventricular receptors and contractility contributes to muscle vasoconstriction; beta-blockade with propranolol reduces forearm vasoconstriction during moderate lower body suction (Ferguson *et al.* 1983). Curiously, propranolol does not reduce vasoconstrictor responses to lower body suction in hypertensive patients with left ventricular hypertrophy (Trimarco *et al.* 1986). The minimal vasoconstrictor responses of heart transplant patients to lower body suction (Ferguson *et al.* 1983) argues against important mediation of responses to lower body suction by visceral receptors (Sarnoff and Yamada 1959; Andrews *et al.* 1972).

Intense (McNamara *et al.* 1969; Beiser *et al.* 1970; Rowell *et al.* 1973; Skagen *et al.* 1981) but not minor (Tripathi and Nadel 1986; Baily *et al.* 1990) lower body suction also reduces skin blood flow. Reduced skin flow is presumed to be mediated by arterial baroreceptors (Skagen *et al.* 1981); however, the possibility that non-specific arousal (Delius *et al.* 1972*c*) by noise, altered respiration, and the sensation of being drawn into the chamber may also mediate skin vasoconstriction has not been excluded.

Not surprisingly, lower body suction increases concentrations of circulating neurohumoral substances. Vasopressin levels do not increase with non-hypotensive lower body suction (Goldsmith *et al.* 1982). Plasma renin activity and aldosterone levels increase gradually (Kiowski and Julius 1978; Grassi *et al.* 1988; Mohanty *et al.* 1988). The early and elegant study of Egan *et al.* (1983) shows that increased renin activity levels result from increased sympathetic nerve traffic to the kidneys; renin activity does not increase during lower body suction in patients with (denervated) kidney transplants, and in healthy subjects given the beta-blocker propranolol (Kiowski and Julius 1978). Arterial noradrenaline levels increase; however, this appears to be due to reduced clearance rather than increased spillover (Baily *et al.* 1990).

Abrupt termination of lower body suction reverses the autonomic changes described above, and provokes responses that tend to mimic changes occurring after release of Valsalva straining (Chapter 4). There is a substantial increase of arterial pressure and slowing of heart rate, and an increase of forearm blood flow to above baseline levels (Stevens and Lamb 1965; E. Brown *et al.* 1966). As with the Valsalva manoeuvre, these changes result from a sudden increase of the level of venous return; they are reduced greatly if thigh cuffs impede venous return from the legs (E. Brown *et al.* 1966).

Lower body suction has been performed for decades; however, the method is not totally safe. We are aware of four published *untoward reactions*. Stevens and Lamb (1965) used intense lower body suction in a prospective study of syncope. One subject experienced a vasovagal reaction marked by 6 s of asystole and tonic–clonic convulsions. Wolthuis *et al.* (1974) discussed a subject who experienced asystole for 30 s during lower body suction; this episode was terminated (successfully) by a blow to the chest. Scott *et al.* (1988) provoked a profound vasovagal reaction with only 10 mmHg lower body suction in a diabetic patient in whom hyperinsulinaemia with constant blood sugar levels was maintained by infusions of insulin and glucose. The mechanism for this patient's response is unknown. Finally, F. Hilton *et al.* (1989) reported that a healthy young man experienced 9 s of asystole during sustained 70 mmHg lower body suction.

The technique of lower body suction is used routinely to provoke *syncope* or presyncope. In well-hydrated healthy subjects, syncope usually does not occur with suction levels less than 40 mmHg. If suction is sufficiently intense (to 80 mmHg), syncope occurs in all people (Stevens and Lamb 1965). (The same can be said of phlebotomy; if sufficient quantities of blood are withdrawn, all subjects faint (Barcroft *et al.* 1944).) Syncope occurs at lower levels of lower body suction if blood is withdrawn also (R.H. Murray *et al.* 1967).

Fascinating physiological changes occur on the way to syncope (produced by lower body suction or vasodilator infusions). Central venous pressure increases prior to syncope, and heart rate, arterial pressure, stroke volume, cardiac output, and systemic vascular resistance decrease (R.H. Murray *et al.* 1968). Mechanisms responsible for divergent changes of central venous pressure and systemic vascular resistance are unknown. Conceivably, central venous pressure increases because of intense reflex venoconstriction, and arterial pressure falls because of stimulation of cardiopulmonary receptors. Blood pressure and heart rate fluctuate widely immediately prior to the onset of hypotension (Stevens and Lamb 1965). We are aware of no mechanism to explain these changes. Atropine prevents heart rate slowing during vasovagal attacks, but does not prevent hypotension or syncope (R.H. Murray and Shropshire 1970).

Several groups have recorded abrupt reductions of muscle sympathetic nerve activity prior to syncope (Wallin and Sundlöf 1982; Sanders and Ferguson 1989; Scherrer *et al.* 1990*b*; van Lieshout *et al.* 1991). One of these records is depicted in Fig. 7.19. This record (van Lieshout *et al.* 1991) shows increasing muscle sympathetic nerve activity and decreasing R–R intervals during nitroprusside infusion rates of 1 and 2 μg/kg/min. Although nitroprusside infusion was stopped when arterial pressure had fallen to 100/40 mmHg, pressure continued to fall, sympathetic activity declined dramatically, and R–R intervals lengthened. The most widely accepted mechanism for vasovagal syncope is that it is triggered by strong left ventricular contractions, beginning at low end-diastolic volumes during periods of heightened sympathetic stimulation (Gauer and Henry 1964; Öberg and White 1970; Öberg and Thorén 1972). A case report by Scherrer *et al.* (1990*b*), however, shows that ventricular receptors are not *necessary* for vasovagal reactions to occur; they reported a classic vasovagal reaction

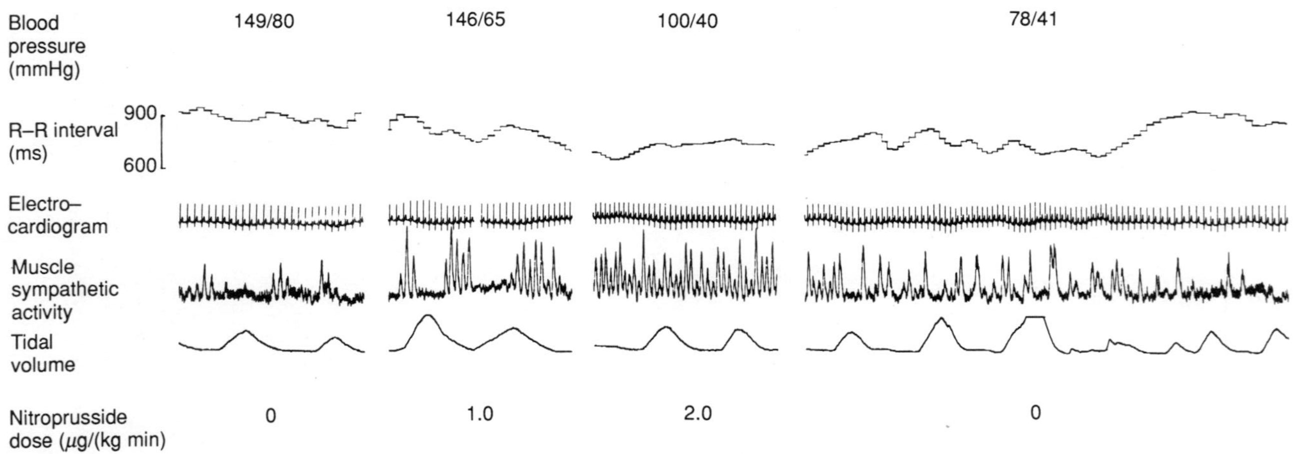

FIG. 7.19. A vasovagal reaction recorded during infusion of the vasodilator sodium nitroprusside. The infusion was stopped when the arterial pressure was 100/40 mmHg. Reproduced with permission (van Lieshout *et al.* 1991).

during vasodilator infusion in a recipient of an orthotopic heart transplant (with ventricular denervation). (The possibility that arterial baroreceptors may contribute to vasovagal reactions is discussed in Chapter 6.)

Patients with cardiovascular diseases may have quantitatively or even qualitatively abnormal responses to lower body suction. Data on the influence of hypertension on responses to lower body suction are contradictory. Trimarco *et al.* (1986) found that patients with hypertension and left ventricular hypertrophy have normal increases of forearm vascular resistance during lower body suction. Mark and Kerber (1982) found that young men with blood pressures sometimes above and sometimes below 150/90 mmHg ('borderline' hypertensives) have supranormal forearm vasoconstrictor responses to non-hypotensive lower body suction. Ookuwa *et al.* (1987) found that young subjects with normal blood pressures but with positive family histories of hypertension have *subnormal* vasoconstrictor responses to minor lower body suction. Some patients with severe heart failure have qualitatively abnormal responses to lower body suction; in such patients, suction may increase (not reduce) cardiac output and left ventricular stroke volume and reduce (not increase) systemic vascular resistance (R.H. Murray *et al.* 1969).

Arterial and cardiopulmonary baroreceptor interactions

We applaud efforts, discussed below, of workers trying to sort out complex baroreflex interactions in conscious human subjects. We bring to the nettlesome issue of human baroreflex interactions a critical questioning of both methods and conclusions. We suggest that before data characterizing human baroreflex 'interactions' be accepted as trustworthy, several contingencies should be satisfied. We recognize that, given constraints inherent in human research, these requirements may be unattainable; nevertheless, the exercise may have value in its identification of goals to be upheld.

1. *The reflex stimulus should be as 'physiological' as possible.*

2. *The baroreceptor stimulus–response relation should be defined completely.* Protocols that involve use of a single intensity of baroreceptor stimulation and measurement of a single response are least informative. Studies that involve different intensities of stimuli and define *slopes* are better than those involving single stimuli. However, if inferences are drawn from changes of slopes after interventions, a tacit (or explicit) assumption is made that both the range and set (or equilibrium) point of the reflex response did not change.

3. *Interventions should be brief.* Long interventions are difficult to interpret.

4. *Baroreflex stimuli should be applied repeatedly and averaged.* This is necessary because of the great variability of baroreflex responses during experimental sessions in conscious humans (Eckberg 1980*a*; Parati *et al.* 1983; Eckberg *et al.*, 1992).

5. *The intervention suspected of 'interacting' with baroreflex responses should be selective and should not alter arterial baroreceptor input.*

6. *For valid comparisons between carotid and aortic baroreflex responses to be made, stimuli to both regions should be identical.*

7. *In situations where data from humans are lacking, responses should be interpreted in terms of what is known from more invasive, often more definitive animal research.*

8. *Interpretation of results should be economical.*

Several groups have modified cardiopulmonary baroreceptor activity with low levels of lower body suction or pressure and measured responses to carotid baroreceptor stimulation. In general, these studies define a reflex 'interaction' as a change of responses to neck pressure or suction occurring during changes of pressure around the lower body.

Bevegård *et al.* (1977*c*) applied sustained −40 or 40 mmHg lower body pressure and measured arterial pressure and heart rate responses to sinusoidally varying neck suction (−10 to −40 mmHg with a period of about 33 s). They found that peak-to-peak changes of systolic and mean arterial pressures are increased by moderate lower body suction and that heart rate is not changed. Bevegård concluded that changes of input from intrathoracic receptors modify blood pressure responses to carotid stimulation. An original record from this study is depicted in Fig. 7.20.

In a second study (Bevegård *et al.* 1977*d*), this same group stimulated carotid baroreceptors with 40 mmHg neck suction for about 5 min, with or

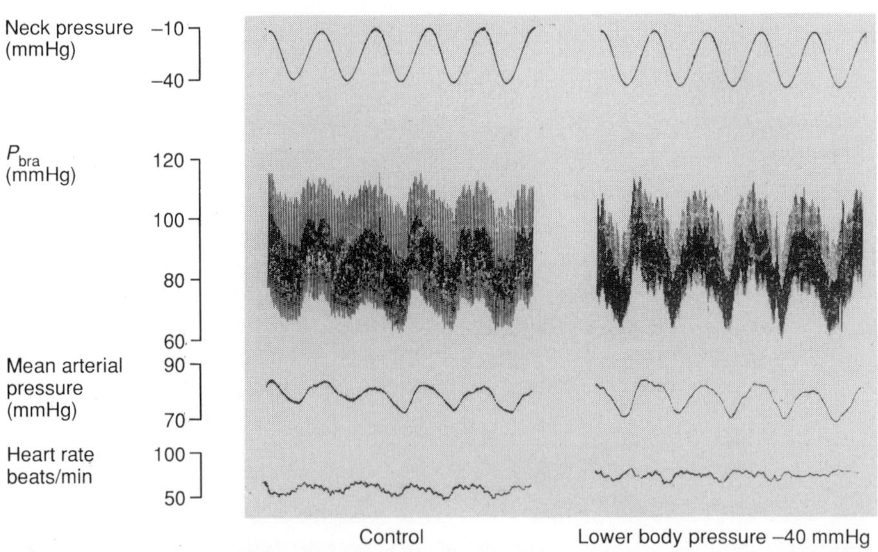

FIG. 7.20. Arterial pressure responses to sinusoidal neck suction. Adapted with permission (Bevegård *et al.* 1977*c*).

without disengagement of cardiopulmonary and arterial baroreceptors provoked by sustained lower body suction of 40 mmHg. They found that carotid stimulation leads to greater reduction of mean arterial pressure and forearm vascular resistance when it is applied during than before lower body suction. They concluded that haemodynamic responses to carotid baroreceptor stimulation are modified by changes of blood volume distribution and vascular resistance.

The advantage of Bevegård's first study (1977*c*) is that *sinusoidal* neck suction was used; this reduced the baroreceptor adaptation that must have occurred in the second study during 5 min of continuous neck suction (Bevegård *et al.* 1977*d*). The disadvantages of both studies are that the neck pressure range did not encompass the entire arterial pressure–effector response relation, and the intensity of lower body pressure change clearly affected both cardiopulmonary and arterial baroreceptors. Therefore, although a change in responses to neck suction occurred, it is impossible to know if the results reflect a true interaction, and if the differences were due exclusively to cardiopulmonary baroreceptors, or if it involved changes of arterial baroreceptor activity as well. Conceivably these data might indicate that lower body suction shifted subjects' resting arterial pressures onto steeper portions of their reverse sigmoid carotid pressure–sympathetic response relations (Fig. 10.4), and the same excursions of carotid distending provoked greater fluctuations of sympathetic outflow. This economical interpretation of Bevegård's data does not require an interaction between arterial and cardiopulmonary baroreceptors.

Takeshita *et al.* (1979) measured R–R interval responses to brief (0.6 s) carotid baroreceptor stimulation with 30 mmHg neck suction, or carotid and aortic baroreceptor stimulation with arterial pressure ramps produced by bolus phenylephrine injections. They found that non-hypotensive lower body suction does not alter R–R interval responses to baroreceptor stimulation. Although Takeshita's study had the advantage that neck suction was applied briefly and R–R interval responses were measured immediately, it had the disadvantage that only one intensity of stimulation was used.

A study by Parati *et al.* (1987) also failed to identify any influence of cardiopulmonary receptors on carotid baroreflex control of R–R intervals. These authors *stimulated* cardiopulmonary mechanoreceptors massively (with head-out immersion, which increased average intrathoracic venous pressure from a low (1.5) to a supranormal (12.0) mmHg level). Parati and co-workers also measured R–R interval changes immediately, and added a new and distinctive feature: they stimulated carotid baroreceptors over a wide range of neck pressures (from −6 to −57 mmHg). Although this wide range did not encompass the entire arterial pressure–sinus node response range, it was large enough to permit the authors to conclude that immersion (which raised average baseline R–R intervals from 760 to 939 ms) reset baroreflex responses (that is, they occurred over a higher R–R interval range). In this study, immersion lowered arterial pressure; therefore Parati's conclusions must be tempered by knowledge that reduced *arterial* baroreceptor afferent activity may have influenced results.

Abboud *et al.* (1979) stimulated carotid baroreceptors with two levels of sustained (1 to 2 min) neck suction with or without lower body suction of 20 or 40 mmHg. They found that forearm vasoconstriction caused by lower body suction is not affected by neck suction, but that splanchnic vasoconstriction and cardioacceleration are largely abolished. They concluded that cardiopulmonary receptors exert a greater effect on muscle resistance vessels than carotid baroreceptors, but that carotid baroreceptors exert a greater effect on splanchnic resistance and heart rate than cardiopulmonary receptors. One of these results is puzzling: the finding which indicates that neck suction does not alter forearm vascular resistance during lower body suction is at variance with findings of Bevegård *et al.* (1977*d*), described above, and T.J. Ebert (1983), described below. The conclusion that cardiopulmonary receptors do not influence heart rate importantly is accepted.

T.J. Ebert (1983) measured arterial pressure and heart rate responses to intense gradually increasing neck suction (50 mmHg) or pressure before and during lower body suction of 12 mmHg. He found that when neck suction was applied during lower body suction, reductions of forearm vascular resistance are greater and more sustained than they are when neck suction is applied alone. Ebert advanced the concept that withdrawal of sympathetic activity secondary to baroreceptor stimulation may be slight and evanescent when sympathetic traffic is small (as it may be in healthy young volunteers (Sundlöf and Wallin 1978*b*; Wallin and Eckberg 1982)), and large and protracted when sympathetic traffic is augmented (by reduced levels of cardiopulmonary inhibition). A similar argument was made subsequently by Victor and Mark (1985). Ebert did not define the entire baroreceptor stimulus–cardiac response relation; therefore it is impossible to know if the greater reduction of vascular resistance that occurred when neck suction was applied during lower body suction reflects a true reflex interaction.

Victor and Mark (1985) studied the effects of low-level (10 mmHg) lower body suction on responses to carotid baroreceptor disengagement with sustained neck pressure (20 and 30 mmHg). They found that lower body suction causes greater than twofold increases of forearm vasoconstrictor responses to neck pressure. They concluded that their results provide the first clear evidence for an inhibitory interaction between carotid and cardiopulmonary baroreceptors. The validity of this conclusion (perhaps incorrect — see above) rests upon the assumption, discussed above, that low levels of lower body suction do not affect afferent arterial baroreceptor activity.

Finally, Pawelczyk and Raven (1989) reinvestigated this physiology with tools used by others, but added two new refinements. First, they disengaged cardiopulmonary receptors with varying levels of lower body suction and then attempted to define *entire* carotid baroreceptor–cardiac reflex relations with a variant on the brief neck pressure–suction profile shown in Fig. 5.13. Second, they modelled R–R interval and mean arterial pressure responses to neck pressure changes with a four-parameter logistic function to derive a mathematical expression of reflex gain. Their results are summarized in Fig. 7.21. Modest lower body suction did not alter baseline R–R intervals and modified baroreflex relations in subtle ways; the gain of both R–R interval and mean arterial pressure responses derived from logistic modelling increased significantly. Pawelczyk concluded that disengagement of cardiopulmonary receptors by reductions of central blood volume augments both vagal and sympathetic carotid baroreflex adjustments.

Most of the studies quoted above show some change of arterial baroreflex responses during lower body suction. (We have not discussed responses to lower body *pressure*, which tend to be in the opposite direction, and of minor consequence.) The most impressive changes of arterial baroreflex responses occur when intense, hypotensive lower body suction is used.

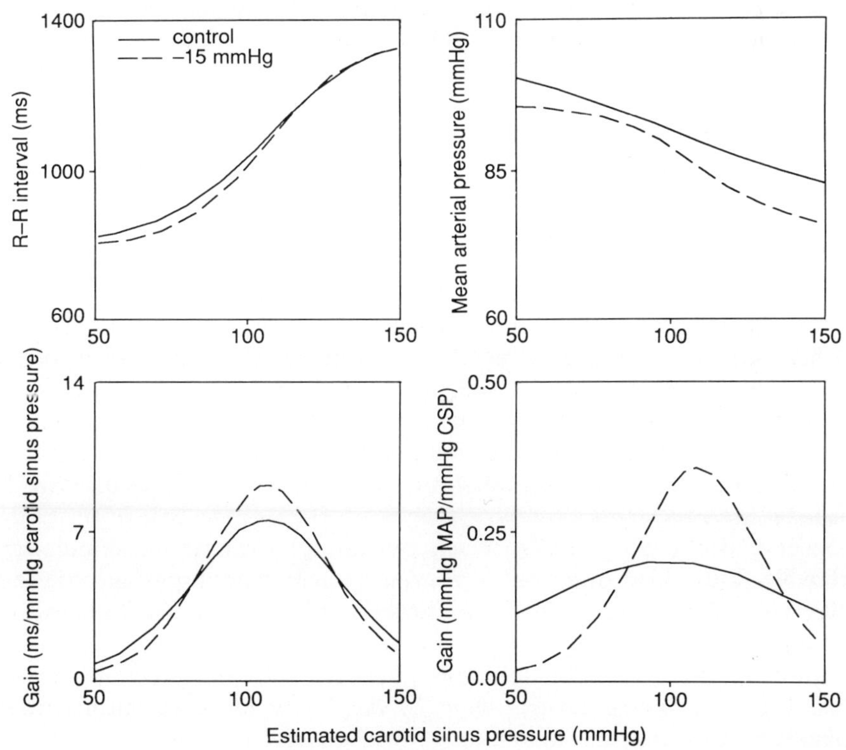

FIG. 7.21. R–R interval and arterial pressure responses to graded positive and negative neck pressure changes with four-parameter logistic modelling of gain. Adapted with permission (Pawelczyk and Raven 1989).

Arguably, the most sophisticated research approach used is that of Pawelczyk and Raven (1989). These authors delineated entire baroreflex stimulus–response relations and documented increases of gain. We suggest that this study and most of the others discussed above show that lower body suction alters responses to altered arterial baroreceptor input. To the extent that lower body suction disengages cardiopulmonary baroreceptors *selectively*, these data point to an interaction between arterial and cardiopulmonary baroreceptors.

CAROTID AND AORTIC BAROREFLEX INTERACTIONS

Static neck suction provokes short-lived blood pressure reductions (Mancia *et al.* 1977). Lindblad *et al.* (1982) showed that the return of blood pressure to normal during sustained neck suction is due in part to return of vascular resistance to normal, and Båth *et al.* (1981) showed that the return of vascular resistance to normal is due in part to return of (muscle) sympathetic nerve activity to normal. Thus abrupt changes of carotid distending pressure alter sympathetic outflow to the important muscle vascular bed, but only transiently. Therefore the question becomes: why is the sympathetic response to neck suction so evanescent?

Escape from sympathoinhibition is probably due to several mechanisms. First, in many studies, neck suction is applied continuously at one intensity; such stimuli are highly unphysiological. Since baroreceptor firing (Landgren 1952*a*) and baroreflex sinus node responses (Eckberg 1977*a*) decay as power functions during sustained carotid sinus pressure elevations, increases of carotid sinus nerve activity after minutes of neck suction are probably minuscule. As noted in Chapter 6, viscoelastic creep occurs also when tension on baroreceptors is reduced; therefore it is likely that decay of responses occurs with neck pressure as well as neck suction. However, decay of sympathetic responses to baroreflex forcings is probably not due entirely to adaptation of baroreceptors, since sympathetic activity also returns towards normal in humans during electrical carotid sinus nerve stimulation (Wallin *et al.* 1975).

Second, decay of sympathetic activity during baroreceptor stimulation probably results also from central nervous system adaptation (Gabriel and Seller 1970); however, there may be no certain way to test this possibility in humans.

Third, decay of sympathoinhibition during sustained neck suction is probably due importantly to *opposition by aortic receptors*. (Evidence from animals suggests that cardiopulmonary mechanoreceptors play a minor role in modulation of sympathetic and vagal responses to brief changes of baroreceptor input (Guo *et al.* 1982).) *The question of aortic baroreceptor opposition to changes of carotid transmural pressure is of central importance to baroreflex research in humans*. Stated in another way, how can responses to carotid baroreflex forcings be interpreted in a closed-loop preparation in which carotid baroreflex effects are opposed by aortic baroreceptors?

It may be that aortic baroreceptor opposition to carotid sinus stimulation has been rediscovered by researchers whose subjects are humans. Schneyer (1935*a,b*) and C.V. Winder (1937) showed clearly in anaesthetized dogs that carotid sinus pressure provokes biphasic heart rate and blood pressure responses; cardiac slowing is followed by speeding, and arterial pressure reduction is followed by pressure elevation. Since late paradoxical responses do not occur after section of aortic and vagus nerves (C.V. Winder 1937), they are most likely due to opposition of responses to carotid stimulation by aortic baroreceptors sensing reflex arterial pressure reductions.

Valiant attempts have been made to explore possible interactions between aortic and carotid baroreceptors in humans. This task is daunting, for at least two reasons. First, the sequential baroreceptor denervation used in animals cannot be used in humans. Second, there is probably no way to alter human aortic baroreceptor input precisely and selectively. Therefore research done in this area with humans as subjects is perforce a highly deductive process. That such research can be attempted is based on the availability of a method (involving neck chambers) for stretching or compressing *carotid* baroreceptors selectively, and a method (involving vasoactive drug infusions) for increasing or decreasing activity of *all* baroreceptors simultaneously. In the paragraphs that follow, we summarize published studies and evaluate their results and conclusions critically, according to principles enumerated above.

Ferguson *et al.* (1985) measured vagally mediated R–R interval lengthening caused by neck suction during sustained or brief arterial pressure elevations caused by infusions or bolus injections of phenylephrine. During phenylephrine infusions, when an increment of mean arterial pressure of 12.5 mmHg was countered by a neck pressure increase (perceived as carotid pressure reduction) of 15.5 mmHg, R–R interval responses were reduced by 28 per cent. After bolus phenylephrine injections, carotid pressure was increased as a gradual ramp to neutralize increases of mean arterial pressure (the linear regression correlation coefficient between increases of systemic artery and neck pressure was 0.996!). Neck pressure applied during arterial pressure ramps reduced baroreflex slopes by 30 per cent. Ferguson concluded that extracarotid baroreceptors participate in control of heart rate in humans, and play a greater role in heart rate control than carotid baroreceptors during dynamic arterial pressure changes.

Sanders *et al.* (1988) used a related experimental paradigm to evaluate baroreflex modulation of human *sympathetic nerve* activity. They raised arterial pressure with phenylephrine infusions and measured the effects of neck pressure on sympathetic traffic. Neck pressure began abruptly, and was increased gradually to counter additional reflex pressure elevations resulting from carotid compression. Sanders found that combined stimulation of aortic and carotid baroreceptors (phenylephrine infusion) leads to sustained, important reductions of muscle sympathetic nerve activity, and that stimulation of aortic receptors alone (phenylephrine infusion plus neck pressure) produces similar degrees of sympathoinhibition. In a second protocol, carotid baroreceptor stimulation alone with neck suction failed to inhibit sympathetic activity. (This finding is puzzling because both Båth *et al.* (1981) and Wallin and Eckberg (1982) showed that neck suction leads to transient sympathoinhibition.) Sanders concluded that aortic baroreceptors are more important than carotid baroreceptors in regulation of human muscle sympathetic nerve activity.

The studies of Ferguson (1985) and Sanders (1988) and their co-workers show clearly that aortic baroreceptors modulate human muscle sympathetic nerve activity. These studies are less clear in proving that aortic baroreceptors are *more important* than carotid baroreceptors in control of sympathetic traffic. As discussed in Chapter 4, *measured* arterial pressure changes during vasoactive drug infusions may not reflect the stimulus driving aortic baroreceptors; the measured pressure change is the *error signal.* Reflex adjustments during vasoactive drug infusions are ongoing, and the actual baroreceptor driving stimulus is much greater than the registered pressure changes indicate (Evans *et al.* 1979). Thus the concern with these studies is that the relative importance of each area cannot be known unless transmural pressure changes applied to aortic and carotid baroreceptors are known with certainty to be identical.

In a third study by the same group, Sanders *et al.* (1989*a*) used different combinations of arterial and carotid pressure changes to examine the role of aortic baroreceptors in mediating responses to *falling* arterial pressure. They recorded muscle sympathetic nerve activity during deactivation of aortic and carotid baroreceptors (with nitroprusside infusions) and deactivation of aortic receptors alone (nitroprusside infusion plus carotid baroreceptor stimulation with sustained (for minutes) neck suction). In a second protocol, they applied sustained neck pressure and found no significant change of sympathetic traffic. (This finding is also puzzling, since Wallin and Eckberg (1982) and Rea and Eckberg (1987) found major but brief increases of muscle sympathetic nerve activity during application of neck pressure.) Sanders found that increases of sympathetic traffic are similar with both interventions, and concluded that aortic baroreceptors exert a greater influence on human sympathetic activity than carotid baroreceptors.

We find this study more difficult to interpret than the ones described above (Ferguson *et al.* 1985; Sanders *et al.* 1988). The same criticism, that measured arterial pressure reductions indicate the error signal after reflex adjustments have been made, applies to the study of Sanders *et al.* (1989*a*). In addition, the criticism can be made that because of baroreflex adaptation (Eckberg 1977*a*), neck suction sustained over 3 min must result in extremely small increments of carotid baroreceptor afferent activity. Sanders's assertion that aortic baroreceptors are *more important* than carotid baroreceptors depends upon the exact equivalence of provoked changes of afferent activity. It is likely that changes of baroreceptor nerve traffic in this study were not equivalent; increases of aortic afferent traffic were much larger, and increases of carotid afferent traffic were much smaller, than the authors imagined.

Fritsch *et al.* (1991) used a different approach to study the way carotid and aortic baroreceptors modulate sympathetic activity. They examined sympathetic responses to brief, stereotyped, neck pressure–suction sequences (Fig. 5.13) which provoked serial, alternating, sometimes opposite changes of transmural pressure in aortic and carotid areas. They found that abrupt reductions of carotid dimensions (provoked by changes of neck chamber pressure from 0 to 40, or from −65 to 0 mmHg) triggered immediate large bursts of sympathetic activity. On the basis of these observations, they concluded that carotid baroreceptors are important in regulation of human muscle sympathetic nerve activity. However, they also found that when neck pressure is reduced progressively (by R-wave-triggered 15 mmHg decrements), huge increases of carotid afferent baroreceptor activity (to −65 mmHg neck pressure) are nullified and reversed by minor reductions of aortic baroreceptor afferent activity (average diastolic pressure fell to only 9.0 mmHg below baseline). Fritsch concluded that both carotid and aortic baroreceptors are important for regulation of human muscle sympathetic nerve activity. Since she did not employ any strategy to alter aortic baroreceptor activity selectively, she drew no conclusions on the relative importance of either receptor group in control of sympathetic activity. These data can be interpreted as showing that aortic baroreceptors are more important than carotid baroreceptors in control of muscle sympathetic nerve activity in humans.

However, human responses to Valsalva's manoeuvre (discussed in Chapter 4) support the opposite conclusion. Valsalva straining provokes large reductions of aortic transmural pressure and small reductions of intraarterial pressure. Increases of sympathetic activity early during straining lead to increases of arterial pressure, and these apparently lead to reductions of muscle sympathetic nerve activity, notwithstanding reductions of aortic distending pressure. These changes are depicted in Fig. 4.2. Such patterns support the notion that increases of carotid arterial distending pressure override reductions of aortic distending pressure because carotid baroreceptors are more important than aortic baroreceptors in modulation of human sympathetic activity.

We postulate that the apparent primacy of aortic baroreceptors in some human studies (Ferguson *et al.* 1985; Sanders *et al.* 1988, 1989*a*; Fritsch *et al.* 1991) is artefactual, and arises as a consequence of the limited repertory of techniques available for human research. This hypothesis (for it is no more than that) is supported by studies conducted in dogs. Glick and Covell measured heart rate responses of anaesthetized dogs to opposite pressure changes delivered to aortic and carotid circulations. They found that carotid pressure elevation and aortic pressure reduction provoked cardiac slowing, and aortic pressure elevation and carotid pressure reduction also provoked bradycardia. This study would have supported the conclusion that aortic baroreceptors are prepotent in regulation of heart rate, if the only combination of pressure changes applied had been pressure elevation in the aorta and pressure reduction in the carotid circulation.

Guo *et al.* (1982) also compared the influence of carotid and aortic baroreceptors on arterial pressure. They found that section of either aortic or carotid baroreceptor afferent nerves causes modest arterial pressure elevations. Transection of aortic baroreceptor nerves *after* transection of carotid baroreceptor nerves causes a large increase of arterial pressure. Similarly, transection of carotid baroreceptor nerves after transection of aortic baroreceptor nerves causes a large increase of arterial pressure. Therefore, as suggested by A.S. Greene and Brunner (1988), when aortic and carotid baroreceptor nerves are sectioned serially, *the afferent limb sectioned later appears to be the more powerful of the two*. We conclude that in the context of the human studies discussed above, aortic baroreceptors may appear to be more important than carotid baroreceptors because methods have not been devised that permit delivery of selective pressure changes to the aorta.

POSTURE

Assumption of upright posture is a complex manoeuvre which appears to involve arterial baroreceptors as well as other physiological sensors. There are several excellent reviews on physiological responses to standing (Gauer and Thron 1965; Blomqvist and Stone 1983; Rowell 1986; van Lieshout 1989; Sprangers 1990; Dambrink 1991). Standing involves physical exertion (and therefore, 'central command' (Freyschuss 1970; C. Borst *et al.* 1982*b*)), changes of respiration, increases of lung volume, changes of sensory input from labyrinths and otoliths, and increases of blood volume in dependent portions of the body and decreases of blood volume in the thorax. The individual importance of each of these factors has not been worked out completely in humans; moreover, there may be no animal model that reproduces adequately the haemodynamic changes that occur when humans stand. There is inferential evidence that altered labyrinthine mechanisms may not be important in mediating human reflex responses to standing (R. Winter *et al.* 1982). Most attention has focused on pooling of blood in dependent portions of the body (and consequent reductions of central blood volumes and pressures) as the primary trigger that initiates the chain of human autonomic responses to standing (Gauer and Thron 1965).

Stimulus for reflex adjustments

Standing increases the amount of blood in dependent portions of the body (Atzler and Herbst 1923; A.H. Turner *et al.* 1930; Waterfield 1931), and reduces right and left ventricular end-diastolic volumes and pressures (Lagerlöf *et al.* 1951; Bevegård *et al.* 1960; Paley *et al.* 1964; Rapaport *et al.* 1966; Stefadouros *et al.* 1975). Not surprisingly, ventricular stroke volumes and cardiac output decline (Bevegård *et al.* 1960; McGregor *et al.* 1961; Poliner *et al.* 1980). Systolic and diastolic pressures fall, rise again early after the onset of tilt, and finally return to or above baseline levels (Holmgren 1956; Gauer and Thron 1965; Ten Harkel *et al.* 1990). At least one report (C. Borst *et al.* 1984) documents a brief early rise of pressures before the onset of pressure reduction. Systemic vascular resistance falls during active (but not passive) standing (Sprangers *et al.* 1991).

Baroreflex modulation of standing responses

The evidence that autonomic fluxes during standing are mediated by arterial (and possibly cardiopulmonary) baroreceptors takes several forms. First, arterial pressure and heart rate changes are nearly mirror images of each other (C. Borst *et al.* 1982*b*, 1984; Rossberg and Peňáz 1988). This provides circumstantial evidence for mediation of heart rate responses by baroreceptors. More direct evidence comes from studies involving baroreceptor denervation. Orthostatic hypotension was observed transiently by Holton and Wood (1965) in one of two patients treated with bilateral carotid sinus denervation for asthma, and by Kochar *et al.* (1984) in a patient who had extremely depressed arterial baroreflex responses. Orthostatic hypotension, associated with supine hypertension, was reported in three patients who had had radiation to the neck for tumours (Chamontin *et al.* 1989). More compelling data (but from animals) comes from prospective studies by Cowley (1973) and Persson (1987) and their co-workers, who documented important orthostatic hypotension in conscious dogs after sinoaortic (and in Persson's study, cardiopulmonary) baroreceptor denervation. Bishop *et al.* (1987*b*), however, did not find significant orthostatic hypotension in sinoaortic barodenervated baboons. There is strong evidence that cardiopulmonary receptors are not necessary for orthostatic blood pressure adjustments. Banner (1990) reported that patients with orthotopic heart–lung transplants do not develop orthostatic hypotension during upright tilt.

Additional evidence for baroreflex mediation of postural haemodynamic changes comes from studies involving simulated or actual weightlessness. One of the nagging, as yet not fully explained, problems with space flight is the orthostatic hypotension that astronauts experience after they return to earth (Sandler 1980; Blomqvist and Stone 1983; Bungo 1989; Eckberg 1991). Serial studies before and after American Space Shuttle missions

(Fritsch *et al.* 1990) and weightlessness simulated by prolonged head-down bedrest (Kakurin *et al.* 1976) reduce the gain and range of vagally mediated R–R interval responses to neck pressure changes (Convertino *et al.* 1990; Eckberg and Fritsch 1992). In the bedrest study of Convertino *et al.* (1990), subjects who experienced the greatest baroreflex impairment were also most likely to experience hypotension and subnormal cardioacceleration with standing after bedrest. These data provide inferential evidence that in healthy humans, normal baroreceptor–cardiac reflex function contributes to orthostatic haemodynamic adjustments. Some of Convertino's results are shown in Fig. 7.22. (In this figure, baroreflex responses of non-syncopal and syncopal subjects are shown in the left upper and lower panels.)

In all likelihood, arterial (primarily diastolic (Sundlöf and Wallin 1978*b*)) hypotension mediates increases of sympathetic activity during standing.

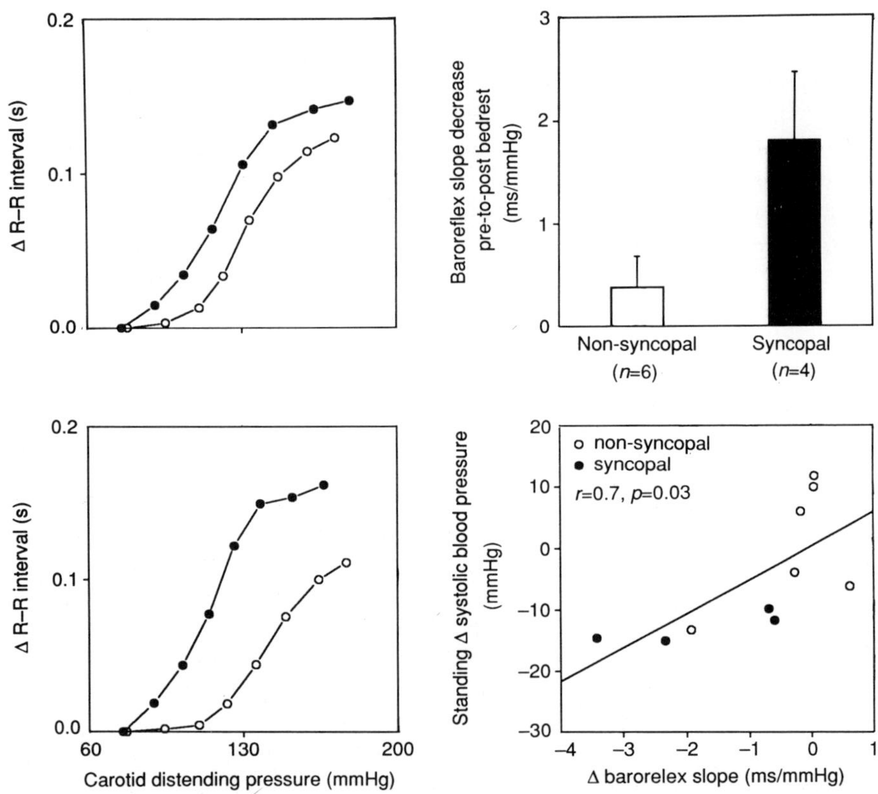

FIG. 7.22. Baroreceptor–cardiac reflex responses of ten healthy subjects to 30 day, 6° head-down bedrest. The left panels show baroreflex responses of nonsyncopal (upper) and syncopal (lower) subjects before (●–●) and after (○–○) bedrest. Adapted with permission (Convertino *et al.* 1990).

Also, there may be more subtle aspects of baroreflex control brought into play by upright posture. Disparate changes of aortic and carotid baroreceptor activity probably alter net autonomic sensory input. In the supine position, pressures in aortic and carotid baroreceptor areas are equal. With standing, however, pressure in the carotid arteries is lower than that in the aortic arch, and levels of neural input from these two sensory areas probably reflect these dissimilar pressures. Reductions of cardiac chamber dimensions and increases of contractility would also be expected to alter the profile of cardiopulmonary sensory input, and reduce levels of atrial natriuretic peptide (Goetz 1988). The relative roles of arterial and cardiopulmonary baroreceptors in mediating postural changes are unclear. It seems clear that small changes of venous return provoke important changes of sympathetically mediated limb vascular resistance (Roddie and Shepherd 1956), as discussed earlier in this chapter.

Arterial baroreceptor function during standing has been studied by several groups. T.G. Pickering *et al.* (1971) measured phenylephrine baroreflex slopes and concluded that reflex control of heart rate is impaired by standing. Harrison *et al.* (1986) found an increase of R–R interval prolongation caused by graded neck suction during head-*down* tilt, which increases central blood volume, as does weightlessness. J.M. Fritsch and D.L. Eckberg (unpublished), however, delivered ramped neck pressure–suction sequences to healthy young volunteers in supine, sitting, and standing positions, and found that vagal baroreflex responses are not influenced by posture.

There is some evidence that sympathetic activity modifies vagal baroreflex responses during standing. Eckberg *et al.* (1976) found *increased* baroreflex gain when the cardioacceleration caused by standing was reduced by the beta-blocker propranolol. The explanation for this is unknown. One possibility is that standing augments baroreflex gain, but that increased gain is not evident because cardiac speeding shifts the time of arrival of inhibitory vagal volleys to a time within the cardiac cycle when they are relatively ineffective (this mechanism is discussed in Chapter 9). After propranolol, increased baroreflex gain becomes evident because heart rate slowing restores the usual temporal relation between the release of acetylcholine and sinus node susceptibility to inhibition (Eckberg 1976). Another possibility is that propranolol exerts its influence by altering reflexes triggered by cardiopulmonary receptor stimulation. Ferguson *et al.* (1232) examined the effects of propranolol on vasoconstrictor reflexes during lower body negative pressure. They found that propranolol reduces vasoconstrictor responses to lower body suction and suggested that this effect results from reduced sympathoinhibitory influences of ventricular receptors (arguably a leap of faith in this minimally invasive human research), and thereby reduces arterial baroreflex responses.

The studies quoted above suggest that human arterial baroreceptors contribute to haemodynamic responses to standing. We now examine those changes.

Vagal responses to standing

The earliest studies of heart rate transients during standing were published by Hoshi *et al.* (1955) and Drischel (1963). C. Borst *et al.* (1982*b*) subsequently carried out particularly painstaking analyses, and compared responses of subjects to active standing, passive tilt, and sustained handgrip. Their findings are summarized in Fig. 7.23. The initial response to active standing (open circles) comprises an immediate large increase in heart rate, which Drischel (1963) called the *Initialkomplex*. As this figure indicates, abrupt handgrip (stippled area) leads to cardioacceleration, whose time course is similar to that of active standing. Conversely, passive tilt (filled circles) provokes minimal initial cardioacceleration. Therefore, it seems likely that initial cardioacceleration with standing results from muscular exercise, possibly associated with 'central command'.

Early transient cardioacceleration is followed, after a delay of about 5 to 12 s, by a reduction of heart rate. This in turn is followed by a secondary rise of heart rate, which probably reflects baroreflex compensation for the fall of blood pressure (see above). The earliest heart rate changes reflect withdrawal of vagal–cardiac nerve activity (Weissler *et al.* 1957; Ewing *et al.* 1980). These changes are illustrated by data of Ewing *et al.* (1980), depicted in Fig. 7.24. The initial R–R interval shortening and subsequent lengthening were not affected importantly by beta-blockade with propranolol, but were nearly abolished when atropine was given. Subsequent (after about 2 min) heart rate speeding results from increased sympathetic activity.

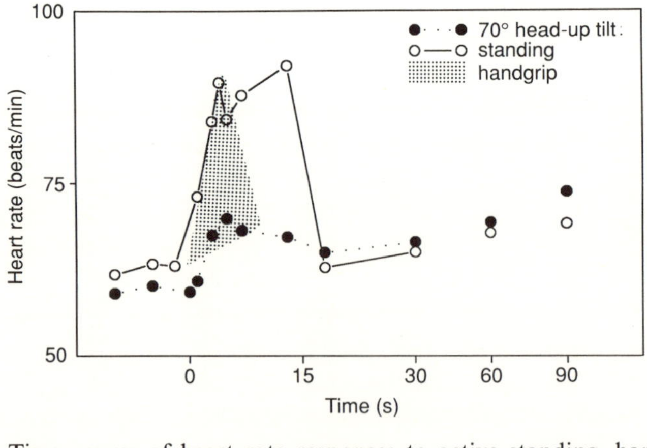

FIG. 7.23. Time course of heart rate responses to active standing, handgrip, and passive upright tilt. Adapted with permission (C. Borst *et al.* 1982*b*).

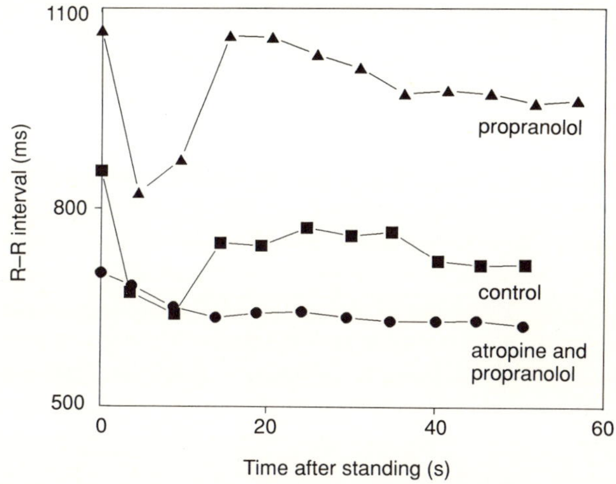

FIG. 7.24. Effect of beta-adrenergic and cholinergic blockade on initial R–R interval transients provoked by active standing. Adapted with permission (Ewing *et al.* 1980).

This is documented by the study of Marin Neto *et al.* (1980); their data are depicted in Fig. 7.25, which shows that heart rate increases during standing, before and after atropine dosage. Combined cholinergic and beta-adrenergic blockade (not-shown) nearly abolished heart rate increases with standing. This is one of the few published studies that has examined heart rate changes after the initial minute of standing.

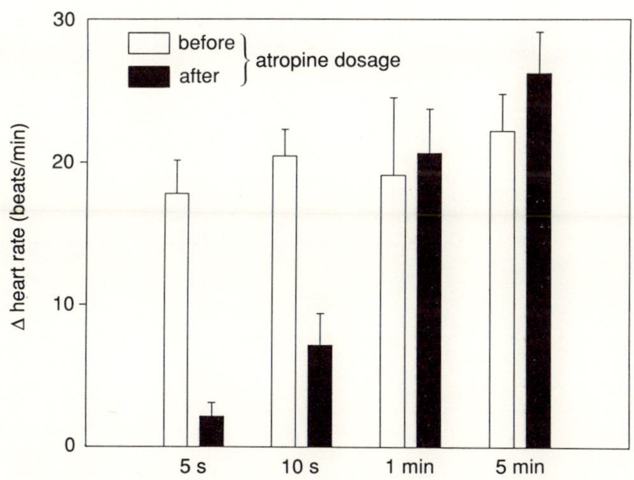

FIG. 7.25. Cardiac-sympathetic contributions to heart rate changes during prolonged standing. Adapted with permission (Marin Neto *et al.* 1980).

Sympathetic responses

Evidence that increased sympathetic activity contributes to cardiovascular adjustments to standing is strong. Sympathetic nervous outflow, as estimated from urinary (von Euler *et al.* 1955) and plasma (Hickler *et al.* 1959; Cryer *et al.* 1974; Lake *et al.* 1976; Fiorica and Kem 1985) noradrenaline concentrations, noradrenaline spillover (Davis *et al.* 1987), and directly recorded muscle sympathetic nerve activity (Burke *et al.* 1977) increases. Postural changes of muscle sympathetic nerve activity, recorded by Iwase *et al.* (1987) are shown in Fig. 7.26. Increased sympathetic activity leads to arteriolar constriction and blood pressure increases (Wallin and Nerhed 1982), reduced forearm (Brigden *et al.* 1950; G. Hildebrandt 1960-1961; Mengesha and Bell 1979), leg (Amery *et al.* 1973), and kidney (Werkö *et al.* 1949; Bakris *et al.* 1986) blood flow, and increased systemic vascular resistance (Werkö *et al.* 1949). Failure of efferent sympathetic activity to increase with standing results in disabling orthostatic hypotension (Ibrahim 1975). Responses of one patient with autonomic failure to arterial pressure reductions caused by Valsalva's manoeuvre are shown in Fig. 4.6.

Importantly, venous tone also increases and thus partly counters orthostatic reductions of venous return to the heart (Rothe 1979). The net reduction of venous return leads to reduction of cardiac output. (Since venous return is difficult to measure, reductions of cardiac output, which are simpler

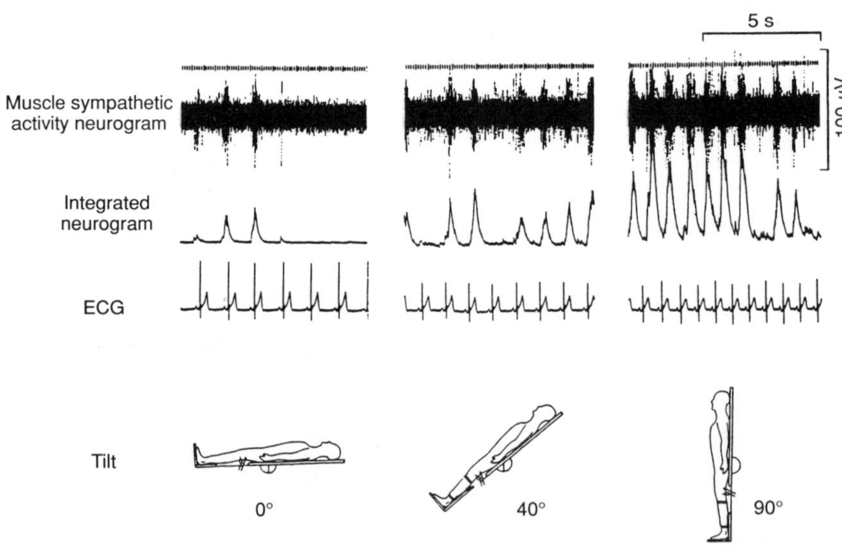

FIG. 7.26. Sympathetic activity in one subject as a function of body position. Reproduced with permission (Iwase *et al.* 1987).

to measure, are used by some authors as indexes of reduced venous return (Ulrych *et al.* 1969; Ibrahim *et al.* 1974)).

Burke *et al.* (1977) closely examined quantitative aspects of sympathetic nerve responses to sitting and standing. They identified three patterns of sympathetic responses to standing:

(1) increased numbers of bursts per heart beat;

(2) increased heart rate with the same number of bursts per beat;

(3) increased burst incidence *and* heart rate (the most common mechanism).

They were unable to evaluate a fourth possibility, increased sympathetic firing per burst (Wallin and Eckberg 1982), because standing changed sympathetic recording sites.

Subjects with a high burst incidence (for example, 90 bursts/100 heart beats) have only a limited ability to increase the number of bursts per heart beat with standing. In such subjects, sympathetic activity increases primarily through heart rate increases (the incidence of sympathetic bursts actually may decline). For such subjects, the ability to increase sympathetic activity with standing is tied closely to their abilities to increase heart rate. Thus individuals may adjust peripheral vascular tone in a predictable manner; some increase vasoconstrictor tone (sympathetic discharge rate) by increasing the bursts for each heart beat, others increase heart rate with the same number of bursts per beat, and others use both mechanisms. The practical importance of these findings is unclear. However, the complex interrelations that Burke identified may explain variability among subjects when muscle sympathetic nerve traffic changes during postural changes are examined in isolation.

SUMMARY

As Chapter 3 indicates, afferent inputs from a diverse array of autonomic sensors converge onto the same neurones in the solitary tract nucleus. Changes of input from one sensor alter the state of these neurones and modulate their susceptibility to influence by inputs from other sensors. Not surprisingly, major changes of the activity of central interneurones provoked by alterations of respiration, chemoreceptor or trigeminal nerve activity, exercise, and posture, alter the responsiveness of those interneurones to subsequent changes of incoming baroreceptor information. This chapter reviews changes of response patterns that result when multiple inputs are altered simultaneously; it emphasizes the enormous complexity of responses to the overtly simple interventions available for use with humans to tease out and study baroreflex 'interactions'.

8

Efferent baroreflex responses

Human baroreflex studies are based necessarily on the experimental 'black box' approach. A major strength of studies conducted with human volunteers — that data can be obtained from intact co-operative subjects without anaesthesia — is also a major weakness. The degree of invasiveness is limited in human studies (however, these limits are being rolled back steadily), and because subjects are intact, much of the information obtained is indirect and inferential. The present chapter focuses on methods available for measuring or estimating efferent neural responses related to baroreceptor activity in humans. Just as Chapters 4 and 5 emphasize that baroreceptor stimuli (input) must be controlled and quantitated precisely, the present chapter emphasizes that efferent responses (output) to changes of baroreceptor activity must also be quantitated accurately. An obvious corollary of this is that when conclusions are based on highly inferential data, the assumptions underlying the inferences must be recognized and dealt with explicitly.

In Chapter 6 we described the highly invasive, almost unique study of Angell-James and Lumley (1974) that involved direct measurements of carotid sinus dimensions and baroreceptor nerve activity in patients undergoing carotid endarterectomy. In that study, the input was the naturally occurring arterial pulse, the output was integrated multifibre carotid sinus nerve activity, and the 'black box' was the carotid sinus baroreceptor mechanism. Other published studies are vastly more complicated than this and involve interposition of a succession of 'black boxes' between baroreceptor input and efferent neural output. This chapter focuses on the ways in which efferent neural responses can be gauged in humans.

SYMPATHETIC MICRONEUROGRAPHY

The simplest, most straightforward approach to evaluation of an efferent limb of human baroreflex responses may be to measure changes of nerve traffic resulting from baroreflex forcings directly. No direct recordings of human *vagal*–cardiac nerve traffic have been made. However, since the innovative early study of Vallbo and Hagbarth in 1967, it has been possible to obtain direct measurements of previously inaccessible postganglionic muscle and skin *sympathetic* nerve activity in conscious, co-operative human volunteers and patients, with safety (Eckberg *et al.* 1989). Vallbo *et al.* (1979) ably summarized the technique of microneurography, as it pertains to research into proprioceptive, somatosensory, and sympathetic functions. However, we are aware of no treatment devoted exclusively to the technique of *sympathetic* microneurography; therefore we describe the technique of sympathetic microneurography in explicit detail.

Neural anatomy

Efferent, postganglionic sympathetic nerve traffic to muscle and skin vascular beds can be measured with electrodes inserted percutaneously into mixed peripheral nerves. Efferent sympathetic activity is carried by unmyelinated 'C' fibres with conduction velocities of about 1 m/s; in a study by Fagius and Wallin (1980*a*), conduction velocities averaged 0.74 for skin and 1.11 m/s for skeletal muscle sympathetic nerves. Electron micrographs of mixed peripheral nerves indicate that unmyelinated fibres (including, presumably, efferent sympathetic fibres) tend to be grouped in bundles (Gasser 1955). Aggregation of these thinnest of all human nerve fibres makes it possible to lodge the tip of an electrode in their vicinity and record their activity. Figure 8.1 is an electron micrograph of a cross-section of a human peroneal nerve, which documents grouping of unmyelinated fibres.

Recordings can be obtained from virtually any human mixed nerve that is accessible to a recording electrode (typical length: about 5 cm). The choice of nerve depends on the goal of the study. For example, if the goal is to measure nerve traffic to a skeletal muscle vascular bed, probably the greatest success will be obtained if the electrode is placed in a nerve supplying a large skeletal muscle mass. The peroneal nerve, impaled at the fibular head, or the tibial nerve, impaled in the popliteal fossa, are used widely for this purpose. These nerves are also suitable for studies of skin sympathetic activity. The position of the peroneal nerve in relation to the fibular head is shown in Fig. 8.2.

Recordings of sympathetic activity can be obtained from either skin or muscle nerve fascicles. In the peroneal nerve, these two types of fascicles usually appear to be separate from each other. Muscle and skin nerve fascicles contain a diverse array of afferent and efferent nerve fibres. Muscle

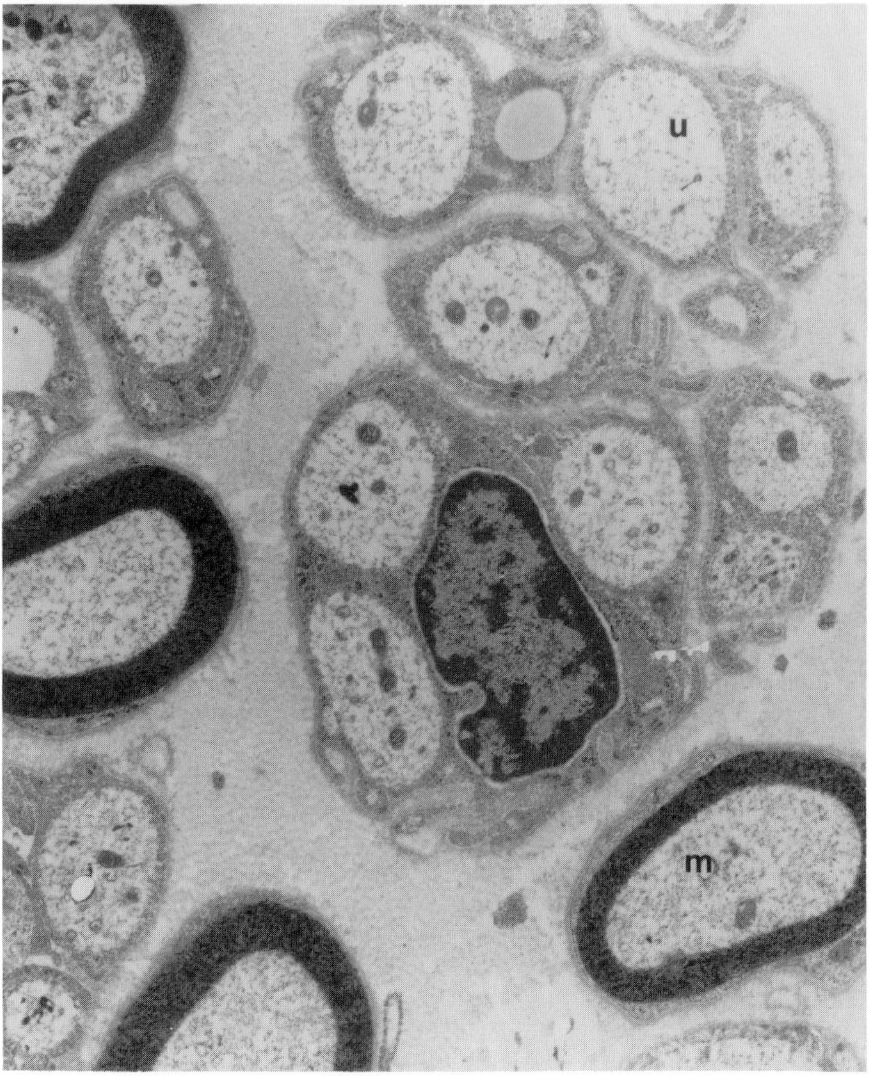

FIG. 8.1. Electron photomicrograph of cross-section of human peroneal nerve. u = unmyelinated fibres; m = myelinated fibres.

fascicles contain sensory nerves from primary or secondary muscle spindles and Golgi tendon organs, and efferent motor as well as sympathetic nerves. Skin fascicles contain afferent nerves subserving mechano-, thermo- and nociceptive functions, as well as efferent vasoconstrictor, piloerector, and sudomotor functions. Figure 8.3 shows a low-magnification light-microscopic section of an *entire* human peroneal nerve at the level of the fibular

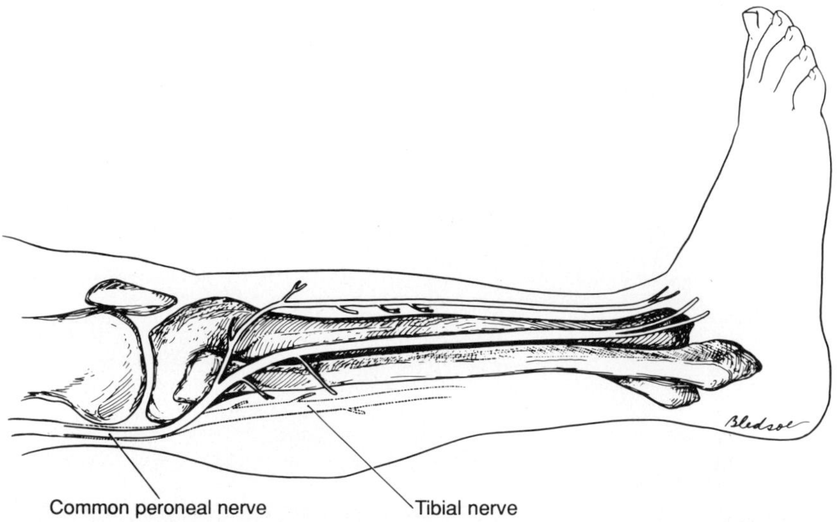

FIG. 8.2. Anatomical detail of the peroneal nerve in relation to the fibular head. The tibial nerve lies medial to the peroneal nerve in the popliteal fossa. Drawing based on a study published in preliminary form (Snyder *et al.* 1989) and photographs kindly supplied by Mark D. Snyder and James J. Gibbons.

head, with a microelectrode used in human neurography overlaid. The diameter of the electrode is 0.225 mm. (On the basis of this measurement, the length and width of the nerve are about 3.0 and 1.6 mm.)

Types of sympathetic activity

Traffic carried by sympathetic nerves can be distinguished from other traffic by its tendency to occur in volleys or 'bursts'. Sundlöf and Wallin (1977) summarized several approaches to validate sympathetic microneurograms and to distinguish muscle from skin sympathetic activity. This neural traffic is *efferent* because it is abolished by proximal but not distal nerve block with local anaesthetic (Hagbarth and Vallbo 1968). It is postganglionic because it is abolished reversibly by administration of a ganglionic blocking drug (Delius *et al.* 1972*b*). Increases of nerve traffic are followed regularly by increases of arterial pressure (Wallin and Nerhed 1982), and reductions of nerve traffic (provoked by electrical carotid sinus nerve stimulation) are followed regularly by increases of limb blood flow (Wallin *et al.* 1975).

It is surprisingly easy to distinguish muscle from skin efferent sympathetic activity. Electrical stimuli delivered through an electrode impaling a muscle fascicle provoke twitches of muscle groups innervated by the fascicle. Conversely, stimuli delivered through an electrode impaling a skin fascicle elicit paraesthesiae. Passive stretch applied to innervated muscle groups provokes

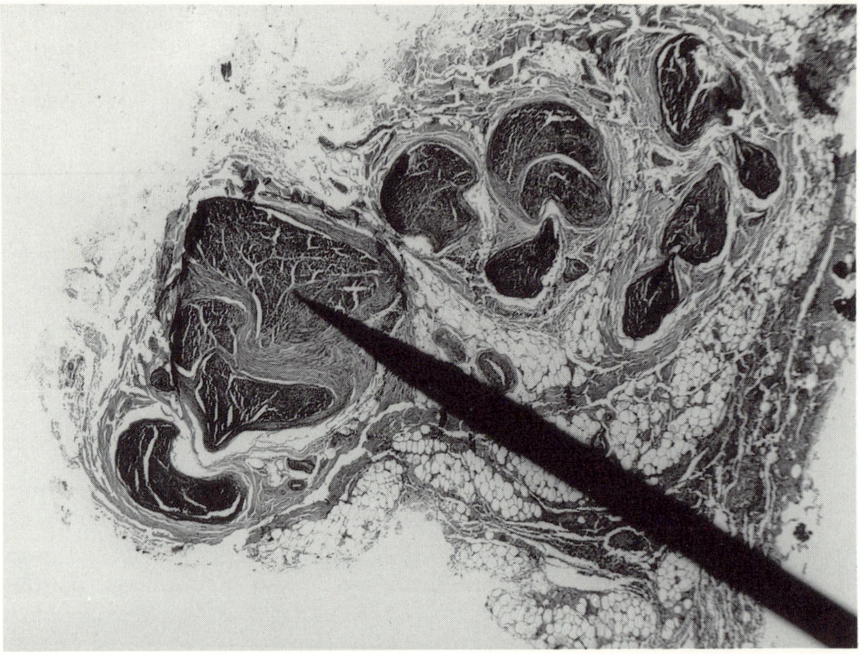

FIG. 8.3. Low-power light-microscopic section of a human peroneal nerve at the level of the fibular head.

volleys of afferent muscle sensory activity, and light stroking of innervated areas of skin provokes volleys of afferent skin traffic. Unexpected sounds, noxious stimuli such as skin pinches, and emotional stress increase skin but usually not muscle sympathetic activity. Finally, muscle and skin sympathetic bursts can be distinguished in part by their sounds; skin sympathetic bursts have longer durations than muscle sympathetic bursts and usually lack the cardiac rhythmicity that characterizes muscle sympathetic nerve activity.

Microneurography hardware

Much of the early effort of Vallbo and Hagbarth (1967) was devoted to developing electrodes suitable for human microneurography. Currently these electrodes are custom-made or produced commercially. (One source is Frederick Haer and Company, Inc., Brunswick, ME, USA.) Typically they are made from 0.2 mm diameter tungsten wire. The flexibility of this wire is thought to dampen the effects of small movements of tissue, and prevent dislodgement of the needle tip from the nerve fascicle (Vallbo *et al.* 1979). Electrodes are coated with epoxy resin and pointed electrolytically. The uninsulated tip of electrodes used for sympathetic recordings is variable but is believed to be about 1 to 5 μm. Figure 8.4 is a photograph of a typical electrode, and Fig. 8.5 is a scanning electronmicrograph of an electrode after it has been used.

A reference electrode with a much larger uninsulated tip is lodged in subcutaneous tissue, within a few centimetres of the recording electrode. (For peroneal nerve recordings, the reference electrode is typically placed directly over the fibular head, where interference from electromyographic potentials is less likely to occur.) The two electrodes are connected by insulated copper wires to a differential preamplifier with a gain of 1000, which is usually positioned close to the recording electrode. The position of reference and recording electrodes and preamplifier for peroneal nerve recordings is illustrated in Fig. 8.6.

Preamplified nerve voltage is amplified further to yield a total gain of 20 000 to 99 000. The amplified nerve signal then is led into a band-pass filter, usually set with low- and high-frequency cutoffs of 700 and 2000 Hz. (We are aware of no prospective study of the propriety of this practice; conceivably, sympathetic firing occurs at rates below or above these limits, and is filtered out and ignored.) The filtered signal is then divided, rectified, and clipped to reduce remaining noise. Several outputs from this amplifier are available; typically, recordings include filtered raw nerve signals and resistance–capacitance integrated nerve activity, which appear as upright bursts. Usually, the time constant for integrated nerve activity is set at 0.1.

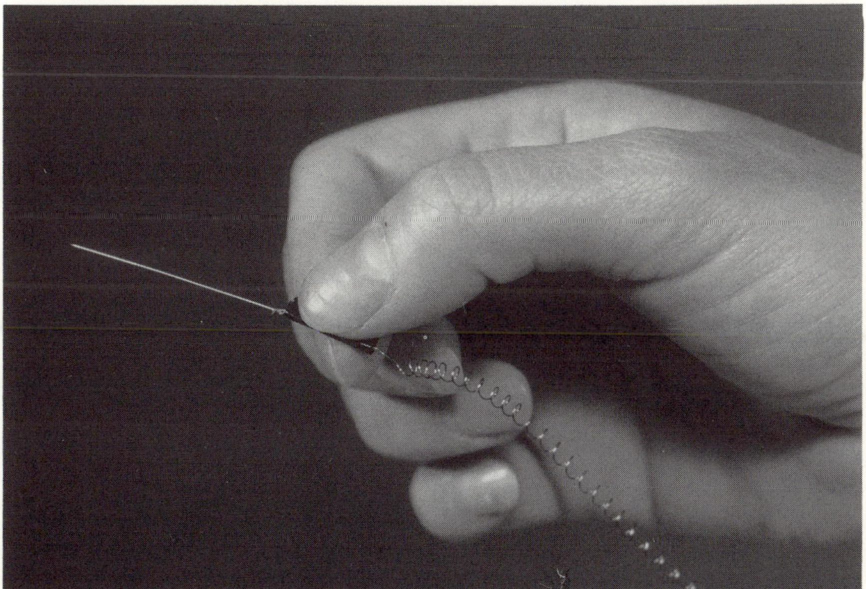

FIG. 8.4. Electrode used for sympathetic microneurography.

FIG. 8.5. Scanning electronmicrograph of electrode after it was used in a human study. Reproduced with permission (Mano 1990).

Placement of recording electrodes

The first task is to outline the position of the nerve to be impaled. (This discussion deals exclusively with the peroneal nerve; however, the principles described here apply also to study of other nerves.) Some Japanese workers (M. Saito *et al.* 1986) prefer the tibial to the peroneal nerve. Tadaaki Mano (personal communication) indicates that this nerve is relatively easy to impale. However, since the tibial nerve lies in the popliteal fossa, either subjects must lie prone rather than supine during studies, or investigators must work from below subjects. Electrical shocks are given transcutaneously with a pencil electrode. These shocks typically are delivered at a rate of 1/s, for 0.1 ms, at 20 to 80 v. Leg or foot twitches occur when the pencil electrode is located over the nerve. After the initial twitches are provoked, fine adjustments are made to find the electrode position that yields the largest twitches. An ink mark is placed at this location, the electrode is moved

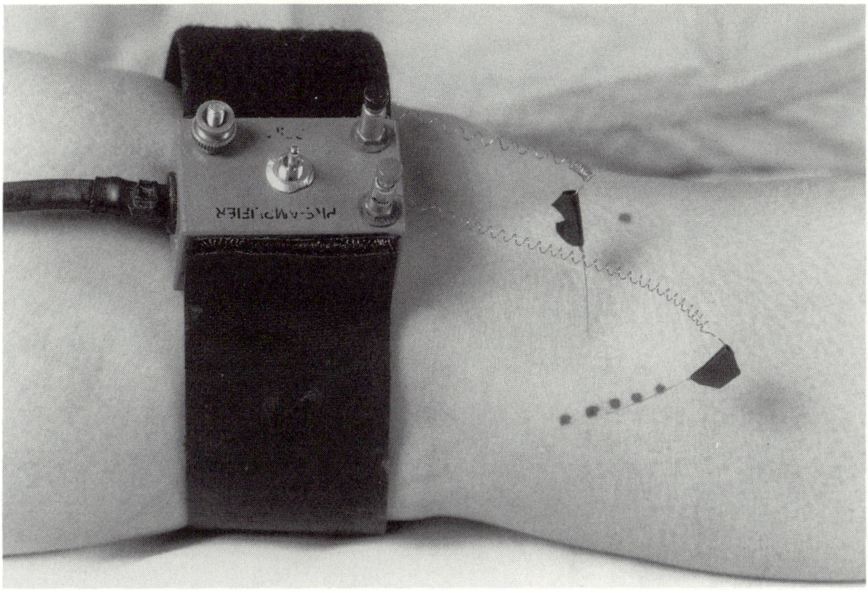

FIG. 8.6. Position of electrodes for recording peroneal nerve sympathetic activity. The superior electrode is a reference electrode. The box near the knee is a preamplifier.

cephalad or caudad, and the process is repeated. By this means, the course of the nerve is outlined by a series of marks.

Next, the reference electrode is inserted over the fibular head, as described above, and the recording electrode is inserted with its tip directed towards the nerve. *At this stage, electrical stimulus parameters are changed*, to render them suitable for intraneural stimulation. Typically, stimulus duration is reduced to 0.02 ms, and voltage is reduced to 5 v or less. As before, shocks are delivered at a rate of 1/s, as the position of the electrode is changed. At this stage, the person inserting the microelectrode and the person controlling the stimulator are prepared to turn switches to stop stimulation; stimuli as low as 5 v can elicit strong reactions if the tip of the electrode happens to be placed exactly within a muscle or skin nerve fascicle.

As the uninsulated electrode tip approaches a nerve fascicle, the subject experiences either paraesthesiae (from skin fascicles) or muscle twitches (from muscle fascicles). Usually muscle twitches are obvious to the person inserting the electrode; however, paraesthesiae are not obvious, and subjects must indicate when they feel paraesthesiae (or any other sensations, including discomfort at the site of needle insertion). Investigators interested in baroreceptor mechanisms usually measure *muscle* sympathetic activity. Therefore, if a subject reports paraesthesiae, the needle is repositioned until twitches are provoked.

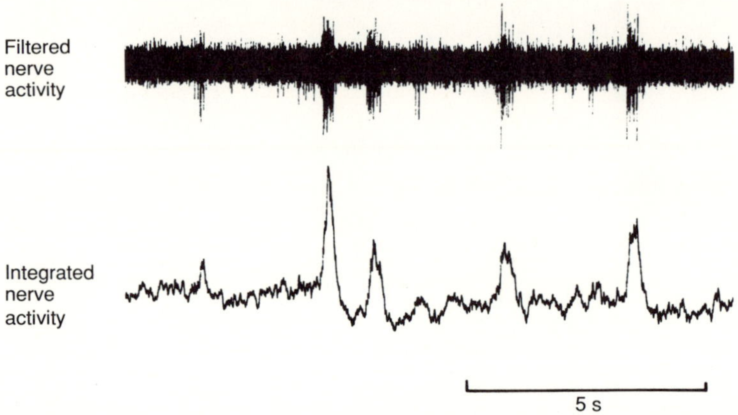

FIG. 8.7. Filtered and integrated multifibre peroneal nerve muscle sympathetic activity. The raw nerve signal was integrated by a resistance–capacitance circuit with a time constant of 0.1.

Then subtle changes of needle position are made to find a site at which twitches occur with a very small voltage (typically, 1 to 2 v). At this stage, the stimulator is turned off, and subsequent needle adjustments, if necessary, are made according to amplified sounds. If characteristic sympathetic bursts are heard, needle adjustments are made to optimize their loudness. More commonly, sympathetic bursts are not heard, and needle adjustments are made during repetitive passive stretches of innervated muscle groups. Passive stretches of innervated muscles elicit barrages of afferent muscle receptor activity; the loudness of these barrages can be used as a guide to home in on the muscle fascicle. When afferent muscle sensory activity is strong, it can be assumed that the microelectrode tip lies within a muscle fascicle. Then, extremely fine needle adjustments are made until sympathetic bursts are heard and optimized.

Rarely it has been possible to record activity in *single* sympathetic nerve fibres (Hallin and Torebjörk 1974). More commonly, electrode tips sense multifibre activity, the amplitude of which depends critically on the position of the electrode tip in relation to nerve fibres. Since tip position cannot be controlled completely, the amplitude of integrated bursts cannot be compared among different subjects. Raw and integrated nerve signals from one subject are shown in Fig. 8.7.

Microneurography software

Several methods have been used to quantitate and otherwise validate sympathetic nerve recordings. The simplest method may be to count the number of bursts, and express the count according to bursts/min or bursts/100 heart beats. This may not be ideal, however, since during large increases of sympathetic traffic, such as those provoked by hypotension, increases of total sympathetic activity may occur through increased firing rates of fibres as well as through increased numbers of bursts. Figure 8.8 illustrates several measures of sympathetic bursts during changes of traffic induced by intravenous infusions of sodium nitroprusside and phenylephrine hydrochloride. These data indicate that simple measures of burst *numbers* do not characterize total sympathetic activity completely.

Another way to deal with changes of sympathetic activity is to multiply burst *dimensions* by burst numbers. The simplest approach may be to determine the average height of bursts and multiply burst numbers by this amount. (One study (Eckberg *et al.* 1988) showed that at baseline blood pressures, burst height, which is simple to measure, is a highly linear function of burst area, which is more difficult to measure.) This yields an excellent estimate of total sympathetic activity, which can be used to express changes of sympathetic outflow *within individual subjects*.

It is more difficult to compare burst heights *among different subjects*. One approach to burst normalization was developed by Wallin and Eckberg (1982). In their study, the largest sympathetic burst occurring during a rest period was assigned a value of 1000, and all other bursts occurring in that subject were calibrated against that standard. This method (or variations

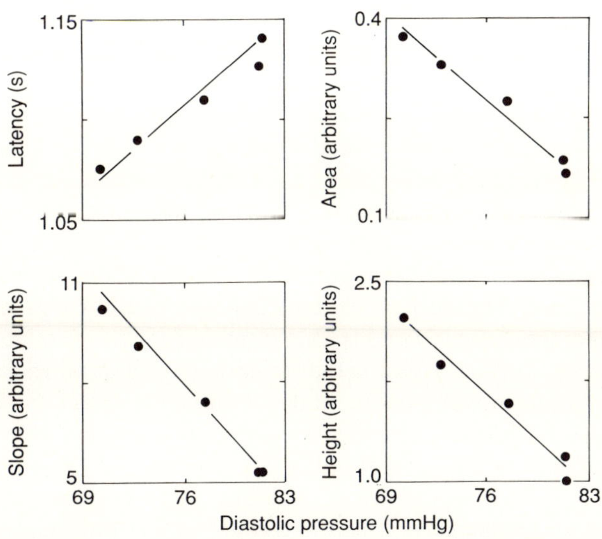

FIG. 8.8. Four different characterizations of muscle sympathetic nerve activity. Average data obtained from healthy young men during intravenous infusions of nitroprusside and phenylephrine. (Data derived from retrospective analysis of data published by Eckberg *et al.* (1988).)

based on it) quantitates sympathetic activity with units that can be compared among subjects.

The methods described above are based on the assumption that firing of human sympathetic nerves occurs synchronously as *'bursts'*. They are also based on the assumption (to our knowledge never explicitly owned in studies of human sympathetic activity) that either sympathetic activity does not occur tonically between bursts, or that if it does, it rises and falls in parallel with the frequency or size of bursts. This assumption may not be valid, since the occurrence of irregular firing of postganglionic sympathetic fibres in experimental animals is well documented (Adrian and Bronk 1932; J.H. Green and Heffron 1968*a*). As illustrated in Chapter 13, in certain circumstances including rapid dysrhythmias, typical sympathetic bursting patterns may be absent. Under such conditions, the raw rather than the electronically integrated nerve signal must be used for analysis. Figure 8.9 shows two approaches to the analysis of raw nerve signals. For these analyses, an electronic threshold crossing was set arbitrarily at a level of about 75 per cent of the 'noise' level, and the counter was reset either at a fixed number of crossings, or at a fixed time. (For such analyses, the threshold is set at the beginning of the record and not changed until all activity has been analysed.)

MEASUREMENTS OF NEUROTRANSMITTERS

Thus far, the only direct recordings made of human sympathetic nerve activity are of outflows to skeletal muscle and skin vascular beds. Microneurography has not been used to record traffic over other important sympathetic outflows, including those to the renal, cardiac, and splanchnic vascular beds. Therefore if sympathetic activity is to be characterized comprehensively, it may be necessary to supplement direct sympathetic recordings with other types of information. One alternative approach to direct sympathetic recordings is estimation of levels of neural traffic from plasma concentrations of neurotransmitters. The earliest example of this method was published by S.H. Grossman *et al.* (1982) and is illustrated in Fig. 5.7. A related approach is to give a drug such as tyramine that releases catecholamines, and measure changes of plasma noradrenaline and adrenaline levels (D.S. Goldstein *et al.* 1983; Jie *et al.* 1987).

This approach holds several advantages over the direct sympathetic recordings described above. First, accurate and sensitive techniques for measuring neurotransmitters, including especially noradrenaline, are much more widely available than techniques for measuring nerve traffic directly. Second, microneurography may not be suited well to the study of patients with autonomic neuropathy, including those with diabetes mellitus, in whom satisfactory recordings of sympathetic activity may not be obtainable (Fagius

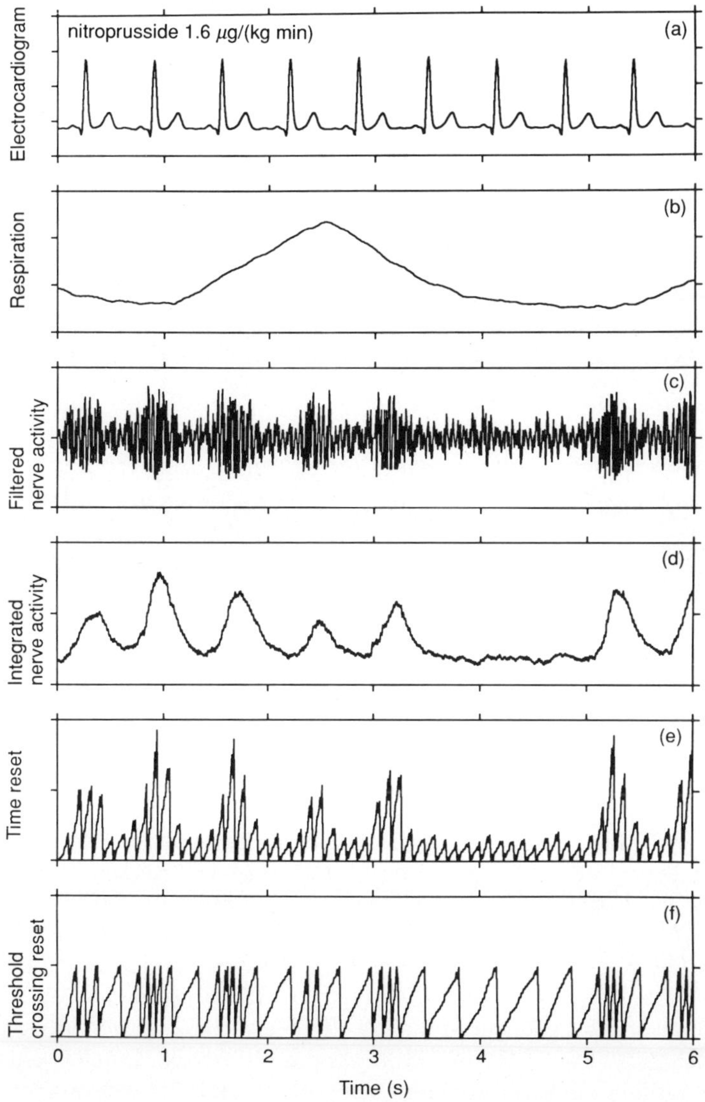

FIG. 8.9. Analyses performed by setting a threshold on raw nerve signal, displayed as number of threshold crossings (e) or number of resets (f) per unit time.

and Wallin 1980*b*). Third, many organs contribute to plasma pools of neurotransmitters; therefore measurements of (arterial) plasma neurotransmitter concentrations may reflect whole body levels of autonomic traffic better than measurements of nerve traffic confined to the outflow of a single organ. Fourth, kinetic analyses of neurotransmitters can be made which provide information about neuronal spillover and re-uptake from individual organs, and presynaptic modulation of transmitter release (Hasking *et al.* 1986; Esler *et al.* 1988,1990; D.S. Goldstein *et al.* 1988). Such information about autonomic function cannot be derived from simple sympathetic microneurograms. Fifth, some organs in human subjects are inaccessible to microelectrodes but can be reached with catheters. Finally, histochemical studies indicate that autonomic nerve endings contain a variety of neurotransmitters that may influence human cardiovascular function (Pernow 1988). Measurements of these diverse neurotransmitters may characterize autonomic outflow more comprehensively than measurements of the nerve traffic that releases them.

However, use of neurotransmitter measurements to study autonomic responses also holds disadvantages which are recognized widely. First, this approach may not be applicable for all neurotransmitters; some, such as acetylcholine, probably disappear too rapidly from the circulation to be measured in humans (Kilbinger and Löffelholz 1976). Second, experimental interventions, such as exercise or drug infusions, which alter rates of release of neurotransmitters, also alter blood flow. Usually, only *concentrations* of neurotransmitters are measured, without regard to concomitant changes of blood flow; changes of blood flow may alter concentrations independently, by simple dilution, or by modification of neurotransmitter re-uptake (Carlsson *et al.* 1964; Esler *et al.* 1985; Chang *et al.* 1986). Third, simple measurements of concentrations of neurotransmitters may not reveal changes of the kinetics of transmitter release and re-uptake, and may provide no information regarding sympathetic outflow to individual vascular beds. Fourth, most published studies of human plasma noradrenaline levels are based on samples obtained from antecubital veins; these measurements may be dominated by spillover from forearm muscles and may not be representative of measurements obtained from samples collected in arteries or other veins (Folkow *et al.* 1983; Esler *et al.* 1984). Moreover, antecubital vein blood also drains the cutaneous circulation. Since sympathetic nerve traffic to (and, presumably, noradrenaline spillover from) skin is determined importantly by ambient temperatures (Bini *et al.* 1980), room temperatures may influence antecubital vein plasma noradrenaline concentrations. In most human studies, room temperature is neither controlled nor reported. Finally, the relation between plasma noradrenaline and sympathetic traffic is determined importantly by the timing of blood sampling after interventions (Rea *et al.* 1990); it may be wrong to assume that sympathetic traffic and transmitter change proportionally, as square waves after interventions.

Correlations between sympathetic nerve traffic and plasma neurotransmitter concentrations

Several studies have examined the relation between baseline or experimentally altered levels of plasma noradrenaline and directly measured postganglionic muscle sympathetic nerve activity. Notwithstanding the numerous theoretical concerns enumerated above, very good correlations between *baseline* levels of antecubital vein plasma noradrenaline and muscle sympathetic nerve traffic have been found in normal subjects (Wallin *et al.* 1981) and patients with hypertension (Mörlin *et al.* 1983) or congestive heart failure (Leimbach *et al.* 1986; Porter *et al.* 1990). Moreover, plasma noradrenaline and muscle sympathetic nerve activity correlate closely during *interventions* such as the cold pressor test (Victor *et al.* 1987*a*), isometric exercise (Wallin *et al.* 1987), and infusions of pressor and depressor drugs (Eckberg *et al.* 1988). (However, such linear relations must not be interpreted as signifying linearity between noradrenaline spillover and sympathetic nerve traffic; direct correlations indicate that the relation between human sympathetic nerve traffic and noradrenaline spillover is complex (Baily *et al.* 1990; Pryor *et al.* 1990)). Importantly, during mental stress, neither antecubital vein plasma noradrenaline concentration (D.S. Goldstein *et al.* 1987) nor muscle sympathetic nerve activity (Wallin *et al.* 1973*a*; Hjemdahl *et al.* 1984) changes appreciably. Under this circumstance, antecubital vein plasma noradrenaline does not reflect generalized sympathetic activity. There is one known exception to the rule that antecubital vein plasma noradrenaline reflects sympathetic activity: hypoxaemia increases human muscle sympathetic nerve activity but does not increase venous plasma noradrenaline (Chase *et al.* 1988). As indicated, however, simple plasma catecholamine measurements mask great complexities of adrenergic kinetics.

The case is less well developed for other neurotransmitters. Neuropeptide Y is located with noradrenaline in central and peripheral sympathetic nerve endings, in humans as well as other species (Lundberg *et al.* 1982, 1984). Limited evidence from healthy human volunteers suggests that, unlike plasma noradrenaline, release of neuropeptide Y into the circulation is not a simple linear function of muscle sympathetic nerve activity. Eckberg *et al.* (1988) found that neuropeptide Y levels rise in proportion to muscle sympathetic nerve activity and plasma noradrenaline levels during infusions of sodium nitroprusside but do not fall during infusions of phenylephrine. Although neuropeptide Y is released by stimulation of cardiac sympathetic nerves in experimental animals (M.R. Warner and Levy 1989), this possibility has not been studied in humans.

As mentioned, plasma levels of the parasympathetic neurotransmitter acetylcholine have not been measured in humans. However, another neurotransmitter released by electrical vagus nerve stimulation, pancreatic polypeptide, has been measured by two groups. Tansey *et al.* (1981) found no correlation between the degree of bradycardia and the level of pancreatic polypeptide in patients with acute inferior myocardial infarction. However, Sander-Jensen and co-workers (1986*a,b*) found that in three out of 20 patients with haemorrhagic shock, and in six healthy subjects during passive

60 degree head-up tilt, reductions of heart rate were associated with elevations of pancreatic polypeptide levels.

MEASUREMENTS OF END ORGAN RESPONSES

Probably the most common approach used to gauge the effects of changes of human baroreceptor afferent activity in humans is to measure responses of innervated organs. This approach may be expedient, but it adds complexity to the research by interposing additional 'black boxes' in series. Typically in this research, neural input to an organ and the responsiveness of that organ to changes of neural input are unmeasured. Moreover, during autonomic interventions, all end organs are susceptible to influences by circulating hormones and neurotransmitters, as well as to changes of neural traffic.

Individual end organs bring unique complexities to this research. The *sinus node* responds to changes of right atrial stretch as well as neural input (Tansey *et al.* 1981), and it is likely that some heart rate variability is due to non-autonomic mechanical influences. Patients with orthotopic cardiac transplants (whose donor sinus nodes are denervated) maintain extremely small but measurable respiration-related heart rate changes (Zbilut *et al.* 1986; Bernardi *et al.* 1989). Presumably, sinus arrhythmia in these patients is provoked by respiration-related fluctuations of venous return and right atrial distension. In addition, because the sinus node functions as an oscillator, responses to changes of neural input depend on the *timing* as well as the *amount* of neurotransmitter released (Eckberg 1976).

Use of heart rate changes as simple indices of changing autonomic neural input to the sinus node may be complicated by interactions between sympathetic and parasympathetic inputs (Löffelholz and Muscholl 1970; Levy 1971; Carrier and Bishop 1972; Levy and Blattberg 1976; Chassaing *et al.* 1983); pre- and post-junctional interactions provide for highly non-linear responses to changing neural inputs. The complexity of autonomic interactions was underscored by H. Inoue and Zipes (1987*a*), who showed in anaesthetized dogs that different combinations of electrical sympathetic and vagus nerve stimulation frequencies could be chosen to produce identical heart rates but grossly dissimilar atrioventricular nodal conduction times and ventricular refractory periods.

Atrioventricular node function is modulated closely by changes of neural input; however, most neurally mediated dromotropic changes are masked by concurrent changes of sinus rate (Mancia *et al.* 1979*a*; Welch *et al.* 1986). The *left ventricle* may be influenced by changes of sympathetic and vagal neural input, but these changes may be confounded by concurrent changes of contractility, heart rate, preload, and afterload.

Additional complexities are added when healthy subjects are compared with patients. For example, in patients with hypertension, changes of vascular architecture alter vascular responses to changes of neural input (Folkow 1978; Korner 1982). The situation in these patients (see Chapter 11) is complex; hypertensive patients have supranormal pressor responses to infusions of both an adrenergic agonist, noradrenaline, and a non-adrenergic agent, angiotensin-II (Philipp *et al.* 1978). Since hypertensive patients also have supranormal forearm vascular resistance during reactive hyperaemia (Conway 1963), abnormal responses to pressor infusions may reflect abnormalities of vessels as well as abnormalities of reflex control. Thus, although in hypertensive patients unmistakable differences in responses to baroreceptor interventions may be found (Mancia *et al.* 1978*b*), it may be difficult to know whether these measured differences are due to alterations of baroreceptor function or to end organ responsiveness (Folkow 1982). Altered end organ responsiveness may result also from changes of autonomic receptors. Patients with congestive heart failure (see Chapter 14) appear to have supranormal levels of sympathetic–cardiac nerve traffic, as reflected by increased noradrenaline spillover (Hasking *et al.* 1986), and these are associated with subnormal levels of adrenergic receptor density (M.R. Bristow *et al.* 1982). They also may have abnormal responsiveness of peripheral vessels to vasodilator stimuli (Zelis *et al.* 1968).

Perhaps the end organ most exhaustively studied in baroreflex research is the *sinus node* (or more properly, 'atrial pacemaker complex' (Boineau *et al.* 1978, 1988, 1989)). Its pacemaker function can be evaluated non-invasively from surface electrocardiograms with great accuracy. It is responsive to changes of both sympathetic and parasympathetic neural input. Importantly, thus far, measurements of sinus node function constitute the most attractive approach available for quantitating levels of vagal–cardiac nerve traffic in humans. What follows is a review of this usage and a consideration of both its strengths and its weaknesses.

Estimates of human vagal–cardiac nerve activity

Human vagal–cardiac nerve traffic has not been measured directly; therefore, in humans, methods used to quantitate vagal influences are based on indirect approaches. Almost all of these methods are based on information extracted from the electrocardiogram. Among electrocardiographic parameters, R–R intervals (and their fluctuations) appear to provide more linear indications of vagal–cardiac activity than does heart rate. Validation of approaches used in humans has come from studies conducted in experimental animals and humans.

Jewett (1964) and Katona *et al.* (1970) correlated R–R intervals with directly measured vagal–cardiac nerve traffic in anaesthetized dogs, and concluded that during moderate changes of arterial pressure, R–R intervals are linear functions of vagal–cardiac nerve activity. In 1985, Koizumi *et al.* (1985) measured cardiac–vagal and –sympathetic traffic directly and simultaneously during spontaneously occurring blood pressure waves in chloralose-anaesthetized dogs. Their results, depicted in Fig. 8.10, document linear relations between sympathetic as well as vagal activities and R–R intervals over a range of intervals that occurs physiologically. The correlation coefficient for the vagal–heart period relation was an astonishing 0.98.

In a subsequent study, Parker *et al.* (1984) measured R–R intervals over a very wide range of electrical vagus nerve stimulation frequencies in anaesthetized dogs. We have replotted R–R interval responses shown in their Fig. 1 (right vagus stimulation) with corresponding calculated heart rate responses; see Fig. 8.11. The heart rate plot (dashed line) emphasizes an obvious mathematical relation: that the reciprocal of a linear function cannot be linear. The practical importance of this is that the magnitude of changes of heart rate provoked by changes of vagus nerve traffic depends critically on the preceding heart rate. Changes of heart rate after interventions such as exercise or hypoxic ventilation that change resting heart rates cannot be considered simply as linear functions of vagal traffic. On the other hand, over a wide range of R–R intervals, as studied by Parker and associates (1984), changes of R–R intervals provoked by changes of vagus nerve stimulation are independent of baseline R–R interval.

There are comparable but extremely limited data on human sinus node responses to electrical vagus stimulation. Carlsten *et al.* (1957) stimulated the left or right vagus nerve electrically in six patients undergoing radical neck surgery for cancer, and in whom the proximal vagus nerve was blocked by lignocaine. Their results are adapted in Fig. 8.12. These data document

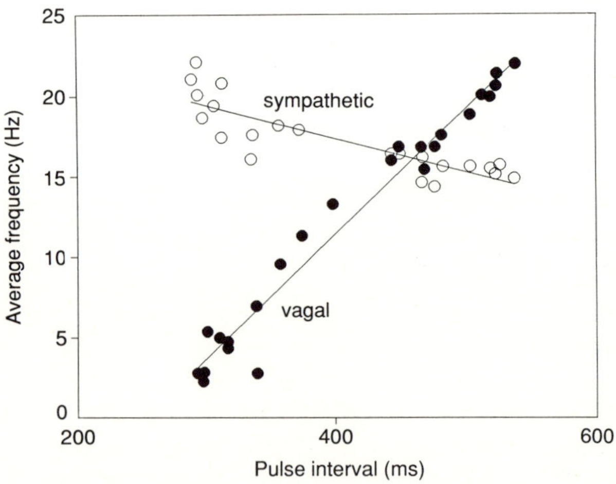

FIG. 8.10. R–R interval plotted as a function of directly recorded vagal– and sympathetic–cardiac nerve activities in an anaesthetized dog. Adapted with permission (Koizumi *et al.* 1985).

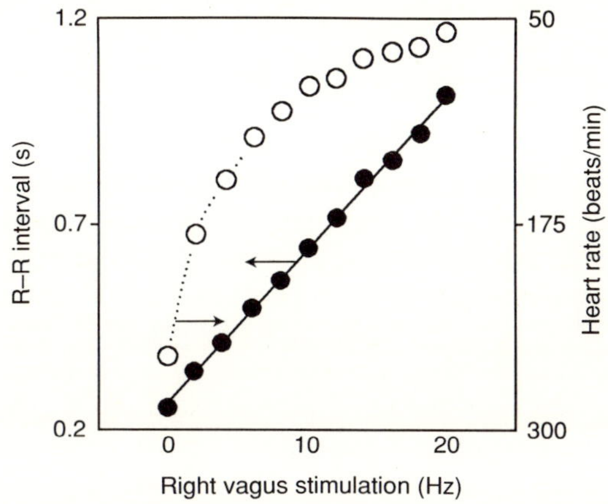

FIG. 8.11. Measured R–R intervals and calculated heart rates resulting from vagus nerve stimulation in anaesthetized dogs. Adapted with permission (Parker *et al.* 1984).

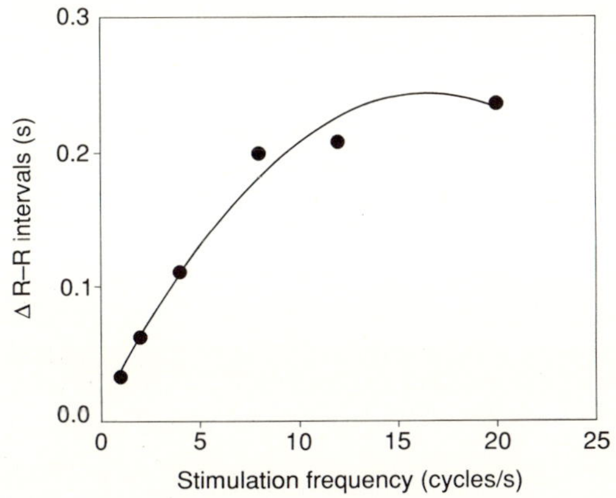

FIG. 8.12. R–R interval changes from six patients provoked by electrical vagus nerve stimulation during radical neck surgery. Adapted with permission (Carlsten *et al.* 1957).

a linear range and a plateau, as the frequency of vagus nerve stimulation increases.

We make a strong case for use of R–R intervals rather than heart rate in studies that involve estimation of vagal–cardiac nerve activity. Figure 8.11 illustrates the mathematical impropriety of regarding heart rate as a linear function of vagal traffic. However, heart rate (which is a calculated reciprocal of R–R interval and cannot be measured directly) is probably more important than R–R interval as a physiological variable. This issue is also discussed by Rowell (1986).

It does not follow necessarily from the above discussion that average R–R intervals accurately reflect average vagal activity at rest. Baseline R–R intervals are determined by a host of factors, including levels of sympathetic and vagal efferent activity, levels of noradrenaline (Izzo 1983) and angiotensin (Akselrod *et al.* 1981), the degree of right atrial distension (Bussan *et al.* 1957), intrinsic sinus node function (Jose and Taylor 1969), body core temperature (Kitney 1975), and the level of physical fitness (Steinhaus 1933; Katona *et al.* 1982). Thus, although the average R–R intervals are related to efferent vagal activity, they are influenced importantly by other factors, and probably cannot be considered specific indices of vagal activity.

Because of this, attempts to quantitate resting levels of human vagal–cardiac activity have focused on *changes* rather than baseline levels of R–R intervals. Three types of changes (and variations thereupon) have been examined; spontaneous fluctuations of R–R intervals related to respiration (respiratory sinus arrhythmia); shortening of R–R intervals provoked by large doses of muscarinic receptor blocking drugs; and changes of R–R intervals resulting from altered afferent (particularly baroreceptor) sensory activity. The first two types of measurements may have utility as estimates of resting levels of vagal–cardiac activity.

Rhythmic, respiration-related changes of R–R intervals reflect the ebb and flow of vagal–cardiac traffic, since they can be abolished by bilateral vagotomy, vagal cooling, or large doses of atropine. At rest, respiration-related changes of R–R intervals are thought to occur as linear functions of changes of vagal–cardiac traffic. Katona and Jih (1975) and Fouad *et al.* (1984) made the empirical observation that the level of respiratory sinus arrhythmia is related quantitatively to the level of ongoing vagal–cardiac traffic, as estimated from R–R interval shortening provoked by vagal cooling or large doses of atropine. On the basis of this observation, these authors proposed that measurements of peak minus valley changes of R–R intervals during quiet breathing may be used as surrogates for measurements of actual levels of vagal–cardiac nerve traffic.

Katona and Jih (1975) based their use of peak minus valley R–R interval changes on the observations that R–R intervals are linear functions of vagal–cardiac nerve traffic (as discussed above), and during inspiration, vagal–cardiac neurons are *silent* (Iriuchijima and Kumada 1964; Jewett 1964; Katona *et al.* 1970). Thus, during quiet respiration, activity of vagal–cardiac motoneurones increases from zero during inspiration to some higher level during expiration; this higher level is gauged by the expiratory R–R interval, which is a linear function thereof. The relation between average peak and valley P–P intervals during frequency and tidal volume-controlled breathing for five subjects is depicted in Fig. 8.13 (Eckberg 1983). This figure shows that both peak and valley P–P intervals are determined by respiratory

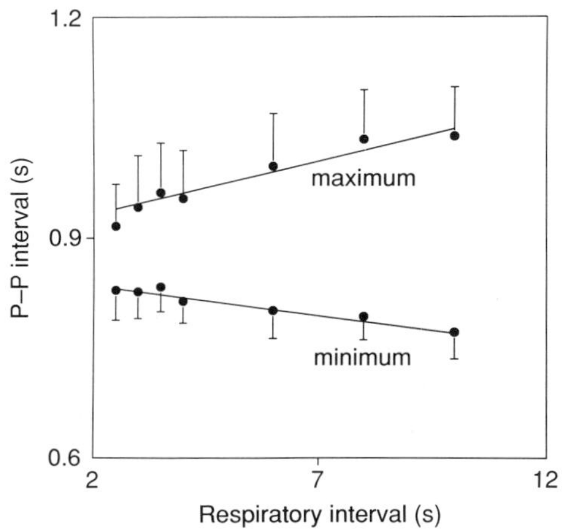

FIG. 8.13. Average maximum and minimum P–P intervals during frequency and tidal volume-controlled breathing from five healthy subjects. Reproduced with permission (Eckberg 1983).

rate, and that the longer the respiratory interval (the slower the breathing), the lower the minimum P–P intervals. The implication of these data is that in humans, vagal–cardiac motoneurone firing is not minimal, and therefore, these neurones are not silent at usual breathing rates.

Kollai and Mizsei (1990) tested this possibility more directly; they found that minimum R–R intervals at different breathing rates are always higher than those measured during acute interventions known to withdraw vagal–cardiac nerve traffic (hand-grip, deep breathing, and standing) or block it (large dose atropine). These authors concluded that human vagal–cardiac motoneurone activity does not fall to zero during inspiration, as it does in dogs, and that therefore, simple measurements of respiratory sinus arrhythmia do not necessarily provide quantitative indices of vagal–cardiac nerve traffic. Negoescu and Csiki (1989) drew a similar conclusion from the results of their computer modelling of human respiratory sinus arrhythmia.

Eckberg and co-workers (1988) studied the relation between the degree of baroreceptor stimulation of vagal–cardiac motoneurones and respiratory sinus arrhythmia in human volunteers with a stratagem that did not involve use of autonomic blocking drugs. They measured the effects of several levels of baroreceptor afferent activity on respiratory sinus arrhythmia in healthy young men during supine rest. Arterial pressure was altered with graded intravenous infusions of sodium nitroprusside and phenylephrine hydrochloride. Subjects breathed at a rate of 12 breaths/min. (Choice of this respiratory rate may have been unfortunate, since the breathing interval, 5 s, is a divisor of the 10 s interval that contains important sympathetic periodicity (Pagani *et al.* 1986).) Breathing rate was controlled because respiratory rate strongly influences respiratory sinus arrhythmia (see above), and sodium nitroprusside infusions provoke changes of respiration (possibly mediated by nasal stuffiness). Tidal volume was not controlled. Respiratory sinus arrhythmia was considered to be the average longest minus the shortest R–R intervals (peak minus valley changes) during ten single breaths.

Results from this analysis are summarized in Fig. 8.14. The lowest average peak minus valley R–R interval change, 0.048 s (found during infusions of the highest dose of nitroprusside, extreme left), was significantly less than all other averages. Peak minus valley R–R interval changes were comparable ($p > 0.05$) at the highest seven pressures studied. These data support the view that in healthy young adults, respiratory sinus arrhythmia is proportional to vagal–cardiac activity, but they suggest also that proportionality exists only when levels of vagal traffic are *below* baseline levels, a situation that usually does not obtain in resting young healthy adults.

Gilbey *et al.* (1984) studied the electrophysiological basis of respiratory sinus arrhythmia with microelectrodes positioned in or near vagal–cardiac motonuclei in the nucleus ambiguus of anaesthetized cats. They concluded that the instantaneous rate of vagal firing reflects a dynamic balance among the competing influences of respiration-related changes of cell membrane properties and incoming autonomic sensory information. This conclusion is in accord with other experimental evidence which shows that sinus arrhythmia is determined both by respiration (Hirsch and Bishop 1981; Eckberg 1983) and by autonomic inputs, including those from baroreceptors (Anrep *et al.* 1936), chemoreceptors (Eckberg *et al.* 1982), and trigeminal receptors (Eckberg *et al.* 1984).

The largest dose of phenylephrine that was infused by Eckberg *et al.*

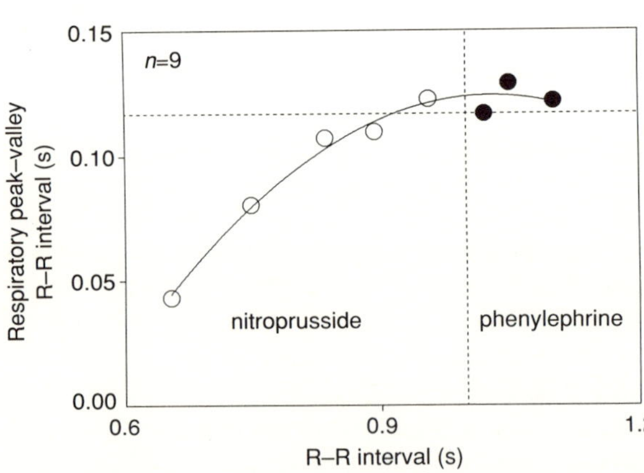

FIG. 8.14. Average respiration related peak minus valley R–R intervals during infusions of nitroprusside and phenylephrine. Adapted with permission (Eckberg *et al.* 1988).

(1988) provoked only modest increases of diastolic pressure (average rise: 11 mmHg). It is likely that if larger doses of phenylephrine had been used, respiratory peak minus valley R–R interval changes would have *diminished.* Anrep *et al.* (1936) found that large elevations of arterial pressure abolish respiratory sinus arrhythmia in anaesthetized dogs. Similarly, intense carotid baroreceptor stimulation with neck suction in human volunteers abolishes the inspiratory–expiratory differential responsiveness that is found with less intense stimuli (Eckberg and Orshan, 1977). These studies suggest that respiration does not function as a 'gate,' which either allows or does not allow stimulatory inputs to influence vagal–cardiac motonuclei (Lopes and Palmer 1976). Rather, respiratory influences are measurable and finite; they are competitive, and as such, they are overcome by intense levels of stimulatory autonomic inputs (Fig. 8.14, far right), and perhaps, even baseline levels (dashed vertical line).

Thus it may be unwarranted to extrapolate from the data published by Katona and Jih (1975) and Fouad *et al.* (1984) to conclude that proportionality between sinus arrhythmia and vagal activity exists for all levels of vagal activity. Although these authors did not comment on it, the interventions they used focused primarily on *reductions*, not elevations, of vagal–cardiac activity from baseline levels. Katona and Jih (1975) measured respiratory sinus arrhythmia during elevations as well as reductions of arterial pressure, but the dogs they studied were anaesthetized, and probably had subnormal baseline levels of vagal–cardiac activity (in these dogs, control heart rate averaged 103 beats/min). Therefore in their study it appears that very few measurements were obtained when vagal–cardiac activity was at or above levels that exist in resting conscious dogs. Fouad and associates (1984) measured resting levels of respiratory sinus arrhythmia, and then *reduced* the sinus node effects of those levels with incremental injections of large doses of atropine.

A second extrapolation from the respiratory sinus arrhythmia–vagal activity relation is that differences among resting peak minus valley R–R intervals obtained from different groups of subjects (such as diabetic patients and nondiabetic subjects (T. Bennett *et al.* 1978), and patients with and without heart failure (Saul *et al.* 1988*b*; Porter *et al.* 1990; Sopher *et al.* 1990)) reflect differences of vagal–cardiac activity. This is intuitively attractive; however, definitive experiments to support this extrapolation have not yet been performed in experimental animals or humans. Moreover, there are pitfalls in attempts to compare levels of sinus arrhythmia in patients and healthy subjects. One obvious problem is that breathing is different in heart failure patients than in healthy subjects; they have reduced compliance (which probably translates into augmented respiratory motoneurone activity per breath), and they breathe more rapidly (and possibly more deeply). The important influence of breathing parameters on respiratory sinus arrhythmia is discussed below.

Also, there may be subtleties that are peculiar to the specific diseases of patients. One is that the influence of breathing on autonomic sensory inputs may be different in patients than in healthy subjects. For example, inferior vena cava (and by inference, atrial) diameter does not change as much during quiet breathing in heart failure patients than in healthy subjects. Patients with structural heart disease present different problems. R.R. Taylor and Hopkins (1972) showed that respiratory sinus arrhythmia exerts strong influences on the degree of valvular aortic regurgitation. On the basis of this observation, we predict that the breathing pattern in patients with aortic regurgitation influences the degree of sinus arrhythmia through respiration-related variations of left ventricular stroke volume.

QUANTITATION OF HEART RATE FLUCTUATIONS

A variety of related measurements has been used to estimate autonomic nerve traffic to the sinus node from surface electrocardiograms. These include simple averages of R–R intervals or heart rates, measurements of the clustering of R–R intervals around mean values or beat-to-beat changes of R–R intervals, correlations of R–R intervals with other physiological variables such as respiration (as discussed above), and analysis of rhythms with frequency-domain techniques. A detailed discussion of each of these approaches, ranked in ascending order of complexity, follows.

All of these methods typically require that subjects and patients be in normal sinus rhythm. Commonly, patients with cardiovascular diseases have frequent premature beats; it may be that researchers have paid insufficient attention to technical problems posed by dysrhythmias of patients. One way to deal with premature beats is to exclude the premature beat, and the next two or three beats, and reconstitute the data stream with exclusively sinus beats. This approach was used by Kleiger *et al.* (1987) who studied R–R interval variability in patients recovering from acute myocardial infarction. Another approach is to scan electrocardiogram records for segments that are free of premature beats and restrict analyses to these segments. This may not be ideal if the goal of a study is to estimate fluxes of autonomic neural activity associated with dysrhythmias; autonomic traffic may be different during periods free of premature beats than when they are present (Lovelace and Knoebel 1986).

The simplest measure of autonomic traffic to the sinus node is average heart rate or R–R interval. Average heart rate was used to estimate levels of autonomic outflow by Webb *et al.* (1972); they considered that patients with acute myocardial infarction who had heart rates above 100 beats/min had excess sympathetic activity, and patients who had heart rates below 60 beats/min had excess vagal activity. We know of no experimental validation of this intuitively attractive approach. Although average R–R intervals are related significantly to other measures of autonomic outflow, as indicated above, R–R intervals are determined by a host of other factors and probably cannot be considered specific indicators of neural traffic. Moreover, modest rate fluctuations occur in the denervated atria of heart transplant patients (Alexopoulos *et al.* 1988). Therefore several permutations of varying complexity, based on R–R intervals or heart rate measurements, have been developed.

Figure 8.15, adapted from the study of T. Bennett *et al.* (1978), suggests that, at least in populations that include diabetic patients, peak minus valley R–R interval changes during quiet respiration correlate closely with baroreceptor–cardiac reflex slopes measured after bolus injections of phenylephrine. Peak minus valley R–R intervals were used by Katona and Jih (1975) and Fouad *et al.* (1984), as described above. The basis for this association is that rapid fluctuations of R–R intervals appear to be mediated entirely by fluctuations of vagal–cardiac nerve activity (Akselrod *et al.* 1981; Eckberg 1983; Fouad *et al.* 1984). Changes of sympathetic traffic to the sinus node probably occur as frequently as changes of vagal activity, but these provoke *gradual* changes of R–R intervals; the sinus node functions as a 'low pass' filter for sympathetic traffic, and frequencies greater than about six per minute are dampened considerably (R.D. Berger and Cohen 1988). As mentioned, peak minus valley changes of R–R interval correlate closely with total vagal–cardiac activity (Katona and Jih 1975). One issue that is usually not addressed is that use of respiratory peak minus valley measurements requires that signals be stationary (Porges 1986). Another is

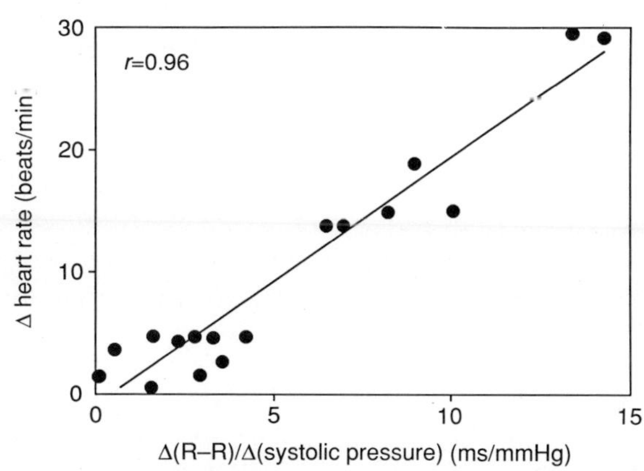

FIG. 8.15. Slopes of R–R interval–systolic pressure relations after bolus phenylephrine injections plotted as functions of respiration-related peak minus valley R–R interval changes in healthy subjects and diabetic patients. Adapted with permission (T. Bennett *et al.* 1978).

that measurements of peak minus valley R–R interval is that they require accurate registration and control of respiratory activity.

Another approach, which does not require knowledge of respiration, is to measure variability or standard deviation of R–R (or P–P) intervals. Frequently occurring vagally mediated fluctuations of R–R intervals increase standard deviations. Accordingly, if the R–R interval standard deviation is subnormal, it is likely that vagal–cardiac nerve traffic is also subnormal. However, if the standard deviation is normal, the interpretation becomes more complex. If the segment of time chosen for analysis includes large changes of state, such as changes of posture or activity, these will contribute to and increase the standard deviation. The contribution of such factors to measurements of R–R interval standard deviation has not been determined. De Boer (1985) discussed several measures of R–R interval variability in his thesis. He points out that standard deviations provide no information about the wave forms from which they are derived. He developed a series of wave forms that have identical standard deviations; these are shown in Fig. 8.16.

To our knowledge, there has been no prospective comparison of the various methods used to estimate human vagal–cardiac nerve activity. In Eckberg's study (1983), an excellent ($r=0.97$) linear correlation was documented between R–R interval standard deviations and peak minus valley R–R interval changes. It should be emphasized, however, that this correlation was obtained under rigidly controlled conditions, with subjects lying

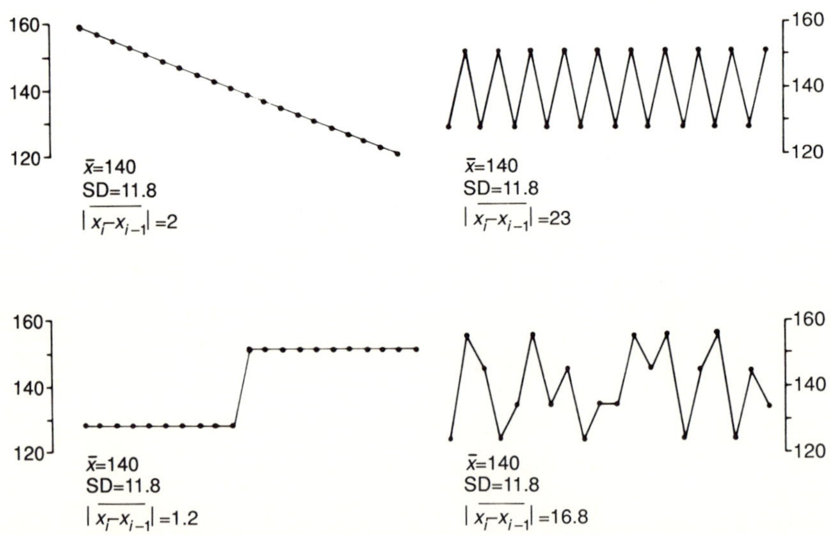

FIG. 8.16. Theoretical argument against use of R–R interval standard deviations (see text). Reproduced with permission (de Boer *et al.* 1985*a*).

supine at rest and breathing at constant rates and tidal volumes. We do not know if such a strong correlation would be found if comparisons were made under less controlled conditions. Also, although there is a growing literature involving use of standard deviations to characterize R–R interval variability, it should not be assumed that all studies employ the same method to calculate standard deviation. Thus 'standard deviation' may refer to mathematical standard deviations calculated from all R–R intervals during a test period (Dibner-Dunlap *et al.* 1985), but it may also refer to an average of standard deviations derived from all R–R intervals during fixed periods of measurement (Magid *et al.* 1985), or to standard deviations of *average* R–R intervals (Kleiger *et al.* 1987). Other approaches based on the differences between successive R–R intervals (large abrupt R–R interval changes are mediated by vagal–cardiac activity changes, and are taken as indications of the level of vagal control) have been described (Ewing *et al.* 1981); these methods may be less influenced by baseline R–R intervals and slow changes of R–R intervals over time.

More sophisticated mathematical treatments of R–R interval and heart rate, including power spectral density, cross-correlation, and cross-spectra, are used widely; for many, they have become non-invasive 'probes' for estimation of levels of sympathetic and parasympathetic nerve traffic to the sinus node. Kitney and Rompelman (1987) have edited a splendid review of such usages. These methods have several theoretical advantages over more simple measures based on the variability of R–R intervals. The magnitude of R–R interval standard deviations increases in proportion to average R–R intervals, and tends to become larger as the period of recording becomes longer, because of unavoidable non-stationarities (Saul *et al.* 1988*a*). Moreover, these measurements do not indicate whether observed changes of standard deviations are due to high- or low-frequency changes of heart rate, and they do not allow for correlations to be made between R–R intervals (or heart rates) and other parameters, including blood pressure, respiration, and sympathetic nerve activity.

Spectral analyses of R–R intervals

There is a burgeoning literature on mathematical analyses of heart rate, R–R interval, or arterial pressure waveforms as a means to detect and quantify underlying fluxes of autonomic neural traffic. The electrocardiogram is a complicated waveform, which must be changed before it can be analysed by spectral techniques. The initial approach is usually to transform the electrocardiogram into a point series of events, such as R–R intervals. Since spectral analyses are based on Fourier transformations, R–R interval point series must subsequently be transformed into some type of continuous waveform. The easiest approach is to subject continuous analogue R–R interval recordings, offset by one beat, to spectral analysis. This widely used approach has two shortcomings, as discussed by de Boer *et al.* (1985*a*). First, such tracings give undue weight to short intervals preceding long intervals. Second, although such a signal can be used for autospectral analyses, it may not be appropriate for correlations between R–R intervals and other variables, such as respiration. This usage mandates a phase lag of one heart beat (plus the P–R interval). The alternative, and the one advocated by de Boer and co-workers (1985*a*), is to shift R–R intervals to the left by one heart beat, and render the duration of the signal connecting R–R intervals equal to those intervals.

Power spectral analyses of changes of sinus node frequency have been used in an attempt to quantify autonomic neural fluxes underlying heart rate changes. As will be seen, the major problem with this exciting new approach is that it may not have been validated adequately; there is a substantial probability that technological improvements have outstripped growth of understanding of the underlying physiology. We suggest that the 'gold standard' for validation of indirect means for measuring neural traffic is nerve traffic itself. We are aware of no direct comparisons between sympathetic– or vagal–cardiac nerve traffic and estimates of such traffic derived from spectral analyses. Therefore, as with many other techniques used for human research, validation of power spectral analyses is based on inferential proofs.

There appears to be universal agreement that rapid changes of R–R intervals are mediated almost exclusively by changes of vagal–cardiac activity (Akselrod *et al.* 1981; Eckberg 1983; Fouad *et al.* 1984). Thus absolute or relative R–R interval power at respiratory frequencies has been used as an indirect means to quantify vagal–cardiac traffic. (Changes of sympathetic neural input may contribute to respiratory sinus arrhythmia at slow breathing frequencies.) The arguments given above, that R–R interval variability is proportional to vagal–cardiac nerve activity, should apply also to variability determined by spectral analyses. However, limited data suggest that for R–R interval recordings made at rest with subjects rigorously controlling their breathing, peak minus valley, respiratory R–R interval changes, standard deviations of R–R intervals, and R–R interval spectral power at respiratory frequencies convey similar information, and the superiority of one method over another has not been proven. Spectral methods may be superior to more simple methods when breathing rates vary (Saul *et al.* 1989).

The most important potential advantage of spectral analyses of R–R interval power may lie in the possibility that this approach can be used to estimate levels of *sympathetic*–cardiac nerve traffic. As discussed earlier, although changes of both sympathetic (J.H. Green and Heffron 1968*b*; Ninomiya *et al.* 1971; Wallin and Eckberg 1982) and parasympathetic (Eckberg 1976, 1980*a*) outflows occur rapidly after baroreceptor stimuli and have similar periodicities (Eckberg *et al.* 1985; Saul *et al.* 1988*b*), sinus node *responses* to these changes are vastly different. The sinus node can follow changes of vagal activity up to about 0.5 Hz (that is, changes occurring every 2 s), but can follow changes of *sympathetic* activity only up to about 0.15 Hz (changes occurring every 7 s) (H.R. Warner and Russell 1969; Akselrod *et al.* 1985; R.D. Berger *et al.* 1989). From these considerations, it would be expected that R–R interval power in high-frequency (respiratory) bands is not, but that R–R interval power at low frequencies is, modulated importantly by changes of sympathetic nerve traffic.

As mentioned, in this controversy there is no gold standard. Although muscle and skin sympathetic activity can be measured readily in humans, cardiac sympathetic activity has not been measured. Therefore, the issue devolves to one of competing indirect proofs. We see at least two problems with using low-frequency R–R interval power as an index of cardiac–sympathetic activity. First, as with the more simple measures of R–R interval fluctuations discussed above, this usage makes the assumption that *fluctuations* are related directly to *absolute* levels of sympathetic traffic. There are circumstances in which this simple assumption does not hold: although heart failure patients have high levels of sympathetic traffic to the heart (Hasking *et al.* 1986), some may have extremely low levels of low-frequency R–R interval power (Saul *et al.* 1988*b*). Healthy people increase their sympathetic activity during exercise, but reduce their low-frequency R–R interval power (Arai *et al.* 1989). Low-frequency R–R interval power may not reflect sympathetic outflow in either of these circumstances. Second, this usage is based on the assumption that vagal–cardiac traffic does not contribute to low-frequency R–R interval power; although this view is supported by some data (Rimoldi *et al.* 1990), there is substantial evidence that refutes it (Akselrod *et al.* 1981, 1985; Pomeranz *et al.* 1985; Saul *et al.* 1988*b*; R.D. Berger *et al.* 1989). If both sympathetic and vagal traffic contribute to low-frequency R–R interval variations, a simple frequency analysis of heart rate variability may be insufficient to determine from one resting measurement the contribution of each.

Since human sympathetic–cardiac nerve activity has not been measured, the evidence that low-frequency power is mediated nearly exclusively by sympathetic input is circumstantial. First, upright tilt, an intervention known to increase (muscle) sympathetic nerve activity (Burke *et al.* 1977), increases low-frequency R–R interval power and reduces high-frequency power. Second, beta-adrenergic blocking drugs appear to reduce low-frequency power selectively (Akselrod *et al.* 1985; Pagani *et al.* 1986). Third, sympathectomy reduces low-frequency R–R interval power (Pagani *et al.* 1986; Rimoldi *et al.* 1990). Fourth, patients with traumatic cervical spinal cord transections lack R–R interval spectral power at low frequencies (K. Inoue *et al.* 1990).

The weaknesses of these proofs may be that upright posture is a complex intervention which assuredly also reduces vagal–, as it increases sympathetic–cardiac traffic, and beta-blocking drugs exert complex effects that are understood incompletely (see below).

Objections to use of R–R interval power as an indirect registration of sympathetic–cardiac nerve traffic are theoretical and practical. The principal theoretical objection is that this usage requires that vagal–cardiac nerve traffic changes occur rapidly, but not slowly; obviously if vagal fluxes occur slowly, there may be no way to know what proportion of low-frequency fluctuations is attributable to vagal influences. Saul *et al.* (1988*b*) compared R–R interval power and human muscle sympathetic nerve power at low and high-frequencies. They found no significant correlation between these two measures at rest. Subjects with higher levels of resting muscle sympathetic nerve activity did not necessarily have higher levels of R–R interval power at low frequencies, and subjects with high vagal traffic (R–R interval power spectrum at respiratory frequencies) did not necessarily have low levels of low-frequency R–R interval power. The main weakness of this study is that its authors measured *muscle*, not *cardiac*, sympathetic activity. The authors make the assumption (supported by data from experimental animals (Ninomiya *et al.* 1971; Kollai and Koizumi 1989)) that during infusions of vasoactive drugs, nerve traffic to cardiac and muscular vascular beds increases and decreases in parallel. This group did find, however, that when sympathetic nerve traffic increases (and presumably, vagal traffic decreases) during pharmacological arterial pressure reductions, low-frequency R–R power and muscle sympathetic nerve activity increase in proportion to one another. However, when sympathetic nerve traffic decreases and vagal traffic increases during phenylephrine infusions, there is no correlation between muscle sympathetic nerve activity and low-frequency power. To our knowledge, correlations between cardiac noradrenaline spillover and low-frequency R–R interval power at different arterial pressures have not been sought; such correlations might help to validate (or refute) use of low-frequency R–R interval power to estimate sympathetic–cardiac nerve traffic.

At the time of writing, it is not clear how this controversy can be resolved. It may not be sufficient to measure sympathetic traffic to the muscle vascular bed directly and assume that parallel changes of sympathetic outflow to the heart occur also. Studies conducted in experimental animals indicate clearly that sympathetic outflow to various vascular beds is differentiated (E. Simon and Riedel 1975; Ninomiya and Fujita 1976). Moreover, studies conducted in humans indicate that some interventions which increase sympathetic outflow to the muscle vascular bed, such as immersion of the face (Fagius and Sundlöf 1986) or arm (Victor *et al.* 1987*a*) in ice water, or mental stress (E.A. Anderson *et al.* 1987) provoke bradycardia. In these circumstances, bradycardia is probably due to reflex increases of cardiac–vagal nerve traffic; however, the occurrence of sympathetic withdrawal cannot be excluded on the basis of present evidence.

Another important use for spectral analyses of human electrocardiograms is to identify *correlations* between different physiological signals (Peňáz *et al.* 1978). Thus de Boer *et al.* (1986) correlated changes of R–R intervals with changes of systolic pressures within the frequency band occupied by respiration. Their method is illustrated in Figs. 5.22 and 5.23. As is evident, these two parameters were related to each other closely (least-squares linear regression $r=0.90$) and had a slope, 10 ms/mmHg, that is comparable to baroreflex slopes derived from R–R interval responses to bolus injections of phenylephrine (Smyth *et al.* 1969). It could be argued that methods of gauging baroreflex responses based on spectral analysis do not hold any great advantage over other methods, including the phenylephrine method, since both methods require arterial cannulation. This argument may no longer be valid, because the non-invasive approach to beat-by-beat registration of human arterial pressure developed by Peňáz (1962) has become available commercially (Finapres, Ohmeda, Englewood, CO, USA; USM-801, Ueda, Tokyo, Japan).

There is a practical problem associated with analyses of R–R interval fluctuations as indirect indices of efferent vagal–cardiac nerve activity. As discussed in in this chapter and Chapter 7, vagal outflow is determined importantly by respiratory activity. Kitney and co-workers in important articles that are largely ignored (Selman *et al.* 1982; Kitney *et al.* 1982) showed convincingly that high-frequency (that is, respiratory) R–R interval power is determined by the frequency of breathing and tidal volume. One implication of these results is that sinus arrhythmia cannot be compared in groups of subjects whose breathing rates (or, perhaps breathing mechanics) are dissimilar. This conclusion is underscored by a study by Drummond (1990). He found that respiratory sinus arrhythmia is subnormal in patients with borderline blood pressure elevation because they breathe more rapidly than normotensive subjects.

The importance of breathing frequency for R–R interval spectral power (at a constant, controlled tidal volume) in one healthy subject is illustrated in Fig. 8.17 (T.E. Brown *et al.* 1991). This record documents a critically important relation between breathing rate and R–R interval spectral power. Spectral power declines precipitously as breathing frequency increases from very low frequencies (left) to usual frequencies (middle). Moreover, it is highly likely that at low breathing frequencies, vagal– and sympathetic–cardiac traffic both contribute to R–R interval spectral power at the breathing frequency. Kitney's and Brown's observations have practical importance, since most workers who use indirect methods to gauge vagal–cardiac nerve activity control do not measure or control respiratory frequency and tidal volume. These human data support the dictum Neil (1979) derived from animal research, that '. . . *the influence of the discharge of central respiratory neurons should not be ignored in experiments which purport to examine cardiac vagal reflex responses*.'

Spectral analysis techniques have their greatest utility in study of *periodic*

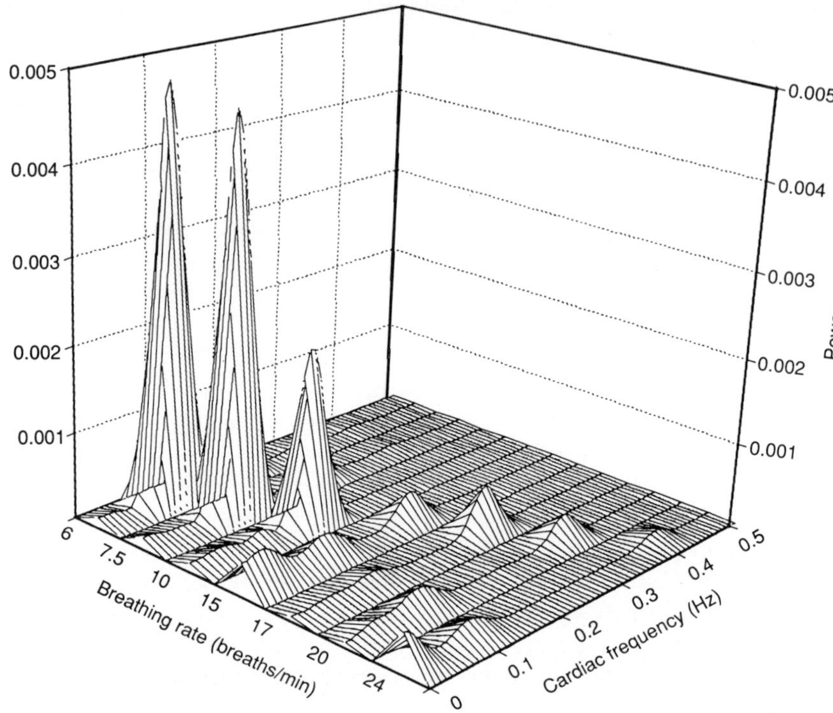

FIG. 8.17. R–R interval spectral power at one tidal volume (1.0 litre) and different respiratory frequencies (T.E. Brown *et al.* 1991). Note that the scale for respiratory frequency is not linear.

changes of autonomic outflow, such as those related to cardiac or respiratory activity. These methods are less useful for study of non-periodic changes, including degree of arousal or baroreflex responsiveness, body position, and physical activity. Such non-linear influences are better characterized by methods that display instantaneous changes (Abraham and Shaw 1983). One such method, the Poincaré plot, as adapted for Holter analyses by Woo *et al.* (1990), is illustrated in Fig. 8.18. These graphs were made by plotting each R–R interval against the next R–R interval during a 24 h period of Holter recording. The qualitatively different plot from the heart failure patient may reflect disruption of normal baroreflex control, as discussed in Chapter 14.

PHARMACOLOGICAL BLOCKADE

Another indirect means used to characterize human baroreflex function, which has several variations, is to measure autonomic outflow at rest or in

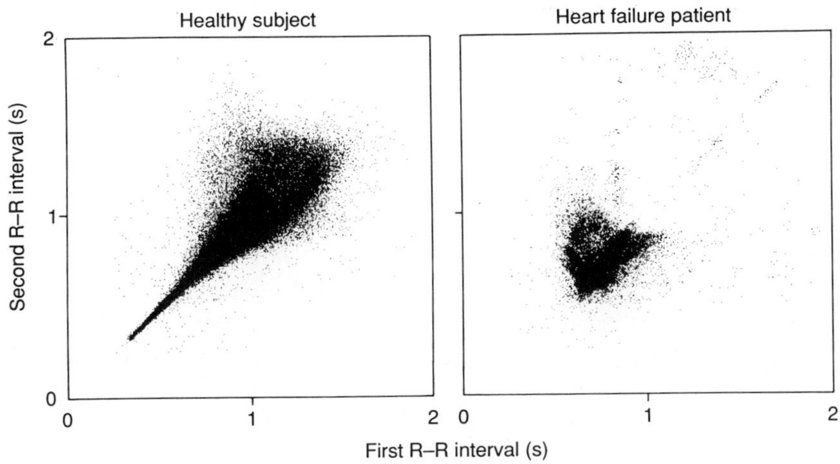

FIG. 8.18. Poincaré plot of each R–R interval plotted as a function of the next R–R interval during a 24 h Holter recording. Figure kindly supplied by Mary A. Woo (Woo *et al.* 1990).

response to baroreflex forcings, before and after administration of drugs that influence autonomic function. According to this usage, changes occurring after an autonomic blocking drug provide some quantitative estimate of the neural traffic whose effects were blocked. A great theoretical advantage of this approach is that a variety of drugs are available to block agonist effects; therefore this method is limited only by the availability of specific blocking drugs. Pharmacological approaches open the possibility that otherwise obscure mechanisms might be defined; however, as with most of the approaches described earlier, pharmacological methods yield indirect evidence bearing on baroreflex function.

Therefore, as with other methods, it is important that underlying assumptions be owned and dealt with explicitly. We are particularly concerned about possible taxonomic traps wherein the assumption is made uncritically that when a prefix such as 'adrenergic-' or 'cholinergic-' is attached to a suffix, such as '-blocker' or '-agonist', the drug used does only what its name implies, or that it acts on only one isolated receptor population. Autonomic blocking drugs have complex peripheral and central modes of action (Katona *et al.* 1977; Patel and Turner 1981; Lowensohn 1986). In all likelihood, responses to adrenergic agents are modulated importantly by cholinergic mechanisms, and responses to cholinergic agents are modulated importantly by adrenergic mechanisms. Levy (1971) has catalogued pre- and post-synaptic adrenergic–cholinergic interactions in experimental animals, and Leveston *et al.* (1979) have shown that administration of the anticholinesterase drug edrophonium increases plasma noradrenaline levels in humans.

Another pitfall is the assumption that after administration of autonomic blocking drugs in large doses, the remaining functional status of an organ reflects simply its function deprived of neural influences. This approach was used first by Jose and Taylor (1969), who characterized heart rate after large intravenous doses of the beta-blocker propranolol and the muscarinic cholinergic blocker atropine sulphate as the 'intrinsic heart rate'. The difficulties with this construct are that efferent autonomic neurones contain many informational substances (Hökfelt *et al.* 1980) whose influences may not be blocked by the drugs given, and, as discussed below, blocking drugs may do more than their names imply. This problem is underscored by the study of Evans *et al.* (1990) conducted in sedated dogs. They found that heart rate after surgical denervation is significantly slower than after ganglionic blockade or 'total' autonomic blockade with muscarinic cholinergic and beta- and alpha-adrenergic blocking drugs. The implication of this study is that efferent nerves continue to influence sinus node function after blocking drugs are given.

Cholinergic blockade

A time-honoured approach used to estimate baseline levels of vagal–cardiac activity is to measure the R–R interval shortening that occurs after large intravenous doses of atropine sulphate. These methods are based on the assumptions that shortening of R–R intervals after atropine is mediated solely by blockade of sinus node muscarinic receptors, and that such changes are related quantitatively to the level of efferent vagal–cardiac traffic. There are several published variations on this method. The simplest is to measure average R–R intervals or heart rate before and after a large intravenous dose of atropine sulphate (Jose and Taylor 1969). In such studies, R–R interval shortening or cardioacceleration after atropine is taken as an index of the amount of vagal–cardiac traffic before atropine. Another is to measure R–R interval shortening when atropine is given after acetylcholinesterase blockade (Birnbaum *et al.* 1988).

A.C. Simon *et al.* (1977) conceived a more sophisticated variation of this approach. They measured slopes of vagally mediated R–R interval increases after each of several cumulative intravenous atropine doses. Slopes of this relation declined with each dose. Regression of baroreflex slope on the cumulative atropine dose, extrapolated to zero, yielded the 'atropine blocking dose,' which was taken as a function of vagal–cardiac nerve traffic. These authors found that in a group of 95 normotensive and hypertensive men, the atropine blocking dose varied inversely with diastolic pressure. This inverse relation was significant in normotensive but not in hypertensive subjects. The method is illustrated in Fig. 8.19. We are not aware of any other published use of this method for quantifying levels of human vagal–cardiac activity.

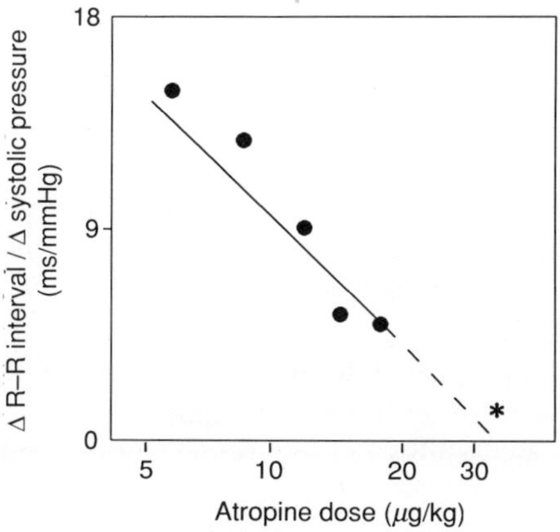

FIG. 8.19. Method developed to judge amount of baseline vagal–cardiac nerve activity from changes of phenylephrine baroreflex slopes after incremental atropine doses. Asterisk indicates 'atropine blocking dose'. Adapted with permission (A.C. Simon *et al.* 1977).

Large-dose atropine methods tend to ignore potential contributions of baseline sympathetic stimulation to cardioacceleration after atropine. This issue is complex. Patients with high spinal cord transections and absent cardiac–sympathetic innervation have subnormal heart rate speeding after high-dose atropine (Lehmann *et al.* 1987). However, patients with severe heart failure, who have high levels of muscle sympathetic nerve activity (Leimbach *et al.* 1986; Porter *et al.* 1990) and noradrenaline (J.A. Thomas and Marks 1978; Cohn *et al.* 1984) also may have only minimal R–R interval shortening with high-dose atropine (Eckberg *et al.* 1971). Such patients may be relatively insensitive to adrenergic stimulation by virtue of down-regulation of beta-adrenergic cardiac receptors (M.R. Bristow *et al.* 1982).

There are theoretical objections to the use of atropine to estimate human vagal–cardiac nerve activity. First, it is clear that responses to large doses of atropine do not result simply from competitive blockade of sinus node cholinergic receptors; in experimental animals, R–R interval shortening is greater after large doses of atropine than after bilateral vagotomy or vagal cooling (Brunsting *et al.* 1979; Rigel *et al.* 1984). Cardioacceleration after atropine is mediated by vagus nerves; Brouha and Nowak (1939) showed that although vagotomy prevents the cardioacceleration that occurs after atropine, marked heart rate speeding occurs if vagal traffic is restored by electrical stimulation after atropine. Also, intravenous atropine produces opposing central and peripheral effects: in sympathectomized or adrenergically blocked experimental animals, it causes a dose-related augmentation of vagal–cardiac activity, even at high doses (Brouha and Nowak 1939;

D.E. Donald *et al.* 1967; Chassaing *et al.* 1976; Katona *et al.* 1977; Brunsting *et al.* 1979), as well as peripheral muscarinic receptor blockade. R–R interval shortening after intravenous atropine probably reflects, in some way, algebraic summation of these opposing influences. However, atropine sulphate exerts negligible direct effects on denervated sinus nodes, and it is likely that very large doses of atropine (0.04 mg/kg (Jose and Taylor 1969)) effectively block the restraining influence of central vagal–cardiac activity, however large. Results of an atropine dose-ranging study in patients with cardiac transplants by A.E. Epstein *et al.* (1990) are depicted in Fig. 8.20.

None the less, the degree of R–R interval shortening after intravenous atropine correlates with other indices of vagal–cardiac activity. Patients with organic heart diseases exhibit several poorly understood abnormalities of parasympathetic control. The oldest observation is that patients experience less R–R interval shortening after large doses of atropine than do healthy subjects (Müller 1891; Crawford 1923; Eckberg *et al.* 1971). In patients with congestive heart failure (Porter *et al.* 1990) and primary hypertension (Alicandri *et al.* 1985), reduced R–R interval shortening after large doses of atropine is associated with reduced R–R interval lengthening after small doses of atropine, and reduced vagally mediated baroreceptor–cardiac reflex responses (J.D. Bristow *et al.* 1969*b*; Eckberg *et al.* 1971). These diverse observations provide inferential evidence that responsiveness of central vagal–cardiac motoneurones to increased levels of baroreceptor stimulation or cholinergic blockade (Gilbey *et al.* 1984) is tied to resting levels of efferent vagal–cardiac activity. Within this context, one indirect technique tends to validate another.

However, as far as we know, there have been no attempts to correlate

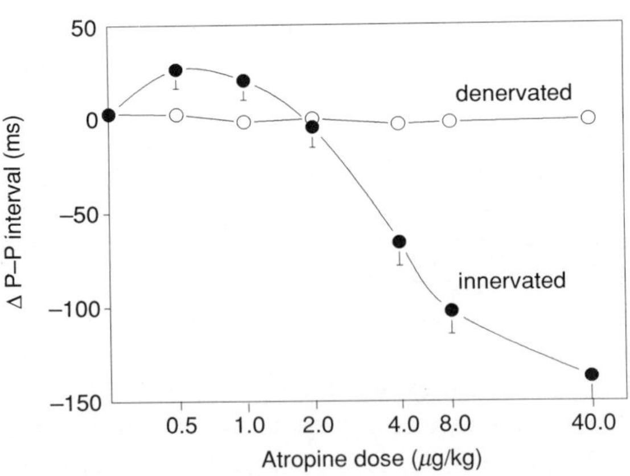

FIG. 8.20. Average P–P intervals of denervated and innervated sinoatrial nodes during sequential atropine doses measured in heart transplant patients. Adapted with permission (A.E. Epstein *et al.* 1990).

direct measurements of resting levels of vagal–cardiac activity with R–R interval shortening after large doses of atropine in different populations of experimental animals or humans. T. Bennett *et al.* (1978) studied diabetic subjects and found a close linear relation between baroreceptor responsiveness (the R–R interval–systolic pressure relation after bolus injections of phenylephrine), and peak minus valley R–R interval changes (Fig. 8.15). Raczkowska *et al.* (1983) made similar observations in a group of healthy subjects; however, the correlation they found was not as close as that described by Bennett and co-workers, whose study population included diabetic patients. These observations suggest that baseline levels of vagal activity are proportional to responsiveness of vagal nuclei to stimulatory inputs, and thus provide indirect validation for the use of respiratory sinus arrhythmia as an index of vagal–cardiac activity.

Adrenergic blockade

Arterial baroreceptor stimulation slows heart rate through increased vagal– and decreased sympathetic–cardiac stimulation in humans (Kilbinger and Löffelholz 1976). Therefore, when beta-adrenergic blocking drugs became available, these agents were used to quantitate sympathetic withdrawal during baroreceptor stimulation. The earliest study of the effects of beta-blockade on baroreflex responses was that of T.G. Pickering *et al.* (1972*b*). They measured R–R interval prolongation following bolus injections of phenylephrine, before and after large intravenous doses of propranolol, and found that slopes of these relations are *increased*, not decreased, by beta-blockade.

These findings were surprising. If sympathetic withdrawal contributes to R–R interval prolongation during arterial pressure elevations, baroreflex slopes should be less, not more, after beta-adrenergic blockade. If sympathetic withdrawal during pressure elevations develops so slowly that it occurs after the period of measurement (Samaan 1935), the slope of the R–R interval–systolic pressure relation should be unchanged, not increased, after beta-adrenergic blockade. Moreover, since the review by Levy (1966) showed that vagal volleys inhibit the sinus node more when sympathetic activity is high, when the effects of sympathetic stimulation are blocked by drugs, responses to vagal volleys should be less, not more.

The observations of Pickering, that acute administration of beta-blockers augment vagally mediated baroreflex responses, have been confirmed by subsequent studies (Eckberg *et al.* 1976; Takeshita *et al.* 1979; Volpe *et al.* 1982). Other indirect evidence suggests that beta-blockers may increase baseline levels of vagal–cardiac restraint. Chess *et al.* (1975) measured heart period power spectra in decerebrate cats and found that propranolol augments power in vagally mediated heart period frequencies. S.H. Gottlieb and Engel (1979) found that propranolol enhances the vagally mediated ability of operantly conditioned conscious monkeys to slow their heart rates. Finally, many (Fouad *et al.* 1984; Coker *et al.* 1984; Zemaityte *et al.* 1984; Alicandri *et al.* 1985; Bittiner and Smith 1988; Kollai and Mizsei 1990) but not all (P. Grossman *et al.* 1990) studies conducted in resting humans show that peak minus valley, respiration-related changes of R–R intervals are augmented after beta-blockade.

The explanation for *increased* vagal effects after beta-blockade is not clear. One possibility is that beta-blockers increase baroreceptor gain. Angell-James and Bobik (1978) in a preliminary communication found that propranolol augments carotid baroreceptor afferent responses to pressure increases in renal hypertensive rabbits. However, Dorward and Korner (1978) conducted similar experiments in renal hypertensive rabbits, and found that beta-blockade leads to small reductions of aortic baroreceptor afferent activity. Another possibility is that beta-adrenergic blockade alters the effects of efferent vagal activity on sinus node function. As mentioned, the degree of inhibition resulting from a volley of efferent vagal–cardiac activity depends on the timing of the arrival of that volley, as well as its strength. Therefore, one potential mechanism for the enhancement of vagal–cardiac inhibition by beta-blockers is that they slow the heart rate and move the timing of phasic vagal volleys to a more favourable position within the cardiac cycle (Stuesse *et al.* 1981). Is not known if enhancement of vagal effects after beta-blockade is mediated centrally or peripherally. Such enhancement occurs both with non-lipophilic beta-blockers such as atenolol that do not enter the brain readily (Neil-Dwyer *et al.* 1981; Kostis and Rosen 1987) and with lipophilic beta-blockers such as metoprolol that do (Bittiner and Smith 1988). We are aware of no recordings of vagal–cardiac nerve traffic before and after beta-blockade.

Whatever the mechanism, the above observations complicate interpretation of the effects of beta-blockers on human baroreflex responses. They indicate that these agents cannot be regarded simply as sinus node post-synaptic beta-adrenergic receptor blockers. It is assumed that bradycardia after beta-blockers is due to beta-blockade. This seems reasonable, since beta-blockers shift sinus node isoproterenol dose–response relations in humans to the right (Cleaveland *et al.* 1972) and reduce the effects of stellate ganglion stimulation in dogs (Jose and Taylor 1969). The studies quoted above, however, suggest that some cardiac slowing after beta-blockade may be due to increased vagal inhibition.

There are other complexities in interpretations of responses to beta-blockade. Large intravenous doses of beta-adrenergic blocking drugs increase splanchnic vascular resistance and plasma noradrenaline levels (J.F. Hansen *et al.* 1978). Although the studies of Best and Halter (1985) with tritiated noradrenaline suggest that increases of plasma noradrenaline are due to reduced clearance rather than to increased spillover, direct measurements of (muscle) sympathetic activity document increased levels of sympathetic traffic after intravenous metoprolol (Sundlöf *et al.* 1983). As discussed in Chapter 6, one mechanism whereby beta-blockade increases muscle sympathetic nerve activity may be that prolongation of diastole increases the period of afferent baroreceptor silence.

Although beta-blockers increase plasma noradrenaline levels in heart failure patients (Binkley *et al.* 1988), they may reduce noradrenaline levels in patients with myocardial infarction. Mueller and Ayres (1980) showed that patients given propranolol within about 10 h of the onset of infarction symptoms experience large reductions of their elevated plasma noradrenaline levels. Patients with congestive heart failure may have extremely high levels of circulating noradrenaline (J.A. Thomas and Marks 1978; Cohn *et al.* 1984; Porter *et al.* 1990), large amounts of muscle sympathetic nerve traffic (Leimbach *et al.* 1986; Porter *et al.* 1990), and increased cardiac noradrenaline spillover (Hasking *et al.* 1986). Surprisingly, as indicated by the limited literature available on this subject, such patients do not have greater cardiac slowing after beta-blockade than do healthy subjects (Eckberg *et al.* 1971).

The above considerations apply to *acute* administration of beta-blockers. The evidence suggests that responses to *chronic* beta-blocker therapy are different from those to acute beta-blockers. Acutely, beta-blockers increase plasma noradrenaline concentrations (J.F. Hansen *et al.* 1978) and raise resting muscle sympathetic nerve activity (Sundlöf *et al.* 1983); chronically (in hypertensive patients), beta-blockers lower muscle sympathetic nerve activity (Wallin *et al.* 1984*a*). Also, augmentation of phenylephrine-induced increases of R–R interval (in hypertensive patients) occur in some (R.D.S. Watson *et al.* 1979) but not all (Takata *et al.* 1986) studies that employed chronic beta-blocker therapy. The reasons for the disparate responses to acute and chronic beta-adrenergic blockade, and for disparate responses among studies involving chronic beta-blockade, are unknown.

Autonomic responses to *clonidine* also illustrate complexities involved with the use of blockers and agonists to gauge efferent autonomic outflow. Clonidine is a pre- and post-synaptic alpha-adrenergic agonist which is thought to lower arterial pressure primarily by reducing sympathetic vasoconstrictor activity (J.V.O. Reid *et al.* 1977; Kobinger 1978). D.S. Goldstein and co-workers (1985) found that single doses of clonidine reduce arterial pressure in proportion to baseline noradrenaline levels, and in proportion to the reduction of plasma noradrenaline levels in hypertensive patients. The mechanism whereby clonidine lowers arterial pressure is not entirely clear, however. Guthrie and Kotchen (1983) showed that a striking (78 per cent) increase of the slope of phenylephrine-induced baroreflex R–R interval prolongation occurs after clonidine. This suggests that augmentation of baroreflex responsiveness contributes to the hypotensive effects of clonidine. Clonidine reduces plasma noradrenaline responses of patients with high cervical spinal cord transections to bladder compression (Mathias *et al.* 1979); this suggests that clonidine also influences spinal–sympathetic mechanisms. However, Wallin and Frisk-Holmberg (1981) measured muscle sympathetic activity directly and correlated reductions of arterial pressure with plasma levels of clonidine. They found that whereas clonidine at all concentrations lowered arterial pressure, it increased muscle sympathetic nerve activity at low concentrations, and reduced sympathetic traffic at high concentrations.

SUMMARY

This chapter catalogues means used for gauging human efferent neural responses to baroreflex forcings. It emphasizes the complexities of the measurements that can be made and it evaluates critically the assumptions underlying their use. The methods discussed have great diversity and power; their availability does much to strengthen the scientific credibility of human baroreflex research.

9

Effector responses to baroreceptor forcings

With the exception of studies in which sympathetic nerve traffic or levels of plasma neurotransmitters are measured, most human baroreflex research concentrates on *effector organ* responses to baroreflex forcings. In such studies, the effector organ is at the end of a cascade of changes imposed at earlier stations in the reflex arc. In some instances, effector responses betray changes imposed earlier; certain effector response patterns may offer the only glimpse available into contributions of earlier segments in the human baroreflex arc. For example, sinus node responses must reflect, in some way, characteristics of baroreceptor transduction. It seems likely that proportionality between sinus node responses and induced changes of arterial distending pressure reflects proportionality of baroreceptor responses. An intuitively attractive extrapolation of this is that thresholds and saturations of sigmoid baroreflex response relations measured from electrocardiograms (Eckberg 1980*a*; Kasting *et al.* 1987) and muscle sympathetic nerve recordings (Rea and Eckberg 1987) reflect thresholds and saturations of baroreceptor neurones. Similarly, as discussed in Chapter 6, rate sensitivity, adaptation, and resetting seen in sinus node responses must in some way reflect rate sensitivity, adaptation, and resetting of baroreceptors.

Notwithstanding these considerations, certain aspects of end organ responses to baroreflex interventions are probably conditioned primarily by characteristics of end organs themselves. These are discussed next.

SINUS NODE

Baroreflex latencies

The response times of individual components of the baroreflex arc have been worked out well in animals. The time required for transduction of abrupt pressure changes and conduction over afferent (myelinated) baroreceptor nerves to the central nervous system in anaesthetized dogs is about 18 ms (Katona *et al.* 1968). The duration of central transmission is variable, and probably depends on the number of synapses and the lengths of neurones interposed in central pathways. Kunze found that in anaesthetized cats, the shortest central latency was 26 ms (Kunze 1972). In animals, vagal–cardiac nerve traffic is carried over small myelinated ('A-delta') fibres whose conduction velocities average about 8 m/s (Iriuchijima and Kumada 1964; Jewett 1964). If the distance from human vagal–cardiac motonuclei to the heart is 25 cm, the contribution of conduction time over vagus nerves to the total baroreflex latency is only about 3 ms. The final latency — the time required for release of acetylcholine, its diffusion across the synaptic cleft, and its influence on sinus node pacemaker cells — is about 120 ms (G.L. Brown and Eccles 1934). Thus in animals the minimum total reflex latency, derived by adding individual component latencies, is (in ms)

arterial transduction (Katona *et al.* 1968)	18
central transmission (Kunze 1972)	26
efferent vagus conduction (G.L. Brown and Eccles 1934)	3
effector response (G.L. Brown and Eccles 1934)	120
Total	167

Data on the durations of *segments* of the human baroreceptor–cardiac reflex arc are limited to responses of ambulatory patients to electrical carotid sinus nerve stimulation (C. Borst and Karemaker 1983) and responses of anaesthetized patients to electrical vagus nerve stimulation. Figure 9.1 (a) and (b) show *vagal*–cardiac reflex responses measured in one dog by G.L. Brown and Eccles (1934), and in two patients during neck surgery by D.L. Eckberg (unpublished); Fig. 9.1 (c) shows *baroreceptor*–cardiac reflex responses measured after abrupt intense neck suction in conscious young volunteers (Eckberg 1976). In these three studies, stimuli were applied at different times in the cardiac cycle, and P–P interval prolongation was plotted as a function of time from onset of stimuli until successive P waves. The data from the dog and the (very limited) data from the two patients suggest that the majority of delay in the baroreceptor–cardiac reflex arc is in the efferent segment. The data obtained with neck suction indicate that the latency of the entire baroreceptor–cardiac reflex in humans is about 240 ms.

Total baroreceptor–cardiac reflex arc duration, as shown in Fig. 9.1

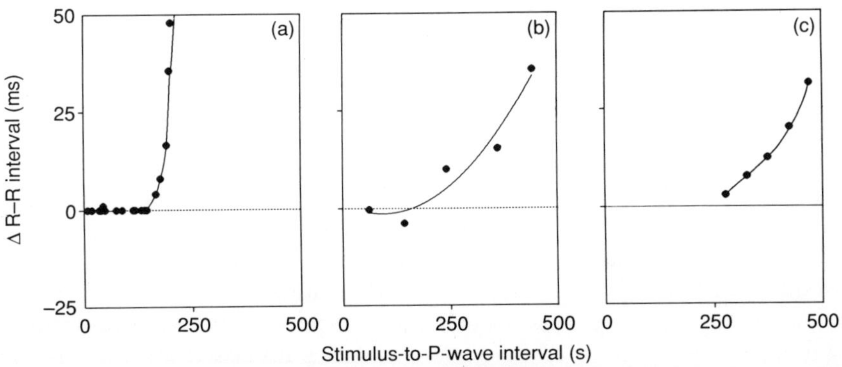

FIG. 9.1. P–P intervals from the beginning of electrical cervical vagus nerve stimulation in (a) one dog and (b) two humans, and (c) neck suction in five humans. Graphs (a) and (c) adapted with permission (G.L. Brown and Eccles 1934; Eckberg 1976).

(right), has been measured by several groups. The first study of this latency was published by Ashman and Gouaux (1937). They examined the electrocardiogram of a patient with complete atrioventricular block and a parasystolic pacemaker, and, with methods not described explicitly, concluded that the latency between baroreceptor stimulation and cardiac responses was about 370 ms. Carlsten and Heyman (1965) applied carotid massage for about 0.5 s in three subjects and measured successive P–P intervals. They recorded latencies of 200 and 400 ms in two of these subjects. Bevegård *et al.* (1967) related P–P intervals to preceding arterial pressure pulses in patients with complete atrioventricular block and found that in one patient, the baroreceptor–cardiac reflex latency was 400 ms.

Later studies proposed somewhat longer reflex latencies. T.G. Pickering and Davies (1973) studied responses of subjects to baroreceptor stimulation provoked by bolus phenylephrine injections and estimated baroreceptor–cardiac reflex latency to be about 475 ms. This value is close to the 500-600 ms that was obtained by C. Borst and Karemaker (1983) who measured delays between the onset of electrical carotid sinus nerve stimulation and sinus node responses in patients with surgically implanted carotid sinus nerve stimulators (see Chapter 5). Reflex latencies measured by Eckberg and co-workers in a succession of articles (Eckberg 1976; Eckberg 1980*a*; Sopher *et al.* 1990) are somewhat shorter, about 240-300 ms, than latencies measured by Pickering and Davies and by Borst and Karemaker.

Reasons for the 200-300 ms disparities among published human baroreceptor–cardiac reflex latencies are not readily apparent. It might be expected that intense neck suction beginning extremely rapidly (with maximum rates of pressure change exceeding 2000 mmHg/s), as used by Eckberg and co-workers (Eckberg *et al.* 1975; Eckberg 1976, 1980*a*) would stimulate more baroreceptor fibres earlier, and define the reflex latency more precisely (Dong and Reitz 1970), than more gradual pharmacological increases of arterial pressure following bolus pressor injections, as used by T.G. Pickering and Davies (1973).

On the other hand, instantaneous mass stimulation of baroreceptor (and probably chemoreceptor) fibres by electrical carotid sinus nerve stimulation, as used by C. Borst and Karemaker (1983), should yield the shortest reflex latencies of all. Probably differences among studies cannot be explained on the basis of the age or clinical diagnoses of subjects studied, because Sopher *et al.* (1990) found baroreceptor–cardiac reflex latencies of about 300 ms in middle-aged patients with congestive heart failure.

We emphasize two points regarding the disparities among measurements of human baroreceptor–cardiac reflex latencies discussed above. First, the magnitude of sinus node inhibition occurring between 240 (the shortest latency recorded) and 500 ms is very small, as shown in Fig. 9.1 (right). Second, *all* studies indicate that reflex delays are sufficiently short in humans for baroreflex stimuli to influence the sinus node in the same heart period in which they occur.

Baroreceptor–*sympathetic* latencies are longer than baroreceptor–*cardiac* latencies. Wallin and Eckberg (1982) delivered abrupt neck suction to healthy young adults and recorded changes of peroneal nerve muscle sympathetic activity, with latencies of about 1.0 to 1.5 s. It is likely that most of this delay is due to slow conduction from sympathetic ganglia over unmyelinated fibres to the recording site at the peroneal nerve; Fagius and Wallin (1980*a*) found that peroneal conduction velocity averages 1.1 m/s in healthy subjects. Fagius *et al.* (1987) measured intervals between electrocardiographic R waves and the peaks of sympathetic bursts occurring in the next cardiac cycle, and found latencies similar to those recorded by Wallin and Eckberg. Interestingly, they documented variability of reflex latencies during autonomic perturbations, including the Valsalva manoeuvre and simulated diving. Animals, whose postganglionic nerves are not as long as those of humans, have much shorter latencies. Kezdi and Geller (1968) found that latencies between carotid sinus pressure changes and renal and splanchnic sympathetic nerve responses lay between 150 and 360 ms in anaesthetized cats.

Latencies between baroreceptor (or electrical carotid sinus nerve) stimuli and systolic pressure changes are about 2 to 3 s (Eckberg 1980*a*; C. Borst and Karemaker 1983). Latencies between baroreceptor stimuli and *diastolic* pressure changes are shorter because diastolic pressures are influenced importantly by the duration of diastole. Since baroreceptor stimuli can easily prolong the same cardiac cycle in which they are applied (see above), they can also influence the diastolic pressure of that interval.

The human sinoatrial and atrioventricular nodes, atrial muscle, atrioventricular node bypass tracts, and Purkinje networks are richly innervated by cholinergic and adrenergic fibres (James and Spence 1966; K.M. Kent *et al.*

1974). As the following discussion indicates, there is a large body of data regarding neurally mediated cardiac responses to baroreflex forcings.

Sinus node responses to single baroreceptor stimuli

Human sinus node potentials cannot be seen on conventional electrocardiograms (West 1955; Paes de Carvalho *et al.* 1959). Therefore baroreflex influences on the sinus node are determined on the basis of changes observed in the morphology of (atrial) P waves. In some subjects (D.L. Eckberg, unpublished), changes of baroreceptor input lead to changes of P wave morphology; these probably betray shifts of cardiac pacemaker location. Although animal research indicates clearly that changes of autonomic input shift the location of atrial pacemaker regions (Meek and Eyster 1913; J.M. Goldberg *et al.* 1973; Schuessler *et al.* 1988), we are unaware of any attempts to map changes of pacemaker location after baroreflex forcings in humans. It seems likely that such shifts occur and go unrecognized.

The landmark study by G.L. Brown and Eccles (1934) was the first to demonstrate that sinus node responses to changes of vagal (and therefore baroreceptor) input are determined importantly by the timing of stimuli within the cardiac cycle. The pattern of sinus node responses of one anaesthetized dog to single vagal volleys is shown in Fig. 9.2 (a) left (Iano *et al.* 1973). This figure illustrates several aspects of sinus node responses to vagal input. First, as discussed above, there is an absolute latency, during which there is no apparent influence of vagal activity on sinus node function (this

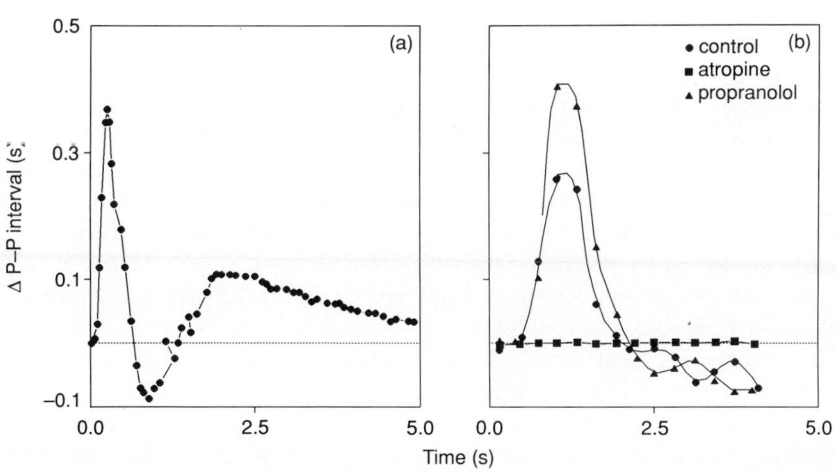

FIG. 9.2. Changes of P–P intervals provoked by (a) electrical vagus nerve stimuli in one anaesthetized dog and (b) brief (0.6 s) intense (60 mmHg) neck suction in seven conscious human subjects. Adapted with permission (Iano *et al.* 1973; Eckberg 1978).

latency is shown also in Fig. 9.1). Second, after the absolute latency is exceeded, sinus node inhibition increases abruptly; as the timing of intense stimuli is changed, there may be no points between absence of responses (during a period of absolute refractoriness) and maximum responses. Third, after a vagal stimulus, the level of sinus node inhibition declines rapidly. Finally, post-inhibitory cardiac acceleration may occur.

Average sinus node responses of seven human volunteers to brief intense neck suction delivered at different times in the cardiac cycle are depicted in Fig. 9.2 (b) (Eckberg 1978). The great similarity between these measurements and those recorded in dogs after electrical vagus nerve stimulation (left), suggests that both response patterns are triggered by local release of acetylcholine. The right panel illustrates the major components of *human* responses to brief baroreceptor stimulation: a latent period, a period of sinus node inhibition, and a period of sinus node facilitation (cardioacceleration). In animals a second period of sinus node inhibition, mediated by reduction of the slope of phase-4 diastolic depolarization (Jalife and Moe 1979) may occur (G.L. Brown and Eccles 1934; Levy *et al.* 1970). Prystowsky and Zipes (1985) observed secondary sinus node inhibition after brief neck suction in one of the subjects they studied.

The right panel of Fig. 9.2 also shows that post-inhibitory R–R interval shortening is not influenced by beta-adrenergic blockade, and therefore is not mediated by adrenergic mechanisms. (This observation was made earlier in animals (G.L. Brown and Eccles 1934; Iano *et al.* 1973; Jalife and Moe 1979) and was also confirmed by Prystowsky and Zipes (1985) in humans.) Cardioacceleration may result in part from post-excitatory depression of baroreceptor firing (Landgren 1952*a*); however, this mechanism would not explain the occurrence of cardioacceleration after electrical carotid sinus (McCloskey and Potter 1981) or vagus nerve stimulation (G.L. Brown and Eccles 1934). Other factors must participate, including complex bidirectional changes of sinus node membrane responses to hyperpolarization (Jalife and Moe 1979). The degree of cardioacceleration is proportional to the degree of prior cardiac inhibition, as described below. Moreover, all R–R interval changes (increases and decreases) after *brief* baroreceptor stimulation are nearly abolished by pretreatment with large atropine doses. Thus the entire sinus node response pattern to brief baroreceptor stimuli is mediated by muscarinic cholinergic receptors.

The time course of decay of sinus node inhibition and post-inhibitory cardioacceleration has relevance to the question of which R–R interval should be paired with systolic pressure pulses during arterial pressure ramps (Chapter 5). For the responses shown in Fig. 9.2 (right), sinus node inhibition lasted until about 2 s after the onset of 0.6 s stimuli. In these subjects therefore, single baroreceptor stimuli exerted inhibitory effects on more than one heart beat, and an offset of either 0 or 1 (Fig. 5.5) would yield a positive relation between baroreceptor stimulus intensity and P–P (or R–R) interval responses. When brief baroreceptor stimuli are delivered to subjects with extremely long baseline R–R intervals, however, an offset of zero may yield an appropriate positive baroreflex slope, but an offset of 1 may yield a negative slope (Eckberg and Eckberg 1982).

Rosenbaum and Lepeschkin (1955) identified clinical counterparts of these experimentally observed phenomena. In patients with high-grade atrioventricular block, phasic inhibition or facilitation of sinus node function occurs, as the timing of ventricular contraction (and arterial pulsation) shifts in relation to sinus node firing.

Dong and Reitz (1970) also studied sinus node responses to electrical vagus nerve stimulation, but used a new analytic method. The plot developed by G.L. Brown and Eccles (1934), shown in Fig. 9.2, does not indicate the timing of sinus node *responsiveness* to vagal inhibition precisely, because delays in the appearance of P waves are included in stimulus-to-P-wave intervals. Dong and Reitz (1970) modified the method of G.L. Brown and Eccles by plotting delays of the appearance of P waves according to when stimuli were delivered before the P waves would have occurred, had no stimuli been delivered. (In the anaesthetized dogs they studied, P–P intervals were nearly constant in the absence of stimuli, and therefore the timing of the occurrence of P waves could be predicted with accuracy.) Figure 9.3 shows plots of (a) responses of one dog to multiple electrical vagus nerve stimuli (Dong and Reitz 1970), and (b) responses of one human volunteer to multiple applications of brief neck suction (Eckberg 1976).

The pattern but not the time course of responses in the two experiments is similar. Responses of the human subject (which were representative of

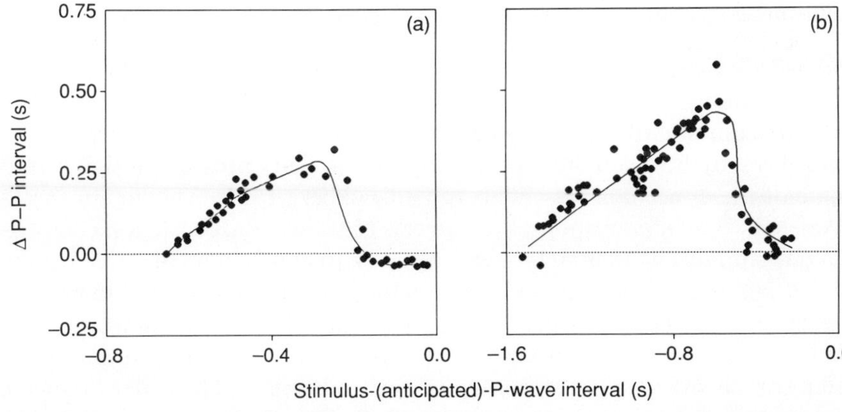

FIG. 9.3. Changes of P–P intervals provoked (a) in one anaesthetized dog by electrical vagus nerve stimuli and (b) in one conscious human subject with brief (0.58 s) intense (60 mmHg) neck suction. Adapted with permission (Dong and Reitz 1970; Eckberg 1976).

those of other subjects) indicate that the sinus node is maximally responsive to vagal inhibition when baroreceptor stimuli occur about 0.8 s before the expected appearance of the next P wave. In the study of Dong and Reitz (1934), delivery of vagal stimuli of great intensity sharpened the definition of the latency (at about −0.2 s), but did not reduce the latency. In both animal and human studies, the difference between the timing of stimuli that yielded the smallest and the largest responses was very small.

The right panel of Fig. 9.3 also illustrates an intriguing aspect of vagal control of the human sinus node. This subject had an extremely long resting P–P interval. Because of this, it was possible to time the delivery of intense baroreceptor stimuli so that the large pulse of acetylcholine released arrived too soon to influence the earliest sinus node depolarization, and too late to influence the next sinus node depolarization (presumably because acetylcholine had been hydrolysed completely by that time). Thus, in subjects with very slow resting heart rates (who may be very responsive to baroreflex inhibition), intense, brief, critically timed baroreceptor stimuli may elicit no sinus node response whatever.

The baroreceptor stimulus–response relation was discussed in Chapter 6. It seems likely that the sigmoid shape of this relation reflects primarily properties of the baroreceptors themselves, rather than properties of the sinus node. Sinus node responses to a very wide range of electrical vagus nerve stimulus frequencies are linear rather than sigmoid functions (Parker *et al.* 1984). Moreover, during spontaneous (Kasting *et al.* 1987) or pharmacologically induced pressure changes (Fritsch *et al.* 1989), this sigmoid relation shifts widely on the R–R interval and pressure axes. This is illustrated in Fig. 6.16.

Sinus node responses to repetitive baroreceptor stimuli

Published evidence indicates that sinus node responses to *repetitive* baroreceptor stimuli are conditioned by the timing of stimuli within the cardiac cycle, the amount of acetylcholine released, and the kinetics of acetylcholine hydrolysis by acetylcholinesterase. Figure 9.4 (left) provides insights into some of these mechanisms (Eckberg and Eckberg 1982). This figure shows average P–P interval responses of subjects to trains of neck suction comprising one stimulus of 10 mmHg, two stimuli of 10 and 20 mmHg, three stimuli of 10, 20, and 30 mmHg, and four stimuli of 10, 20, 30, and 40 mmHg. These data suggest that when baroreceptor stimuli occur as ramps (as they do in healthy humans (Fritsch *et al.* 1986; Parati *et al.* 1988)), responses to later stimuli build upon responses to earlier stimuli. (These data indicate also that post-vagal cardioacceleration, P–P interval shortening at the right of the figure, is proportional to the level of preceding vagal inhibition.)

Figure 9.4 (right) also shows the integral of P–P interval prolongation divided by the integral of neck chamber pressure change (upper panel), and

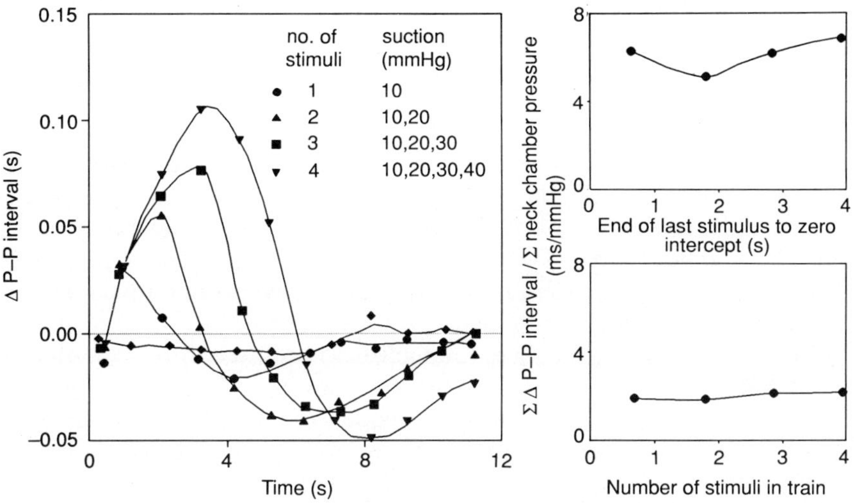

FIG. 9.4. Average P–P interval responses of 18 healthy subjects to trains of neck suction stimuli. See text for explanation. Adapted with permission (Eckberg and Eckberg 1982).

the time from the end of the last neck suction stimulus until the response crossed the zero intercept, as calculated with linear regression (lower panel). These data suggest that for such baroreceptor stimuli (which probably fell on the 'linear' portion of the stimulus–response relation), sinus node responses are simple functions of the amount of baroreceptor stimulation. These data also indicate that in humans, the rate of hydrolysis of acetylcholine is independent of the amount of acetylcholine liberated. In this study, the calculated time required for return of R–R intervals to baseline (not shown) averaged 2.0 s. On the basis of data from animals, Dexter *et al.* (1989) calculated a half-life of 2.7 s for acetylcholine after cessation of electrical vagus nerve stimulation.

Baroreflex hysteresis

In the original article that described the bolus pressor injection technique for stimulating baroreceptors, Smyth *et al.* (1969) recognized that the relation between systolic pressure and R–R intervals is not the same during pressure elevations and pressure reductions. Part of the apparent hysteresis reported in this study is probably due to the multiple effects of the pressor angiotensin-II, as discussed in Chapters 5 and 7.

Similar asymmetries have been observed when responses to rising pressures are compared with responses to falling pressures. Figure 5.4, published by T.G. Pickering *et al.* (1972*c*), shows responses of one subject to a ramped pressure elevation after an intravenous injection of phenylephrine and to a ramped pressure reduction after an injection of nitroglycerine. Based on data derived from responses to neck pressure changes (Eckberg 1980*a*), it seems likely that some of the differences between R–R interval–systolic pressure slopes during pharmacological pressure elevations and reductions results because responses to pressure reductions occur in the threshold region of the stimulus–response relation. Figure 9.5 shows R–R interval responses of one healthy subject to boluses of phenylephrine followed by nitroprusside (T.J. Ebert, unpublished). This loop indicates clearly that instantaneous R–R interval responses to arterial pressure changes depend upon the direction of those changes.

J.M. Fritsch and D.L. Eckberg (unpublished) studied baroreflex hysteresis by applying ramped neck pressure–suction sequences (shown in Fig. 5.13) and mirror-image suction–pressure sequences in healthy volunteers. Responses of two subjects to the pressure–suction sequence (open circles) and its mirror image (closed circles) are shown in Fig. 9.6. In these subjects, R–R intervals during neck pressure reductions tended to be lower than those during neck pressure increases.

Hysteresis of integrated baroreflex responses cannot be ascribed simply to changes occurring at the sinus node, because related changes occur at the level of the receptors themselves (Angell James 1971*a*). Thus, although the occurrence of hysteresis may be apparent in baroreflex responses triggered by vasoactive drugs (Figs. 5.4 and 9.5) and neck pressure changes

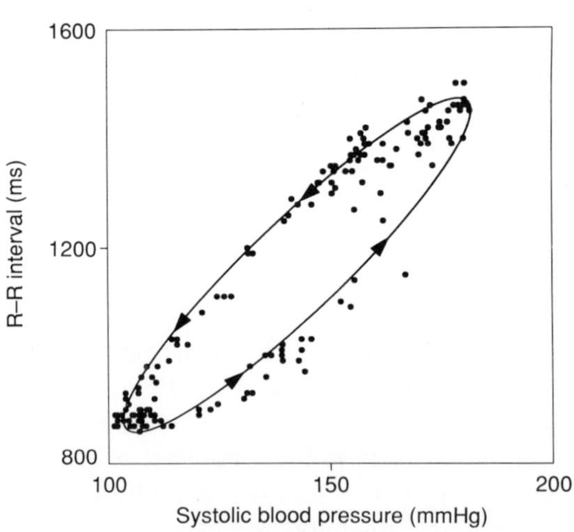

FIG. 9.5. R–R interval changes of one subject after sequential intravenous bolus injections of phenylephrine and nitroprusside. Figure kindly provided by Thomas J. Ebert.

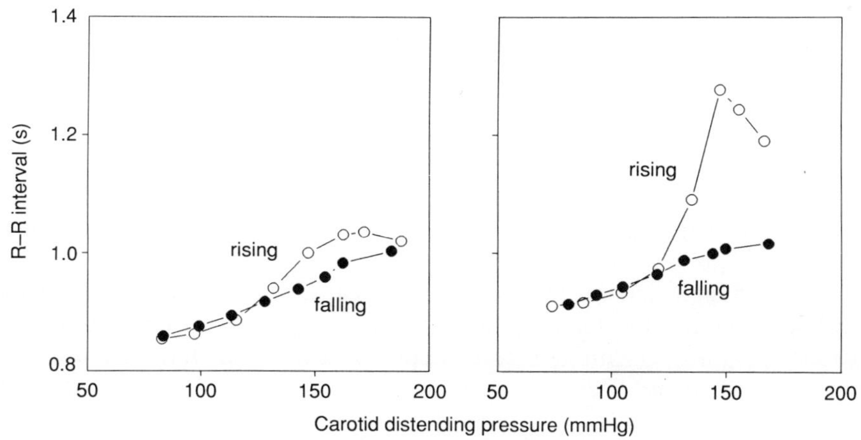

FIG. 9.6. Responses of two subjects to ramped neck pressure sequences shown in Fig. 5.13 (J.M. Fritsch and D.L. Eckberg, unpublished).

(Fig. 9.6), the proportion of hysteresis attributable to baroreceptors, the central nervous system, sinus node, or other factors is unknown.

ATRIOVENTRICULAR NODE

Several studies (Rossi 1954; James 1961; K.M. Kent *et al.* 1974) point clearly to rich sympathetic and cholinergic innervation of human atrioventricular nodes. Data regarding baroreceptor influences on atrioventricular node function in humans comes from responses to carotid massage, pharmacological alterations of arterial pressure, the Valsalva manoeuvre, neck chamber pressure changes, and electrical carotid sinus nerve stimulation.

In humans as well as animals, the most important influence of baroreceptors on atrioventricular node function is mediated by changes of heart rate. Since atrioventricular node conduction time is related inversely to heart rate (Damato *et al.* 1969; Ferrier and Dresel 1973; Loeb *et al.* 1985), baroreceptor mediated changes of heart rate influence atrioventricular node conduction in ways that oppose those related to release of neurotransmitters. Although carotid sinus massage, which massively stimulates baroreceptors, may provoke atrioventricular block (Lown and Levine 1961; Castellanos *et al.* 1974; Glasser *et al.* 1974), the effects of changes of baroreceptor activity on atrioventricular node function are seen most clearly when concomitant fluctuations of sinus node cycle length are prevented by atrial pacing. However, in animals, electrical vagus nerve volleys can be timed critically so that stimuli occur too late in the cardiac cycle to influence sinus node firing, but in time to slow atrioventricular node conduction (P. Martin 1977;

Jalife and Moe 1983). In many cases in which carotid sinus massage prolongs human atrioventricular conduction, sinus arrest occurs also.

An early important study of baroreflex modulation of the human atrioventricular node was published by Mancia *et al.* (1979*a*), who measured the effects of phenylephrine and nitroglycerine on sinus and atrioventricular nodes in patients during diagnostic cardiac catheterization. They found that elevations and reductions of arterial pressure provoke only minimal changes of atrioventricular conduction (atrial-to-bundle-of-His, or A–H intervals) when sinus node cycle length is allowed to fluctuate. During atrial pacing however, similar arterial pressure changes cause substantial A–H interval prolongation and shortening. Figure 9.7 depicts responses of one subject to arterial pressure elevations before (upper tracings) and during (lower tracings) atrial pacing. Without atrial pacing, an arterial pressure elevation of 21 mmHg prolonged the A–A (sinus) intervals by 250 ms, but did not alter atrioventricular node conduction. During atrial pacing however, an arterial

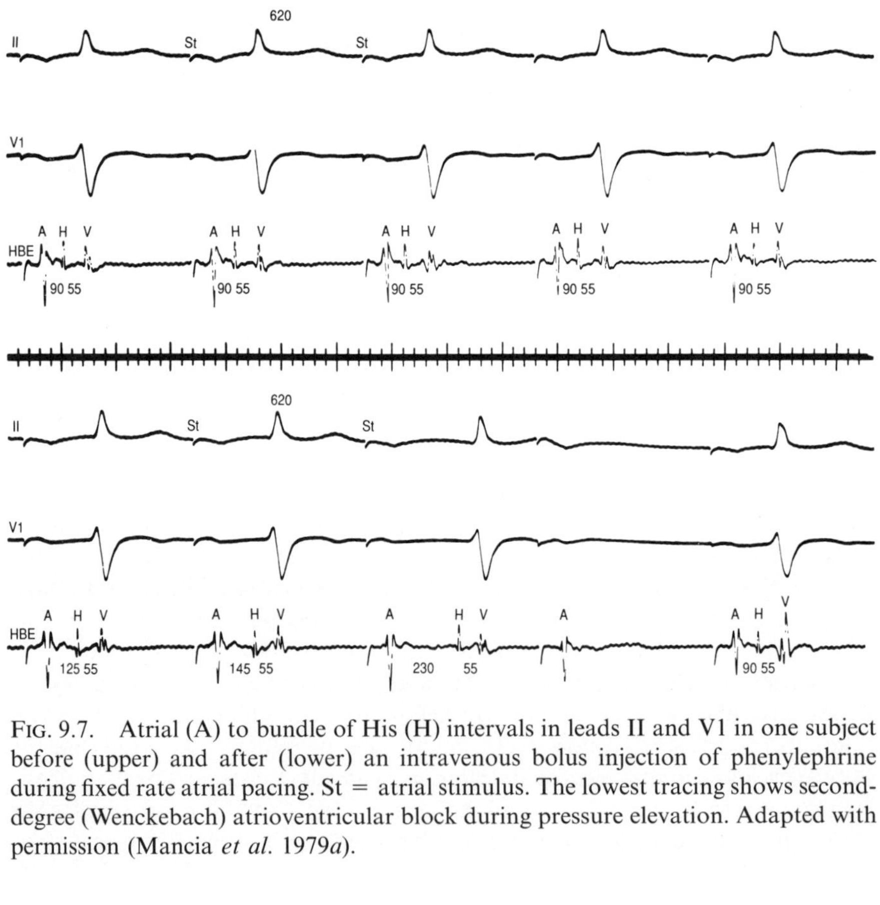

FIG. 9.7. Atrial (A) to bundle of His (H) intervals in leads II and V1 in one subject before (upper) and after (lower) an intravenous bolus injection of phenylephrine during fixed rate atrial pacing. St = atrial stimulus. The lowest tracing shows second-degree (Wenckebach) atrioventricular block during pressure elevation. Adapted with permission (Mancia *et al.* 1979*a*).

pressure elevation of 28 mmHg led to progressive (Wenckebach, or Mobitz type I) atrioventricular block.

Mancia *et al.* (1979*a*) found no change of His-Purkinje-to-ventricle (H–V) intervals during changes of arterial pressure. However, Castelanos *et al.* (1974) found that intraventricular conduction may be slowed by carotid massage, and Markel *et al.* (1988) found in patients with Mobitz type II atrioventricular block, muscarinic blockade with atropine shortens the atrial cycle length at which block occurs. These results would be predicted on the basis of the large subendocardial concentrations of acetylcholinesterase present in human ventricular myocardium (see below).

Mancia *et al.* (1979*a*) also studied *autonomic mediation* of the changes of atrioventricular node conduction they measured. In three patients, large intravenous doses of atropine sulphate shortened baseline R–R intervals and virtually abolished baroreflex-induced changes of sinoatrial rate and atrioventricular conduction. This finding suggests that in humans, changes of atrioventricular node conduction provoked by brief arterial pressure changes are mediated primarily by muscarinic cholinergic mechanisms.

Parasympathetic modulation of atrioventricular node conduction was studied also by Das *et al.* (1975), who measured P–R interval changes after graded doses of intravenous atropine sulphate, with and without atrial pacing. Without pacing, small doses of atropine prolonged P–P intervals (by increasing vagal–cardiac nerve traffic — see Chapter 8) and shortened P–R intervals. In this circumstance, P–R intervals probably shortened because the slower heart rate allowed more time for recovery of the atrioventricular node. Large doses of atropine, with or without atrial pacing, also led to shortening of P–R intervals, probably by blocking vagal inhibitory effects on the atrioventricular node.

Others found evidence for sympathetic as well as parasympathetic modulation of human atrioventricular conduction. C. Borst *et al.* (1982*a*) suggested that some prolongation of atrioventricular node conduction caused by electrical carotid sinus nerve stimulation is due to sympathetic withdrawal, since it is lessened by beta-blockade. Support for this conclusion comes from the studies of Prystowsky (1981) and Cinca (1985) and their co-workers, who showed that intravenous beta-adrenergic blockade or left stellate ganglion anaesthesia slows atrioventricular conduction.

Ferrari *et al.* (1981*a*) measured atrioventricular node conduction times during isometric handgrip before and after beta-adrenergic or cholinergic blockade, with and without fixed-rate atrial pacing. They found that shortening of atrioventricular conduction time during handgrip is reduced by both beta-adrenergic and cholinergic blockade. They concluded that in humans, the effects of sympathetic activity on the atrioventricular node are small at rest, but are much larger when sympathetic traffic is increased above baseline levels.

Two human studies examined the proportionality between atrioventricular node conduction changes and baroreceptor stimuli. Ferrari *et al.* (1983) raised or lowered arterial pressure with bolus injections of phenylephrine or nitroglycerine, and documented linear proportionality between atrioventricular conduction time and systolic pressure. Butrous *et al.* (1987) made similar measurements, but stimulated baroreceptors physiologically with Valsalva manoeuvres. In some patients in both studies, elevations of arterial pressure led to second degree (Wenckebach or Mobitz type I) block.

These studies indicate that sympathetic as well as vagal traffic modulates conduction in the human atrioventricular node, and that in resting humans, arterial pressures lie above the baroreceptor stimulus–atrioventricular node sympathetic and vagal response relation thresholds.

Several studies have estimated human baroreceptor–atrioventricular node *latencies*. As mentioned, Ferrari *et al.* plotted changes of atrioventricular node conduction with latencies of one heart beat (that is, with an offset of 1 (T.G. Pickering and Davies 1973)), and concluded that baroreceptor stimuli influence atrioventricular node function in the next cardiac interval. A subsequent study from the same group confirmed this finding, but showed also that latencies vary among subjects (Ferrari *et al.* 1987). C. Borst and Karemaker (1983) measured atrioventricular node conduction changes provoked by electrical carotid sinus nerve stimuli, and found latencies of 0.8 to 1.4 s in one subject, and 0.9 to 1.6 s in the other. They also found that in one subject, a large atropine sulphate dose prolonged the latency to about 3.0 s.

Welch *et al.* (1986) measured carotid baroreceptor–atrial and –atrioventricular node latencies during abrupt 5 s applications of neck suction in patients undergoing electrophysiological study. Figure 9.8 shows average sinus node

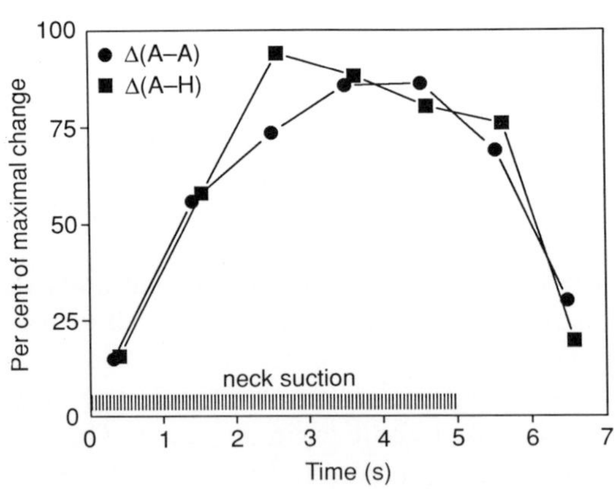

FIG. 9.8. Average changes of atrial intervals without atrial pacing (A–A), and atrial-to-bundle-of-His intervals during fixed rate atrial pacing (A–H) provoked by 5 s of 60 mmHg neck suction. Figure kindly supplied by William J. Welch.

responses (A–A intervals) without atrial pacing and atrioventricular node conduction time (atrial-to-bundle-of-His, or A–H intervals) during atrial pacing. (As discussed above, autonomic effects on atrioventricular node conduction are most apparent when changes of atrial cycle length are prevented.) In this figure, responses are normalized and plotted as percentages of maximum changes. These unique data suggest that in humans, the time course of baroreflex influences on sinus and atrioventricular nodes is similar. Calculated latencies averaged 0.26 s for baroreceptor–sinus node effects, and 0.32 s for baroreceptor–atrioventricular node effects. Pace and Martin (1985) found similar short latencies from the onset of electrical carotid sinus nerve stimulation until P–P or P–R interval prolongation in anaesthetized dogs. These data indicate that in animals and humans, changes of baroreceptor afferent activity can influence the atrioventricular node in the *same* cardiac cycle in which the change of arterial pressure occurs.

As discussed above, the timing of baroreceptor stimuli within the cardiac cycle determines the magnitude of sinus node responses. Studies in animals indicate that timing of vagal stimuli also determines atrioventricular node responses (Mazgalev *et al.* 1986; Salata *et al.* 1986). We are aware of no studies of the importance of timing of baroreceptor stimuli on atrioventricular nodal conduction in humans.

Variability of baroreceptor–atrioventricular node latencies has been studied in animals but not in humans. Jalife and Moe (1983) showed that atrioventricular node latency is a function of baseline sinus cycle length, and can be less than 100 ms when underlying sinus node cycle lengths are long. Respiratory modulation of atrioventricular node conduction, documented in animals (M.R. Warner and Loeb 1986; M.R. Warner *et al.* 1986), has been studied once in humans. Rawles *et al.* (1989) used cosinor analysis (Halberg *et al.* 1972) to identify respiratory effects on P–R intervals (without atrial pacing) in 70 patients and healthy subjects during quiet uncontrolled breathing. They found respiration-related variations of P–R intervals in 39 per cent of subjects, and found that the time course of P–R interval changes paralleled that of P–P interval changes. The longest P–R intervals occurred at the end of expiration, and the shortest occurred at the end of inspiration. Baroreflex modulation of dysrhythmias involving the atrioventricular node is discussed in Chapter 13.

Normal baroreflex control of atrioventricular node conduction can masquerade as pathology. O'Toole and co-workers (1984) showed that in anaesthetized dogs during fixed-rate atrial pacing, changes of arterial pressure can provoke atrioventricular node block in the absence of pathology. Butrous *et al.* (1987) speculate that in some patients with sinus node disease, loss of autonomically mediated fluctuations of P–P intervals may unmask autonomic effects on the atrioventricular node. Such patients may have normal autonomic modulation of atrioventricular conduction which presents as atrioventricular block.

MUSCLE SYMPATHETIC NEURONES

The strictures that apply to research into human *vagal* mechanisms do not apply to human *sympathetic* mechanisms. Although changes of human vagal activity must be inferred from electrocardiographic changes, changes of sympathetic traffic to skin and muscle vascular beds can be measured directly. This capability reduces the need to infer changes of sympathetic activity from changes of blood pressure or peripheral vascular resistance, neither of which is interchangeable with sympathetic nerve activity. Fluctuations of blood pressure reflect changes of both cardiac output and vascular resistance. Since cardiac output is determined by heart rate as well as stroke volume, interventions that lead to transient vagally mediated changes of heart rate may alter blood pressure independent of changes of sympathetic activity. Moreover, changes of vascular resistance may not reflect changes of sympathetic traffic accurately because vascular responsiveness is determined by vessel wall architecture (Folkow 1982) as well as sympathetic traffic.

In part because of methodological difficulties, some earlier studies suggested that arterial baroreceptors exert little or no influence on vascular resistance (Ernsting and Parry 1957; Bevegård and Shepherd 1966*a*; Beiser *et al.* 1970; Abboud *et al.* 1979). This conclusion was contradicted by the earliest studies that employed sympathetic recordings (Delius *et al.* 1972*a*; Wallin *et al.* 1975; Sundlöf and Wallin 1978*b*), which documented clear and significant inverse relations between blood pressure and muscle sympathetic nerve traffic. (*Skin* sympathetic traffic is a separate case and is discussed below.)

Direct recordings of muscle sympathetic nerve activity and R–R intervals provide strong inferential evidence that both sympathetic and vagal motonuclei receive beat-by-beat information regarding arterial pressure. This was shown convincingly by Wallin and Nerhed (1982) who triggered a signal-averaging circuit on muscle sympathetic bursts and recorded arterial pressure and the electrocardiogram. Changes of diastolic pressure after sympathetic bursts in three subjects are shown in Fig. 9.9, and changes of R–R intervals in Fig. 9.10. Wallin and Nerhed suggested that blood pressure and R–R intervals change in opposite directions after a sympathetic burst because both are responding to reduction of arterial baroreceptor input.

Changes of muscle sympathetic traffic are inversely proportional to changes of blood pressure

Inverse proportionality between arterial pressure and sympathetic activity is easiest to demonstrate when pressure changes are seen by all baroreceptor areas and are short-lived, and most difficult to demonstrate when pressure changes are seen by only one baroreceptor area and are sustained.

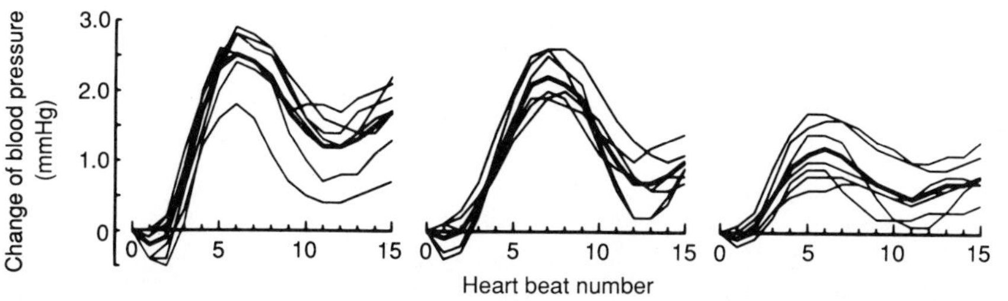

FIG. 9.9. Changes of diastolic pressures after the peak of a sympathetic burst, in three subjects. Reproduced with permission (Wallin and Nerhed 1982).

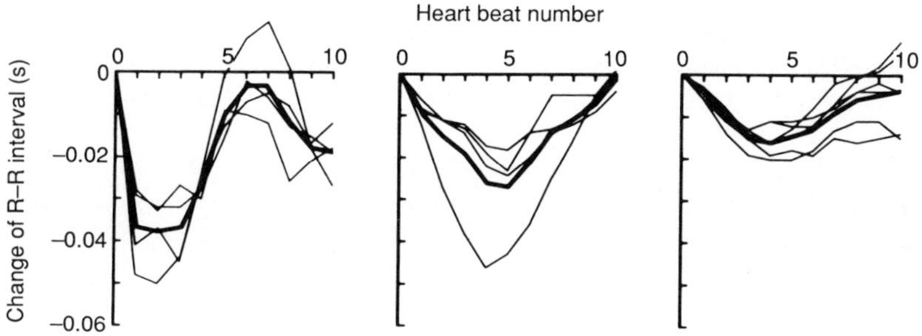

FIG. 9.10. Changes of R–R intervals after the peak of a sympathetic burst, in three subjects. Reproduced with permission (Wallin and Nerhed 1982).

The earliest study to map out the relation between arterial pressure and sympathetic traffic was that of Wallin *et al.* (1974*b*), who measured muscle sympathetic nerve activity in subjects with large variability of diastolic pressure secondary to respiratory sinus arrhythmia or atrial fibrillation. Wallin elaborated baroreceptor stimulus–sympathetic nerve response relations, which he called 'threshold variability diagrams', in which the percentage of diastolic pressures associated with sympathetic bursts was plotted as a function of diastolic pressure. Reverse sigmoid relations between diastolic pressure and muscle sympathetic nerve activity from two patients with atrial fibrillation are shown in Fig. 9.11. In these patients there appeared to be a pressure threshold, above which no sympathetic bursts occurred, a linear range, and a saturation pressure, below which no additional increases of sympathetic bursts could occur.

Sundlöf and Wallin (1978*b*) extended these observations by determining the percentages of diastolic, mean, systolic, and pulse pressures that were associated with sympathetic bursts. These findings, summarized in Fig. 9.12, documented inverse relations between diastolic and mean arterial pressures and sympathetic bursts, and a *direct* relation between pulse pressure and sympathetic activity. Sundlöf and Wallin postulated that a positive relation

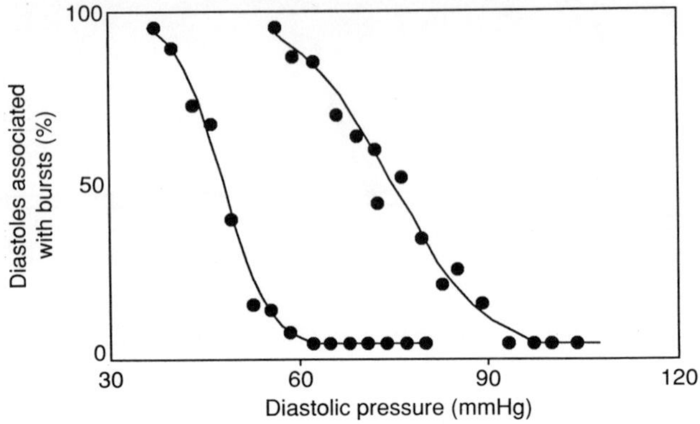

FIG. 9.11. Percentage of heart beats associated with muscle sympathetic bursts in two patients with atrial fibrillation. Adapted with permission (Wallin *et al.* 1974*b*).

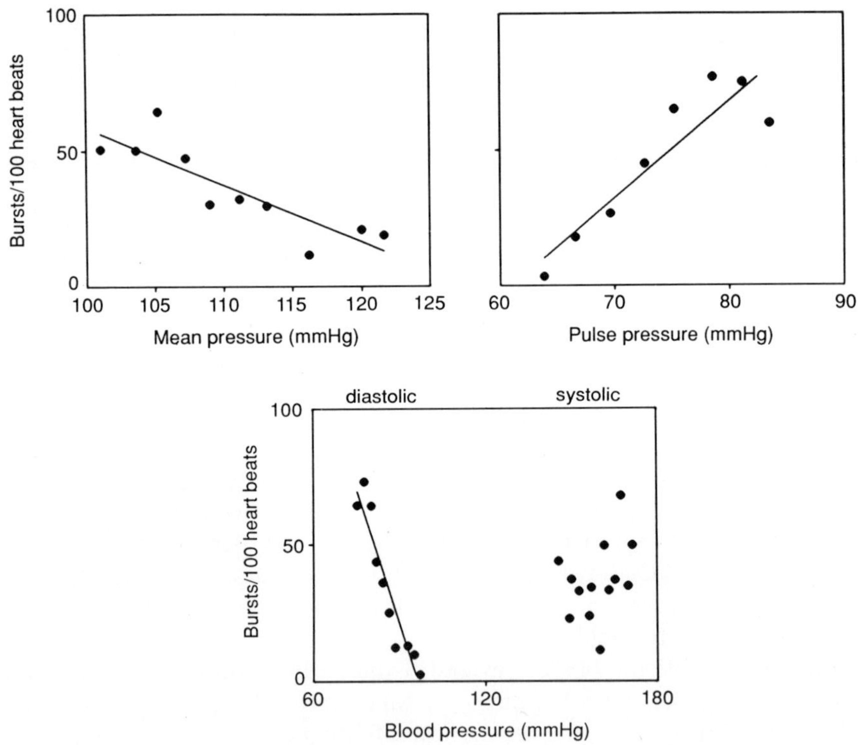

FIG. 9.12. Percentage of heart beats associated with muscle sympathetic bursts in 16 subjects, plotted according to preceding arterial pressures. Adapted with permission (Sundlöf and Wallin 1978*b*).

exists between pulse pressure and sympathetic activity because heart beats with the largest pulse pressures also have the lowest diastolic pressures. Importantly, they found *no* correlation between *systolic* pressure and sympathetic activity. They explained this by saying that afferent baroreceptor input suppresses muscle sympathetic nerve activity completely; sympathetic firing occurs during late diastole, when baroreceptor activity is absent (Chapter 6); and proportionality between sympathetic firing and diastolic pressure exists because the duration of baroreceptor silence is longer when diastolic pressure is lower.

Other studies document inverse sigmoid relations between afferent baroreceptor and efferent sympathetic activities. The relation between antecubital vein plasma noradrenaline, used as an indirect measure of sympathetic activity, and diastolic pressure is depicted in Fig. 10.5. To the extent that antecubital vein plasma noradrenaline levels reflect sympathetic activity (this controversial issue is discussed in Chapter 8), this study suggests that sympathetic activity does not simply remain at a plateau as diastolic pressure falls. After peak levels are attained, plasma noradrenaline levels *decline* as diastolic pressure falls further. Although this may the first suggestion in humans of *reductions* of sympathetic activity as diastolic pressure is reduced progressively, the phenomenon has been remarked (but with actual sympathetic nerve recordings) in conscious dogs during non-hypotensive haemorrhage by Morita and Vatner (1985), and in conscious rabbits during perivascular occlusion by Dorward *et al.* (1985).

Finally, Rea and Eckberg (1987) studied sympathetic responses to changes of pressure seen by only the carotid baroreceptor areas; they applied brief steps of neck suction and pressure and found an inverse sigmoid relation between muscle sympathetic nerve activity and carotid distending pressure (systolic pressure less neck chamber pressure). These results are depicted in Fig. 10.4 (upper left panel) and are very similar to the results of Wallin *et al.* (1974*b*) shown in Fig. 9.11. This study and the others quoted above, indicate clearly that *brief changes* of arterial pressure elicit reciprocal changes of muscle sympathetic nerve activity. These results do not address the question of how *sustained* changes of pressure influence sympathetic activity. This issue is discussed next.

Relation between sustained changes of carotid baroreceptor input and sympathetic activity

Wallin and Eckberg (1982) studied the time course of sympathetic responses during continuous 5 s step changes of afferent carotid baroreceptor activity. All responses of one subject to ten applications of 30 mmHg neck pressure are shown in Fig. 9.13. (Fig. 5.15 shows a similar plot of all responses of one subject to neck suction.) For this figure, data sweeps were triggered by the onset of neck pressure (which was in turn triggered by electrocardio-

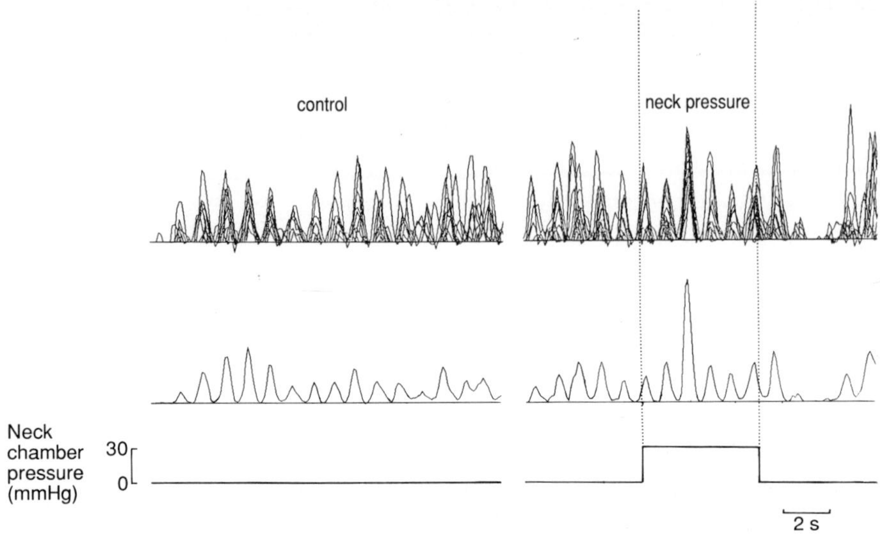

FIG. 9.13. Muscle sympathetic nerve activity in one subject before and during ten applications of neck pressure. Reproduced with permission (Wallin and Eckberg 1982).

graphic R waves). The upper tracing shows all muscle sympathetic bursts during ten applications of neck pressure superimposed upon one another, and the middle tracing shows computer-averaged responses. Although neck pressure increased muscle sympathetic nerve activity, the increase was limited to one burst. Sympathetic activity returned to baseline levels *during* neck pressure (perceived as hypotension), and was suppressed completely when neck pressure returned to ambient levels (far right).

Although these recordings were made in one peroneal nerve, the changes are probably representative of those occurring in other muscle sympathetic nerves (Sundlöf and Wallin 1977). The sympathetic silence that followed sudden stretches of carotid sinus baroreceptors at the release of neck pressure (*no* bursts occurred during ten periods of observation) suggests that *all* muscle sympathetic motoneurones are inhibited by arterial baroreceptor input from one receptive field (in this case, the carotid sinuses). Sympathetic silence also constitutes inferential evidence that aortic and carotid baroreceptors influence the same muscle sympathetic motoneurone pool; if aortic baroreceptors controlled different motoneurones, all sympathetic traffic would not have been suppressed by sudden increases of carotid sinus baroreceptor input. There was no evidence from this study that some muscle sympathetic motoneurones were baroreflex-insensitive (Gebber *et al.* 1973).

Vagally mediated *R–R interval* responses are not shown in Fig. 9.13 (or Fig. 5.15). As expected, neck suction increased, and neck pressure decreased,

R–R intervals. In contrast to sympathetic responses, however, R–R interval responses were *sustained* (Wallin and Eckberg 1982). These results may point in part to differences in central processing of baroreceptor information. The rapid decay of sympathetic responses suggests that sympathetic motoneurones respond more to changes than to absolute levels of arterial pressure. The persistence of vagal responses suggests that vagal–cardiac motoneurones respond simply to the level of afferent baroreceptor input. These conclusions reached from human studies are remarkably similar to conclusions drawn from more direct data by Kollai and Koizumi (1979), who measured sympathetic–cardiac (not sympathetic–*muscle*) and vagal–cardiac nerve traffic directly in anaesthetized dogs. Intravenous injection of a bolus of noradrenaline led to a sustained, slowly adapting increase of vagal–cardiac traffic and an evanescent reduction of sympathetic–cardiac traffic. A sample tracing from this study is shown in Fig. 9.14.

The most sustained blood pressure changes studied in humans are those associated with hypertension. In a series of articles (Wallin *et al.* 1973*a*;

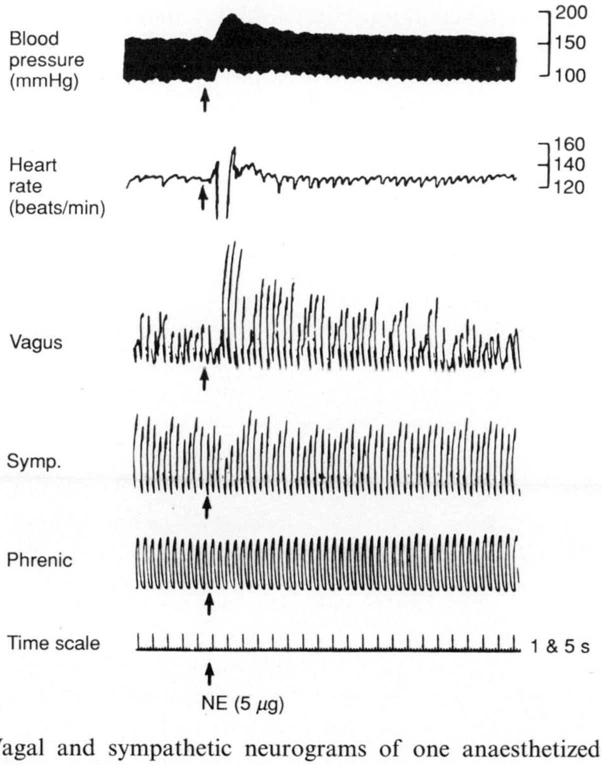

FIG. 9.14. Vagal and sympathetic neurograms of one anaesthetized dog during bolus injection of noradrenaline (arrow). Reproduced with permission (Kollai and Koizumi 1979).

Sundlöf and Wallin 1978*c*; Mörlin *et al.* 1983), Wallin and co-workers reported no significant relation between resting arterial pressure and muscle sympathetic nerve activity. Their findings support the conclusion of Cowley and associates (1973) (discussed and disputed in Chapter 11) that baroreceptors respond to changes, but do not establish absolute levels, of arterial pressure. More recent evidence, also discussed in Chapter 11, indicates that in some hypertensive subjects, particularly those with 'borderline' hypertension, baseline levels of muscle sympathetic nerve activity are supranormal (Y. Yamada *et al.* 1988; E.A. Anderson *et al.* 1989).

Baroreceptor–sympathetic responses depend on the past history of sympathetic activity

The possibility that past blood pressures influence ongoing sympathetic outflow was studied elegantly by Sundlöf and Wallin (1977). They evaluated the relation between diastolic pressure and sympathetic bursts according to whether diastolic pressure was falling or rising. Sundlöf found that for all diastolic pressures, the likelihood that a sympathetic burst would occur was greater when pressure was falling than when it was rising.

Other evidence suggests that pre-existing levels of sympathetic traffic determine sympathetic responses to changes of baroreceptor input. Båth *et al.* (1981) measured muscle sympathetic nerve activity during sinusoidally varying neck pressures between ambient and −40 mmHg. Since neck suction is perceived as hypertension, sinusoidal neck suction would be expected to *reduce* muscle sympathetic nerve activity. Båth found that volunteers with high levels of sympathetic activity reduced sympathetic traffic during sinusoidal neck suction, as expected. However, she also found that volunteers with *low* levels of sympathetic activity *increased* their sympathetic traffic. Båth explained her findings on the basis that muscle sympathetic motoneurones follow trends rather than absolute levels of arterial pressure. During sinusoidal neck suction, afferent baroreceptor activity is increasing as neck pressure falls, but is *decreasing* as neck pressure rises to ambient levels. Subjects with high levels of baseline sympathetic activity are able to reduce their sympathetic outflow substantially as carotid distending pressure increases. However, subjects with low levels of baseline sympathetic activity may not be able to reduce their sympathetic activity further, but can increase their sympathetic traffic importantly as carotid distending pressure declines (that is, as neck suction is reduced).

Sympathetic oscillators are entrained by arterial baroreceptors

There is substantial evidence from animals that sympathetic motoneurones do not fire in simple inverse relation to arterial pressure; rather, spinal sympathetic motoneurone pools are seen as independent oscillators that under usual conditions are *entrained* by baroreceptors (M.I. Cohen and Gootman 1970; D.G. Taylor and Gebber 1975; Gebber and Barman 1980). Several studies suggest that human sympathetic motoneurones are also entrained by baroreceptor inputs. In healthy resting subjects, termination of each sympathetic burst is thought to result from a barrage of afferent baroreceptor input caused by the preceding arterial pulse (Fagius 1988). Permanent loss of baroreceptor entrainment is seen in patients with high complete spinal cord transections. In such patients, muscle sympathetic bursts have the appearance of skin sympathetic bursts (which at usual temperatures are not related to arterial pressure). In intact subjects, muscle sympathetic bursts are short, do not increase after arousal stimuli, occur without temporal relation to skin sympathetic bursts, and have clear cardiac periodicities. In spinal patients, muscle sympathetic bursts are long, increase during arousal, may occur in synchrony with skin sympathetic bursts, and are not time-locked to cardiac activity (Stjernberg *et al.* 1986).

Fagius *et al.* (1985) showed that baroreceptor entrainment of muscle sympathetic nerve activity can be interdicted experimentally in healthy subjects. These authors injected local anaesthetic at the base of the skull in two volunteers and interrupted afferent baroreceptor activity and afferent and efferent vagal activity. After these injections, muscle sympathetic bursts looked and behaved like skin sympathetic bursts; this is shown in Fig. 9.15.

Loss of entrainment of sympathetic motoneurones by baroreceptors is thought to occur also during abrupt reductions of systemic arterial pressure, such as those that occur after premature ventricular beats (Wallin *et al.* 1974*b*; Fagius 1988; Welch *et al.* 1989). Muscle sympathetic bursts following premature beats bear no fixed temporal relation to the R wave of the

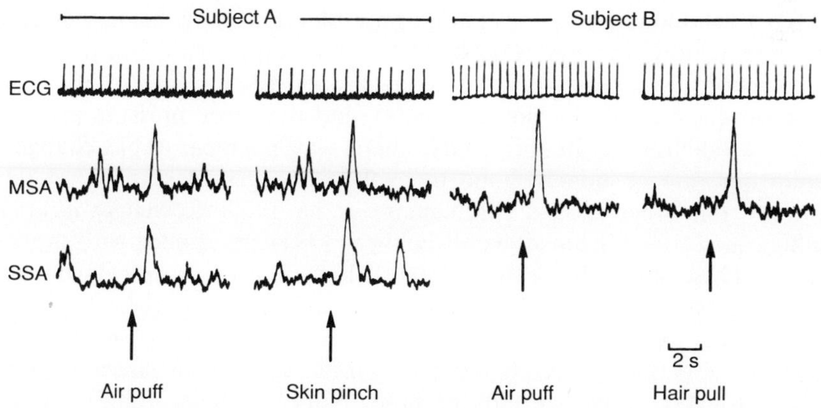

FIG. 9.15. Muscle (MSA) and skin (SSA) sympathetic nerve activity in two subjects after blockade of afferent carotid sinus nerves with injection of local anaesthetic at the base of the skull. Reproduced with permission (Fagius *et al.* 1985).

preceding beat (Fagius 1988), tend to be larger and longer than bursts occurring during sinus rhythm (Welch *et al.* 1989), and have the appearance of skin sympathetic bursts. With profound hypotension, such as that which occurs during rapid ventricular tachycardia, muscle sympathetic nerve activity may lose its bursting pattern altogether. This is shown in Fig. 13.6 (middle and right panels).

SKIN SYMPATHETIC NEURONES

Human skin sympathetic nerve activity is thought to be uninfluenced by baroreceptors; however, this issue is complex, and published studies are contradictory. Several studies in animals suggest that skin sympathetic outflow may be modulated by baroreceptors under unusual circumstances. Horwitz *et al.* (1979) reported that in anaesthetized cats, elevation of carotid sinus pressure reduces sweat gland potentials. Jänig and co-workers (Gregor *et al.* 1977; Jänig and Kummel 1977) identified cardiac periodicities in vasoconstrictor fibres supplying cats' hindpaws and hindlimbs. As discussed above, cardiac periodicity in muscle sympathetic outflow is thought to represent entrainment of sympathetic motonuclei by baroreceptors. (However, a baroreceptor influence may exist without cardiac periodicity; Kollai and Koizumi (1980) showed that postganglionic cardiac or vertebral sympathetic nerves which do *not* have cardiac periodicity may nevertheless be inhibited by increased carotid sinus pressure.)

There are several studies in which changes of skin sympathetic activity during arterial pressure changes were inferred from indirect measurements. With the exception noted, these studies suggest that skin blood flow is related inversely to arterial pressure (Crossley *et al.* 1966). Duprez *et al.* (1987) measured finger plethysmographic blood flow in healthy humans during reductions and elevations of neck chamber pressure. They found that neck suction provokes consistent finger vasoconstriction (a response which was reproducible and did not habituate), and that neck pressure provokes finger vasodilatation. In this study, there was no measurable change of respiratory rate or tidal volume to explain changes of skin blood flow. Results from other groups (McNamara *et al.* 1969; Rowell *et al.* 1973; Kellogg *et al.* 1990; Schmedtje and Eckberg 1991) are at odds with those of Duprez. These data indicate that moderate lower body suction (which lowers arterial pressure, and therefore reduces arterial baroreceptor traffic) transiently reduces hand or forearm skin blood flow. The study of Kellog *et al.* (1990) suggests that reduced baroreceptor input causes skin vasoconstriction by withdrawing active vasodilator tone. None of these studies involved measurement of responses to lower body *positive* pressure on skin flow; therefore they have may not excluded a non-specific arousal response to the unusual intervention of lower body suction. Figure 9.16 shows average

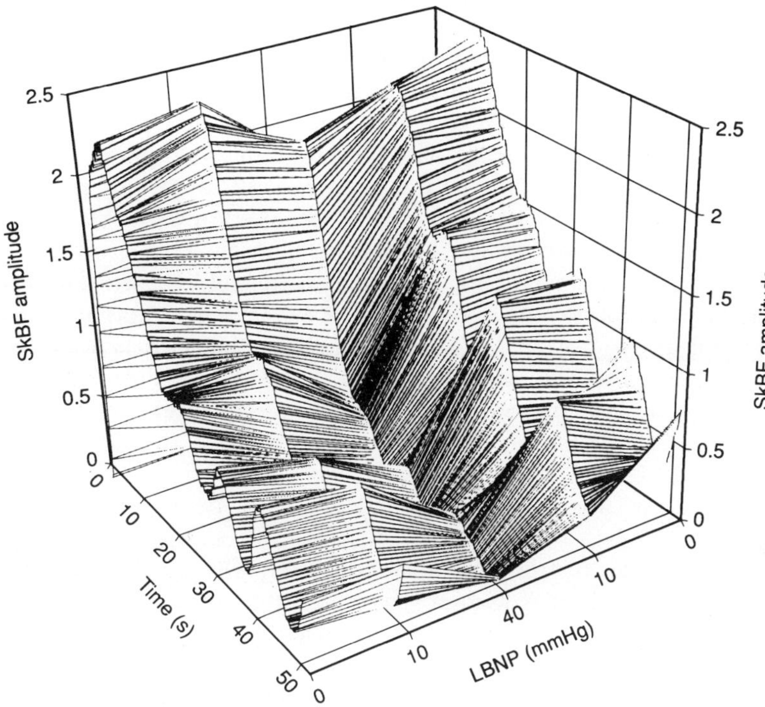

FIG. 9.16. Average responses of healthy subjects to onset and offset of ramped lower body suction. Skin blood flow (SkBF) declined in proportion to the intensity of lower body suction (LBNP). Postocclusive reactive hyperaemic skin blood flow oscillations were preserved. Reproduced with permission (Schmedtje and Eckberg 1991).

postocclusive skin reactive hyperaemia of nine healthy subjects to the onset and offset of sequential, ramped lower body suction. This study documented reductions of skin blood flow that were proportional to the intensity of lower body suction, and preservation of the periodicity of skin blood flow that occurs after release of tourniquets.

Data from direct microneurographic recordings has generally failed to document baroreceptor control of human skin sympathetic traffic. In an early study, there was no clear relation between skin sympathetic activity and blood pressure (Hagbarth *et al.* 1972). However, it may be necessary to distinguish between skin vasoconstrictor and sudomotor outflows in studies of possible baroreceptor influences. In multifibre skin sympathetic nerve recordings, reductions of ambient temperatures increase vasoconstrictor and decrease sudomotor traffic (Hagbarth *et al.* 1972). Bini *et al.* (1981) showed that although vasoconstrictor skin sympathetic activity elicited by cooling does not have cardiac rhythmicity, sudomotor skin sympathetic activity elicited by body heating has clear cardiac rhythmicity. Since lower body suction did not increase skin sudomotor sympathetic activity, this study suggests baroreceptors entrain skin sudomotor neurones at high body temperature, but do not control the level of traffic. Moreover, although in this study, Valsalva manoeuvres increased skin sudomotor sympathetic activity, bursts were not necessarily related inversely to blood pressure changes.

Finally, Wallin *et al.* (1975) found that electrical carotid sinus nerve stimulation (which increases traffic over baroreceptor and chemoreceptor nerves) usually did not change skin sympathetic activity significantly. In some instances, skin sympathetic traffic increased, and in others it decreased. In individual subjects, habituation of both increases and decreases of skin sympathetic traffic occurred. Thus, although heating increases sudomotor skin sympathetic activity and confers on it clear cardiac rhythmicity, the evidence linking this change to baroreceptors is equivocal.

MYOCARDIUM

Anatomical studies of Chidsey (1965) and K.M. Kent (1974) and their co-workers show clearly that human ventricular myocardium is richly innervated by sympathetic and cholinergic neurones. Although the left ventricle has rich cholinergic innervation (K.M. Kent *et al.* 1974), acetylcholinesterase staining is limited almost entirely to endocardial muscle and Purkinje fibres and subjacent myocardium; there is very sparse cholinergic innervation of the remaining ventricular myocardium. Cholinergic innervation of human ventricular endocardium is illustrated in Fig. 9.17. In the discussion that follows, baroreflex influences on myocardial function are divided into effects on electrophysiological properties and contractility.

Myocardial electrophysiological properties

Direct data on baroreflex influences on human myocardial electrophysiological properties are extremely limited. This stands in contrast with data on the effects of baroreceptor stimulation on dysrhythmias, which are substantial (see Chapter 13). In the case of re-entrant dysrhythmias involving the atrioventricular node, baroreceptor stimulation blocks conduction through the atrioventricular node and interrupts re-entrant pathways. Mechanisms underlying baroreceptor termination of re-entrant dysrhythmias whose pathways do not include the atrioventricular node (Waxman and Wald 1977*b*) are not well understood.

There are relatively few studies on the influence of baroreflexes on myocardial refractoriness in *animals*. Kowey *et al.* (1983*a*) raised arterial pressure with methoxamine or phenylephrine infusions in anaesthetized dogs and measured ventricular effective refractory periods and repetitive extrasys-

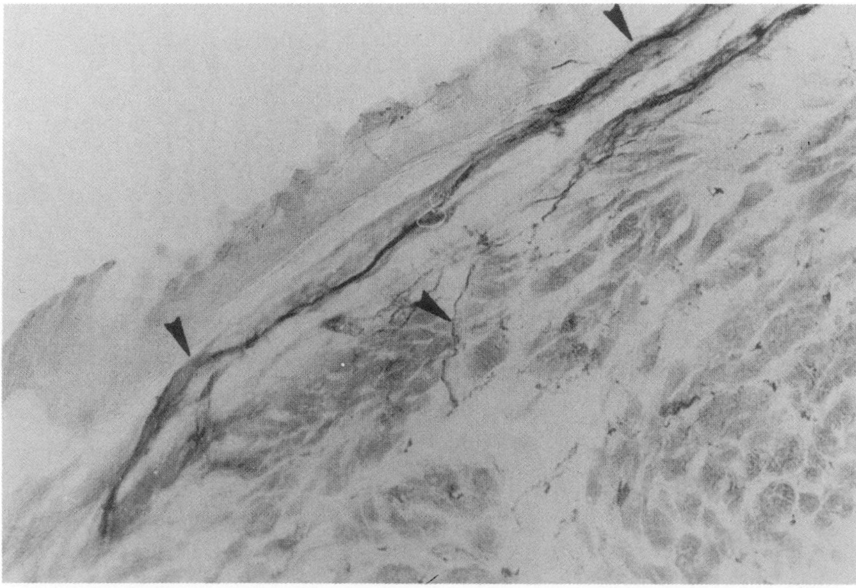

FIG. 9.17. Acetylcholinesterase activity in human left ventricular myocardium. Reproduced with permission (K.M. Kent *et al.* 1974).

tole thresholds (putative indexes of vulnerability to ventricular fibrillation (Kowey *et al.* 1983*b*)). They found that baroreceptor stimulation increases the repetitive extrasystole threshold (an effect that appears to be mediated by baroreceptors, because it is not present after sinoaortic denervation), but does not alter refractory periods. Kamiya *et al.* (1983) studied baroreflex influences indirectly. They measured the effects of coronary occlusion on refractoriness of non-ischaemic ventricular myocardium and found that coronary occlusion causes a small but significant reduction of refractory periods, which is absent after sinoaortic denervation. L.A. Turner (1985) showed that in anaesthetized dogs, sympathetic nerve stimulation shortens Purkinje fibre refractoriness, but that vagal stimulation does not alter it.

Baroreflex influences on *human* myocardial refractoriness have been studied by several groups. This literature suggests that increased vagal activity reduces atrial and increases ventricular refractoriness. Prystowsky and co-workers (1983) measured vagally mediated changes of *atrial* refractory periods provoked by intense neck suction in patients undergoing diagnostic electrophysiological testing. They found that atrial refractory periods are reduced during baroreceptor stimulation; in three patients, reflex reductions of atrial refractoriness were associated with repetitive atrial responses after premature atrial stimuli. Coumel *et al.* 1978, 1979) provide strong evidence (discussed in Chapter 13) that increased vagal activity promotes the occurrence of atrial fibrillation. Reductions of atrial refractory periods in response to electrical vagus nerve stimulation have also been documented in dogs (Zipes *et al.* 1974).

These findings may be at variance with those of Dhingra and Hordof and their co-workers. Dhingra *et al.* (1976) found that large doses of atropine shorten atrial refractoriness in patients. The implication of this is that vagal activity lengthens atrial refractoriness. Other evidence for vagal inhibition of atrial myocardial electrophysiological comes from the unique study of Hordof *et al.* (1978). They measured electrophysiological properties of atrial tissue removed from patients during heart surgery and found that acetylcholine superfusion increases maximum diastolic potential, reduces the slope of phase 4 depolarization, and slows the spontaneous rate of pacemaker firing. These effects were ascribed to acetylcholine because they were prevented by pretreatment with atropine.

There is a consensus that vagal activity inhibits electrophysiological properties of *ventricular* myocardium. The most direct evidence comes from the study of Ellenbogen and associates (1990*a*), who showed that intense carotid baroreceptor stimulation or moderate pharmacological elevation of arterial pressure provokes significant vagally mediated increases of right ventricular refractory periods. Prolongation of ventricular refractoriness was prevented by pretreatment with atropine. Prystowsky *et al.* (1981) and Morady *et al.* (1988) and their co-workers showed that large dose atropine reduces human ventricular refractory periods. Prystowsky found that the inhibitory effects of atropine occur with and without prior beta-adrenergic blockade; their observation may be the first evidence of vagal inhibition of ventricular electrophysiological properties in humans. Morady found that reductions of ventricular refractory periods after atropine are greater during beta-receptor stimulation with isoproterenol than after beta-blockade with propranolol. Thus, part of the putative prolongation of ventricular refractoriness by vagal activity may result because of adrenergic enhancement of the effects of acetylcholine.

Studies in dogs also suggest that vagal influences are mediated partly by antagonism of sympathetic effects (Bailey *et al.* 1979; Martins and Zipes 1980; Wendt and Martins 1990). Martins *et al.* (1988) studied the effects of reduced baroreceptor activity on ventricular refractoriness. They found that during intense lower body suction, ventricular refractoriness is shortened. This effect was largely prevented by pretreatment with a beta-blocker. Thus, it appears that increased baroreceptor activity prolongs ventricular refractoriness by stimulating release of acetylcholine, and reduced baroreceptor activity shortens ventricular refractoriness by stimulating release of noradrenaline.

It is important to underscore one aspect of all of the above studies on human atrial and ventricular refractoriness: changes provoked by altered baroreceptor activity are extremely small. None the less, there is considerable evidence (discussed in Chapter 13) that these small changes are involved in baroreflex interruption of re-entrant atrial (Gould and Lyon 1967) and possibly ventricular (Waxman and Wald 1977*b*) dysrhythmias.

Baroreceptor influences on myocardial contractility

Baroreflex influences on myocardial contractility are difficult to study in humans, because changes of baroreceptor afferent activity alter other haemodynamic determinants of contraction, including heart rate, diastolic filling periods, and arterial pressure. Ernsting and Parry (1957), Bevegård and Shepherd (1966*a*), and D.S. Goldstein and Keiser (1984) found no change of cardiac output with neck suction or pressure. T.J. Ebert *et al.* (1986*b*), however, in a preliminary communication, reported that left ventricular stroke volume measured with impedance cardiography during neck suction or pressure varies directly with changes of diastolic filling period. (Impedance cardiography is a poorly understood non-invasive method that provides accurate estimates of changes of left ventricular stroke volumes, but not of absolute levels (Secher *et al.* 1977; T.J. Ebert *et al.* 1984*a*; D.S. Goldstein *et al.* 1986)). In an unpublished study, T.J. Ebert estimated stroke volume changes after most R–R interval responses to neck pressure changes had been prevented by pretreatment with large-dose atropine. His results suggest that in the absence of significant changes of heart rate, abrupt changes of afferent carotid baroreceptor input provoke small but significant (presumably sympathetically mediated) opposite changes of ventricular function.

S.F. Vatner *et al.* (1972) electrically stimulated carotid sinus nerves in conscious dogs and concluded that baroreceptors exert negligible effects on myocardial function. However, subsequent studies by Yoran *et al.* (1981) and Aylward *et al.* (1983) indicate that when afferent inputs of *all* baroreceptor areas are altered, important vagally and sympathetically mediated effects on ventricular function can be demonstrated.

Yoran and co-workers (1981) lowered arterial (and cardiopulmonary) pressures by occluding the inferior vena cava in conscious dogs, and estimated ventricular contractility with measurements of the rate of left ventricular pressure change (dP/dt) after release of the occlusion. After occlusion, arterial pressure and left ventricular contractility increased in proportion to preceding reductions of arterial pressure. Augmented contractility seemed to be mediated by increased sympathetic stimulation, because pretreatment with a beta-blocker prevented its occurrence.

The study of Aylward *et al.* (1983) complements that of Yoran *et al.* (1981). Aylward occluded the aorta (to raise pressure) as well as the inferior vena cava in conscious rabbits, and documented important baroreflex modulation of ventricular function. In Aylward's study however, most of the changes of ventricular contractility seemed to be mediated by vagal–cardiac activity, because pretreatment with atropine largely prevented their occurrence. (However, as discussed in Chapter 8, atropine does not function simply as a peripheral muscarinic cholinergic blocker.)

Information regarding baroreflex modulation of *human* myocardial function is sparse and indirect. Both W.C. Smith *et al.* (1929) and McMichael and Sharpey-Schafer (1944) found that large doses of intravenous atropine modestly increase cardiac output. Carlsten *et al.* (1967), in a preliminary communication, reported that direct electrical stimulation of the vagus nerves in patients undergoing neck surgery reduces cardiac output. Another report (Hendrickson 1969) indicates that large-dose atropine prevents hypotension during surgical manipulation of the vagus nerves (which also may involve the carotid sinuses).

The more contemporary study of Stratton *et al.* (1987) explored this physiology in ways that had not been possible earlier. These workers infused adrenaline and noradrenaline intravenously before and after large dose atropine in healthy human volunteers and measured stroke volumes with radionuclide ventriculography. They found that the increases of ejection fraction, stroke volume, and cardiac output provoked by catecholamine infusion were augmented by atropine, and suggested that atropine unmasked baseline cholinergic ventricular depression. An alternative or complementary hypothesis is that left ventricular stroke volume increased because of a sympathetic-parasympathetic interaction resulting from the experimentally-induced elevation of catecholamine levels (Levy 1971; Levy and Zieske 1969). The study of Stein *et al.* (1966) renders it unlikely that atropine increases cardiac output merely by increasing heart rate; Stein showed that during atrial pacing, cardiac output is constant over a wide range of heart rates.

Pretreatment with large dose atropine also augments arterial pressure increases after release of Valsalva straining (Elisberg *et al.* 1953) or injections of vasoactive drugs (Pickering *et al.* 1972b; Evans *et al.* 1979). Moreover, atropine abolishes the reductions of left ventricular ejection, as estimated with pulsed-wave aortic Doppler ultrasound, that occur after bolus injections of phenylephrine (Casadei *et al.* 1992).

There is also one remarkable direct documentation of negative inotropic effects of acetylcholine. Sieber *et al.* (1973) measured the force of isometrically contracting human papillary muscles removed during heart surgery. After pretreatment with guanethedine, high voltage stimulation led to modest reductions of contractability. Since this effect was not observed in muscles pretreated with atropine, it was attributed to acetylcholine.

CENTRAL NERVOUS SYSTEM

In Chapter 2 it was noted that the Greek root verb καρο, from which the name of the carotid arteries was derived, means to plunge into a deep, heavy sleep. This name appears to have been chosen because people become sleepy during carotid massage. Since it is clear that there are monosynaptic connections between the termination of afferent baroreceptor neurones in the solitary tract nucleus and higher central nervous system structures (Chapter 3), it is reasonable to expect that ever-changing baroreceptor inputs in some way influence behaviour. This speculation has engendered a very lively discussion, especially in the psychophysiology literature.

One of the earliest links between baroreceptor activity and behaviour was discovered by Koch (1932) (and subsequently reported by Schweitzer (1937)), who found that sudden elevation of pressure in a blind carotid sac produces muscle inhibition similar to that which occurs during sleep. Many studies published subsequently support the notion that baroreceptor input *inhibits* central nervous system activity. A much quoted study by Bonvallet *et al.* (1953) showed that under certain circumstances, carotid sinus distention reduces electroencephalogram activity.

These observations have been codified in what has come to be known as the 'Lacey hypothesis'. John and Beatrice Lacey first noted an association between the attention human subjects paid to external stimuli and cardiac deceleration (they characterized these observations as 'data in search of a theory' (Lacey *et al.* 1963)). Subsequently the Laceys evolved a theory which, broadly stated, is that the degree of cortical excitation is proportional to arterial pressure (Lacey and Lacey 1962).

Considerable evidence has been marshalled to support this theory. Dworkin and colleagues (1979) measured the distance that conscious rats run to avoid aversive stimuli (electrical stimulation of trigeminal nuclei) during saline or phenylephrine infusions. They found that rats ran less during hypertension (blood pressure elevations of 20 to 50 mmHg) than during normotension. Rats subjected earlier to sinoaortic denervation ran the same distance during phenylephrine and saline infusions. The authors interpreted these findings as showing that high blood pressure levels influence behaviour by reducing central responses to aversive stimuli.

Other evidence supports a link between baroreceptor activity and behaviour. Gellhorn (1941) showed that the increased sympathetic activity (as reflected by nictitating membrane contraction) that occurs during hypothalamic stimulation is less during than before adrenaline-induced blood pressure elevations. He and his co-workers (1942) showed subsequently that the frequency of convulsive activity is inversely related to blood pressure; that pharmacological blood pressure elevations increase electroencephalographic slow potentials and reduce fast potentials; and that these effects are abolished by sinoaortic denervation (Nakao *et al.* 1956). More recently, Coleridge *et al.* (1976) showed that carotid sinus distention inhibits individual pyramidal tract neurone firing.

Elbert *et al.* (1988*b*) and co-workers tested the Lacey hypothesis in human subjects. They measured pain thresholds in normotensive and borderline hypertensive subjects during moderate neck suction (and an auditory tone) or mild neck pressure (and a different auditory tone). Under baseline conditions, tolerance to pain was less in normotensive than in borderline hypertensive subjects. Moreover, in normotensive subjects, baroreceptor stimulation lowered pain thresholds, but in borderline hypertensive subjects, baroreceptor stimulation raised pain thresholds. These results were interpreted as showing that in certain subjects, blood pressure elevations render painful stimuli more tolerable, and therefore lead to further blood pressure elevations (Dworkin 1988).

It should be noted that the theories discussed above are highly controversial. In particular, J. Green (1980) discussed problems with the chain of reasoning associated with the Lacey hypothesis. He questioned the validity of studies of central nervous system function that involve use of general anaesthetics. He also quoted several studies which show that baroreceptor stimulation results in central *excitation* rather than inhibition. It is difficult for us to ascribe the increased tolerance for pain found in borderline hypertensive subjects to their higher levels of baroreceptor input; on the basis of numerous studies in animals with hypertension (see Chapter 11), we would expect afferent baroreceptor traffic to be similar in borderline hypertensive and normotensive groups.

KIDNEY

Renin is a proteolytic enzyme elaborated primarily (but not exclusively) by juxtaglomerular cells surrounding afferent renal arterioles (Taugner *et al.* 1984) and macula densa cells of adjacent distal convoluted tubules, where it is stored as granules. Renin acts on the hepatic globulin angiotensinogen to form the decapeptide angiotensin-I (Braun-Menendez *et al.* 1940; Page and Helmer 1940). Angiotensin-I is converted to the active octapeptide angiotensin-II by converting enzyme. Angiotensin-II is a powerful vasoconstrictor that (among other actions) stimulates central thirst centres and increases water intake, and stimulates adrenal production of the sodium-retaining hormone aldosterone (Severs and Daniels-Severs 1973).

At least four interrelated mechanisms control renin release:

(1) juxtaglomerular cells sense the degree of arteriolar stretch and initiate intrarenal pressure-driven adjustments of plasma renin activity;

(2) macula densa cells sense changes of plasma sodium concentration and modulate renin release to maintain osmolarity;

(3) efferent sympathetic impulses act on juxtaglomerular adrenergic receptors (Atlas *et al.* 1977) and release renin;

(4) circulating hormones and neurotransmitters modulate renin release.

Baroreceptor reflexes have the potential to influence renin secretion at several levels, in extraordinarily complex ways.

Renal baroreceptor mechanisms

Skinner *et al.* (1964) discovered renal baroreflex control of renin release. In their studies, aortic constriction above the level of the renal arteries in anaesthetized dogs led to reductions of renal artery pulse and mean pressures without reductions of renal blood flow, and increases of plasma renin activity. Guazzi *et al.* (1981) showed that partial renal artery occlusion also increases renal vein plasma renin activity in humans.

Subsequent studies quantified renal baroreflex mechanisms in terms of the relation between arterial pressure and plasma renin activity (Farhi *et al.* 1982, 1987), renin release (Eide *et al.* 1973; Finke *et al.* 1983; Kirchheim *et al.* 1985), or angiotensin-I production (Cowley and Guyton 1972). These data indicate that there is a threshold for renin release, which in normotensive animals lies slightly below baseline arterial pressures. Pressure elevations above this threshold do not reduce plasma renin activity appreciably, and pressure reductions below this threshold increase plasma renin activity linearly. Pressure reductions below a certain level (about 60 mmHg in the conscious dogs studied by Kirchheim *et al.* (1985)), do not increase renin activity further. (Therefore, although a plateau region is usually not plotted, the arterial pressure–plasma renin activity relation seems to be a reverse sigmoid function remarkably similar to the baroreceptor–muscle sympathetic nerve activity relation (Rea and Eckberg 1987).)

The relation between angiotensin-I levels (used as an index of plasma renin activity) and the distribution of mean arterial pressure is depicted in Fig. 9.18 (Ehmke *et al.* 1987*a*). In this figure, reductions of pressure below the threshold for renin release lead to sharp increases of angiotensin-I levels. In the study of Finke *et al.* (1983) a reduction of mean arterial pressure of only 2.5 mmHg increased plasma renin activity 2.7-fold. This figure suggests also that a small increase of the threshold for renin release, such as might occur during sympathetic stimulation, may increase renin release (and consequently, arterial pressure) dramatically.

Although the relation between renal arterial pressure and plasma renin activity is influenced importantly by renal sympathetic nerve activity (and serum sodium concentrations — see below), renal baroreflex control of renin activity also exists independent of renal sympathetic activity. Large doses of the diuretic furosemide increase plasma renin activity in patients with renal transplants and in quadriplegic patients, who have sympathetic denervation (Horvath *et al.* 1976). Moreover, when healthy subjects are made hypotensive by vasodilator infusions, only part of the elevation of plasma renin activity is prevented by pretreatment with ganglionic blockade (Kaneko *et al.* 1970).

The practical consequences of the arterial pressure–plasma renin activity relation have been worked out elegantly by Ehmke *et al.* (1987*a,b*). In conscious dogs on normal sodium diets, the average mean arterial pressure

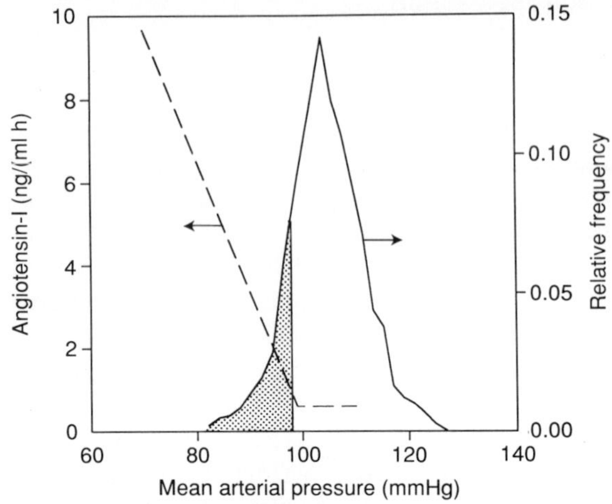

FIG. 9.18. Relation between mean arterial pressure and angiotensin-I (used as an index of plasma renin activity), and the frequency distribution of mean arterial pressure. The stippled area signifies pressures below the threshold for renin release. Adapted with permission (Ehmke *et al.* 1987*a*).

is 12 mmHg above threshold for renin release. Variability of blood pressure is greater in dogs with lower thresholds and more shallow slopes for renin release. Dogs with higher thresholds for renin release also have higher resting arterial pressures. Emhke's data describe an on–off relation between arterial pressure and renin release, such that renin release occurs intermittently during daily activities (with as much as 300 per cent increases above baseline levels), and contributes, perhaps importantly, to the stabilization of arterial pressure. It should be noted that in their animals, sympathetic pathways were intact, and therefore the relation between arterial pressure and renin release was likely to be influenced by ongoing fluctuations of sympathetic outflow.

Sympathetic control of renin release

Although, as mentioned, some of the effects of sympathetic stimulation may be secondary to renal baroreflex mechanisms, sympathetic mechanisms may also stimulate renin release directly. Haemorrhage (Buñag *et al.* 1966) and renal sympathetic nerve stimulation (La Grange *et al.* 1973) increase plasma renin activity independent of changes of arterial pressure and renal blood flow. Ganglionic blockade or local renal nerve anaesthesia prevents increases of renin activity during haemorrhage (Buñag *et al.* 1966). Beta-adrenergic blockade prevents elevations of plasma renin activity during bilateral carotid occlusion with controlled renal artery pressure (R. Gross *et al.* 1981*a*) and the elevation of renin activity that occurs with head-up tilt (Winer *et al.* 1969). Moreover, baseline plasma renin activity declines after intravenous propranolol (R. Gross *et al.* 1981*a*), and renal vein renin activity is lower in denervated than in innervated kidneys (Grandjean *et al.* 1978). These results indicate that baseline plasma renin activity is determined in part by baseline sympathetic activity.

It is not clear if *alpha*-adrenergic receptors participate in sympathetic control of renin release. M.L. Blair (1983) showed that intrarenal infusions of the alpha-adrenergic agonist phenylephrine, which do not produce measurable changes of total renal blood flow, increase renin release. However, since phenylephrine reduces renin release from cultured endothelial (Tang *et al.* 1990) cells *in vitro*, the meaning of Blair's findings is unclear. Conceivably, phenylephrine modulates renin release in the intact kidney by subtly redistributing renal flow or modifying glomerular filtration rate or sodium excretion (Kirchheim *et al.* 1985).

Circulating hormones and neurotransmitters

There are several negative feedback loops that govern renin release. Increased angiotensin-II levels inhibit subsequent renin production through (1) a direct effect on juxtaglomerular cells in the vascular component of the juxtaglomerular apparatus, (2) an indirect effect mediated by increases of renal artery pressure secondary to vasoconstriction, and (3) an indirect effect mediated by increased levels of aldosterone, vasoconstriction, and sodium and water retention. Other negative feedback loops are initiated by atrial stretch caused by pressure elevations. Atrial stretch reduces plasma renin activity and aldosterone levels, and increases atrial natriuretic peptide levels (Annat *et al.* 1976; Burnett *et al.* 1984). Atrial natriuretic peptide opposes extrarenal actions of angiotensin-II and causes direct vasodilatation and sodium and water excretion. (However, atrial natriuretic peptide does not appear to oppose efferent arteriolar constriction caused by angiotensin-II (Burnett *et al.* 1984; Maack *et al.* 1984).)

Baroreceptors involved in modulation of renin secretion

The considerable confusion in the literature regarding which baroreceptors (arterial or cardiopulmonary) are responsible for modulation of plasma renin activity in humans may have arisen from the complexity of this physiology, and from experimenters' failure or inability to control all necessary experimental variables. Literature for animals is difficult, and literature for human subjects may be impossible to interpret.

There is however substantial agreement on one issue: most workers believe that mechanoreceptors in the cardiopulmonary region sense changes of blood volume and modulate plasma renin activity (D.E. Donald 1979). This view is based on studies which show that plasma renin activity increases in healthy or hypertensive human subjects when venous return to the right heart is reduced by lower body suction (Fasola and Martz 1972; Mark *et al.* 1978) or thigh tourniquets (Kiowski and Julius 1978; Egan *et al.* 1983). Some (Kiowski and Julius 1978; Egan *et al.* 1983; Grassi *et al.* 1988) but not all (Mark *et al.* 1978) studies, show that renin activity levels increase with very small reductions of central venous pressure that do not alter arterial pressure or heart rate. To the extent that these venous pressure reductions alter activity of cardiopulmonary receptors selectively (this issue is joined in Chapter 7), these data support a role for cardiopulmonary receptor modulation of renin release in humans.

Arguments arise over the role of sinoaortic baroreceptors in regulation of renin secretion. Thames *et al.* (1978*a*) and D.E. Donald (1979) lucidly discuss the confounding influences of baroreceptor areas whose activities are not controlled in studies of renin release. First, there is a genre of research that defines the role of a receptor area in terms of the effects of denervation of that area on the variable being measured. Denervation of cardiopulmonary receptors provokes increases of plasma renin activity, but only transiently (Stella *et al.* 1978). Evanescence of renin activity increases probably results from the influence of remaining afferent carotid and aortic baroreceptor inputs. When cardiopulmonary baroreceptors are denervated selectively, plasma renin levels remain normal, but when all (including cardiopulmonary) baroreceptor areas are denervated, plasma renin activity is elevated permanently (Persson *et al.* 1987, 1988). Thus only one functioning baroreceptor area is necessary to maintain normal levels of plasma renin activity.

Second, many studies fail to consider that changes of pressure in one baroreceptor area alter sympathetic outflow and haemodynamic factors, and thereby modify intrarenal mechanisms responsible for control of plasma renin activity. We suggest that because of this problem, it may be impossible to sort out this physiology in humans. An example of the difficulties of such research can be found in a report by Mancia *et al.* (1981). These authors studied the effects of neck pressure changes on plasma renin activity in hypertensive subjects, and found little evidence for carotid baroreceptor control of renin release. Although these findings may point away from a significant role for carotid baroreceptors in modulation of renin activity, they can be explained more economically. In Mancia's study, neck suction stretched carotid baroreceptors (Kober and Arndt 1970) and reduced mean arterial pressure by an average of 16 mmHg. In all likelihood, reductions of renin secretion secondary to baroreceptor stimulation were opposed by increases of renin secretion secondary to arterial pressure and renal blood flow reductions.

The study of Finke *et al.* (1983) indicates that pressure reductions of this magnitude produce huge increases of plasma renin activity in conscious dogs. Rocchini and Barger (1979) and R. Gross *et al.* (1981*b*) showed that influences of carotid baroreceptors (as defined by responses to carotid occlusion) can be demonstrated *only* if increases of renal artery pressure are prevented. Thus, notwithstanding elegant attempts to sort out this physiology (Egan *et al.* 1983), methods available for study of humans may not adequately isolate one system for study.

HYPOTHALAMUS

Arginine vasopressin is a peptide neurotransmitter and hormone released by the neurohypophysis (Brownstein *et al.* 1980) in response to a variety of stimuli, including hypertonicity (Verney 1947; Skowsky *et al.* 1974), blood volume reduction (Pullan *et al.* 1980), hypotension (G.L. Robertson and Athar 1976), and hypoxia and hypercapnia (Forsling and Rees 1975). The principal sensors responsible for vasopressin release are central osmoreceptors; small increases of plasma osmolality predictably trigger proportional increases of vasopressin concentrations in plasma (Verney 1947; G.L. Robertson 1984) and cerebrospinal fluid (B.C. Wang *et al.* 1985). Blood volume sensors and baroreceptors also contribute importantly to vasopressin release, since major reductions of blood volume lead to proportional increases of plasma vasopressin concentrations (Pullan *et al.* 1980).

A partial explanation for separate influences of diverse physiological stimuli on vasopressin secretion may be that supraoptic nuclei involved in vasopressin release receive separate inputs from osmoreptors and baroreceptors (Kannon and Yagi 1978). Moreover, osmoreceptors and baroreceptors have different thresholds for vasopressin release; subtle increases of osmolarity provoke proportional increases of vasopressin levels, but only marked reductions of blood volume lead to vasopressin increases (F.L. Dunn *et al.* 1973). The relation between plasma vasopressin and osmolality in healthy subjects is shown in Fig. 9.19, and the relations between vasopressin and blood volume and osmolality in rats are shown in Fig. 9.20.

Baroreflex control of vasopressin release

Evidence that baroreceptors modulate release of vasopressin seems incontrovertible. Figure 9.21 depicts the loose but significant relation that exists between blood pressure and vasopressin levels in rats, obtained (at time of sacrifice) after subcutaneous injections of nitroprusside. These results complement the data relating blood volume to vasopressin levels depicted in Fig. 9.20; they suggest that reductions of arterial pressure must be substantial before vasopressin defense mechanisms are invoked.

Because of ethical considerations, data obtained from humans are not as comprehensive as these from rats; however, human data also point towards

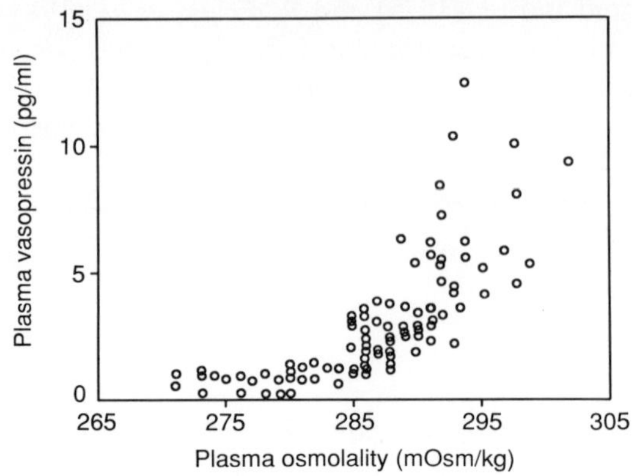

FIG. 9.19. Regulation of vasopressin levels by osmolar receptors in 25 healthy subjects. Adapted with permission (G.L. Robertson 1977).

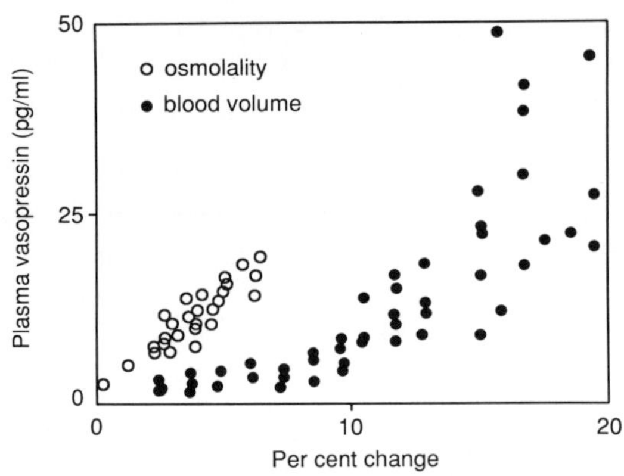

FIG. 9.20. Comparison of the influence of osmolality and blood volume on vasopressin levels in conscious rats. Adapted with permission (F.L. Dunn *et al.* 1973).

a blood pressure threshold for vasopressin release. In healthy subjects, neither reduction of blood volume by a phlebotomy of about 10 per cent (Goetz *et al.* 1974) nor upright posture (G.L. Robertson and Athar 1976; Bie *et al.* 1986), which reduces central blood volume by about 10 to 15 per cent, importantly increases plasma vasopressin concentrations. However, the combination of phlebotomy and upright posture provokes large increases of vasopressin levels (G.L. Robertson and Athar 1976).

A substantial number of experiments has focused on which baroreceptors (cardiopulmonary or arterial) are primarily responsible for modulating vasopressin release. It is clear that reduction of baroreceptor input leads to increases of plasma vasopressin concentrations. Share (1967) studied vaso-

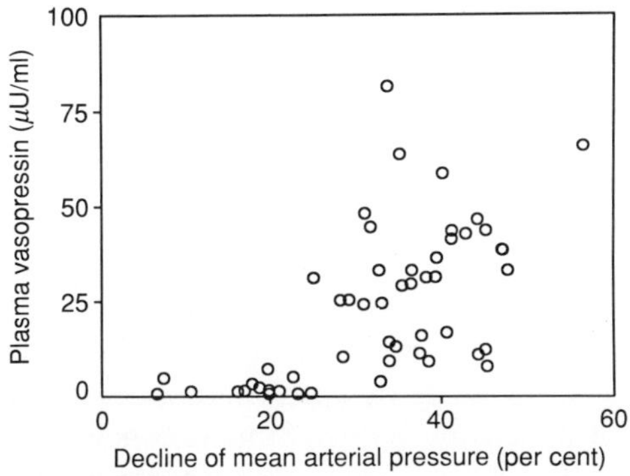

FIG. 9.21. Vasopressin levels after subcutaneous injections of sodium nitroprusside in conscious rats. Adapted with permission (G.L. Robertson 1977).

pressin release in anaesthetized vagotomized dogs during graded haemorrhage. He found that when carotid baroreceptor areas were perfused at constant normal pressures, haemorrhage of up to 40 per cent of blood volume did not provoke important increases of plasma vasopressin levels. However, when carotid sinus pressure was allowed to fall with systemic pressure during haemorrhage, vasopressin levels increased substantially. These results suggest that in the absence of cardiopulmonary receptors, arterial baroreceptors modulate vasopressin secretion.

M. Morris and Alexander (1989) found that sinoaortic denervation in conscious rats increases oxytocin levels slightly, and increases osmotically provoked oxytocin and vasopressin levels markedly. Thames and Schmid (1979) also studied the effects of sinoaortic denervation or vagotomy on plasma vasopressin concentrations in anaesthetized dogs. They found that when vagus nerves were intact, sinoaortic denervation led to minimal elevations of vasopressin concentrations. Subsequent vagotomy, however, provoked large increases of vasopressin levels (notwithstanding major elevations of arterial pressure). This result suggests that cardiopulmonary receptors are more important than arterial baroreceptors in modulating vasopressin release. However, in other dogs the order of sensory nerve section was changed. In these experiments, section of aortic and vagal nerves led to (moderate) increases of vasopressin levels, but subsequent carotid sinus nerve transection led to substantial additional increases of vasopressin levels. These results suggest that total afferent information from all baroreceptors may be more important in determining vasopressin release than input from individual receptor areas.

Several studies in humans focused on the role of arterial baroreceptors in control of vasopressin release. One group (Norsk *et al.* 1987) found that extremely prolonged (20 min) carotid baroreceptor stimulation with neck suction does not alter plasma vasopressin levels. (However, in three out of eight subjects, vasopressin levels rose substantially after release of neck suction.) The relevance of this observation is unknown, because as discussed in Chapter 5, prolonged neck pressure changes provoke complex physiological responses that are extraordinarily difficult to interpret.

Other studies from animals suggest that arterial baroreceptors may not play an important role in modulating vasopressin secretion. B.C. Wang *et al.* (1983) suggested that cardiac receptors bear a primary responsibility for changes of vasopressin levels, since in conscious dogs, cardiac denervation nearly abolishes the increases of vasopressin that occur during slow haemorrhage or infusion of hypertonic saline. Similar conclusions were reached by Quail *et al.* (1987), who measured vasopressin responses of conscious rabbits to slow haemorrhage before and after denervation of arterial baroreceptors or pharmacological denervation of cardiac receptors (with intrapericardial procaine). One difference between these two studies is that B.C. Wang *et al.* (1983) found that haemorrhage increases vasopressin levels after cardiac denervation, but Quail *et al.* (1987) found no change. The former study left open the possibility that arterial baroreceptors make some (albeit small) contribution to vasopressin release during haemorrhage. J.R. Oliver *et al.* (1990) studied the effects of blood volume reduction on plasma vasopressin concentration in conscious rabbits. They found that during the early 'non-hypotensive' phase of haemorrhage, plasma vasopressin (and renin activity) levels remain low. However, after total adrenergic and cholinergic blockade, vasopressin levels rise rapidly during minor haemorrhage. These results suggest that ongoing baroreflex adjustments during minor haemorrhage minimize secretion of vasopressin.

Certain observations suggest that there may be species differences in the non-osmotic control of vasopressin release. Arnauld *et al.* (1977) and Gilmore *et al.* (1982) found that in anaesthetized monkeys, haemorrhage led to elevations of vasopressin which were related to reductions of arterial rather than left ventricular pressure.

This controversy has led to attempts to study *cardiopulmonary* baroreceptor control of vasopressin release in humans. Goldsmith *et al.* (1982) applied two levels of prolonged lower body suction to healthy normally hydrated human volunteers. Eight out of nine subjects experienced no increase of vasopressin levels. However, one subject who experienced a vasovagal reaction had an increase of plasma vasopressin concentration from 3 to 81 pg/ml. Presumably, since arterial pressure fell in this subject, this effect was mediated by arterial baroreceptors. These results are consistent with animal studies (Fig. 9.21) which suggest that vasopressin release occurs only when reduction of blood pressure is substantial.

Leimbach *et al.* (1984) conducted similar experiments, but in subjects who were dehydrated. They found that low levels of lower body suction do not alter plasma vasopressin concentration. High levels of lower body suction, however, which reduce arterial pressure (and therefore reduce arterial, as well as cardiopulmonary baroreceptor input) significantly (but modestly) increase vasopressin levels (from 4 to 15 μU/ml). Subsequently, Goldsmith *et al.* (1987) documented a similar interaction between osmoreceptor and baroreceptor control of vasopressin. Taken together, this evidence from human studies suggests that in healthy normally hydrated subjects, moderate reductions of baroreceptor afferent activity do not increase plasma vasopressin concentrations, but that in dehydrated subjects, reductions of baroreceptor input lead to small increases of vasopressin concentrations.

Vasopressin control of baroreflex function

The above discussion indicates that baroreceptors may influence vasopressin release in healthy humans. A separate but related issue is whether released vasopressin influences baroreflex function. The studies of Altura and Altura (1977) indicate that on a molecular basis, vasopressin is one of the most powerful naturally occurring vasoconstrictors. This finding stands in contrast to abundant evidence that experimental elevations of vasopressin levels within or even above the physiological range may not increase arterial pressure. There is strong evidence that under such circumstances, arterial pressure fails to rise because of baroreflex buffering. It is now apparent that although such increases of vasopressin levels may not raise arterial pressure, they are associated with reductions of heart rate and cardiac output and increases of peripheral vascular resistance, in animals (Szczepanska-Sadowska 1973; Montani *et al.* 1980) and humans (T.J. Ebert *et al.* 1986*a*).

The evidence that baroreflexes mediate these haemodynamic changes is compelling. Interference with baroreflex responses by baroreceptor denervation (Montani *et al.* 1980), atropine (Liard 1984), or ganglionic blockade (H.N. Wagner and Braunwald 1956) largely prevents cardiac slowing and reductions of cardiac output, and augments pressor responses to vasopressin. Arguments also have been made that vasopressin uniquely modifies baroreflex function in ways that further diminish pressor responses to its infusion. The evidence for this comes from a variety of sources, but may not be definitive (T. Bennett and Gardiner 1986).

The first type of proof comes from comparisons among haemodynamic responses to vasopressin and other pressors, including particularly, the alpha-agonists phenylephrine and methoxamine. Cowley *et al.* (1974) studied responses of unanaesthetized dogs to vasopressin and noradrenaline infusions before and after sinoaortic denervation. They found that after baroreceptor denervation, pressor responses to vasopressin were augmented much more than pressor responses to noradrenaline. B.F. Cox *et al.* (1990) studied rabbits with infusions of phenylephrine and vasopressin. They found that slopes of heart rate were greater during vasopressin than during phenylephrine infusions, but that after placement of lesions in the area postrema, slopes were similar. Heyndrickx *et al.* (1976) and Matsuguchi and Schmid (1982) drew similar conclusions from studies conducted in conscious dogs and rats. Imai *et al.* (1983) measured R–R interval prolongation during infusions of phenylephrine given to Brattleboro rats (which lack the ability to synthesize vasopressin). They found that the slope of the R–R interval–systolic pressure relation was significantly lower in Brattleboro rats than in normal rats. Administration of a subpressor (but nevertheless large) intravenous dose of vasopressin to Brattleboro rats increased their baroreflex slopes towards those of control rats. Thus data from different species and from different types of experiments seem to indicate that baroreceptors modulate autonomic responses to vasopressin differently than they do to other pressors.

A subsequent study by Courtice *et al.* (1984) provided direct support for this conclusion. They studied conscious sheep and found that baroreflex slopes after vasopressin injections were steeper than those after phenylephrine. They then studied dogs and measured activity directly in single vagal–cardiac fibres during elevations of arterial pressure provoked by vasopressin or phenylephrine, or aortic occlusion. They found that vagal activity always rises more after vasopressin than after other pressor stimuli. They also found that single fibre carotid sinus nerve responses to pressure elevations are comparable after vasopressin and phenylephrine. By exclusion, this suggests that baroreflex augmentation by vasopressin occurs within the central nervous system. The influence of vasopressin on baroreceptor afferent activity was also studied by Abboud *et al.* (1986), who obtained conflicting results. They found that in anaesthetized rabbits, single unit baroreceptor nerve firing per unit of pressure elevation was increased after vasopressin. Curiously, baseline firing rates were not increased by vasopressin, even though baseline R–R intervals were prolonged.

Others failed to find augmentation of vagally mediated responses to baroreceptor stimuli after vasopressin infusion. Cowley *et al.* (1984) increased vasopressin levels of anaesthetized hypophysectomized dogs to 455 pg/ml and measured heart rate responses to changes of pressure in an isolated carotid sinus. They found no change of heart rate responses to rising pressure. Elliott *et al.* (1985) studied conscious rabbits and similarly failed to document an influence of elevated plasma vasopressin levels (to 200 pg/ml) on R–R interval responses to pressure elevations. O'Callaghan and Sheehan (as quoted by T. Bennett and Gardiner (1986)) studied human responses to carotid stimulation with neck suction and saw no change after infusion of vasopressin at an extremely low rate (5 pg/(kg min)). Thus the evidence that vasopressin augments vagally mediated baroreflex responses is conflicting.

The evidence that vasopressin augments sympathetic responses is less controversial. Undesser *et al.* (1985) studied conscious rabbits and found that vasopressin provokes larger reductions of renal sympathetic nerve activity per unit pressure increase than does phenylephrine. The study of Cowley *et al.* (1984) quoted above, shows that gain of the baroreflex control of arterial pressure is increased by vasopressin, but only when pressures in the isolated carotid sinus are lowered below baseline levels. This conclusion is supported by the human study of T.J. Ebert *et al.* (1986*a*). They measured increases of peripheral and forearm vascular resistance provoked by low levels (10 mmHg) of lower body suction before and after two doses of intravenous vasopressin, which raised average plasma vasopressin concentrations from 5 to 18 and 36 pg/ml. The lower dose increased average diastolic pressure by 2 mmHg, but did not increase systolic or mean arterial pressures, or reduce plasma noradrenaline levels significantly. Ebert found that increases of peripheral vascular resistance provoked by lower body suction are 60 per cent greater after than before vasopressin administration, and concluded that vasopressin increases baroreflex gain. (There may be an alternative or complementary explanation for this result, however; vasopressin infusions probably increased the baseline wall-to-lumen ratio of resistance vessels, and this change of geometry may have amplified subsequent vasoconstrictor responses to lower body suction (Folkow 1978).)

Although, as discussed above, published evidence from animals and humans shows that baroreceptor reflexes modulate release of vasopressin, and are in turn modulated by vasopressin, the subject is not free of controversy. In most studies, vasopressin is treated as a circulating hormone rather than a neurotransmitter. It is well recognized that vasopressin-containing neurones are located in important areas of the medulla concerned with cardiovascular regulation, including solitary tract and ambiguus nuclei (Buijs 1978).

In all likelihood, responses to intravenous infusions of vasopressin do not duplicate responses to physiological release of vasopressin from nerve endings. Inferential support for this contention comes from studies in which haemodynamic responses to intravenous and intracerebral artery injections of vasopressin were compared. Varma *et al.* (1969) found in anaesthetized dogs that intracerebroventricular vasopressin injections caused much more bradycardia than intravenous injections. Liard *et al.* (1981) found that although infusions of vasopressin into the inferior vena cava and vertebral artery lead to similar increases of plasma vasopressin levels, intravertebral infusions provoke smaller reductions of arterial pressure but greater reductions of heart rate than do intravenous infusions. This result suggests that the route of administration importantly determines the response to exogenous vasopressin; however, it falls short of proving that intravertebral artery (or, for that matter, intracerebroventricular (Imaizumi and Thames 1986)) administration of vasopressin reproduces the effects of physiological, neuronal release of vasopressin.

If it be accepted that intravenous vasopressin infusions reproduce responses to physiological vasopressin release, the question remains: what plasma concentration is necessary to reproduce physiological vasopressin effects? Cowley (1982) summarized published measurements of plasma vasopressin concentrations made in healthy animals and humans. He found that human plasma vasopressin concentrations average 4.3 pg/ml during normal hydration, and 1.5 and 15.8 pg/ml during overhydration and dehydration. Under extreme experimental conditions, vasopressin levels may rise much higher. Baylis *et al.* (1978) recorded vasopressin levels as high as 136 pg/ml in healthy subjects during hypotensive lower body suction. Goldsmith *et al.* (1982) found that in one healthy volunteer who developed a vasovagal reaction during lower body suction, plasma vasopressin levels rose from 2.6 to 81 pg/ml, and Sander-Jensen *et al.* (1986*a*) found that in healthy young adults during prolonged tilt and the development of presyncope, average vasopressin levels rose from 4 to 126 pg/ml. These measurements may bracket the physiological range of plasma vasopressin concentrations in healthy humans.

There has been an ongoing debate about whether changes of plasma vasopressin concentrations within this rather wide physiological range produce measurable haemodynamic effects. Malayan *et al.* (1980), who studied conscious dogs, and T.J. Ebert *et al.* (1986*a*), who studied conscious humans, found that elevations of plasma vasopressin levels within the range which occur during dehydration or non-hypotensive haemorrhage lead to measurable haemodynamic changes. Others observed only negligible haemodynamic changes during infusions of small doses of vasopressin.

Because of this, in many studies, infusion rates have been extremely large, and have led to increases of plasma vasopressin levels that are in excess of those reported to occur in all but the most extreme clinical circumstances. For example, Aylward *et al.* (986) found that a vasopressin infusion at a rate of 0.4 ng/(kg min), which raised plasma vasopressin concentrations to 24 pg/ml (by any reckoning, a major increase), produced only slight (but significant) reductions of heart rate (2 beats/min) and increases of central venous pressure (0.7 mmHg). Because of these modest results, the authors repeated their measurements with an infusion rate ten times that used in their first group of subjects. The higher infusion rate provoked substantial haemodynamic changes, but led also to an average increase of plasma vasopressin concentration to 290 pg/ml. In the study of conscious rabbits by Elliott *et al.* (1985), average vasopressin levels rose to 10 698 pg/ml! It is not clear how responses to 'industrial doses' (T. Bennett and Gardiner 1986) of vasopressin relate to physiology. Large infusion rates may lead to unexpected results, such as forearm dilatation (Aylward *et al.* 1986) and cardioacceleration (Möhring *et al.* 1980). However, we do not rule out their physiological relevance, since, as with noradrenaline (see above), concentrations of vasopressin are probably much greater in nerve synapses than in plasma. Thus research employing high doses of exogenously infused vasopressin may be necessary to replicate the high synaptic concentrations that occur physiologically.

Another way to determine if physiological vasopressin concentrations exert haemodynamic effects has been to measure the effects of vasopressin blocking drugs on resting blood pressure in normally hydrated human subjects. The rationale for this type of study is that if physiological concentrations of vasopressin exert haemodynamic effects, vasopressin blockade should lower blood pressure. In two studies (Bussien *et al.* 1984; Liard and Spadone 1984), administration of doses of a vasopressin blocker shown to prevent the effects of infused vasopressin did not lower arterial pressure. These studies may not provide definitive insights into the question, however, because, as with other types of antagonist drugs, vasopressin blocking drugs may have agonist properties (Creager *et al.* 1986). Conceivably, agonist vasoconstrictor properties could oppose antagonist properties and blood pressure might remain unchanged. Studies in animals suggest that vasopressin blocking drugs lower arterial pressure only when plasma vasopressin levels are elevated (Aisenbrey *et al.* 1981). A variant on this is that vasopressin blocking drugs which do not alter arterial pressure under control conditions may convert non-hypotensive haemorrhage into hypotensive haemorrhage (J. Schwartz and Reid 1981).

There are other problems with methods available for study of vasopressin physiology in humans. As far as we know, there has been no direct correlation made between the nerve activity that leads to vasopressin release and resulting plasma vasopressin levels (although bioelectrical activity of secretory cells and vasopressin release has been studied *in vitro* (Nordmann *et al.* 1985)). This stands in contrast to published correlations between muscle sympathetic nerve traffic and antecubital vein plasma concentrations of the released neurotransmitters noradrenaline and neuropeptide Y (Eckberg *et al.* 1988). Thus, although plasma vasopressin levels probably reflect activity in vasopressin secretory neurones, the nature of the relation between plasma vasopressin levels and neural activity is unknown. A related problem is that many studies treat vasopressin release as a steady-state constant; limited data from humans undergoing neck surgery points towards pulsed rather than continuous vasopressin release from the central nervous system (Hammer and Engell 1982).

SUMMARY

This chapter reviews many (but not all) end-organ responses to changes of arterial baroreceptor input. It documents a ubiquitous influence of arterial pressure on the sinoatrial and atrioventricular nodes and myocardium, as well as skeletal muscle and skin. Baroreflex influences on the kidney and hypothalamus provide additional means whereby arterial pressure modulates neurohumoral mechanisms and haemodynamic function.

PART D
Pathophysiology

10

Baroreflex malfunction in cardiovascular patients: general principles

The medical literature indicates clearly that baroreceptor reflex mechanisms are distorted in patients with cardiovascular diseases. Abnormalities of *parasympathetic* cardiac control have been recognized since the studies of Müller in 1891, and Crawford in 1923. These physicians found that their patients with heart diseases experienced less cardioacceleration after blocking doses of atropine sulphate than did healthy people. Similar observations were made in 1971 by Eckberg *et al.*, whose studies indicate also that patients with subnormal cardioacceleration after intravenous atropine also have subnormal vagally mediated R–R interval prolongation during arterial pressure elevations; in this study, the extent of both abnormalities was proportional to the functional severity of the underlying cardiovascular disease. These results suggest that patients with cardiovascular diseases have subnormal baseline levels of vagal–cardiac activity and subnormal increases of vagal–cardiac activity during baroreceptor stimulation.

Abnormalities of *sympathetic* cardiovascular control are also well recognized. Chidsey *et al.* (1965) reported histochemical evidence for profoundly reduced catecholamine concentrations in myocardium obtained from patients undergoing open heart surgery for valvular heart disease. Subsequently, Hasking *et al.* (1986) measured the kinetics of ^{3}H-noradrenaline turnover in patients with congestive heart failure and documented a fivefold increase of cardiac noradrenaline spillover, which was interpreted as signifying increased sympathetic neural traffic to the heart. The analyses of M.R. Bristow *et al.* (1982) indicate that increased sympathetic traffic to the heart is associated with decreased beta-adrenergic density. They found that biopsy specimens from failing hearts have decreased beta-adrenergic receptor densities, and subsensitivity to adrenergic stimulation. Brush *et al.* (1989) document a more subtle abnormality of sympathetic control; they found that hypertrophic cardiomyopathy patients with normal arterial plasma noradrenaline levels have impaired cardiac neuronal noradrenaline uptake.

Other evidence points towards more generalized increases of sympathetic activity in cardiac patients. In patients with heart failure, elevated levels of antecubital venous plasma noradrenaline (J.A. Thomas and Marks 1978; Cohn *et al.* 1984; Leimbach *et al.* 1986; Porter *et al.* 1990) and muscle sympathetic nerve traffic (Leimbach *et al.* 1986; Porter *et al.* 1990) tend to correlate with the severity of cardiac disease. Thus in heart failure patients there appears to be excessive sympathetic traffic to the heart associated with down-regulation of cardiac beta-receptors, and increased sympathetic traffic to other vascular beds, including especially that of skeletal muscle, associated with increased peripheral vascular resistance.

Although most patients reported in the studies cited above had congestive heart failure, some did not, and those with failure suffered from a variety of diseases. Since these patients and their diseases were heterogeneous, autonomic involvement in their pathophysiological processes may not be simple. Nevertheless, there may be broad principles that apply to baroreflex malfunction across disease categories. In the next five chapters we discuss baroreceptor reflex involvement in specific human disease processes. In this chapter we deal with baroreflex abnormalities found in cardiovascular patients in general terms. We try to develop some understanding of this pathophysiology with a consideration of the abnormalities that occur, from a variety of perspectives.

REPRODUCIBILITY OF BASELINE AUTONOMIC MEASUREMENTS

There is substantial published evidence which indicates that average human baroreflex responses and resting levels of autonomic efferent activity are reproducible when they are studied serially under exacting laboratory conditions, in the absence of interventions. (If experimental baroreflex forcings are dissimilar (D.S. Goldstein *et al.* 1982), or if physiological or pharmacological interventions are made between measurements, baroreflex responses and autonomic outflow may not be reproducible.) In healthy subjects (Lake *et al.* 1976) and patients with congestive heart failure (T.B. Levine *et al.* 1983), average antecubital vein plasma noradrenaline levels measured during supine rest on different days are reproducible. Moreover, in healthy volunteers, directly measured levels of muscle sympathetic nerve traffic are reproducible (and proportional to plasma noradrenaline levels), even when they are measured over intervals as long as months to years (Wallin *et al.* 1981).

So far there are no reports of serial measurements of resting muscle sympathetic nerve activity in patients with congestive heart failure.

Indirect estimates of resting levels of *vagal–cardiac* activity have also been found to be reproducible, at least in healthy volunteers. Kenney (1985) found that measurements of peak minus valley R–R interval changes during carefully controlled respiration are highly reproducible in healthy subjects studied on separate days. Gribbin *et al.* (1971) and Eckberg (1977*a*) found that in small groups of healthy volunteers studied twice over intervals up to 15 months, average R–R interval responses to phenylephrine injections or graded neck suction were virtually identical. Similarly, Kioski *et al.* (1986) found that baroreceptor–cardiac reflex slopes obtained in 13 hypertensive men after bolus injections of phenylephrine were comparable before and after a 6 weeks course of oral nifedipine therapy. (In this study, reproducibility of responses speaks against a significant long-term influence of nifedipine on baroreceptor–cardiac reflexes.)

Kasting *et al.* (1987) measured R–R interval responses of 11 healthy subjects to ramped neck pressure changes serially every 3 h, for 24 h. They found that although the position of the sigmoid relation between carotid distending pressures and R–R intervals varied on arterial pressure and R–R interval axes (according to circadian variations of those parameters), the characteristics of those responses were unchanged. The eleven subjects studied by Kasting also were evaluated on different days, 7 to 10 days apart. Average responses during such serial tests are depicted in Fig. 6.16.

The most exhaustive study of reproducibility of human vagal–cardiac reflex function is by Eckberg *et al.* (1992). These authors measured R–R interval interval responses of 34 healthy subjects to ramped neck pressure changes (depicted in Fig. 5.13) serially 7 to 10 days apart (a) or 10 weeks apart (b). These data, summarized in Fig. 10.1, indicate that human baroreflex responses are highly reproducible.

RECIPROCAL SYMPATHETIC AND VAGAL–CARDIAC RELATIONS

There are several studies of patients with cardiovascular diseases which suggest that pathological distortion of baroreflex responses tends to involve *reciprocal* changes of sympathetic and parasympathetic activity. Cross-sectional studies of hypertensive patients suggest that autonomic cardiovascular responses to tilt change reciprocally with advancing age: during upright tilt, hypertensive patients have smaller increases of heart rate and larger increases of systemic vascular resistance than younger people (London *et al.* 1987).

Jose and Taylor (1969) measured heart rate responses of healthy subjects and patients with cardiovascular diseases to large intravenous doses of the beta-adrenergic and muscarinic cholinergic blockers propranolol hydrochloride and atropine sulphate. In healthy subjects, heart rate increased

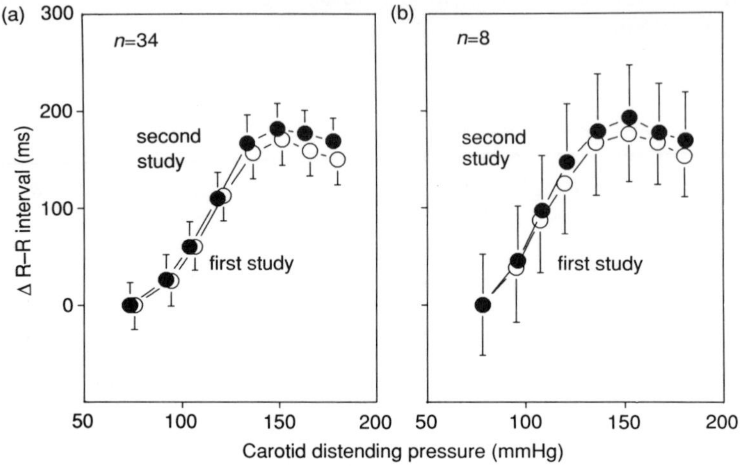

FIG. 10.1. Carotid distending pressure (systolic minus neck chamber pressure) for 34 subjects studied twice, (a) 7 to 10 days apart, or (b) 10 weeks apart. Vertical bars encompass one standard error of measurement. Reproduced with permission (Eckberg *et al.* 1992).

after injection of these drugs (the resulting heart rate was termed the 'intrinsic heart rate'). The increase of heart rate after combined blockade suggested that in these subjects, baseline levels of vagal inhibition were greater than baseline levels of sympathetic stimulation. In patients, heart rate increased also, but to a lesser extent than in normal subjects. In a few patients with severe cardiovascular diseases, after combined beta-adrenergic and muscarinic cholinergic blockade, heart rate slowed. The authors interpreted these findings as showing that as cardiovascular disease advances, there is a reciprocal reduction of vagal–cardiac inhibition, and increase of sympathetic–cardiac stimulation.

Julius *et al.* (1971) found that subjects with borderline hypertension (defined as diastolic pressures above and below 90 mmHg during serial testing) experience greater reductions of cardiac output after large intravenous doses of propranolol, and less acceleration of heart rate after large intravenous doses of atropine than do healthy subjects. This response pattern was interpreted as signifying reciprocal elevation of sympathetic, and depression of parasympathetic activity in these subjects.

D.S. Goldstein subsequently (1983*b*) obtained more direct evidence for such a relation. He studied 29 normotensive volunteers and 25 patients with hypertension and found that across all subjects, sympathetic activity, as estimated from antecubital vein plasma noradrenaline concentrations, and vagal–cardiac baroreflex responsiveness, as estimated from R–R interval interval prolongations after intravenous bolus injections of phenylephrine, are loosely but significantly related ($r = -0.58$, $p < 0.01$, linear regression). His data are shown in Fig. 10.2.

Porter and co-workers (1990) subsequently studied patients with severe congestive heart failure and age-matched healthy subjects. In this study, sympathetic activity was measured directly with percutaneously inserted peroneal microelectrodes and indirectly with antecubital vein plasma noradrenaline levels, and vagal traffic was estimated with standard deviations of R–R intervals during controlled breathing. Plasma noradrenaline levels were much higher than in the hypertensive patients studied by D.S. Goldstein (1983*b*). Porter's data point towards a reciprocal relation between sympathetic and vagal activities in heart failure patients. Patients with the highest plasma noradrenaline concentrations also had the lowest R–R interval standard deviations. However, a reciprocal relation was not apparent in data obtained from healthy subjects. D.S. Goldstein (1983*b*) also failed to identify a significant relation between plasma noradrenaline levels and vagal baroreflex slopes in healthy subjects. These observations suggest that under resting conditions, it is only when vagal–cardiac activity is at very low levels that sympathetic activity is elevated. That is, reciprocal depression of vagal outflow and elevation of sympathetic outflow may occur only under pathological conditions. (Indeed, Lindblad (1980) found in a group of eight healthy subjects that resting muscle sympathetic nerve activity may be related *directly* to vagal traffic, reflected by baseline R–R intervals.) Data from Porter's study are depicted in Fig. 10.3.

It should be noted that the reciprocal patterns of autonomic abnormalities

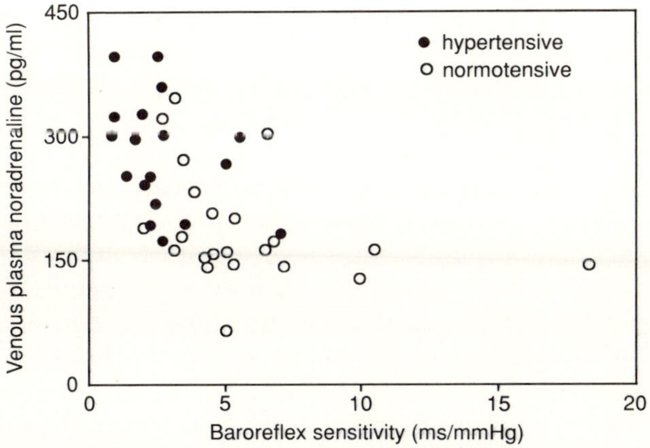

FIG. 10.2. Relation between slopes of baroreceptor–cardiac reflex responses (ratio of changes of R–R intervals to changes of systolic pressure after bolus phenylephrine injections) and antecubital vein plasma noradrenaline concentration, in 25 patients with hypertension and 29 healthy subjects. Adapted with permission (D.S. Goldstein 1983*b*).

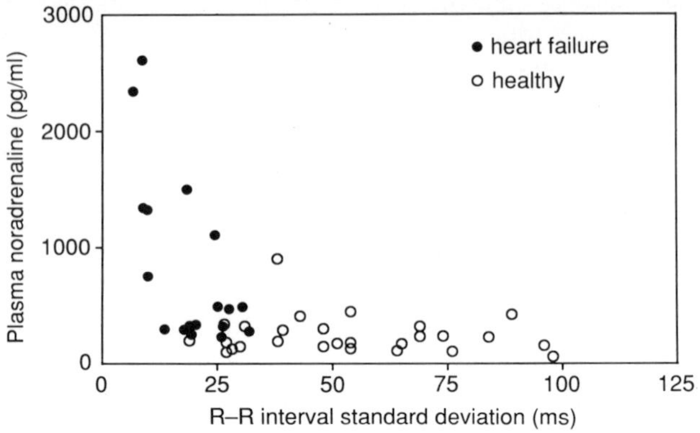

FIG. 10.3. Relation between R–R interval standard deviations and antecubital vein plasma noradrenaline levels in 18 patients with heart failure and 46 healthy subjects. Adapted with permission (Porter *et al.* 1990).

described above may *not* be found in certain specific diseases. For example, patients with diabetes mellitus may have parallel loss of sympathetic and parasympathetic reflex control due to peripheral autonomic neuropathy (B.F. Clarke *et al.* 1979). Patients with Chagas' disease may have degeneration of vagal–cardiac nerves and impairment of vagal–cardiac reflexes without corresponding abnormalities of sympathetic activity (Marin Neto *et al.* 1975).

THE LINK BETWEEN SYMPATHETIC AND VAGAL LIMBS OF THE BAROREFLEX ARC

Sympathetic–parasympathetic relations in healthy subjects

Most studies of baroreflex function tend to concentrate on either vagal–cardiac or sympathetic (or more likely, blood pressure) responses to experimental interventions. Typically, measurements of vagally mediated cardiac slowing are made, and the results are taken as representative of all aspects of baroreflex function. Not surprisingly, questions have been raised (Ludbrook *et al.* 1980) regarding which is the more important aspect of baroreflex regulation: heart rate or blood pressure control? The answer to this question may depend upon the goals of a study. If the goal is to evaluate sympathetic and vagal baroreflex responses, it may not be helpful to focus strictly upon heart rate or arterial pressure, since both parameters may change in response to changes of sympathetic or vagal activity. Moreover, although abrupt changes of heart rate seem to be mediated exclusively by changes of vagal–cardiac efferent activity (Katona and Jih 1975; Akselrod *et al.* 1981; Eckberg

1983), changes of blood pressure are complex and are not necessarily synonymous with changes of sympathetic activity (Båth *et al.* 1981).

Also, implicit in this question is the assumption that changes of heart rate play no role in regulation of arterial pressure. Data from conscious rabbits (Faris *et al.* 1980*b*) support this view. However, studies conducted in other animals and patients point towards an important role for heart rate changes in blood pressure regulation. Akselrod *et al.* (1985) found that fixed-rate atrial pacing or high-dose atropine in conscious dogs nearly abolishes blood pressure variability. Data published in preliminary form (Stephenson and Tagett 1984) from conscious dogs whose resting heart rates are close to those of humans suggest that vagally mediated heart rate changes may be very important in arterial pressure regulation.

Dogs given chronic (28 day) infusions of noradrenaline sufficient to raise their average venous plasma concentrations to over 4000 pg/ml remain normotensive, apparently entirely through vagal bradycardia (King *et al.* 1987). Healthy people whose muscle sympathetic nerve bursts occur at nearly one per heart beat cannot increase the incidence of bursts per heart beat, and consequently increase their sympathetic traffic during standing by increasing their heart rates (Burke *et al.* 1977). In humans, changes of heart rate may be particularly important to maintenance of haemodynamic function. Several studies (R.D.S. Watson *et al.* 1980; Conway *et al.* 1984; Mancia *et al.* 1986; Floras *et al.* 1988*a*) document an inverse relation between vagally mediated baroreflex responses and blood pressure variability. Båth *et al.* (1981) documented a significant reciprocal relation between between muscle sympathetic nerve activity and R–R interval responses to sinusoidal neck suction. Subjects whose net response was to increase the number of sympathetic bursts per 100 heart beats also had large increases of R–R interval, but subjects who reduced the number of sympathetic bursts had only small increases of R–R interval.

The suggestion that changes of heart rate are not important in blood pressure control also makes the tacit assumption that sympathetic and vagal baroreflex stimulus–response relations are distinct. This is an intriguing possibility which may not have been researched adequately in animals or humans.

The approach to the study of human baroreceptor reflexes has been dictated in part by methods available for study. In most studies, baseline arterial pressure has been treated as a watershed; reflex responses to pressures above resting levels have been studied with one type of intervention (such as pressor injections or neck suction), and reflex responses to pressures below resting levels have been studied with another (such as intravenous nitroglycerine or neck pressure). The principal problem with studies that employ successive increases and decreases of arterial pressure to characterize baroreflex responses is that they ignore hysteresis; R–R intervals and sympathetic firing respond differently to rising and falling pressures (Smyth *et al.* 1969; Sundlöf and Wallin 1977).

Nevertheless, this approach has some physiological justification. One study employing neck pressure changes suggested that vagal mechanisms are engaged primarily by pressure changes *above* resting levels (Eckberg 1980*a*), and another employing pharmacological arterial pressure changes suggested that sympathetic mechanisms are engaged primarily by pressure changes *below* resting levels (Eckberg *et al.* 1988). There is yet another experiment that examined both limbs of baroreflex responses in the same subjects. This study by Rea and Eckberg (1987) suggests that in healthy human volunteers, small-to-moderate variations of carotid afferent traffic above and below resting levels provoked by neck pressure changes modify *both* sympathetic and vagal cardiovascular outflows. Because there is some overlap of working areas, it may be reasonable to expect that for pressures immediately above and below resting levels, the sigmoid vagal baroreceptor–cardiac reflex relation may bear some mathematical relation to the inverse sigmoid baroreceptor–sympathetic relation. If this be correct, it may be reasonable to extrapolate from vagal to sympathetic reflex responses. Figure 10.4 documents a remarkable similarity between *human* baroreceptor–cardiac vagal and –muscle sympathetic nerve relations (upper panels (Rea and Eckberg 1987)) and *canine* baroreceptor–vagal and –sympathetic nerve relations (lower panels (Kollai and Koizumi 1989)). (The solid lines indicate reflex gain.) The importance of this comparison may be not so much that response patterns are similar in human and canine species, but that they are similar in human *muscle* and canine *cardiac* sympathetic outflows.

One study conducted in healthy chloralose–urethane-anaesthetized rabbits (Guo *et al.* 1982) suggests that baroreceptor–vagal and –sympathetic relations are not necessarily related. Responses to pharmacological arterial pressure changes were studied during serial sections of aortic and carotid baroreceptor nerves. Section of either aortic or carotid sinus nerves impaired baroreflex modulation of heart rate, but did not alter baroreflex modulation of resistance in an isolated perfused hindlimb. The authors extrapolate from these data and suggest that measurements of baroreflex control of heart rate in healthy or unhealthy humans may not convey accurate information about baroreflex control of arterial pressure.

Data obtained from healthy human subjects are contradictory. Mancia *et al.* (1979*b*) measured R–R interval and cardiac output responses to static neck suction and concluded that the magnitudes of the two types of responses are unrelated to each other. Båth *et al.* (1981) used sinusoidal neck suction and found that changes of R–R intervals correlate closely with changes of directly measured muscle sympathetic nerve activity. We suspect that since changes of sympathetic and vagal cardiovascular outflow are driven by the same stimulus — afferent baroreceptor activity — they must

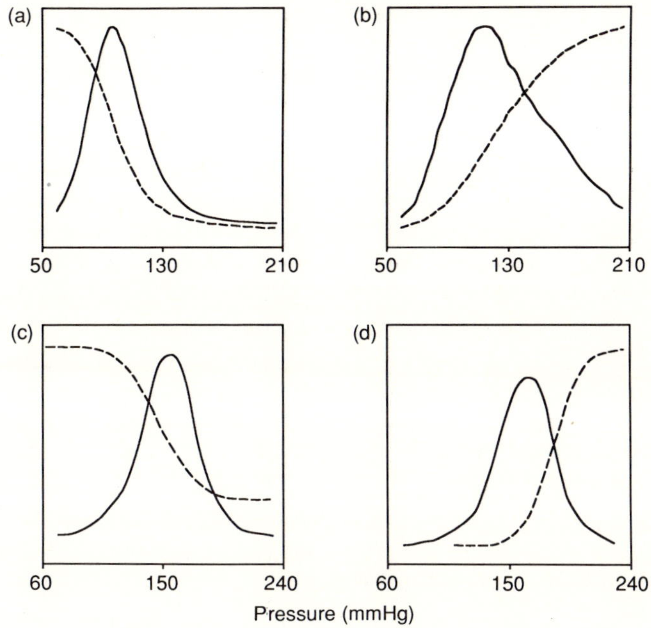

FIG. 10.4. Upper panels: relation between carotid distending pressure (systolic minus neck chamber pressure) and (a) peroneal nerve muscle sympathetic activity and (b) changes of R–R intervals in nine healthy subjects. Adapted with permission (Rea and Eckberg 1987). Lower panels: relation between systolic arterial pressure and (c) cardiac sympathetic and (d) vagal nerve activity in anaesthetized dogs. Adapted with permission (Kollai and Koizumi 1989). Dashed lines indicate nerve or sinus node responses and solid lines indicate calculated reflex gain.

be related mathematically. However, this question is controversial, and merits further study.

Sympathetic–parasympathetic relations in patients with cardiovascular diseases

Data from patients suggest (but fall short of proving) that measurements of vagal baroreflex function provide insights into sympathetic baroreflex function. Kiowski and co-workers (1986) found in 13 hypertensive patients that the slopes of R–R interval–systolic pressure relations after intravenous bolus injections of phenylephrine were related inversely to subsequent reductions of arterial pressure provoked by 10 mg doses of sublingual nifedipine. Similarly, Shepherd *et al.* (1981) found in six hypertensive patients that the degree of R–R interval shortening during arterial pressure reduction with nitroprusside (which is thought to be mediated partly by vagal withdrawal

(S.F. Vatner *et al.* 1974; Eckberg 1980*a*)) was proportional to the elevation of plasma noradrenaline levels.

A similar conclusion was drawn in a study published by Packer and co-workers (Packer *et al.* 1979). They gave intravenous sodium nitroprusside to 20 patients with severe congestive heart failure and studied rebound hemodynamic responses (which included increases of systemic vascular resistance) to cessation of infusions. In these patients, the magnitude of the increase of systemic vascular resistance was related inversely to the magnitude of the tachycardia that occurred in response to the initial infusion to nitroprusside. If tachycardia in these patients resulted from vagal withdrawal, these results point to a relation between vagally and sympathetically mediated baroreflex responses. However, if tachycardia during hypotension in these patients was mediated by increased sympathetic rather than by decreased vagal activity, these data may indicate merely that sympathetic tachycardia is proportional to sympathetic vasoconstriction. Studies conducted in conscious dogs (Higgins *et al.* 1972*a*; S.F. Vatner *et al.* 1974) and humans (R.E. Goldstein *et al.* 1975) with heart failure suggest that cardioacceleration responses to hypotension are mediated more by sympathetic augmentation than vagal withdrawal.

MECHANISMS UNDERLYING ABNORMAL BAROREFLEX FUNCTION

Mechanisms underlying baroreflex abnormalities in patients with cardiovascular diseases are understood poorly. It is not yet known whether there is a common abnormality that transcends disease categories, or whether individual disease processes have individual pathophysiological features. For purposes of discussion, we have taken the approach that a broad range of cardiovascular diseases share pathophysiological features. We recognize that certain pathological conditions, such as Chagas' disease (a parasitic infestation that may damage vagal–cardiac nerves directly (S.A. Morris *et al.* 1990)) may constitute special cases, and that subsequent research may allow finer discriminations to be made.

Abnormalities of baroreceptors

There is inferential evidence which suggests that baroreflex malfunction may arise because of abnormalities in baroreceptors. Patients recovering from acute myocardial infarction may have subnormal R–R interval prolongation in response to boluses of phenylephrine (Imaizumi *et al.* 1984*b*). However, such patients may also have supranormal heart rate slowing in response to facial immersion in cold water (a simulated diving response (Gooden *et al.* 1978)). Although the afferent pathways for the simulated diving response and the baroreflex are different (Kumada *et al.* 1977), the efferent portions of the reflexes are the same. Therefore an abnormality of baroreflex mediated bradycardia without an abnormality in simulated diving bradycardia points towards an abnormality of baroreceptors (or central pathways). Studies by Niebauer and Zucker (1985) and White (1981) document abnormal carotid baroreceptor responses in dogs with aortocaval fistulae, and a magnetic resonance imaging study by Mohiaddin *et al.* (1989) documents subnormal aortic compliance (which should impair baroreceptor transduction) in patients with coronary artery disease.

Paradoxical baroreflex responses in cardiovascular patients

There are a number of baroreflex response patterns of patients with cardiovascular diseases that appear to be *qualitatively* different from those of healthy people. Phenylephrine, which raises blood pressure and slows heart rate in healthy subjects, may cause *heart rate speeding* in some patients with heart failure (Eckberg *et al.* 1971) or diabetic autonomic cardiac neuropathy (Lloyd-Mostyn and Watkins 1976). Carotid baroreceptor stimulation, which reduces left ventricular stroke volume in healthy subjects (T.J. Ebert, unpublished) increases stroke volume in patients with severe heart failure (Osterziel *et al.* 1990*a*). Exercise, which raises blood pressure in healthy subjects, may lower pressure in aortic stenosis patients (L.S. Schwartz *et al.* 1969; Mark *et al.* 1973). Digitalis, which increases forearm vascular resistance in healthy subjects, may reduce resistance (D.T. Mason and Braunwald 1964) and muscle sympathetic nerve activity (Ferguson *et al.* 1989) in heart failure patients. Upright tilt or standing, which increases noradrenaline spillover and causes vasoconstriction in healthy people, may reduce noradrenaline spillover (Davis *et al.* 1987) and cause vasodilatation in heart failure patients (Brigden and Sharpey-Schafer 1950; Goldsmith *et al.* 1983*a*; Ferguson *et al.* 1984; Kassis *et al.* 1986). Lower body suction, which in healthy people increases vascular resistance and plasma noradrenaline levels, may provoke vasodilatation and reduction of plasma noradrenaline levels in some patients with severe heart failure (Ferguson *et al.* 1984; Mohanty *et al.* 1989). The second phase of the Valsalva manoeuvre, which is associated with reductions of pulse pressure in healthy people, may increase pulse pressure in heart failure patients (Sharpey-Schafer 1955). Nitroglycerine (Faxon *et al.* 1980*b*), hydralazine (Fitchett *et al.* 1979), and nitroprusside (Packer *et al.* 1979), which lower arterial pressure and speed heart rate in healthy subjects, commonly slow the heart rate in patients with heart failure. Sodium nitroprusside, which increases plasma noradrenaline levels in healthy subjects, may *lower* plasma noradrenaline levels in patients with heart failure (Olivari *et al.* 1983). Interventions which cause venous pooling in the legs, such as lower body negative pressure and upright posture, increase sympathetic activity, and provoke vasoconstriction in healthy subjects may reduce plasma noradrenaline levels and cause vasodilatation in patients with heart failure (R.H. Murray *et al.* 1969; Kubo and Cody 1983; T.B. Levine *et al.* 1983).

Another example of qualitatively abnormal responses in heart failure comes from a study of conscious dogs by LeWinter *et al.* (1978). They found that prior to the placement of an arteriovenous fistula, progressive haemorrhage provokes cardioacceleration in proportion to the amount of blood withdrawn. After the creation of high output failure, however, heart rate fails to increase during progressive haemorrhage and then abruptly declines. A closely related observation was made by Branham in 1890; he compressed a traumatic femoral arteriovenous fistula in a patient (and presumably reduced blood pressure) and provoked striking bradycardia. Come and Pitt (1976) described five patients with acute myocardial infarction who experienced vasovagal reactions in response to small doses of intravenous nitroprusside or sublingual nitroglycerine.

On the surface, these untoward response patterns of patients with cardiovascular diseases seem to be drawn from a different repertory from that available to healthy people. This issue warrants further study. However, we suggest that some qualitatively abnormal response patterns may have counterparts in normal human physiology. For example, extreme hypotensive stresses provoke vasovagal reactions in healthy subjects. If healthy volunteers are bled sufficiently (Barcroft *et al.* 1944; Stevens and Lamb 1965), or subjected to prolonged, intense lower body negative pressure (R.H. Murray *et al.* 1969), or passively tilted and maintained at 60 degrees (Sander-Jensen *et al.* 1986*a*), they may experience abrupt reductions of arterial pressure and slowing of heart rate. Healthy people may reduce muscle sympathetic nerve activity during upright tilt if their pre-tilt levels of sympathetic activity are high (Burke *et al.* 1977).

Other evidence from healthy animals and humans suggests that similar reflex changes may occur in response to less severe stresses. In conscious dogs (Morita and Vatner 1985) and rabbits (Dorward *et al.* 1985), changes of renal sympathetic nerve traffic during moderate progressive haemorrhage are bimodal; sympathetic activity rises first and then falls. In healthy humans, stepwise infusions of sodium nitroprusside lead to increases, then reductions of antecubital vein plasma noradrenaline levels (Eckberg *et al.* 1986). In the human study, the possibility that reductions of plasma noradrenaline levels at higher rates of nitroprusside infusion reflect dilution secondary to increased forearm blood flow was not excluded. Also, another study (Eckberg *et al.* 1988) failed to document reductions of directly measured muscle sympathetic nerve traffic during progressive stepwise nitroprusside infusions. However, the study of Roddie and Shepherd (1957) suggests that reductions of human sympathetic traffic may occur during progressive reductions of afferent baroreceptor activity; they documented forearm vasodilatation during large reductions of carotid artery pressure during manual bilateral carotid artery occlusions. Data from the study of Eckberg *et al.* (1986) are depicted in Fig. 10.5. These three studies point towards late reductions of sympathetic activity during progressive, mild-to-moderate blood pressure reductions. Such responses, seen in healthy animals and humans provide reason to suspect that some abnormal baroreflex mediated responses of patients with cardiovascular diseases may be merely quantitatively, not qualitatively, different from normal.

Thus, for reasons that are as yet unclear, minor hypotensive stimuli which in healthy subjects increase sympathetic activity reduce sympathetic activity (and increase vagal activity) in some patients with cardiovascular diseases. We suspect that much could be learned about human pathophysiology if factors underlying vasovagal reactions were understood better.

Vasovagal reactions, and their early less dramatic manifestations, appear to be mediated by paradoxical increases of afferent autonomic sensory input. As discussed in Chapter 6, studies by Öberg and White (1970) and Öberg and Thorén (1972) suggest that vasovagal reactions result from increases of afferent traffic carried over unmyelinated vagal fibres from the left ventricle (Sleight 1964; Sleight and Widdicombe 1965). During blood volume reductions, contractility is augmented by exaggerated levels of sympathetic input, augmented contractility leads to obliteration of the left ventricular cavity, and this leads to increased firing of ventricular receptors (Gauer and Henry 1964). Curiously, ventricular receptors are also implicated in the the syncope that occurs in patients with aortic stenosis (L.S. Schwartz *et al.* 1969; Mark *et al.* 1973). However, in this circumstance firing of ventricular receptors is thought to result from the very high intraventricular pressures that may be generated (usually during exercise).

Morita and Vatner (1985) suggested that such paradoxical responses may

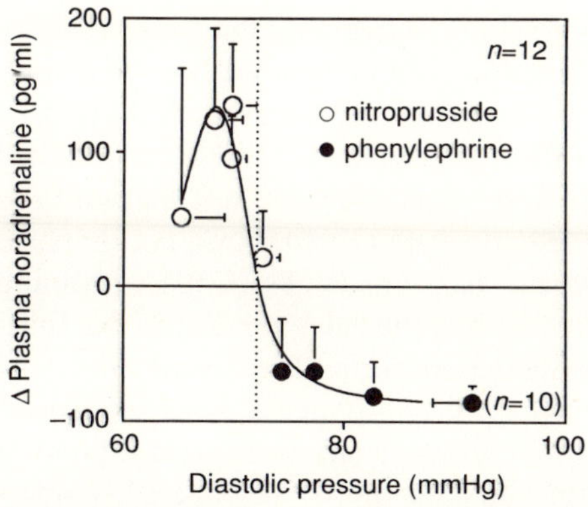

FIG. 10.5. Plasma noradrenaline levels of healthy subjects during arterial pressure reductions with nitroprusside and pressure elevations with phenylephrine. Dashed line indicates control diastolic pressure. Adapted with permission (Eckberg *et al.* 1986).

not be mediated exclusively by left ventricular receptors; they showed in conscious dogs that progressive haemorrhage may lead to reductions of renal sympathetic nerve traffic *after vagotomy* (which denervates cardiac receptors). A human counterpart to this observation may be the reduction of plasma renin activity that occurs in healthy subjects with vasovagal reactions triggered by upright tilt (Oparil *et al.* 1970). Other studies indicate that paradoxical augmentation of firing in response to pressure reductions occurs in arterial (Heymans *et al.* 1931; Landgren 1952*a*; McCubbin *et al.* 1956) as well as ventricular baroreceptors. Also, as discussed in Chapter 6, one case report suggests that innervated ventricles are not *necessary* for vasovagal reactions to occur. Scherrer *et al.* (1990*b*) reported the occurrence of a vasovagal reaction, with near cessation of efferent muscle sympathetic nerve activity during infusion of the vasodilator nitroprusside in a patient with a heart transplant. (Of interest, published recordings of human muscle sympathetic nerve activity during vasovagal reactions (Wallin and Sundlöf 1982; Sanders and Ferguson 1989; Scherrer *et al.* 1990*b*; van Lieshout *et al.* 1991) all show near *cessation* of activity. Similarly, norepinephrine spillover to cardiac and renal vascular beds declines strikingly (Esler *et al.* 1990).) If vasovagal reactions are due to active vasodilatation, as suggested by Barcroft and Edholm (1945), sympathetic traffic and norepinephrine spillover should increase, not decrease.)

Baroreflex modulation of autonomic outflow

The next chapter indicates that the role of arterial baroreceptors in hypertension is highly controversial. A related issue which has received much less attention is: what role, if any, do baroreflexes play in establishing ongoing levels of sympathetic and vagal efferent activity? We consider this to be an open question. The possibility that ongoing baroreflex responses determine *chronic* levels of autonomic outflow has not been explored in depth; there are reasons to think that this might be the case. Although Persson *et al.* (1988) showed that denervation of sinoaortic and cardiopulmonary baroreceptors leads to sustained hypertension, they did not measure sympathetic traffic or noradrenaline levels. Shade *et al.* (1990) found that hypertension after chronic sinoatrial denervation in baboons is associated with increased plasma levels of noradrenaline. This finding provides some evidence that denervation hypertension results from increased sympathetic outflow.

As mentioned, heart failure patients have increased sympathetic and reduced vagal outflows. When heart transplantation is used as therapy for heart failure, patients' vagal baroreflex responses become steeper (Ellenbogen *et al.* 1989), their plasma noradrenaline levels return towards normal (Olivari *et al.* 1987), and they increase their resting respiratory sinus arrhythmia (of the innervated atrium) (Smith *et al.* 1989, 1990). Although these data document an association between improvement of baroreflex function and chronic changes of autonomic cardiovascular nervous outflow, they do not prove a cause-and-effect relation.

There is evidence for substantial ongoing reciprocally changing levels of sympathetic and vagal outflows, at least in resting healthy volunteers. In almost all healthy subjects (Eckberg *et al.* 1985), there are strong cardiac and respiratory periodicities in muscle sympathetic nerve recordings. Båth *et al.* (1981) and Wallin and Nerhed (1982) showed that episodic sympathetic bursts are followed by elevations of arterial pressure, and Fritsch *et al.* (1986) showed that these pressure elevations are associated with proportional, parallel prolongations of R–R intervals. These data suggest that changes of muscle sympathetic nerve activity provoke changes of arterial pressure, which in turn provoke changes of vagal–cardiac activity. Conversely, abrupt prolongations of R–R intervals provoke abrupt increases of muscle sympathetic nerve activity (Fagius 1988; Welch *et al.* 1989). Physiological alternation of sympathetic and vagal outflows in response to recurring small changes of arterial pressure in a healthy, resting subject is shown in Fig. 10.6.

It is possible that vagally mediated baroreflex responses in some way determine absolute levels of sympathetic nerve traffic. At least three studies (R.D.S. Watson *et al.* 1980; Ogawa *et al.* 1981; Mancia *et al.* 1986) show that arterial pressure variability is greater in hypertensive patients than in healthy subjects. In 24 hour recordings of intraarterial pressure, the extent of arterial

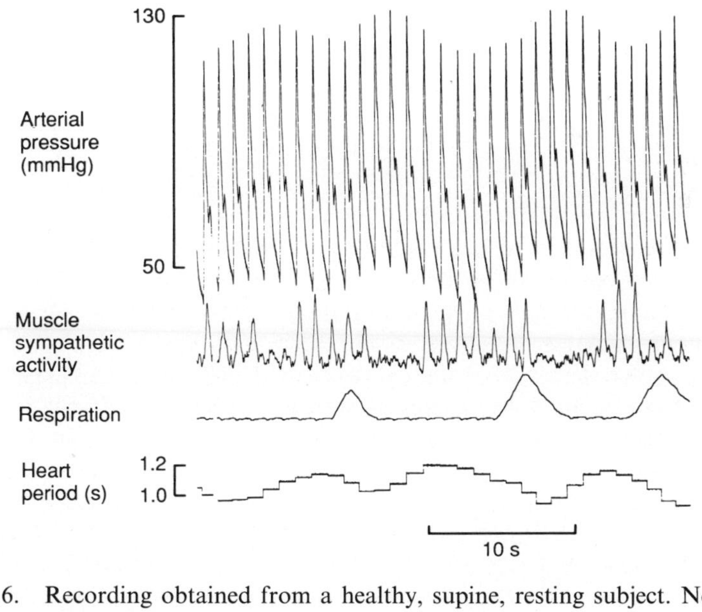

FIG. 10.6. Recording obtained from a healthy, supine, resting subject. Notice the reciprocal relation between sympathetic activity and R–R intervals. Adapted with permission (Eckberg *et al.* 1985).

pressure variations (presumably mediated primarily by sympathetic nerve traffic variations) is inversely related to slopes of *vagal* baroreceptor–cardiac reflex responses. It may be that in such patients, impairment of vagally mediated cardiac slowing and speeding reduces buffering of blood pressure by changes of heart rate, and thereby promotes greater than normal fluctuations of arterial pressure.

How might such impairment of *vagal*–cardiac reflex changes be translated into increased levels of sympathetic nerve traffic? The studies of Wallin and co-workers have provided critical insights into the relation between arterial pressure and muscle sympathetic nerve activity. Their data indicate clearly that average levels of arterial pressure are unrelated to average levels of muscle sympathetic nerve activity (Wallin *et al.* 1973*a*; Sundlöf and Wallin 1978*b*; Wallin and Sundlöf 1979). However, this is not to say that there is no relation between arterial pressure and sympathetic activity. In subjects at supine rest, there is a highly significant relation between fluctuations of arterial pressure and fluctuations of muscle sympathetic activity, such that sympathetic activity is inversely related to *diastolic* pressure, but is unrelated to *systolic* pressure (Sundlöf and Wallin 1977). Thus these data link changes of diastolic pressure with changes of muscle sympathetic activity in resting humans.

Is there a link between vagal–cardiac baroreflex responses and diastolic pressure? De Boer *et al.* (1985*b*) used power spectral analyses to study the relation between respiration and systolic and diastolic pressures. They found a strong respiratory periodicity in systolic pressures, but *no* respiratory periodicity in diastolic pressures. They interpreted their findings as follows. Changes of systolic pressures evoke parallel changes of R–R intervals with short latency. During brief elevations of systolic pressure, prolongations of R–R intervals lead to prolongations of diastolic runoff, which reduce increases of diastolic pressure that might be expected to follow increases of systolic pressure.

From this, it would be expected that subjects with very responsive vagal–cardiac baroreflexes have less variability of diastolic pressures than subjects with depressed vagal–cardiac baroreflexes. Since changes of diastolic pressure critically determine levels of muscle sympathetic nerve activity (Sundlöf and Wallin 1977), greater variability of diastolic pressure should be associated with augmented levels of sympathetic activity. We are not aware of any analysis of arterial pressure variability in patients with congestive heart failure. However, such patients have grossly impaired vagal responses to arterial pressure changes (Eckberg *et al.* 1971), and elevated levels of muscle sympathetic nerve activity (Leimbach *et al.* 1986; Porter *et al.* 1990), which are strongly related to respiration (Porter *et al.* 1990). Therefore we speculate that impaired vagal–rdiac baroreflex responses to respiration-related fluctuations of arterial pressure contribute to exaggerated fluctuations of diastolic pressure, and that these in turn contribute to exaggerated levels of muscle sympathetic activity. This speculation could be tested experimentally.

Appropriateness of animal models of human baroreflex malfunction

Much current understanding of baroreflex abnormalities comes from research involving baroreceptor *denervation*. As far as we can determine, the notion that the physiological role of baroreceptors can be understood best when measurements are compared before and after baroreceptor denervation has not been challenged.

Support for the view that denervation is the most extreme form of baroreceptor malfunction in humans comes from the small but important literature on the effects of local anaesthesia of baroreceptor areas (Lampen *et al.* 1949; Guz *et al.* 1966; Fagius *et al.* 1985). This research suggests that baroreceptor mechanisms are important indeed: after injection of lignocaine into the regions of the carotid sinuses or base of the skull, blood pressure may rise to great heights. (In the extraordinary study of Lampen *et al.* (1949) conducted in patients with malignant hypertension, *average* pressures in six patients reached levels of 321/187 mmHg!) It is likely that in most of these studies, local anaesthesia involved afferent and efferent *vagal* fibres as well as afferent carotid baroreceptor (and chemoreceptor) fibres. Therefore the relative contributions of afferent carotid sinus nerve traffic and afferent and efferent vagus nerve traffic cannot be determined.

Although these studies suggest that baroreceptors regulate baseline levels of autonomic cardiovascular outflow over minutes to hours, they do not prove that baroreceptors regulate autonomic outflow chronically. Blood pressure and plasma noradrenaline concentrations rise substantially in dogs (Cowley *et al.* 1973) and rats (N. Alexander and DeCuir 1966) immediately following baroreceptor denervation. However, in such studies, average blood pressures (or plasma noradrenaline levels) eventually return to near normal levels. Such studies raise doubts about the role of baroreceptors in determining chronic levels of efferent autonomic outflow.

There are no studies of humans with complete arterial and cardiopulmonary denervation. However, Aksamit *et al.* (1987) reported a patient with paroxysmal elevations of arterial pressure and plasma noradrenaline levels, whose extensive mediastinal and neck irradiation was thought to have destroyed all arterial baroreceptors. Holton and Wood (1965) studied two patients before and after bilateral carotid sinus denervation for asthma. Both patients had hypertension, tachycardia, and impaired blood pressure responses to tilt in the post-operative period. Sixty-one and 43 weeks after surgery, heart rate and responses to tilt returned to normal. Blood pressure remained elevated in one patient and returned to normal in the other.

In their seminal study of the effects of sinoaortic denervation in conscious dogs, Cowley *et al.* (1973) showed that after denervation, blood pressure is extremely labile and does not increase appropriately after blood pressure reductions, such as those provoked by standing. In their study, mean arterial pressure averaged only 11 mmHg above that in sham-operated dogs. Cowley's conclusion, that arterial baroreceptors are not necessary to maintain pressure at normal levels, is challenged by Scher and Ito (1980), who found that arterial baroreceptor denervation leads to increased resting blood pressure levels. The study of Persson *et al.* (1988) may have resolved this issue. These authors went to great lengths to be certain that *cardiopulmonary* as well as arterial baroreceptors were denervated in their dogs. They found that average mean arterial pressure after complete denervation of all afferent baroreceptors was substantially elevated (by 22 mmHg). Frequency distributions of blood pressures in Cowley's normal and baroreceptor denervated dogs are depicted in Fig. 10.7.

In all studies of baroreceptor denervation, including the one involving carotid sinus denervation in humans (Holton and Wood 1965), blood pressure was elevated in the early days and weeks after surgery. Anatomical or functional changes (neural plasticity) may take place in medullary synapses in response to distortion of afferent input. Early reversal of hypertension following section of one set of baroreceptor afferent nerves (O'Leary and Scher 1990) may reflect this adaptive process. Thus partial baroreceptor denervation may increase the importance of remaining baroreceptor inputs, such that one group of receptors is able to maintain pressure at normal or near normal levels. In the case of dogs with sinoaortic denervation, that group appears to be cardiopulmonary baroreceptors. However, cardiopulmonary receptors seem unable to respond to *abrupt* changes of pressure. This has several consequences: blood pressure variability is increased, blood pressure is not supported during standing, and R–R intervals may not lengthen after bolus injections of pressors. Cardiopulmonary receptors may provide information on chronic levels of pressure, however, and this input restores blood pressure towards normal. An important corollary of this conclusion is that baroreceptors *'remember'* and function to establish chronic blood pressure levels.

In a provocative but rarely cited study, Burstyn *et al.* (1972) suggested

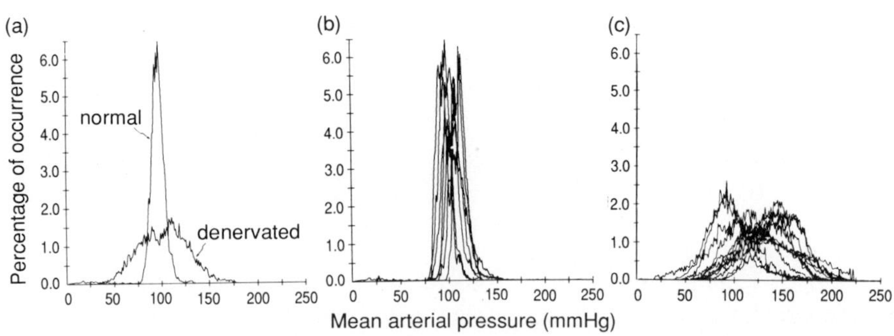

FIG. 10.7. Frequency distribution of mean arterial pressure in conscious dogs with all baroreceptors intact and with sinoaortic denervation: (a) mean data; (b), (c) individual data from (b) intact and (c) sinoaortic denervated dogs. Adapted with permission (Cowley *et al.* 1973).

malfunction. These authors placed rigid epoxy casts around both carotid sinuses of eight rabbits and around the carotid artery above the carotid sinuses in four rabbits. Subsequent reflex testing (bilateral carotid clamping and intravenous angiotensin injections) disclosed that two rabbits had absent and one rabbit had diminished baroreflex responses. Baroreflex responses were intact in five other rabbits with carotid sinus casts and four sham-operated rabbits. At the end of 12 weeks, sham-operated rabbits and rabbits with putative baroreceptor denervation had normal blood pressures, but rabbits with carotid sinus casts and *functioning* carotid sinus baroreflexes had significantly elevated pressures (averaging 38 per cent above preoperative levels). The authors interpreted their results as indicating that *abnormal* baroreceptor input may increase arterial pressure more than *absent* baroreceptor input.

It seems likely that levels of sympathetic and vagal cardiovascular outflow are determined importantly by second-to-second changes of the *profile* of all autonomic sensory inputs. It may be overly simplistic to assume that changes of autonomic output that occur after changes of a sensory input reflect merely the loss of that input. For example, although minor arterial pressure reductions provoke increases of muscle sympathetic nerve activity that are sustained (Eckberg *et al.* 1988), large increases of neck pressure provoke increases of muscle sympathetic nerve activity that last for only one or two heart beats (Wallin and Eckberg 1982). Sanders and co-workers (1988) suggested that this disparity in the duration of responses indicates that carotid baroreceptors are not important for regulation of arterial pressure.

There are at least three alternative interpretations, however. First (and obviously), although afferent carotid baroreceptor activity is reduced by neck pressure, aortic baroreceptor activity is increased (by reflex pressure elevations). Second, servo-regulatory responses to abrupt changes of only a part of arterial baroreceptor input may be subtle; under such circumstances, information from carotid baroreceptors, which conflicts with information from aortic baroreceptors, may be discounted. Third, changes of input from one barosensory area may alter pressure transduction by another barosensory area. Behnia *et al.* (1987) found that during maintenance of constant carotid sinus pressures, lignocaine blockade of one carotid sinus augments firing rates of contralateral carotid baroreceptors.

PRACTICAL CONSEQUENCES OF BAROREFLEXES

Influence of baroreflexes upon therapy

There is substantial evidence that baroreceptor reflexes alter responses of patients to drugs and that these influences have practical importance. It is well recognized that the antihypertensive effects of hydralazine are limited by baroreflex responses to induced hypotension (Koch-Weser 1976), and that this adverse influence can be prevented by beta-adrenergic blockade (Zacest *et al.* 1972). Thus baroreflex mediated cardioacceleration opposes the ability of hydralazine to lower blood pressure. Similarly, as mentioned, the magnitude of the acute hypotensive response to the calcium-channel blocker nifedipine is inversely related to the vagal bradycardia that occurs in response to bolus phenylephrine injections (Kioski *et al.* 1986); therefore, antihypertensive effects of nifedipine may be diminished in patients with sensitive baroreflexes. Patients with heart failure who develop cardioacceleration after sublingual nitroglycerine usually also experience reductions of cardiac output; patients who develop minimal cardioacceleration (or who develop bradycardia) do not experience reductions of cardiac output (Faxon *et al.* 1980*b*). Myocardial infarction patients who increase their heart rates in response to vasodilators do not have as great an increase of cardiac output as those who do not increase their heart rates (Chatterjee *et al.* 1973).

Influence of baroreflexes upon prognosis

There is compelling if circumstantial evidence that baroreflex function is related inversely to prognosis. The most direct evidence comes from studies conducted in conscious dogs by Billman *et al.* (1982). They measured slopes of vagally mediated increases of R–R intervals after boluses of phenylephrine in dogs who had recovered from anterior myocardial infarction, and related these slopes to the occurrence of ventricular dysrhythmias during and following treadmill exercise and circumflex coronary artery occlusion (leading to inferior left ventricular ischaemia). Dogs with diminished baroreflex slopes were much more likely to develop ventricular fibrillation during exercise and myocardial ischaemia than dogs with high slopes. One dog improved its baroreflex slope spontaneously (with no therapeutic intervention) and changed from being susceptible to being resistant to ventricular fibrillation. Other dogs placed upon a programme of regular exercise also increased their baroreflex slopes and became resistant to exercise-induced ventricular fibrillation. Dogs whose baroreflex responses remained depressed continued to be vulnerable to ventricular fibrillation.

Data from humans are less extensive but tend to support similar conclusions. In patients, the risk of death after acute myocardial infarction is not distributed uniformly over time; the risk is greatest during the earliest weeks after hospital discharge (Gilpin *et al.* 1983). Two groups have documented transient reductions of baroreflex function during this period. Imaizumi *et al.* (1984*b*) found a temporary impairment of forearm vasoconstriction in response to intense lower body negative pressure during 11 patients' early convalescence from uncomplicated acute myocardial infarctions. This group and that of P.J. Schwartz *et al.* (1988*b*) also documented temporary reductions of baroreflex slopes provoked by phenylephrine injections. Whether transient impairment of baroreflex function in any way contributes to the approximately parallel (Gilpin *et al.* 1983) early mortality in these patients is unknown. However, in an extensive, prospective study of baroreceptor–cardiac reflex responses of patients after their first myocardial infarctions, LaRovere and her associates (1988) found that post-infarction mortality appears to be limited to patients with minimal baroreflex slopes. These important results are summarized in Fig. 10.8.

It may be noteworthy that the only therapeutic intervention which has been shown to reduce mortality in patients after acute myocardial infarction is beta-adrenergic blockade. In published trials, prevention of sudden cardiac death is a prominent feature of the action of beta-blockers. Moreover, subgroups of patients most likely to benefit from beta-blockers include elderly and physically inactive people (Beta-Blocker Heart Attack Trial Research Group 1982) and diabetic patients (Norwegian Multicenter Study Group 1981), who would be expected to have small baroreflex responses. The mechanism whereby this weakly antidysrhythmic (Winkle *et al.* 1978) class of drugs prolongs life is unclear. Beta-blockers may prevent dysrhythmias by preventing repeat myocardial infarctions (Ahlmark *et al.* 1974; Gundersen *et al.* 1982). Other beneficial effects of beta-blockers may be related to their effects upon baroreflex responses (Eckberg 1984); acute administration of beta-blockers increases vagally mediated R–R interval prolongation after bolus injections of phenylephrine (T.G. Pickering *et al.* 1972*b*; Eckberg *et al.* 1976) and respiratory sinus arrhythmia (Coker *et al.* 1984; Bittiner and Smith 1988).

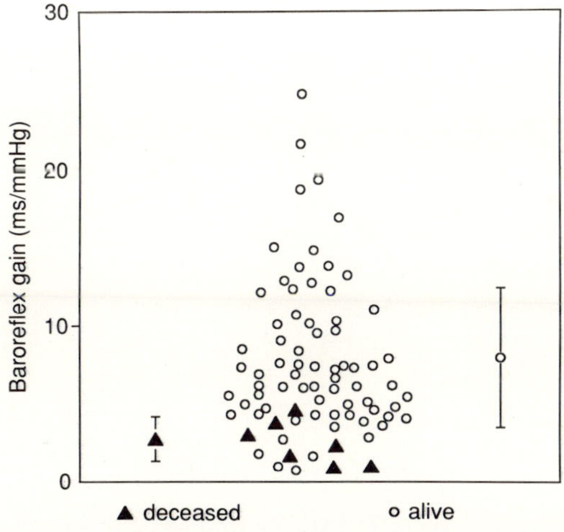

FIG. 10.8. Relation between baroreflex slopes (after bolus phenylephrine injections) and mortality in 78 patients after their first myocardial infarctions. Vertical bars encompass one standard error of measurement. These data are part of an ongoing study (LaRovere *et al.* 1988) and were kindly supplied by Peter J. Schwartz.

There is also substantial evidence that links diminished vagal–cardiac and exaggerated sympathetic outflows (as measured by indirect means) with poor prognoses. The prognosis of patients early (M.M. Wolf *et al.* 1978) and late after (Kleiger *et al.* 1987) acute myocardial infarction is related inversely to the magnitude of their respiratory sinus arrhythmia (which is thought to reflect vagal traffic to the heart — see Chapter 8). Loss of sinus arrhythmia may be in part an epiphenomenon; that is, patients with severe heart disease are more likely to have both a poor prognosis and reduced sinus arrhythmia. However, the study by Kleiger and associates (1987) suggests that reduced R–R interval variability has predictive value that is independent of ventricular function.

Similarly, high resting noradrenaline levels (Cohn *et al.* 1984) and failure of noradrenaline levels to rise during nitroprusside infusions carry an extremely poor prognosis (Olivari *et al.* 1983). A life table analysis of the influence of the presence or absence of reflex increases of noradrenaline concentrations on survival is shown in Fig. 10.9. The extent to which baroreflex function controls resting levels of efferent vagal–cardiac and sympathetic activity is unknown. However, the studies cited above provide in part strong justification for the pursuit of possible interrelations, because baroreflex function is susceptible to therapeutic manipulation.

THERAPEUTIC IMPROVEMENT OF BAROREFLEX FUNCTION

It is well recognized that baroreceptor–cardiac reflex responses may vary according the physiological circumstances of the moment. There is general

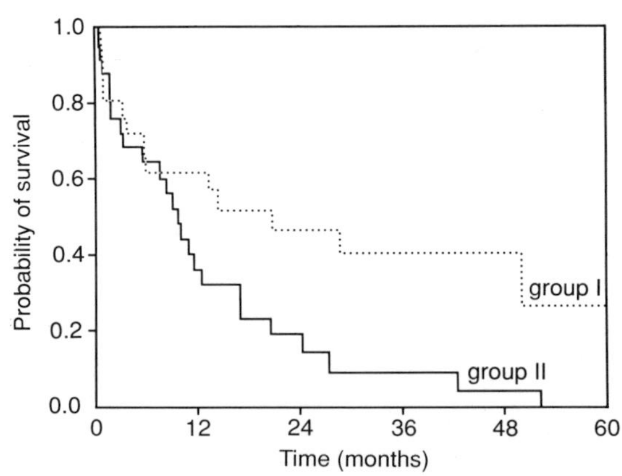

FIG. 10.9. Life table analysis of 21 patients with heart failure whose plasma noradrenaline levels increased by more than 17 per cent during nitroprusside infusion (group I), and 25 patients whose plasma noradrenaline did not increase, or even decreased, during nitroprusside infusions (group II). Adapted from Olivari *et al.* (1983).

agreement that in healthy subjects, slopes of baroreceptor–cardiac reflex responses increase during sleep (J.D. Bristow *et al.* 1969*a*) and decline during mental arousal or physical exercise (J.D. Bristow *et al.* 1971*b*). Moreover, there is a variety of circumstances in which patently abnormal baroreceptor reflex responses can become normal or improve. First, subnormal baroreflex responses may improve spontaneously. Studies documenting recovery of sympathetically and vagally mediated baroreflex function have been cited (Imaizumi *et al.* 1984*b*; M.L. Smith *et al.* 1989).

Second, a variety of pharmacological interventions alter baroreceptor reflex function acutely. In healthy subjects low doses of atropine or scopolamine increase vagal–cardiac outflow and baroreceptor–cardiac reflex responsiveness (Raczkowska *et al.* 1983; Dibner-Dunlap *et al.* 1985). Patients with diabetes mellitus who have minimal respiratory R–R interval variation (which has been taken as evidence for structural disease of vagus nerves (A. Murray *et al.* 1975)) may prolong their R–R intervals and increase their standard deviations of R–R intervals during arterial pressure elevations provoked by phenylephrine infusions (Eckberg *et al.* 1986). Acute administration of other drugs, including propranolol (T.G. Pickering *et al.* 1972*b*; Eckberg *et al.* 1976) and digitalis (Cokkinos *et al.* 1979; Ferrari *et al.* 1981*b*), may also increase the slopes of baroreceptor–cardiac reflex responses.

Moreover, there is evidence that chronic pharmacological or physiological interventions may improve baroreceptor reflex function. Chronic administration of the calcium-channel blocker felodipine to hypertensive patients augments baroreceptor–cardiac responses (Timmis *et al.* 1984). Chronic oral clonidine, which is thought to exert its effects primarily through stimulation of central alpha-adrenergic receptors (Kobinger 1978), augments baroreceptor–cardiac slopes and reduces plasma noradrenaline concentrations (Korner *et al.* 1974; Sleight *et al.* 1975; Guthrie and Kotchen 1983), and in some patients reduces muscle sympathetic nerve traffic (Wallin and Frisk-Holmberg 1981). Endurance training (which reduces plasma noradrenaline concentrations (W.W. Winder *et al.* 1979) and slows resting heart rates (Sedgwick *et al.* 1974)) increases the slopes of baroreceptor–cardiac reflex relations (LaRovere *et al.* 1986). The variability of baroreflex responses suggests that abnormal baroreflex function in patients does not result necessarily from structural or anatomical abnormalities, and holds the promise that baroreflex function may be improved by therapeutic interventions.

SUMMARY

This chapter makes several points. First, in the absence of interventions, baseline baroreflex function (or baseline levels of autonomic outflow) tends to be reproducible in healthy subjects and patients. Second, abnormalities of autonomic outflow described in patients may be related reciprocally; that is, some patients have exaggerated sympathetic and diminished vagal activity. Third, although the issue is controversial, abnormalities in one efferent baroreflex limb, sympathetic or vagal, tend to be associated with abnormalities of the other. Fourth, mechanisms underlying baroreflex abnormalities in cardiovascular patients are understood poorly. Fifth, the evidence, much of it circumstantial, points towards a close link between baroreflex function and prognosis. Finally, baroreflex function can be improved by a variety of therapeutic interventions.

11

Hypertension

The discovery of the powerful pressure sensing and regulating role of the carotid sinuses by Heinrich Hering in 1923 led to the intuitively attractive hypothesis that baroreceptor malfunction *causes* hypertension. This hypothesis was supported by his studies conducted in anaesthetized dogs, which showed that acute carotid sinus baroreceptor denervation provokes hypertension (Hering 1924; Koch and Mies 1928). However, it was not long before evidence against this hypothesis began to appear. As early as 1934, Koch and Mattonet showed that baroreceptor denervation does not necessarily lead to hypertension. In 1936, other arguments were raised by the study of G.W. Pickering *et al.*, which showed that unilateral carotid artery occlusion in hypertensive patients increases blood pressure. Persistence of this reflex response in hypertensive patients suggests that carotid baroreceptors are sensing pressure changes and initiating appropriate reflex adjustments.

Further questions were raised by the dangerous study of Lampen *et al.* (1949), who injected procaine hydrochloride into the regions of the carotid sinuses of 33 volunteers. In subjects with malignant hypertension, the *average* maximum systolic and diastolic pressures after nerve block were 321 and 187 mmHg! (It should be noted that injections of local anaesthetics into the region of the carotid sinuses also block afferent and efferent vagus nerve traffic.)

These studies provide strong evidence that baroreceptors are functional in human hypertension; however, they do not answer the questions: (1) if the baroreceptors are working, why is the blood pressure so high?; (2) since the blood pressure is so high and the baroreceptors are working, why is not the heart rate much slower than it is?

BARORECEPTOR RESETTING

An important partial answer to these questions came with the landmark experiments of McCubbin *et al.* (1956), who studied normotensive dogs and dogs with hypertension induced by cellophane wrapping of both kidneys. In the first part of their experiment, one carotid sinus nerve and both aortic nerves were cut, and the remaining innervated carotid sinus was explanted in a tube of skin so that it could be compressed manually (to lower carotid pressure without anaesthesia). Carotid hypotension provoked increases of blood pressure before and during the development of renal hypertension (one of the two records published suggests that reflex blood pressure elevation actually is *greater* after than before the development of hypertension). This portion of the experiment corroborated the earlier study conducted by Pickering *et al.* (1936) in hypertensive patients.

Next, multifibre recordings of carotid sinus nerve traffic were made in normotensive dogs and in dogs after renal hypertension had been established for at least 6 weeks. These recordings did much to clarify the role of baroreceptors in hypertension. First, carotid sinus nerve activity in the hypertensive dogs was intermittent, not continuous. Although in normotensive dogs, traffic was also intermittent at usual blood pressure levels, it was continuous when their pressures were raised acutely to the same levels as those of the renal hypertensive dogs. Second, carotid sinus responses to sinusoidal fluctuations of pressures about mean values of 60, 120, and 240 mmHg differed in hypertensive and normotensive dogs. In normotensive dogs, nerve activity was intermittent at 60 and continuous at 240 mmHg. In renal hypertensive dogs, traffic was absent at 60 and intermittent at 240 mmHg. The authors concluded that in hypertension, baroreceptors are *reset* to function at higher than normal pressure levels. That is, they respond appropriately to blood pressure changes occurring within their new hypertensive range, and buffer reductions as well as elevations of pressure. Carotid sinus nerve recordings from McCubbin's study are depicted in Fig. 6.14.

Subsequently, Sleight *et al.* (1977) confirmed McCubbin's observations and showed that the relation between *single fibre* baroreceptor activity and blood pressure is shifted to the right, such that the maximum gain of the reflex occurs at higher pressures in hypertensive than in normotensive dogs. When recordings are made at dogs' usual (normotensive or hypertensive) pressures, afferent input to the central nervous system is identical in the two populations. However, gain of baroreceptor fibres during stimulation by pulsatile pressure is subnormal in hypertensive dogs.

Baroreflex resetting was discussed in Chapter 6. The process seems to be mediated primarily by pressure changes; it does not occur when baroreceptor areas are protected from high pressures (Kezdi *et al.* 1973). The earliest studies suggested that resetting is due to mechanical splinting of receptors by hypertrophied (McCubbin *et al.* 1956) or oedematous (Tobian *et al.* 1961) arterial wall elements. Subsequent research showed that increased stiffness of baroreceptor arteries is not necessary for resetting to occur. Although reduced compliance has been found in baroreceptor areas in chronically hypertensive animals (Aars 1968; Wolinsky 1971; Angell-James 1973, 1974; Andresen *et al.* 1978), compliance of baroreceptor areas may be normal, and the resetting process may also involve other segments of the baroreflex arc (A.M. Brown *et al.* 1976; Sapru and Krieger 1979; Koushanpour and Behnia 1987). The study by Andresen and Yang (1990) suggests that acute resetting is not a laboratory artefact, secondary to use of non-physiological steady-state pressure changes; they documented similar shifts of pressure–single fibre aortic nerve response relations after both steady and pulsatile pressure changes.

BARORECEPTOR DENERVATION

The hypothesis that baroreceptor malfunction causes hypertension became nearly untenable when Heymans and Bouckaert (1934) found that hypertension following baroreceptor denervation occurs primarily when denervated dogs are awake and aroused; when barodenervated dogs are sleeping or resting quietly, blood pressure is normal.

Cowley *et al.* (1973) confirmed this early work and extended it in important ways. In their study, 27 young adult dogs were instrumented chronically to allow physiological evaluation without anaesthesia; all recordings were made while dogs were isolated in a quiet room. Fifteen of these dogs underwent baroreceptor denervation, which included the following: stripping of nerves and adventitia from the aortic arch and carotid sinus areas, application of phenol solution and isopropyl alcohol to these areas, and transection of the left, and medial bundle of the right, vagosympathetic trunks. Evidence that arterial baroreceptor denervation was complete included absence of bradycardia during bolus injections of angiotensin, noradrenaline, and vasopressin (indeed, phenylephrine caused modest *cardioacceleration*); exaggeration of orthostatic blood pressure reductions; and only minimal blood pressure elevation during bilateral carotid artery occlusion.

Baroreceptor-denervated dogs differed from sham-operated dogs in several important respects. First, their blood pressure variability was much greater. Second, although their average blood pressures were significantly higher, the differences were much smaller than the authors expected (for example, average 24 h mean arterial pressure was 113 mmHg in denervated dogs, and 102 mmHg in sham-operated dogs). Third, average heart rate and heart rate variability were greater. The authors concluded that the primary function of baroreceptors is to minimize variations rather than to set absolute levels of blood pressure.

The studies of Cowley *et al.* (1973) are widely misquoted as indicating that baroreceptor denervation does not lead to elevation of blood pressure. In their study, as they point out, average mean, systolic and diastolic blood pressures all were significantly higher in denervated than in sham-operated dogs, by 11, 13, and 18 mmHg. Moreover, *some of the dogs in this study had unequivocal severe hypertension; a subset of six of the twelve hypertensive dogs had average mean pressures of almost 150 mmHg.*

The same issue was joined subsequently by Persson et al. (1987), who also studied conscious dogs after baroreceptor denervation. These authors chose surgical procedures which enabled them to delineate individual roles of aortic, carotid, and cardiopulmonary baroreceptors in blood pressure control. Their preparations may have differed from those of Cowley *et al.* (1973) in that they used an elaborate approach to assure total denervation of *cardiopulmonary* as well as arterial baroreceptors. They found appreciable elevations of mean arterial pressure (from 100 to 120 mmHg) after total denervation. Blood pressure frequency distributions from their study are shown in Fig. 11.1.

The substantial literature on the influence of baroreceptor denervation on blood pressure suggests the following:

1. Acute *arterial* baroreceptor denervation in anaesthetized animals provokes hypertension, which is usually transient (Kezdi *et al.* 1980; Walgenbach 1984; Geer *et al.* 1986).

2. Persistence of hypertension after denervation depends importantly on the extent of denervation (Ito and Scher 1979); blood pressure may return

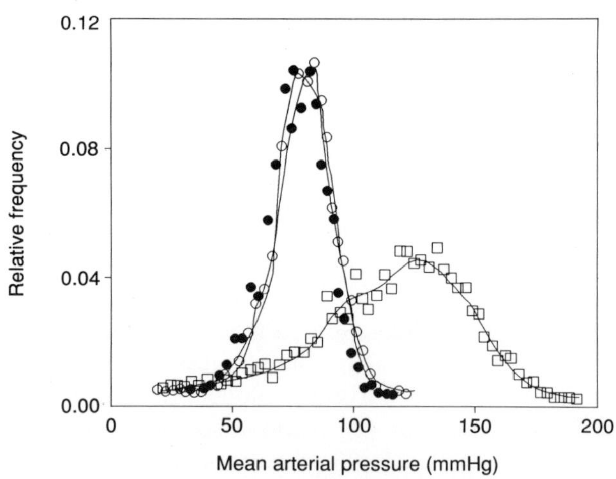

FIG. 11.1. Frequency distribution of mean arterial pressure in intact dogs (○), dogs with cardiopulmonary denervation (●), and dogs with total cardiopulmonary and arterial baroreceptor denervation (□). Adapted with permission (Persson *et al.* 1987).

to normal if only a small percentage of afferent baroreceptor fibres are intact (Cowley *et al.* 1980; Guo *et al.* 1983; Persson *et al.* 1987, 1988). It is likely that the reduction of arterial pressure to normal levels that occurs after denervation of one baroreceptor area is due to central plasticity and enhancement of the influence of other baroreceptor areas (O'Leary and Scher 1990). This conclusion supports two important corollaries. The first is that failure to document arterial pressure elevation after baroreceptor denervation constitutes indirect evidence that denervation was incomplete. The second, which is of fundamental importance to understanding of baroreceptor physiology, is that *baroreceptor afferent activity importantly determines resting arterial pressure levels.*

It is likely that central nervous system plasticity exaggerates the importance of small numbers of afferent baroreceptor fibres that remain after baroreceptor denervation procedures. The elegant study of Seller and Illert (1969) provides a basis for such a possibility. These authors stimulated the carotid sinus nerves electrically and recorded evoked potential responses in solitary tract nuclei. They found that evoked potentials produced by contralateral carotid sinus nerve stimulation are similar to those produced by bilateral carotid sinus nerve stimulation. They suggested that when both carotid sinus nerves are stimulated simultaneously, impulses from the contralateral side arrive *after* those from the ipsilateral side, at times when solitary tract interneurones are unresponsive. However, when ipsilateral carotid sinus input is absent, contralateral firing assumes much greater importance. These results may help explain the strong depressor reflexes that are elicited when only a fraction of afferent baroreceptor traffic is preserved (Koch 1931; Stegemann and Müller-Bütow 1966).

There are no human counterparts for the controlled prospective studies conducted by Cowley *et al.* (1973) and Persson *et al.* (1987). However, there are scattered reports of hypertensive patients with altered baroreceptor function that may be related to denervation studies. Lampen (1949) described ten patients in whom transient hypertension was associated with neuritis involving baroreceptor nerves. Tanaka *et al.* (1975) described three patients who developed hypertension during exacerbations of Takayasu's aortitis, an inflammatory disease that involves large arteries, including the aortic arch and carotid arteries. In these patients, hypertension disappeared after inflammation was treated successfully with steroids.

Sleight (1979) followed a patient who had surgical denervation of his carotid sinuses as treatment for asthma. This patient's pressure rose immediately after surgery (from about 140/75 to as high as about 200/100 mmHg), declined to near preoperative levels within the next few months, rose again to high levels (above 200/100 mmHg) during the next 5 years, and thereafter remained at high levels. Kochar *et al.* (1984) studied a patient who had striking hypertension in the supine position (220/120 mmHg) and hypotension in the standing position. The only autonomic abnormalities found in this patient were markedly depressed heart rate responses to changes of arterial baroreceptor afferent activity provoked by blood pressure changes, neck suction, and carotid massage.

Aksamit *et al.* (1987) studied a patient in whom earlier irradiation of the chest and neck appeared to have caused sinoaortic denervation. This patient had paroxysmal but not sustained hypertension. His responses to lower body negative pressure suggested that low-pressure baroreceptor control of blood pressure was preserved, or perhaps even heightened. Langford *et al.* (1987) reported two young adult patients in whom surgery for haemangioblastoma of the obex had damaged the brainstem in the region of the solitary tract nucleus. Both patients were consistently hypertensive after surgery, and both had sustained but minor elevations of plasma noradrenaline levels. In one patient, hypertension began during surgery. In the other, hypertension was most marked in the supine position, and was supplanted by hypotension and absence of cardioacceleration in the upright position. Chamontin *et al.* (1989) reported three patients treated with cervical irradiation who had supine hypertension and orthostatic hypotension. (Both Cowley *et al.* (1973) and Persson *et al.* (1987) noted orthostatic hypotension in their conscious baroreceptor-denervated dogs.) Although these isolated human cases provide intriguing clues, they are not substitutes for prospective controlled research into the possible role of baroreceptor malfunction in causation of human hypertension.

BAROREFLEX FUNCTION IN HYPERTENSIVE PATIENTS

Hypertensive patients may have degeneration of nerves in baroreceptor areas (Hilgenberg 1958). This allows for the possibility that hypertension may impair baroreceptor function. The following discussion addresses two questions: (1) is baroreflex function normal in patients with hypertension?; (2) if it is not, is there any evidence that baroreflex abnormalities are implicated as factors that promote or perpetuate hypertensive processes?

Baroreflex control of vagal–cardiac outflow

There is nearly universal agreement that baroreflex control of heart rate (or R–R interval) is impaired in patients with hypertension, and that the degree of impairment is proportional to the degree of blood pressure elevation. Impaired vagal control is evident in R–R interval responses to acute arterial pressure elevations, and R–R interval fluctuations during quiet breathing.

J.D. Bristow *et al.* (1969*b*) measured R–R interval prolongations during the systolic pressure elevations that follow bolus intravenous injections of angiotensin-II or phenylephrine (Smyth *et al.* 1969) in subjects with resting mean blood pressures less than 100 mmHg, and subjects with resting mean pressures greater than 100 mmHg. They found that the R–R interval–systolic pressure relation operates in a higher pressure range in hypertensive than in normotensive subjects (that is, it is 'reset'), and that the slope of this relation is not as steep as it is in the normotensive group (that is, it is also depressed).

The concept of 'resetting' is central to an understanding of baroreflex mechanisms in hypertension. Figure 11.2 published by Korner *et al.* (1974) depicts this concept as it relates to baroreflex control of R–R interval in groups of normotensive and moderately and severely hypertensive humans. As discussed in Chapter 6, the relation between pressure in arterial baroreceptor areas and R–R intervals is sigmoid (Koch 1931) and comprises threshold (the lowest pressure that elicits a response), 'linear', and saturation (the pressure that elicits a maximum response) ranges. A normal relation is shown in Fig. 11.2, left. Pure resetting, as in early hypertension (centre), shifts this relation to the right on the pressure axis, but does not alter its slope. Resetting with reduction of sensitivity (or gain), shifts the relation and also diminishes the slope of the linear portion (right). A diminished *range* of baroreceptor–cardiac R–R interval responses is also present in hypertensive patients, as it is in heart failure patients (discussed in Chapter 14). In both hypertensive and heart failure patients, reduction of baroreflex range appears to be due entirely to reduction of the upper R–R interval plateau (Korner *et al.* 1974; Sopher *et al.* 1990).

Resetting of baroreceptor–cardiac reflexes in hypertensive patients has been documented by different groups in different ways. It is present in middle-aged patients who have had sustained hypertension for long periods

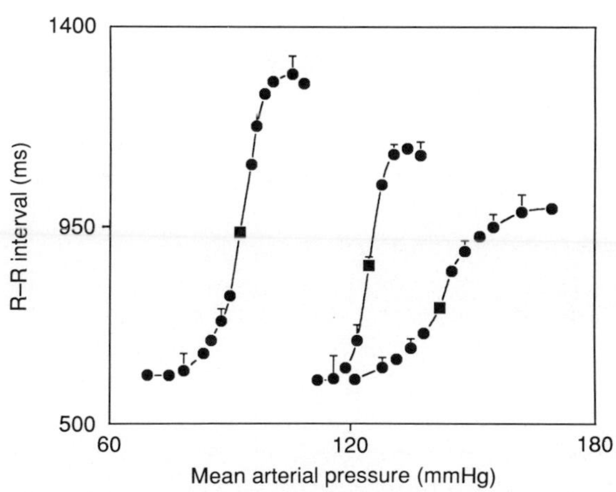

FIG. 11.2. Average R–R interval responses of normotensive (left) and moderately and severely hypertensive patients (centre and right) to infusions of phenylephrine and nitroprusside. Adapted with permission (Korner *et al.* 1974).

(Gribbin *et al.* 1971), and it is present also in young people (not yet patients) who have blood pressures that fluctuate above and below levels that might be considered hypertensive ('borderline hypertension' (Takeshita *et al.* 1975; Eckberg 1979; Volpe *et al.* 1982)). Impaired baroreflex control of R–R intervals is present even in *normotensive* relatives of hypertensive patients (Takata *et al.* 1985). Moreover, it is found when baroreceptors are stimulated by neck suction (Eckberg 1979), by intravenous injections of pressors as boluses (Gribbin *et al.* 1971; Mancia *et al.* 1978*b*) or infusions (Korner *et al.* 1974; Ishizuka *et al.* 1987), and even by noradrenaline released by spontaneously occurring bursts of muscle sympathetic nerve activity (Fritsch *et al.* 1986).

The cumulative experience (by 1979) of the Oxford group (Sleight 1979) with phenylephrine injections with 114 subjects is summarized in Fig. 11.3. These data demonstrate that baroreflex slopes are related inversely to blood pressure, and that this effect exists independent of age-related reductions of baroreceptor–cardiac reflex slopes. This figure shows that although the influence of ageing is strong, diminished baroreceptor–cardiac reflex responsiveness in hypertensive patients probably cannot be explained entirely on the basis of age-related baroreflex impairment.

Subnormal levels of *resting vagal–cardiac nerve outflow* (which are related to subnormal gain of baroreceptor–cardiac reflexes) have also been documented in hypertensive patients. Hypertensive patients may have higher resting heart rates (Gribbin *et al.* 1971; Julius *et al.* 1971; Korner *et al.* 1973; Alicandri *et al.* 1985), and less respiratory sinus arrhythmia, R–R interval

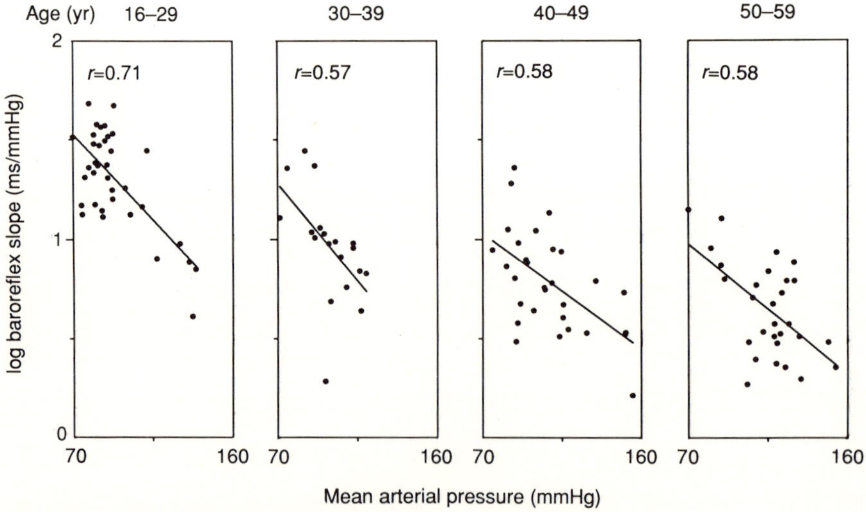

FIG. 11.3. Phenylephrine baroreflex slopes as functions of mean arterial pressure in four age groups. Adapted with permission (Sleight 1979).

prolongation with low-dose atropine, and R–R interval shortening (cardioacceleration) with high-dose atropine than have normotensive subjects (Julius *et al.* 1971; A.C. Simon *et al.* 1977; Alicandri *et al.* 1985). This issue is controversial, however, and many studies document normal heart rates in hypertensive patients.

Mechanisms responsible for impaired baroreflex control of vagal activity in hypertension are not well understood. One possibility is that baroreceptor transduction is abnormal. This may be the case in older hypertensive patients who have subnormal pulsatile common carotid artery expansion (von Ungern-Sternberg *et al.* 1975). Another is that it may be related (in poorly understood ways) to disordered sympathetic outflow. In hypertensive or borderline hypertensive subjects, beta-adrenergic blockade alone with propranolol (Floras *et al.* 1988*c*) or beta-blockade combined with acetylcholinesterase blockade with neostigmine may restore vagal baroreflex slopes towards normal (Volpe *et al.* 1982). (We note that this issue too is controversial; not all studies show that acute or chronic beta-adrenergic blockade improves baroreceptor–cardiac reflex control (G. Simon *et al.* 1977; Krediet and Dunning 1979).)

Control of sympathetic outflow

It is clear that there is an important relation between sympathetic activity and blood pressure. Figure 11.4 is from a retrospective study of men with spinal injuries that documents a strong relation (the linear regression correla-

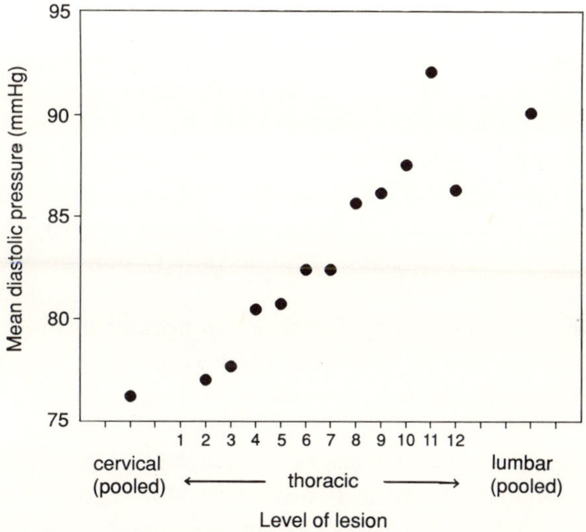

FIG. 11.4. Diastolic pressure of patients with complete spinal cord injuries. Adapted with permission (Frankel *et al.* 1972).

tion coefficient for the average data is 0.92) between the level of complete spinal cord lesions and diastolic pressure (Frankel *et al.* 1972). On the basis of such data, it is reasonable to expect that there may be some direct relation between blood pressure and sympathetic activity in hypertensive patients.

A variety of attempts has been made to determine if arterial baroreflex control of *blood pressure* is normal in hypertension. Vasoactive drugs cannot be used for such research because they influence the variable being measured. Therefore most workers use neck suction–pressure techniques to perturb baroreceptors. As discussed in Chapter 5, responses to 'steady-state' neck pressure and suction are difficult to interpret because of ongoing opposite changes of afferent baroreceptor activity in other baroreceptor areas. Nevertheless, Mancia *et al.* (1978*b*) studied blood pressure and heart rate responses to neck suction and pressure in a group of hypertensive patients and compared results with those obtained earlier from a group of normotensive subjects (Mancia *et al.* 1977). They found that the shape of the stimulus–response relation was different in hypertensive than in normotensive subjects; in hypertensive patients, the hypotensive response to neck suction was greater than the hypertensive response to neck pressure. In both studies, this group found that gain of blood pressure responses to changes of carotid transmural pressure is normal in hypertensive subjects.

These results are at variance with those of Angell-James and George (1980) and Guo *et al.* (1983), who studied open-loop baroreflex control of vascular resistance in anaesthetized renal hypertensive rabbits. Both groups found that the gain of carotid baroreflex control of vascular resistance is augmented in hypertension. Angell-James and George speculated that increased gain results from amplification of responses by vascular smooth muscle receptors secondary to pathologically increased arterial wall-to-lumen ratios (Folkow 1978, 1982; Jie *et al.* 1986). Since it is clear that hypertensive patients have exaggerated constrictor responses to pressors (see below), *normal* blood pressure increases with reduced carotid baroreceptor input (as with neck pressure) may signify *subnormal* baroreflex regulation of blood pressure. An emerging literature focuses directly on sympathetic pathophysiology in human hypertension, rather than on secondary end-organ responsiveness.

Resting levels of plasma *noradrenaline* are normal in most hypertensive patients (Lake *et al.* 1977). However, there is growing evidence that in a subset of young people with hypertension, resting antecubital vein plasma noradrenaline concentrations are slightly higher than those of age-matched normotensive people (Sever *et al.* 1977; Kotchen *et al.* 1982; Esler *et al.* 1986). Of great interest, Esler and co-workers (1986) studied noradrenaline kinetics across renal and cardiac vascular beds and found that in young hypertensive subjects, renal noradrenaline spillover is 2.4 times normal. This allows for the possibility that in some patients, increased renal sympathetic outflow may increase renin release by raising renal baroreceptor thresholds (see below).

D.S. Goldstein (1983*a*) analysed 64 studies that reported plasma catecholamine levels in hypertensive patients. Average plasma noradrenaline levels were higher in hypertensive patients than in normotensive subjects in 81 per cent of these studies; in 39 per cent, this disparity was statistically significant. In 74 per cent of 31 of these studies, average *adrenaline* levels were also higher in hypertensive patients than in normotensive subjects (in 48 per cent, the difference was statistically significant).

Sever's survey showing the range of plasma noradrenaline levels in recumbent and standing normotensive and hypertensive subjects is summarized in Fig. 11.5. This figure indicates that the majority of hypertensive patients have normal plasma noradrenaline levels, and that differences between average baseline levels of hypertensive patients and normotensive subjects are small (80 pg/ml, in this study, and 55 pg/ml in Goldstein's survey of 64 articles (1983*a*)). Studies by Silverberg *et al.* (1978), Izzo (1983), and Chang *et al.* (1988) suggest that such small increments of circulating plasma noradrenaline should have negligible haemodynamic consequences. However, plasma levels provide only indirect information on noradrenaline concentrations at the receptor level; therefore there is still some likelihood that excessive sympathetic activity plays an aetiological role in some patients with hypertension.

Other data suggest that sympathetic abnormalities may be more widespread than Goldstein's survey indicates, and that mere measurement of plasma noradrenaline levels may not be sufficiently discriminating to identify a sympathetic contribution to hypertension. Numerous studies show that patients with hypertension or borderline hypertension have exaggerated pressor responses to infused noradrenaline or phenylephrine (Doyle and

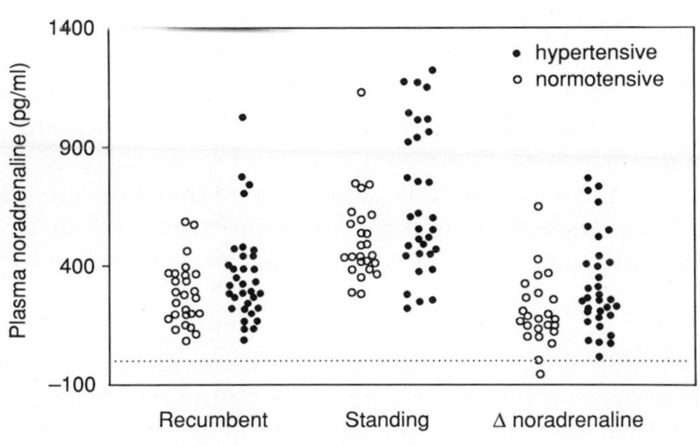

FIG. 11.5. Plasma noradrenaline levels in normotensive and hypertensive subjects. Adapted with permission (Sever *et al.* 1977).

Black 1955; Philipp *et al.* 1978; Weidmann *et al.* 1980; Meier *et al.* 1981) or to reflexly increased (with lower body suction (Mark and Kerber 1982)) sympathetic activity. One or more of several mechanisms may explain exaggerated pressor responses in hypertensive patients. First, they may result because of defective buffering of pressure by baroreceptors; similar hyperresponsiveness to pressors occurs in other disease conditions associated with defective baroreflex control, including autonomic failure (H.N. Wagner and Braunwald 1956; Möhring *et al.* 1980) and diabetes mellitus (Eckberg *et al.* 1986). Second, adrenergic hyperresponsiveness may be due exclusively to up-regulation of adrenergic receptors; however, this cannot be the only mechanism, since responses to the non-adrenergic constrictor angiotensin-II also are exaggerated (Philipp *et al.* 1978). Third, increased vascular wall-to-lumen ratios of resistance vessels may render hypertensive patients hyperresponsive to constrictor stimuli (Korner *et al.* 1973; Folkow 1978). Fourth, altered vascular responses in hypertensive patients may result in part from disturbed endothelial mechanisms (Linder *et al.* 1990).

Philipp *et al.* (1978) found a significant hyperbolic relation between resting blood pressure, pressor responses to infused noradrenaline, and plasma noradrenaline increases after 2 min of bicycle exercise. They studied two groups of subjects: 29 normotensive men, and 28 hypertensive men and one woman. Hypertensive patients were subdivided into groups with average mean pressures between 100 and 135 mmHg, and average mean pressures above 135 mmHg. Figure 11.6 shows the relation between noradrenaline reactivity (reciprocal of the dose of noradrenaline necessary to raise mean pressure by 20 mmHg) and plasma noradrenaline levels at the end of 2 min of bicycle exercise. In each cohort, subjects with the smallest noradrenaline reactivity required the largest amount of noradrenaline to increase their mean arterial pressure by 20 mmHg, and experienced the largest increase of noradrenaline release with exercise.

A multiple regression analysis of these data was used to derive a formula to calculate mean arterial pressure from noradrenaline data. Use of this formula identified a close correlation between pressure measured directly and pressure calculated from noradrenaline reactivity and exercise-induced noradrenaline increases. These data document highly significant relations among blood pressure, exercise-induced noradrenaline elevations, and pressor responsiveness to infused noradrenaline; they point towards a complicated but important role for sympathetic mechanisms in hypertensive patients.

In the studies quoted above, plasma noradrenaline was used as an indirect index of sympathetic activity. More direct evidence for sympathetic involvement in hypertension has come from percutaneous microneurograms. As discussed in Chapter 8, two sympathetic neural outflows can be recorded in conscious human volunteers: to skin and to muscle (Vallbo *et al.* 1979). Of these, muscle sympathetic nerve activity may be particularly important in

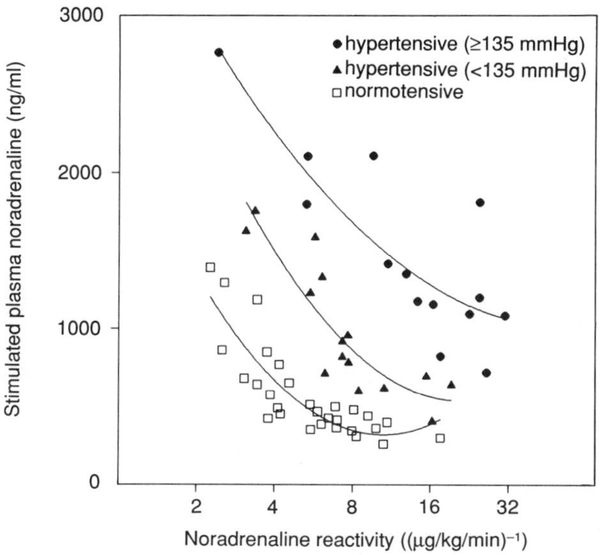

FIG. 11.6. Relation between the amount of administered noradrenaline necessary to increase mean arterial pressure by 20 mmHg ('noradrenaline reactivity') and noradrenaline levels during bicycle exercise. Adapted with permission (Philipp *et al.* 1978).

hypertension, because skeletal muscle comprises about 42 per cent of human body mass (Rein and Schneider 1956), and sympathetic outflow to the muscle (but usually not skin) vascular bed is under arterial baroreflex control (Wallin and Eckberg 1982; Eckberg *et al.* 1988).

The earliest direct recordings showed that resting levels of muscle sympathetic nerve activity are comparable in normotensive and hypertensive patients (Wallin *et al.* 1973*b*; Wallin and Sundlöf 1979; Mörlin *et al.* 1983). (However, sympathetic control is reset, such that sympathetic bursts are regulated over a higher arterial pressure range in hypertensive than normotensive patients.) These data tend to explain results of other studies which show that forearm (primarily muscle) blood flow is normal (or even supranormal) in hypertensive patients (Brod *et al.* 1962; Kiowski *et al.* 1979; Jie *et al.* 1986). They suggest further that increased peripheral vascular resistance in hypertensive patients does not result primarily from increased muscle vascular resistance.

Figure 11.7 shows a microneurogram from a *normotensive* subject which documents exquisitely sensitive, ongoing, iterative, presumably baroreflex-mediated modulation of arterial pressure (and R–R intervals). (Evidence that sympathetic bursts are triggered by baroreceptor mechanisms includes the constant latency between an abrupt drop of pressure and the sympathetic response, and the sudden inhibition of sympathetic activity at the onset of

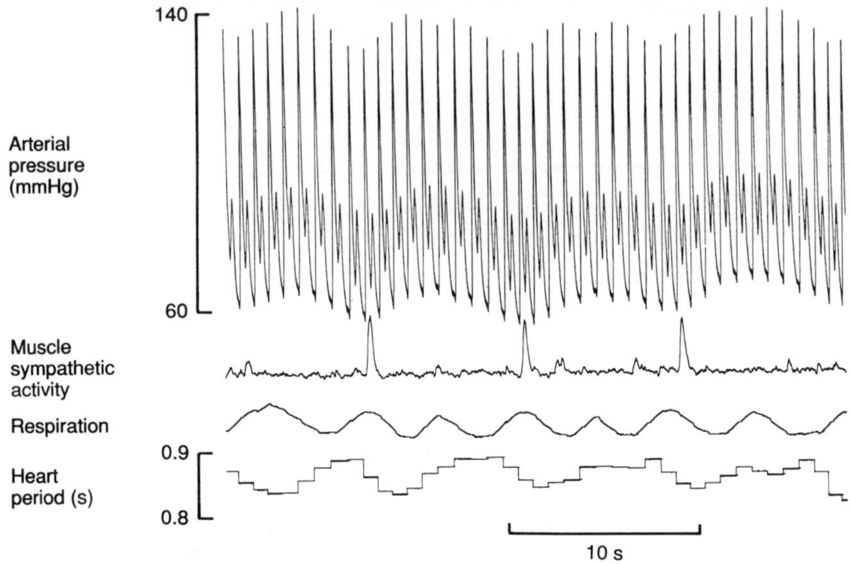

FIG. 11.7. Relations among arterial pressure, muscle sympathetic activity, and vagally mediated R–R interval changes during supine rest, in a normotensive volunteer. Reproduced with permission (Eckberg *et al.* 1985).

electrical carotid sinus nerve stimulation (Wallin *et al.* 1975) or neck suction (Wallin and Eckberg 1982).) This ongoing baroreflex activity might be overlooked when resting sympathetic activity is characterized merely in terms of average numbers of sympathetic bursts, or average R–R intervals.

Attempts to document increased sympathetic activity in hypertensive patients may be criticized on the basis that they tend to be based on a view of structure and function as static and fixed in time. Such research asks questions that have binary answers (is sympathetic activity high or normal in hypertensive patients?) and may conceivably ignore qualitative non-stationary abnormalities in hypertensive subjects. It is clear from serial studies performed during the development of experimental hypertension that hypertensive pathophysiology is evolutionary (J.G.G. Borst and Borst-De Geus 1963; Krieger 1970; Barger 1979).

Studies of Wallin and Sundlöf disclose subtle differences between the muscle sympathetic nerve activities of normotensive and hypertensive subjects, and may explain in part why muscle sympathetic nerve activity is normal (not subnormal) in hypertensive patients despite their high pressure levels. In two elegant studies (Sundlöf and Wallin 1978*b*; Wallin and Sundlöf 1979), they suggest that ongoing variability of autonomic cardiovascular outflow is related to the tightness of arterial pressure control. They plotted relations between diastolic pressures and the subsequent occurrence of muscle sympathetic bursts. With their method, steep slopes resulted if small reductions of diastolic pressure were met by large increases of the number and amplitude of sympathetic bursts. They found that variability of diastolic pressure is related inversely to the steepness of these diastolic pressure–muscle sympathetic burst relations. That is, subjects who respond to small reductions of diastolic pressure with large increases of sympathetic activity have less variability of diastolic pressure than subjects who respond to falling pressure with small increases of sympathetic activity. The distribution of slopes in hypertensive and borderline hypertensive patients was skewed to the right. These findings are summarized in Fig. 11.8.

Studies published after those quoted above tend to show that muscle sympathetic nerve activity levels are *higher* in young hypertensive than in normotensive subjects. Y. Yamada *et al.* (1988) showed that resting muscle sympathetic nerve traffic is higher in hypertensive *and normotensive* adolescents with family histories of hypertension than in normotensive subjects without such histories. E.A. Anderson *et al.* (1989) also found slightly higher levels of muscle sympathetic nerve activity in subjects with borderline hypertension than in age-matched normotensive subjects. In Anderson's study, these disparities were not altered by variations of dietary sodium content. Another study by Y. Yamada *et al.* (1989) is unique in that it shows supranormal levels of muscle sympathetic nerve activity in hypertensive subjects *regardless of age*; the reasons why these conclusions are different

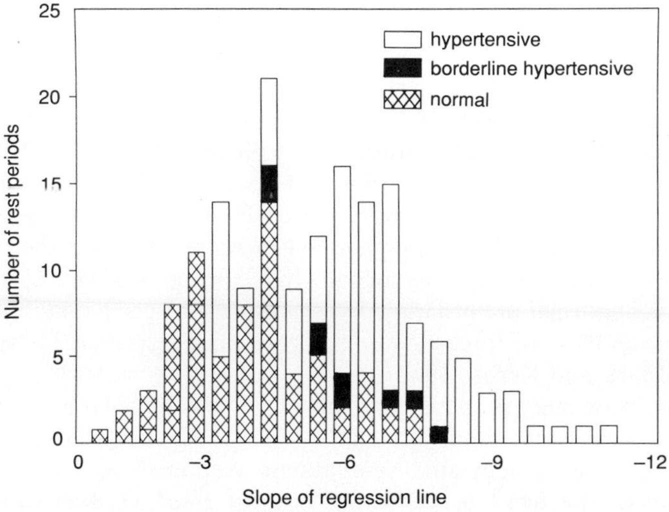

FIG. 11.8. Frequency distribution (number of periods of observation) of slopes of the relation between diastolic pressure and muscle sympathetic nerve activity in subjects with normal blood pressures and borderline hypertensive and hypertensive subjects. Adapted with permission (Wallin and Sundlöf 1979).

from those of Wallin and co-workers cited above (Wallin *et al.* 1973*b*; Sundlöf and Wallin 1978*b*; Wallin and Sundlöf 1979) are unclear.

One implication of elevated plasma catecholamine levels and sympathetic activity in hypertensive subjects is that they in some way *cause* hypertension. Indirect support for this notion comes from studies which gauge the effects of circulating noradrenaline levels by the reductions of blood pressure that occur after blocking drugs. There is a proportionality between pre-existing noradrenaline levels and reductions of blood pressure after administration of drugs such as bethanidine and debrisoquine that reduce neuronal release of noradrenaline (Weidmann *et al.* 1980; Kakembo 1982), drugs such as atenolol that block noradrenaline's actions on beta-receptors (Myers and de Champlain 1983), and drugs such as the centrally acting alpha-adrenergic agonist clonidine that reduce efferent sympathetic outflow (Wallin and Frisk-Holmberg 1981; D.S. Goldstein *et al.* 1985). However, this is not to say that *all* of the increased peripheral vascular resistance found in hypertensive patients is due to augmented neurohumoral vasoconstriction; the study of hypertensive patients' responses to autonomic blocking drugs by Korner *et al.* (1973) suggests that only 20 to 40 per cent of resistance is attributable to autonomic factors. Presumably the other 80 to 60 per cent of resistance is due to structural vascular abnormalities.

The reasons for elevated levels of sympathetic activity in hypertension are not well understood. Perhaps the simplest explanation is that sympathetic outflow is high because it is not buffered normally by arterial baroreceptors. Two groups have studied this question in subjects with borderline hypertension. In a study reported in preliminary form, Miyajima *et al.* (1989) reported that reductions of muscle sympathetic nerve activity during phenylephrine infusions are subnormal in borderline hypertensives. A very similar study by Rea and Hamdan (1990) drew opposite conclusions; these workers found that sympathetic responses to infusions of a phenylephrine and nitroprusside are similar in borderline hypertensive and normotensive subjects. However, their study showed that borderline hypertensive subjects have higher central venous pressures at rest and exaggerated increases of sympathetic activity during lower body suction. This suggests that cardiopulmonary baroreceptor function is abnormal in borderline hypertensive subjects, a possibility raised earlier on the basis of forearm vascular resistance responses to lower body suction (Mark and Kerber 1982).

Another possibility is that hypertensive subjects have larger sympathetic responses to external stimuli. Miyajima *et al.* (1988) showed that hypertensive subjects have exaggerated sympathetic responses to pain caused by immersion of the hand in ice water. Somers *et al.* (1988*a*) showed that borderline hypertensive subjects have exaggerated muscle sympathetic nerve responses to hypoxia. An intriguing variant on this theme is that afferent baroreceptor input raises pain thresholds, and that patients operantly raise their blood pressures to minimize pain (Dworkin *et al.* 1979). Several psychophysiological studies into hypertensive mechanisms are discussed in a monograph edited by Elbert *et al.* (1988*a*).

Another explanation for elevated sympathetic activity in hypertensive subjects is that their baroreflex control of sympathetic traffic is impaired. Results obtained with direct muscle sympathetic nerve activity recordings support this conclusion. T. Matsukawa *et al.* (1988) found that the slopes of the relations between arterial pressure and muscle sympathetic nerve activity after phenylephrine injections are lower in hypertensive than in normotensive subjects. Miyajima *et al.* (1989) found that in borderline hypertensive subjects, increased levels of muscle sympathetic nerve activity are related inversely to vagal baroreflex slopes. Ishizuka *et al.* (1987) showed that when arterial pressure is lowered with nitroprusside, increases of plasma noradrenaline are related inversely to R–R interval changes. Similarly, Meguro *et al.* (1989) showed that both R–R interval and plasma noradrenaline changes during infusions of pressor and depressor drugs are subnormal in hypertensive patients. These data establish a link between impaired baroreflex function and blood pressure and sympathetic outflow in some hypertensive subjects.

The above discussion focuses upon sympathetic outflow in hypertensive patients; one unstated assumption is that sympathetic activity influences blood pressure primarily by increasing cardiac contractility (Broughton and Korner 1983) and output (Widimisky *et al.* 1957), or by causing vasoconstriction. Sympathetic activity may influence pressure through other mechanisms, including importantly by stimulating the renin–angiotensin–aldosterone system. Chronic intrarenal infusions of adrenaline (Farhi *et al.* 1982) or noradrenaline (Katholi *et al.* 1977) increase the pressure threshold for renin release, and thereby increase renin release and blood pressure (see Chapter 9). Ehmke *et al.* (1987*a*) documented significant positive correlations between renal artery pressure thresholds for renin release and mean arterial pressures; dogs in which renin was released at higher renal artery pressures maintained higher mean arterial pressures. Thus baroreflex-mediated increases of sympathetic outflow to the kidney may increase blood pressure by increasing the pressure threshold for renin release.

R.D.S. Watson *et al.* (1980) showed that in hypertensive patients, the logarithm of plasma renin activity is related directly to systolic pressure variability. Since blood pressure variability is related inversely to vagal baroreceptor–cardiac reflex gain (R.D.S. Watson *et al.* 1980; Ogawa *et al.* 1981; Conway *et al.* 1984; Mancia *et al.* 1986), these results establish a link between impaired baroreflex function and supranormal renin activity in humans. Grassi *et al.* (1988) showed that patients with severe hypertension who have subnormal forearm vascular resistance increases and decreases during mild lower body negative pressure and leg raising have also subnormal (cardiopulmonary) baroreflex control of plasma renin activity.

It is highly unlikely that the abnormal baroreceptor–vagal and –sympathetic reflex function documented by studies quoted above arises from experimental artefact. However, a large question remains: does abnormal regulation of vagal and sympathetic outflows in human hypertension have any practical consequences? We believe that it does.

RELATION BETWEEN IMPAIRED BAROREFLEXES AND BLOOD PRESSURE CONTROL

Evidence is accumulating which suggests that baroreceptor reflex control of arterial pressure is defective in hypertensive patients, and that abnormal regulation of pressure is tied closely to the abnormalities of baroreflex control of R–R interval. The gain of R–R interval responses to pressure elevations is related inversely to blood pressure elevations caused by bolus injections or infusions of pressor drugs (D.S. Goldstein 1983*b*; Conway *et al.* 1984; Eckberg *et al.* 1986). Baroreflex gain is also related inversely to the degree of hypotension caused by sublingual administration of the calcium-channel blocker nifedipine (Kioski *et al.* 1986), or by intravenous infusion of the vasodilator nitroprusside (Shepherd *et al.* 1981). These data provide highly suggestive evidence that baroreflex gain is an important determinant of *short-term* blood pressure fluctuations.

Some studies document parallel impairment of vagal–cardiac reflexes and blood pressure control. Lindblad *et al.* (1981) stimulated carotid baroreceptors with sinusoidal neck suction (−40 mmHg, with 30 s periods) in normotensive and moderately hypertensive subjects (whose average mean arterial pressure was 30 mmHg greater than that of control subjects). Sinusoidal neck suction provoked sinusoidal R–R interval and arterial pressure changes, which the authors related mathematically to stimulus intensity to derive expressions of gain for each response. Their results point towards a loose but significant inverse relation between baseline pressure and carotid baroreflex control of both heart rate and blood pressure. Ookuwa *et al.* (1987) found that normotensive subjects with hypertensive family members have impairment of both R–R interval responses to phenylephrine and forearm vascular resistance responses to lower body suction.

These results support conclusions drawn from animals with experimental hypertension (Angell-James and George 1980), but are at variance with studies performed with static neck suction in man (J. Wagner *et al.* 1968; Bevegård *et al.* 1977*a*; Mancia *et al.* 1978*b*). Lindblad *et al.* (1981) suggested that their use of sine wave neck suction avoided problems of interpretation associated with sustained neck suction. As discussed in Chapter 5, prolonged static neck suction is an extraordinarily complex intervention whose net effects are mediated by a multiplicity of factors (which include only trivial reductions of muscle sympathetic nerve activity).

However, other evidence tends to support the conclusion of Lindblad *et al.* (1981) that baroreflex abnormalities in hypertensive patients impair their ability to control sympathetic activity, blood pressure, and heart rate. At least five studies, involving over 250 patients, show that intra-arterial pressure variability during ambulatory recordings is inversely related to slopes of vagally mediated R–R interval prolongations provoked by bolus injections of phenylephrine (R.D.S. Watson *et al.* 1980; Ogawa *et al.* 1981; Conway *et al.* 1984; Mancia *et al.* 1986; Floras *et al.* 1988*a*). These studies are in keeping with studies of Cowley *et al.* (1973) and Persson *et al.* (1987) in denervated dogs, and support the notion that baroreflex malfunction leads to increased intermediate-range blood pressure variability.

A study by Floras *et al.* (1988*b*) may tie together several of the findings described above. They evaluated hypertensive patients (average mean systolic and diastolic pressures 173 and 107 mmHg) and correlated phenylephrine baroreflex slopes with increases of blood pressure and plasma noradrenaline occurring during bicycle exercise. Multiple linear regression analyses documented significant inverse correlations between maximum mean arterial pressures during exercise and ambulatory blood pressure variability, and logarithm of phenylephrine baroreflex slopes. This study suggests that baroreflexes work actively to minimize blood pressure elevation and contain sympathetic outflow during exertion.

SUMMARY

The evidence that human hypertension is associated with impairment of baroreflex control of vagal–cardiac outflow is inarguable. The evidence that human hypertension is associated with corresponding impairment of sympathetic outflow is less strong. The subtlety of baroreceptor–sympathetic abnormalities taxes the ability of available research techniques to discriminate between normotensive and hypertensive humans. We suggest that subtle abnormalities of baroreflex control of sympathetic outflow exist in human hypertensives; however, differences from normal tend to be quantitative rather than qualitative. We suggest further that the possibility that abnormal baroreflex control contributes to the development of human hypertension cannot be dismissed altogether. The great deficiency of the data base generated from studies in humans (where the hypertension model cannot be criticized) is that all data are *cross-sectional*; longitudinal data obtained before, during, and after the development of human hypertension have not been (and probably will not be) published. Nevertheless, a substantial amount of published evidence indicates that baroreflex function is abnormal in populations of young people from whom most fixed hypertensive patients will come; this supports the possibility (which has never gone away entirely) that in poorly understood ways, baroreflex abnormalities *cause* hypertension.

12

Myocardial ischaemia

Baroreflexes may be involved in the pathophysiology of human myocardial ischaemia in a variety of ways. Since ischaemia alters arterial pressure, and since arterial baroreceptors sense and respond to extremely small changes of pressure, it is likely that baroreceptors sense all pressure changes provoked by ischaemia and help shape autonomic responses. In patients with ischaemic heart disease, baroreceptors seem to be involved intimately with prognosis; they may modify responses to therapy; and there is even the possibility that under certain circumstances, baroreflex mechanisms modulate coronary vascular resistance and provoke (or relieve) myocardial ischaemia.

Understanding of human baroreflex involvement during myocardial ischaemia is surprisingly well developed. However, in some cases this understanding derives from oblique rather than frontal approaches to myocardial ischaemia. For example, more is known about *models of ischaemia*, such as reflex changes provoked by coronary arteriography, than about *ischaemia itself*. Therefore, understanding about baroreflex mechanisms in ischaemia relies heavily on animal studies. However, there is gratifying congruence between observations made in humans and those made under more controlled circumstances in animals. Data from humans on the pathophysiology of myocardial ischaemia come from patients studied during ischaemia due to coronary artery spasm or acute myocardial infarction, and from patients studied during clinical events thought to share mechanisms with myocardial ischaemia, including post-thrombolytic reperfusion of occluded coronary arteries and coronary arteriography.

None the less, there are important gaps in knowledge derived from humans. For example, as far as we can determine, there have been no quantitative assessments of baroreflex function in patients *before* as well as *after* myocardial infarction. Again, data obtained from patients during myocardial ischaemia point towards the existence of important *interactions* between cardiac receptors and arterial baroreceptors; however, as far as we can determine, no one has been ingenious enough to sort out these interactions in humans, and determine the precise contribution of different receptor groups to net autonomic responses.

HUMAN MODELS OF MYOCARDIAL ISCHAEMIA

Coronary artery spasm

Perez-Gomez *et al.* (1979) studied heart rate responses of 14 patients during attacks of variant angina (due to coronary artery spasm (Prinzmetal *et al.* 1960)) and added 63 case reports of electrocardiograms before and during attacks published by others. Their results were similar to those Webb *et al.* (1972) obtained in patients during acute myocardial infarction (see below). In patients with anterior ischaemia, average heart rate *increased* significantly from 68 to 95 beats/min, but in patients with inferior or lateral ischaemia, average heart rate *decreased* insignificantly from 91 to 77 beats/min. Similar but significant trends were found when these authors added responses of patients culled from the literature to those of their own patients. In 38 patients with anterior ischaemia, average heart rate increased from 69 to 86, and in 37 patients with inferior ischaemia, heart rate decreased from 77 to 65 beats/min. These data are consistent with a notion supported by research in animals (Thames *et al.* 1978*b*; Thames and Abboud 1979), that myocardial receptors which trigger bradycardia are located preferentially (but not exclusively) in the inferior wall of the left ventricle.

Perez-Gomez *et al.* (1979) also adduced evidence that this distribution of receptors plays some role in dysrhythmias in patients with coronary artery spasm. Thirty-five episodes of ventricular tachycardia or fibrillation were reported in 196 patients. Of the 35 patients with potentially catastrophic ventricular dysrhythmias, 30 had anterior and five had inferior or lateral ischaemia. (Premature ventricular beats also were more prevalent in anterior than inferior ischaemia (22 vs. 7).) These data obtained during myocardial ischaemia induced by spontaneous coronary artery spasm suggest that inferior ischaemia is more likely to trigger inhibitory cardiac reflexes and less likely to provoke ventricular dysrhythmias than is anterior ischaemia.

This suggests that anterior ischaemia predisposes to, and inferior ischaemia protects against, ventricular dysrhythmias. However, this conclusion is difficult to reconcile with data published by M.J. Davies and Thomas (1984) which show that in patients dying suddenly, thrombi (not necessarily occlusive) are more common in the *right* coronary artery (which supplies the inferior wall of the left ventricle in about 90 per cent of humans (James 1961)) than in the left anterior descending or circumflex coronary arteries.

Coronary reperfusion

Patients with acute myocardial infarction treated with thrombolytic agents commonly have dysrhythmias at the time of lysis of thrombi and restoration of coronary flow (S. Goldberg *et al.* 1983); their responses may provide additional insights into the pathophysiology of ischaemia. Wei *et al.* (1983) measured haemodynamic changes of 41 myocardial infarction patients treated with intra-coronary thrombolysis. Flow was re-established in infarct coronary vessels in 27 patients. In patients with anterior infarctions, average systolic pressure fell (by 7 mmHg) and average heart rate fell (by 6 beats/min) at the time of reperfusion. In patients with inferior infarctions, both the average fall of systolic pressure (36 mmHg) and the average reduction of heart rate (21 beats/min) were significantly greater than the changes measured in patients with anterior infarctions. Linear regression analysis of these changes indicated that in both groups, heart rate and blood pressure were related positively (that is, reductions of blood pressure were associated with *slowing* of heart rate), and that for any given change of pressure, slowing of heart rate was greater for inferior than anterior left ventricular reperfusion. These data are summarized in Fig. 12.1.

These results suggest that reperfusion of the inferior left ventricle interferes with usual arterial baroreflex modulation of heart rate. Presumably, the downward displacement of the heart rate–systolic pressure relation in patients with inferior reperfusion reflects greater prevalence of inhibitory

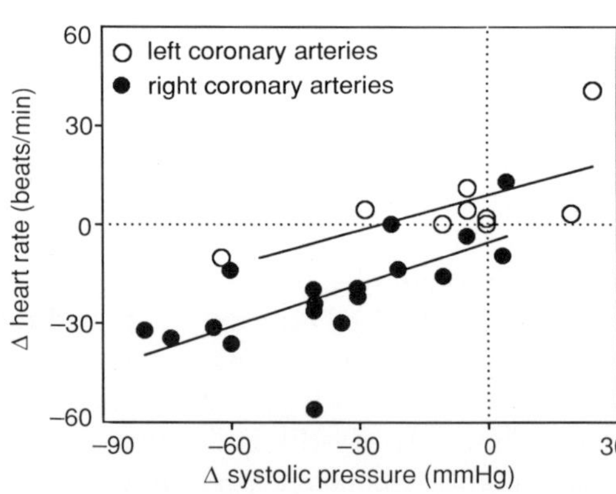

FIG. 12.1. Reflex changes occurring at the time of lysis of coronary thrombi in 37 patients treated successfully with streptokinase. Closed circles depict responses after reperfusion of the right coronary artery, which in about 90 per cent of humans supplies the left ventricular inferior wall. Adapted with permission (Wei *et al.* 1983).

receptors in the inferior than in the anterior wall of the left ventricle. The study of Wei and co-workers (1983) does not indicate what the stimulus to these receptors is. However, a very similar study of 92 patients was published in the same year by Esente *et al.* (1983). Although these authors do not provide data, they indicate that in patients with inhibitory reflex changes, left ventriculography failed to disclose inferior systolic bulging. Their data provide inferential evidence that during reperfusion, left ventricular receptors are stimulated chemically or metabolically rather than mechanically.

Coronary arteriography

Coronary arteriography triggers changes strikingly similar to those described during myocardial ischaemia or reperfusion. Injection of contrast medium into a coronary artery leads to bradycardia which can be prevented by pretreatment with atropine injected intravenously (Eckberg *et al.* 1974) or into the coronary artery supplying the sinoatrial node (Frink *et al.* 1975). As with ischaemia provoked by coronary spasm (Perez-Gomez *et al.* 1979) and reperfusion (Esente *et al.* 1983; Wei *et al.* 1983), receptors seem to be distributed preferentially to the inferior left ventricle. The human response to intracoronary injections of contrast material appears to be a counterpart of the experimental Bezold–Jarisch reflex (Bezold and Hirt 1867; Krayer 1961; Mark 1983). Animal studies provide one candidate, prostaglandins, for the chemical mediator of these reactions (Hintze 1987).

Data from the study of Perez-Gomez and Garcia-Aguada (1977) are depicted in Fig. 12.2. These authors plotted changes of R–R intervals during intracoronary injections of contrast medium according to which coronary

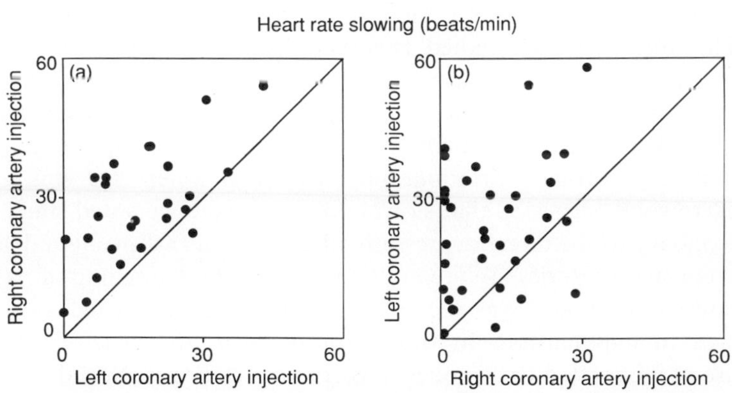

FIG. 12.2. Heart rate slowing provoked by myocardial receptor stimulation during diagnostic coronary arteriography, in patients whose left ventricular inferior wall is supplied by (a) the right coronary artery or (b) the left coronary artery. Adapted with permission (Perez-Gomez and Garcia-Aguado 1977).

artery (left or right) supplied the inferior left ventricle. Figure 12.2 (a) shows responses of 25 patients in whom the inferior left ventricle was perfused by the right coronary artery, and Fig. 12.2 (b) shows responses of 37 patients in whom the inferior left ventricle was perfused by the left coronary artery. These data provide additional proof that stimulation of myocardial receptors, located primarily in the inferior left ventricular wall, initiates reflex changes which override arterial baroreflex mechanisms.

Other studies have worked out the nature of these reflex changes. As indicated, the bradycardia is mediated by muscarinic cholinergic mechanisms, since it can be prevented by atropine (Eckberg *et al.* 1974; Frink *et al.* 1975). The receptors are not necessarily located in the proximal coronary arteries, because reflex changes can be triggered by contrast injections into saphenous vein grafts which bypass proximal coronary arteries (Eckberg *et al.* 1974). Haemodynamic changes do not appear to be caused by increases of coronary artery pressure during injections, because reflex responses are also triggered by slow injections of contrast medium, which presumably do not increase coronary pressure (Eckberg *et al.* 1974). Hypotension does not result primarily from bradycardia, because it is not prevented by atrial pacing (Eckberg *et al.* 1974). The reflex may be mediated in part by sympathetic cholinergic mechanisms; hypotension can be prevented by pretreatment with atropine injected intravenously (White *et al.* 1973*b*), and forearm vasodilatation can be prevented by atropine injected into the brachial artery (Zelis *et al.* 1976). Presumably, sympathetic *withdrawal* also occurs during coronary arteriography; however, we are not aware of any data on this issue from human studies. The study of Zucker and Cornish (1981) conducted in conscious dogs supports such a possibility.

The occurrence of bradycardia during hypotension suggests that arterial baroreflex mechanisms are in some way interdicted by coronary arteriography. This question was studied creatively by Arrowood *et al.* (1989) in patients with heart transplants. Coronary arteriography provoked hypotension, probably secondary to the direct depressant effects of contrast medium on myocardial function. Patients' denervated (donor heart) sinus nodes slowed during coronary arteriography, but their innervated sinus nodes sped. Arrowood suggested that slowing of patients' denervated sinus nodes was secondary to the direct depressant effects of contrast medium delivered via the coronary arteries (White *et al.* 1976), and that speeding of patients' innervated sinus nodes was secondary to baroreflex-mediated autonomic responses to hypotension. In transplant patients, the usual bradycardic response of innervated sinus nodes to coronary arteriography did not occur, because the afferent pathways from ventricular receptors that mediate this reflex (Sleight 1964) had been interrupted by transplantation. These results suggest that arterial baroreceptors oppose hypotension during coronary arteriography, but that in intact patients baroreflex responses are overridden by more powerful reflexes triggered by stimulation of left ventricular receptors.

MYOCARDIAL INFARCTION

Acute responses to myocardial infarction

The greatest mortality during acute myocardial infarction occurs during the first hour (Kuller *et al.* 1966; Fulton *et al.* 1969). The most complete information regarding events during this critical period comes from Pantridge and his group in Belfast (summarized by Pantridge *et al.* (1975)). Of 89 patients seen within the first 30 min of the onset of symptoms (who did not have ventricular fibrillation), 21 per cent had hypertension (systolic pressure of 160/100 mmHg or greater) and 39 per cent had hypotension (systolic pressure of 100 mmHg or less). Of the 89 patients, 22 per cent had systolic pressures below 80 mmHg, and 4 per cent had blood pressures that were unobtainable. Thus 60 per cent of patients seen during the first 30 min after the onset of acute myocardial infarction had abnormal blood pressures. Electrocardiogram changes were also prevalent. Sinus tachycardia (heart rate of 100 beats/min or more) was present in 24 per cent of patients, sinus bradycardia (heart rate of 60 or less) was present in 40 per cent, and complete atrioventricular block (associated with profound hypotension) was present in 8 per cent. Only 17 per cent of patients seen during the first 30 min after the onset of acute myocardial infarction had normal blood pressures and heart rates.

Of 294 patients reported by Pantridge *et al.* (1975) who came under treatment within one hour, 55 experienced ventricular fibrillation. Twenty-three of these had ventricular fibrillation when the mobile team arrived, and 18 more developed fibrillation during the next 2 h. Thus 14 per cent of all patients, or 75 per cent of patients who fibrillated, did so within 2 h of the onset of symptoms. The picture painted by these data obtained very early in the course of acute myocardial infarction is one of major changes of haemodynamic function associated with great cardiac risk. These data are summarized in Fig. 12.3.

Blood pressure and electrocardiogram abnormalities were much less prevalent in the second half hour after the onset of symptoms. Blood pressure was abnormal in 33 per cent of patients (down from 61 per cent during the first half hour) and heart rate was abnormal in 32 per cent (down from 64 per cent). Blood pressure and heart rate were normal in 44 per cent (up from 17 per cent). The evanescence of reflex changes suggests that receptors (or other elements in the reflex arc) mediating blood pressure and heart rate responses adapt rapidly during the course of myocardial ischaemia. This possibility is borne out by the study of Thorén in anaesthetized cats (1976),

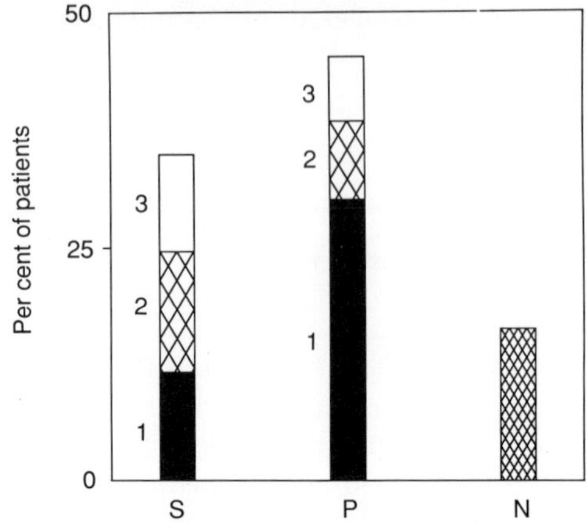

FIG. 12.3. Inferential evidence for autonomic dysfunction in 89 patients studied within 30 min of the onset of acute myocardial infarction symptoms. S = sympathetic overactivity with sinus tachycardia and hypertension (1), sinus tachycardia (2), or hypertension (3). P = parasympathetic overactivity with bradydysrhythmia with hypotension (1), bradydysrhythmia (2), or hypotension (3). N = neither sympathetic nor parasympathetic overactivity. Adapted with permission (Pantridge *et al.* 1973).

which showed that afferent traffic recorded from unmyelinated left ventricular fibres increases abruptly at the beginning of coronary occlusion, but returns to baseline levels within minutes, even though coronary occlusion is maintained. Rapid adaptation of reflex responses to myocardial ischaemia was also documented by Lombardi *et al.* (1983), who showed that transient increases of preganglionic cardiac sympathetic traffic occur during coronary occlusion in anaesthetized dogs and are associated with reductions of ventricular fibrillation thresholds.

The extent to which these early changes of blood pressure and heart rate and rhythm reflect or are modified by changes of baroreflex-mediated autonomic outflow is unknown. It is not known what is stimulus and what is response; in all likelihood, stimuli and responses are intermingled. Pantridge and his co-workers (Webb *et al.* 1972) assumed that in their patients sinus tachycardia or hypertension reflected increased sympathetic efferent activity, and sinus bradycardia, atrioventricular block, or hypotension reflected increased vagal efferent activity. According to this construct, sympathetic overactivity was present in 36 per cent and vagal overactivity was present in 55 per cent. In patients with anterior infarctions, sympathetic overactivity was present in 54, and vagal overactivity was present in 32 per cent. In patients with inferior or posterior infarctions, sympathetic overactivity was present in 20 per cent, and vagal overactivity was present in 77 per cent. As these authors recognize, such a scheme breaks down when sympathetic and vagal activities increase together, as they may during myocardial ischaemia

(Gillis 1971; Koizumi and Kollai 1981). There was indirect evidence for this in their own study: 8 per cent of patients given atropine for bradycardia or hypotension subsequently developed tachycardia or hypertension. Nevertheless, the different pressures and heart rates recorded in patients with anterior and inferior myocardial infarctions suggest that receptors within the myocardium play a role in determining net changes of autonomic outflow.

There is more direct (but still inferential) evidence on changes of human autonomic activity during acute myocardial infarction. Several groups measured plasma noradrenaline levels serially during acute myocardial infarction. In an early study by L. McDonald *et al.* (1969), plasma adrenaline and noradrenaline levels were measured in 50 patients within 24 h of the onset of acute myocardial infarction, 50 age-matched patients admitted for non-cardiovascular problems, and (curiously) three automobile racing drivers within 10 min of the completion of a race. Adrenaline levels were comparable in infarction and control patients, but noradrenaline levels were much higher in patients with infarctions (600 vs. 280 pg/ml). (Since the three racing drivers had levels of 6810, 1360, and 5060 pg/ml, the authors concluded that high noradrenaline levels do not usually provoke serious ventricular dysrhythmias in people with healthy hearts.)

Mueller and Ayres (1980) extended these data. They measured plasma catecholamines in 70 patients earlier (at an average of 10 h after the onset of myocardial infarction), and analysed their results according to the site of infarction. Plasma noradrenaline levels were comparable in patients with anterior and inferior or posterior infarctions (1950 vs. 2020 pg/ml), even though patients with anterior infarctions had significantly higher resting heart rates than those with inferior infarctions (84 vs. 69 beats/min). Blood pressures were not different in the two groups. These data suggest that elevations of sympathetic activity, estimated from plasma noradrenaline levels, are common in acute myocardial infarction, and occur with equal frequency in patients with anterior and inferior infarctions.

Some animal studies indicate that increased catecholamine levels post-infarction may arise from reflexes initiated directly by ischaemic myocardium (A.M. Brown 1967; Staszewska-Barczak 1971). An alternative or complementary possibility is that sympathetic activity increases because of haemodynamic deterioration, and may result in part from baroreflex adjustments. Karlsberg *et al.* (1979) measured plasma noradrenaline levels serially after acute coronary artery occlusion in conscious dogs. They found that the extent of plasma noradrenaline elevation is proportional to infarct size and to decrements of cardiac output and arterial pressure. Thus in this study, elevation of noradrenaline levels may have resulted from baroreceptor responses to falling pressure. Karlsberg also found that there is no difference between catecholamine levels in matched anterior and inferior infarctions.

A study reported in preliminary form by Schömig *et al.* (1984) documented an inverse relation between angiographically determined left ventricular ejection fraction and plasma noradrenaline and adrenaline levels in post-infarction patients. It seems likely that increased circulating noradrenaline levels in post-infarction patients reflect generalized augmentation of systemic activity rather than selective augmentation of sympathetic traffic to the heart. Esler *et al.* (1984) showed that under normal conditions, contributions of noradrenaline spillover from the heart are small. Animal studies indicate that sympathetic traffic to the heart increases during acute myocardial infarction. Increased sympathetic traffic coupled with ischaemia sets in motion a complex cascade of events that may end with myocardial damage secondary to accumulation of catecholamines (Schömig 1990).

Zipes and his co-workers (Barber *et al.* 1985; H. Inoue *et al.* 1988; H. Inoue and Zipes 1988) and others (Martins *et al.* 1989*b*) have studied chronic changes of cardiac innervation that occur in animals after transmural infarctions. These studies show that infarctions create areas of focal afferent and efferent vagal and sympathetic denervation. Minardo *et al.* (1988) showed that although reinnervation may occur in some animals, denervation supersensitivity may persist. The practical consequences of these changes are unknown. It can be said however that focal areas of supersensitivity to catecholamines (H. Inoue and Zipes 1987*b*; Kammerling *et al.* 1987) have the potential to participate in dysrhythmogenesis (Kimura *et al.* 1987; H. Inoue *et al.* 1988), and may be present in the hearts of post-infarction patients (Stanton *et al.* 1989). The presence of pre-existing myocardial disease adds further complexity; Martins *et al.* (1989*a*) showed that chronic hypertension and left ventricular hypertrophy increase the likelihood that ventricular tachycardia will be induced by coronary artery occlusion in dogs.

Post-infarction changes of autonomic outflow

The consequences and natural history of autonomic disturbances in patients with acute myocardial infarctions have been studied by several groups. Pantridge and his co-workers (1975) assumed (reasonably enough) that tachycardia and hypertension, or bradycardia and hypotension, were harmful, and they treated patients with these changes of blood pressure and heart rate with large doses of intravenous practolol (an early, subsequently withdrawn beta-blocker) or atropine.

M.M. Wolf *et al.* (1978) estimated vagal–cardiac activity with R–R interval standard deviations squared and considered that patients with variances above 1000 (standard deviations of about 32 ms) had sinus arrhythmia, and those with variances below 1000 did not. According to this scheme most patients (59 per cent) did *not* have sinus arrhythmia. Thirty-seven per cent of patients with anterior infarctions, and 47 per cent of patients with inferior infarctions, had sinus arrhythmia. (Wolf does not indicate whether the increased likelihood of sinus arrhythmia found in patients with inferior infarction was significant.) Mortality was 4 per cent in patients with sinus arrhythmia and 16 per cent in patients without sinus arrhythmia ($P < 0.05$). McAreavey *et al.* (1989) measured abrupt, vagally mediated R–R interval changes in patients during the first 24 hours after acute myocardial infarction; they found that such abrupt changes occur in fewer patients with anterior than with inferior infarctions.

Lombardi *et al.* (1987) analysed electrocardiograms with an autoregressive power spectral technique (Pagani *et al.* 1986) to estimate vagal and sympathetic outflow in patients after acute myocardial infarction. They found that the high-frequency (respiratory) component, which is accepted widely as an index of vagal–cardiac outflow, was subnormal at 2 weeks, and less subnormal by 12 months post-infarction. They found also that the low-frequency component (0.10 Hz), which they consider to be an an index of sympathetic cardiac activity (see Chapter 8), was supranormal at 2 weeks, and normal at 12 months. The ratio between high- and low-frequency power, which averaged 9 in early post-infarction patients and 2 in healthy subjects, further underscored the reciprocal nature of these changes. There was no difference between power spectra of patients with anterior and inferior infarctions. These data provide strong but inferential evidence for transient depression of vagal– and augmentation of sympathetic–cardiac nervous outflow in the early weeks following acute myocardial infarction. The time course of changes of efferent autonomic outflow documented by Lombardi is approximately similar to those of post-infarction baroreflex depression (P.J. Schwartz *et al.* 1988*b*) and mortality (Gilpin *et al.* 1983).

The mechanism mediating post-infarction reductions of vagal–cardiac nervous outflow is not known with certainty. One possibility is that it results from abnormal sensory input from damaged myocardium or from adjacent normal myocardium. Another is that it results simply from physical deconditioning. Musch *et al.* (1989) showed that subnormal heart rate responses to exercise in post-infarction rats can be restored to normal by endurance training.

Post-infarction baroreflex function

Come *et al.* (1975) gave nitroglycerine to patients with acute myocardial infarctions and found that 20 mmHg reductions of mean arterial pressure did not alter heart rate. This points to impaired baroreflex function early in the course of infarctions. Chatterjee *et al.* (1973) also studied responses of post-infarction patients to vasodilators. They gave the alpha-adrenergic blocking drug phentolamine or sodium nitroprusside to patients with acute myocardial infarction, and categorized haemodynamic responses according the severity of left ventricular dysfunction. Patients with nearly normal left ventricular end-diastolic pressures and stroke volumes increased their heart rates, reduced their stroke volumes, and did not benefit from vasodilator therapy. In contrast, patients with elevated end-diastolic pressures and reduced stroke volumes did *not* increase their heart rates and improved their left ventricular function. These data suggest that patients with severe left ventricular dysfunction lose baroreflex mediated cardioacceleration responses to hypotension, and that baroreflex *impairment* improves responses to vasodilator therapy. The authors did not focus on the putative baroreflex malfunction in their patients and they did not indicate whether baroreceptor reflex responses improved after patients' recoveries from myocardial infarction.

Martens *et al.* (1981) measured heart rate responses of post-infarction patients to carotid sinus massage; they found 'hyperactive' responses in 29 per cent on the first day after infarction, but in only 10 per cent one month after infarction. Imaizumi *et al.* (1984*b*) measured vagally mediated R–R interval prolongation after intravenous boluses of phenylephrine, and forearm vasoconstriction during lower body suction serially in patients following acute myocardial infarction. They found that baroreflex slopes were markedly depressed (compared with responses of healthy subjects reported earlier) during the early convalescent period (average time of study: seven days) but were close to normal by the late convalescent period (41 days). Imaizumi also documented subnormal early, but nearly normal late, adrenergically mediated reflex forearm vasoconstriction after acute infarction. Baroreceptor–cardiac reflex responses of one patient studied early and late after infarction are shown in Fig. 12.4.

P.J. Schwartz and co-workers (1988*b*) also studied baroreflex function in post-infarction patients. They showed that phenylephrine baroreflex slopes are lower in post-infarction patients than healthy volunteers 18 days but not 3 months after the event. Interestingly, post-infarction patients whose

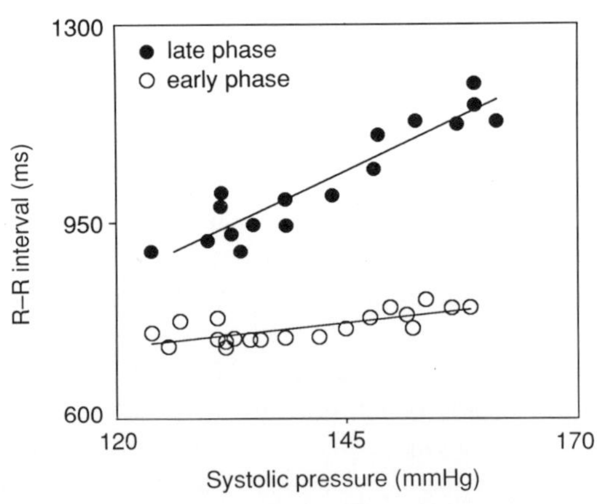

FIG. 12.4. Phenylephrine-induced R–R interval prolongation in one post-infarction patient studied early (about 7 days) and late (about 41 days) after the onset of infarction symptoms. Adapted with permission (Imaizumi *et al.* 1984*b*).

infarctions had been complicated by ventricular fibrillation had significantly lower baroreflex slopes than post-infarction patients who had not had this dysrhythmia.

The study of Bigger *et al.* (1989) indicates that the site of infarction influences baroreflex slopes; surprisingly, average baroreflex slopes were slightly but significantly lower in patients with *inferior* than in patients with *anterior* infarctions. One interpretation of these data is that in healthy subjects, inferior ventricular receptors contribute to vagal–cardiac inhibition during transient arterial pressures elevations. (This notion is not supported by Persson *et al.* (1987), whose studies in conscious dogs suggest that cardiopulmonary receptors do not mediate responses to abrupt transient pressure changes.) Bigger's results are supported by a study by Billman and Marsh (1989) conducted in dogs. They found that haemodynamic responses to bilateral carotid artery occlusion are impaired by acute inferior ischaemia, but are not altered by acute anterior ischaemia. In their study, haemodynamic abnormalities were prevented by prior vagotomy; this further supports the notion that inferior ischaemia alters arterial baroreflex function by altering afferent traffic from the heart.

Data quoted above suggest that acute myocardial infarction depresses arterial baroreflex responses, and that reflex impairment may not be due entirely to structural damage to autonomic nerves, since it is in part reversible. The time course of baroreflex impairment is roughly similar to that of mortality of infarction patients after discharge from the hospital (Gilpin *et al.* 1983). Whether increased risk during this period is related to disturbed baroreflex function is unknown. The possibility is discussed below.

Some studies of post-infarction prognosis use estimates of baseline vagal–cardiac activity rather than direct measures of baroreflex responses. As mentioned in Chapter 8, T. Bennett *et al.* (1978) found that the slope of phenylephrine-induced R–R interval changes in diabetic patients is related significantly to their respiratory sinus arrhythmia (in his study, peak minus valley R–R intervals during quiet breathing). Similarly, Bigger *et al.* (1989) found loose but statistically significant correlations between phenylephrine baroreflex slopes and other measures of R–R interval variation (the percentage of R–R interval changes exceeding 50 ms; the root mean square difference of normal R–R intervals; and the R–R interval power spectral density at high, respiratory frequencies). These observations provide some theoretical support for use of R–R interval variability as surrogates for baroreflex slopes, determined invasively.

In humans as in animals, myocardial infarction appears to reduce vagal–cardiac nervous outflow. In a preliminary report, Schneider and Costiloe (1965) found that a measure of heart rate variability (acceleration or deceleration of heart rate of 8 or more beats/min within 10 s) was reduced in young and middle-aged post-infarction patients. Similarly, Airaksinenn *et al.* (1987) found that heart rate changes during deep breathing are significantly less in patients with coronary artery disease than in healthy subjects; in patients, reductions of sinus arrhythmia were unrelated to cardiac function and the extent of coronary artery disease.

If this usage is justified, the observations by Kleiger *et al.* (1987) in patients, may be related to those of Billman and associates (1982) in dogs. Studies in 808 post-infarction patients showed that R–R interval variability on 24 h Holter electrocardiogram recordings is related inversely to survival. It is not known if the inverse relation between R–R interval variability and prognosis is related to the apparent protection from ventricular fibrillation conferred by baroreceptor–cardiac reflex responses (Volpe *et al.* 1985).

Post-infarction prognosis

Billman *et al.* (1982) studied dogs with healed anterior myocardial infarctions and occluded their circumflex coronary arteries (to produce acute *inferior* left ventricular ischaemia) during treadmill exercise. They showed that the likelihood of exercise-induced ventricular fibrillation was related inversely to the slopes of their vagally mediated baroreceptor–cardiac reflex responses. Subsequently this same group (P.J. Schwartz *et al.* 1988*a*) reported results from 307 dogs (only 192 were studied fully!). This study showed that anterior myocardial infarctions produced early in the protocol reduced baroreflex slopes. Curiously, post-infarction dogs most likely to experience ventricular fibrillation during exercise and inferior ischaemia had lower *pre-infarction* baroreflex slopes than dogs resistant to post-infarction fibrillation. Thus these dogs appeared to be at the low end of the spectrum of baroreflex slopes before infarction, reduced their slopes further after infarction, and moved into a high-risk category. After infarction, 91 per cent of ventricular fibrillation episodes occurred in dogs with baroreflex slopes less than 9, and only 12 per cent occurred in dogs with slopes above 20 ms/mmHg. These data are summarized in Fig. 12.5.

This work raises at the question: why does infarction impair vagal arterial baroreflex responses? One possibility is that impaired vagal reflexes are due to increased firing of left ventricular receptors secondary to abnormal contraction patterns. P.J. Schwartz *et al.* (1973) documented increased firing of left ventricular sympathetic afferent nerves after intracoronary injections of a veratrum alkaloid in anaesthetized cats and showed that increased afferent sympathetic firing was associated with reduced efferent vagal–cardiac nerve firing. Another study reported in preliminary form also supports the possibility that increased afferent sympathetic traffic impairs vagal–cardiac control; Cerati and Schwartz (1990) showed that increases of vagal–cardiac nerve traffic after phenylephrine injections are greater after than before left stellectomy.

A large study of *patients* after their first myocardial infarctions is under way, and preliminary results have been reported (LaRovere *et al.* 1988). In

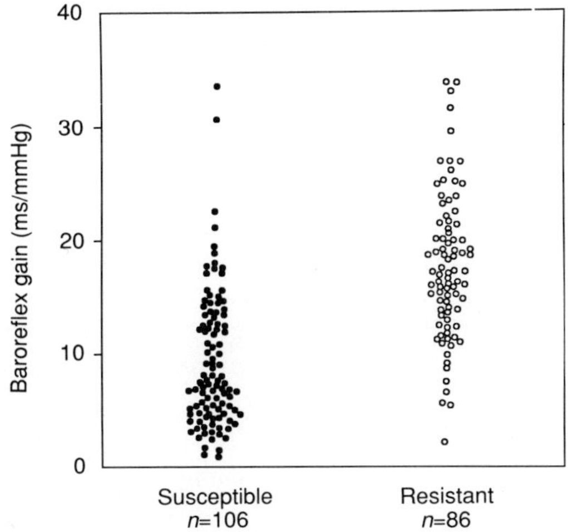

FIG. 12.5. Slopes of R–R interval–systolic pressure relations after phenylephrine boluses in dogs susceptible, or resistant to ventricular fibrillation (see text). Adapted with permission (P.J. Schwartz *et al.* 1988*a*).

studies of 78 patients, during an average follow-up of 24 months, baroreceptor–cardiac reflex slopes 30 days after infarction averaged 8.3 in 71 patients who survived, and 2.5 ms/mmHg in seven patients who died. Mortality was 50 per cent in post-infarction patients with baroreflex slopes below 3, but only 3 per cent in patients with baroreflex slopes above 3 ms/mmHg. The distribution of baroreflex slopes in all post-infarction patients is depicted in Fig. 10.8.

The mechanism whereby preservation of arterial baroreflex responses may protect against ventricular fibrillation is not well understood. In Billman's study (1982), dogs with steep arterial baroreflex slopes were *also* more likely to develop cardiac slowing during inferior ischaemia than dogs with shallow slopes. One interpretation of these data is that preservation of arterial baroreflex responses merely identifies dogs who also have preserved reflex responses to inferior left ventricular ischaemia, and that it is the preservation of inhibitory *ventricular* reflexes that confers protection against ventricular fibrillation. Alternatively, in these dogs, increased *arterial* baroreflex gain may be a marker for augmented central gain of all inhibitory reflexes, including those arising from the left ventricle.

Another explanation for the better prognosis of post-infarction patients with preserved baroreflex responses is that these patients have less severe heart disease than patients with impaired baroreflex responses. In the study of LaRovere *et al.* (1988), baroreflex impairment was related directly to the number of coronary arteries diseased, and of the seven patients who died, all but one had left ventricular ejection fractions below 40 per cent. However, of those with subnormal ejection fractions, patients were more likely to die if they had depressed than if they had normal baroreflex slopes. The study of Kleiger *et al.* (1987) showed that subnormal levels of another index of vagal–cardiac inhibition — heart rate variability — carried an adverse prognosis in post-infarction patients that was independent of ventricular function.

Another study (Billman *et al.* 1984) suggests that the apparent protective effect of sensitive baroreflexes is not due simply to structural anatomic differences. In that study, post-infarction dogs placed on an aerobic exercise programme increased their baroreflex slopes and became resistant to exercise and ischaemic sudden death; of great, but probably anecdotal, interest is that one dog not enrolled in the exercise programme spontaneously increased its baroreflex slope and also became resistant to ventricular fibrillation.

If it be accepted that arterial baroreflexes protect against sudden cardiac death, the mechanism is far from clear. One possibility is that normal baroreflex function is protective because normal vagal–cardiac nerve traffic is protective. The protective role of beta-adrenergic blockade in post-infarction patients is established beyond doubt (May *et al.* 1982). One possible mechanism, suggested originally by Eckberg (1984), is that beta-blockers prolong life by increasing vagal–cardiac nerve traffic. Numerous studies provide inferential evidence (discussed in Chapter 8) that beta-adrenergic blockers increase vagal–cardiac nerve activity in healthy humans, and a preliminary report (Guidera *et al.* 1990) suggests that similar effects occur in post-infarction patients. (A different approach was reported in preliminary form by Vybiral *et al.* (1990) and Casadei *et al.* (1991) who gave transdermal scopolamine to post-infarction patients; this treatment is thought to increase vagal–cardiac nerve traffic in humans by stimulating inhibitory central cholinergic vagal motonuclei (Dibner-Dunlap *et al.* 1985).)

However, a study by Verrier *et al.* (1974) suggests that acute blood pressure elevation (and baroreceptor stimulation) protects dogs against ventricular fibrillation, by reducing *sympathetic* rather than augmenting *vagal* outflow. This opens the possibility that abnormalities of baroreceptor–vagal reflexes referred to above may be markers for abnormalities of baroreceptor–sympathetic reflexes. This possibility is supported by the study of Schneider and Costiloe (1966), which showed that variability of systolic and diastolic pressures is greater than normal in post-infarction patients. Thus, impaired vagally mediated R–R interval control was related to impaired adrenergically mediated blood pressure control. Myocardial spillover of noradrenaline is normal in patients who have angina but not congestive heart failure (Swedberg *et al.* 1984). However, acute myocardial ischaemia selectively increases myocardial noradrenaline spillover (McCance and Forfar 1989). This important finding supports the possibility that acute augmentation of cardiac sympathetic activity during myocardial ischaemia contributes to sudden death. Acutely induced ventricular bigeminy increases coronary sinus noradrenaline levels (Geschwind *et al.* 1984); and premature ventricular beats elicit large increases of muscle and cardiac sympathetic nerve activity (see Chapter 13 (Delius *et al.* 1972*a*; Morrison *et al.* 1984; Herre and Thames 1987; Fagius 1988; Welch *et al.* 1989)).

Other evidence, however, supports the notion that protection against fibrillation is conferred in some way by reflexly augmented *vagal* activity. Cerati and Schwartz (1989) showed that baseline vagal–cardiac efferent activity is similar in cats who fibrillate during coronary artery occlusion and cats who do not fibrillate; however, cats who do not fibrillate have greater increases of vagal–cardiac nerve responses to coronary occlusion than cats who fibrillate. Moreover, electrical vagus nerve stimulation protects against exercise-induced ventricular fibrillation (De Ferrari *et al.* 1987). Some protection may result from lower heart rates and lower myocardial oxygen consumption. Another possibility is that protection results from some as yet to be defined sympathetic–cholinergic interaction.

ARTERIAL AND CARDIAC BAROREFLEX INTERACTIONS

As mentioned, there are at least two major deficiencies in human studies of baroreceptor reflexes in myocardial ischaemia; quantitative studies of baroreflex function have not been done in patients before and after infarctions, and the quantitative importance of interactions among autonomic receptor groups in patients has not been sorted out. Therefore great reliance must be placed on research conducted in animals which defines important reflex *interactions* in the pathophysiology of myocardial ischaemia. Most workers who have investigated the role of arterial baroreceptors in integrated responses to myocardial ischaemia have obtained measurements before and after baroreceptor denervation, and have considered that the differences in responses with and without baroreceptors define the contributions of baroreflexes.

Peterson *et al.* (1973) measured haemodynamic responses of conscious dogs to circumflex coronary artery occlusion before and after sinoaortic baroreceptor denervation. Their main results are summarized in Fig. 12.6. Before denervation, coronary occlusion led to an average reduction of mean arterial pressure of 10 mmHg and an increase of heart rate of 34 beats/min. After baroreceptor denervation, circumflex occlusion led to an average reduction of mean arterial pressure of 35 mmHg (from levels 40 mmHg higher than those of the dogs studied with baroreceptors intact), and an increase of heart rate of 18 beats/min. These results were confirmed by a subsequent study (Peterson and Bishop 1974); they suggest that arterial baroreceptors oppose the reduction of arterial pressure provoked by inferior myocardial ischaemia by increasing heart rate and vascular resistance. Of interest in that later study is that engagement of arterial baroreceptors

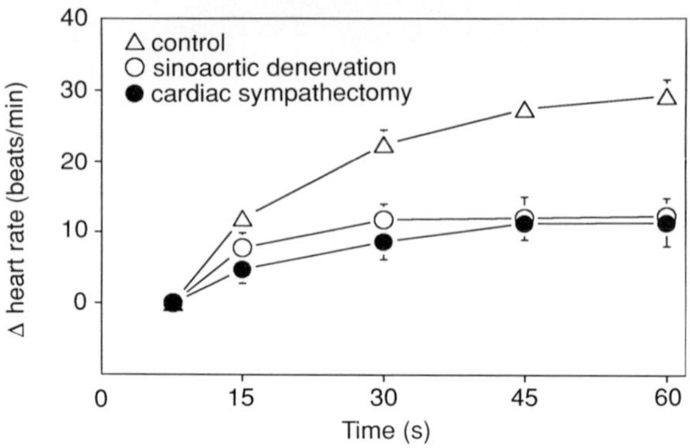

FIG. 12.6. Heart rate changes after occlusion of the left circumflex coronary artery in conscious dogs with or without arterial baroreceptor or cardiac sympathetic innervation. Adapted with permission (Peterson *et al.* 1973).

appeared to begin before a measurable reduction of arterial pressure occurred.

Thorén (1973) found that left anterior descending or right coronary artery occlusions in anaesthetized cats lead to moderate reductions of heart rate and blood pressure and slight increases of limb vascular resistance. When coronary occlusion is performed during bilateral carotid artery occlusions, reductions of heart rate and blood pressure are more pronounced and vasoconstriction is replaced by vasodilatation. These findings also suggest that arterial baroreceptors oppose haemodynamic deterioration caused by myocardial ischaemia.

Felder and Thames (1979) extended these findings by measuring the effects of ischaemia on multifibre efferent sympathetic–cardiac nerve activity, before and after sinoaortic denervation in dogs. They found that with baroreceptors intact, myocardial ischaemia leads to small increases of cardiac sympathetic traffic. However, with baroreceptors denervated, inferior myocardial ischaemia leads to moderate *reductions* of cardiac sympathetic traffic (and much larger reductions of arterial pressure). (In this experiment, after denervation, *anterior* ischaemia did not alter cardiac sympathetic nerve activity, and provoked a small reduction of arterial pressure.) The principal conclusions from this study are that inferior left ventricular ischaemia stimulates myocardial receptors, and increased firing from cardiac receptors leads to reductions of cardiac and systemic sympathetic outflow. In intact dogs, hypotension is opposed by arterial baroreflexes which act to increase sympathetic efferent activity in response to falling arterial pressure.

H.I. Chen (1979) studied arterial baroreceptor and cardiac reflex interactions with intravenous injections of veratridine (the intervention used originally by Bezold and Hirt (1867)) given to anaesthetized rabbits during controlled carotid sinus pressure changes. These studies showed that arterial baroreceptors partially buffer haemodynamic responses to veratridine, and that conversely, reflexes triggered by veratridine (presumably mediated by cardiac receptors in his study) buffer haemodynamic responses to altered baroreceptor activity. Chen's findings are summarized in Fig. 12.7.

Interactions between ventricular receptors and arterial baroreceptors were studied further by Holmberg *et al.* (1983), who showed that intracoronary injections of veratrine reduce the gain and R–R interval range of baroreceptor–cardiac reflex responses. In this study, the reduced range of vagally mediated R–R interval changes was found to result primarily from reduced vagal withdrawal; thus ventricular receptor stimulation shifted baroreceptor stimulus–sinus node response relations towards the threshold region (Chapter 6).

In our view, the question of how *healed* myocardial infarctions modify arterial baroreceptor and cardiac reflex responses to subsequent myocardial ischaemia has not been studied adequately in humans. Autonomic responses

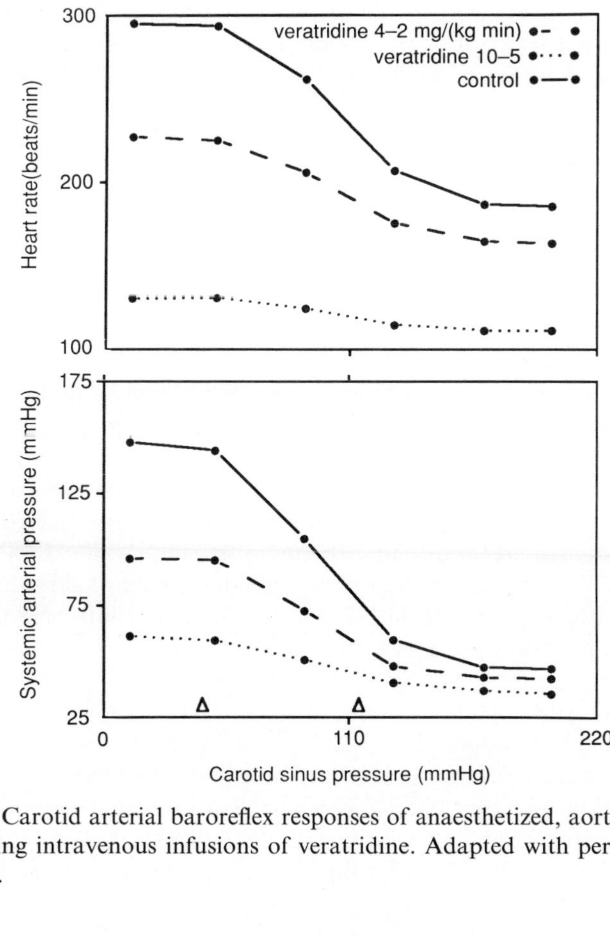

FIG. 12.7. Carotid arterial baroreflex responses of anaesthetized, aortic denervated rabbits during intravenous infusions of veratridine. Adapted with permission (H.I. Chen 1979).

during myocardial ischaemia that result from increased activity of myocardial receptors should be reduced in proportion to the extent of myocardial necrosis or scar. Some evidence in support of this comes from a study of inhibitory reflexes following successful reperfusion of infarct arteries. Koren *et al.* (1986) found that the occurrence of typical bradycardia and hypotension following reperfusion of the inferior left ventricle was inversely related to ejection fraction and non-invasive indices of the extent of infarction. Pantridge *et al.* (1975) found that when patients with previous infarctions are excluded, evidence for increased vagal activity during myocardial infarction is found almost universally. The implication of this observation, to which they call attention, is that prior infarctions interfere with usual reflex responses to myocardial ischaemia.

The observations of Wei *et al.* (1983) conflict with this conclusion. They found that six out of 27 patients who had had prior myocardial infarctions experienced autonomic changes during reperfusion that were similar to those of patients who had not had prior infarctions. Esente *et al.* (1983) report that of three patients with prior inferior infarctions who reperfused their right coronary arteries after thrombolytic therapy, hypotension and bradycardia occurred in two. Since longevity after myocardial infarction is clearly related inversely to the extent of myocardial damage (Geltman *et al.* 1979), it may be very important to understand better how earlier infarctions modify reflex responses to subsequent ischaemia. As mentioned, reinnervation of left ventricular receptors occurs in animals recovering from myocardial infarctions (Minardo *et al.* 1988).

MODULATION OF REFLEX RESPONSES BY DRUGS

Other perspectives can be obtained regarding myocardial–baroreceptor reflex interactions from the responses of patients to drugs. Responses of myocardial infarction patients to nitrates may result in part from interactions between reflexes triggered by arterial baroreceptors and myocardial receptors. As mentioned in Chapter 8, Come and Pitt (1976) reported the occurrence of vasovagal reactions in patients with acute myocardial infarction treated with intravenous sodium nitroprusside or sublingual nitroglycerine.

Subsequently Bossi *et al.* (1977) studied this reaction prospectively by giving 100 consecutive patients with acute myocardial infarction 5 mg doses of isosorbide dinitrate sublingually. In their series, 14 patients developed severe hypotension, associated with absolute or relative bradycardia. These reactions were not influenced by the site of infarction; seven patients had inferior or lateral infarctions, and six had anterior infarctions. This article did not indicate whether these patients also had right ventricular infarctions; conceivably, impaired right ventricular function may have predisposed the patients to greater than normal blood pressure reductions and vasovagal responses after nitrates. Also, Bossi and co-workers assumed that such hypotensive reactions would not have occurred in patients with coronary artery disease who had *not* had myocardial infarctions; however, no control group was studied. The mechanism for these responses is unknown; one possibility is that altered afferent input from ischaemic myocardial receptors modifies responses to changes of arterial baroreceptor input centrally. Presumably the mechanism for hypotension and bradycardia after nitrates is different from that for ischaemia or coronary arteriography, since it occurs equally in patients with anterior and inferior infarctions (Bossi *et al.* 1977).

BAROREFLEX REGULATION OF CORONARY RESISTANCE

One mechanism that might explain the relation between baroreflex function and prognosis, but which has not been explored fully in humans, is that arterial baroreflexes modulate the level of coronary vasoconstriction. Volpe *et al.* (1985) found that neck pressure (which compresses carotid sinus and simulates hypotension) significantly increases coronary vascular resistance (in normal subjects, but not in hypertensive patients).

A more direct relation between baroreflex (but probably cardiopulmonary rather than arterial) function and myocardial ischaemia was described by this same group. Trimarco *et al.* (1990) studied responses of patients to low-level lower body suction who had spontaneous episodes of myocardial ischaemia. They arbitrarily divided subjects into those who experienced increases of calculated coronary vascular resistance of more than or less than 20 per cent. The group with less coronary constriction during lower body suction had significantly longer periods of myocardial ischaemia. Both groups experienced coronary vasoconstriction during intense neck pressure. Responses to neck pressure indicated that efferent reflex coronary vasoconstrictor mechanisms were intact in both groups. By exclusion, impaired responses to lower body suction in the group with more prolonged ischaemic episodes were probably due to impairment of the afferent *cardiac* portion of the reflex arc. The implication of these observations is that intact reflex control of coronary vasomotion may terminate ischaemic episodes, presumably by promoting coronary vasodilatation. These data support the notion that apparent protective effects of normal autonomic reflexes are related to reflex control of coronary artery calibre.

It has long been recognized that carotid massage terminates anginal attacks (S.A. Levine 1962). Disappearance of symptoms may result from reduced myocardial oxygen demand secondary to heart rate slowing and reduced afterload and contractility. An alternative mechanism, and one that has not been sought, is that carotid massage improves oxygen delivery in part by dilating epicardial coronary arteries.

There is inferential evidence that changes of sympathetic nerve traffic precede the onset and termination of spontaneous angina. Bernardi *et al.* (1987) reported preliminarily that increases of low-frequency heart rate spectral power precede asymptomatic nocturnal angina attacks. Ricci *et al.* (1979) found that QT interval prolongation may precede spontaneous or ergonovine-induced coronary spasms. Whether these putative increases of sympathetic coronary vasoconstriction are related to baroreceptor reflex function is unknown.

However, a provocative study published by Lange *et al.* (1972) supports the possibility that under certain circumstances, baroreflexes may *promote* myocardial ischaemia. They reported cases of nine munitions workers who developed typical angina (one of whom died suddenly) after brief absences from their work, which involved handling nitroglycerine. In five of these workers, coronary arteriography failed to disclose significant atherosclerosis. One interpretation (the exact mechanism is not known) of the data of Lange and co-workers (1972) is that when workers were exposed regularly to nitroglycerine at work, their blood pressures were low, and consequently their efferent, baroreflex mediated levels of sympathetic activity were high. On work days, their coronary arteries were dilated by the direct effects of nitroglycerine (Feldman *et al.* 1982). However, after days off work, their nitrate levels and coronary dilatation declined more than their sympathetic activity, and they developed coronary artery spasm.

SUMMARY

Myocardial ischaemia alters haemodynamic function directly and reflexly and thereby modifies baroreflex mechanisms. There is abundant evidence that anterior and inferior infarctions affect baroreflex function differently. Acute inferior ischaemia, especially in patients without prior infarctions, provokes cholinergically mediated bradycardia and hypotension and partially interdicts appropriate arterial baroreflex responses. Inferior ischaemia does not have as much potential for development of catastrophic ventricular dysrhythmias as does anterior ischaemia. Although it is widely believed that inferior infarction increases sympathetic activity less than does anterior infarction, the evidence for this is not compelling. Healed inferior myocardial infarctions appear to impair arterial baroreflex function more than do anterior infarctions. There is an intriguing direct association between arterial baroreflex gain and survival post-infarction; the mechanism for this apparent protective effect is far from certain.

13

Dysrhythmias

Although organic heart disease, including particularly coronary artery disease, forms an essential substrate for the occurrence of most catastrophic cardiac dysrhythmias (Cobb *et al.* 1975), other factors also may be important. Several articles (Corr and Gillis 1978; Verrier and Lown 1984; P.J. Schwartz and Zaza 1986) criticize static views of cardiac anatomy and function and emphasize the role of dynamic changes of efferent autonomic activity in the pathophysiology of cardiac dysrhythmias. This chapter describes a wealth of published information which directly or indirectly implicates arterial baroreflexes in the pathophysiology of dysrhythmias. In some situations such as the provocation (Porus and Marcus 1963; M.V. Cohen 1972) or termination (Waxman *et al.* 1980*c*) of rhythm disturbances with carotid sinus massage, involvement of baroreflexes is inarguable. In other situations however, involvement of baroreflexes seems likely but is far from proven.

Preceding discussions of baroreceptor reflex physiology (Chapters 8 and 9) provide a basis for the prediction that arterial baroreflex mechanisms play major roles in all stages of cardiac dysrhythmias. Arterial baroreceptors are exquisitely sensitive and respond to extremely small changes of arterial pressure. In animals, pressure elevations as small as 1 mmHg trigger bursts of afferent baroreceptor activity (Landgren 1952*a*); this is illustrated in Fig. 6.7. In humans, arterial pressure changes as small as 5 mmHg have been shown to trigger changes of efferent vagal–cardiac (Eckberg 1980*a*;

Fritsch *et al.* 1986) and sympathetic–muscle (Sundlöf and Wallin 1978*b*) nerve activity. Also in humans, ventricular bigeminy increases coronary sinus and antecubital vein plasma catecholamine concentrations even though it does not alter mean arterial pressure (Geschwind *et al.* 1984). In animals, changes of arterial pressure that are too small to measure seem to provoke baroreceptor reflex-mediated changes of autonomic neural outflow (Peterson and Bishop 1974; Stinnett *et al.* 1976). Also, small reductions of blood volume that do not provoke detectable reductions of blood pressure reduce baroreceptor afferent activity (Hakumäki *et al.* 1985). Therefore *all* changes of arterial pressure associated with changes of cardiac rhythm are likely to alter autonomic neural outflow. (This assertion follows on another, that *all* rhythm changes perturb arterial pressure.)

Since most healthy humans function on or near the linear portions of their baroreceptor–sympathetic (Rea and Eckberg 1987) and –vagal (Eckberg 1980*a*) response relations, all *small* changes of arterial pressure resulting from dysrhythmias are likely to trigger changes of both sympathetic and vagal neural cardiovascular outflows. However, since resting arterial pressure is poised asymmetrically on arterial pressure–autonomic neural response relations (Rea and Eckberg 1987), responses to *large* changes of arterial pressure are probably mediated nearly exclusively by *either* sympathetic *or* vagal systems. For example, it is likely that a precipitous reduction of arterial pressure at the onset of a rapid dysrhythmia removes a patient from the pressure range over which both autonomic divisions respond, into a range where responses are exclusively sympathetic.

Under such circumstances, during large pressure reductions, engagement of autonomic responses is *sequential* as arterial pressure moves through vagal and sympathetic response relations. The very earliest pressure reduction simultaneously increases sympathetic and decreases vagal outflows. However, as pressure declines further, it may fall below the range for vagal responses, and vagal neurones will fall silent. Within the new pressure range, defences against the dysrhythmia are marshalled exclusively by increased sympathetic activity.

Conversely, if pressure subsequently rises above baseline levels in response to increased sympathetic activity, final responses are probably mediated exclusively by increased vagal activity. Evidence for this comes from studies of patients with paroxysmal supraventricular tachycardia; increases of heart rate during the tachycardia are nearly abolished by beta-blockade, and slowing of heart rate before restoration of rhythm to normal is abolished by hysocine butylbromide, an atropine analogue, which does not prevent hypotension or pressure recovery during the tachycardia, but prevents vagally mediated heart rate slowing and termination of the tachycardia (Waxman *et al.* 1982).

The time constants for baroreceptor mediated changes of *neural* activity are short (Wallin and Eckberg 1982). Moreover, at least in the case of vagally mediated baroreflex responses, the time constant of cardiac *effector* responses also is short; cardiac responses to abrupt changes of carotid baroreceptor activity occur in less than 0.5 s in healthy subjects (Eckberg 1976; Eckberg 1980*a*) and patients with congestive heart failure (Sopher *et al.* 1990) (see Chapter 9). Thus baroreflex mediated changes of autonomic neural activity may occur during dysrhythmias on a beat-by-beat basis. (However, based on experiments conducted in animals, the time constant for sympathetic–cardiac responses is likely to be longer (Spear and Moore 1973). We are aware of no measurements of the latency of sympathetically mediated baroreceptor–cardiac reflex responses in humans.)

On the basis of these considerations, it is unlikely that *any* dysrhythmia occurs that does not involve changes of baroreceptor reflex function. From this follows the notion that since autonomic neural outflow alters cardiac electrophysiological properties, baroreceptors probably participate in the pathophysiology of all dysrhythmias. This participation is dynamic, however, and as discussed in Chapter 9, the rules of engagement are probably changing rapidly. For example, sustained changes of arterial pressure during prolonged dysrhythmias probably lead to shifts of the baroreflex operating range ('resetting' (McCubbin *et al.* 1956; M.L. Smith *et al.* 1989*b*)) as discussed in Chapter 6; a possible example of resetting of responses to ventricular premature beats is shown in Fig. 13.1.

In this patient, premature ventricular beats were introduced serially and timed to occur at exactly the same time in the cardiac cycle during a progressive reduction of arterial pressure after a phenylephrine infusion was stopped. The leftmost panel shows that such a premature beat does not

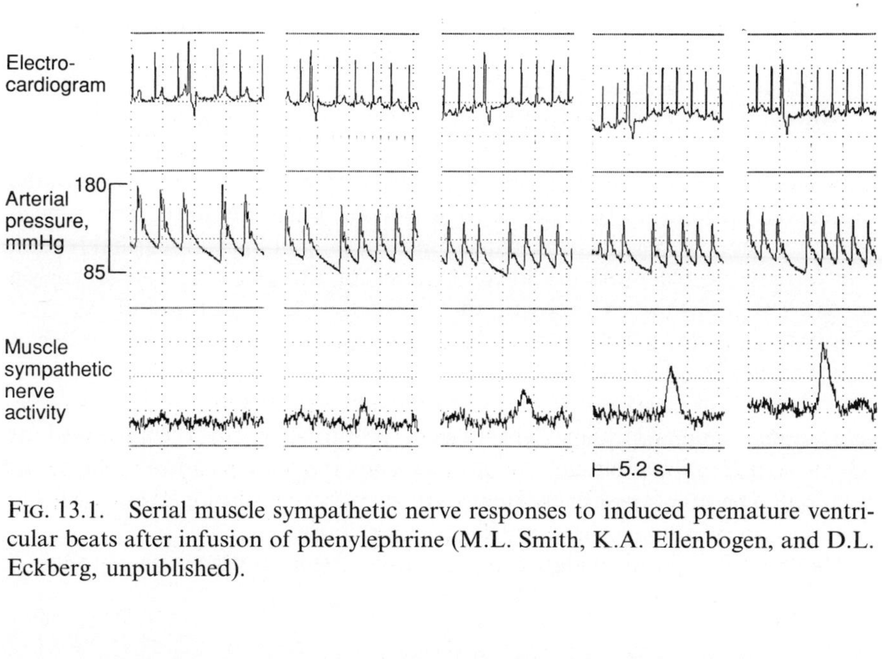

FIG. 13.1. Serial muscle sympathetic nerve responses to induced premature ventricular beats after infusion of phenylephrine (M.L. Smith, K.A. Ellenbogen, and D.L. Eckberg, unpublished).

elicit a sympathetic burst when it occurs during pressure elevation. However, the last three panels show that with no apparent further reduction of pressure, premature beats elicited increasingly large sympathetic bursts. These data suggest that the return of sympathetic activity (over a period of about 4 min) reflects gradual resetting of the usual relation between arterial pressure and sympathetic activity.

This chapter reviews evidence that changes of baroreceptor activity may (1) set the stage for or even provoke dysrhythmias, (2) modulate haemodynamic responses to dysrhythmias, and (3) terminate dysrhythmias or transform them into more or less threatening dysrhythmias. It is not necessarily our intention to prove that baroreflexes are prepotent in causing rhythm disturbances, but merely that baroreflexes are involved importantly, and in some patients critically, in the pathophysiology of dysrhythmias.

BAROREFLEXES MODULATE CARDIAC ELECTROPHYSIOLOGICAL PROPERTIES

There is unequivocal histological and functional evidence that changes of sympathetic and vagal efferent inflow importantly influence function of *human* sinoatrial and atrioventricular nodes and atrial and ventricular myocardium. Older studies employed histochemical staining and documented the presence of noradrenaline (Chidsey *et al.* 1965) and acetylcholinesterase (James and Spence 1966; K.M. Kent *et al.* 1974) in all of these structures. Figure 9.17 shows cholinergic innervation of the human ventricle. More recent studies employ highly specific radioligand binding and document the presence of alpha- and beta-adrenergic and muscarinic–cholinergic receptors in all human specialized cardiac pacemaking and conducting tissues (Stiles *et al.* 1983; Syrota *et al.* 1985).

However, in healthy hearts the distribution of adrenergic and cholinergic receptors is not uniform, and in diseased hearts additional non-uniformities may be present. Cholinergic and adrenergic receptors are not distributed uniformly throughout the left ventricular wall. Barber *et al.* (1985) showed that cholinergic receptors are located preferentially in the endocardial region, and adrenergic receptors are located preferentially in the epicardial region of the left ventricle. Moreover, the function and quantities of receptor populations is not fixed. Myocardial ischaemia may increase the number of beta-adrenergic receptors (Mukherjee *et al.* 1979). Myocardial infarction may damage or destroy these receptors; 'subendocardial' or 'non-Q-wave' infarctions damage primarily cholinergic receptors, whereas 'transmural' or 'Q-wave' infarctions damage or destroy both types of receptors (Barber *et al.* 1985). The functional consequences of such focal pathological myocardial denervation in humans have not been worked out.

There is also solid evidence from humans that traffic over autonomic nerves to the heart leads to release of neurotransmitters. Older measurements involving arterial plasma levels (Chidsey *et al.* 1962; Swedberg *et al.* 1984) and more recent kinetic studies with radiolabelled noradrenaline (Hasking *et al.* 1986) establish that the heart releases noradrenaline into the systemic circulation. (As far as we know, acetylcholine, which is hydrolysed extremely rapidly by acetylcholinesterase, has not been measured in human coronary sinus blood samples.) Thus there are receptors capable of responding to released neurotransmitters, and there is direct evidence that at least one neurotransmitter, noradrenaline, is released by nerve traffic to the human heart.

Evidence that the heart responds to released neurotransmitters takes several forms. First, human hearts respond to direct electrical stimulation of efferent autonomic nerves. Although we are aware of no published studies of responses of patients to direct electrical stimulation of *sympathetic*–cardiac nerves, there is at least one study of responses to direct stimulation of *vagal*–cardiac nerves in humans (Carlsten *et al.* 1957). This study indicates that, as expected, vagus stimulation slows the heart rate, and it establishes (although this probably was not needed) a link between the presence of receptors and responses to nerve stimulation. Figure 9.1 (centre) depicts unpublished human sinoatrial node responses to vagus nerve stimulation.

Other less direct functional evidence also exists. Carotid baroreceptor stimulation with abrupt neck suction increases right ventricular refractory periods by a small but highly significant amount (Ellenbogen *et al.* 1990*a*). Electrical field stimulation of excised human papillary muscles after beta-adrenergic blockade releases a substance (presumed to be acetylcholine) which inhibits contraction (Sieber *et al.* 1973). In anaesthetized dogs, electrical vagus nerve stimulation slows ventricular pacemakers when complete atrioventricular block is present (Eliakim *et al.* 1961; Spear and Moore 1973) and reduces left ventricular systolic pressure in isovolumetric preparations (Levy *et al.* 1966). Finally, functional evidence for important autonomic influences on cardiac structures comes from studies which employ autonomic blocking drugs (see Chapter 8).

Much of the foregoing considerations catalogue evidence that cardiac structures involved importantly in dysrhythmias are susceptible to autonomic influences in humans. From this follows the notion that changes of efferent autonomic traffic to the heart may potentially play necessary or permissive roles in the genesis of dysrhythmias and modify their outcomes.

ACUTE CHANGES OF AUTONOMIC OUTFLOW MAY PRECEDE DYSRHYTHMIAS

Heart rate changes may precede the onset of *atrial* dysrhythmias. Studies of Coumel and his associates (1978, 1979) and Coumel (1990) elegantly describe changes leading to atrial fibrillation in patients in whom excess vagal activity seems to contribute to the dysrhythmia. First, they document the occurrence of cardiac slowing prior to the onset of atrial fibrillation. (Changes of heart rate point strongly toward changes of autonomic nervous traffic to the heart, since at rest heart rate changes are extremely small in patients with transplanted, and therefore denervated, hearts (Beck *et al.* 1969; M.L. Smith *et al.* 1989).) When the cardiac interval reaches a critical lengthening, atrial fibrillation begins. In Coumel's experience, the last cycles before the onset of tachycardia are longer than preceding cycles in 70 to 80 per cent of patients. In their study, interventions known (or suspected) to increase vagal–cardiac activity, such as digitalis (Ferrari *et al.* 1981*b*) or beta-blocker administration (Eckberg 1984), physical inactivity, or sleep, increase the likelihood that atrial fibrillation will develop, and interventions known to reduce vagal (and increase sympathetic) activity, such as exercise, prevent their occurrence. Thus, in the patients studied, increased vagal (and probably reduced sympathetic) activity seems to precede the development of atrial fibrillation.

The mechanisms whereby increases of vagal activity predispose to atrial rhythm disturbances have been studied in animals and humans. Vagus nerve stimulation provokes non-uniform shortening of atrial (Alessi *et al.* 1958; Prystowsky *et al.* 1983) and lengthening of ventricular (Kolman *et al.* 1975; Martins 1984; Ellenbogen *et al.* 1990*a*) refractoriness and produces intra-atrial conduction delays (Spear *et al.* 1979). These electrophysiological changes predispose to re-entrant atrial dysrhythmias (Moe 1975). Also, vagal stimulation may lead to atrial enlargement through bradycardia (with longer atrial diastole) and a negative inotropic effect (Boyden and Hoffman 1981). Waxman *et al.* (1983) speculate that the beneficial effects of quinidine, procainamide, and disopyramide in preventing recurrence of atrial fibrillation are mediated in part by their weak but unmistakable anticholinergic properties.

Prystowsky *et al.* (1983) found that abrupt increases of vagal–cardiac activity provoked by carotid baroreceptor stimulation with neck suction lead to small but significant reductions of atrial refractory periods. Insight into the functional importance of this observation came from three patients who had repetitive depolarizations in response to single atrial stimuli. In these patients repetitive responses occurred only during neck suction when vagal activity was higher than usual. In patients with acute severe high cervical spinal cord injuries, vagal traffic is unopposed by sympathetic traffic and supraventricular dysrhythmias occur commonly (Lehmann *et al.* 1987). These studies may implicate vagal traffic directly the in pathogenesis of human supraventricular dysrhythmias.

Heart rate changes are also known to precede *ventricular* dysrhythmias. In a series of 48 patients with acute myocardial infarction, Adgey *et al.* (1982) recorded electrocardiographic changes leading to the onset of ventricular fibrillation. In two-thirds of patients the sinus rate was faster immediately before the onset of ventricular fibrillation than it was at the time when electrocardiographic monitoring began. Cardioacceleration exerts different effects in healthy and diseased hearts. In healthy hearts tachycardia reduces the dispersion of ventricular refractory periods. In ischaemic hearts however, tachycardia increases the dispersion of refractory periods and reduces the threshold for ventricular fibrillation (K.M. Kent *et al.* 1973). In the remarkable, celebrated case reported by Wellens *et al.* (1972), a 14-year-old girl with the long QT interval syndrome (P.J. Schwartz 1985; P.J. Schwartz *et al.* 1991) experienced syncope in response to startling sounds. Such sounds initiated an electrocardiographic sequence which began predictably with sinus tachycardia and culminated with ventricular tachycardia. The electrocardiogram in this patient is depicted in Fig. 13.2.

Speeding of the sinus rate prior to the onset of ventricular tachycardia or fibrillation has been remarked commonly in other studies of Holter recordings obtained from people who died suddenly. Kempf and Josephson (1984) examined Holter recordings from 20 patients who died of tachydysrhythmic sudden cardiac death. In 11 of the 20 there was an increase of the sinus rate of greater than 20 per cent in the hour preceding death. Similarly, Leclercq *et al.* (1988) found that in 46 patients during the hour before the onset of ventricular tachycardia, heart rate increased in 33, decreased in six, and did not change in seven patients.

Zimmermann *et al.* (1986) studied a group of young patients with idiopathic ventricular tachycardia and made two observations regarding the sinus rate preceding the ventricular tachycardia. First, they found that the sinus rate was faster prior to ventricular tachycardia than before isolated premature ventricular beats. Second, they found that there was an extremely close linear correlation ($r = 0.98$) between the sinus rate before the tachycar-

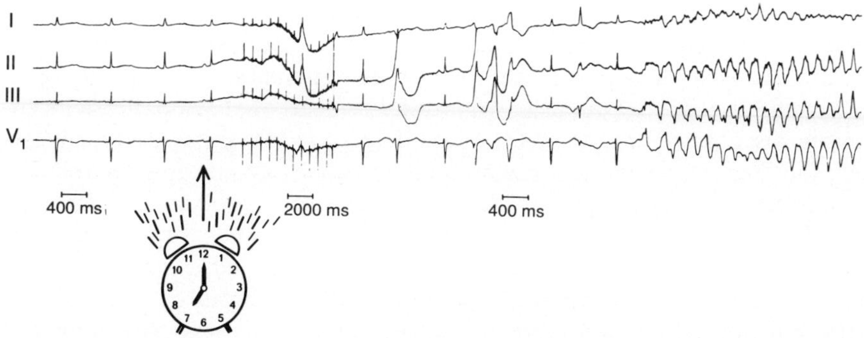

FIG. 13.2. Electrocardiogram of the patient of Wellens *et al.* (1972) who developed ventricular fibrillation after startling sounds, including those of her alarm clock. Reproduced with permission.

dia began and the duration of that tachycardia. Zimmermann's findings provide strong if inferential evidence that changes of autonomic outflow to the heart, possibly related to baroreflex function, predispose to the occurrence of ventricular dysrhythmias and partly determine their duration. Yusuf *et al.* (1980) found a significant correlation between the heart rate preceding accelerated idioventricular rhythms and the rate of those rhythms in patients with acute myocardial infarction.

The nature of the change of autonomic activity giving rise to heart rate speeding is not known. Although it is generally thought to be secondary to increased sympathetic activity (Adgey *et al.* 1982), sinoatrial node speeding may be due also to withdrawal of vagal restraint or to simultaneous increase of sympathetic and reduction of vagal activity. Any of these combinations could predispose to ventricular dysrhythmias. (Moreover, H. Inoue and Zipes (1987*a*) showed that levels of simultaneous electrical vagal and sympathetic cardiac stimulation can be adjusted in a way that substantial changes of electrophysiological properties occur with no changes of R–R interval.)

The preceding paragraphs describe what appear to be spontaneous changes of autonomic outflow prior to the onset of rhythm disturbances. In other situations, changes of autonomic outflow may be provoked by definable events. *Exercise* is a complex intervention which, among many other effects, may modify baroreflex control of sympathetic (Eckberg and Wallin 1987) and vagal (Bristow *et al.* 1971*b*; Cunningham *et al.* 1972*a*; Eckberg *et al.* 1972) autonomic outflow. As mentioned, Coumel found that in some patients, exercise seems to prevent the occurrence of paroxysmal atrial fibrillation (Coumel *et al.* 1979). Exercise may also provoke dysrhythmias. Patients with severe aortic stenosis may experience sudden death during exercise. L.S. Schwartz *et al.* (1969) showed that in some of these patients hypotension (probably reflex (Mark *et al.* 1973)) is followed by ventricular flutter or fibrillation. Frenneaux *et al.* (1990) suggested that exercise-induced hypotension may be a marker for risk of sudden death in patients with hypertrophic cardiomyopathy. Some patients experience ventricular tachycardia primarily during exercise (Zimmermann *et al.* 1986).

Leriche *et al.* (1932) published anecdotal evidence that sympathetic mechanisms are important in the pathophysiology of paroxysmal atrial tachycardia; they reported a patient in whom bilateral sympathetic ganglia ablation prevented recurrence of paroxysmal atrial tachycardia. Waxman *et al.* (1983) found that in five of 23 patients with paroxysmal supraventricular dysrhythmias, the rhythm disturbance could be induced by exercise or isoproterenol infusion. Similarly, ventricular tachycardia can be induced after but not before isoproterenol infusion in some patients (Freedman *et al.* 1984; Olshansky and Martins 1987). The response to isoproterenol suggests that in these patients, exercise provokes dysrhythmias primarily by increasing sympathetic activity rather than by reducing vagal activity. (However, since isoproterenol reduces arterial pressure, vagal withdrawal probably occurs.) This conclusion is strengthened by the finding of Olshansky and Martins (1987) that some patients in whom ventricular tachycardia can be induced only after isoproterenol are not inducible after isoproterenol and a beta-blocker. However, Waxman and co-workers (1983) also found that in seven of 14 patients with paroxysmal supraventricular dysrhythmias confined to the atrioventricular node, sustained tachycardia could be induced only after muscarinic blockade with atropine. This observation points towards a role for vagal as well as sympathetic outflow in the pathogenesis of this rhythm disturbance.

Other factors are known to modify baroreflex responses. B. Liang and co-workers (1979) showed that conscious dogs placed in an aversive sling experience a striking tachycardia and elevation of plasma catecholamine levels, and a reduction of their repetitive extrasystole thresholds (used as an index of vulnerability to ventricular fibrillation (Kowey *et al.* 1983*b*)). The study of Conway *et al.* (1983) indicates that the slopes of phenylephrine baroreflex responses are inversely related to the state of *arousal*. Emotion is a well-known antecedent of ventricular fibrillation. Conceivably, emotion may provoke rhythm disturbances in part by opposing vagal baroreflex mechanisms.

BAROREFLEX STIMULI THAT CAUSE DYSRHYTHMIAS

The medical literature contains numerous examples of dysrhythmias *caused* by baroreceptor stimulation. The study of Killip and Gault (1965) showed that atrial fibrillation may be initiated by premature atrial stimuli. In one of the 14 patients they reported, atrial fibrillation was provoked by baroreceptor stimulation with carotid massage. Waxman *et al.* (1983) also documented a case in which atrial fibrillation was provoked by carotid massage. Cases are reported in which carotid massage applied during atrial flutter increased the rate of flutter (F.N. Wilson 1924), or converted atrial flutter to atrial fibrillation (Bussan *et al.* 1957). There is also a literature which indicates that carotid massage may lead to ventricular fibrillation (usually in patients receiving digitalis (Porus and Marcus 1963; M.V. Cohen 1972; Strasberg *et al.* 1989)). Carotid massage is employed routinely for diagnosis of dysrhythmias and treatment of supraventricular tachycardias (Waxman and Wald 1977*b*; Waxman *et al.* 1980*c*); although ventricular fibrillation is reported, its occurrence must be very rare. Valsalva straining and disengagement of arterial baroreflexes may precipitate ventricular tachycardia in patients with the long QT interval syndrome (Mitsutake *et al.* 1981).

BAROREFLEX RESPONSES TO DYSRHYTHMIAS

Since baroreceptors sense every arterial pulse, it is almost inescapable that they continuously modify efferent autonomic traffic during dysrhythmias. Responses of arterial baroreflexes to dysrhythmias have been worked out in surprisingly complete detail.

Premature beats

In the great majority of instances, sustained dysrhythmias do not arise simply in association with speeding or slowing of heart rate, as described above; usually there are one or more *premature beats* as well. As mentioned, it has long been recognized (Killip and Gault 1965; Wellens and Durrer 1975) that supraventricular dysrhythmias may be initiated by premature atrial beats. Moreover, a rapidly expanding literature indicates that the onset of ventricular fibrillation, as captured on Holter recordings, is also usually preceded by one or more premature ventricular beats (Nikolic *et al.* 1982; Weaver *et al.* 1982; Leclercq *et al.* 1988). Premature atrial or ventricular beats may lead directly to dysrhythmias with no intervening rhythm disturbance. In such instances, baroreflex mechanisms may be brought into play simply by the fall of arterial pressure that occurs in response to the rhythm disturbance (Schlepper *et al.* 1978; Curry 1979; Bauman *et al.* 1984; Crozier *et al.* 1989).

Often however, premature beats prior to rhythm disturbances occur in sequences that are punctuated by sinus beats. In such instances premature and sinus beats may conspire to provoke complex haemodynamic and autonomic changes which may set the stage for subsequent rhythm disturbances. It may not be accidental that both supraventricular (Coumel *et al.* 1979) and ventricular (Bauman *et al.* 1984; Crozier *et al.* 1989) tachycardias are commonly preceded, or can be induced (Denker *et al.* 1984), by long–short sequences of cardiac cycle length; that is, a premature beat is followed by a sinus beat with a compensatory pause, and this in turn is followed by a second premature beat. In the case of the patient reported by Wellens *et al.* (1972), ventricular tachycardia triggered by loud noises was preceded by long–short cardiac cycle sequences. This sequence is shown in Fig. 13.2.

Bauman *et al.* (1984) retrospectively collected data on 31 patients taking quinidine who had experienced polymorphic ventricular tachycardia (*torsades de pointes*). In 11 patients, the electrocardiogram had been recorded at the onset of tachycardia. In ten of these 11 patients, ventricular tachycardia was preceded by a long–short cycle. A schematic illustration of these electrocardiograms is depicted in Fig. 13.3. In all 13 patients with *torsades de pointes* described by Leclercq *et al.* (1988), the dysrhythmia was initiated by long–short cardiac cycles.

Zimmerman (1986) found in a series of patients with idiopathic ventricular

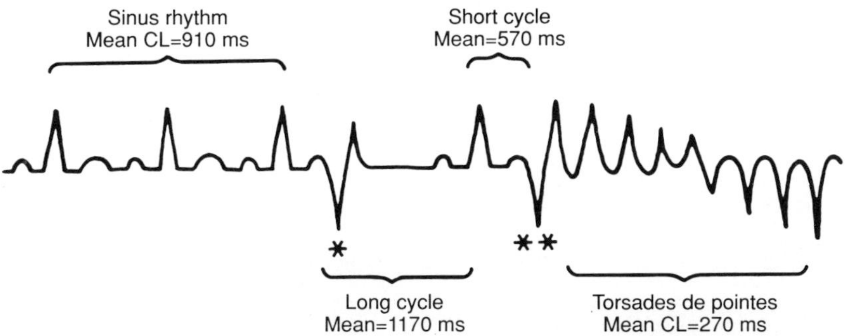

FIG. 13.3. Schematic drawing of common electrocardiographic sequence leading to polymorphic ventricular tachycardia (*torsades de pointes*) in ten out of eleven patients treated with quinidine. Adapted with permission (Bauman *et al.* 1984).

tachycardia that fully 77 per cent of episodes were preceded by a long R–R interval (which in most cases followed an isolated premature beat). Kempf and Josephson (1984) found that ventricular tachycardia or fibrillation was initiated by a long–short cycle in two of 27 patients with sudden cardiac death. Long–short sequences may alter susceptibility to ventricular dysrhythmias by provoking large sympathetic bursts (Welch *et al.* 1989); an additional, perhaps complementary mechanism is that long–short sequences may modify ventricular refractoriness (Marchlinski 1987).

Some studies of human sympathetic mechanisms in dysrhythmia are based on a critical assumption: that sympathetic nerve traffic recorded in a mixed peripheral nerve reflects in some way *cardiac* sympathetic nerve activity. This seems eminently reasonable. First, Ninomiya and co-workers (1971) showed that baroreflex forcings change sympathetic traffic to the heart, spleen, and kidneys in qualitatively similar ways (there are differences in gain). Second, the pattern of change of sympathetic nerve traffic after premature beats is similar in cardiac preganglionic outflow (Lombardi *et al.* 1989), and in cardiac (Morrison *et al.* 1984; Herre and Thames 1987; Neely *et al.* 1990), renal (Herre and Thames 1987), and muscle (Fagius 1988; Welch *et al.* 1989) postganglionic sympathetic outflows. Third, measurements of coronary sinus plasma noradrenaline levels (Crozier *et al.* 1989) and renal artery blood flow (Benchimol and Desser 1971) in humans suggest that ventricular tachycardia increases sympathetic traffic to these beds as it does to the muscle vascular bed (Ellenbogen *et al.* 1990*b*; M.L. Smith *et al.* 1991).

Autonomic responses to isolated premature beats have been examined in detail in animals (Morrison *et al.* 1984; Herre and Thames 1987; Lombardi *et al.* 1989) and humans (Geschwind *et al.* 1984; Fagius 1988; Welch *et al.* 1989). Muscle sympathetic nerve responses of one patient to two induced premature ventricular beats are shown in Fig. 13.4. A variety of haemody-

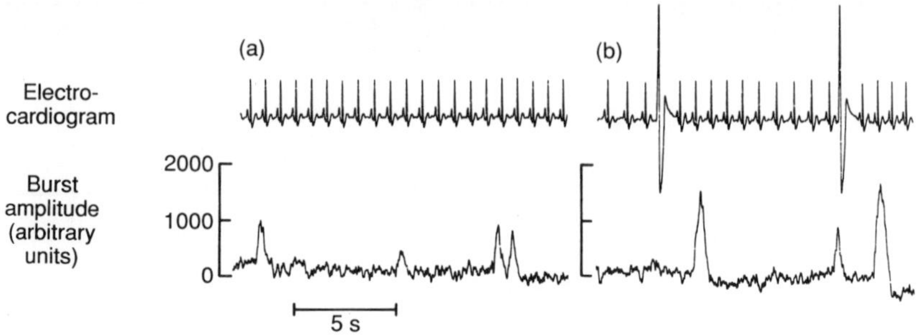

FIG. 13.4. Muscle sympathetic nerve activity (a) during sinus rhythm and (b) during sinus rhythm with two induced premature ventricular beats. Adapted with permission (Welch *et al.* 1989).

namic and autonomic changes occur following a premature beat which might help set the stage for subsequent rhythm disturbances.

First, since the beat is premature, the time available for diastolic pressure runoff of the preceding beat is reduced and the diastolic pressure for that beat is higher than usual. Since the magnitude of sympathetic bursts is determined importantly by diastolic arterial pressure (Sundlöf and Wallin 1978*b*), the sympathetic burst related to that arterial pulse is smaller than that following non-premature beats, or is absent altogether.

Second, since the period of diastolic ventricular filling is shortened, left ventricular end-diastolic volume and therefore force of contraction is reduced. Also, since the diastolic pressure against which the ventricle ejects is higher than usual, the systolic arterial pressure and stroke volume resulting from the premature beat are lower than usual. If the beat is very premature, left ventricular pressure may not exceed aortic pressure, the aortic valve will not open, and there will be no arterial pulse.

Third, several consequences issue from the post-premature *pause* that occurs before the next (sinus) beat. During such a pause the period of diastolic arterial runoff is lengthened and diastolic pressure falls to unusually low levels. Thus the higher than usual diastolic pressure of the cardiac cycle terminated by the premature beat is followed by the lower than usual diastolic pressure of the cycle initiated by the premature beat. This combination elicits a burst of sympathetic traffic that is greater than most bursts that occur during sinus rhythm (Wallin *et al.* 1974; Fagius 1988; Welch *et al.* 1989). Importantly, such bursts last longer than those occurring during sinus rhythm and may reflect transient interruption of baroreflex entrainment of sympathetic oscillators (Morrison *et al.* 1984; Fagius *et al.* 1985; Welch *et al.* 1989). (Entrainment is thought to occur normally because afferent baroreceptor input terminates each burst of sympathetic firing; this situation may be analogous to termination of inspiratory motoneurone firing by afferent input from pulmonary and thoracic stretch receptors (M.I. Cohen 1975; J.L. Feldman and Gautier 1976).) Such sympathetic surges are triggered by abruptly reduced baroreceptor input; they do not occur during bilateral carotid artery occlusion or after sinoaortic denervation (Herre and Thames 1987; Lombardi *et al.* 1989). Also, because of the longer diastolic filling period (Sung *et al.* 1980) and post-extrasystolic potentiation (Lendrum *et al.* 1960; Palmero and Caeiro 1976), the force of the next sinus beat is unusually large, and the resulting systolic and diastolic arterial, and left atrial (Lombardi *et al.* 1989) pressures are higher than usual. A higher systolic pressure elicits an increase of vagal–cardiac activity, and a higher diastolic pressure in the post-premature beat leads to sympathetic silence (Welch *et al.* 1989). Afferent traffic from cardiopulmonary receptors appears to modulate sympathetic responses to premature beats, since cardiac preganglionic sympathetic responses are greater after than before vagotomy (Lombardi *et al.* 1989). Alicandri *et al.* (1978) documented large reductions of arterial pressure during venous cannon waves provoked by ventricular pacing; these authors suggested that the hypotensive response may be modulated in part by increased afferent cardiopulmonary receptor activity. Fluctuations of (muscle) sympathetic nerve activity surrounding an induced premature beat are depicted in Fig. 13.5.

The above discussion suggests that isolated premature beats may provoke major temporally dispersed fluctuations of sympathetic and vagal outflow to the heart. Importantly, fluctuations of sympathetic activity are mediated by arterial baroreceptors, since they do not occur following sinoaortic

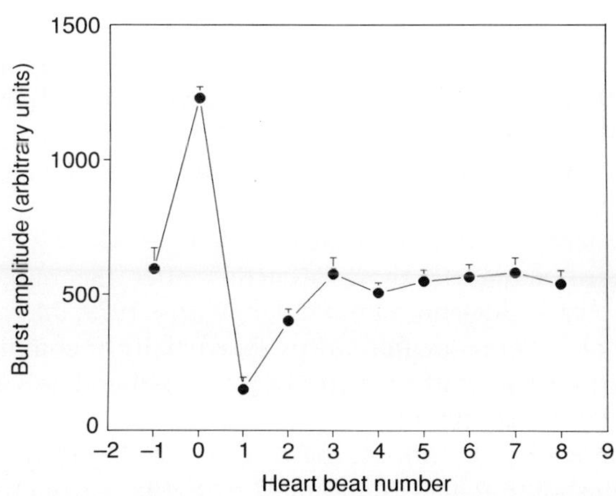

FIG. 13.5. Average amplitude of peroneal nerve muscle sympathetic nerve activity recorded before and after an induced ventricular premature beat in nine patients. The premature beat is designated 0 on the abscissa. Adapted with permission (Welch *et al.* 1989).

denervation (Herre and Thames 1987; Lombardi *et al.* 1989). Since the time constant for *sympathetic* sinoatrial node responses is long (Spear and Moore 1973) and the time constant for *vagal* sinoatrial node responses is short (Eckberg 1976; Eckberg 1980*a*), vagal and sympathetic effects may be temporally dispersed. Such temporal dispersion may modify myocardial electrophysiological properties in a way that favours the occurrence of dysrhythmias. (However, this is only speculation; we are aware of no studies in animals or humans that address this possibility.)

Baroreflex responses to sustained rhythm disturbances

In Chapter 8 we indicate that some research with arterial baroreceptor denervated dogs (Cowley *et al.* 1973) suggests that the primary physiological role of arterial baroreflexes is not to establish absolute levels of arterial pressure, but to buffer pressure transients. On the basis of this construct, it seems likely that baroreceptors are active during abrupt changes of afferent input and that they initiate haemodynamic responses to these changes with short latency. *It may be that mediation of responses to dysrhythmias constitutes a prime role of arterial baroreflexes.* In what follows, we analyse extant data on baroreflex *responses* to dysrhythmias in terms of principles elaborated above.

Baroreflex mediated responses to *sustained* rhythm disturbances are determined by a host of factors. These include the autonomic health of the patient, the patient's baseline level of autonomic efferent activity, the baseline position of the patient on the baroreceptor–sympathetic and –vagal response relations, the magnitude and time course of changes of baroreceptor afferent activity, the rate and extent of baroreflex resetting during pressure reductions, central translation of changes of afferent baroreceptor input into efferent sympathetic and vagal activity, effector receptor activity, and haemodynamic changes that result from receptor responses. A key question asked by this exercise is: why are some rhythms sustained and others non-sustained?

In general terms, arterial pressure changes during non-sustained dysrhythmias are similar to those occurring after the onset of Valsalva straining (Chapter 4 (Schlepper *et al.* 1978; Curry 1979; M.L. Smith *et al.*, unpublished)): blood pressure falls abruptly, partially or completely recovers during the dysrhythmia, and then rises above baseline levels by the end of and following the dysrhythmia.

During sustained dysrhythmias, the early fall of blood pressure initiates an autonomic response that is conditioned by the rate, extent, and duration of pressure reduction (Curry 1979). Arterial pressure falls, and (muscle) sympathetic nerve activity increases in proportion to the speeding of heart rate (M.L. Smith *et al.* 1991). Reductions of left ventricular systolic pressure, rate of pressure change (dP/dt), stroke volume, and cardiac output occur in parallel with reductions of arterial pressure (Schlepper *et al.* 1978). Sinus rate increases, in part secondary to beta-adrenergic stimulation (Huikuri *et al.* 1989). Atrial dimensions and afferent cardiac nerve activity decrease (Linden 1979; Zucker and Gilmore 1981). These changes are associated with early reductions of renal artery flow velocity (Benchimol and Desser 1971), and increases of muscle sympathetic nerve traffic (Delius *et al.* 1972*b*; M.L. Smith *et al.* 1991), of plasma noradrenaline (Nakano *et al.* 1961; Morady *et al.* 1987*b*), atrial natriuretic factor (Nilsson *et al.* 1987), vasopressin, aldosterone, and vasopressin (Crozier *et al.* 1989) levels, and of plasma renin activity.

Figure 13.6 illustrates several aspects of human responses to induced ventricular tachycardia (in this patient at a rate of about 150/min). Figure 13.6 (a) shows that during sinus rhythm, sympathetic activity occurred as discrete bursts related to cardiac activity. Rapid ventricular pacing abruptly reduced systolic, diastolic, and pulse pressures. The sympathetic response to these pressure changes was marked by an increased level of activity with bursts that tended to coalesce. Figure 13.6 (b) shows that during sustained ventricular tachycardia, systolic and pulse pressures were below, and diastolic pressure was above, baseline levels. Sympathetic activity tended to be *below* baseline levels; however, large surges occurred in response to respiration-related reductions of arterial pressure. Figure 13.6 (c) shows sympathetic responses to the rapid ventricular pacing that was used to terminate ventricular tachycardia. There was an abrupt reduction of arterial pressure and a reciprocal sympathetic surge. At this time, sympathetic firing became tonic and bursts were absent. When sinus rhythm was reinstated, arterial pressure was above baseline levels and sympathetic activity was absent.

Changes of blood pressure triggered by dysrhythmias shift the position of patients on the baroreceptor stimulus–response relations. The time course of this change (resetting) has not been worked out completely in humans; however, the available evidence suggests that it may occur within tens of seconds of a sudden pressure change (Fritsch *et al.* 1989; M.L. Smith *et al.* 1991). A direct consequence of a possible shift on the baroreceptor–sympathetic reflex relation (Fig. 10.4) is that if baroreflex resetting does not occur rapidly, the extent of the increase of sympathetic activity that can occur as a result of further pressure reductions may be limited.

The studies quoted above suggest that patients' positions on baroreflex stimulus–autonomic response relations shift into the pressure range for sympathetic responses, and out of the pressure range for vagal responses during abrupt reductions of blood pressure provoked by tachydysrhythmias (Rea and Eckberg 1987). A corollary of this, and one which has not been worked out well in animals or humans, is that at least during brief dysrhythmias, important baroreflex resetting may not occur.

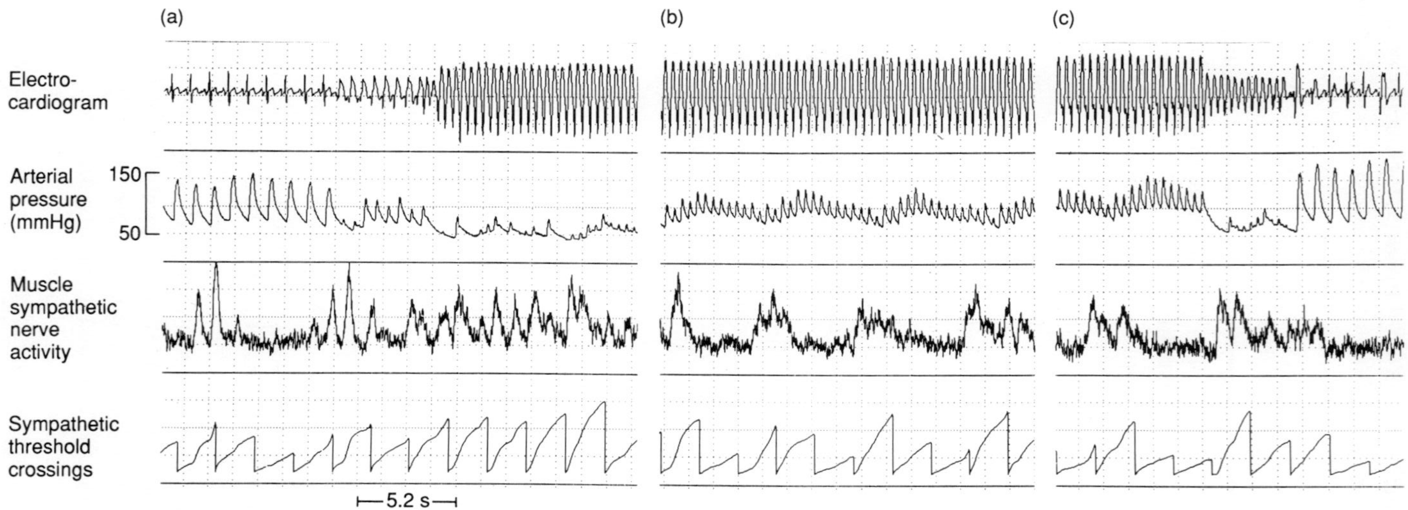

FIG. 13.6. Muscle sympathetic nerve traffic in one patient (a) before, (b) during, and (c) after induced ventricular tachycardia. (M.L. Smith, K.A. Ellenbogen, and D.L. Eckberg, unpublished)

As with the Valsalva manoeuvre (Chapter 4), autonomic responses to ongoing dysrhythmias modify usual physiological relations. An example of this is the blood pressure response to respiration. Under usual circumstances, during inspiration left ventricular stroke volume and arterial pressure decrease (Karam *et al.* 1984; Buda *et al.* 1979). During expiration, the reverse occurs. During tachydysrhthmias, however, respiratory activity exerts qualitatively different effects on arterial pressure than during sinus rhythm (Waxman *et al.* 1980*a*). Presumably, the normal increase of venous return during inspiration occurs. However, because of tachycardia, the transit time of this increased volume of blood from the right to left ventricles is shorter. Thus, during inspiration, venous return to the left ventricle is increased. This leads to increased ventricular ejection, by virtue of the Frank–Starling relation (Braunwald *et al.* 1961). This larger ventricular ejection occurs in the setting of increased peripheral vascular resistance (as after Valsalva straining). The net result of these changes is that during tachydysrhythmias inspiration may be associated with large *increases* rather than decreases of arterial pressure; inspiration-related increases of pressure stimulate baroreceptors and may be sufficient to terminate paroxysmal supraventricular tachycardia (Waxman *et al.* 1980*a*).

Published evidence suggests that different rhythm disorders trigger subtly different autonomic responses. For example, ventricular tachycardia leads to greater depressions of ventricular function and reductions of blood pressure, and to greater increases of atrial pressure than atrial tachycardia (Nakano 1964). (However, ventricular pacing and ventricular tachycardia at the same rate appear to alter haemodynamic parameters and sympathetic traffic in the same way (M.L. Smith *et al.* 1991).) This factor may be important, since the abruptness of blood pressure reduction determines the autonomic response to the dysrhythmia. It would be expected that a greater reduction of blood pressure with ventricular tachycardias triggers a greater initial surge of sympathetic activity. This in turn may lead to spontaneous termination of the dysrhythmia, or to more effective haemodynamic compensation if the dysrhythmia is not terminated (J.L. Feldman *et al.* 1988). More gradual reductions of blood pressure occurring with supraventricular dysrhythmias may provoke smaller initial surges of sympathetic activity, and therefore may be more likely to be sustained. Baroreflex resetting during more modest reductions of pressure might further contribute to maintenance of the dysrhythmia.

Additional complexities might arise if atrioventricular dissociation is present. Loss of synchrony between atrial and ventricular contractions may lead to striking variability of arterial pressure and hence sympathetic and vagal outflows. As far as we know, the influence of atrioventricular asynchrony and the resulting variability of arterial pressures on dysrhythmias has not been studied; in all likelihood, during atrioventricular dissociation, levels of autonomic outflow are changing profoundly and rapidly.

Insight into what happens during a tachydysrhythmia also can be gained by inspection of haemodynamic changes occurring immediately *after* the rhythm disturbance. The arterial pressure elevation seen soon after termination of dysrhythmias (M.L. Smith *et al.* 1991) is associated with increased left ventricular pressure and rate of pressure change, and cardiac output (Schlepper *et al.* 1978). Post-tachycardia pressure elevation and possibly baroreceptor–sympathetic resetting lead to sympathetic silence (M.L. Smith *et al.* 1991). In patients with ventricular dysrhythmias and atrioventricular dissociation, the earliest sinus rate is faster than that which preceded the dysrhythmia (Curry *et al.* 1977). This feature is shown in Fig. 13.7, far right. Immediate post-tachydysrhythmia changes provide additional evidence that sympathetic activity was increased during the dysrhythmia (vagal withdrawal may also contribute to post-dysrhythmia tachycardia).

Figure 13.7 shows clearly normal sympathetic bursting prior to the onset of tachycardia (left), abrupt reduction of arterial pressure and increase of sympathetic activity (without a clearly defined bursting pattern) during the tachycardia, and pressure elevation, sinus node speeding, and sympathetic silence after the tachycardia (right). Of interest, the premature ventricular beat that occurred after termination of ventricular tachycardia did *not* provoke a sympathetic response. We assume that a post-premature beat sympathetic surge was absent because post-tachycardia arterial pressure was above the threshold for sympathetic activation.

Figure 13.7 also documents another feature of sympathetic responses to tachydysrhythmias. During profound hypotension (right portion of the left panel), the usual sympathetic bursting pattern is absent. Figure 13.8 depicts autocorrelation and power spectral analyses of this patient's sympathetic traffic during sinus rhythm (a,b) and ventricular tachycardia (c,d). There was a large peak of sympathetic power at the cardiac frequency, 1.6 Hz (96 beats/min), during sinus rhythm (b). During ventricular tachycardia (c,d), periodicity was absent.

The data depicted in Fig. 13.4 and Figs. 13.6-8 suggest that human (muscle) sympathetic nerve activity may increase during tachydysrhythmias through three more or less sequential mechanisms.

1. *Minor* reductions of pressure increase sympathetic traffic through reductions of the level of baroreceptor inhibition of sympathetic motonuclei — the normal arterial baroreflex mechanism.

2. *Moderate* reductions of pressure increase sympathetic activity through loss of baroreflex entrainment of sympathetic motonuclei. The usual sympathetic bursting pattern persists, but is no longer time locked to cardiac activity.

3. *Large* reductions of pressure increase sympathetic traffic through mechanisms yet to be worked out. Sympathetic activity loses its normal bursting pattern and becomes tonic. This pattern is shown well at the termination of

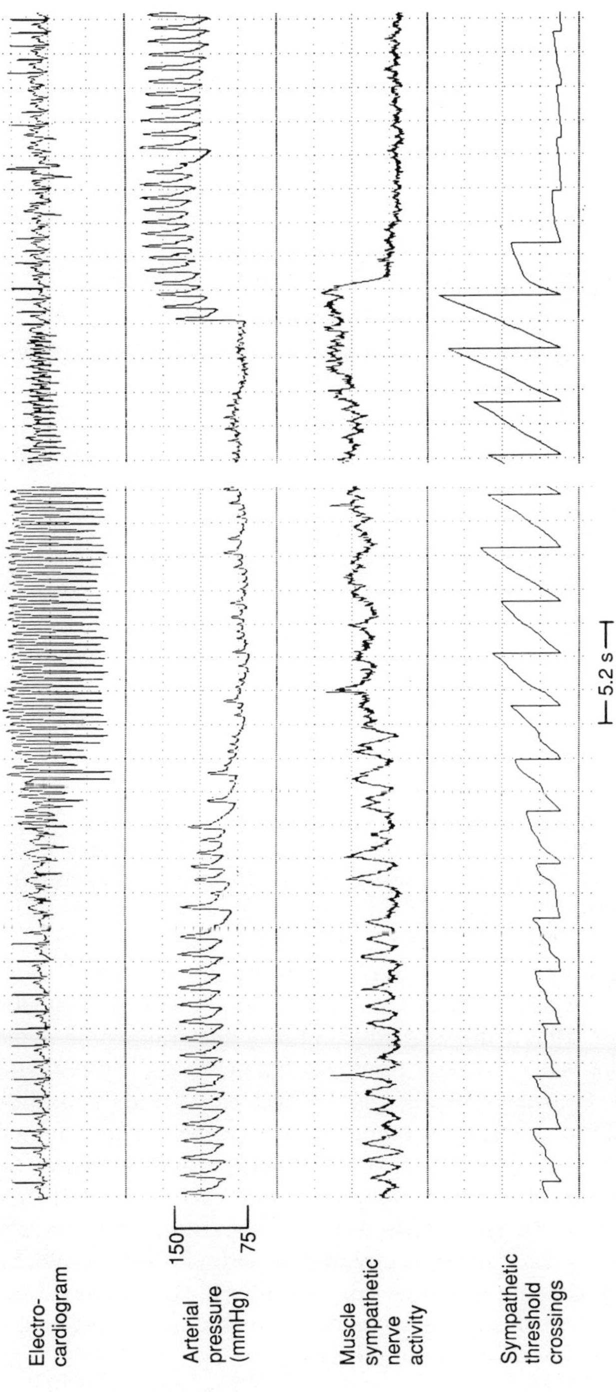

FIG. 13.7. Onset and termination of induced episode of ventricular tachycardia in another patient. (M.L. Smith, K.A. Ellenbogen, and D.L. Eckberg, unpublished)

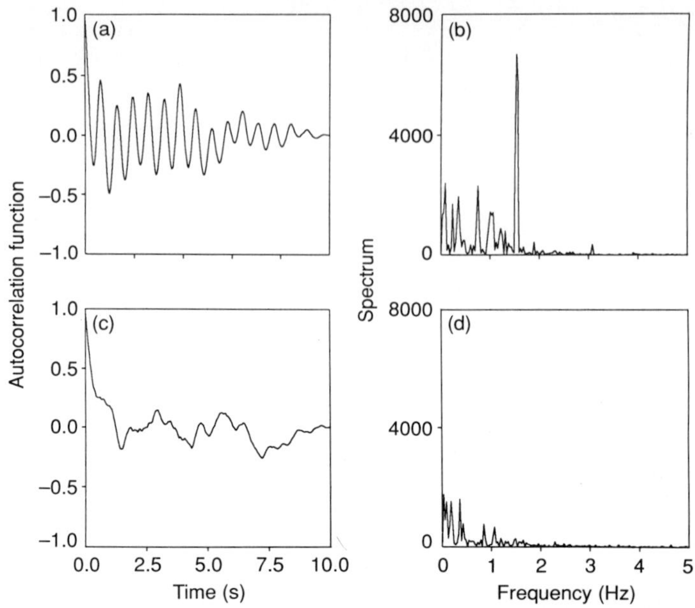

FIG. 13.8 Autocorrelation and power spectral analyses of muscle sympathetic nerve activity from the patient whose record is depicted in Fig. 13.7; (a,b) sinus rhythm; (c,d) ventricular tachycardia. Reproduced with permission (M.L. Smith *et al.* 1991*a*).

ventricular tachycardia, in Fig. 13.7, right. This may represent a cerebral ischaemic response, as documented first by Koscis *et al.* (1989).

Yet another way to understand autonomic responses to tachydysrhythmias is to determine how haemodynamic changes during the dysrhythmia are modified by autonomic blocking drugs. Sinus node speeding during tachycardia probably reflects vagal withdrawal and sympathetic augmentation; although beta-blockade does not prevent heart rate speeding during the first 20 s of ventricular tachycardia, it reduces subsequent speeding (Huikuri *et al.* 1989). Tachycardia after cessation of the dysrhythmia does not occur in patients pretreated with beta-blocking drugs (Waxman *et al.* 1982); this provides indirect but compelling evidence that the tachycardia was provoked primarily by increased sympathetic–cardiac nerve activity. Curiously, there is less blood pressure reduction during the tachydysrhythmia, and less overshoot, after beta-blockade than before. This provides some evidence that tachycardia (and perhaps increased left ventricular contractility) may play a more important role in the post-dysrhythmia blood pressure overshoot than does increased peripheral vascular resistance (which must be present also). Pretreatment with cholinergic blocking drugs (such as hyoscine methylbromide) does not alter the blood pressure reduction occurring paroxysmal supraventricular tachycardia, but prevents its spontaneous termination (Waxman *et al.* 1982).

AUTONOMIC RESPONSES TO DYSRHYTHMIAS ARE DETERMINED BY INITIAL CONDITIONS

The preceding paragraphs suggest that abrupt blood pressure reductions at the onset of dysrhythmias shift the position of the patient on the autonomic stimulus–response relations. From this logically follows the notion that the position of the patient on these relations at the onset of dysrhythmias importantly determines autonomic responses to the dysrhythmia. Baseline autonomic activity at the onset of dysrhythmias may be distorted pathologically by the presence of disease, or physiologically.

Chronic pathological changes of autonomic outflow

There is a great deal of published information regarding baroreflex mediated autonomic responses of patients with cardiovascular diseases. While this evidence points towards important baroreflex abnormalities in patients, the implications of these abnormalities are understood poorly. Questions that need to be addressed by future research regarding the importance of differences in patients with cardiovascular diseases include the following. Are arterial pressure reductions at the onset of dysrhythmias greater (or less) than those of healthy subjects? With abrupt reductions of pressure, do baroreflex responses reset more (or less) rapidly in patients than healthy subjects? Does hypotension trigger the same change of autonomic sensory input profile in patients as in healthy subjects? (For example, does the presence of fibrosis modify contributions of changes of left ventricular baroreceptors to hypotension?) Since vagal mechanisms appear to be grossly deranged in cardiovascular patients (see below), do such patients respond to hypotension with the same mix of changes of sympathetic and vagal outputs as healthy people? Does the presence of disease in effector organs, such as myocardial or vascular hypertrophy or myocardial fibrosis, modify autonomic responses to hypotension? Does distortion of pre- and postsynaptic sympathetic and parasympathetic interactions present in cardiovascular patients (Porter *et al.* 1990) modify their responses to hypotension?

An answer to one of these questions came from a study of exercising dogs with healed anterior myocardial infarctions. In such dogs Billman and Hoskins (1989) produced acute inferior left ventricular ischaemia during graded treadmill exercise and correlated changes of vagal–cardiac activity, as estimated from R–R interval spectral power at respiratory frequencies, with the occurrence or non-occurrence of ventricular fibrillation. During exercise, vagal–cardiac activity declined more rapidly and to a greater extent in dogs susceptible than in dogs resistant to ventricular fibrillation. In a preliminary report, Cerati and Schwartz (1989) measured vagal–cardiac efferent nerve activity directly in anaesthetized cats during coronary occlusion; they found that vagal traffic increased more in cats which did not fibrillate than in those which fibrillated. Collins and Billman (1989) inferred that dogs susceptible to exercise-induced ventricular fibrillation also have greater than normal increases of sympathetic activity.

Figure 13.9 shows the influence of exercise training on phenylephrine baroreflex slopes and susceptibility to sudden cardiac death in dogs with healed myocardial infarctions and inferior ischaemia during exercise (Billman *et al.* 1984). Some dogs (left) had steep baroreflex slopes and were resistant to ventricular fibrillation before and after exercise training. Other dogs (centre) had shallow baroreflex slopes and were susceptible to ventricular fibrillation before exercise training, but had steep baroreflex slopes and were resistant to ventricular fibrillation after exercise training. Some dogs were not given exercise training (right). All but one remained susceptible to ventricular fibrillation. The exception was a dog who for unknown reasons increased its baroreflex slope. These data document an inverse relation between vagal baroreflex responsiveness and sudden cardiac death and indicate that an intervention that increases baroreflex slopes also reduces dysrhythmia susceptibility.

Presumably, the presence of diseases that chronically alter autonomic outflow place patients at additional risk for the occurrence of dysrhythmias. Infants at risk for sudden infant death syndrome have subnormal baseline

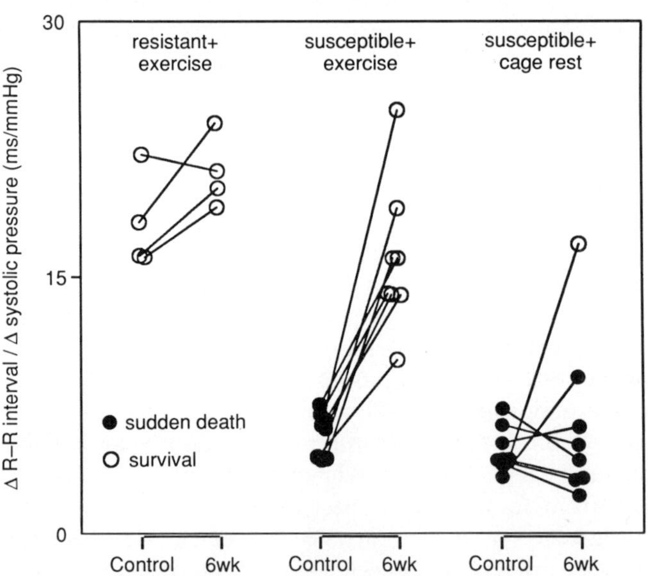

FIG. 13.9. Phenylephrine baroreflex slopes of dogs with healed anterior myocardial infarctions measured before and after 6 week exercise training programs (left and centre) or before and after cage rest (right). Adapted with permission (P.J. Schwartz and Stone 1985).

levels of vagal–cardiac traffic, as judged by spectral analysis of heart rate (Kitney 1984; Kluge *et al.* 1988) and abnormal heart rate responses to changes of posture (Fox and Matthews 1989). In patients with known cardiovascular diseases, acute physiological changes of autonomic outflow may be superimposed upon chronic derangements.

A variety of chronic autonomic abnormalities has been defined in cardiovascular patients. Probably the earliest recognized *vagal* abnormality was impairment of muscarinic–cholinergic control. First, patients with heart diseases have subnormal heart rate speeding in response to large intravenous doses of atropine (Müller 1891; Crawford 1923; Eckberg *et al.* 1971; Porter *et al.* 1990). This suggests that resting levels of efferent vagal–cardiac traffic are subnormal (see Chapter 8). This defect is associated with others that point towards diminished vagal–cardiac outflow: patients with heart diseases tend to have subnormal levels of respiration-related fluctuations of heart rates (Bigger *et al.* 1988; Saul *et al.* 1988*b*; Porter *et al.* 1990); they do not develop heart rate slowing in response to very low doses of intravenous atropine (Porter *et al.* 1990), which are believed to increase vagal–cardiac motoneurone firing rates through a central mechanism (Katona *et al.* 1977; Raczkowska *et al.* 1983; Gilbey *et al.* 1984); and they have diminished R–R interval prolongation in response to bolus injections of pressor drugs or neck suction (Eckberg *et al.* 1971; Amorim *et al.* 1981; Osterziel *et al.* 1988; Sopher *et al.* 1990). Finally, heart failure patients seem to operate closer to the threshold for vagal activation than healthy people (Sopher *et al.* 1990). Therefore cardioacceleration during even small reductions of pressure in such patients probably develops slowly and is mediated almost exclusively by increased sympathetic activity, without important vagal withdrawal. These abnormalities probably diminish patients' vagal responses to sustained dysrhythmias.

Second, patients with heart disease may have disordered *sympathetic* activity and sympathetic responses to changes of baroreceptor activity. (Sympathetic as well as vagal abnormalities have been worked out best for patients with hypertension and congestive heart failure; less is known regarding baroreflex function in patients with other types of cardiovascular diseases.) Possibly the oldest observation is that the myocardium of patients with heart failure is depleted of catecholamines (Chidsey *et al.* 1962). Levels of sympathetic traffic to the heart (Hasking *et al.* 1986) and skeletal muscle (Leimbach *et al.* 1986; Porter *et al.* 1990) are supranormal, and increased sympathetic traffic is associated with excessive plasma noradrenaline levels (J.A. Thomas and Marks 1978; Cohn *et al.* 1984; Leimbach *et al.* 1986; Porter *et al.* 1990). Sympathetic responses to changes of baroreceptor input have not been studied in heart failure patients; however, as mentioned in Chapter 10, plasma noradrenaline levels may decline paradoxically when arterial pressure is lowered in heart failure patients (Olivari *et al.* 1983). As indicated above, surges of sympathetic activity at the the outset of dysrhythmias oppose and reduce the extent of blood pressure reductions, and may even increase blood pressure above baseline levels and terminate the dysrhythmia. Conceivably, in some heart failure patients, surges of sympathetic activity do not occur at the onset of dysrhythmias, or they are blunted. This mechanism could substantially impair patients' abilities to cope with dysrhythmias by increasing arterial pressure.

There are certain groups, including particularly patients with heart failure, with extraordinarily high probabilities for the occurrence of ventricular rhythm disturbances, who appear to have deranged baseline levels of autonomic cardiovascular outflow. In such patients, high levels of sympathetic activity are associated with low levels of vagal activity, as reflected by standard deviations of R–R intervals or R–R interval shortening after large doses of atropine (Eckberg *et al.* 1971; Porter *et al.* 1990). Recordings from a patient with heart failure (left) and a healthy subject (right) are depicted in Fig. 14.1. As discussed by Corr and Gillis (1978), this combination of changes of autonomic outflow should favour the occurrence of ventricular dysrhythmias. The incidence of sudden cardiac death in heart failure patients is as high as 40 to 70 per cent yearly (Dargie *et al.* 1987), and an astonishing 50 per cent of heart failure patients have ventricular tachycardia during a single 24 h period of Holter monitoring (Bigger 1987).

Vagal–cardiac nervous outflow, as reflected by standard deviations of R–R intervals during Holter recordings, may also be reduced in patients who have recovered from acute myocardial infarction (Kleiger *et al.* 1987). In such patients, survival is related directly to the magnitude of estimated vagal activity. In the study published by Kleiger *et al.* (1987), patients with standard deviations below 50 ms were about six times as likely to die during the year after their infarctions than were patients with standard deviations above 100 ms. Other types of data support this conclusion indirectly. Standard deviations of R–R intervals were found to be very low in five patients who were wearing Holter monitors at the time of sudden death (G.J. Martin *et al.* 1987). In another study published in preliminary form (Magid *et al.* 1985), patients resuscitated outside the hospital in whom sustained ventricular tachycardia could be induced by electrophysiological means were compared with age-matched patients with frequent, high-grade ventricular dysrhythmias in whom ventricular tachycardia could not be induced. R–R interval standard deviations were significantly lower in the group of patients in whom ventricular tachycardia could be induced than in those in whom it could not. As discussed in Chapter 8, standard deviations of R–R intervals are related directly (but loosely) to baroreflex responsiveness (T. Bennett *et al.* 1978; Raczkowska *et al.* 1983). Moreover, in hypertensive (D.S. Goldstein 1983*b*) and heart failure (Porter *et al.* 1990) patients, this measure is related inversely to plasma noradrenaline levels. Therefore there are interrelated data from humans that seem to link vagal–cardiac activity, sympathetic activity, and baroreflex function to the occurrence of dysrhythmias.

Finally, LaRovere *et al.* (1988) have shown that reduced vagally mediated arterial baroreceptor–cardiac reflex responses identify post-myocardial infarction patients at risk for sudden cardiac death. They measured slopes of R–R interval–systolic pressure relations after bolus intravenous injections of phenylephrine in 78 patients under age 65 who were recovering from their first myocardial infarctions. In this study, seven patients died during an average 24 month follow-up period. Baroreflex slopes were significantly ($p = 0.004$) lower in these patients than in patients who survived (2.5 vs. 8.3 ms/mmHg). Mortality was 50 per cent in patients with baroreflex slopes less than 3, but only 3 per cent in patients with baroreflex slopes greater than 3 ms/mmHg. In a related study, Farrell and co-workers (1991) found that phenylephrine baroreflex slopes are lower in post-infarction patients in whom ventricular tachycardia can be induced than in those in whom it cannot be induced. Farrell's prospective study may provide the most compelling evidence to date for a link (as yet unexplained) between depressed vagally mediated arterial baroreflex responses and sudden cardiac death.

Baseline autonomic outflow may also be altered by acute *myocardial ischaemia*. Although the mechanisms by which ischaemia provokes ventricular fibrillation are likely to be multiple, one may be interference with usual baroreflex mechanisms. Chapter 10 discusses experimental analogues of coronary ischaemia used for research in humans. One of these is coronary arteriography. During intracoronary injections of contrast medium, myocardial receptors (probably chemosensitive) provoke reflex bradycardia despite concomitant hypotension (Eckberg *et al.* 1974). Two groups (Takeshita *et al.* 1980; Trimarco *et al.* 1987) studied the influence of acute coronary occlusion on carotid baroreceptor cardiac reflexes in animals. Both found that baroreflex mediated cardiac slowing is suppressed during coronary occlusion. One group (Pomidossi *et al.* 1989) reported preliminarily that R–R interval shortening after intravenous nitroglycerine is subnormal during coronary angioplasty. These results suggest that myocardial ischaemia may also interfere with baroreflex adjustments. If this inference is correct, this mechanism could contribute to ischaemia mediated sudden cardiac death. As mentioned above, the studies of P.J. Schwartz and his group indicate that depressed baroreceptor–cardiac responses identify post-infarction dogs (Billman *et al.* 1982) and humans (LaRovere *et al.* 1988) at risk for sudden cardiac death.

Physiological changes of initial conditions

Unusual baseline levels of autonomic outflow may also result simply from physiological changes. For example, standing patients are likely to begin dysrhythmias on a different part of the baroreceptor stimulus-autonomic response relation than are supine patients. In the standing position, increases of sympathetic activity probably occur earlier after the onset of the dysrhythmias, but blood pressure reductions are greater than they are in the recumbent position (Saunders and Ord 1962). The reasons for these differences are not known with certainty. It may be that sympathetic activity increases more rapidly because the patient is closer to the pressure at which maximum sympathetic responses occur (Rea and Eckberg 1987). A related possibility is that the greater initial reduction of pressure may result in part because the patient is closer to the threshold for vagal responses (that is, there is less vagal restraint to withdraw, and hence there is less immediate cardioacceleration resulting from vagal withdrawal). This mechanism could also explain why some dysrhythmias which terminate spontaneously in the recumbent position are sustained in the standing position (Saunders and Ord 1962). In the standing position, the patient may be below the position where maximum vagal responses occur, and compensatory increases of sympathetic activity are insufficient to restore pressure to the range in which vagal–cardiac responsiveness is maximal.

Another physiological change that influences the occurrence of dysrhythmias is respiration. In five patients with frequent premature ventricular beats, Patrick *et al.* (1984) documented clear respiratory periodicity; the frequency of premature beats varied nearly sinusoidally during respiration, with the greatest frequency occurring at about 2 s after the onset of inspiration.

BAROREFLEX RESPONSES MODIFY THE OUTCOME OF DYSRHYTHMIAS

The preceding discussion suggests that baroreflexes are active during all stages of dysrhythmias, and in many patients during the critical time before the onset of the rhythm change. What follows is a consideration of evidence that baroreflexes play an important, even pivotal, role in determining the outcome of some dysrhythmias. During typical dysrhythmias, baroreflexes trigger corrective changes of autonomic outflow that restore blood pressure towards normal. In this process, baroreflexes have the possibility to transform the dysrhythmia into a different type of rhythm. The new rhythm may be normal sinus, some other dysrhythmia, or, given a nearly unique set of circumstances, ventricular fibrillation.

Probably the vast majority of dysrhythmias are transformed into sinus rhythm. As mentioned earlier, the sequence of events occurring after the onset of a tachydysrhythmia is initiated by hypotension. This leads to withdrawal of vagal restraint; augmentation of sympathetic traffic to the heart (Godin *et al.* 1985), kidneys (Benchimol and Desser 1971), and skeletal muscle (Fig. 13.7); and elaboration of vasopressor substances (Nilsson *et al.* 1987; Crozier *et al.* 1989). Vascular resistance, the rate of the dysrhythmia, and probably left ventricular contractility increase, and arterial pressure begins to return towards normal. If arterial pressure rises sufficiently, vagal–cardiac neurones are stimulated, sympathetic firing is reduced or silenced, the rate of the dysrhythmia slows, and the dysrhythmia abruptly converts to sinus rhythm.

Waxman and co-workers (1982) evaluated factors that predispose to the conversion of paroxysmal *supra*ventricular tachycardia to normal sinus rhythm in 20 patients. In nine, dysrhythmias were self-terminating. Compared with dysrhythmia episodes that were sustained, self-terminating episodes were more likely to have had a greater fall of blood pressure and a greater early acceleration of the rate of the dysrhythmia, a more rapid return of blood pressure to normal levels, and a greater elevation of blood pressure above pre-dysrhythmia levels. In other words, self-termination of the dysrhythmia seemed to hinge importantly on the magnitude and the rate of the reduction of baroreceptor afferent activity, and the magnitude and time course of sympathetic and vagal responses.

Waxman *et al.* (1982) then modified sympathetic and vagal responses physiologically and pharmacologically to gain further insight into mechanisms leading to self-termination of the dysrhythmia. Curiously, they found that both head-up and head-down tilt reduced the likelihood of self-termination. Head-down tilt reduced the tachycardia rate and blood pressure reduction, and therefore reduced the sympathetic response to the tachycardia, the sympathetically mediated recovery of blood pressure, and the likelihood of vagal termination. Head-up tilt reduced venous return and thereby blunted the recovery of blood pressure and reduced the likelihood of vagal termination. Abrupt head-down tilt increases baseline levels of vagal–cardiac activity (Convertino *et al.* 1990) and reduces baseline levels of sympathetic neural activity. Abrupt head-up tilt provokes opposite responses.

In the standing position, vagal activity is less than in the supine position, and therefore the patient lies closer to the threshold pressure for activation of vagal responses. Moreover, if the blood pressure fall caused by the dysrhythmia is the same or greater in the standing than in the supine position, the patient will require a greater increase of arterial pressure to reach the threshold for vagal activation. An additional contributing factor may be impairment of the patient's ability to increase venous return because of orthostatic pooling of blood. Also, if the blood pressure fall is great enough, the patient's capacity to augment sympathetic activity may also be reduced; the patient will be closer to (or beyond) the pressure at which maximum sympathetic responses occur, and he or she will also be closer to the pressure at which sympathetic activity paradoxically declines (Fig. 10.5, extreme left).

In a different study, Waxman *et al.* (1980*c*) attempted to quantify the amount of augmentation of vagal–cardiac activity necessary to terminate supraventricular dysrhythmias. He concluded that the likelihood of self-termination of supraventricular dysrhythmias is proportional to the magnitude of baroreceptor stimulation and to vagal responses to such stimulation. Thus unilateral carotid sinus massage was less likely to terminate dysrhythmias than stimulation of all baroreceptor areas by elevations of systemic pressure (caused by phenylephrine injections or the Valsalva manoeuvre). Mehta *et al.* (1988) made similar observations in patients with junctional tachycardia. (It should be remarked that carotid massage does not simply increase baroreceptor afferent activity from one baroreceptive area; it reduces baroreceptor afferent activity from other areas by virtue of the reduction of systemic pressure that it provokes.) The likelihood of self-termination was increased if the hydrolysis of acetylcholine was opposed with edrophonium, and was reduced if the cardiac effects of increased vagal activity were opposed with large doses of atropine. A further, elegant demonstration that vagal activity was responsible for termination of the dysrhythmia came from the observation that termination always occurred during expiration (Waxman *et al.* 1980*a*), when vagal responses to increased baroreceptor activity are greatest (Eckberg and Orshan 1977; Eckberg *et al.* 1980). Finally, a factor that has not been worked out well in humans is baroreflex responsiveness. Patients who do not self-terminate their dysrhythmias seem to require greater pharmacological pressure elevation than those who terminate their dysrhythmias. Conceivably these patients have subnormal baroreflex responsiveness.

Certain data of Waxman *et al.* (1982) suggest that the likelihood of self-termination of paroxysmal supraventricular tachycardias is determined simply by the ratio between sympathetic and vagal cardiovascular outflows. When sympathetic activity is high, more vagal activity is needed to oppose its effects, and when sympathetic activity is low, less vagal activity is needed. This notion is supported by responses to agonist and antagonist drugs. For example, infusion of the beta-agonist isoproterenol increases, and administration of the beta-adrenergic blocker propranolol reduces, the amount of edrophonium necessary to terminate the dysrhythmia (Waxman *et al.* 1980*c*). (These observations run counter to the concept of 'accentuated antagonism' developed in healthy animals (Levy 1971) and applied to humans (Morady *et al.* 1988), that vagal effects are magnified (not reduced) by increased sympathetic activity.)

However, the process of self-termination is likely to be much more complicated than this simple construct suggests. First, isoproterenol probably reduces (Arnold and McDevitt 1986), and, as discussed in Chapter 8, propranolol probably increases (Eckberg 1984), vagal–cardiac nervous activity. Second, beta-blockade provokes haemodynamic changes by slowing the rate of the dysrhythmia. Slower heart rates translate into longer left ventricular diastolic filling periods, increased cardiac dimensions (Waxman *et al.* 1980*b*), larger stroke volumes, greater reductions of diastolic arterial pressure, and, in patients with coronary disease, less ischaemia (Braunwald *et al.* 1967) or less cyclic variation of coronary flow. Some of these changes seem to be independent of blockade of adrenergic receptors; when verapamil is given during dysrhythmias, blood pressure also rises in association with cardiac slowing, notwithstanding verapamil's negative inotropic and vasodilator properties (Heng *et al.* 1975; Curry 1979). Beta-adrenergic blockade may interfere with self-termination of dysrhythmias. Since the initial blood pressure reduction is less, the trigger for increased sympathetic activity is also less. Therefore more time is required for blood pressure to stabilize during the dysrhythmia, and the blood pressure overshoot after the dysrhythmia is less (Waxman *et al.* 1982).

The foregoing discussion of self-termination of dysrhythmias refers primarily to re-entrant supraventricular dysrhythmias involving atrioventricular nodal pathways. In a distinct minority of instances, similar conclusions apply to ventricular dysrhythmias. This seems to be most true for carefully defined subsets of patients, such as those with exercise-induced ventricular tachycardia. Waxman and Wald (1977*b*) and D.S. Hess *et al.* (1982) published examples of ventricular tachycardia terminated by vagal manoeuvres. Waxman *et al.* (1981) also showed that ventricular tachycardia can be provoked by vagal withdrawal (with or without large doses of atropine) during deep inspirations. The study of Dresel and Sutter (1961) conducted in anaesthetized dogs also supports a role for vagal mechanisms in ventricular dysrhythmias; they provoked frequent premature ventricular contractions with adrenaline infusions and found that the likelihood that electrical vagal stimulation would abolish the induced dysrhythmia was proportional to the intensity of vagal stimulation. Subsequently Waxman *et al.* (1989) showed that efferent vagal nerve stimulation increases the doses of catecholamines necessary to provoke ventricular fibrillation in dogs whose hearts are sensitized by halothane. This conclusion is also supported by inferential data from the important study conducted by Zimmerman *et al.* (1986). These authors evaluated 24 hour Holter recordings of 30 patients with repetitive monomorphic idiopathic ventricular tachycardia. They found that slowing of the ventricular rate precedes spontaneous conversion of ventricular tachycardia to sinus rhythm. Presumably, slowing results from reduction of sympathetic or augmentation of vagal efferent activity.

It should be noted that documentation of autonomic influences on ventricular dysrhythmias does not necessarily define mechanisms; in all likelihood, multiple mechanisms are involved (Podrid *et al.* 1990). These include changes of serum electrolytes (especially hypokalaemia and hypomagnesaemia), ischaemia (Morady *et al.* 1987*a*), and alteration of electrophysiological influences of re-entrant pathways (Moe 1975; Waxman and Wald 1977*a*), automatic foci, and triggered activity (Josephson *et al.* 1979; Priori and Corr 1990; Priori *et al.* 1991).

TRANSFORMATION OF ONE DYSRHYTHMIA INTO ANOTHER

Reflex mechanisms initiated in response to hypotension may transform one dysrhythmia into another. As mentioned, increased sympathetic activity may accelerate the rate of paroxysmal supraventricular tachycardia, and as a result of sympathetic stimulation, the tachycardia may be transformed into atrial fibrillation. This transformation can be prevented by pretreatment with beta-adrenergic blocking drugs (Waxman *et al.* 1983). If the patient has a rapidly conducting atrioventricular node bypass tract, conversion of supraventricular tachycardia into atrial fibrillation may lead to ventricular fibrillation (Waxman *et al.* 1983). In all likelihood, baroreflex mediated changes of autonomic outflow contribute to these transformations.

As mentioned, many patients enjoy prolonged useful life after resuscitation for ventricular fibrillation. The longevity of such patients argues that the deaths of many cardiac patients are not due simply to advanced terminal cardiac structural damage. Rather, in many or perhaps most patients, the terminal event is an electrical event made possible by a nearly unique combination of structural, electrophysiological, and neuroendocrine changes (Lown and Verrier 1976; Coumel *et al.* 1986). The evidence is strong that autonomic mechanisms contribute to ventricular fibrillation; since autonomic involvement is transitory, it may be of transcendent importance to understand the nature of autonomic involvement and discover ways to modify it.

The substantial literature on the electrocardiographic antecedents indicates that ventricular fibrillation is usually preceded by less malignant dysrhythmias. Premature ventricular beats occur almost universally before ventricular fibrillation; often they come early (the 'R on T' pattern (Smirk and Palmer 1960)). Roelandt *et al.* (1984) published analyses of Holter recordings made at the time of sudden cardiac death, and added these to 20 other such recordings culled from the medical literature. They noted a common pattern of early premature ventricular beats initiating ventricular fibrillation, which has also been remarked by others (El-Sherif *et al.* 1976; Dhurandhar *et al.* 1971; Adgey *et al.* 1982). Early premature beats, which may fall in the ventricular 'vulnerable' period (Smirk and Palmer 1960), also provoke larger sympathetic responses than do late premature beats (Welch *et al.* 1989). In Roelandt's study, the most consistent feature of the cardiac rhythm before the onset of ventricular fibrillation was a long, post-premature pause before the last sinus beat. However, although ventricular fibrillation was initiated most commonly by early premature ventricular beats, in seven of the 30 patients, premature beats initiating ventricular fibrillation occurred late in the cardiac cycle. (Such a case is illustrated in Fig. 13.4.) He noted also an increase of sinus rate before the onset of ventricular fibrillation. Similar observations were made by Leclercq *et al.* (1988).

It is unclear why the specific premature beats captured by Holter monitors in the patients mentioned above triggered ventricular fibrillation, when multitudinous earlier similar premature beats had no such effect (Leclercq *et al.* 1988). One possible explanation is that the potential for premature ventricular beats to initiate malignant dysrhythmias depends on other permissive factors (Lown and DeSilva 1978). A common (but by no means universal) concomitant of premature beats before ventricular fibrillation is myocardial ischaemia. In Roelandt's series (1984), there was evidence for ischaemia (angina, ST segment depression, or T wave inversion) in 14 of 30 patients. In the meticulous pathology study of M.J. Davies and Thomas (1984), which comprised 100 patients with ischaemic heart disease who died suddenly, there was evidence pointing towards myocardial ischaemia in no fewer than 95 per cent; 74 patients had intracoronary thrombi (not necessarily occlusive), and an additional 21 patients had coronary plaque fissures. D.D.-P. Lee *et al.* (1989) showed that silent myocardial ischaemia, which is associated with sudden cardiac death and dysrhythmias (Carboni *et al.* 1987), increases plasma noradrenaline levels. In such patients, myocardial ischaemia may conspire with increased sympathetic activity to set the stage for ventricular fibrillation. Inferential evidence that increased sympathetic activity may transform a benign premature ventricular beat into a malignant one comes from the study of Leclercq *et al.* (1988), who found that premature beats initiating ventricular tachycardia had similar morphologies to isolated premature beats, but shorter coupling intervals.

Thus it seems likely that ischaemia plus autonomic changes (and additional factors, such as healed myocardial infarctions in coronary heart disease patients) form a common substrate for ventricular fibrillation. In this scheme, autonomic changes form a final common pathway for the terminal event. Autonomic mechanisms may set in motion a cycle of changes: ischaemia impairs cardiac function, reduces blood pressure, increases sympathetic (and reduces vagal) outflow, speeds heart rate, increases myocardial oxygen consumption, and increases ischaemia. Experimental evidence links autonomic mechanisms strongly in this cycle. The study of Skinner and Reed (1981) showed that ventricular fibrillation caused by acute coronary occlusion in psychologically stressed pigs can be prevented by cryogenic blockade of frontocortical–brainstem neural pathways. Extrinsic cardiac denervation also largely (but not completely (J.W. Mason *et al.* 1976)) prevents ventricular tachycardia and fibrillation provoked by acute coronary artery occlusion in anaesthetized dogs (Schaal *et al.* 1969; P.A. Ebert *et al.* 1970).

The above suggests that when a unique set of circumstances comes together, baroreflex mechanisms may contribute to the transformation of less malignant dysrhythmias into ventricular fibrillation. The question, how can normally protective baroreflexes contribute to sudden cardiac death? is one of no small importance. One of the curious and incompletely explained features of ventricular fibrillation is its predilection for patients with *anterior* infarction. The choice of the dog model of sudden cardiac death developed by Billman *et al.* (1982) may have been extremely fortunate. They created *anterior* infarctions with left anterior descending coronary artery occlusions, and subsequently studied the effects of *inferior* myocardial ischaemia produced by acute left circumflex coronary artery occlusion during treadmill exercise. Dogs with active arterial baroreceptor–cardiac reflexes (as stimulated by bolus phenylephrine injections) were resistant to ventricular fibrillation. These relations are illustrated in Fig. 13.9.

The explanation for these observations is not entirely clear. One possibility is that the degree of bradycardia provoked by arterial pressure elevations parallels the degree of bradycardia provoked by myocardial ischaemia. Since reflexes triggered by inferior left ventricular ischaemia appear to override arterial baroreflexes (bradycardia occurs in association with sympathetic withdrawal and hypotension), resistance of dogs to ventricular fibrillation may signify that in these dogs, myocardial reflexes are able to overcome harmful arterial baroreflex responses. In the study of Billman *et al.* (1982), dogs that were resistant to ventricular fibrillation experienced cardiac slowing rather than speeding when inferior ischaemia was added to exercise stress.

We suspect that valuable insights into mechanisms underlying sudden cardiac death might come from a better understanding of how reflexes arising from the heart interact with reflexes arising from the arterial baroreceptors during myocardial ischaemia. Also, clues to such interactions might come from a better understanding of why anterior ischaemia or infarction differs from inferior ischaemia or infarction. Several differences have been remarked. In the prospective study of LaRovere and co-workers (1988), arterial baroreflex responses to phenylephrine injections were greater in patients with *anterior* than with inferior infarctions. Nevertheless, patients with anterior infarctions are much more likely to suffer ventricular fibrillation than patients with inferior infarctions (Adgey and Webb 1979; P.J. Schwartz and Zaza 1986).

The studies of Corr and Gillis (1974) and Corr *et al.* (1976) suggest that autonomic reflex function figures importantly in electrophysiological responses to myocardial ischaemia. For reasons that remain obscure, large-dose atropine or vagotomy increase the incidence of ventricular fibrillation during anterior ischaemia, but do not increase the incidence of ventricular fibrillation during inferior ischaemia in anaesthetized cats.

SUMMARY

This chapter reviews participation of arterial baroreflexes in human dysrhythmias. Baroreceptors are exquisitely sensitive to changes of arterial pressure, and it is likely that baroreflex mechanisms are invoked in all dysrhythmias. Our discussion suggests that baroreflex responses to dysrhythmias are extremely important; indeed, such responses may even constitute the most important physiological role of baroreflex mechanisms.

14

Congestive heart failure

In no category of patients with cardiovascular problems is abnormal baroreflex function more apparent than it is in those with congestive heart failure. Starling as early as 1897 recognized the possibility that tachycardia and vasoconstriction in heart failure patients might be due to reflex mechanisms. Baroreflex abnormalities are well documented in patients with heart failure resulting from a variety of underlying diseases, including particularly hypertension, coronary artery disease, and cardiomyopathy. Abnormalities are present (or have been studied) most commonly when disease involves the left ventricle; however, abnormal baroreflex-mediated responses may also be present in patients with processes that do not involve the left ventricle, including mitral stenosis (McIntosh Set al. 1954) and obstructive lung disease (Patakas *et al.* 1982). Although abnormal baroreflex function is most striking in patients with advanced heart diseases, it also occurs in patients (Eckberg *et al.* 1971) and animals (Higgins *et al.* 1972*a*) with mild heart disease and even in patients whose heart disease has not yet produced symptoms (Amorim *et al.* 1981; Gallo *et al.* 1987). (Patients with left ventricular dysfunction may also have elevated levels of plasma noradrenaline, vasopressin, and atrial natriuretic factor before they become symptomatic (Francis *et al.* 1990).) There is some evidence, reviewed below, that abnormal baroreflex function contributes to symptoms in heart failure patients. In addition, there are unquestioned, if unexplained, direct correlations between autonomic, possibly baroreflex-mediated abnormalities, and prognosis. Of great interest is the possibility that therapy of heart failure, which improves symptoms and life expectancy, may do so in part through improvement of autonomic function.

As discussed in Chapter 6, baroreceptors sense every arterial pulse and orchestrate complex, ongoing neurohumoral adjustments. Thus baroreflex influences on the circulation are pervasive; there may be no important alterations of state, such as changes of physical or mental activity, posture, or respiration, that do not in some way alter baroreflex function. In heart failure patients, symptoms that may be related to baroreflex dysfunction include increased or decreased orthostatic tolerance, reduced exercise capacity, and sudden cardiac death (discussed in Chapter 13).

Sympathetic abnormalities in heart failure patients include: increased traffic to cardiac and muscle vascular beds (Hasking *et al.* 1986; Leimbach *et al.* 1986; Ferguson *et al.* 1989; Porter 1990) and probably reduced traffic to splanchnic and renal vascular beds (Faxon *et al.* 1984); reduced responsiveness and numbers of cardiac adrenergic receptors (D.E. Vatner *et al.* 1985; Fowler *et al.* 1986), depletion of cardiac noradrenaline stores (Chidsey *et al.* 1965; Kramer *et al.* 1968); and reduced tachycardia and hypotension during infusion of the beta-adrenergic agonist isoproterenol (Erne *et al.* 1988). Vagal abnormalities include: reduced traffic to the heart, as manifested by short, nearly monotonic resting R–R intervals (Saul *et al.* 1988*b*; Arai *et al.* 1989; Porter *et al.* 1990), reduced R–R interval shortening with large atropine doses (Müller 1891; Crawford 1923; Eckberg *et al.* 1971; Porter *et al.* 1990), and diminished R–R interval lengthening after small doses of atropine (Porter *et al.* 1990), bolus pressor injections (Amorim *et al.* 1968; Eckberg *et al.* 1971; Amorim *et al.* 1981), and neck suction (Osterziel *et al.* 1988; Sopher *et al.* 1990); and reduced muscarinic cholinergic receptor density and agonist binding (D.E. Vatner *et al.* 1988).

Biochemical abnormalities include: elevations of plasma renin activity (Kluger *et al.* 1982; T.B. Levine *et al.* 1983) and levels of antidiuretic hormone (M. Stein *et al.* 1954) or arginine vasopressin (Yamane 1968; Szatalowicz *et al.* 1981; Uretsky *et al.* 1985; Creager *et al.* 1986), noradrenaline (J.A. Thomas and Marks 1978; Francis *et al.* 1982*a*; T.B. Levine *et al.* 1983; Cohn *et al.* 1984), neuropeptide Y (Maisel *et al.* 1989), growth hormone (Anand *et al.* 1989), cortisol (C.-s. Liang *et al.* 1987; Anand *et al.* 1989), beta-endorphin (C.-s. Liang *et al.* 1987; Sakamoto and Liang 1989), atrial natriuretic factor (Tikkanen *et al.* 1985), and prostaglandins (Dzau *et al.* 1984); and reductions of serum sodium concentration (Lilly *et al.* 1984; W.H. Lee and Packer 1986). Since *acute* changes of these hormones, neurotransmitters, and electrolytes can be provoked by changes of afferent baroreceptor activity, it may be reasonable to assume that *chronic* changes may also be related to baroreceptor dysfunction. Chapter 8 discusses the evidence that chronic baroreceptor input determines arterial pressure levels. The study of Shade *et al.* (1990) showed that baroreceptor denervation leads to persistent hypertension and chronic elevation of circulating catecholamine levels in baboons.

SUBNORMAL BAROREFLEX INHIBITION

As mentioned in Chapter 10, the earliest study pointing towards autonomic abnormalities in heart failure patients showed that they have exaggerated plasma catecholamine responses to exercise (Chidsey *et al.* 1962). (Catecholamine responses were greater in patients than healthy subjects, at comparable *absolute* work loads; the complex issue of whether work loads should be expressed in absolute or relative terms for such comparisons is discussed below.) Many subsequent studies of possible baroreflex abnormalities have worked counterclockwise in the baroreflex arc, from effect to cause, to suggest that elevated sympathetic (and reduced vagal–cardiac) outflow results from subnormal inhibitory input (that is, from an impaired negative feedback mechanism). Figure 14.1 depicts experimental records from a patient with severe heart failure (a) and a healthy subject (b). The patient had nearly constant, short R–R intervals and increased levels of muscle sympathetic nerve activity (which, interestingly, were strongly influenced by respiration). The healthy subject had longer, more variable R–R intervals and modest levels of sympathetic activity. Baseline antecubital vein plasma noradrenaline levels in the patient and healthy subject were 2345 and 180 pg/ml. The observation that heart failure patients have reciprocal reductions of vagal and increases of sympathetic outflows is supported by data from conscious dogs with pacing-induced failure (J.-S. Chen *et al.* 1991).

The simplest baroreflex-related explanation for altered autonomic neural output in heart failure patients is that arterial pressure is subnormal. Harris (1987) argues that autonomic abnormalities in heart failure can be explained entirely on the basis of purely physiological responses by patients to support

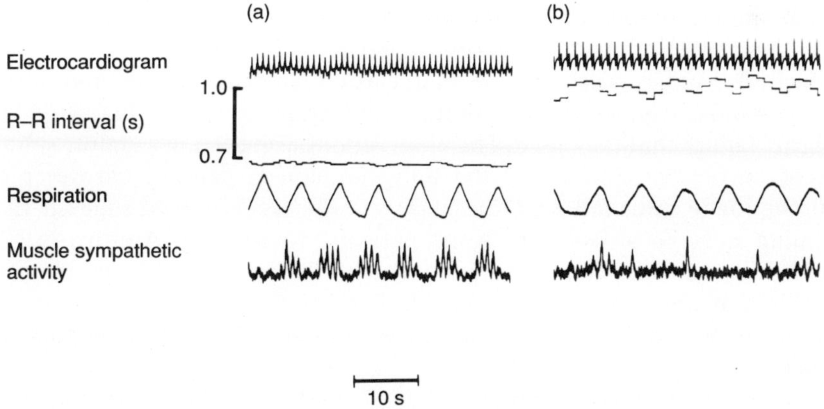

FIG. 14.1. Recordings obtained from (a) a patient with heart failure and (b) a healthy subject. Adapted with permission (Porter *et al.* 1990).

their blood pressures in the face of subnormal cardiac outputs. (Implicit in this suggestion are the notions 1) that baroreceptors set absolute levels of arterial pressure and efferent autonomic neural outflow, and do not merely oppose acute changes of pressure, and (2) that sympathetic outflow is related inversely to arterial pressure in heart failure, as it is in health. These issues are joined in Chapter 10.) This suggestion should not be dismissed altogether; many end-stage heart failure patients have subnormal blood pressures, and in some studies of autonomic mechanisms in heart failure, patients' blood pressures are significantly lower than those of healthy volunteers (Swedberg *et al.* 1984) or those of stable heart failure patients (Dzau *et al.* 1981). Moreover, most studies in this genre involve comparisons only of average systolic, diastolic, and mean pressures, and not other possible determinants of sympathetic activity, such as arterial pressure variability (de Boer *et al.* 1985*a*), pulse pressure, and heart rate (Sundlöf and Wallin 1977).

In heart failure patients whose arterial pressure is normal, baro*reflexes* may be abnormal because baro*receptors* respond abnormally to arterial pressure changes. Evidence from humans on this question (discussed below) is indirect. Evidence from animals is more direct, however, and provides strong support for the view that the afferent limb of baroreflexes is abnormal in experimental heart failure. In a series of studies, Zucker and co-workers (Niebauer and Zucker 1985; Niebauer *et al.* 1986; W. Wang *et al.* 1990) documented impaired baroreceptor responses in dogs with heart failure caused by aortocaval fistulae or rapid ventricular pacing. The study by W. Wang *et al.* (1990) showed that in dogs with pacing-induced failure, the threshold for baroreceptor activation is higher, and maximum baroreceptor firing rates and gain are lower than in healthy dogs. These abnormalities are reversed by ouabain, and therefore may result (in incompletely understood ways) from augmented sodium, potassium ATPase activity. It is not known if the observation of Ferguson *et al.* (1989), that cardiac glycosides reduce muscle sympathetic nerve activity in heart failure patients, results from augmented afferent baroreceptor activity.

Zucker's studies also showed that carotid arterial sodium and water contents and calculated aortic wall stress and strain are normal in the models of heart failure that they used. His observations argue against simple splinting of baroreceptor areas as the only mechanism causing baroreceptor malfunction in heart failure. Other indirect evidence, however, suggests that splinting may be a factor in some patients. Kassis and Amtorp (1987) showed that chronic therapy of patients with felodipine, a calcium channel blocker that appears to be highly selective for vascular tissues, improves reflex responses to upright tilt, and increases echographic aortic root pulsatile stretch.

A study by Dibner-Dunlap and Thames (1989) also showed that afferent (multifibre aortic nerve) baroreceptor responses to pharmacological pressure elevations and reductions are impaired in dogs with pacing-induced heart failure. They also found, however, that efferent renal sympathetic nerve responses are normal. This observation points towards augmented gain of the *central* portion of the baroreflex arc. The mechanism is far from clear; presumably, increased central gain (of sympathetic, but not vagal) baroreflex responses in heart failure reflects in some way plasticity secondary to chronically reduced arterial baroreceptor inputs.

The majority of studies of baroreflex abnormalities in heart failure *patients* has focused on the *vagal* efferent limb of the reflex. The earliest evidence for abnormal baroreflex function in cardiovascular patients came from the study of Amorim *et al.* (1968). These Brazilian investigators evaluated 15 patients with Chagas' disease, a chronic multiorgan *Trypanosoma cruzi* parasitic infestation that leads to widespread degeneration of autonomic nerves, including particularly vagal–cardiac nerves and ganglia (S.A. Morris *et al.* 1990). Amorim used a unique method to study vagal–cardiac function: he gave an intravenous injection of metaraminal (a sympathomimetic amine) in doses sufficient to raise systolic pressure by about 50 mmHg, and followed it with a 1 mg atropine sulphate injection, given at the time of peak pressure elevation. Eight patients with clinical evidence of Chagas' myocardial disease experienced less bradycardia during blood pressure elevation, and less tachycardia after (partial) muscarinic cholinergic blockade than seven patients without evidence of myocardial disease. Amorim suggested that in these special patients, abnormalities result from degeneration of parasympathetic nerves supplying the sinoatrial node.

In 1971, Eckberg *et al.* (1971) reported studies involving patients with a variety of heart diseases not known to involve autonomic nerves directly. They compared responses of 22 patients with coronary heart disease, idiopathic cardiomyopathy, congenital heart disease, and hypertension to baroreceptor stimulation and autonomic blockade, with those of 23 age-matched healthy subjects. They found that patients experienced the same degree of bradycardia as control subjects after intravenous propranolol, but less tachycardia after intravenous atropine sulphate. Average slopes of post-phenylephrine R–R interval lengthening were much less in patients than healthy subjects (3.7 vs. 16.0 ms/mmHg), and the degree of baroreflex impairment was related directly to the severity of cardiac symptoms. A subsequent prospective experiment conducted in conscious dogs with experimental heart failure (Higgins *et al.* 1972*a*) confirmed the relation between severity of failure and the degree of baroreflex impairment. Thus, these studies document abnormal reflex responses when *all* baroreceptors are stimulated.

Several studies document abnormal reflex responses when only *carotid* baroreceptors are stimulated. Osterziel *et al.* (1988) showed that slopes of R–R interval responses to graded neck suction are less in heart failure patients than in normal subjects. A second study from this group (Osterziel *et al.* 1990*b*) documented loose but significant direct correlations between depression of R–R interval responses to neck suction and elevations of plasma renin activity and noradrenaline concentrations, and an inverse correlation between baroreflex responses and stroke volume. This study (and an earlier one (Vogt *et al.* 1987)) also documented increased gain of baroreceptor–cardiac improvement after angiotensin converting enzyme inhibition.

Sopher *et al.* (1990) evaluated sigmoid carotid baroreceptor–cardiac reflex relations in heart failure patients with neck pressure changes that were larger and more complex than those used by Osterziel. Figure 14.2 depicts average responses of heart failure patients and age-matched control subjects to such forcings. Compared with healthy subjects, patients had lower thresholds, significantly less baroreflex gain (2.0 vs. 3.5 ms/mmHg) and smaller R–R interval ranges during neck pressure sequences. There is a striking similarity between the pattern of patients' abnormal vagal responses to neck suction and the pattern of heart failure dogs' abnormal carotid sinus baroreceptor nerve responses to pressure changes documented by W. Wang *et al.* (1990). The extent to which abnormal afferent baroreceptor nerve responses in

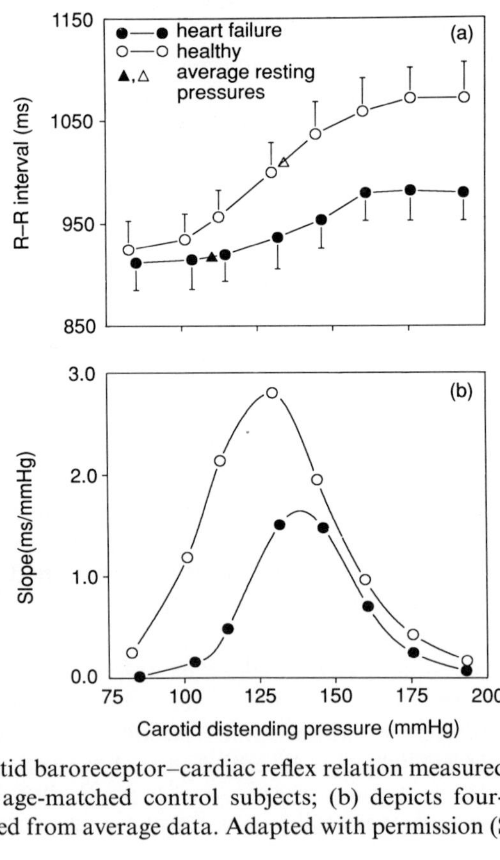

FIG. 14.2. Carotid baroreceptor–cardiac reflex relation measured in 14 heart failure patients and 14 age-matched control subjects; (b) depicts four-parameter logistic function calculated from average data. Adapted with permission (Sopher *et al.* 1990).

patients contribute to abnormal cardiac responses is unknown. Of interest is that patients' average resting pressures ('operational points', shown as triangles in Fig. 14.2 (a)) lay in the threshold region, whereas those of healthy subjects lay about midway in the 'linear' portion of the stimulus–response relation. Conceivably, patients' positions on this relation are related causally to their relative lack of R–R interval shortening with large doses of atropine; that is, they do not have much vagal–cardiac efferent activity to block.

The studies discussed above document impairment of vagally mediated baroreflex responses. They support but do not prove the notion that levels of inhibitory afferent baroreceptor input to the central nervous system are subnormal. It is not known if such a mechanism explains abnormalities of sympathetic outflow, which are much more difficult to study.

Although the above discussion focuses on *arterial* baroreflex mechanisms, it is likely that *cardiopulmonary* baroreflex mechanisms are also abnormal in heart failure patients. Greenberg *et al.* (1973) and Zucker *et al.* (1977) documented abnormal atrial mechanoreceptor firing and reduced compliance in heart failure, and Niebauer *et al.* (1986) and Niebauer and Zucker (1985) showed that the same high-output heart failure model used by Zucker *et al.* (1977) also leads to arterial baroreceptor abnormalities. Also, Kappagoda *et al.* (1987) and Kappagoda and Ravi (1989) showed that afferent traffic from *pulmonary* receptors may contribute to reflex abnormalities during pulmonary congestion. Thus in this chapter we do not attempt to discriminate between the possible roles of either cardiopulmonary or arterial baroreceptors in mediating cardiovascular abnormalities in heart failure patients. Indeed, one possible mechanism for return of plasma catecholamine levels to normal after sinoaortic denervation (N. Alexander *et al.* 1980) is that absence of sinoaortic baroreceptor inputs renders cardiopulmonary mechanoreceptors prepotent, a mechanism that may be related to central nervous system plasticity (Milgram *et al.* 1987).

In this regard, Abboud *et al.* (1981) speculate that activity of cardiopulmonary receptors may explain supranormal baseline levels of sympathetic outflow, and paradoxical sympathetic responses to arterial pressure changes (see Chapter 10) in heart failure patients. They reason that increased blood volume (Samet *et al.* 1957), chronic stretch, and diminished compliance (Zucker *et al.* 1977) of cardiac chambers lead to diminished contractions, and therefore diminished phasic inhibitory activity of cardiac receptors. When blood volume is reduced by treatment, normal contractions are restored, inhibitory traffic from cardiac receptors increases, and sympathetic outflow declines.

An alternative (or complementary) mechanism to explain supranormal sympathetic outflow in heart failure patients is that they have augmented sympatho*stimulatory* reflexes. Ruscone *et al.* (1987) showed that cardiac stretch may increase afferent traffic over sympathetic nerves to the spinal cord, and that such activity may increase efferent cardiac–sympathetic nerve activity. A study by Malliani and Pagani (1983) showed that sympathoexcitatory reflexes may inhibit baroreceptor–cardiac reflex responses. To our knowledge, there are no data from humans on the role, if any, of sympathostimulatory reflexes in heart failure.

Several studies show that baroreflex abnormalities in heart failure are *reversible*. A study of baroreflex responses in dogs with high-output failure by White (1981) showed that R–R interval prolongation after phenylephrine injections is partially restored to normal by surgical closure of aortocaval fistulae. This finding is supported by studies in humans by Ellenbogen *et al.* (1989) and Marin-Neto *et al.* (1991) which show that heart failure patients partly increase the steepness of their phenylephrine-induced baroreflex slopes after cardiac transplantation or successful medical therapy. Thus, although partial reversal of abnormal baroreflex responses does not exclude structural damage to baroreceptors (Zucker *et al.* 1977; Earle *et al.* 1980), it suggests that some of the abnormality is functional. The cause of functional baroreflex impairment is unknown; one theoretical possibility is that baroreceptors may function abnormally because in some patients with severe heart failure, they are bathed by hyponatraemic plasma (Kunze and Brown 1978; Andresen *et al.* 1979).

CENTRAL MECHANISMS

Studies of heart failure patients that employ experimentally induced arterial or neck pressure changes do not indicate where within the baroreflex arc abnormalities lie. It is probably impossible to gauge afferent baroreceptor input in humans. However, there are some published data that provide indirect information on central mechanisms. Porter *et al.* (1990) measured coherence between between spectral densities of the electrocardiogram and muscle sympathetic nerve activity in heart failure patients. This approach was based on the observation that activity of muscle sympathetic motoneurones is time-locked to cardiac activity, in all likelihood because of baroreflex entrainment (Fagius *et al.* 1985). Because of this relation, significant coherence between the two signals could be taken as evidence that *some* baroreceptor input is present within the central nervous system to influence sympathetic oscillators. Porter found significant coherence between muscle sympathetic nerve and cardiac activities in seven out of eight patients. However, his study says little about the *normality* of baroreceptor inputs.

Porter's study also provides some information on the sharply diminished respiratory sinus arrhythmia found in heart failure patients (Schlomka 1937; Saul *et al.* 1988*b*). The strong correlation between breathing and sympathetic activity in these patients (Fig. 14.1, left) provides evidence that near-absence or reduction of sinus arrhythmia is not due to absence of respiratory rhythms within the central nervous system.

Porter *et al.* (1990) also showed that heart failure patients do not prolong their R–R intervals after injections of low intravenous doses of atropine. Since low dose atropine increases firing of vagal–cardiac nuclei (Katona *et al.* 1977), Porter's finding points towards a *central* abnormality within the baroreflex arc in patients. However, it does not indicate whether the defect results from subnormal baroreceptor stimulation of vagal motoneurones or from intrinsic abnormalities of the neurones themselves.

CIRCULATING VASOCONSTRICTORS

Circulating vasoconstrictor substances, including noradrenaline, adrenaline, arginine vasopressin, renin and angiotensin-II, and neuropeptide Y may contribute to increased peripheral resistance in heart failure patients. Some patients, including particularly those with high renin levels, experience reductions of forearm vascular resistance after acute administration of angiotensin-converting enzyme inhibitors (Curtiss *et al.* 1978; Gavras *et al.* 1978; Kluger *et al.* 1982). Others have vasodilatation with vasopressin-blocking drugs (Creager *et al.* 1986). These responses however do not occur in the majority of patients with heart failure. On the other hand, intra-arterial *adrenergic* blocking drugs, such as phentolamine, reduce vascular resistance in a majority of patients (Zelis *et al.* 1968; Creager *et al.* 1985, 1986).

Noradrenaline

Chidsey *et al.* (1962) first reported abnormal regulation of plasma noradrenaline levels in patients with congestive heart failure. Subsequent studies (J.A. Thomas and Marks 1978; Moskowitz *et al.* 1979; Francis *et al.* 1982*a*; Cohn *et al.* 1984) confirmed and extended these early findings. Evidence obtained from healthy animals in their basal states suggests that the great majority (about 95 per cent) of circulating plasma noradrenaline comes from adrenergic nerve terminals, and only a small amount comes from the adrenal medulla (Price 1957). We are not aware of any study that has determined the source of circulating plasma noradrenaline in patients with heart failure. However, some heart failure patients are known to have elevated plasma levels of adrenal-derived adrenaline (Hasking *et al.* 1986). It is likely that a substantial portion of the excess circulating plasma noradrenaline found in heart failure patients derives from sympathetic nerve endings, because three studies in patients (Leimbach *et al.* 1986; Ferguson *et al.* 1989; Porter *et al.* 1990) document supranormal levels of nerve traffic to the large (about 42 per cent of body mass (Rein and Schneider 1956)) skeletal muscle vascular bed.

Plasma noradrenaline levels are related inversely (and weakly) to cardiac index (Cody *et al.* 1982*b*), and directly (and more strongly) to the severity of heart disease (J.A. Thomas and Marks 1978; Francis *et al.* 1982*a*; Kluger *et al.* 1982), and tend to increase with time (Francis *et al.* 1988). Most studies involving measurements of plasma noradrenaline levels have been conducted with patients who have systolic left ventricular dysfunction. We are not aware of any studies in which the independent contributions of systolic and diastolic left ventricular dysfunction to elevated sympathetic outflow have been assessed. Indirect evidence that diastolic dysfunction contributes to supranormal sympathetic outflow comes from the study of heart failure patients by Leimbach *et al.* (1986). They documented a statistically significant direct proportionality between left ventricular end-diastolic pressure and muscle sympathetic nerve activity. Figure 14.3 depicts data from the seminal study of J.A. Thomas and Marks (1978) on the relation between plasma noradrenaline levels and severity of heart disease.

Only about half of the supranormal circulating plasma noradrenaline levels found in heart failure patients results from increased noradrenaline spillover; the other half results from reduced noradrenaline clearance (Hasking *et al.* 1986; Davis *et al.* 1988). Most released noradrenaline undergoes reuptake, which in healthy subjects appears to be mediated more by neuronal than by non-neuronal mechanisms (Cryer 1980). Since heart failure patients appear to have normal *neuronal* noradrenaline reuptake (Clemson *et al.* 1990), reduced clearance must result importantly from impaired *non-neuronal* mechanisms (Leuenberger *et al.* 1989). Presumably, impaired non-

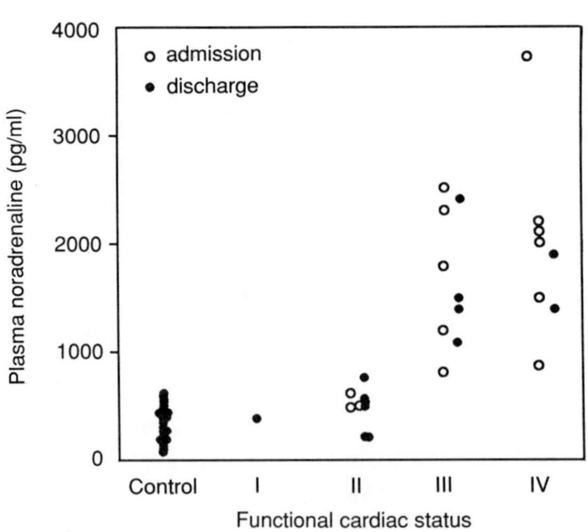

FIG. 14.3. Relation between plasma noradrenaline levels and the functional class of heart failure patients. Adapted with permission (J.A. Thomas and Marks 1978).

neuronal noradrenaline reuptake in heart failure patients results primarily from low tissue blood flow.

Direct microneurographic recordings indicate that sympathetic traffic to skeletal muscle is supranormal in patients with heart failure (Fig. 14.1 (Leimbach *et al.* 1986; Ferguson *et al.* 1989; Porter *et al.* 1990)). Studies of noradrenaline kinetics suggest that nerve traffic to the heart is also augmented. Swedberg *et al.* (1984) found that in heart failure patients, myocardial noradrenaline release, estimated from arterial–coronary sinus differences, averaged 20 times that of patients with angina without failure. Hasking *et al.* (1986) found large (540 and 206 per cent) increases of noradrenaline spillover from the heart and kidneys (this finding is consistent with increased sympathetic nerve traffic to those organs) in patients with left ventricular failure, but normal spillover from the lungs.

The consequences of chronically increased levels of plasma noradrenaline in heart failure patients are unclear, in part because mechanisms involving the effects of high levels of circulating noradrenaline are complex. One complexity is that released adrenaline exerts a positive feedback, and increases neuronal release of additional noradrenaline (M.D. Johnson *et al.* 1983). Some workers dispute the notion that the levels of noradrenaline found in heart failure patients exert measurable haemodynamic effects (Silverberg *et al.* 1978). Others consider that such concentrations exert subtle but unmistakable haemodynamic effects (Izzo 1983; Chang *et al.* 1986). A third issue is that even though heart failure patients have high circulating noradrenaline levels, down-regulation of beta-adrenergic receptors blunts cardiac inotropic responses (M.R. Bristow *et al.* 1982; Goldsmith *et al.* 1985; Fowler *et al.* 1986). (However, heart failure patients appear to have normal alpha-adrenergic mediated vasoconstriction (Kubo *et al.* 1989).)

The study of Ferguson *et al.* (1984) points away from significant down-regulation of adrenergic *vascular* receptors. They showed that heart failure patients experience normal increases of forearm vascular resistance after intra-arterial noradrenaline injections. King *et al.* (1987) studied the effects of chronic (28 day) infusions of noradrenaline in healthy dogs, and found that elevated noradrenaline levels do not change arterial pressure, cause slight cardiac slowing, and provoke reductions of baroreflex mediated bradycardia. In this study, large-dose atropine administration provoked substantial tachycardia and hypertension; this suggests that absence of hypertension is secondary to muscarinic cholinergic reflex buffering of the effects of noradrenaline.

Two studies document a strong inverse relation between plasma noradrenaline levels and survival in patients with heart failure (Creager *et al.* 1982; Cohn *et al.* 1984). Actuarial data from the study of Cohn *et al.* (1984) are shown in Fig. 14.4. A subsequent analysis from this group (Rector *et al.* 1987) suggests that the direct relation between mortality and plasma noradrenaline in heart failure patients is better described by a threshold function

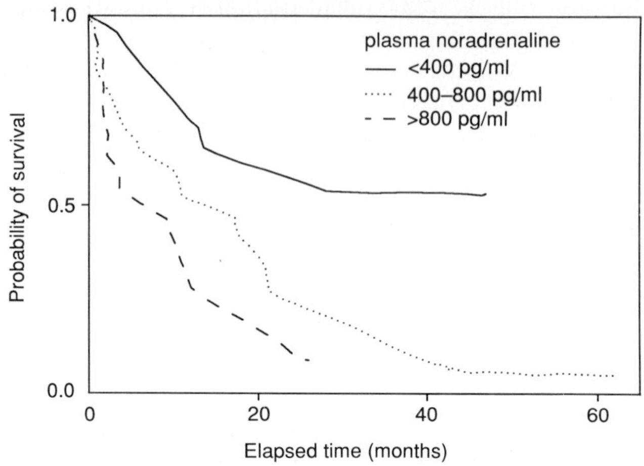

FIG. 14.4. Actuarial analysis of survival of heart failure patients plotted in tertiles according to plasma noradrenaline levels. Adapted with permission (Cohn *et al.* 1984).

(at about 600 pg/ml) than by a continuous variable. There is much that is not known about this intriguing relation. For example, it is not known if elevation of plasma noradrenaline levels contributes directly to mortality, or is merely an epiphenomenon that provides a biochemical index of the severity of heart disease (Massie 1988). Indirect support for the former view is provided by data which show that myocardial infarction patients with heart failure are more likely to benefit from beta-adrenergic blockade than patients without failure (Chadda *et al.* 1986). Also, one therapy of heart failure patients that lowers plasma noradrenaline levels — administration of the angiotensin converting enzyme inhibitor enalapril — improves survival (CONSENSUS Trial Study Group 1987) in part by reducing sudden death (SOLVD Investigators 1991).

Renin, angiotensin, and aldosterone

Chapter 9 indicates that baroreflex mechanisms are involved importantly in the physiological production of renin (and angiotensin-II and aldosterone). Baroreflexes have the potential to modulate plasma renin activity through several interrelated mechanisms. First, baroreflexes modulate arterial and therefore renal artery pressures, and influence plasma renin activity through direct actions on pressure-sensitive juxtaglomerular cells. Second, baroreflexes modulate levels of efferent sympathetic traffic to the kidneys, which influence renin activity by stimulating juxtaglomerular cell beta-adrenergic receptors. Third, baroreflexes in part regulate levels of circulating neurohumoral substances, such as adrenaline and vasopressin, which in turn modulate renin release. Finally, baroreflex mediated beta-adrenergic receptor stimulation modifies the threshold for renal baroreflex mediated release of renin, and thereby modulates levels of angiotensin-II and aldosterone levels, and influences sodium and water balance (Kirchheim *et al.* 1988).

Merrill *et al.* (1946) were the first to document elevated renin activity (as measured by an early bioassay) in heart failure patients. In their study, elevated renal vein renin activity was present in eight out of 11 patients. Subsequently, Merrill *et al.* (1946) and Anand *et al.* (1989) showed that reduction of renal blood flow in heart failure patients is much greater than reduction of cardiac output, but that glomerular filtration tends to be preserved. Preservation of nearly normal glomerular filtration in the face of reduced renal blood flow is probably secondary to renin-provoked constriction of efferent renal arterioles. The finding of Creager *et al.* (1981) and T.B. Levine *et al.* (1984) that converting enzyme blockers increase renal blood flow much more than they increase cardiac output supports this view.

Kirchheim *et al.* (1988) suggest that high plasma renin activity in heart failure results importantly from resetting of the arterial pressure–renin relation (see Chapter 9). According to this construct, in early heart failure, impaired cardiac function leads to subtle changes of the arterial pulse contour, modifies baroreceptor input, and increases sympathetic nerve traffic to the kidneys. Increased sympathetic traffic raises the pressure threshold for renin release (that is, episodic renin release occurs at higher pressures than before), and plasma renin activity increases. According to Kirchheim, an increase of renal baroreflex threshold can occur without a reduction of arterial pressure. If blood pressure actually falls in some heart failure patients, as it did in studies cited below, this mechanism seems even more likely. Kirchheim's presumption is that augmented sympathetic outflow to the kidneys is in some way mediated by baroreflexes. Thus, in early heart failure, abnormally high levels of plasma renin activity may result from *normal* functioning of arterial baroreflexes.

Others besides Merrill *et al.* (1946) showed that venous or arterial plasma renin activity is increased in heart failure patients, but in smaller percentages than they found. J.J. Brown *et al.* (1970) studied a group of 21 untreated heart failure patients and found that plasma renin activity was subnormal in three, normal in 12, and supranormal in six (29 per cent of the total). In heart failure patients, elevated levels of plasma renin activity tend to correlate directly with the severity of heart failure (Dzau *et al.* 1981) and plasma noradrenaline (Creager *et al.* 1986) and blood urea nitrogen levels, and inversely with serum sodium levels (J.J. Brown *et al.* 1970).

One explanation for the *normal* levels of plasma renin activity found in many heart failure patients is that plasma renin levels (and, by inference, angiotensin-II) fluctuate during the course of clinical heart failure, and may be either elevated or normal at the time blood sampling occurs. This possibility was predicted in a low output heart failure model in conscious dogs by

Watkins *et al.* (1976). They found that soon after constriction of the inferior vena cava or pulmonary artery, blood pressure fell to subnormal levels and plasma renin activity and aldosterone levels rose. Although urine production remained normal, blood volume increased, because of increased water intake and drastic reduction of sodium excretion. Days later, after the development of ascites and oedema, plasma renin activity, aldosterone, and blood pressure returned to normal. In dogs whose blood pressure did not return to normal, plasma renin activity and aldosterone levels remained elevated. Converting enzyme blockade caused *no* reduction of arterial pressure in normal dogs and in heart failure dogs whose blood pressure and plasma renin activity had returned to normal, but caused hypotension in dogs whose renin activity remained elevated. Conversely, converting enzyme blockade begun concurrently with vena cava or pulmonary artery constriction prevented compensatory haemodynamic responses, and the heart failure dogs remained severely hypotensive. This study points strongly towards hypotension, and by implication, reduced afferent baroreceptor activity, as an important mediator of some neurohumoral responses to heart failure.

A related study in humans also supports a role for baroreflexes in pathophysiological augmentation of renin activity. Dzau *et al.* (1981) divided 23 heart failure patients into three groups: five with recent onset of severe heart failure and pulmonary oedema, nine with recent worsening of heart failure, and nine with stable chronic heart failure. They found that patients with recent onset of failure are more likely to have elevated renin levels than patients with stable heart failure. They postulate that, as in the dogs studied by Watkins *et al.* (1976), plasma renin activity increases early in the course of clinical heart failure to maintain acceptable blood pressure in the face of falling cardiac output. As blood volume increases (secondary to the actions of angiotensin-II and aldosterone and sodium and water retention), blood pressure and plasma renin activity levels return towards normal.

Other work suggests that augmentation of renin activity represents a compensatory mechanism to support blood pressure. Curtiss *et al.* (1978) and Dzau *et al.* (1981) found that in groups of patients whose severity of heart failure varied, blood pressure and plasma renin activity are inversely related. Indirect support for the view that elevated levels of plasma renin activity result from attempts to maintain blood pressure come from responses to converting enzyme inhibition. Curtiss *et al.* (1978) found that post-treatment reductions of blood pressure and peripheral vascular resistance are related inversely to pre-treatment levels of plasma renin activity.

Renin-related vasoconstriction is not found in all heart failure patients. Although in some patients, increased plasma renin activity contributes to increased vascular resistance (Curtiss *et al.* 1978; Gavras *et al.* 1978; Faxon *et al.* 1980*a*), not all patients experience reductions of resistance with converting enzyme inhibition. Faxon *et al.* (1980*a*) found that converting enzyme inhibitors reduce peripheral vascular resistance in only 19 per cent of heart failure patients. This contrasts with the finding (Creager *et al.* 1986) that alpha-adrenergic blockade with phentolamine reduces systemic vascular resistance in virtually all heart failure patients. Moreover, reductions of resistance following converting-enzyme inhibition do not necessarily prove that excess vasoconstriction is *caused* by angiotensin; one action of converting-enzyme blockers is to retard degradation of vasodilating kinins (McCaa *et al.* 1978).

The above discussion focuses on *normal* baroreflex responses to cardiac failure; it is not entirely clear how patients with *deranged* baroreflex mechanisms cope with diminished cardiac output. A study by Lilly *et al.* (1984) supplies a link between baroreflex malfunction and impaired renal function in heart failure. They found that heart failure patients who are hyponatraemic also have greatly impaired haemodynamic responses to standing and severely reduced renal blood flow. There is some evidence that baroreflex malfunction impairs control of plasma renin activity in heart failure patients; some patients may not experience a normal reduction of renin activity levels with upright tilt or lower body negative pressure (Kubo and Cody 1983; T.B. Levine *et al.* 1983; Creager *et al.* 1985).

Vasopressin

The relation between baroreflex function and plasma arginine vasopressin in healthy subjects is discussed in Chapter 9. Plasma vasopressin levels, which average about 4.3 pg/ml in normally hydrated healthy adults (Cowley 1982), may be elevated in heart failure patients in proportion to the degree of their cardiac disability (Yamane 1968). This relation is illustrated in Fig. 14.5. Although striking elevations of vasopressin concentrations may be found in some heart failure patients, levels of vasopressin are normal in most untreated patients (Anand *et al.* 1989).

However, as indicated in Chapter 9, the normality of plasma vasopressin levels cannot be determined without knowledge of plasma osmolality. This relation is shown in Fig. 9.19. In the study of Szatalowicz *et al.* (1981), only four out of 37 patients with heart failure had vasopressin levels above the normal range; however, since in the patients they studied, serum sodium concentration averaged only 131 mmol/l, and plasma osmolality averaged only 272 mOsm/kg of water, their finding of *normal* vasopressin concentrations in 30 out of 37 patients points towards abnormalities of vasopressin release. Levels of sodium and osmolality found in heart failure patients are below the threshold for vasopressin release in normal subjects (as shown in Fig. 9.19), about 285 mOsm/kg water (G.L. Robertson *et al.* 1973).

Water retention in heart failure patients may be secondary to a variety of factors, including reductions of fluid delivered to the distal diluting segments of the nephron, as well as to increased vasopressin levels. A study by Kortas *et al.* (1986) stresses the importance of augmented vasopressin levels to water

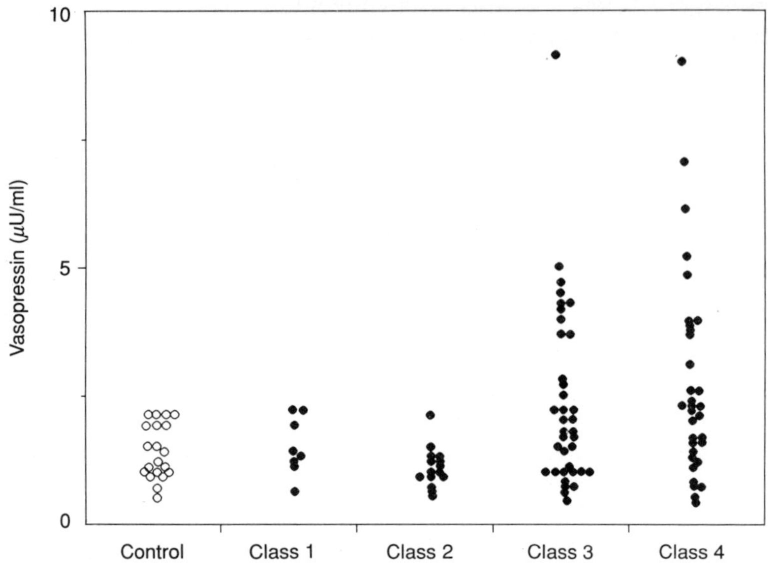

FIG. 14.5. Plasma arginine vasopressin levels in heart failure patients, plotted according to functional class. Most patients have normal vasopressin levels. Adapted with permission (Yamane 1968).

retention. They found that increases of water excretion and reductions of plasma vasopressin levels after vasodilator therapy are not associated with increases of glomerular filtration rates. Szatalowicz *et al.* (1981) suggested that in their patients, vasopressin levels were sufficient to enhance tubular reabsorption of water, and therefore to contribute to hyponatraemia. Their study does not indicate whether the vasopressin levels found in heart failure patients were sufficient to cause vasoconstriction. Probably in most patients they were not. G.L. Robertson (1983) found that maximum water absorption occurs with vasopressin levels of about 5 pg/ml, and that higher levels result primarily in increased vasoconstriction. However, other studies have shown that if vasopressin levels are above normal, administration of a vascular vasopressin receptor blocker reduces systemic vascular resistance (Nicod *et al.* 1985; Creager *et al.* 1986).

Thus the 'normal' vasopressin concentrations found in many heart failure patients are inappropriate for their serum osmolalities, and probably contribute to abnormal sodium and water retention and circulatory congestion. Frankly elevated vasopressin levels are found in a small percentage of heart failure patients, and in this subset, vasopressin probably contributes to abnormally elevated vascular resistance. However, vasoconstriction due to vasopressin is not as prevalent as vasoconstriction due to other neurohumoral mechanisms, since in heart failure patients, converting enzyme inhibitors and alpha-adrenergic blocking drugs are more likely to reduce vascular resistance than are vasopressin blocking drugs (Creager *et al.* 1986).

Why are vasopressin concentrations abnormal in heart failure patients? As discussed in Chapter 9, in healthy subjects, vasopressin levels appear to be determined primarily by osmolality and blood volume and pressure. It is highly unlikely that abnormal osmoreception leads to elevation of vasopressin levels in heart failure patients. First, as indicated, osmolality is subnormal in many heart failure patients; subnormal osmolality should be associated with undetectable, not normal or elevated levels of plasma vasopressin. Second, heart failure patients respond appropriately to water loading with reductions of serum osmolality and vasopressin levels (Yamane 1968; Kortas *et al.* 1986). Moreover, many heart failure patients respond to osmolar loads with supranormal increases of vasopressin levels (Uretsky *et al.* 1985). In some patients, elevations of vasopressin levels occur even though experimentally induced increases of serum osmolality raise their osmolality to levels below those necessary to trigger vasopressin release in healthy subjects.

Thus, inappropriate vasopressin levels found in heart failure patients cannot be explained on the basis of insensitivity of osmoreceptors to changes of osmolality. Therefore, by exclusion, normal or elevated vasopressin levels in patients probably result from abnormal regulation of vasopressin secretion by blood volume or pressure. A series of studies (G.L. Robertson and Athar 1976; G.L. Robertson *et al.* 1976) suggests that the relation between osmolality and blood volume and pressure may be described by a family of relations, such that the magnitude of vasopressin changes provoked by changes of osmolality is determined by blood volume or pressure. This is illustrated in Fig. 14.6; increasing hypotension or hypovolaemia (lines to the left) reduce osmolar thresholds for vasopressin release and increase the steepness of the relation between osmolality and vasopressin. Such a scheme might explain increased steepness of vasopressin responses to osmolar loads in heart failure patients (Uretsky *et al.* 1985), if these patients' blood volumes or pressures were perceived to be subnormal. This issue is not simple, however, since heart failure patients usually have increased central blood volume (as reflected by cardiac chamber dilatation), and usually do not have hypotension. However, the range of 'normal' arterial pressure is wide, and, as mentioned, in some studies (Dzau *et al.* 1981; Swedberg *et al.* 1984) arterial pressure was significantly lower in heart failure patients than in healthy subjects.

It is not known if the elevated levels of plasma renin activity, angiotensin II, and aldosterone found in some patients with heart failure reflect some abnormality of autonomic sensory input that is interpreted as reduced blood volume. Although it is likely that abnormal pressure or volume sensing contributes to inappropriate vasopressin levels in humans with heart failure, it is by no means clear what that autonomic sensor is. It may be that

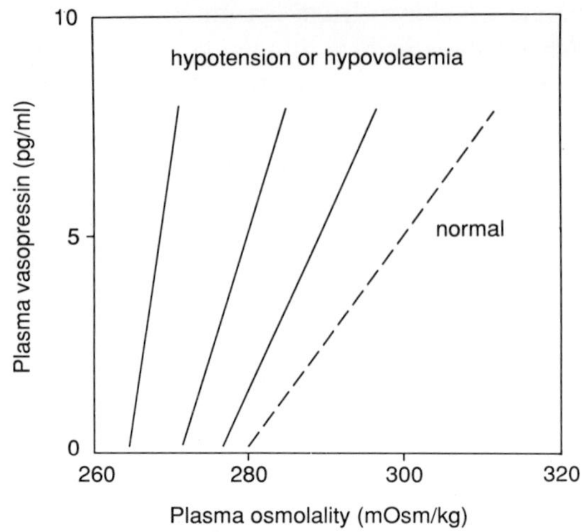

FIG. 14.6. Interaction between plasma osmolality and blood volume or hypovolaemia. Lines to the left show increasing hypotension or hypovolaemia. Adapted with permission (G.L. Robertson *et al.* 1976).

techniques available for use in humans are not sufficiently sensitive to detect changes of sensory input profiles that trigger inappropriate vasopressin responses, and are incapable of discriminating among various mechanisms which might be responsible for these changes. Nevertheless, some indirect information is available regarding arterial and cardiopulmonary baroreflex modulation of vasopressin secretion in humans with heart failure.

Although simple measurements of arterial pressure do not predict which heart failure patients have elevated vasopressin levels (Szatalowicz *et al.* 1981; Riegger *et al.* 1982), certain data are compatible with the notion that abnormal arterial baroreflex function contributes to the inappropriate vasopressin secretion that occurs. Intravenous infusions of nitroprusside lower arterial pressure and trigger baroreflex mediated increases of sympathetic activity (Eckberg *et al.* 1988). Presumably nitroprusside also raises plasma vasopressin levels in healthy people. Although we are not aware that this possibility has been tested in humans, data from animals (Fumoux *et al.* 1978) suggest that this should be the case. Nitroprusside does not raise plasma vasopressin levels in heart failure patients, however, and this could reflect an abnormality of arterial baroreflex control of vasopressin secretion (Francis *et al.* 1983).

A second indication that abnormal arterial baroreflexes may contribute to inappropriate vasopressin levels in heart failure patients comes from their responses to tilt. As discussed in this and Chapter 7, upright tilt alters barosensory input and invokes reflexes that are thought to be mediated in part by reduced arterial baroreceptor activity. Upright tilt leads to small elevations of plasma vasopressin levels in most healthy people (R. Davies *et al.* 1977; Goldsmith *et al.* 1983*b*; Sander-Jensen *et al.* 1986*a*), and large increases of vasopressin levels in the minority of healthy people who develop vasovagal reactions during tilt (Goldsmith *et al.* 1983*b*; Sander-Jensen *et al.* 1986*a*). Upright posture does not lead to elevations of vasopressin levels in heart failure patients (Goldsmith *et al.* 1983*b*). Absence of increases of vasopressin levels during upright posture correlates with absence of increases of plasma catecholamines and renin activity in these patients (Bito *et al.* 1980; Goldsmith *et al.* 1983*a*; Creager *et al.* 1985). All of these abnormal response patterns could be mediated in part by abnormal baroreflex function in patients.

More attention has focused on the role of cardiopulmonary than on that of arterial baroreflexes in modulation of vasopressin secretion. There is a substantial literature regarding the role of atrial pressure in modulation of vasopressin levels (for example, Goetz *et al.* 1975; Ledsome *et al.* 1983; M.-E. Lee *et al.* 1986). Several studies suggest that left and right atria are differentially involved in modulation of vasopressin secretion. Brennan *et al.* (1971) and DeTorrente *et al.* (1975) report that left atrial distension lowers plasma vasopressin concentrations, but right atrial distension does not (Brennan *et al.* 1971).

There is some circumstantial evidence that differential atrial control of vasopressin secretion also occurs in patients with heart failure. Yamane's study (1968) provides indirect support for the notion that elevated vasopressin levels correlate more closely with right than with left atrial distension. She found that 15 out of 48 patients deemed to have 'pure' left heart failure had normal vasopressin levels. Many of the remaining patients with right, or right and left, heart failure had significantly elevated vasopressin levels.

Additional indirect evidence for differential left and right atrial control of vasopressin release comes from the study of F.H. Epstein *et al.* (1953). These authors reported the effects of manual compression of chronic traumatic arteriovenous fistulae. They found that closure of fistulae raised arterial (and presumably, left heart) pressures, and usually reduced right heart pressures. Fistula closure usually led to increased renal sodium excretion, an effect which may have been mediated by reduced vasopressin release secondary to left atrial distension.

Another example may be found in patients with acute myocardial infarction. Patients with engorgement of vessels in the upper lung fields (presumably associated with left atrial dilatation) experience acute increases of glomerular filtration rates and urine output (E.D. Bennett *et al.* 1977). The possible role of atrial natriuretic factor in mediating these responses is discussed below.

Although it seems likely that cardiopulmonary receptors modulate vasopressin secretion in heart failure patients, responses of patients cannot necessarily be predicted from responses of healthy subjects. The quality of responses of patients may depend on the chronicity of disease. Indirect evidence suggests that acute haemodynamic changes may trigger normal vasopressin responses. For example, although most patients with paroxysmal dysrhythmias experience a diuresis, patients with chronic left atrial dilatation due to mitral stenosis do not (P. Wood 1963). Presumably, in patients without chronic atrial disease, tachydysrhythmias lead to left atrial stretch which reduces vasopressin levels; in patients with chronic mitral stenosis, atrial disease prevents this reflex reaction. (The role of atrial natriuretic factor in mediating diuretic responses to paroxysmal atrial tachycardias is discussed below.)

Patients with chronic heart failure seem to have *qualitative* abnormalities of vasopressin regulation. For example, increases of heart volume that occur when healthy dehydrated people lie supine lead to natriuresis (Hulet and Smith 1961), presumably secondary to reductions of vasopressin levels. In patients, similar chronic increases of central blood volume may be associated with elevated vasopressin levels and sodium and water retention. Moreover, in heart failure patients, therapeutic interventions that reduce blood volume, such as haemofiltration (Riegger *et al.* 1982) or inotropic or vasodilator therapy (Uretsky *et al.* 1985; Kortas *et al.* 1986), reduce plasma vasopressin levels. In one study (Uretsky *et al.* 1985), changes of vasopressin levels after therapy were related inversely to changes of arterial pressure.

Levels of other vasoconstrictor substances may be supranormal in heart failure patients. These include the noradrenaline precursor dopamine (Francis *et al.* 1984); the non-adrenergic vasoconstrictor neuropeptide Y (Maisel *et al.* 1989), which is co-released with noradrenaline from sympathetic nerve terminals, and which (in healthy subjects) appears to be under baroreflex control (Eckberg *et al.* 1988); and endothelin (R.M. Robertson *et al.* 1990), a potent vasoconstrictor elaborated by endothelial cells, which appears to influence central baroreflex modulation (Mosqueda-Garcia *et al.* 1990).

CIRCULATING VASODILATORS

Atrial natriuretic factor

Atrial natriuretic factor (ANF, atrial natriuretic peptide — ANP, or atriopeptin) is a family of low molecular weight peptides (de Bold *et al.* 1981; Flynn and Davies 1985) with important natriuretic and vasodilator properties, stored as parts of a 126 amino-acid prohormone in perinuclear regions of cardiac (primarily atrial) myocytes (de Bold 1979; de Bold *et al.* 1981; Currie *et al.* 1983). Specific atrial natriuretic factor receptors are abundant in vascular tissues (Napier *et al.* 1984).

Increases of atrial natriuretic factor appear to be mediated primarily by mechanical atrial stretch, rather than by stretch-induced acetylcholine or noradrenaline release (Schiebinger and Linden 1986). Several groups (Schiffrin *et al.* 1985; Tikkanen *et al.* 1985; Yamaji *et al.* 1985; Nilsson *et al.* 1987) showed that supraventricular tachycardia (which increases atrial pressures (P. Wood 1963)) increases atrial natriuretic factor levels. Although it seems likely that increased atrial natriuretic factor levels *cause* the diuresis that occurs in patients with tachycardia, the mechanism may be complex. Singer *et al.* (1987) showed that after acute saline infusions, atrial natriuretic factor levels fall to normal long before sodium excretion returns to normal. *Reductions* of atrial natriuretic factor levels are difficult to provoke; in healthy humans, levels tend to be low, and they may not fall further during reductions of atrial dimensions caused by upright posture or haemorrhage (Hodsman *et al.* 1986).

There is good evidence that the changes of atrial natriuretic factor levels that occur during changes of arterial pressure are mediated in part by *baroreflexes*. Interventions that increase arterial pressure, including infusions of noradrenaline, angiotensin-II, and vasopressin, increase atrial natriuretic factor levels (Sonnenberg and Veress 1984; Lang *et al.* 1985; Manning *et al.* 1985). Conversely, interventions that reduce arterial pressure may reduce atrial natriuretic factor levels (see below). Since changes of arterial pressure and atrial pressure (and dimensions) tend to occur in parallel, baroreflex influences on atrial natriuretic factor release may be mediated indirectly through changes of atrial dimensions.

Erne *et al.* (1987) ingeniously evaluated this issue in patients undergoing electrophysiological study. They paced the right ventricle at rates slightly above the naturally occurring rate, and divided subjects according to their arterial pressure responses to ventricular pacing (and loss of atrioventricular synchronization). Patients who maintained their usual arterial pressures (but who probably increased atrial pressures) increased atrial natriuretic factor levels. Patients who reduced their arterial pressures (who also increased atrial pressures) *reduced* atrial natriuretic factor levels. These data suggest that arterial hypotension prevents atrial natriuretic factor release by stretched atrial myocytes. Similarly, neck pressure, which reduces carotid artery dimensions (Kober and Arndt 1970) and is perceived as hypotension, and may not alter atrial pressure, provokes proportional atrial natriuretic factor reductions (Volpe *et al.* 1988*a*).

The mechanism whereby reduced baroreceptor afferent input overrides the effects of mechanical atrial stretch is unclear. This interaction seems to be mediated neurohumorally because it is prevented by pretreatment with alpha- and beta-adrenergic blocking drugs (Volpe *et al.* 1988*a*). Sonnenberg and Veress (1984) showed that noradrenaline (and vasopressin) increase atrial natriuretic factor release by cultured atrial myocytes, and suggested that this effect is mediated by activation of the cellular phosphoinositide system. Whatever the mechanism, these findings make teleological sense;

increased levels of a natriuretic and diuretic hormone would be inappropriate during hypotension.

The literature on the influence of *infused* atrial natriuretic factor in humans is confusing and contradictory. In general, published data indicate that large pharmacological doses increase sodium excretion, glomerular filtration fraction, osmolar and free water clearance, and haematocrit, and reduce plasma renin activity, aldosterone levels, and arterial and pulmonary capillary wedge pressures (Cody *et al.* 1986). Volpe *et al.* (1990) showed that infusions of large doses of atrial natriuretic peptide increase R–R interval lengthening provoked by phenylephrine; since this effect does not occur after converting enzyme inhibition, atrial natriuretic factor may influence human baroreflexes in part by blocking the opposing influence of angiotensin-II. Small, physiological doses of atrial natriuretic factor consistently increase urinary sodium and water excretion but do not lower arterial pressure (J.V. Anderson *et al.* 1987).

A major problem with human studies is that many involve administration of pharmacological or even *industrial* doses of atrial natriuretic factor. A related problem is that the method of administering atrial natriuretic factor varies among studies. Some studies (Tikkanen *et al.* 1985; Volpe *et al.* 1987, 1988*b*; T.J. Ebert *et al.* 1988) involve loading doses of atrial natriuretic factor followed by constant intravenous infusions; others (Takeshita *et al.* 1987) begin simply with infusions. This methodological detail is important, because T.J. Ebert *et al.* (1988) showed that transient reductions of arterial pressure and systemic vascular resistance occur after loading boluses of atrial natriuretic factor, but do not necessarily occur during continuous infusions not preceded by bolus loading doses.

The study of Hirooka *et al.* (1988) in anaesthetized rabbits may explain some complexities involved with studies of the effects of infused atrial natriuretic factor on baroreceptors. They found that arterial hypotension resulting from atrial natriuretic factor infusions does not reduce afferent aortic nerve activity; they postulated that nerve activity does not change, because aortic dilatation also occurs. Thus, reduced tension on baroreceptors secondary to hypotension may be offset exactly by increased tension secondary to direct vascular smooth muscle relaxation. This finding may help explain that of Holtz *et al.* (1987), that in conscious dogs, hypotension provoked by atrial natriuretic peptide does not elicit tachycardia.

T.J. Ebert (1988) studied some physiological consequences of atrial natriuretic factor infusions. He showed that reductions of arterial pressure provoked by large intravenous infusions increase muscle sympathetic nerve activity. He felt that this effect is mediated by cardiopulmonary mechanoreceptors because prevention of reductions of central venous pressure with concurrent head-down tilt or lower body positive pressure prevents increases of sympathetic traffic. Atrial natriuretic factor infusions at rates that do not reduce arterial pressure or increase heart rate also blunt tachycardic responses to carotid compression by neck pressure (T.J. Ebert and Cowley 1988).

Studies of the influence of atrial natriuretic factor on human responses to low levels of lower body negative pressure (thought to reduce cardiopulmonary receptor activity without altering arterial baroreceptor activity — see Chapter 7) are contradictory. Takeshita *et al.* (1987) found that infusions of 30 ng/(kg min) reduce forearm vasoconstrictor responses to lower body suction. Conversely, Volpe *et al.* (1988*b*) gave much larger doses of atrial natriuretic factor (500 ng/kg boluses followed by 50 ng/(kg min) infusions) and found that forearm vascular responses to lower body suction are enhanced. Large-dose infusions of atrial natriuretic factor in patients with severe heart failure provoke paradoxical reductions of plasma noradrenaline levels (Molina *et al.* 1988).

Heart failure patients, who may have elevated noradrenaline, vasopressin, and angiotensin-II levels (see above), also have elevated atrial natriuretic factor levels (Shenker *et al.* 1985; Tikkanen *et al.* 1985). Indeed, elevations of atrial natriuretic factor levels, which may be more than five times normal (Tikkanen *et al.* 1985; Cody *et al.* 1986; Y. Saito *et al.* 1987), may be the most striking neuroendocrine abnormality in heart failure patients (Anand *et al.* 1989). There is a step-up of atrial natriuretic factor concentrations between peripheral veins and the right ventricle; this provides direct evidence in humans that cardiac muscle produces atrial natriuretic factor. Reductions of cardiac pressures with lower body suction reduce atrial natriuretic factor levels in heart failure patients (Creager *et al.* 1988), but volume expansion in dogs with experimental failure does not lead to further increases (Redfield *et al.* 1989). Forearm (Hirooka *et al.* 1990) and renal (K. Tsunoda *et al.* 1988) vasodilator responses to atrial natriuretic factor infusions are subnormal in heart failure, possibly due to chronically elevated levels and receptor down-regulation. An additional mechanism may be that vasodilatory responses to atrial natriuretic factor are determined importantly by circulating vasoconstrictors (Lappe *et al.* 1987), whose levels vary markedly among patients.

There is a direct relation between the degree of cardiac dilatation and the amount of immunoreactive atrial natriuretic factor in human ventricular muscle granulocytes (Takemura *et al.* 1989); a proportionality between the severity of heart failure symptoms and the plasma level of atrial natriuretic factor (Tikkanen *et al.* 1985; Burnett *et al.* 1986); and an inverse relation between blood levels of atrial natriuretic factor and prognosis (S.S. Gottlieb *et al.* 1989; Swedberg *et al.* 1990). (Prognosis is also related to levels of other circulating neurohumoral substances that are exaggerated in heart failure patients (Cohn *et al.* 1984; Swedberg *et al.* 1990). It seems likely that the inverse relation between prognosis and plasma levels of atrial natriuretic factor results because atrial natriuretic factor is a marker for the severity of heart failure; atrial natriuretic factor's actions should be advantageous rather than harmful to such patients.)

Since salt and water retention, subnormal urine output, and vasoconstriction characterize many patients with advanced heart failure, the role of high concentrations of atrial natriuretic factor seems paradoxical. Intravenous infusions of atrial natriuretic factor in heart failure patients may not increase renal sodium and water excretion and reduce vasoconstrictor levels as they do in healthy subjects (Cody *et al.* 1986; Y. Saito *et al.* 1987), or such infusions may improve function only slightly (Fifer *et al.* 1990).

Thus, although supranormal levels of atrial natriuretic factor should be helpful, in some patients reduced renal responsiveness to atrial natriuretic factor may contribute to salt and water retention. Cody *et al.* (1986) could not determine from their data whether unresponsiveness to infused atrial natriuretic factor results from down-regulation of atrial natriuretic factor receptors (which occurs (Schiffrin 1988)) or from the abnormal renal haemodynamics found in heart failure patients. Three patients who increased their glomerular filtration rates during atrial natriuretic factor infusions also excreted more sodium. This provides some evidence that the subnormal renal responses to atrial natriuretic factor of heart failure patients result in part from deranged renal haemodynamics.

Some haemodynamic improvement may occur in response to atrial natriuretic peptide infusions in heart failure states. Münzel *et al.* (1991) documented reductions of preload and increases of cardiac output and urine production during 20 h infusions of atrial natriuretic peptide in heart failure patients. These effects were short-lived, however, in part secondary to suppression of endogenous atrial natriuretic peptide production. Drexler *et al.* (1980) showed that injections of purified antibodies to atrial natriuretic factor increase right heart pressures and systemic vascular resistance in post-infarction with severe heart failure, in proportion to pre-injection levels of atrial natriuretic peptide. This result suggests that extremely high levels of atrial natriuretic peptide do provoke some vasodilatation. Evidence for vasodilatation was not found in rats with moderate heart failure. Drexler suggests that the translocation of fluid from the intra- to the extravascular spaces that occurs normally with the peptide (Almeida *et al.* 1986; V.H. Huxley *et al.* 1987) and that may contribute to afterload reduction (Münzel *et al.* 1991) may have opposed direct vasodilatation by augmenting cardiac output and thereby reducing vascular resistance.

Prostaglandins

Prostaglandins, particularly I2 and E2, form a second endogenous vasodilator system that opposes vasoconstriction in some patients with heart failure. Dzau *et al.* (1984) measured plasma prostaglandin metabolites in patients whose heart failure had been characterized biochemically and haemodynamically. They found striking prostaglandin elevations in many patients, and noted close inverse correlations between prostaglandin E metabolites and serum sodium levels and plasma renin activities. They divided their patients into groups with and without hyponatraemia. The group with hyponatraemia was more likely to have elevated levels of prostaglandins and angiotensin II, and was more likely to experience deleterious increases of blood pressure, systemic vascular resistance, and pulmonary capillary wedge pressures and reductions of cardiac output after inhibition of prostaglandin synthesis with indomethacin than was the group with normonatraemia. Although baroreflex function was not studied directly in these patients, there was indirect evidence for baroreflex malfunction; indomethacin increased mean arterial pressure from 86 to 97 mmHg in hyponatraemic patients, but did not slow heart rate.

The specific causes of supranormal levels of prostaglandin metabolites are not known with certainty. It is known that renal and cardiac ischaemia provokes increases of prostaglandin synthesis (Needleman *et al.* 1978; J.A. Oliver *et al.* 1981). Also, it is known that vasoconstrictor substances elevated in heart failure patients, including angiotensin-I and -II and noradrenaline, increase prostaglandin levels (Needleman *et al.* 1973; Blumberg *et al.* 1977). Whatever the cause, it is likely that augmented prostaglandin production provides protection of kidneys when levels of circulating vasoconstrictors are high (J.A. Oliver *et al.* 1981).

Levels of other vasodilating substances are also elevated in heart failure patients. C.-s. Liang and co-workers showed that beta-endorphin levels are elevated in conscious dogs with heart failure caused by pulmonic stenosis and tricuspid regurgitation (C.-s. Liang *et al.* 1987; Sakamoto and Liang 1989). In these dogs, opiate receptor antagonists increased aortic pressure and cardiac output, and arterial baroreflex gain.

UPRIGHT POSTURE

Physiological responses to upright posture were reviewed in Chapter 7. It has long been recognized that *patients with heart failure* may not respond normally to upright posture (Brigden and Sharpey-Schafer 1950; Howard and Leathart 1951; Taquini *et al.* 1961; Abelmann and Fareeduddin 1967). In some patients, responses are directionally normal but quantitatively abnormal. In these patients, standing may provoke less reduction of left ventricular volume (Stefadouros *et al.* 1975), cardiac output (Rapaport *et al.* 1966), stroke volume (Rapaport *et al.* 1966), systolic pressure (Stefadouros *et al.* 1975), and blood flow to the kidneys (Werkö *et al.* 1949) and liver (Goldsmith *et al.* 1983*a*), and less cardioacceleration (Howard and Leathart 1951; R.E. Goldstein *et al.* 1975; Kassis 1989) than it does in healthy people. Also, in some heart failure patients, systemic or forearm vascular resistance may not increase normally (Goldsmith *et al.* 1983*a*; Creager *et al.* 1985), or may fall (Brigden and Sharpey-Schafer 1950; Kassis *et al.* 1986) during standing.

The combination of arterial hypotension, absence of cardioacceleration, and vasodilatation during standing points strongly towards abnormal baroreflex function in heart failure patients. This may be related to subnormal increases (or even paradoxical decreases — see Chapter 10) of sympathetic activity, as reflected by urinary noradrenaline concentrations (Bito *et al.* 1980) and noradrenaline spillover (Davis *et al.* 1987), or of renin activity (Cody *et al.* 1982*a*; Goldsmith *et al.* 1983*a*; Creager *et al.* 1985) and vasopressin concentrations (Rogge and Moore 1968; Goldsmith *et al.* 1982) during standing or 'simulated' orthostatic stress with lower body negative pressure. The degree of orthostatic venous pooling is related inversely to the severity of heart failure (Kubo and Cody 1983); some patients with mild or successfully treated heart failure may have normal venous pooling during standing, and appropriate reflex responses (Zelis 1974; Kubo and Cody 1983).

Conversely, some patients with advanced heart failure may have *qualitatively abnormal* responses to standing (see Chapter 10). Such patients may experience vasodilatation rather than vasoconstriction (Brigden and Sharpey-Schafer 1950; Goldsmith *et al.* 1983*a*; Kassis *et al.* 1986); increased aortic pulsation (Kassis 1987); augmentation, rather than reduction, or no change of cardiac output (Taquini *et al.* 1961; Kassis 1989); and no change of heart rate, or slowing rather than heart rate speeding (Howard and Leathart 1951). The mechanism for vasodilatation is unclear; Kassis *et al.* (1986) postulate that it is mediated by adrenaline-induced beta-2-adrenergic vascular mechanisms.

Some factors responsible for the increased tolerance of heart failure patients to standing are understood well, and others are understood poorly. The simplest explanation, which obtains in patients with severe heart failure (Kubo and Cody 1983), is that the reflex stimulus resulting from standing is not the same when heart failure is present as when it is absent. Thus in some patients, venous tone is abnormally elevated (D.T. Mason and Braunwald 1964; Zelis 1974) and standing does not lead to venous pooling and reflex responses to venous pooling (J.E. Wood *et al.* 1956; Jeffrey *et al.* 1970; Cody *et al.* 1982*a*; Kubo and Cody 1983). As a consequence, cardiac dimensions and arterial pressures do not decrease, and presumably there is little or no reduction of afferent input from most baroreceptor areas. Although such patients may also have major derangements of autonomic reflexes, baroreflex abnormalities need not be invoked to explain their increased orthostatic tolerance.

There is a variety of explanations for absence of orthostatic venous pooling in patients with heart failure. Increased resting venous tone (Zelis 1974*b*) places heart failure patients on a steep portion of the pressure–volume relation. As a consequence, small changes of volume during standing increase venous pressure in the legs greatly, reduce venous distensibility (J.E. Wood *et al.* 1956; Habak *et al.* 1974), and oppose additional venous distension. Experimental evidence that this mechanism may be operative comes from studies in healthy subjects in whom increases of blood volume secondary to saline infusions reduce orthostatic venous pooling (Jeffrey *et al.* 1970). Venous distensibility may be reduced also by sympathetically mediated venoconstriction (J.E. Wood *et al.* 1956), and possibly by increased sodium and water in vein walls and perivascular tissues (Zelis *et al.* 1974*a*; Magrini and Niarchos 1983).

In part because of increased sympathetic activity, systemic vascular resistance is increased (Kramer *et al.* 1968). Heart failure patients have reduced abilities to respond to vasodilator stimuli, such as ischaemia, local heating, and hand exercise (Zelis *et al.* 1968). Some but not all of this vasodilatory impairment is due to increased sympathetic activity (Zelis *et al.* 1968; J.R. Wilson *et al.* 1985).

There is a subset of heart failure patients, however, who experience venous pooling and reduction of left ventricular end-diastolic volume (T.B. Levine *et al.* 1983; Timmis *et al.* 1984) or right atrial pressure (Kassis *et al.* 1986) with standing, but who do not have compensatory reflex vasoconstriction. In contrast to patients with mild failure who may have normal reflex responses, and patients with severe failure who may have no reflex responses because the stimulus of venous pooling is absent, such patients are particularly prone to develop orthostatic hypotension. Afferent and efferent mechanisms responsible for absence of orthostatic reflex responses in these patients are understood incompletely. One favoured afferent mechanism is that cardiovascular disease impairs responses of cardiac (particularly left ventricular) receptors, which contribute importantly to normal orthostatic reflex control (Ferguson *et al.* 1984).

Dysfunction of left ventricular receptors cannot explain subnormal orthostatic reflex responses in heart failure patients fully, however, since reflex function may be restored towards normal in these patients by orthotopic heart transplantation (T.B. Levine *et al.* 1986). Although patients with heart transplants have preservation of some atrial innervation (since donor atria are sutured to remnants of native atria), both *ventricles* are denervated. In such patients, ventricular receptors cannot be responsible for normal orthostatic reflex responses. By exclusion, this suggests that normal orthostatic reflex responses may be initiated importantly by arterial baroreceptors.

The study of Ferguson *et al.* (1984) tends to rule out unresponsiveness of vascular receptors to noradrenaline as an explanation of impaired sympathetic reflex responsiveness to simulated standing (lower body suction). They found that patients who had little vasoconstriction (or even vasodilatation) during lower body suction experienced normal increases of peripheral vascular resistance in response to intra-arterial injections of noradrenaline.

Provocative studies summarized by Kassis (1989) explored efferent mechanisms in heart failure patients who have paradoxical responses to upright tilt. Patients' mean arterial pressures fell significantly, and possibly as a consequence of afterload reduction, their left ventricular stroke volumes increased. Pulsatile changes of aortic diameter increased significantly. (In contrast, healthy subjects experienced no significant change of arterial pressure, stroke volume, or aortic pulsation during upright tilt.) Kassis suggests that augmented aortic pulsations during tilt in heart failure patients probably augment baroreceptor afferent traffic. Such putative augmentation of aortic baroreceptor input, however, is not sufficient to explain vasodilatation. This was shown by other experiments.

In studies conducted with the same patients on different days, Kassis *et al.* (1986) found that vasodilatation during tilt is converted to vasoconstriction by prior intra-arterial infusions of the non-selective beta-adrenergic blocker propranolol. This response suggests that muscle vasodilatation is mediated by beta-2-adrenergic receptors; when these receptors are blocked by propranolol, alpha-adrenergic vasoconstriction is unmasked. Studies conducted on other days indicated that some of the vasodilatation during tilt is mediated by increases (not decreases) of efferent neural traffic; proximal nerve blockade prevented vasodilatation, and even converted the response to vasoconstriction. As mentioned, in heart failure patients, tilt increased aortic pulsations, and presumably afferent baroreceptor activity. Since increased baroreceptor stimulation would not explain increased efferent sympathetic outflow, Kassis suggests that his data point to impaired afferent baroreceptor mechanisms. The occurrence of some vasoconstriction during tilt after nerve blockade points towards the action of a humoral vasoconstrictor. This elegant series of studies suggests that paradoxical muscle vasodilatation during upright tilt is mediated by impaired baroreceptor responses, augmented sympathetic neural traffic, and neural and humoral beta-adrenergic mechanisms.

We are not aware of any direct measurements of muscle sympathetic nerve activity in heart failure patients during upright tilt. Therefore it is not known if sympathetic nerve traffic increases normally, increases excessively, or decreases in heart failure patients during upright tilt. There is some reason to suspect that heart failure patients may not experience normal increases of muscle sympathetic nerve activity when they stand. In humans, directly recorded sympathetic traffic is usually expressed in terms of bursts per unit time or bursts per 100 heart beats. Healthy subjects can increase muscle sympathetic nerve activity by three mechanisms: they can increase the number of bursts per 100 heart beats, the number of heart beats per minute, or the sympathetic activity per burst. Burke *et al.* (1977) showed that healthy subjects who have high incidences of sympathetic bursts may not increase burst incidence further during standing; these subjects increase total sympathetic activity by increasing heart rate. Such mechanisms may not be helpful in patients with heart failure who may have an unfortunate combination of an extremely high incidence of sympathetic bursts (Leimbach *et al.* 1986; Ferguson *et al.* 1989; Porter *et al.* 1990), and an inability to increase heart rate during standing (Howard and Leathart 1951). Such patients may be unable to increase their sympathetic activity during standing by increasing either burst incidence or heart rate.

REDUCED EXERCISE CAPACITY

Exercise intolerance is one of the important hallmarks of congestive heart failure. Since there is strong evidence that normal arterial baroreflex mechanisms modulate haemodynamic responses of healthy people to exercise (see Chapter 7), it is likely that abnormal baroreflex mechanisms contribute to pathophysiological responses of patients with heart failure. This possibility in no way denies participation of other mechanisms, including disordered skeletal muscle energetics (Massie *et al.* 1988). The following discussion reviews evidence that may implicate baroreflexes in the reduced exercise tolerance of heart failure patients.

Heart rate responses

Abnormal baroreflex function may impair heart rate responses to exercise. Healthy people may increase their cardiac output during exercise more by increasing heart rate than by increasing stroke volume. The contributions of heart rate and stroke volume changes in healthy subjects to the increase of cardiac index during exercise are shown in Fig. 14.7. Stroke volume reaches its maximum level after about 40 per cent of maximum oxygen consumption has been achieved; heart rate, however, continues to increase until maximum oxygen consumption has been attained (Åstrand *et al.* 1964). Thus in healthy people, exercise-induced elevations of cardiac output, particularly at high levels of exercise, result primarily from cardioacceleration.

As with healthy subjects, heart failure patients can increase their cardiac outputs during exercise only by increasing their heart rates and stroke volumes. (However, patients may compensate partly for failure of cardiac output to increase adequately during exercise by increasing oxygen extraction and resorting to anaerobic metabolism earlier than healthy people (Zelis *et al.* 1974*b*; Longhurst *et al.* 1976; J.R. Wilson *et al.* 1984).) Heart rate speeding may be particularly important to these patients, however, because they may have a reduced ability to increase their stroke volumes adequately during exercise (K.T. Weber *et al.* 1982).

In some patients with heart diseases, maximal heart rates during standard exercise tests are lower than those in healthy subjects (Bruce *et al.* 1973; Litchfield *et al.* 1982; Colucci *et al.* 1989). In other patients, normal heart rate responses may be preserved, and exercise may be normal (Litchfield

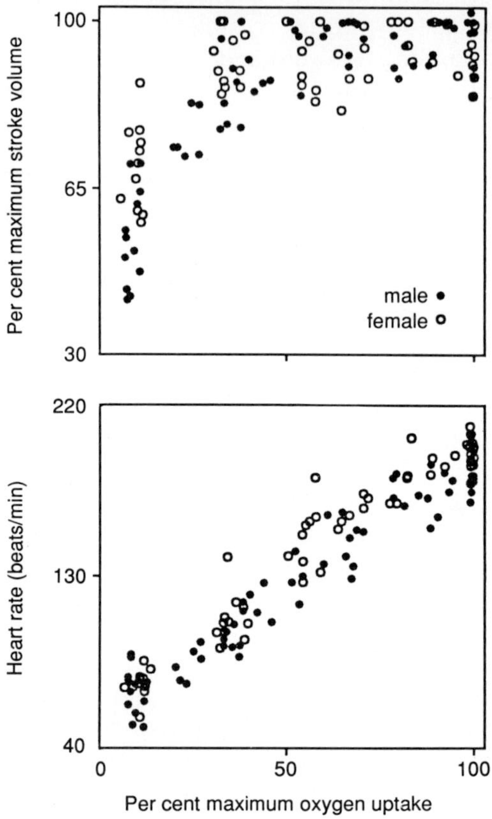

FIG. 14.7. Relation between oxygen uptake and maximum stroke volume and heart rate plotted from the beginning of aerobic exercise in healthy subjects. Adapted with permission (Åstrand *et al.* 1964).

et al. 1982). There is some evidence that heart rate responses to lesser degrees of exertion are impaired in heart failure patients. This issue is controversial, however, and the practical importance of patients' impaired heart rate responses to exercise is ill-defined. Interpretation of published evidence depends in part on how heart rate changes are characterized. If responses are portrayed in terms of the change of heart rate for each standard workload, heart rate increases during exercise may be exaggerated in heart failure patients (Francis *et al.* 1985; McElroy *et al.* 1988). However, standard workloads may present inordinately large challenges to heart failure patients; therefore, it may be more appropriate to characterize heart rate responses of patients according to the challenge presented to them (expressed as per cent of maximum workload) rather than according to the standard workload. Such comparisons document impaired heart rate responses to exercise in heart failure patients (Francis *et al.* 1985).

One way in which baroreflex abnormalities might impair cardioacceleration during exercise is through establishment of subnormal baseline levels of vagal–cardiac nervous outflow (see Chapter 8). The earliest heart rate speeding that occurs in healthy people during exercise results from vagal withdrawal, since it is not associated with increases of plasma catecholamine levels (Christensen and Brandsborg 1973) or muscle sympathetic nerve activity (Mark *et al.* 1985), and is abolished by pretreatment with large doses of atropine (Freyschuss 1970; C.E. Martin *et al.* 1974). Patients with cardiovascular diseases appear to have subnormal resting levels of vagal–cardiac nerve activity. Their resting blood pressures lie in the threshold region of the baroreceptor stimulus–sinus node response relation (Fig. 14.2 (Sopher *et al.* 1990)), they have lower resting mean R–R intervals and standard deviations, and they experience subnormal cardioacceleration after large doses of atropine (Müller 1891; Crawford 1923; Eckberg *et al.* 1971; Porter *et al.* 1990). Obviously, if vagal restraint is less before exercise, there will be less vagal withdrawal during exercise, even though mechanisms mediating vagal withdrawal might be intact. Impairment of heart rate speeding at the beginning of exercise may be particularly important to patients with severe heart failure who may be unable to progress beyond the earliest stages of exercise.

Abnormal baroreflex function may impair cardioacceleration during exercise in patients with normal levels of vagal–cardiac activity if mechanisms mediating vagal withdrawal are faulty. Although it is not known with certainty whether changes of baroreflex function contribute to early cardioacceleration during exercise in healthy subjects, some evidence supports the possibility. At resting arterial pressures, baroreflexes contribute to levels of vagal–cardiac activity, since intravenous nitroglycerine (T.G. Pickering *et al.* 1972*c*) and brief neck pressure (which is perceived as reduction of arterial pressure (Eckberg 1980*a*)) cause cardioacceleration. If resting levels of vagal–cardiac activity are determined by baroreceptor input, subnormal heart rate responses to exercise could result in part from impaired baroreflex mediated vagal withdrawal. In healthy people, exercise reduces R–R interval prolongations triggered by arterial pressure elevations (J.D. Bristow *et al.* 1971*b*), electrical carotid sinus nerve stimulation (Eckberg *et al.* 1972), and neck suction (Eckberg and Wallin 1987). Since changes of R–R intervals are related linearly to changes of vagal–cardiac nerve activity (Katona *et al.* 1970; Parker *et al.* 1984; Kollai and Koizumi 1989), exercise may impair baroreflex control of vagal activity. Although we are not aware that the possibility has been examined, heart failure may interfere with normal exercise-induced suppression of vagal–cardiac outflow.

Abnormal heart rate responses to exercise could result also from deranged pre- and post-synaptic sympathetic–parasympathetic interactions. Subnormal levels of vagal–cardiac nerve traffic, with presumed subnormal levels of acetylcholine in the region of the sinus node, may enhance noradrenaline release from sympathetic nerve terminals through a presynaptic mechanism (Löffelholz and Muscholl 1970; Lavallee *et al.* 1978; Vanhoutte and Levy 1980). This interaction may work also in the opposite direction: high levels of alpha-1-adrenergic agonists may reduce acetylcholine release (Wetzel and Brown 1985; Wetzel *et al.* 1985) and vagally mediated bradycardia (McGrattan *et al.* 1987). Finally, chronic high levels of sympathetic stimulation associated with elevated heart rates may shift the time of arrival of phasic vagal volleys to a position within the cardiac cycle where their cardioinhibitory actions are less effective (Stuesse *et al.* 1981).

As noted in Chapter 7, there is not universal agreement that exercise interferes with baroreflex mediated sinus node control in healthy subjects. Some investigators (Walgenbach and Donald 1983; Rowell 1986) argue that heart rate, not R–R interval, is the physiologically relevant variable, and that when changes of sinus node function are expressed as heart rate changes, baroreflex control of sinus node function during exercise is normal. The reasons for choosing R–R interval over heart rate as an index of vagal–cardiac nerve activity have been discussed by J.D. Bristow *et al.* (1971*b*), Cunningham *et al.* (1972*a*), and Eckberg and Wallin (1987). The question may not be of overriding importance, however, because if exercise is intense enough, baroreceptor stimulation provokes neither R–R interval nor heart rate changes (Bevegård and Shepherd 1966*a*; McRitchie *et al.* 1976). Moreover, peak cardiac responses of heart failure patients may be abnormal whether they are expressed as R–R interval or heart rate changes (Colucci *et al.* 1989).

Cardioacceleration occurring later during exercise appears to be due in part to increased sympathetic activity (C.E. Martin *et al.* 1974; Hume *et al.* 1979). Supranormal levels of sympathetic outflow may lead to down-regulation of beta-adrenergic cardiac receptors (Fowler *et al.* 1986), reductions of regulatory guanine nucleotide regulatory proteins (Horn *et al.* 1988), and impairment of ionotropic (Colucci *et al.* 1988) and chronotropic (K.T. Weber *et al.* 1982; Francis *et al.* 1985; Colucci *et al.* 1989) responses to beta-agonists. The strikingly supranormal resting levels of sympathetic activity found in some heart failure patients may arise in part from baroreflex dysfunction (see Chapter 8). It is not known whether high levels of sympathetic outflow in heart failure patients are as susceptible to baroreflex modulation as much lower levels are in healthy people.

Blood pressure changes

As indicated in Chapter 9, reflex control of blood pressure (which some authors consider as interchangeable with reflex control of sympathetic activity) is complex, and has not been studied as exhaustively as reflex control of heart rate. In healthy people, blood pressure increases moderately during exercise (Åstrand *et al.* 1964; Bruce *et al.* 1973). In most patients with heart disease, blood pressure also increases during exercise, but not to the same extent as it does in healthy people (Bruce *et al.* 1973). However, in some patients, including (but not limited to) those with severe aortic stenosis (L.S. Schwartz *et al.* 1969; Mark *et al.* 1973), coronary artery disease (Levites *et al.* 1978; Sanmarco *et al.* 1980; Weiner *et al.* 1982), and hyponatraemia (W.H. Lee and Packer 1986), blood pressure may *fall* during exercise. In these patients, blood pressure falls partly because increases of cardiac output are limited by heart disease. Since blood pressure = flow x resistance, inadequate flow or resistance (or both) may lead to blood pressure reductions during exercise in these patients. If adequate perfusion pressure is not maintained during exercise, pressure-dependent flow to the brain may fall to unacceptably low levels, and syncope may occur. Thus in some patients high vascular resistance during exercise may have a salutary effect and prevent hypotension, despite inadequate augmentation of cardiac output.

However, most patients with heart failure do not develop hypotension or syncope during exercise. In the majority of patients with mild-to-moderate heart failure, the primary exercise-related symptom resulting from their heart failure is *fatigue*, which results in part because increases of blood flow fail to keep pace with increases of oxygen requirements of exercising muscle (J.R. Wilson and Ferraro 1983; J.R. Wilson *et al.* 1984, 1989*b*). As discussed above, inadequate increases of muscle blood flow during exercise may result from subnormal increases of stroke volume and heart rate in response to exercise, and these deficiencies may be countered only partly by increased tissue extraction of oxygen (Zelis *et al.* 1974*b*). Blood flow may also be reduced by reductions of arterial pressure; however, one excellent study of stable heart failure patients (Sullivan *et al.* 1989) documented normal pressure responses to exercise.

An additional mechanism that may contribute to exercise intolerance in heart failure patients is impairment of muscle blood flow secondary to excessive muscle vascular resistance. The potential importance of vascular resistance to blood flow in exercising muscle can be appreciated when the relation: blood pressure = flow × resistance is rearranged to: flow = blood pressure/resistance.

Vascular resistance

Inordinately large vascular resistance may contribute to impaired muscle blood flow in heart failure patients. (In healthy people, neurohumoral vasoconstriction may actually override metabolic vasodilatation in working muscles during maximum exercise (Rowell and Blackmon 1987; Sinoway *et al.* 1988).) In patients with severe heart failure, limb vascular resistance may be abnormally high and limb blood flow may be abnormally low at rest

(Hewlett and van Zwaluwenburg 1909; K.W. Donald *et al.* 1955; Zelis *et al.* 1968; Longhurst *et al.* 1976; Leithe *et al.* 1984). In patients with mild heart failure, however, limb vascular resistance may be normal at rest but may rise excessively with exercise (K.W. Donald *et al.* 1955; J.R. Wilson *et al.* 1989*a*). Thus abnormal resistance may contribute to symptoms in patients with either mild or severe heart failure. A related explanation for the inordinate fatigue that heart failure patients experience may be that some (but not all (K.W. Donald *et al.* 1955)) patients do not redistribute blood flow from non-working to working muscles, as healthy people do (J.R. Wilson *et al.* 1989*a*; Sullivan *et al.* 1989).

Heart failure patients also have neurally mediated increases of vascular resistance; such increased resistance may have salutary effects if it prevents exercise-induced hypotension. However, neurogenically mediated excess vascular resistance may contribute to exercise intolerance if it limits muscle blood flow inappropriately. During exercise, vessels in working muscles dilate (and thereby reduce resistance and increase flow) because of accumulation of metabolic products, including adenosine (Tominaga *et al.* 1980). As mentioned, during maximal exertion or during prolonged ischaemia, powerful forces promoting metabolic vasodilatation tend to overcome neural vasoconstrictor activity (Zelis *et al.* 1973). However, with lesser degrees of exertion, a competitive balance exists between neurohumoral vasoconstriction and metabolic vasodilatation (Remensnyder *et al.* 1962; Strandell and Shepherd 1967).

Thus during severe exertion, neural vasoconstriction in exercising muscles is largely overcome by metabolic factors, and does not limit muscle blood flow. (However, in heart failure patients, fixed changes, including increased vascular stiffness due to increased sodium content (Zelis *et al.* 1974*a*) are not overcome by metabolic vasodilatation, and therefore continue to limit muscle blood flow (Zelis *et al.* 1973).) J.R. Wilson *et al.* (1985) studied the effects of intra-arterial alpha-adrenergic blockade on leg blood flow during bicycle exercise in patients with heart failure. They concluded, as did Zelis *et al.* (1973) before them, that neurogenic vasoconstrictor activity does not limit blood flow to exercising muscles in heart failure patients. However, both groups studied only the effects of maximal exercise on neurogenic vasoconstriction. Since the importance of local metabolic factors is related directly to the intensity of exercise, Wilson's data do not exclude the possibility that neurogenic vasoconstriction may limit muscle blood flow during mild exertion. As far as we know, this possibility has not been evaluated. Such a mechanism may have great clinical importance, because in patients with severe heart failure, symptoms may preclude all but only mild or moderate exertion.

These considerations raise the possibility that supranormal levels of sympathetic vasoconstrictor activity limit muscle blood flow during exertion (at least during mild exertion). We next consider the evidence that baroreflex mechanisms modulate sympathetic outflow during exercise, and that abnormalities of reflex control might contribute to exercise intolerance in patients with heart failure.

In dogs with baroreceptor denervation (but not dogs with intact reflexes), blood pressure falls transiently at the beginning of exercise (Higgins *et al.* 1972*b*; Melcher and Donald 1981); this would be expected to increase sympathetic outflow. Melcher and Donald (1981) studied responses of instrumented conscious dogs to steady-state carotid arterial pressure changes during dynamic exercise. They showed that although the inverse sigmoid carotid pressure–blood pressure relation was shifted on its pressure axis during exercise (to operate in a higher pressure range), it was quantitatively unaltered. Others showed that exercise reduces blood pressure responses to acute changes of carotid baroreceptor activity. Faris *et al.* (1982) documented exercise-induced reductions of blood pressure responses to carotid pressure changes in conscious rabbits, and T.J. Ebert (1986) documented reduction of blood pressure responses to changes of carotid distending pressure (neck pressure and suction) in conscious human volunteers.

In one study, carotid baroreflex control of directly recorded muscle sympathetic nerve activity was evaluated during handgrip in conscious healthy human subjects. Eckberg and Wallin (1987) showed that increments of sympathetic activity resulting from carotid compression with neck pressure are smaller, and decrements of sympathetic activity resulting from carotid stretch with neck suction are greater, during exercise than during rest. Taken together, these observations (the only extant measurements of baroreflex control of directly measured muscle sympathetic activity during voluntary exercise) suggest that in healthy people, baroreflex mechanisms work in ways that limit increases of sympathetic activity during exercise.

In healthy subjects, plasma noradrenaline concentrations (Christensen and Brandsborg 1973; Lake *et al.* 1976; Vecht *et al.* 1978; Blomqvist *et al.* 1981) and muscle sympathetic nerve activity (Mark *et al.* 1985; M. Saito *et al.* 1986) begin to increase about 1 min after the onset of isometric exercise. As with heart rate responses (see above), conclusions drawn regarding the normality of sympathetic responses of heart failure patients to exercise depends in part on how data are interpreted. Early studies suggested that plasma noradrenaline concentrations rise excessively in heart failure patients during exercise (Chidsey *et al.* 1962). Subsequent studies expressed plasma noradrenaline concentrations in relation to the maximal amount of exercise that could be performed. When the amount of exercise is expressed in terms of absolute oxygen consumption, heart failure patients experience inordinately large increases of plasma noradrenaline concentrations (Francis *et al.* 1982*b*, 1985). However, since these patients cannot perform as much exercise as healthy people, their absolute noradrenaline concentrations may not rise as high during exercise as those of healthy people. It is possible that heart failure patients are challenged more than healthy subjects by a given level of exercise (although the 'challenge' posed by an exercise load may be difficult to quantify). If this be true, it may be appropriate to characterize exercise responses in relative, rather than absolute terms.

Figure 14.8 shows plasma noradrenaline responses of healthy subjects and patients with heart failure during exercise, plotted according to absolute and relative workloads. In this figure, group 1 comprises subjects who are healthy and groups 2 and 3 comprise subjects with moderate and severe heart failure. Figure 14.8 (a) documents an excessive increase of plasma noradrenaline per unit of exercise in heart failure patients, and supports the conclusion of Chidsey *et al.* (1962). Since levels of plasma noradrenaline during exercise parallel levels of directly recorded muscle sympathetic activity (Wallin *et al.* 1987), excessive increases of noradrenaline levels in heart failure patients during exercise probably point towards excessive increases of sympathetic vasoconstrictor activity.

If baroreflex mechanisms operate to limit vasoconstrictor outflow to the

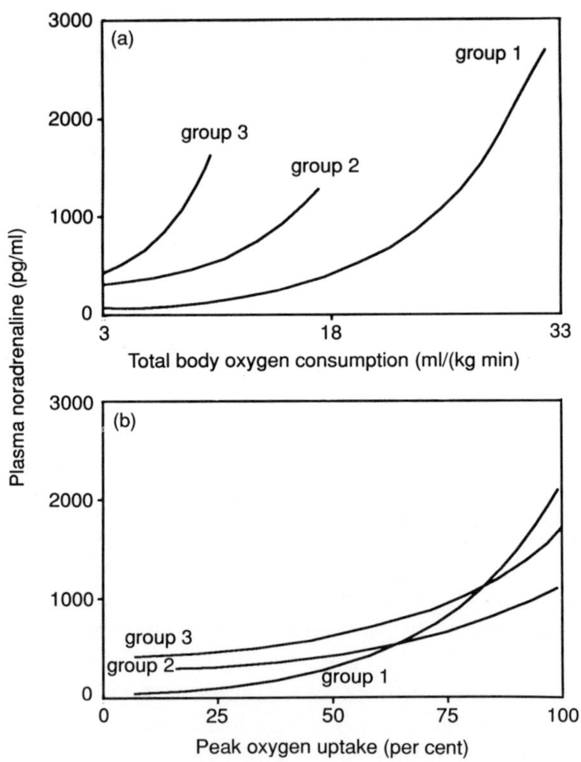

FIG. 14.8. Plasma noradrenaline responses to exercise, plotted as functions of (a) absolute and (b) relative work intensities. Adapted with permission (Francis *et al.* 1985).

muscle vascular bed during exercise (as Fig. 7.16 suggests), and if baroreflex mediated, exercise-related sympathetic reflexes are abnormal in heart failure patients (we are not aware that they have been studied), then abnormal baroreflexes may contribute to excess sympathetic vasoconstriction during exercise.

Neurohumoral factors which may be subject to baroreflex control (see below) contribute to this abnormal resistance but do not explain it fully. There are also fixed non-adrenergic factors that increase vascular resistance and importantly limit muscle blood flow in heart failure patients. Zelis *et al.* (1968) studied vasodilatation resulting from exercise, ischaemia (reactive hyperaemia), and other interventions in healthy subjects and patients with heart failure. The degree of vasodilatation after these interventions was not affected by intra-arterial adrenergic blockade or local nerve anaesthesia, and therefore was not neurally mediated. In this study, non-neural forearm dilatation was reduced sharply in patients with heart failure. The authors concluded that factors other than neural vasoconstriction are partly responsible for the reduced maximum vasodilatation seen in heart failure. The study of J.R. Wilson *et al.* (1989*a*) suggests that since heart failure patients may experience huge increases of plasma renin activity during exercise, some of their exercise-induced, non-adrenergic vasoconstriction may be caused by angiotensin-II.

Other factors may also contribute to impairment of non-neural vasodilatation in heart failure patients. First, vessels may have increased stiffness due to increased sodium and water content. This possibility is supported indirectly by the improvement of vasodilator responses that occurs in heart failure patients after diuresis (Zelis *et al.* 1968). This possibility is supported also by studies conducted in dogs with heart failure due to prolonged rapid ventricular pacing, which were found to have increased sodium content of large and small arteries (Zelis *et al.* 1970), and by studies conducted in healthy human subjects given salt loads and mineralocorticoids, who increased their vascular stiffness and reduced their reactive hyperaemia responses (Zelis and Mason 1970). Finally, a study reported in preliminary form (Drexler *et al.* 1991), points towards impaired endothelial vasodilatory mechanisms in heart failure patients.

SUMMARY

Clinical and even preclinical heart failure is associated with major abnormalities of autonomic cardiovascular control. These include subnormal levels of baseline vagal–cardiac nervous traffic and responsiveness to baroreceptor stimulation, and supranormal levels of baseline sympathetic nervous traffic. Other biochemical evidence for deranged autonomic control includes elevations of the circulating vasoconstrictors angiotensin, aldosterone, vasopressin, neuropeptide Y, and endothelin, and of the circulating vasodilators atrial natriuretic factor and prostaglandins. The extent to which baroreflex abnormalities is responsible for altered neurohumoral cardiovascular control in heart failure is uncertain. The mechanisms responsible for baroreflex impairment are not well understood; it is clear, however, that impairment of baroreceptor transduction occurs. Baroreflex function can be restored towards normal by medical therapy or cardiac transplantation; this suggests that baroreflex impairment in heart failure is not simply due to irreversible structural abnormalities.

15

Carotid sinus reflex hypersensitivity

I didn't know sharks had such handsome, beautifully formed tails.
Ernest Hemingway, *The Old Man and the Sea*

CLINICAL ASPECTS

The final chapter of this book deals with a much discussed but nevertheless poorly understood condition described originally by Roskam (1930), and popularized by Weiss and co-workers (Weiss and Baker 1933; Ferris *et al.* 1935) and Sigler (1941, 1942), who reported the occurrence of asystole or high-grade atrioventricular block, arterial hypotension, and symptoms of presyncope or syncope during unilateral carotid sinus massage. Several excellent reviews of this enigmatic medical problem have been published (Marcel 1956; Franke 1963; J.E. Thomas 1972; Morley and Sutton 1984; Strasberg *et al.* 1989). The review of Morley and Sutton (1984) is particularly commended for its discussion of pathophysiology and natural history, and for its survey of modern treatment methods.

Subjects with carotid sinus reflex hypersensitivity are distinguished from patients with carotid sinus syndrome. Although both groups have abnormal responses to carotid sinus massage, patients with carotid sinus *syndrome* also have spontaneous attacks of presyncope or syncope.

Carotid sinus reflex *hypersensitivity* is classified arbitrarily according to its predominant haemodynamic manifestation. *Cardioinhibitory* carotid sinus hypersensitivity is the most common type (accounting for as many as 80 per cent of cases), and is usually defined by asystole lasting three or more seconds during no more than five seconds of carotid massage. Asystole is usually due to arrest of all cardiac pacemakers; although concurrent atrioventricular block is common, asystole is rarely due to isolated high-grade atrioventricular block. Indirect sinoatrial node recordings suggest that what appears to be sinus arrest is in fact sinoatrial exit block (Reiffel and Bigger 1983). *Vasodepressor* carotid sinus reflex hypersensitivity (without cardioinhibition), however, occurs uncommonly, in perhaps 5 to 10 per cent of cases, and is defined by systolic pressure reductions of at least 50 mmHg in asymptomatic subjects, or 30 mmHg in patients with presyncope or syncope during carotid massage. A third category is defined by the occurrence of both asystole and systolic pressure reduction during carotid massage. A fourth, *cerebral*, category — presyncope or syncope without asystole or systolic pressure reduction — has been abandoned; in this situation, neurological symptoms are thought to result from interruption of cerebral arterial flow in patients with coexistent obstruction of other cerebral arteries.

The above categorization of carotid hypersensitivity already betrays ambiguities that permeate this literature. First, published records made during carotid sinus massage show that subjects who have primarily vasodepression also have bradycardia, if not outright asystole. Therefore 'cardioinhibition' is a quantitative rather than qualitative descriptor. Second, it is difficult to imagine that carotid massage sufficient to cause asystole does not also trigger blood pressure reductions. Therefore the diagnosis of vasodepression is also quantitative rather than qualitative; it is made when blood pressure falls below some arbitrarily defined limit. Third, since asystole lowers blood pressure, the diagnosis of vasodepression must be made independently of the confounding influence of asystole. The use of large doses of atropine to abolish heart rate slowing during diagnostic carotid massage has largely been abandoned. Electrical cardiac pacing is a more widely accepted method to prevent confounding effects of asystole. Since atrioventricular asynchrony may lower blood pressure, however, atria and ventricles should be paced sequentially (Morley *et al.* 1982). Obviously, sequential atrioventricular pacing adds greatly to the complexity of what is regarded as a non-invasive bedside diagnostic procedure.

A major problem with carotid sinus syndrome as a clinical entity is that it is extremely difficult to document the occurrence of spontaneous attacks. Because of this, diagnosis depends on use of provocative manoeuvres. There is general agreement on the *method* of carotid massage. (However, this is not to say that the degree of baroreceptor stimulation during carotid massage is reproducible or even quantifiable from subject to subject.) Cerebral embolization is a complication of carotid massage (see below); therefore, a history of transient ischaemic attacks, or physical findings for partial or total carotid occlusion contraindicate the procedure. Carotid massage is performed during continuous electrocardiogram monitoring. Most carotid massage is performed with patients supine; however, some workers perform carotid massage with patients sitting (Sigler 1941) or even standing (the position in which most attacks occur (Hammill *et al.* 1984)). Massage is performed at the level of the cricothyroid cartilage (over the carotid pulsation) for 5 s (Hudson *et al.* 1985) at a time, on only one side. Since some patients may be exquisitely sensitive, carotid massage should be initiated with very gentle pressure. Care should be taken to avoid carotid occlusion; this may provoke cerebral symptoms in patients with occlusion of other cerebral vessels, and moreover this will reduce rather than increase carotid sinus nerve afferent traffic (Roddie and Shepherd 1957). (Carotid occlusion does not elicit symptoms or signs in patients who have the carotid sinus syndrome (Ferris *et al.* 1935).) The simplest way to determine whether carotid occlusion occurs during carotid massage is to palpate the temporal artery; a more involved method is to record supraorbital artery flow with a Doppler device.

Although carotid massage is generally considered safe, major complications have been reported; these include cerebral embolization and stroke (Beal *et al.* 1981; Bastulli and Orlowski 1985), and atrial (Strasberg *et al.* 1989) and ventricular fibrillation and death (Lown and Levine 1961; Greenwood and Dupler 1962; Bastulli and Orlowski 1985). Of interest is that most patients who have had ventricular fibrillation during carotid massage have been taking digoxin. A link between cholinergic outflow and digoxin toxicity has also been documented in experimental animals; atropine pretreatment prevents digoxin-toxic ventricular fibrillation in conscious stressed guinea pigs (Cagin and Natelson 1981).

Figure 15.1 depicts responses of a 57-year-old man with otherwise unexplained syncopal attacks brought on by shaving or unusual head movements, who had profound haemodynamic responses to very gentle right carotid massage. We believe this to be the only extant recording of muscle sympathetic nerve activity during carotid massage performed in a patient who fulfils

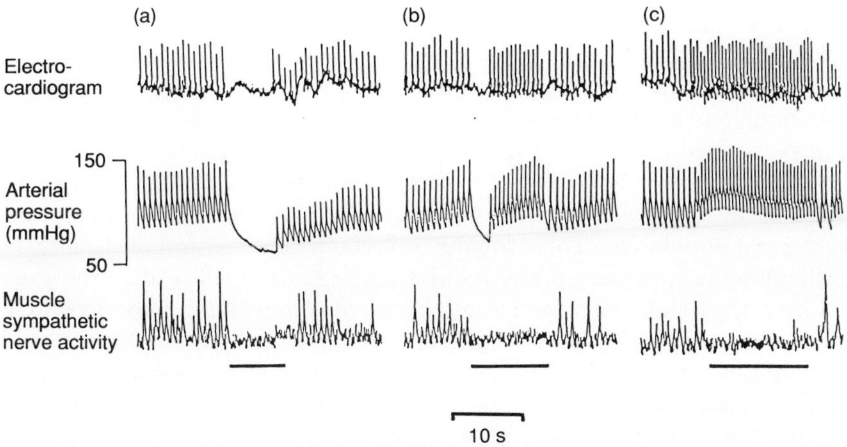

FIG. 15.1. Responses of one patient with carotid sinus syndrome (a) to carotid massage, and to right ventricular pacing begun (b) during and (c) before and during carotid massage. (M.L. Smith, K.A. Ellenbogen, and D.L. Eckberg, unpublished).

diagnostic criteria for carotid sinus syndrome. The record depicts responses to carotid massage performed three times. Figure 15.1 (a) shows this patient's responses when carotid massage was performed with no other interventions. During carotid massage, the patient had asystole for the duration of massage, profound arterial hypotension, and interruption of muscle sympathetic nerve activity. Figure 15.1 (b) shows responses when carotid massage was interrupted after about 4 s by right ventricular pacing. Under this circumstance, arterial pressure returned promptly to normal, notwithstanding continued absence of muscle sympathetic nerve activity. Figure 15.1 (c) shows responses when rapid ventricular pacing began before and continued during carotid massage. During pacing, arterial pressure did not fall, even though sympathetic traffic was absent. This patient had *'cardioinhibitory'* carotid sinus reflex hypersensitivity.

The incidence of carotid sinus syndrome and reflex hypersensitivity is not known with certainty. Carotid sinus *syndrome*, with presyncope or syncope triggered by head movement, shaving, or tight collars is rare. In one prospective study of 204 patients with syncope (Kapoor *et al.* 1983), only one patient was diagnosed as having carotid sinus syndrome. Morley *et al.* (1982) estimate that the number of new diagnoses of carotid sinus syndrome is 35 per million population per year. (Leatham (1982) suggests that this estimate may be high because it includes patients with simple sinoatrial node disease.)

Carotid sinus reflex *hypersensitivity* is much more common. As with many cardiovascular diagnostic procedures, the incidence of abnormal responses to carotid sinus massage depends very much on the population studied. As early as 1932, Mandelstamm and Lifschitz (1932) showed that carotid sinus reflex hypersensitivity is much more common in older than in younger subjects; they reported average reductions of systolic pressure of 37 mmHg in aged Russian pensioners, but only 5 mmHg in young Russian soldiers. Sigler (1942) suggests that as many as 10 per cent of the adult population have carotid sinus reflex hypersensitivity!

Although series have been published in which carotid sinus massage has been performed in hundreds or even thousands of subjects (by 1941, Sigler had reported responses of 2312 patients (Sigler 1936, 1941)), we are not aware of any studies that define the sensitivity and specificity of this test in the general population. This probably reflects the imprecision of the method itself, and the absence of any 'gold standard'. It is clear that the great majority of patients diagnosed as having carotid sinus reflex hypersensitivity do not experience presyncope or syncope, and most people who have presyncope or syncope do not have carotid sinus reflex hypersensitivity. This is underscored by the prospective study of K.A. Brown *et al.* (1980) conducted in patients undergoing coronary arteriography. They diagnosed carotid sinus reflex hypersensitivity in 32 per cent of patients; however, the incidence of abnormal responses was the same in patients with and without histories of presyncope or syncope. Although as many of 63 per cent of patients diagnosed as having carotid sinus syndrome experience spontaneous remissions (Sugrue *et al.* 1986), some patients continue to have symptoms during long-term follow-up (Walter *et al.* 1978; Sugrue *et al.* 1986).

In earlier times, given the vagaries of diagnosis and the variable natural history, available *treatment* of carotid sinus syndrome was generally unsatisfactory. Use of anticholinergic drugs has been abandoned. Since most patients fall into the 'cardioinhibitory' category, excellent results have been obtained with electronic cardiac pacemakers. Simple atrial pacing should not be used, because of the presence of atrioventricular node disease in many elderly patients and the occurrence of atrioventricular block in patients with carotid sinus reflex hypersensitivity, whose sinus rate is controlled. The ideal treatment involves use of demand sequential atrioventricular pacemakers. Morley *et al.* (1982) reported use of pacemakers in a large series of patients. Fully 89 per cent of patients were asymptomatic with appropriate cardiac pacing; vasodepression was common in patients who remained symptomatic. The treatment of residual vasodepression in carotid sinus syndrome may not be satisfactory (Almquist *et al.* 1985); mineralocorticoids, ergotamines, and adrenergic agonists have been used with mixed results and high incidences of side-effects (Almquist *et al.* 1985; McLaran *et al.* 1986).

PATHOPHYSIOLOGY

Perhaps the most compelling evidence that exaggerated baroreceptor input can provoke pathological baroreflex responses comes from patients with glossopharyngeal neuralgia who have throat pain when they yawn or swallow that appears to trigger bradycardia and hypotension (Riley *et al.* 1942). Wallin *et al.* (1984*b*) published the only extant recording of muscle sympathetic nerve activity during a spontaneous episode of glossopharyngeal neuralgia. This record, shown in Fig. 15.2, documents cardiac slowing, cessation of sympathetic activity, and extreme hypotension during one attack. The mechanism responsible for this condition is unclear; the authors postulate that a surge of afferent nerve activity from a trigger zone in the pharynx spreads by means of an ephapse (a point of lateral, non-synaptic contact between neurones) and thereby augments traffic carried over afferent baroreceptor fibres. This report establishes the likelihood that augmentation of afferent carotid baroreceptor activity may occur pathologically. However, this case does not necessarily define a mechanism for carotid sinus syndrome, which, as discussed below, may be associated with *subnormal*, not exaggerated, afferent baroreceptor input.

Two prospective studies from one group implicate abnormal baroreflexes in the pathogenesis of the carotid sinus syndrome. Dehn *et al.* (1984) used graded neck suction to evaluate ten patients with spontaneous syncope and three or more seconds of asystole during carotid massage. Patients' responses

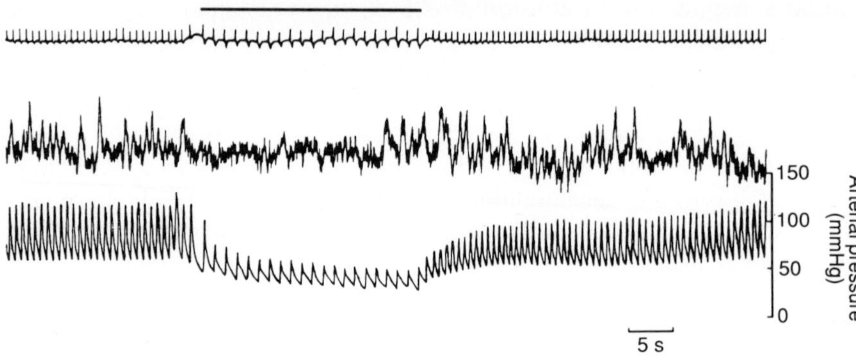

FIG. 15.2. Measurements obtained in one patient during a spontaneous attack of glossopharyngeal neuralgia. Reproduced with permission (Wallin *et al.* 1984*b*).

were compared with responses of a group of 16 other patients, which included seven patients with syncope but normal responses to carotid massage. Patients with carotid sinus reflex hypersensitivity had average baroreflex slopes (R–R interval prolongation/neck pressure change) of 12 ms/mmHg (vastly higher than responses of healthy subjects reported in the literature), and control subjects had slopes of 4 ms/mmHg ($P < 0.02$). In a second study, Morley *et al.* (1984) reported similarly exaggerated responses to baroreceptor stimulation by bolus intravenous pressor injections. These results document supranormal responsiveness to baroreceptor stimulation in patients diagnosed as having carotid sinus syndrome, but they do not delineate where within the reflex arc the abnormality arises.

Several studies implicate pathology in the region of the carotid sinus. Three out of 15 patients described by Weiss and Baker (1933) had tumours 'pressing' on the carotid sinus. Other patients have had locally infiltrating malignant neck tumours (Cicogna *et al.* 1985; Onrot *et al.* 1987). It has been apparent since the earliest reports (Mehrmann 1925; Mandelstamm and Lifschitz 1932; Weiss and Baker 1933; Ferris *et al.* 1935) that subjects with carotid sinus reflex hypersensitivity have an unusually high incidence of generalized atherosclerosis. K.A. Brown *et al.* (1980) reported highly significant correlations between the severity of coronary artery disease or anginal symptoms and heart rate and blood pressure responses to carotid massage. The report of Martens *et al.* (1981) is of great interest. These authors measured responses to carotid massage in a group of patients early and late after acute myocardial infarction. Abnormal responses were present in 29 per cent during the first 48 h, and in 10 per cent at 4 weeks. Carotid sinus reflex hypersensitivity was four times as likely in patients with inferior as in those with anterior infarction. These results suggest that structures other than the carotid sinus may participate in the pathophysiology of carotid sinus reflex hypersensitivity. The belief that the carotid sinus syndrome arises from the carotid sinus itself prompted local (largely abandoned) therapies, including irradiation (Greeley *et al.* 1955) and carotid sinus nerve transection (Carlsten *et al.* 1959).

Mechanisms responsible for carotid sinus reflex hypersensitivity are not known with certainty. Is there a common denominator in this disparate patient group? It seems unlikely that tumours *stimulate* baroreceptors mechanically (even though they are described as 'pressing' on the carotid sinus). Similarly, carotid sinus rigidity in patients with atherosclerosis would be unlikely to increase afferent baroreceptor firing. We suspect that *subnormal* afferent baroreceptor activity is more likely than supranormal activity in these conditions. Mandel *et al.* (1972) and Leatham (1982) propose that in many patients, carotid sinus syndrome is a manifestation of sinoatrial node disease, which is common in elderly patients. Morley *et al.* (1983) dispel this notion in a study reported in preliminary form. They found that electronically paced patients with the diagnosis of carotid sinus syndrome had significantly shorter sinoatrial node recovery times, a smaller incidence of atrial tachydysrhythmias, and a faster average heart rate after large dose atropine and propranolol than had electronically paced patients with sinoatrial node disease.

We offer the following speculation to explain the pathophysiology found in the disparate group of patients who carry the diagnosis of carotid sinus syndrome. We suggest that this condition is due to dynamic factors rather than to discrete anatomical lesions. The variable incidence of carotid sinus reflex hypersensitivity in the same patients studied serially after acute myocardial infarction, and in the different groups of patients studied after inferior and anterior infarctions (Martens *et al.* 1981), speaks against variability of carotid sinus anatomy. We propose that *carotid sinus reflex hypersensitivity may be due to abnormal central nervous system processing of afferent baroreceptor information.* In patients with local carotid sinus disease (including those with tumours and atherosclerosis), baroreceptors may be splinted and may be relatively unresponsive to normal arterial pulsations. In these patients, baroreceptors are not destroyed, however, and can respond to unphysiological manual deformation. Chronically subnormal levels of carotid baroreceptor afferent input lead to exaggerated importance of artificially induced barrages of baroreceptor activity, secondary to central nervous system plasticity (Milgram *et al.* 1987). The mechanism of carotid sinus reflex hypersensitivity may be slightly different in patients with acute myocardial infarction. In these patients, afferent input from the heart may in some way magnify responses to carotid baroreceptor input.

SUMMARY

The above discussion of carotid sinus reflex hypersensitivity is meant to convey our scepticism about this clinical entity. None the less, we suggest that the extensive literature on patients with syncope and exquisitely sensitive carotid sinus reflexes cannot be discounted altogether; we believe that the conclusion that *pathophysiologically exaggerated baroreceptor reflex responses occur in some human subjects* is inescapable. As our discussion indicates, the proper diagnosis of this condition is difficult. In those patients in whom carotid sinus reflex hypersensitivity is documented unambiguously, especially those with primarily severe bradycardia, therapy may be gratifying.

> Nonumque prematur in annum,
> Membranis intus positis: delere licebit
> Quod non edideris; nescit vox missa reverti.
> (Let it be kept till the ninth year, the manuscript put away at home: you may destroy whatever you haven't published; once out, what you've said can't be stopped.)
>
> Horace

References

Aars, H. (1968). Aortic baroreceptor activity in normal and hypertensive rabbits. *Acta Physiologica Scandinavica*, **72**, 298-309.

Abboud, F.M. and Chapleau, M.W. (1988). Effects of pulse frequency on single-unit baroreceptor activity during sine-wave and natural pulses in dogs. *The Journal of Physiology*, **401**, 295-308.

Abboud, F.M., Eckberg, D.L., Johannsen, U.J., and Mark, A.L. (1979). Carotid and cardiopulmonary baroreceptor control of splanchnic and forearm vascular resistance during venous pooling in man. *The Journal of Physiology*, **286**, 173-184.

Abboud, F.M., Thames, M.D., and Mark, A.L. (1981). Role of cardiac afferent nerves in regulation of circulation during coronary occlusion and heart failure. In *Disturbances in neurogenic control of the circulation*, ed. Abboud, F.M., Fozzard, H.A., Gilmore, J.P., and Reis, D.J., pp. 65-86. American Physiological Society, Bethesda, Maryland.

Abboud, F.M., Aylward, P.E., Floras, J.S., and Gupta, B.N. (1986). Sensitization of aortic and cardiac baroreceptors by arginine vasopressin in mammals. *The Journal of Physiology*, **377**, 251-265.

Abelmann, W.H. and Fareeduddin, K. (1967). Circulatory response to upright tilt in patients with heart disease. *Clinical Aviation and Aerospace Medicine*, **38**, 60-65.

Ábrahám, A. (1953). Die innervation der Blutgefässe. *Acta Biologica Academiae Scientiarum Hungaricae*, **4**, 69-160.

Ábrahám, A. (1967). The structure of baroreceptors in pathological conditions in man. In *Baroreceptors and hypertension*, ed. Kezdi, P., pp. 273-291. Pergamon, Oxford.

Ábrahám, A. (1969). *Microscopic innervation of the heart and blood vessels in vertebrates including man*, pp. 174-318. Pergamon, Oxford.

Abraham, R.H. and Shaw, C.D. (1983). *Dynamics — the geometry of behavior — Part 2: chaotic behavior*, Aerial Press, Santa Cruz, California.

Abramson, D.I., Landt, H., and Benjamin, J.E. (1943). Peripheral vascular response to acute anoxia. *Archives of Internal Medicine*, **71**, 583-593.

Adams, L., Guz, A., Innes, J.A., and Murphy, K. (1987). The early circulatory and ventilatory response to voluntary and electrically induced exercise in man. *The Journal of Physiology*, **383**, 19-30.

Adams, W.E. (1958). *The comparative morphology of the carotid body and carotid sinus*, pp. 1-272. Charles C. Thomas, Springfield, Illinois.

Adgey, A.A.J. and Webb, S.W. (1979). The treatment of ventricular arrhythmias in acute myocardial infarction. *British Journal of Hospital Medicine*, **21**, 356-379.

Adgey, A.A.J., Devlin, J.E., Webb, S.W., and Mulholland, H.C. (1982). Initiation of ventricular fibrillation outside hospital in patients with acute ischaemic heart disease. *British Heart Journal*, **47**, 55-61.

Adrian, E.D. (1926). The impulses produced by sensory nerve endings. Part I. *The Journal of Physiology*, **61**, 49-72.

Adrian, E.D. and Bronk, D.W. (1932). Discharges in mammalian sympathetic nerves. *The Journal of Physiology*, **74**, 115-133.

Agarwal, S.K. and Calaresu, F.R. (1990). Reciprocal connections between nucleus tractus solitarii and rostral ventrolateral medulla. *Brain Research*, **523**, 305-308.

Ahlmark, G., Saetre, H., and Korsgren, M. (1974). Reduction of sudden deaths after myocardial infarction (Letter to the editor). *The Lancet*, **2**, 1563.

Ahlquist, R.P. (1948). A study of the adrenotropic receptors. *American Journal of Physiology*, **153**, 586-600.

Airaksinen, K.E.J., Ikäheimo, M.J., Linnaluoto, M.K., Niemelä, M., and Takkunen, J.T. (1987). Impaired vagal heart rate control in coronary artery disease. *British Heart Journal*, **58**, 592-597.

Aisenbrey, G.A., Handelman, W.A., Arnold, P., Manning, M., and Schrier, R.W. (1981). Vascular effects of arginine vasopressin during fluid deprivation in the rat. *The Journal of Clinical Investigation*, **67**, 961-968.

Akins, V.F. and Bealer, S.L. (1991). Central nervous system histamine regulates peripheral sympathetic activity. *American Journal of Physiology*, **260**, H218-H224.

Aksamit, T.R., Floras, J.S., Victor, R.G., and Aylward, P.E. (1987). Paroxysmal hypertension due to sinoaortic baroreceptor denervation in humans. *Hypertension*, **9**, 309-314.

Akselrod, S., Gordon, D., Ubel, F.A., Shannon, D.C., Barger, A.C., and Cohen, R.J. (1981). Power spectrum analysis of heart rate fluctuation: a quantitative probe of beat-to-beat cardiovascular control. *Science*, **213**, 220-222.

Akselrod, S., Gordon, D., Madwed, J.B., Snidman, N.C., Shannon, D.C., and Cohen, R.J. (1985). Hemodynamic regulation: investigation by spectral analysis. *American Journal of Physiology*, **249**, H867-H875.

Alam, M. and Smirk, F.H. (1937). Observations in man upon a blood pressure raising reflex arising from the voluntary muscles. *The Journal of Physiology*, **89**, 372-383.

Alam, M. and Smirk, F.H. (1938). Observations in man on a pulse-accelerating reflex from the voluntary muscles of the legs. *The Journal of Physiology*, **92**, 167-177.

Alessi, R., Nusynowitz, M., Abildskov, J.A., and Moe, G.K. (1958). Nonuniform distribution of vagal effects on the atrial refractory period. *American Journal of Physiology*, **194**, 406-410.

Alexander, N. and DeCuir, M. (1966). Loss of baroreflex bradycardia in renal hypertensive rabbits. *Circulation Research*, **19**, 18-25.

Alexander, N., Velasquez, M.T., DeCuir, M., and Maronde, R.F. (1980). Indices of sympathetic activity in the sinoaortic-denervated hypertensive rat. *American Journal of Physiology*, **238**, H521-H526.

Alexander, R.S. (1946). Tonic and reflex functions of medullary sympathetic cardiovascular centers. *Journal of Neurophysiology*, **9**, 205-217.

Alexopoulos, D., Yusuf, S., Johnston, J.A., Bostock, J., Sleight, P., and Yacoub, M.H. (1988). The 24-hour heart rate behavior in long-term survivors of cardiac transplantation. *The American Journal of Cardiology*, **61**, 880-884.

Alicandri, C., Fouad, F.M., Tarazi, R.C., Castle, L., and Morant, V. (1978). Three cases of hypotension and syncope with ventricular pacing: possible role of atrial reflexes. *The American Journal of Cardiology*, **42**, 137-142.

Alicandri, C. *et al.* (1985). Autonomic nervous system control of heart rate in essential hypertension. *Journal of Hypertension*, **3** (Suppl. 3), S117-S119.

Allen, A.M., Chai, S.Y., Clevers, J., McKinley, M.J., Paxinos, G., and Mendelsohn, F.A.O. (1988). Localization and characterization of angiotensin II receptor binding and angiotensin converting enzyme in the human medulla oblongata. *The Journal of Comparative Neurology*, **269**, 249-264.

Almeida, F.A., Suzuki, M., and Maack, T. (1986). Atrial natriuretic factor increases hematocrit and decreases plasma volume in nephrectomized rats. *Life Sciences*, **39**, 1193-1199.

Almquist, A., Gornick, C., Benson, D.W., Jr., Dunnigan, A., and Benditt, D.G. (1985). Carotid sinus hypersensitivity: evaluation of the vasodepressor component. *Circulation*, **71**, 927-936.

Al-Timman, J.K.A., Hainsworth, R., and Vukasovic, J.L. (1988). A modified neck chamber for the study of carotid baroreceptors in humans (Abstract). *The Journal of Physiology*, **399**, 14P.

Altura, B.M. and Altura, B.T. (1977). Vascular smooth muscle and neurohypophyseal hormones. *Federation Proceedings*, **36**, 1853-1860.

Alzamora-Castro, V., Battilana, G., Garrido-Lecca, G., Rubio, C., Abugattas, R., and Bouroncle, J. (1955). Acute left ventricular failure and carotid sinus stimulation. *The Journal of the American Medical Association*, **157**, 226-229.

Amendt, K., Czachurski, J., Dembowsky, K., and Seller, H. (1979). Bulbospinal projections to the intermediolateral cell column; a neuroanatomical study. *Journal of the Autonomic Nervous System*, **1**, 103-117.

Amery, A., Bossaert, H., Deruyttere, M., Vanderlinden, L., and Verstraete, M. (1973). Influence of body posture on leg blood flow. *Scandinavian Journal of Clinical and Laboratory Investigation*, **31**, 29-36.

Amorim, D.S., Godoy, R.A., Manço, J.C., Tanaka, A., and Gallo, L., Jr. (1968). Effects of acute elevation in blood pressure and of atropine on heart rate in Chagas' disease. A preliminary report. *Circulation*, **38**, 289-294.

Amorim, D.S. *et al.* (1981). Is there autonomic impairment in congestive (dilated) cardiomyopathy? *The Lancet*, **1**, 525-527.

Anand, I.S., Ferrari, R., Kalra, G.S., Wahi, P.L., Poole-Wilson, P.A., and Harris, P.C. (1989). Edema of cardiac origin. Studies of body water and sodium, renal function, hemodynamic indexes, and plasma hormones in untreated congestive cardiac failure. *Circulation*, **80**, 299-305.

Andersen, H.T. (1963). The reflex nature of the physiological adjustments to diving and their afferent pathway. *Acta Physiologica Scandinavica*, **58**, 263-273.

Andersen, H.T. (1966). Physiological adaptations in diving vertebrates. *Physiological Reviews*, **46**, 212-243.

Andersen, H.T. and Blix, A.S. (1974). Pharmacological exposure of components in the autonomic control of the diving reflex. *Acta Physiologica Scandinavica*, **90**, 381-386.

Andersen, P. and Saltin, B. (1985). Maximal perfusion of skeletal muscle in man. *The Journal of Physiology*, **366**, 233-249.

Anderson, E.A., Wallin, B.G., and Mark, A.L. (1987). Dissociation of sympathetic nerve activity in arm and leg muscle during mental stress. *Hypertension*, **9** (Suppl. III), III-114–III-119.

Anderson, E.A., Sinkey, C.A., Lawton, W.J., and Mark, A.L. (1989). Elevated sympathetic nerve activity in borderline hypertensive humans. Evidence from direct intraneural recordings. *Hypertension*, **14**, 177-183.

Anderson, F.D. and Berry, C.M. (1956). An oscillographic study of the central pathways of the vagus nerve in the cat. *The Journal of Comparative Neurology*, **106**, 163-181.

Anderson, J., Struthers, A., Christofides, N., and Bloom, S. (1986). Atrial natriuretic peptide: an endogenous factor enhancing sodium excretion in man. *Clinical Science*, **70**, 327-331.

Anderson, J.V., Donckier, J., Payne, N.N., Beacham, J., Slater, J.D.H., and Bloom, S.R. (1987). Atrial natriuretic peptide: evidence of action as a natriuretic hormone at physiological plasma concentrations in man. *Clinical Science*, **72**, 305-312.

Andresen, M.C. and Yang, M. (1990). Dynamic and static conditioning pressures evoke equivalent rapid resetting in rat aortic baroreceptors. *Circulation Research*, **67**, 303-311.

Andresen, M.C. and Yang, M. (1990). Non-NMDA receptors mediate sensory afferent synaptic transmission in medial nucleus tractus solitarius. *American Journal of Physiology*, **259**, H1307-H1311.

Andresen, M.C., Krauhs, J.M., and Brown, A.M. (1978). Relationship of aortic wall and baroreceptor properties during development in normotensive and spontaneously hypertensive rats. *Circulation Research*, **43**, 728-738.

Andresen, M.C., Kuraoka, S., and Brown, A.M. (1979). Individual and combined actions of calcium, sodium, and potassium ions on baroreceptors in the rat. *Circulation Research*, **45**, 757-763.

Andrews, C.J.H., Andrews, W.H.H., and Orbach, J. (1972). A sympathetic reflex elicited by distension of the mesenteric venous bed. *The Journal of Physiology*, **226**, 119-131.

Andrezik, J.A., Chan-Palay, V., and Palay, S.L. (1981). The nucleus paragigantocellularis lateralis in the rat. Conformation and cytology. *Anatomy and Embryology*, **161**, 355-371.

Angell James, J.E. (1971*a*). The effects of altering mean pressure, pulse pressure and pulse frequency on the impulse activity in baroreceptor fibres from the aortic arch and right subclavian artery in the rabbit. *The Journal of Physiology*, **214**, 65-88.

Angell James, J.E. (1971*b*). The effects of changes of extramural, 'intrathoracic', pressure on aortic arch baroreceptors. *The Journal of Physiology*, **214**, 89-103.

Angell James, J.E. (1971*c*). The responses of aortic arch and right subclavian baroreceptors to changes of non-pulsatile pressure and their modification by hypothermia. *The Journal of Physiology*, **214**, 201-223.

Angell-James, J.E. (1973). Characteristics of single aortic and right subclavian baroreceptor fiber activity in rabbits with chronic renal hypertension. *Circulation Research*, **32**, 149-161.

Angell-James, J.E. (1974). Pathophysiology of aortic baroreceptors in rabbits with vitamin D sclerosis and hypertension. *Circulation Research*, **34**, 327-338.

Angell-James, J.E. and Bobik, A. (1978). Modification of blood pressure and aortic baroreceptor activity in propranolol-treated hypertensive rabbits (Abstract). *The Journal of Physiology*, **278**, 16P.

Angell James, J.E. and Daly, M.deB. (1969). Cardiovascular responses in apnoeic asphyxia: role of arterial chemoreceptors and the modification of their effects by a pulmonary vagal inflation reflex. *The Journal of Physiology*, **201**, 87-104.

Angell James, J.E. and Daly, M.deB. (1970). Comparison of the reflex vasomotor responses to separate and combined stimulation of the carotid sinus and aortic arch baroreceptors by pulsatile and non-pulsatile pressures in the dog. *The Journal of Physiology*, **209**, 257-293.

Angell James, J.E. and Daly, M.deB. (1972). Some mechanisms involved in the cardiovascular adaptations to diving. *Symposia of the Society for Experimental Biology*, **26**, 313-341.

Angell-James, J.E. and George, M.J. (1980). Carotid sinus baroreceptor reflex control of the circulation in medial sclerotic and renal hypertensive rabbits and its modification by the aortic baroreceptors. *Circulation Research*, **47**, 890-901.

Angell-James, J.E. and Lumley, J.S.P. (1974). The effects of carotid endarterectomy on the mechanical properties of the carotid sinus and carotid sinus nerve activity in atherosclerotic patients. *British Journal of Surgery*, **61**, 805-810.

Angell-James, J.E., Daly, M.deB., and Elsner, R. (1978). Arterial baroreceptor reflexes in the seal and their modification during experimental dives. *American Journal of Physiology*, **234**, H730-H739.

Angell-James, J.E., Elsner, R., and Daly, M.deB. (1981). Lung inflation: effects on heart rate, respiration, and vagal afferent activity in seals. *American Journal of Physiology*, **240**, H190-H198.

Annat, G., Grandjean, B., Vincent, M., Jarsaillon, E., and Sassard, J. (1976). Effects of right atrial stretch on plasma renin activity. *Archives Internationales de Physiologie et de Biochimie*, **84**, 311-315.

Anrep, G.V., Pascual, W., and Rössler, R. (1936). Respiratory variations of the heart rate. I - The reflex mechanism of the respiratory arrhythmia. *Proceedings of the Royal Society of London. Series B — Biological Sciences*, **119B**, 191-230.

Appian of Alexandria (1958). *Appian's Roman history*, Volume III, pp. 408-409. Harvard University Press, Cambridge, Massachusetts.

Applegate, R.J., Hasser, E.M., and Bishop, V.S. (1987). Vagal cold block in area postrema-lesioned dogs: interaction of vasopressin and sympathetic nervous system. *American Journal of Physiology*, **252**, H135-H141.

Arai, Y. *et al.* (1989). Modulation of cardiac autonomic activity during and immediately after exercise. *American Journal of Physiology*, **256**, H132-H141.

Arango, V., Ruggiero, D.A., Callaway, J.L., Anwar, M., Mann, J.J., and Reis, D.J. (1988). Catecholaminergic neurons in the ventrolateral medulla and nucleus of the solitary tract in the human. *The Journal of Comparative Neurology*, **273**, 224-240.

Arnauld, E., Czernichow, P., Fumoux, F., and Vincent, J-D. (1977). The effects of hypotension and hypovolaemia on the liberation of vasopressin during haemorrhage in the unanaesthetized monkey (*Macaca Mulatta*). *Pflügers Archiv für die Gesamte Physiologie des Menschen und der Tiere*, **371**, 193-200.

Arndt, J.O., Dörrenhaus, A., and Wiecken, H. (1975). The aortic arch baroreceptor response to static and dynamic stretches in an isolated aorta-depressor nerve preparation of cats *in vitro*. *The Journal of Physiology*, **252**, 59-78.

Arndt, J.O., Morgenstern, J., and Samodelov, L. (1977). The physiologically relevant information regarding systemic blood pressure encoded in the carotid sinus baroreceptor discharge pattern. *The Journal of Physiology*, **268**, 775-791.

Arnold, J.M.O. and McDevitt, D.G. (1986). Reflex vagal withdrawal and the hemodynamic response to intravenous isoproterenol in the presence of beta-antagonists. *Clinical Pharmacology and Therapeutics*, **40**, 199-208.

Arrowood, J.A., Mohanty, P.K., Hodgson, J.McB., Dibner-Dunlap, M.E., and Thames, M.D. (1989). Ventricular sensory endings mediate reflex bradycardia during coronary arteriography in humans. *Circulation*, **80**, 1293-1300.

Ashman, R. and Gouaux, J.L. (1937-1938). Reflex inhibition of the human heart: complete A–V block and parasystole. *Proceedings of the Society for Experimental Biology and Medicine*, **37**, 25-27.

Asmussen, E. and Chiodi, H. (1941). The effect of hypoxemia on ventilation and circulation in man. *American Journal of Physiology*, **132**, 426-436.

Åstrand, P.-O., Cuddy, T.E., Saltin, B., and Stenberg, J. (1964). Cardiac output during submaximal and maximal work. *Journal of Applied Physiology*, **19**, 268-274.

Atlas, D., Melamed, E., and Lahav, M. (1977). β-Adrenergic receptors in rat kidney. Direct localization by a fluorescent β-blocker. *Laboratory Investigation*, **36**, 465-468.

Atweh, S.F. and Kuhar, M.J. (1977). Autoradiographic localization of opiate receptors in the rat brain. I. Spinal cord and lower medulla. *Brain Research*, **124**, 53-67.

Atzler, E. and Herbst, R. (1923). Die Schwankungen des Fussvolumens und deren Beeinflussung. *Zeitschrift für die gesamte Experimentelle Medizin*, **38**, 137-152.

Aviado, D.M., Jr. and Schmidt, C.F. (1955). Reflexes from stretch receptors in blood vessels, heart and lungs. *Physiological Reviews*, **35**, 247-300.

Aylward, P.E., McRitchie, R.J., Chalmers, J.P., and West, M.J. (1983). Baroreflex control of myocardial contractility in conscious normotensive and renal hypertensive rabbits. *Hypertension*, **5**, 916-926.

Aylward, P.E., Floras, J.S., Leimbach, W.N., and Abboud, F.M. (1986). Effects of vasopressin on the circulation and its baroreflex control in healthy men. *Circulation*, **73**, 1145-1154.

Baertschi, A.J., Munzner, R.F., Ward, D.G., Johnson, R.N., and Gann, D.S. (1975). Right and left atrial B-fiber input to the medulla of the cat. *Brain Research*, **98**, 189-193.

Bailey, J.C., Watanabe, A.M., Besch, H.R., Jr., and Lathrop, D.A. (1979). Acetylcholine antagonism of the electrophysiological effects of isoproterenol on canine cardiac Purkinje fibers. *Circulation Research*, **44**, 378-383.

Baily, R.G., Prophet, S.A., Shenberger, J.S., Zelis, R., and Sinoway, L.I. (1990). Direct neurohumoral evidence for isolated sympathetic nervous system activation to skeletal muscle in response to cardiopulmonary baroreceptor unloading. *Circulation Research*, **66**, 1720-1728.

Bakris, G.L., Wilson, D.M., and Burnett, J.C., Jr. (1986). The renal, forearm, and hormonal responses to standing in the presence and absence of propranolol. *Circulation*, **74**, 1061-1065.

Baldwa, V.S. and Ewing, D.J. (1977). Heart rate response to Valsalva manoeuvre. Reproducibility in normals, and relation to variation in resting heart rate in diabetics. *British Heart Journal*, **39**, 641-644.

Bamford, O.S. and Jones, D.R. (1976). Respiratory and cardiovascular interactions in ducks: the effect of lung denervation on the initiation of and recovery from some cardiovascular responses to submergence. *The Journal of Physiology*, **259**, 575-596.

Bandler, R. and McCulloch, T. (1984). Afferents to a midbrain periaqueductal grey region involved in the 'defense reaction' in the cat as revealed by horseradish peroxidase. II. The diencephalon. *Behavioural Brain Research*, **13**, 279-285.

Banner, N.R., Williams, M., Patel, N., Chalmers, J., Lightman, S.L., and Yacoub, M.H. (1990). Altered cardiovascular and neurohumoral responses to head-up tilt after heart–lung transplantation. *Circulation*, **82**, 863-871.

Barber, M.J., Mueller, T.M., Davies, B.G., Gill, R.M., and Zipes, D.P. (1985). Interruption of sympathetic and vagal-mediated afferent responses by transmural myocardial infarction. *Circulation*, **72**, 623-631.

Barcroft, H. and Edholm, O.G. (1945). On the vasodilatation in human skeletal muscle during post-haemorrhagic fainting. *The Journal of Physiology*, **104**, 161-175.

Barcroft, H., Edholm, O.G., McMichael, J., and Sharpey-Schafer, E.P. (1944). Posthaemorrhagic fainting. Study by cardiac output and forearm flow. *The Lancet*, **1**, 489-491.

Barger, A.C. (1979). The Goldblatt Memorial Lecture. Part I: Experimental renovascular hypertension. *Hypertension*, **1**, 447-455.

Barillot, J.-C. (1970). Dépolarisation présynaptique des fibres sensitives vagales et laryngées. *Journal de Physiologie*, **62**, 273-294.

Barker, J.L., Crayton, J.W., and Nicoll, R.A. (1970). Supraoptic neurosecretory cells: autonomic modulation. *Science*, **171**, 206-207.

Barman, S.M. and Gebber, G.L. (1976). Basis for synchronization of sympathetic and phrenic nerve discharges. *American Journal of Physiology*, **231**, 1601-1607.

Barman, S.M. and Gebber, G.L. (1979). Picrotoxin- and bicuculline-sensitive inhibition of cardiac vagal reflexes. *The Journal of Pharmacology and Experimental Therapeutics*, **209**, 67-72.

Barman, S.M. and Gebber, G.L. (1983). Sequence of activation of ventrolateral and dorsal medullary sympathetic neurons. *American Journal of Physiology*, **245**, R438-R447.

Barman, S.M. and Wurster, R.D. (1975). Visceromotor organization within descending spinal sympathetic pathways in the dog. *Circulation Research*, **37**, 209-214.

Barnes, K.L. and Ferrario, C.M. (1981). Anatomical and physiological characterization of the sympatho-facilitative area postrema pathways in the dog. In *Central nervous system mechanisms in hypertension*, ed. Buckley, J.P., and Ferrario, C.M., pp. 25-35. Raven Press, New York.

Baselli, G. *et al.* (1988). Parameter extraction from heart rate and arterial blood pressure variability signals in dogs for the validation of a physiological model. *Computers in Biology and Medicine*, **18**, 1-16.

Baskerville, A.L., Eckberg, D.L., and Thompson, M.A. (1979). Arterial pressure and pulse interval responses to repetitive carotid baroreceptor stimuli in man. *The Journal of Physiology*, **297**, 61-71.

Bastulli, J.A. and Orlowski, J.P. (1985). Stroke as a complication of carotid sinus massage. *Critical Care Medicine*, **13**, 869.

Båth, E., Lindblad, L.E., and Wallin, B.G. (1981). Effects of dynamic and static neck suction on muscle nerve sympathetic activity, heart rate and blood pressure in man. *The Journal of Physiology*, **311**, 551-564.

Batton, R.R., Jayaraman, A., Ruggiero, D., and Carpenter, M.B. (1977). Fastigial efferent projections in the monkey: an autoradiographic study. *The Journal of Comparative Neurology*, **174**, 281-306.

Bauer, D.J. (1939). Vagal reflexes appearing in asphyxia in rabbits at different ages. *The Journal of Physiology*, **95**, 187-202.

Bauernfeind, R.A., Eckberg, D.L., Rea, R.F., and Welch, W.J. (1985). Two-chamber neck collar: effects of right versus left carotid sinus stimulation on heart rate (Abstract). *Clinical Research*, **33**, 864A.

Bauman, J.L., Bauernfeind, R.A., Hoff, J.V., Strasberg, B., Swiryn, S., and Rosen, K.M. (1984). Torsade de pointes due to quinidine: Observations in 31 patients. *American Heart Journal*, **107**, 425-430.

Baylis, P.H., Stockley, R.A., and Heath, D.A. (1978). Influence of lower body negative pressure upon arginine vasopressin release. *Clinical Endocrinology*, **9**, 89-95.

Bayliss, W.M. (1893). On the physiology of the depressor nerve. *The Journal of Physiology*, **14**, 303-325.

Bayliss, W.M. (1902). On the local reactions of the arterial wall to changes of internal pressure. *The Journal of Physiology*, **28**, 220-231.

Beal, M.F., Park, T.S., and Fisher, C.M. (1981). Cerebral atheromatous embolism following carotid sinus pressure. *Archives of Neurology*, **38**, 310-312.

Beau, J.H.S. (1860). Recherches expérimentales sur la mort par submersion. *Archives General Medicine*, **2**, 64-76.

Beck, W., Schrire, V., Vogelpoel, L., Nellen, M., and Swanepoel, A. (1961). Hemodynamic effects of amyl nitrite and phenylephrine on the normal human circulation and their relation to changes in cardiac murmurs. *The American Journal of Cardiology*, **8**, 341-349.

Beck, W., Barnard, C.N., and Schrire, V. (1969). Heart rate after cardiac transplantation. *Circulation*, **40**, 437-445.

Beckstead, R.M., Morse, J.R., and Norgren, R. (1980). The nucleus of the solitary tract in the monkey: projections to the thalamus and brain stem nuclei. *The Journal of Comparative Neurology*, **190**, 259-282.

Behnia, R., Koushanpour, E., Shanks, C.A., and McCarthy, W.J. (1987). Blockade of contralateral carotid sinus nerve enhances ipsilateral nerve action potentials under halothane anaesthesia. *British Journal of Anaesthesia*, **59**, 489-497.

Beiser, G.D., Zelis, R., Epstein, S.E., Mason, D.T., and Braunwald, E. (1970). The role of skin and muscle resistance vessels in reflexes mediated by the baroreceptor system. *The Journal of Clinical Investigation*, **49**, 225-231.

Benchimol, A. and Desser, K.B. (1971). Phasic renal artery blood flow velocity in man during cardiac arrhythmias. *The American Journal of the Medical Sciences*, **261**, 161-166.

Bennett, E.D., Brooks, N.H., Keddie, J., Lis, Y., and Wilson, A. (1977). Increased renal function in patients with acute left ventricular failure: a possible homeostatic mechanism. *Clinical Science and Molecular Medicine*, **52**, 43-50.

Bennett, T. (1987). Cardiovascular responses to central hypovolaemia in man: physiology and pathophysiology. *The Physiologist*, **30** (Suppl.), S-143–S-146.

Bennett, T. and Gardiner, S.M. (1985). Involvement of vasopressin in cardiovascular regulation. *Cardiovascular Research*, **19**, 57-68.

Bennett, T. and Gardiner, S.M. (1986). Influence of exogenous vasopressin on baroreflex mechanisms. *Clinical Science*, **70**, 307-315.

Bennett, T., Farquhar, I.K., Hosking, D.J., and Hampton, J.R. (1978). Assessment of methods for estimating autonomic nervous control of the heart in patients with diabetes mellitus. *Diabetes*, **27**, 1167-1174.

Bennett, T., Wilcox, R.G., and MacDonald, I.A. (1984). Post-exercise reduction of blood pressure in hypertensive men is not due to acute impairment of baroreflex function. *Clinical Science*, **67**, 97-103.

Berger, A.J. (1979). Distribution of carotid sinus nerve afferent fibers to solitary tract nuclei of the cat using transganglionic transport of horseradish peroxidase. *Neuroscience Letters*, **14**, 153-158.

Berger, R.D. and Cohen, R.J. (1988). Analysis of the response of the sino-atrial node to fluctuations in sympathetic and parasympathetic tone using broad band stimulation techniques. *Computers in Cardiology*, 1-4.

Berger, R.D., Saul, J.P., and Cohen, R.J. (1989). Transfer function analysis of autonomic regulation. I. Canine atrial rate response. *American Journal of Physiology*, **256**, H142-H152.

Bergman, S.A., Jr., Campbell, J.K., and Wildenthal, K. (1972). 'Diving reflex' in man: its relation to isometric and dynamic exercise. *Journal of Applied Physiology*, **33**, 27-31.

Berk, J.L. and Levy, M.N. (1977). Profound reflex bradycardia produced by transient hypoxia or hypercapnia in man. *European Surgical Research*, **9**, 75-84.

Bernard, C. (1851). Influence du grand sympathique sur la sensibilité et sur la calorification. *Comptes Rendus des Séances et Mémoires de la Société de Biologie*, **33**, 163-165.

Bernard, C. (1858). *Leçons sur la physiologie et la pathologie du système nerveux.* J.-B. Baillière, Paris.

Bernardi, L., Ferrari, M.R., Ricordi, L., Lumina, C., Vandea, I., Fratino, P., *et al.* (1987). Heart rate variability changes before asymptomatic nocturnal ischemia (Abstract). *Circulation*, **76** (Suppl.IV), IV-349.

Bernardi, L. *et al.* (1989). Respiratory sinus arrhythmia in the denervated human heart. *Journal of Applied Physiology*, **67**, 1447-1455.

Bernthal, T., Greene, W., Jr., and Revzin, A.M. (1951). Role of carotid chemoreceptors in hypoxic cardiac acceleration. *Proceedings of the Society for Experimental Biology and Medicine*, **76**, 121-124.

Bert, P. (1870). *Leçons sur la physiologie comparée de la respiration*, pp. 526-553. J.B. Baillière, Paris.

Bertinieri, G., di Rienzo, M., Cavallazzi, A., Ferrari, A.U., Pedotti, A., and Mancia, G. (1988). Evaluation of baroreceptor reflex by blood pressure monitoring in unanesthetized cats. *American Journal of Physiology*, **254**, H377-H383.

Best, J.D. and Halter, J.B. (1985). Blood pressure and norepinephrine spillover during propranolol infusion in humans. *American Journal of Physiology*, **248**, R400-R406.

β-Blocker Heart Attack Trial Research Group (1982). A randomized trial of propranolol in patients with acute myocardial infarction. I. Mortality results. *The Journal of the American Medical Association*, **247**, 1707-1714.

Bevegård, B.S. and Shepherd, J.T. (1966*a*). Circulatory effects of stimulating the carotid arterial stretch receptors in man at rest and during exercise. *The Journal of Clinical Investigation*, **45**, 132-142.

Bevegård, B.S. and Shepherd, J.T. (1966*b*). Reaction in man of resistance and capacity vessels in forearm and hand to leg exercise. *Journal of Applied Physiology*, **21**, 123-132.

Bevegård, S., Holmgren, A., and Jonsson, B. (1960). The effect of body position on the circulation at rest and during exercise, with special reference to the influence on the stroke volume. *Acta Physiologica Scandinavica*, **49**, 279-298.

Bevegård, S., Jonsson, B., and Karlöf, I. (1967). The instantaneous effect of aortic pressure on atrial rate in complete atrioventricular block. *Acta Medica Scandinavica*, Suppl. **472**, 54-58.

Bevegård, S., Castenfors, J., and Danielson, M. (1977*a*). Carotid baroreceptor function in hypertensive patients. *Scandinavian Journal of Clinical and Laboratory Investigation*, **37**, 495-501.

Bevegård, S., Castenfors, J., and Lindblad, L.E. (1977b). Effect of changes in blood volume distribution on circulatory variables and plasma renin activity in man. *Acta Physiologica Scandinavica*, **99**, 237-245.

Bevegård, S., Castenfors, J., Lindblad, L.E., and Tranesjo, J. (1977*c*). Blood pressure and heart rate regulating capacity of the carotid sinus during changes in blood volume distribution in man. *Acta Physiologica Scandinavica*, **99**, 300-312.

Bevegård, S., Castenfors, J., and Lindblad, L.E. (1977*d*). Effect of carotid sinus stimulation on cardiac output and peripheral vascular resistance during changes in blood volume distribution in man. *Acta Physiologica Scandinavica*, **101**, 50-57.

Bezold, A. and Hirt, L. (1867). Ueber die physiologischen Wirkungen des essigsauren Veratrin's. *Würzburg Universität Physiologisch Laboratorium Untersuchen*, **1-2**, 73-156.

Bhattacharya, A., Knapp, C.F., McCutcheon, E.P., and Evans, J.M. (1979). Cardiac responses of dogs to nonsynchronous and heart synchronous whole-body vibration. *Journal of Applied Physiology*, **46**, 549-555.

Bhattacharya, J., Cunningham, D.J.C., Howson, M.G., Lee, M.J.R., and Sleight, P. (1973). The effects of mild hypoxia with constant PA,CO_2 and ventilation on the setting and sensitivity of the baroreceptor–cardiac-depressor reflex in man (Abstract). *The Journal of Physiology*, **234**, 112P-114P.

Biamino, G. and Kruckenberg, P. (1969). Synchronization and conduction of excitation in the rat aorta. *American Journal of Physiology*, **217**, 376-382.

Bie, P., Secher, N.H., Astrup, A., and Warberg, J. (1986). Cardiovascular and endocrine responses to head-up tilt and vasopressin infusion in humans. *American Journal of Physiology*, **251**, R735-R741.

Bigger, J.T., Jr. (1987). Why patients with congestive heart failure die: arrhythmias and sudden cardiac death. *Circulation*, **75** (Suppl. IV), IV-28–IV-35.

Bigger, J.T., Jr., Kleiger, R.E., Fleiss, J.L., Rolnitzky, L.M., Steinman, R.C., and Miller, J.P. (1988). Components of heart rate variability measured during healing of acute myocardial infarction. *The American Journal of Cardiology*, **61**, 208-215.

Bigger, J.T., Jr. *et al.* (1989). Comparison of baroreflex sensitivity and heart period variability after myocardial infarction. *Journal of the American College of Cardiology*, **14**, 1511-1518.

Billman, G.E. and Hoskins, R.S. (1989). Time-series analysis of heart rate variability during submaximal exercise. Evidence for reduced cardiac vagal tone in animals susceptible to ventricular fibrillation. *Circulation*, **80**, 146-157.

Billman, G.E. and Marsh, D.H. (1989). Effect of myocardial ischemia on hemodynamic response to carotid occlusion. *American Journal of Physiology*, **256**, H672-H680.

Billman, G.E., Schwartz, P.J., and Stone, H.L. (1982). Baroreceptor reflex control of heart rate: A predictor of sudden cardiac death. *Circulation*, **66**, 874-880.

Billman, G.E., Schwartz, P.J., and Stone, H.L. (1984). The effects of daily exercise on susceptibility to sudden cardiac death. *Circulation*, **69**, 1182-1189.

Bini, G., Hagbarth, K.-E., Hynninen, P., and Wallin, B.G. (1980). Thermoregulatory and rhythm-generating mechanisms governing the sudomotor and vasoconstrictor outflow in human cutaneous nerves. *The Journal of Physiology*, **306**, 537-552.

Bini, G., Hagbarth, K-E., and Wallin, B.G. (1981). Cardiac rhythmicity of skin sympathetic activity recorded from peripheral nerves in man. *Journal of the Autonomic Nervous System*, **4**, 17-24.

Binkley, P.F., Lewe, R., Lima, J., Unverferth, D.V., and Leier, C.V. (1988). Neurohumoral profile in congestive heart failure: Response to β-blockade. *The Journal of Laboratory and Clinical Medicine*, **111**, 393-398.

Binswanger, O. (1879). Anatomische Untersuchungen über die Ursprungs-stelle und den Anfangstheil der Carotis interna. *Archiv für Psychiatrie und Nervenkrankheiten*, **9**, 351-368.

Birnbaum, S.G., Richardson, B.C., and Dellinger, J.A. (1988). Quantifying the altered cardiac response to atropine following pyridostigmine in rhesus macaques. *Pharmacology Biochemistry and Behavior*, **31**, 381-386.

Biscoe, T.J. and Sampson, S.R. (1970). Responses of cells in the brain stem of the cat to stimulation of the sinus, glossopharyngeal, aortic and superior laryngeal nerves. *The Journal of Physiology*, **209**, 359-373.

Bishop, V.S., Hasser, E.M., and Nair, U.C. (1987*a*). Baroreflex control of renal nerve activity in conscious animals. *Circulation Research*, **61** (Suppl. I), I-76–I-81.

Bishop, V.S., Shade, R.E., Haywood, J.R., and Hamm, C. (1987*b*). Sinoaortic denervation in the nonhuman primate. *American Journal of Physiology*, **252**, R294-R298.

Bito, K., Kubo, S., and Saimyoji, H. (1980). Role of endocrine factors in chronic congestive heart failure, with emphasis on catecholamines. *Japanese Circulation Journal*, **44**, 117-127.

Bittiner, S.B. and Smith, S.E. (1988). β-adrenergic antagonists increase sinus arrhythmia, a vagotonic effect. *British Journal of Clinical Pharmacology*, **22**, 691-695.

Bjurstedt, H., Rosenhamer, G., and Tydén, G. (1975). Cardiovascular responses to changes in carotid sinus transmural pressure in man. *Acta Physiologica Scandinavica*, **94**, 497-505.

Bjurstedt, H., Rosenhamer, G., Balldin, U., and Katkov, V. (1983). Orthostatic reactions during recovery from exhaustive exercise of short duration. *Acta Physiologica Scandinavica*, **119**, 25-31.

Black, J.E. and Roddie, I.C. (1958). The mechanism of the changes in forearm vascular resistance during hypoxia. *The Journal of Physiology*, **143**, 226-235.

Blair, D.A., Glover, W.E., and Roddie, I.C. (1961). Vasomotor responses in the human arm during leg exercise. *Circulation Research*, **9**, 264-274.

Blair, M.L. (1983). Stimulation of renin secretion by α-adrenoceptor agonists. *American Journal of Physiology*, **244**, E37-E44.

Blessing, W.W. and Reis, D.J. (1982). Inhibitory cardiovascular function of neurons in the caudal ventrolateral medulla of the rabbit: relationship to the area containing A1 noradrenergic cells. *Brain Research*, **253**, 161-171.

Blessing, W.W., Goodchild, A.K., Dampney, R.A.L., and Chalmers, J.P. (1981). Cell groups in the lower brain stem of the rabbit projecting to the spinal cord, with special reference to catecholamine-containing neurons. *Brain Research*, **221**, 35-55.

Blessing, W.W., Sved, A.F., and Reis, D.J. (1982). Destruction of noradrenergic neurons in rabbit brainstem elevates plasma vasopressin, causing hypertension. *Science*, **217**, 661-663.

Blombery, P.A. and Korner, P.I. (1982). Role of aortic and carotid sinus baroreceptors on Valsalva-like vasoconstrictor and heart rate reflexes in the conscious rabbit. *Journal of the Autonomic Nervous System*, **5**, 303-315.

Blomqvist, C.G. and Stone, H.L. (1983). Cardiovascular adjustments to gravitational stress. In *Handbook of physiology. The cardiovascular system. Peripheral circulation and organ blood flow*, Part 2, ed. Shepherd, J.T., and Abboud, F.M., pp. 1025-1063. American Physiological Society, Bethesda, Maryland.

Blomqvist, C.G., Lewis, S.F., Taylor, W.F., and Graham, R.M. (1981). Similarity of the hemodynamic responses to static and dynamic exercise of small muscle groups. *Circulation Research*, **48** (Suppl. I), I87-I92.

Blumberg, A.L., Denny, S.E., Marshall, G.R., and Needleman, P. (1977). Blood vessel–hormone interactions: angiotensin, bradykinin, and prostaglandins. *American Journal of Physiology*, **232**, H305-H310.

Blumhardt, L.D., Smith, P.E.M., and Owen, L. (1986). Electrocardiographic accompaniments of temporal lobe epileptic seizures. *The Lancet*, **1**, 1051-1056.

Böck, P. and Gorgas, K. (1976). Fine structure of baroreceptor terminals in the carotid sinus of guinea pigs and mice. *Cell and Tissue Research*, **170**, 95-112.

Boineau, J.P. *et al.* (1978). Multicentric origin of the atrial depolarization wave: the pacemaker complex. *Circulation*, **58**, 1036-1048.

Boineau, J.P., Canavan, T.E., Schuessler, R.B., Cain, M.E., Corr, P.B., and Cox, J.L. (1988). Demonstration of a widely distributed atrial pacemaker complex in the human heart. *Circulation*, **77**, 1221-1237.

Boineau, J.P., Schuessler, R.B., Canavan, T.E., Corr, P.B., Cain, M.E., and Cox, J.L. (1989). The human atrial pacemaker complex. *Journal of Electrocardiology*, **22** (Suppl.), 189-197.

Bolme, P., Novotný, J., Uvnäs, B., and Wright, P.G. (1970). Species distribution of sympathetic cholinergic vasodilator nerves in skeletal muscle. *Acta Physiologica Scandinavica*, **78**, 60-64.

Bonvallet, M., Dell, P., and Hiebel, G. (1953). Sinus carotidien et activité électrique cérébrale. *Comptes Rendus des Séances de la Société de Biologie et de ses Filiales*, **147**, 1166-1169.

Borst, C. and Karemaker, J.M. (1983). Time delays in the human baroreceptor reflex. *Journal of the Autonomic Nervous System*, **9**, 399-409.

Borst, C., Karemaker, J.M., and Dunning, A.J. (1982*a*). Prolongation of atrioventricular conduction time by electrical stimulation of the carotid sinus nerves in man. *Circulation*, **65**, 432-444.

Borst, C., Wieling, W., van Brederode, J.F.M., Hond, A., de Rijk, L.G., and Dunning, A.J. (1982*b*). Mechanisms of initial heart rate response to postural change. *American Journal of Physiology*, **243**, H676-H681.

Borst, C., van Brederode, J.F.M., Wieling, W., van Montfrans, G.A., and Dunning, A.J. (1984). Mechanisms of initial blood pressure response to postural change. *Clinical Science*, **67**, 321-327.

Borst, J.G.G. and Borst-De Geus, A. (1963). *Hypertension explained by Starling's theory of circulatory homoeostasis. The Lancet*, **1**, 677-682.

Bossi, M., Cataldo, G., Colombo, A., Fiorista, F., Gentili, D., and Pirelli, S. (1977). Ipotensione, bradicardia e prelipotimia da isosorbide dinitrato sub-linguale nell'infarto miocardico in fase iniziale. *Giornale Italiano di Cardiologia*, **7**, 922-926.

Boyd, J.D. (1937). Observations on the human carotid sinus and its nerve supply. *Anatomischer Anzeiger. Zentralblatt für die gesamte wissenschaftliche Anatomie*, **84**, 386-399.

Boyden, P.A. and Hoffman, B.F. (1981). The effects on atrial electrophysiology and structure of surgically induced right atrial enlargement in dogs. *Circulation Research*, **49**, 1319-1331.

Bradley, D.J., Ghelarducci, B., Paton, J.F.R., and Spyer, K.M. (1987*a*). The cardiovascular responses elicited from the posterior cerebellar cortex in the anaesthetized and decerebrate rabbit. *The Journal of Physiology*, **383**, 537-550.

Bradley, D.J., Paton, J.F.R., and Spyer, K.M. (1987b). Cardiovascular responses evoked from the fastigial region of the cerebellum in anaesthetized and decerebrate rabbits. *The Journal of Physiology*, **392**, 475-491.

Branham, H.H. (1890). Aneurismal varix of the femoral artery and vein following a gunshot wound. *International Journal of Surgery*, **3**, 250-251.

Braun-Menendez, E., Fasciolo, J.C., Leloir, L.F., and Muñoz, J.M. (1940). The substance causing renal hypertension. *The Journal of Physiology*, **98**, 283-298.

Braunwald, E., Frahm, C.J., and Ross, J., Jr. (1961). Studies on Starling's law of the heart. V. Left ventricular function in man. *The Journal of Clinical Investigation*, **40**, 1882-1890.

Braunwald, E., Epstein, S.E., Glick, G., Wechsler, A.S., and Braunwald, N.S. (1967). Relief of angina pectoris by electrical stimulation of the carotid-sinus nerves. *The New England Journal of Medicine*, **277**, 1278-1283.

Braunwald, E., Sobel, B.E., and Braunwald, N.S. (1969). Treatment of paroxysmal supraventricular tachycardia by electrical stimulation of the carotid-sinus nerves. *The New England Journal of Medicine*, **281**, 885-887.

Brennan, L.A., Jr., Malvin, R.L., Jochim, K.E., and Roberts, D.E. (1971). Influence of right and left atrial receptors on plasma concentrations of ADH and renin. *American Journal of Physiology*, **221**, 273-278.

Brest, A.N., Wiener, L., and Bachrach, B. (1972). Bilateral carotid sinus nerve stimulation in the treatment of hypertension. *The American Journal of Cardiology*, **29**, 821-825.

Brigden, W. and Sharpey-Schafer, E.P. (1950). Postural changes in peripheral blood flow in cases with left heart failure. *Clinical Science*, **9**, 93-100.

Brigden, W., Howarth, S., and Sharpey-Schafer, E.P. (1950). Postural changes in the peripheral blood-flow of normal subjects with observations on vasovagal fainting reactions as a result of tilting, the lordotic posture, pregnancy and spinal anaesthesia. *Clinical Science*, **9**, 79-91.

Bristow, J.D., Honour, A.J., Pickering, T.G., and Sleight, P. (1969*a*). Cardiovascular and respiratory changes during sleep in normal and hypertensive subjects. *Cardiovascular Research*, **3**, 476-485.

Bristow, J.D., Honour, A.J., Pickering, G.W., Sleight, P., and Smyth, H.S. (1969*b*). Diminished baroreflex sensitivity in high blood pressure. *Circulation*, **39**, 48-54.

Bristow, J.D., Brown, E.B., Jr., Cunningham, D.J.C., Goode, R.C., Howson, M.G., and Sleight, P. (1971*a*). The effects of hypercapnia, hypoxia and ventilation on the baroreflex regulation of the pulse interval. *The Journal of Physiology*, **216**, 281-302.

Bristow, J.D. *et al.* (1971*b*). Effect of bicycling on the baroreflex regulation of pulse interval. *Circulation Research*, **28**, 582-592.

Bristow, M.R. *et al.* (1982). Decreased catecholamine sensitivity and β-adrenergic-receptor density in failing human hearts. *The New England Journal of Medicine*, **307**, 205-211.

Brizzee, K.R. and Neal, L.M. (1954). A re-evaluation of the cellular morphology of the area postrema in view of recent evidence for a chemoreceptor function. *The Journal of Comparative Neurology*, **100**, 41-59.

Broadwell, R.D. and Brightman, M.W. (1976). Entry of peroxidase into neurons of the central and peripheral nervous systems from extracerebral and cerebral blood. *The Journal of Comparative Neurology*, **166**, 257-283.

Brod, J., Fencl, V., Hejl, Z., Jirka, J., and Ulrych, M. (1962). General and regional haemodynamic pattern underlying essential hypertension. *Clinical Science*, **23**, 339-349.

Brodie, T.G. and Russell, A.E. (1900). On reflex cardiac inhibition. *The Journal of Physiology*, **26**, 92-106.

Brody, M.J. (1986). Central nervous system mechanisms of arterial pressure regulation. *Federation Proceedings*, **45**, 2700-2706.

Bronk, D.W. and Kaltreider, N.L. (1931). The discharge of impulses in the aortic nerve. *American Journal of Physiology*, **97**, 508-508.

Bronk, D.W. and Stella, G. (1932). Afferent impulses in the carotid sinus nerve. I. The relation of the discharge from single end organs to arterial blood pressure. *Journal of Cellular and Comparative Physiology*, **1**, 113-130.

Bronk, D.W. and Stella, G. (1935). The response to steady pressures of single end organs in the isolated carotid sinus. *American Journal of Physiology*, **110**, 708-714.

Brooker, J.Z., Alderman, E.L., and Harrison, D.C. (1974). Alterations in left ventricular volumes induced by Valsalva manoeuvre. *British Heart Journal*, **36**, 713-718.

Brooks, C.McC., Lu, H.-H., Lange, G., Mangi, R., Shaw, R.B., and Geoly, K. (1966). Effects of localized stretch of the sinoatrial node region of the dog heart. *American Journal of Physiology*, **211**, 1197-1202.

Broughton, A. and Korner, P.I. (1983). Basal and maximal inotropic state in renal hypertensive dogs with cardiac hypertrophy. *American Journal of Physiology*, **245**, H33-H41.

Brouha, L. and Nowak, S.J.G. (1939). The role of the vagus in the cardio-accelerator action of atropine in sympathectomized dogs. *The Journal of Physiology*, **95**, 439-453.

Brown, A.M. (1967). Excitation of afferent cardiac sympathetic nerve fibres during myocardial ischaemia. *The Journal of Physiology*, **190**, 35-53.

Brown, A.M. (1980). Receptors under pressure. An update on baroreceptors. *Circulation Research*, **46**, 1-10.

Brown, A.M., Saum, W.R., and Tuley, F.H. (1976). A comparison of aortic baroreceptor discharge in normotensive and spontaneously hypertensive rats. *Circulation Research*, **39**, 488-496.

Brown, D.L. and Guyenet, P.G. (1984). Cardiovascular neurons of brain stem with projections to spinal cord. *American Journal of Physiology*, **247**, R1009-R1016.

Brown, D.L. and Guyenet, P.G. (1985). Electrophysiological study of cardiovascular neurons in the rostral ventrolateral medulla in rats. *Circulation Research*, **56**, 359-369.

Brown, E., Goei, J.S., Greenfield, A.D.M., and Plassaras, G.C. (1966). Circulatory responses to simulated gravitational shifts of blood in man induced by exposure of the body below the iliac crests to sub-atmospheric pressure. *The Journal of Physiology*, **183**, 607-627.

Brown, G.L. and Eccles, J.C. (1934). The action of a single vagal volley on the rhythm of the heart beat. *The Journal of Physiology*, **82**, 211-241.

Brown, J.J., Davies, D.L., Johnson, V.W., Lever, A.F., and Robertson, J.I.S. (1970). Renin relationships in congestive cardiac failure, treated and untreated. *American Heart Journal*, **80**, 329-342.

Brown, K.A., Maloney, J.D., Smith, H.C., Hartzler, G.O., and Ilstrup, D.M. (1980). Carotid sinus reflex in patients undergoing coronary angiography: relationship of degree and location of coronary artery disease to response to carotid sinus massage. *Circulation*, **62**, 697-703.

Brown, T.E., Beightol, L.A., Koh, J., and Eckberg, D.L. (1991). R–R interval variability: respiratory rate and tidal volume strongly influence high frequency components of the power spectrum (Abstract). *The FASEB Journal*, **5**, A1495.

Brownstein, M.J., Russell, J.T., and Gainer, H. (1980). Synthesis, transport, and release of posterior pituitary hormones. *Science*, **207**, 373-378.

Bruce, R.A., Kusumi, F., and Hosmer, D. (1973). Maximal oxygen intake and nomographic assessment of functional aerobic impairment in cardiovascular disease. *American Heart Journal*, **85**, 546-562.

Brunner, M.J., Sussman, M.S., Greene, A.S., Kallman, C.H., and Shoukas, A.A. (1982). Carotid sinus baroreceptor reflex control of respiration. *Circulation Research*, **51**, 624-636.

Brunsting, J.R., Bennekers, J.H., Schuil, H.A., and Zijlstra, W.G. (1979). Incomplete cardiac vagal blockade with atropine in the anesthetized dog. *Pflügers Archiv für die Gesamte Physiologie des Menschen und der Tiere*, **381**, 293-295.

Brush, J.E., Jr. *et al.* (1989). Cardiac norepinephrine kinetics in hypertrophic cardiomyopathy. *Circulation*, **79**, 836-844.

Buda, A.J., Pinsky, M.R., Ingels, N.B. Jr., Daughters, G.T., Stinson, E.B., and

Alderman, E.L. (1979). Effect of intrathoracic pressure on left ventricular performance. *The New England Journal of Medicine*, **301**, 453-459.

Budge,–. (1846), Briefliche Mittheilung über die Herzbewegung. *Archiv für Anatomie, Physiologie und Wissenschaftliche Medicin*, 295-297.

Buijs, R.M. (1978). Intra- and extrahypothalamic vasopressin and oxytocin pathways in the rat. Pathways to the limbic system, medulla oblongata and spinal cord. *Cell Tissue Research*, **192**, 423-435.

Buñag, R.D., Page, I.H., and McCubbin, J.W. (1966). Neural stimulation of release of renin. *Circulation Research*, **19**, 851-858.

Bungo, M.W. (1989). The cardiopulmonary system. In *Space physiology and medicine*, ed. Nicogossian, A.E., Huntoon, C.L., and Pool, S.L., pp. 179-201. Lea and Febiger, Philadelphia.

Burack, B., Pollack, C., Borowiecki, B., and Weitzman, E. (1977). The hypersomnia–sleep apnea syndrome (HSA): a reversible major cardiovascular hazard (Abstract). *Circulation*, **56** (Suppl. III), III-177.

Bürger,–. (1921). Ueber klinische Bedeutung des Valsalvaschen Versuches. *Medizinische Klinik*, **51**, 1568.

Bürger, M. (1926). Ueber die Bedeutung des intrapulmonalen Drucks für den Kreislauf und den Mechanismus des Kollapses bei akuten Anstrengungen. *Klinische Wochenschrift*, **5**, 777-780.

Burke, D., Sundlöf, G., and Wallin, B.G. (1977). Postural effects on muscle nerve sympathetic activity in man. *The Journal of Physiology*, **272**, 399-414.

Burn, J.H. and Rand, M.J. (1965). Acetylcholine in adrenergic transmission. *Annual Review of Pharmacology*, **5**, 163-182.

Burnett, J.C., Jr., Granger, J.P., and Opgenorth, T.J. (1984). Effects of synthetic atrial natriuretic factor on renal function and renin release. *American Journal of Physiology*, **247**, F863-F866.

Burnett, J.C., Jr. *et al.* (1986). Atrial natriuretic peptide elevation in congestive heart failure in the human. *Science*, **231**, 1145-1147.

Burns, A. (1811). *Observations on the surgical anatomy of the head and neck*, pp. 163-164. Thomas Bryce, Edinburgh.

Burnstock, G. (1986). The changing face of autonomic neurotransmission. *Acta Physiologica Scandinavica*, **126**, 67-91.

Burstyn, P.G., Horrobin, D.F., and Lloyd, J.J. (1972). Chronic hypertension in rabbits induced by bilateral placement of rigid casts around the carotid sinus regions. *Cardiovascular Research*, **6**, 54-56.

Bussan, R., Reid, E.R., and Scherf, D. (1957). Conversion of atrial flutter into atrial fibrillation by carotid pressure. *Annals of Internal Medicine*, **46**, 814-818.

Bussien, J.P. *et al.* (1984). Does vasopressin sustain blood pressure of normally hydrated healthy volunteers? *American Journal of Physiology*, **246**, H143-H147.

Butler, P.J. and Jones, D.R. (1968). Onset of and recovery from diving bradycardia in ducks. *The Journal of Physiology*, **196**, 255-272.

Butler, P.J. and Jones, D.R. (1982). The comparative physiology of diving in vertebrates. *Advances in Comparative Physiology and Biochemistry*, **8**, 179-364.

Butrous, G.S., Cochrane, T., and Camm, A.J. (1987). Rapid autonomic tone regulation of atrioventricular nodal conduction in man. *American Heart Journal*, **113**, 934-940.

Cagin, N.A. and Natelson, B.H. (1981). Cholinergic activation produces psychosomatic digitalis toxicity. *The Journal of Pharmacology and Experimental Therapeutics*, **218**, 709-711.

Calaresu, F.R. and Pearce, J.W. (1965). Effects on heart rate of electrical stimulation of medullary vagal structures in the cat. *The Journal of Physiology*, **176**, 241-251.

Calaresu, F.R., Faiers, A.A., and Mogenson, G.J. (1975). Central neural regulation of heart and blood vessels in mammals. *Progress in Neurobiology*, **5**, 1-35.

Calaresu, F.R., Ciriello, J., Caverson, M.M., Cechetto, D.F., and Krukoff, T.L. (1984). Functional neuroanatomy of central pathways controlling the circulation. In *Hypertension and the brain*, ed. Guthrie, G.P., Jr., and Kotchen, T.A., pp. 3-21. Futura, Mount Kisco, New York.

Campagnole-Santos, M.J., Diz, D.I., and Ferrario, C.M. (1988). Baroreceptor reflex modulation by angiotensin II at the nucleus tractus solitarii. *Hypertension*, **11** (Suppl. I), I-167–I-171.

Campbell, L.B., Gooden, B.A., and Horowitz, J.D. (1969). Cardiovascular responses to partial and total immersion in man. *The Journal of Physiology*, **202**, 239-250.

Candel, S. and Ehrlich, D.E. (1953). Venous blood flow during the Valsalva experiment including some clinical applications. *The American Journal of Medicine*, **15**, 307-315.

Cannon, W.B. and Rosenblueth, A. (1933). Studies on conditions of activity in endocrine organs. XXIX. Sympathin E and sympathin I. *American Journal of Physiology*, **104**, 557-574.

Caputa, M. and Cabanac, M. (1979). Bradycardia during face cooling in man may be produced by selective brain cooling. *Journal of Applied Physiology*, **46**, 905-907.

Carboni, G.P., Lahiri, A., Cashman, P.M.M., and Raftery, E.B. (1987). Mechanisms of arrhythmias accompanying ST-segment depression on ambulatory monitoring in stable angina pectoris. *The American Journal of Cardiology*, **60**, 1246-1253.

Carlsson, A., Folkow, B., and Häggendal, J. (1964). Some factors influencing the release of noradrenaline into the blood following sympathetic stimulation. *Life Sciences*, **3**, 1335-1341.

Carlsten, A. and Heyman, F. (1965). Effect of brief carotid-sinus pressure on atrial and ventricular rhythms in complete heart block. *Acta Medica Scandinavica*, **177**, 281-286.

Carlsten, A., Folkow, B., and Hamberger, C.-A. (1957). Cardiovascular effects of direct vagal stimulation in man. *Acta Physiologica Scandinavica*, **41**, 68-76.

Carlsten, A., Folkow, B., Grimby, G., Hamberger, C.-A., and Thulesius, O. (1958). Cardiovascular effects of direct stimulation of the carotid sinus nerve in man. *Acta Physiologica Scandinavica*, **44**, 138-145.

Carlsten, A., Folkow, B., and Hamberger, C.-A. (1959). Physiological analysis and subsequent surgical treatment of a case of carotid sinus syndrome. *Acta Oto-Laryngolica*, **51**, 73-78.

Carlsten, A., Engström, H., Folkow, B., Lindell, S.-F., Nilsson, N.J., and Stenberg, K. (1967). Cardiac output in man during direct vagal stimulation (Abstract). *Proceedings of the International Union of Physiological Sciences*, **2**, 69.

Carrier, G.O. and Bishop, V.S. (1972). The interaction of acetylcholine and norepinephrine on heart rate. *The Journal of Pharmacology and Experimental Therapeutics*, **180**, 31-37.

Carrive, P., Bandler, R., and Dampney, R.A.L. (1988). Anatomical evidence that hypertension associated with the defence reaction in the cat is mediated by a direct projection from a restricted portion of the midbrain periaqueductal grey to the subretrofacial nucleus of the medulla. *Brain Research*, **460**, 339-345.

Casadei, B., Pipilis, A., Sousa, F., Flather, M., Conway, J. & Sleight, P. (1991). Transdermial scopolamine: a potential tool in preventing sudden death in myocardial infarction (Abstract). *European Heart Journal*, **12**, 366.

Casadei, B., Meyer, T.E., Coats, A.J.S., Conway, J., and Sleight, P. (1992). Baroreflex control of stroke volume in man: an effect mediated by the vagus. *The Journal of Physiology*, 1992.

Castellanos, A., Sung, R.J., Cunha, D., and Myerburg, R.J. (1974). His bundle recordings in paroxysmal atrioventricular block produced by carotid sinus massage. *British Heart Journal*, **36**, 487-491.

Caverson, M.M., Ciriello, J., and Calaresu, F.R. (1984). Chemoreceptor and baroreceptor inputs to ventrolateral medullary neurons. *American Journal of Physiology*, **247**, R872-R879.

Cechetto, D.F. and Calaresu, F.R. (1985). Central pathways relaying cardiovascular afferent information to amygdala. *American Journal of Physiology*, **248**, R38-R45.

Cerati, D. and Schwartz, P.J. (1989). Vagal reflexes and survival during acute myocardial ischemia in cats (Abstract). *Circulation*, **80** (Suppl. II), II-552.

Cerati, D. and Schwartz, P.J. (1990). Do inhibitory sympatho-vagal reflexes contribute to sudden death after myocardial infarction? *Circulation*, **82** (Suppl. III), III-634.

Chadda, K., Goldstein, S., Byington, R., and Curb, J.D. (1986). Effect of propranolol after acute myocardial infarction in patients with congestive heart failure. *Circulation*, **73**, 503-510.

Chamontin, B. *et al.* (1989). Hypertension artérielle neurogène de l'homme. *Archives des Maladies du Coeur et des Vaisseaux*, **82**, 1143-1146.

Chan, J.Y.H., Barnes, C.D., and Chan, S.H.H. (1990). Tonic enhancement of the sensitivity of baroreceptor reflex response by endogenous substance P in the rat. *Regulatory Peptides*, **29**, 199-213.

Chang, P.C., Kriek, E., van der Krogt, J.A., Blauw, G.-J., and van Brummelen, P. (1988). Haemodynamic effects of physiological concentrations of circulating noradrenaline in man. *Clinical Science*, **75**, 469-475.

Chang, P.C., van der Krogt, J.A., Vermeij, P., and van Brummelen, P. (1986). Norepinephrine removal and release in the forearm of healthy subjects. *Hypertension*, **8**, 801-809.

Chapleau, M.W. and Abboud, F.M. (1987). Contrasting effects of static and pulsatile pressure on carotid baroreceptor activity in dogs. *Circulation Research*, **61**, 648-658.

Charlton, J.D. and Baertschi, A.J. (1982). Responses of aortic baroreceptors to changes of aortic blood flow and pressure in rat. *American Journal of Physiology*, **242**, H520-H525.

Chase, P.B., Seals, D.R., Johnson, D.G., Comess, K.A., and Rowell, L.B. (1988). Hypoxemia increases sympathetic activity but not catecholamines in humans at rest (Abstract). *The Physiologist*, **31**, A112.

Chassaing, C., Duchene-Marullaz, P., Moins, N., and Combre, A. (1976). Effets paradoxaux de propranolol et d'atropine sur la fréquence cardiaque du chien non narcosé. *Journal de Pharmacologie*, **7**, 609-618.

Chassaing, C., Duchene-Marullaz, P., and Veyrac, M.J. (1983). Effects of catecholamines on cardiac chronotropic response to vagal stimulation in the dog. *American Journal of Physiology*, **245**, H721-H724.

Chatterjee, K. *et al.* (1973). Hemodynamic and metabolic responses to vasodilator therapy in acute myocardial infarction. *Circulation*, **48**, 1183-1193.

Chen, H.I. (1979). Interaction between the baroreceptor and Bezold–Jarisch reflexes. *American Journal of Physiology*, **237**, H655-H661.

Chen, J.-S., Wang, W., Bartholet, T., and Zucker, I.H. (1991). Analysis of baroreflex control of heart rate in conscious dogs with pacing-induced heart failure. *Circulation*, **83**, 260-267.

Chess, G.F., Tam, R.M.K., and Calaresu, F.R. (1975). Influence of cardiac neural inputs on rhythmic variations of heart period in the cat. *American Journal of Physiology*, **228**, 775-780.

Chiang, B.N., Perlman, L.V., and Epstein, F.H. (1969). Overweight and hypertension. A review. *Circulation*, **39**, 403-421.

Chidsey, C.A., Harrison, D.C., and Braunwald, E. (1962). Augmentation of the plasma nor-epinephrine response to exercise in patients with congestive heart failure. *The New England Journal of Medicine*, **267**, 650-654.

Chidsey, C.A., Braunwald, E., and Morrow, A.G. (1965). Catecholamine excretion and cardiac stores of norepinephrine in congestive heart failure. *The American Journal of Medicine*, **39**, 442-451.

Chigr, F. *et al.* (1989). Anatomical distribution of somatostatin immunoreactivity in the infant brainstem. *Neuroscience*, **29**, 615-628.

Christensen, N.J. and Brandsborg, O. (1973). The relationship between plasma catecholamine concentration and pulse rate during exercise and standing. *European Journal of Clinical Investigation*, **3**, 299-306.

Christensen, N.J. and Galbo, H. (1983). Sympathetic nervous activity during exercise. *Annual Review of Physiology*, **45**, 139-153.

Chungcharoen, D., Daly, M.deB., Neil, E., and Schweitzer, A. (1952). The effect of carotid occlusion upon the intrasinusal pressure with special reference to vascular communications between the carotid and vertebral circulations in the dog, cat and rabbit. *The Journal of Physiology*, **117**, 56-76.

Cicogna, R., Curnis, A., Dei Cas, L., and Visioli, O. (1985). Syncope and tumours in the neck: carotid sinus or glossopharyngeal syndrome? European Heart Journal, **6**, 979-984.

Cinca, J. *et al.* (1985). Electrophysiologic effects of unilateral right and left stellate ganglion block on the human heart. *American Heart Journal*, **109**, 46-54.

Ciriello, J. (1983). Brainstem projections of aortic baroreceptor afferent fibers in the rat. *Neuroscience Letters*, **36**, 37-42.

Ciriello, J., Hrycyshyn, A.W., and Calaresu, F.R. (1981). Glossopharyngeal and vagal afferent projections to the brain stem of the cat: a horseradish peroxidase study. *Journal of the Autonomic Nervous System*, **4**, 63-79.

Ciriello, J., Rohlicek, C.V., Poulsen, R.S., and Polosa, C. (1982). Deoxyglucose uptake in the rat thoracolumbar cord during activation of aortic baroreceptor afferent fibers. *Brain Research*, **231**, 240-245.

Ciriello, J., Rohlicek, C.V., and Polosa, C. (1983). Aortic baroreceptor reflex pathway: a functional mapping using [^{3}H]2-deoxyglucose autoradiography in the rat. *Journal of the Autonomic Nervous System*, **8**, 111-128.

Ciriello, J., Caverson, M.M., and Polosa, C. (1986). Function of the ventrolateral medulla in the control of the circulation. *Brain Research Reviews*, **11**, 359-391.

Clarke, B.F., Ewing, D.J., and Campbell, I.W. (1979). Diabetic autonomic neuropathy. *Diabetologia*, **17**, 195-212.

Clarke, R.S.J., Hellon, R.F., and Lind, A.R. (1958). The duration of sustained contractions of the human forearm at different muscle temperatures. *The Journal of Physiology*, **143**, 454-473.

Cleaveland, C.R., Rangno, R.E., and Shand, D.G. (1972). A standardized isoproterenol sensitivity test. The effects of sinus arrhythmia, atropine, and propranolol. *Archives of Internal Medicine*, **130**, 47-52.

Clemente, C.D. and Van Breemen, V.L. (1955). Nerve fibers in the area postrema of cat, rabbit, guinea pig and rat. *The Anatomical Record*, **123**, 65-79.

Clemson, B., Baily, R.G., Davis, D., and Zelis, R. (1990). Effects of desipramine on norepinephrine clearance in congestive heart failure. *American Journal of Physiology*, **259**, E261-E265.

Cobb, L.A., Baum, R.S., Alvarez, H., and Schaffer, W.A. (1975). Resuscitation from out-of-hospital ventricular fibrillation: 4 years follow-up. *Circulation*, **51 and 52** (Suppl. III), III-223–III-235.

Cody, R.J., Franklin, K.W., Kluger, J., and Laragh, J.H. (1982*a*). Mechanisms governing the postural response and baroreceptor abnormalities in chronic congestive heart failure: effects of acute and long-term converting-enzyme inhibition. *Circulation*, **66**, 135-42.

Cody, R.J., Franklin, K.W., Kluger, J., and Laragh, J.H. (1982*b*). Sympathetic responsiveness and plasma norepinephrine during therapy of chronic congestive heart failure with captopril. *The American Journal of Medicine*, **72**, 791-797.

Cody, R.J., Atlas, S.A., Laragh, J.H., Kubo, S.H., and Covit, A.B. (1986). Atrial natriuretic factor in normal subjects and heart failure patients. Plasma levels and renal, hormonal, and hemodynamic responses to peptide infusion. *The Journal of Clinical Investigation*, **78**, 1362-1374.

Cohen, M.I. (1975). Phrenic and recurrent laryngeal discharge patterns and the Hering–Breuer reflex. *American Journal of Physiology*, **228**, 1489-1496.

Cohen, M.I. and Gootman, P.M. (1970). Periodicities in efferent discharge of splanchnic nerve of the cat. *American Journal of Physiology*, **218**, 1092-1101.

Cohen, M.V. (1972). Ventricular fibrillation precipitated by carotid sinus pressure: case report and review of the literature. *American Heart Journal*, **84**, 681-686.

Cohn, J.N. *et al.* (1984). Plasma norepinephrine as a guide to prognosis in patients with chronic congestive heart failure. *The New England Journal of Medicine*, **311**, 819-823.

Coker, R., Koziell, A., Oliver, C., and Smith, S.E. (1984). Does the sympathetic nervous system influence sinus arrhythmia in man? Evidence from combined autonomic blockade. *The Journal of Physiology*, **356**, 459-464.

Cokkinos, D.V. *et al.* (1979). Effect of digitalis on diminished baroreceptor sensitivity in diabetics. *Angiology*, **30**, 549-557.

Coleridge, H.M., Coleridge, J.C.G., and Kidd, C. (1964). Cardiac receptors in the dog, with particular reference to two types of afferent ending in the ventricular wall. *The Journal of Physiology*, **174**, 323-339.

Coleridge, H.M., Coleridge, J.C.G., and Rosenthal, F. (1976). Prolonged inactivation of cortical pyramidal tract neurones in cats by distension of the carotid sinus. *The Journal of Physiology*, **256**, 635-649.

Coleridge, H.M., Coleridge, J.C.G., Poore, E.R., Roberts, A.M., and Schultz, H.D. (1984). Aortic wall properties and baroreceptor behaviour at normal arterial pressure and in acute hypertensive resetting in dogs. *The Journal of Physiology*, **350**, 309-326.

Coleridge, H.M., Coleridge, J.C.G., and Schultz, H.D. (1987). Characteristics of C fibre baroreceptors in the carotid sinus of dogs. *The Journal of Physiology*, **394**, 291-313.

Collins, M.N. and Billman, G.E. (1989). Autonomic response to coronary occlusion in animals susceptible to ventricular fibrillation. *American Journal of Physiology*, **257**, H1886-H1894.

Colucci, W.S. *et al.* (1988). Intracoronary infusion of dobutamine to patients with and without severe congestive heart failure. *The Journal of Clinical Investigation*, **81**, 1103-1110.

Colucci, W.S. *et al.* (1989). Impaired chronotropic response to exercise in patients with congestive heart failure. Role of postsynaptic β-adrenergic desensitization. *Circulation*, **80**, 314-323.

Come, P.C. and Pitt, B. (1976). Nitroglycerin-induced severe hypotension and bradycardia in patients with acute myocardial infarction. *Circulation*, **54**, 624-628.

Come, P.C. *et al.* (1975). Reversal by phenylephrine of the beneficial effects of intravenous nitroglycerin in patients with acute myocardial infarction. *The New England Journal of Medicine*, **293**, 1003-1007.

Concato, L. (1870). Sulla fisioigiologia e fisiopatologia del cuore. *Rivista Clinica*, **9**, 1-9.

CONSENSUS Trial Study Group (1987). Effects of enalapril on mortality in severe congestive heart failure. Results of the Cooperative North Scandinavian Enalapril Survival Study (CONSENSUS). *The New England Journal of Medicine*, **316**, 1429-1435.

Convertino, V.A. and Adams, W.C. (1991). Enhanced vagal baroreflex response during 24 hours after maximal exercise. *American Journal of Physiology*, **260**, R570-R575.

Convertino, V.A., Doerr, D.F., Eckberg, D.L., Fritsch, J.M., and Vernikos-Danellis, J. (1990). Head-down bed rest impairs vagal baroreflex responses and provokes orthostatic hypotension. *Journal of Applied Physiology*, **68**, 1458-1464.

Convertino, V.A., Adams, W.C., Shea, J.D., Thompson, C.A., and Hoffler, G.W. (1991). Impairment of the carotid–cardiac baroreflex in wheelchair-dependent quadriplegics. *American Journal of Physiology*, **260**, R576-R580.

Conway, J. (1963). A vascular abnormality in hypertension. A study of blood flow in the forearm. *Circulation*, **27**, 520-529.

Conway, J., Boon, N., Jones, J.V., and Sleight, P. (1983). Involvement of the baroreceptor reflexes in the changes in blood pressure with sleep and mental arousal. *Hypertension*, **5**, 746-748.

Conway, J., Boon, N., Davies, C., Jones, J.V., and Sleight, P. (1984). Neural and humoral mechanisms involved in blood pressure variability. *Journal of Hypertension*, **2**, 203-208.

Cooper, A. (1836). Some experiments and observations on tying the carotid and vertebral arteries, and the pneumo-gastric, phrenic, sympathetic nerves. *Guy's Hospital Reports*, **1**, 457-475.

Cooper, M.W., Myers, W.D., Stanbaugh, G.H., Deutsch, D.T., and Calvert, J.E. (1980). Echocardiographic evaluation of the Valsalva maneuver in an asymptomatic hemodialysis population. *Transactions of the American Society of Artificial Internal Organs*, **26**, 387-390.

Coote, J.H. (1988). The organization of cardiovascular neurons in the spinal cord. *Reviews of Physiology, Biochemistry and Pharmacology*, **110**, 147-292.

Coote, J.H. and Downman, C.B.B. (1969). Supraspinal control of reflex activity in renal nerves. *The Journal of Physiology*, **202**, 161-170.

Coote, J.H. and Westbury, D.R. (1979). Functional grouping of sympathetic preganglionic neurones in the third thoracic segment of the spinal cord. *Brain Research*, **179**, 367-372.

Cori, C.F. and Buchwald, K.W. (1930). Effect of continuous intravenous injection of epinephrine on the carbohydrate metabolism, basal metabolism and vascular system of normal men. *American Journal of Physiology*, **95**, 71-78.

Corr, P.B. and Gillis, R.A. (1974). Role of the vagus nerves in the cardiovascular changes induced by coronary occlusion. *Circulation*, **49**, 86-97.

Corr, P.B. and Gillis, R.A. (1978). Autonomic neural influences on the dysrhythmias resulting from myocardial infarction. *Circulation Research*, **43**, 1-9.

Corr, P.B., Pearle, D.L., and Gillis, R.A. (1976). Coronary occlusion site as a determinant of the cardiac rhythm effects of atropine and vagotomy. *American Heart Journal*, **92**, 741-749.

Cortes, R., Probst, A., and Palacios, J.M. (1984). Quantitative light microscopic autoradiographic localization of cholinergic muscarinic receptors in the human brain: brainstem. *Neuroscience*, **12**, 1003-1026.

Cortes, R., Probst, A., Tobler, H.-J., and Palacios, J.M. (1986). Muscarinic cholinergic receptor subtypes in the human brain. II. Quantitative autoradiographic studies. *Brain Research*, **362**, 239-253.

Cottle, M.K. (1964). Degeneration studies of primary afferents of IXth and Xth cranial nerves in the cat. *The Journal of Comparative Neurology*, **122**, 329-343.

Cottle, M.K.W. and Calaresu, F.R. (1975). Projections from the nucleus and tractus solitarius in the cat. *The Journal of Comparative Neurology*, **161**, 143-158.

Coumel, P. (1990). Role of the autonomic nervous system in paroxysmal atrial fibrillation. In *Atrial arrhythmias. Current concepts and management*, ed. Touboul, P., and Waldo, A.L., pp. 248-261. Mosby Year Book, St. Louis, Missouri.

Coumel, P., Attuel, P., Lavallée, J., Flammang, D., Leclercq, J.F., and Slama, R. (1978). Syndrome d'arythmie auriculaire d'origine vagale. *Archives des Maladies du Coeur et des Vaisseaux*, **71**, 645-656.

Coumel, P., Leclercq, J.-F., Attuel, P., Lavallée, J.-P., and Flammang, D. (1979). *Autonomic influences in the genesis of atrial arrhythmias: atrial flutter and fibrillation of vagal origin*, pp.243-255. Williams & Wilkins, Baltimore, Maryland.

Coumel, P., Leclercq, J.-F., and Zimmerman, M. (1986). The clinical use of beta-blockers in the prevention of sudden death. *European Heart Journal*, **7** (Suppl. A), 187-201.

Courtice, G.P., Kwong, T.E., Lumbers, E.R., and Potter, E.K. (1984). Excitation of the cardiac vagus by vasopressin in mammals. *The Journal of Physiology*, **354**, 547-556.

Cowan, W.M., Gottlieb, D.I., Hendrickson, A.E., Price, J.L., and Woolsey, T.A. (1972). The autoradiographic demonstration of axonal connections in the central nervous system. *Brain Research*, **37**, 21-51.

Cowley, A.W., Jr. (1982). Vasopressin and cardiovascular regulation. In *Cardiovascular physiology IV*, ed. Guyton, A.C., and Hall, J.E., p. 189. University Park Press, Baltimore, Maryland.

Cowley, A.W., Jr. and Guyton, A.C. (1972). Quantification of intermediate steps in the renin–angiotensin–vasoconstrictor feedback loop in the dog. *Circulation Research*, **30**, 557-566.

Cowley, A.W., Jr., Liard, J.F., and Guyton, A.C. (1973). Role of the baroreceptor reflex in daily control of arterial blood pressure and other variables in dogs. *Circulation Research*, **32**, 564-576.

Cowley, A.W., Jr., Monos, E., and Guyton, A.C. (1974). Interaction of vasopressin and the baroreceptor reflex system in the regulation of arterial blood pressure in the dog. *Circulation Research*, **34**, 505-514.

Cowley, A.W., Jr., Quillen, E.W., and Barber, B.J. (1980). Further evidence for lack of baroreceptor control of long-term level of arterial pressure. In *Arterial Baroreceptors and hypertension*, ed. Sleight, P., pp. 391-398. Oxford University Press, Oxford.

Cowley, A.W., Jr., Merrill, D., Osborn, J., and Barber, B.J. (1984). Influence of vasopressin and angiotension on baroreflexes in the dog. *Circulation Research*, **54**, 163-172.

Cox, B.F., Hay, M., and Bishop, V.S. (1990). Neurons in area postrema mediate vasopressin-induced enhancement of the baroreflex. *American Journal of Physiology*, **258**, H1943-H1946.

Cox, G.E., Jordan, D., Moruzzi, P., Schwaber, J.S., Spyer, K.M., and Turner, S.A. (1986). Amygdaloid influences on brain-stem neurones in the rabbit. *The Journal of Physiology*, **381**, 135-148.

Crawford, J.H. (1923). The influence of the vagus on the heart rate. *The Journal of Pharmacology and Experimental Therapeutics*, **22**, 1-19.

Creager, M.A. *et al.* (1981). Acute regional circulatory and renal hemodynamic effects of converting-enzyme inhibition in patients with congestive heart failure. *Circulation*, **64**, 483-489.

Creager, M.A. *et al.* (1982). Determinants of clinical response and survival in patients with congestive heart failure treated with captopril. *American Heart Journal*, **104**, 1147-1154.

Creager, M.A., Faxon, D.P., Rockwell, S.M., Gavras, H., and Coffman, J.D. (1985). The contribution of the renin–angiotensin system to limb vasoregulation in patients with heart failure: observations during orthostasis and α-adrenergic blockade. *Clinical Science*, **68**, 659-667.

Creager, M.A., Faxon, D.P., Cutler, S.S., Kohlmann, O., Ryan, T.J., and Gavras, H. (1986). Contribution of vasopressin to vasoconstriction in patients with congestive heart failure: comparison with the renin–angiotension system and the sympathetic nervous system. *Journal of the American College of Cardiology*, **7**, 758-765.

Creager, M.A., Hirsch, A.T., Nabel, E.G., Cutler, S.S., Colucci, W.S., and Dzau, V.J. (1988). Responsiveness of atrial natriuretic factor to reduction in right atrial pressure in patients with chronic congestive heart failure. *Journal of the American College of Cardiology*, **11**, 1191-1198.

Criscione, L., Reis, D.J., and Talman, W.T. (1983). Cholinergic mechanisms in the nucleus tractus solitarii and cardiovascular regulation in the rat. *European Journal of Pharmacology*, **88**, 47-55.

Crossley, R.J., Greenfield, A.D.M., Plassaras, G.C., and Stephens, D. (1966). The interrelation of thermoregulatory and baroreceptor reflexes in the control of the blood vessels in the human forearm. *The Journal of Physiology*, **183**, 628-636.

Crozier, I.G., Ikram, H., and Nicholls, M.G. (1989). Hemodynamic and hormone changes during induced ventricular tachycardia secondary to coronary artery disease. *The American Journal of Cardiology*, **63**, 618-621.

Cryer, P.E. (1980). Physiology and pathophysiology of the human sympathoadrenal neuroendocrine system. *The New England Journal of Medicine*, **303**, 436-444.

Cryer, P.E., Santiago, J.V., and Shah, S. (1974). Measurement of norepinephrine and epinephrine in small volumes of human plasma by a single isotope derivative method: response to the upright posture. *Journal of Clinical Endocrinology and Metabolism*, **39**, 1025-1029.

Cunningham, D.J.C., Howson, M.G., Petersen, E.S., Pickering, T.G., and Sleight, P. (1970). Changes in the sensitivity of the baroreflex in muscular exercise (Abstract). *Acta Physiologica Scandinavica*, **79**, 16A-17A.

Cunningham, D.J.C., Petersen, E.S., Peto, R., Pickering, T.G., and Sleight, P. (1972*a*). Comparison of the effect of different types of exercise on the baroreflex regulation of heart rate. *Acta Physiologica Scandinavica*, **86**, 444-455.

Cunningham, D.J.C., Petersen, E.S., Pickering, T.G., and Sleight, P. (1972*b*). The effects of hypoxia, hypercapnia, and asphyxia on the baroreceptor–cardiac reflex at rest and during exercise in man. *Acta Physiologica Scandinavica*, **86**, 456-465.

Currie, M.G., Geller, D.M., Cole, B.R., Boylan, J.G., YuSheng, W., Holmberg, S.W., and Needleman, P. (1983). Bioactive cardiac substances: Potent vasorelaxant activity in mammalian atria. *Science*, **221**, 71-73.

Curry, P.V.L. (1979). The hemodynamic and electrophysiological effects of paroxysmal tachycardia. In *Cardiac arrhythmias: electrophysiology, diagnosis and management*, ed. Narula, O.S., pp. 364-381. Williams & Wilkins, Baltimore, Maryland.

Curry, P.V.L., Evans, T.R., and Krikler, D.M. (1977). Paroxysmal reciprocating sinus tachycardia. *European Journal of Cardiology*, **6**, 199-228.

Curtiss, C., Cohn, J.N., Vrobel, T., and Franciosa, J.A. (1978). Role of the renin–angiotensin system in the systemic vasoconstriction of chronic congestive heart failure. *Circulation*, **58**, 763-770.

Cyon, E. and Ludwig, C. (1866). Die Reflexe eines der sensiblen Nerven des Herzens auf die motorischen der Blutgefässe. *Berichte über die Verhandlungen der Königlich Sächsischen Gesellschaft der Wissenschaften zu Leipzig*, **18**, 307-328.

Czachurski, J., Dembowsky, K., Seller, H., Nobiling, R., and Taugner, R. (1988). Morphology of electrophysiologically identified baroreceptor afferents and second order neurones in the brainstem of the cat. *Archives Italiennes de Biologie*, **126**, 129-144.

Czermak, J. (1866). Ueber mechanische Vagus-Reizung beim Menschen. *Jenaische Zeitschrift für Medicin und Naturwissenschaft*, 384-386.

Dahlström, A. and Fuxe, K. (1964). Evidence for the existence of monoamine-containing neurons in the central nervous system. I. Demonstration of monoamines in the cell bodies of brain stem neurons. *Acta Physiologica Scandinavica*, **62** (Suppl. 232), 1-55.

Dahlström, A. and Fuxe, K. (1965). Evidence for the existence of monoamine neurons in the central nervous system. II. Experimentally induced changes in the intraneuronal amine levels of bulbospinal neuron systems. *Acta Physiologica Scandinavica*, **64** (Suppl. 247), 1-36.

Dale, H.H. (1914). The action of certain esters and ethers of choline, and their relation to muscarine. *The Journal of Pharmacology and Experimental Therapeutics*, **6**, 147-190.

Dale, H.H. and Feldberg, W. (1934). The chemical transmission of secretory impulses to the sweat glands of the cat. *The Journal of Physiology*, **82**, 121-128.

Daly, I.deB. (1927). The localisation of receptors involved in the reflex regulation of the heart rate. *The Journal of Physiology*, **62**, 330-340.

Daly, M.deB. (1986). Interactions between respiration and circulation. In *Handbook of physiology, Section 3: The respiratory system* Volume II, ed. Cherniack, N.S., and Widdicombe, J.G., pp. 529-594. American Physiological Society, Bethesda, Maryland.

Daly, M.deB. and Scott, M.J. (1958). The effects of stimulation of the carotid body chemoreceptors on heart rate in the dog. *The Journal of Physiology*, **144**, 148-166.

Daly, M.deB., Elsner, R., and Angell-James, J.E. (1977). Cardiorespiratory control by carotid chemoreceptors during experimental dives in the seal. *American Journal of Physiology*, **232**, H508-H516.

Damato, A.N., Lau, S.H., Helfant, R.H., Stein, E., Berkowitz, W.D., and Cohen, S.I. (1969). Study of atrioventricular conduction in man using electrode catheter recordings of His bundle activity. *Circulation*, **39**, 287-296.

Dambrink, J.H.A. (1991). Orthostatic regulation of blood pressure in healthy individuals: a comparative study in young and old subjects (Ph.D. Thesis), University of Amsterdam, Amsterdam.

Dampney, R. (1990). The subretrofacial nucleus: its pivotal role in cardiovascular regulation. *News in Physiological Sciences*, **5**, 63-67.

Dampney, R.A.L. (1981*a*). Brain stem mechanisms in the control of arterial pressure. *Clinical and Experimental Hypertension*, **3**, 379-391.

Dampney, R.A.L. (1981*b*). Functional organization of central cardiovascular pathways. *Clinical and Experimental Pharmacology and Physiology*, **8**, 241-259.

Dampney, R.A.L. and McAllen, R.M. (1988). Differential control of sympathetic fibres supplying hindlimb skin and muscle by subretrofacial neurones in the cat. *The Journal of Physiology*, **395**, 41-56.

Dampney, R.A.L. and Moon, E.A. (1980). Role of ventrolateral medulla in vasomotor response to cerebral ischemia. *American Journal of Physiology*, **239**, H349-H358.

Dampney, R.A.L., Goodchild, A.K., Robertson, L.G., and Montgomery, W. (1982). Role of ventrolateral medulla in vasomotor regulation: a correlative anatomical and physiological study. *Brain Research*, **249**, 223-235.

Dampney, R.A.L., Czachurski, J., Dembowsky, K., Goodchild, A.K., and Seller, H. (1987). Afferent connections and spinal projections of the pressor region in the rostral ventrolateral medulla of the cat. *Journal of the Autonomic Nervous System*, **20**, 73-86.

Daremberg, Ch. (1854). *Oeuvres anatomiques, physiologiques et médicales de Galien.* J.B. Baillière, Paris.

Dargie, H.J., Cleland, J.G.F., Leckie, B.J., Inglis, C.G., East, B.W., and Ford, I. (1987). Relation of arrhythmias and electrolyte abnormalities to survival in patients with severe chronic heart failure. *Circulation*, **75** (Suppl. IV), IV-98–IV-107.

Das, G., Talmers, F.N., and Weissler, A.M. (1975). New observations on the effects of atropine on the sinoatrial and atrioventricular nodes in man. *The American Journal of Cardiology*, **36**, 281-285.

Davies, M.J. and Thomas, A. (1984). Thrombosis and acute coronary-artery lesions in sudden cardiac ischemic death. *The New England Journal of Medicine*, **310**, 1137-1140.

Davies, R., Forsling, M.L., and Slater, J.D.H. (1977). The interrelationship between the release of renin and vasopressin as defined by orthostasis and propranolol. *The Journal of Clinical Investigation*, **60**, 1438-1441.

Davis, D. *et al.* (1987). Norepinephrine kinetics during orthostatic stress in congestive heart failure. *Circulation Research*, **61** (Suppl. I), I-87–I-90.

Davis, D., Baily, R., and Zelis, R. (1988). Abnormalities in systemic norepinephrine kinetics in human congestive heart failure. *American Journal of Physiology*, **254**, E760-E766.

Dawson, P.M. (1943). An historical sketch of the Valsalva experiment. *British Institute for the History of Medicine*, **14**, 295-320.

Day, T.A. and Renaud, L.P. (1984). Electrophysiological evidence that noradrenergic afferents selectively facilitate the activity of supraoptic vasopressin neurons. *Brain Research*, **303**, 233-240.

Day, T.A., Ro, A., and Renaud, L.P. (1983). Depressor area within caudal ventrolateral medulla of the rat does not correspond to the A1 catecholamine cell group. *Brain Research*, **279**, 299-302.

De Boer, R.W. (1985). Beat-to-beat blood-pressure fluctuations and heart-rate variability in man: physiological relationships, analysis techniques and a simple model (Ph.D. thesis), University of Amsterdam, Amsterdam.

De Boer, R.W., Karemaker, J.M., and Strackee, J. (1985*a*). Description of heart-rate variability data in accordance with a physiological model for the genesis of heartbeats. *Psychophysiology*, **22**, 147-155.

De Boer, R.W., Karemaker, J.M., and Strackee, J. (1985*b*). Relationships between short-term blood-pressure fluctuations and heart-rate variability in resting subjects. II: a simple model. *Medical and Biological Engineering and Computing*, **23**, 359-364.

De Boer, R.W., Karemaker, J.M., and van Montfrans, G.A. (1986). Determination of baroreflex sensitivity by spectral analysis of spontaneous blood-pressure and heart-rate fluctuations in man. In *Neural mechanisms and cardiovascular disease*, ed. Lown, B., Malliani, A., and Prosdocimi, D., pp. 303-315. Liviana, Padov.

De Boer, R.W., Karemaker, J.M., and Strackee, J. (1987). Hemodynamic fluctuations and baroreflex sensitivity in humans: a beat-to-beat model. *American Journal of Physiology*, **253**, H680-H689.

De Boissezon, P. (1942). *La zone reflexogène carotidienne. Etude anatomique et histophysiologique*. Librairie Marqueste, Toulouse.

De Bold, A.J. (1979). Heart atria granularity effects of changes in water-electrolyte balance. *Proceedings of the Society for Experimental Biology and Medicine*, **161**, 508-511.

De Bold, A.J., Borenstein, H.B., Veress, A.T., and Sonnenberg, H. (1981). A rapid and potent natriuretic response to intravenous injection of atrial myocardial extract in rats. *Life Sciences*, **28**, 89-94.

De Castro, F. (1926). Sur la structure et l'innervation de la glande intercarotidienne (glomus caroticum) de l'homme et des mammifères, et sur un nouveau système d'innervation autonome du nerf glossopharyngien. *Travaux du Laboratoire de Recherches Biologiques de l'Université de Madrid*, **24**, 367-432.

De Castro, F. (1927-1928). Sur la structure et l'innervation du sinus carotidien de l'homme et des mammifères. Nouveaux faits sur l'innervation et la fonction du glomus caroticum. *Travaux du Laboratoire de Recherches Biologiques de l'Université de Madrid*, **25**, 332-380.

De Castro, F. (1940). Nuevas observaciones sobre la inervación de la región carotídea. Los quimio y preso-receptores. *Trabajos del Instituto Cajal de Investigaciones Biológicas*, **32**, 297-384.

De Castro, F. (1951). Sur la structure de la synapse dans les chemocepteurs: leur mécanisme d'excitation et rôle dans la circulation sanguine locale. *Acta Physiologica Scandinavica*, **22**, 14-43.

De Champlain, J., Farley, L., Cousineau, D., and van Ameringen, M.-R. (1976). Circulating catecholamine levels in human and experimental hypertension. *Circulation Research*, **38**, 109-114.

De Ferrari, G.M., Vanoli, E., Stramba-Badiale, M., Foreman, R.D., and Schwartz, P.J. (1987). Vagal stimulation and sudden death in conscious dogs with a healed myocardial infarction (Abstract). *Circulation*, **76** (Suppl. IV), IV-107.

De Torrente, A., Robertson, G.L., McDonald, K.M., and Schrier, R.W. (1975). Mechanism of diuretic response to increased left atrial pressure in the anesthetized dog. *Kidney International*, **8**, 355-361.

Dehn, T.C.B., Morley, C.A., and Sutton, R. (1984). A scientific evaluation of the carotid sinus syndrome. *Cardiovascular Research*, **18**, 746-751.

Delius, W., Hagbarth, K.-E., Hongell, A., and Wallin, B.G. (1972*a*). General characteristics of sympathetic activity in human muscle nerves. *Acta Physiologica Scandinavica*, **84**, 65-81.

Delius, W., Hagbarth, K.-E., Hongell, A., and Wallin, B.G. (1972*b*). Manoeuvres affecting sympathetic outflow in human muscle nerves. *Acta Physiologica Scandinavica*, **84**, 82-94.

Delius, W., Hagbarth, K-E., Hongell, A., and Wallin, B.G. (1972*c*). Manoeuvres affecting sympathetic outflow in human skin nerves. *Acta Physiologica Scandinavica*, **84**, 177-186.

Dembowsky, K. and McAllen, R.M. (1990). Baroreceptor inhibition of subretrofacial neurons: evidence from intracellular recordings in the cat. *Neuroscience Letters*, **111**, 139-143.

Denker, S., Lehmann, M., Mahmud, R., Gilbert, C., and Akhtar, M. (1984). Facilitation of ventricular tachycardia induction with abrupt changes in ventricular cycle length. *The American Journal of Cardiology*, **53**, 508-515.

Dexter, F., Saidel, G.M., Levy, M.N., and Rudy, Y. (1989). Mathematical model of dependence of heart rate on tissue concentration of acetylcholine. *American Journal of Physiology*, **256**, H520-H526.

Dhingra, R.C., Amat-Y-Leon, F., Wyndham, C., Denes, P., Wu, D., Pouget, J.M., and Rosen, K.M. (1976). Electrophysiologic effects of atropine on human sinus node and atrium. *The American Journal of Cardiology*, **38**, 429-434.

Dhurandhar, R.W., MacMillan, R.L., and Brown, K.W.G. (1971). Primary ventricular fibrillation complicating acute myocardial infarction. *The American Journal of Cardiology*, **27**, 347-351.

Dibner-Dunlap, M.E. and Thames, M.D. (1989). Baroreflex control of renal sympathetic nerve activity is preserved in heart failure despite reduced arterial baroreceptor sensitivity. *Circulation Research*, **66**, 1526-1535.

Dibner-Dunlap, M.E., Eckberg, D.L., Magid, N.M., and Cintrón-Trevino, N.M. (1985). The long-term increase of baseline and reflexly augmented levels of human vagal–cardiac nervous activity induced by scopolamine. *Circulation*, **71**, 797-804.

Dietlen, H. (1905-1906). Über Grösse und Lage des normalen Herzens und ihre Abhängigkeit von physiologischen Bedingungen. *Deutscher Archiv für Klinische Medizin*, **88**, 55-122.

Dismukes, R.K. (1979). New concepts of molecular communication among neurons. *The Behavioral and Brain Sciences*, **2**, 409-448.

Dittmar, C. (1870). Ein neuer Beweis für die Reizbarkeit der centripetalen Fasern des Rückenmarks. *Berichte über die Verhandlungen der Königlich Sächsischen Gesellschaft der Wissenschaften zu Leipzig*, **22**, 18-48.

Dittmar, C. (1873). Ueber die Lage des sogenannten Gefässcentrums der Medulla Oblongata. *Berichte über die Verhandlungen der Königlich Sächsischen Gesellschaft der Wissenschaften zu Leipzig*, **25**, 449-469.

Diz, D.I. and Ferrario, C.M. (1988). Bidirectional transport of angiotensin II binding sites in the vagus nerve. *Hypertension*, **11** (Suppl. I), I-139–I-143.

Diz, D.I., Barnes, K.L., and Ferrario, C.M. (1987). Functional characteristics of neuropeptides in the dorsal medulla oblongata and vagus nerve. *Federation Proceedings*, **46**, 30-35.

Djojosugito, A.M., Folkow, B., and Yonce, L.R. (1969). Neurogenic adjustments of muscle blood flow, cutaneous A–V shunt flow and of venous tone during 'diving' in ducks. *Acta Physiologica Scandinavica*, **75**, 377-386.

Doba, N. and Reis, D.J. (1973). Acute fulminating neurogenic hypertension produced by brainstem lesions in the rat. *Circulation Research*, **32**, 584-593.

Dohi, S., Tsuchida, H., and Mayumi, T. (1983). Baroreflex control of heart rate during cardiac sympathetomy by epidural anesthesia in lightly anesthetized humans. *Anesthesia Analgesia*, **62**, 815-820.

Donald, D.E. (1979). Studies on the release of renin by direct and reflex activation of renal sympathetic nerves. *The Physiologist*, **22**, 39-42.

Donald, D.E., Samueloff, S.L., and Ferguson, D. (1967). Mechanisms of tachycardia caused by atropine in conscious dogs. *American Journal of Physiology*, **212**, 901-910.

Donald, K.W., Bishop, J.M., and Wade, O.L. (1955). Changes in the oxygen content of axillary venous blood during leg exercise in patients with rheumatic heart disease. *Clinical Science*, **14**, 531-551.

Donald, K.W., Lind, A.R., McNicol, G.W., Humphreys, P.W., Taylor, S.H., and Staunton, H.P. (1967). Cardiovascular responses to sustained (static) contractions. *Circulation Research*, **20 and 21** (Suppl. I), I-15–I-30.

Dong, E., Jr. and Reitz, B.A. (1970). Effect of timing of vagal stimulation on heart rate in the dog. *Circulation Research*, **27**, 635-646.

Donoghue, S., Fox, R.E., Kidd, C., and Koley, B.N. (1981). The distribution in the cat brain stem of neurones activated by vagal non-myelinated fibres from the heart and lungs. *Quarterly Journal of Experimental Physiology*, **66**, 391-404.

Donoghue, S., Garcia, M., Jordan, D., and Spyer, K.M. (1982). Identification and brain-stem projections of aortic baroreceptor afferent neurones in nodose ganglia of cats and rabbits. *The Journal of Physiology*, **322**, 337-352.

Donoghue, S., Felder, R.B., Jordan, D., and Spyer, K.M. (1984). The central projections of carotid baroreceptors and chemoreceptors in the cat: a neurophysiological study. *The Journal of Physiology*, **347**, 397-409.

Donoghue, S., Felder, R.B., Gilbey, M.P., Jordan, D., and Spyer, K.M. (1985). Post-synaptic activity evoked in the nucleus tractus solitarius by carotid sinus and aortic nerve afferents in the cat. *The Journal of Physiology*, **360**, 261-273.

Dormer, K.J., Person, R.J., Andrezik, J.A., Foreman, R.D., and Braggio, J.P. (1989). Ventrolateral medullary lesions and fastigial cardiovascular response in beagles. *American Journal of Physiology*, **256**, H1200-H1208.

Dornhorst, A.C., Howard, P., and Leathart, G.L. (1952). Respiratory variations in blood pressure. *Circulation*, **6**, 553-558.

Dorward, P.K. and Korner, P.I. (1978). Effect of D,L-propranolol on renal sympathetic baroreflex properties and aortic baroreceptor activity. *European Journal of Pharmacology*, **52**, 61-71.

Dorward, P.K. and Korner, P.I. (1987). Does the brain 'remember' the absolute blood pressure? *News in Physiological Sciences*, **2**, 10-13.

Dorward, P.K. Andresen, M.C., Burke, S.L., Oliver, J.R., and Korner, P.I. (1982). Rapid resetting of the aortic baroreceptors in the rabbit and its implications for short-term and longer term reflex control. *Circulation Research*, **50**, 428-439.

Dorward, P.K., Riedel, W., Burke, S.L., Gipps, J., and Korner, P.I. (1985). The renal sympathetic baroreflex in the rabbit. Arterial and cardiac baroreceptor influences, resetting, and effect of anesthesia. *Circulation Research*, **57**, 618-633.

Douglas, W.W. and Schaumann, W. (1956). A study of the depressor and pressor components of the cat's carotid sinus and aortic nerves using electrical stimuli of different intensities and frequencies. *The Journal of Physiology*, **132**, 173-186.

Douglas, W.W., Ritchie, J.M., and Schaumann, W. (1956*a*). A study of the effect of the pattern of electrical stimulation of the aortic nerve on the reflex depressor responses. *The Journal of Physiology*, **133**, 232-242.

Douglas, W.W., Ritchie, J.M., and Schaumann, W. (1956*b*). Depressor reflexes from medullated and non-medullated fibres in the rabbit's aortic nerve. *The Journal of Physiology*, **132**, 187-198.

Doyle, A.E. and Black, H. (1955). Reactivity to pressor agents in hypertension. *Circulation*, **12**, 974-980.

Dresel, P.E. and Sutter, M.C. (1961). Factors modifying cyclopropane–epinephrine cardiac arrhythmias. *Circulation Research*, **9**, 1284-1290.

Drexler, H., Hirth, C., Stasch, H.-P., Lu, W., Neuser, D., and Just, H. (1980). Vasodilatory action of endogenous atrial natriuretic factor in a rat model of chronic heart failure as determined by monoclonal ANF antibody. *Circulation Research*, **66**, 1371-1380.

Drexler, H. *et al.* (1991). Characterization of endothelial function in patients with chronic heart failure (Abstract). *Journal of the American College of Cardiology*, **17**, 282A.

Dripps, R.D. and Comroe, J.H., Jr. (1947). The effect of the inhalation of high and low oxygen concentrations on respiration, pulse rate, ballistocardiogram and arterial oxygen saturation (oximeter) of normal individuals. *American Journal of Physiology*, **149**, 277-291.

Drischel, H., Fanter, H., Gürtler, H., Labitzke, H., and Priegnitz, F. (1963). Das Verhalten der Herzfrequenz gesunder Menschen beim Übergang vom Liegen zum Stehen. *Archiv für Kreislaufforschung*, **40**, 135-167.

Dropmann, K. (1967). Electron microscopic findings in the carotid sinus of rabbits. In *Baroreceptors and hypertension*, ed. Kezdi, P., pp. 267-271. Pergamon, Oxford.

Drummond, P.D. (1990). Parasympathetic cardiac control in mild hypertension. *Journal of Hypertension*, **8**, 383-387.

Drüner, L. (1925). Ueber die anatomischen Unterlagen der Sinusreflexe Herings. *Deutsche Medizinische Wochenschrift*, **51**, 559-561.

Drysdale, D.B. and Petersen, E.S. (1977). Arterial chemoreceptors, ventilation and heart rate in man. *The Journal of Physiology*, **273**, 109-120.

Duke, P.C., Wade, J.G., Hickey, R.F., and Larson, C.P. (1976). The effects of age on baroreceptor reflex function in man. *Canadian Anaesthetists' Society Journal*, **23**, 111-124.

Dunn, A.J. (1978). Commentary. *Neurosciences Research Program Bulletin*, **16**, 554-555.

Dunn, F.L., Brennan, T.J., Nelson, A.E., and Robertson, G.L. (1973). The role of blood osmolality and volume in regulating vasopressin secretion in the rat. *The Journal of Clinical Investigation*, **52**, 3212-3219.

Dunning, A.J. (1971). *Electrostimulation of the carotid sinus nerve in angina pectoris*, Excerpta Medica, Amsterdam. pp. 1-104.

Duprez, D., De Pue, N., and Clement, D.L. (1987). Peripheral vascular responses during carotid baroreceptor stimulation in normotensive and hypertensive subjects. *Clinical Science*, **73**, 635-640.

Dworkin, B. (1988). Hypertension as a learned response: The baroreceptor reinforcement hypothesis. In *Behavioural medicine in cardiovascular disorders*, ed. Elbert, T., Langosch, W., Steptoe, A., and Vaitl, D., pp. 17-47. Wiley, Chichester.

Dworkin, B.R., Filewich, R.J., Miller, N.E., Craigmyle, N., and Pickering, T.G. (1979). Baroreceptor activation reduces reactivity to noxious stimulation: Implications for hypertension. *Science*, **205**, 1299-1301.

Dzau, V.J., Colucci, W.S., Hollenberg, N.K., and Williams, G.H. (1981). Relation of the renin–angiotensin–aldosterone system to clinical state in congestive heart failure. *Circulation*, **63**, 645-651.

Dzau, V.J., Packer, M., Lilly, L.S., Swartz, S.L., Hollenberg, N.K., and Williams, G.H. (1984). Prostaglandins in severe congestive heart failure. Relation to activation of the renin–angiotensin system and hyponatremia. *The New England Journal of Medicine*, **310**, 347-352.

Ead, H.W., Green, J.H., and Neil, E. (1952). A comparison of the effects of pulsatile and non-pulsatile blood flow through the carotid sinus on the reflexogenic activity of the sinus baroceptors in the cat. *The Journal of Physiology*, **118**, 509-519.

Earle, A.M., Zucker, I.H., Earle, A.C., and Gilmore, J.P. (1980). Morphological changes accompanying physiological changes in atrial stretch receptors. *Basic Research in Cardiology*, **75**, 510-514.

Ebert, P.A., Vanderbeek, R.B., Allgood, R.J., and Sabiston, D.C., Jr. (1970). Effect of chronic cardiac denervation on arrhythmias after coronary artery ligation. *Cardiovascular Research*, **4**, 141-147.

Ebert, T.J. (1983). Carotid baroreceptor reflex regulation of forearm vascular resistance in man. *The Journal of Physiology*, **337**, 655-664.

Ebert, T.J. (1986). Baroreflex responsiveness is maintained during isometric exercise in humans. *Journal of Applied Physiology*, **61**, 797-803.

Ebert, T.J. (1988). Reflex activation of sympathetic nervous system by ANF in humans. *American Journal of Physiology*, **255**, H685-H689.

Ebert, T.J. (1990). Differential effects of nitrous oxide on baroreflex control of heart rate and peripheral sympathetic nerve activity in humans. *Anesthesiology*, **72**, 16-22.

Ebert, T.J. and Cowley, A.W., Jr. (1988). Atrial natriuretic factor attenuates carotid baroreflex-mediated cardioacceleration in humans. *American Journal of Physiology*, **254**, R590-R594.

Ebert, T.J., Eckberg, D.L., Vetrovec, G.M., and Cowley, M.J. (1984*a*). Impedance cardiograms reliably estimate beat-by-beat changes of left ventricular stroke volume in humans. *Cardiovascular Research*, **18**, 354-360.

Ebert, T.J., Hayes, J.J., Ceschi, J., Kotrly, K.J., van Brederode, J., and Smith, J.J. (1984*b*). Repetitive ramped neck suction: a quantitative test of human baroreceptor function. *American Journal of Physiology*, **247**, H1013-H1017.

Ebert, T.J., Cowley, A.W., Jr., and Skelton, M. (1986*a*). Vasopressin reduces cardiac function and augments cardiopulmonary baroreflex resistance increases in man. *The Journal of Clinical Investigation*, **77**, 1136-1142.

Ebert, T.J., Kotrly, K.J., and Kampine, J.P. (1986*b*). Carotid baroreflex regulation of cardiac inotropic state in conscious man (Abstract). *Proceedings of the International Union of Physiological Sciences*, **16**, 169.

Ebert, T.J., Skelton, M.M., and Cowley, A.W., Jr. (1988). Dynamic cardiovascular responses to infusions of atrial natriuretic factor in humans. *Hypertension*, **11**, 537-544.

Eckberg, D.L. (1976). Temporal response patterns of the human sinus node to brief carotid baroreceptor stimuli. *The Journal of Physiology*, **258**, 769-782.

Eckberg, D.L. (1977*a*). Baroreflex inhibition of the human sinus node: importance of stimulus intensity, duration, and rate of pressure change. *The Journal of Physiology*, **269**, 561-577.

Eckberg, D.L. (1977*b*). Adaptation of the human carotid baroreceptor–cardiac reflex. *The Journal of Physiology*, **269**, 579-589.

Eckberg, D.L. (1978). Adrenergic–cholinergic baroreflex interactions in man. In *Neural mechanisms in cardiac arrhythmias*, ed. Schwartz, P.J., Brown, A.M., Malliani, A., and Zanchetti, A., pp. 335-338. Raven Press, New York.

Eckberg, D.L. (1979). Carotid baroreflex function in young men with borderline blood pressure elevation. *Circulation*, **59**, 632-636.

Eckberg, D.L. (1980*a*). Nonlinearities of the human carotid baroreceptor–cardiac reflex. *Circulation Research*, **47**, 208-216.

Eckberg, D.L. (1980*b*). Parasympathetic cardiovascular control in human disease: a critical review of methods and results. *American Journal of Physiology*, **239**, H581-H593.

Eckberg, D.L. (1983). Human sinus arrhythmia as an index of vagal cardiac outflow. *Journal of Applied Physiology*, **54**, 961-966.

Eckberg, D.L. (1984). Beta-adrenergic blockade may prolong life in post-infarction patients in part by increasing vagal cardiac inhibition. *Medical Hypotheses*, **15**, 421-432.

Eckberg, D.L. (1991). Cardiovascular responses to weightlessness. In *The Encyclopedia of Human Biology*, Volume 2. ed. Dulbecco, R., pp. 147-156, Academic Press, New York.

Eckberg, D.L. and Eckberg, M.J. (1982). Human sinus node responses to repetitive, ramped carotid baroreceptor stimuli. *American Journal of Physiology*, **242**, H638-H644.

Eckberg, D.L. and Fritsch, J.M. (1992). Influence of ten day head-down bedrest on human carotid baroreceptor–cardiac reflex function. *Acta Physiologica Scandinavica*.

Eckberg, D.L. and Orshan, C.R. (1977). Respiratory and baroreceptor reflex interactions in man. *The Journal of Clinical Investigation*, **59**, 780-785.

Eckberg, D.L. and Wallin, B.G. (1987). Isometric exercise modifies autonomic baroreflex responses in man. *Journal of Applied Physiology*, **63**, 2325-2330.

Eckberg, D.L., Drabinsky, M., and Braunwald, E. (1971). Defective cardiac parasympathetic control in patients with heart disease. *The New England Journal of Medicine*, **285**, 877-883.

Eckberg, D.L., Fletcher, G.F., and Braunwald, E. (1972). Mechanism of prolongation of the R–R interval with electrical stimulation of the carotid sinus nerves in man. *Circulation Research*, **30**, 131-138.

Eckberg, D.L., White, C.W., Kioschos, J.M., and Abboud, F.M. (1974). Mechanisms mediating bradycardia during coronary arteriography. *The Journal of Clinical Investigation*, **54**, 1455-1461.

Eckberg, D.L., Cavanaugh, M.S., Mark, A.L., and Abboud, F.M. (1975). A simplified neck suction device for activation of carotid baroreceptors. *Journal of Laboratory and Clinical Medicine*, **85**, 167-173.

Eckberg, D.L., Abboud, F.M., and Mark, A.L. (1976). Modulation of carotid baroreflex responsiveness in man: effects of posture and propranolol. *Journal of Applied Physiology*, **41**, 383-387.

Eckberg, D.L., Kifle, Y.T., and Roberts, V.L. (1980). Phase relationship between normal human respiration and baroreflex responsiveness. *The Journal of Physiology*, **304**, 489-502.

Eckberg, D.L., Bastow, H., and Scruby, A.E. (1982). Modulation of human sinus node function by systemic hypoxia. *Journal of Applied Physiology*, **52**, 570-577.

Eckberg, D.L., Mohanty, S.K., and Raczkowska, M. (1984). Trigeminal-baroreceptor reflex interactions modulate human cardiac vagal efferent activity. *The Journal of Physiology*, **347**, 75-83.

Eckberg, D.L., Nerhed, C., and Wallin, B.G. (1985). Respiratory modulation of muscle sympathetic and vagal cardiac outflow in man. *The Journal of Physiology*, **365**, 181-196.

Eckberg, D.L., Harkins, S.W., Fritsch, J.M., Musgrave, G.E., and Gardner, D.F. (1986). Baroreflex control of plasma norepinephrine and heart period in healthy subjects and diabetic patients. *The Journal of Clinical Investigation*, **78**, 366-374.

Eckberg, D.L. *et al.* (1988). Baroreflex modulation of sympathetic activity and sympathetic neurotransmitters in humans. *Acta Physiologica Scandinavica*, **133**, 221-231.

Eckberg, D.L., Wallin, B.G., Fagius, J., Lundberg, L., and Torebjörk, H.E. (1989). Prospective study of symptoms after human microneurography. *Acta Physiologica Scandinavica*, **137**, 567-569.

Eckberg, D.L., Convertino, V.A., Fritsch, J.M., and Doerr, D.F. (1992). Reproducibility of human vagal carotid baroreceptor–cardiac reflex responses. *American Journal of Physiology*, (accepted for publication).

Egan, B.M., Julius, S., Cottier, C., Osterziel, K.J., and Ibsen, H. (1983). Role of cardiovascular receptors on the neural regulation of renin release in normal men. *Hypertension*, **5**, 779-786.

Ehmke, H., Persson, P., and Kirchheim, H. (1987*a*). Pressure-dependent renin release: The kidney factor in long-term control of arterial blood pressure in conscious dogs. *Clinical and Experimental Hypertension. Part A — Theory and Practice*, **A9** (Suppl. 1), 181-195.

Ehmke, H., Persson, P., and Kirchheim, H. (1987*b*). A physiological role for pressure-dependent renin release in long-term blood pressure control. *Pflügers Archiv für die Gesamte Physiologie des Menschen und der Tiere*, **410**, 450-456.

Eide, I., Løyning, E., and Kiil, F. (1973). Evidence for hemodynamic autoregulation of renin release. *Circulation Research*, **32**, 237-245.

Einthoven, W. (1911). Neuere Ergebnisse auf dem Gebiete der tierischen Elektrizität. *Verhandlungen der Gesellschaft Deutscher Naturforscher und Ärzte*, **24**, 80-113.

Einthoven, W., Flohil, A., and Battaerd, P.J.T.A. (1908). On vagus currents examined with the string galvanometer. *Quarterly Journal of Experimental Physiology*, **1**, 243-245.

Eklund, B., Kaijser, L., and Knutsson, E. (1974). Blood flow in resting (contralateral) arm and leg during isometric contraction. *The Journal of Physiology*, **240**, 111-124.

Elbert, T., Langosch, W., Steptoe, A., and Vaitl, D. (1988*a*). *Behavioural medicine in cardiovascular disorders*. Wiley, Chichester.

Elbert, T., Rockstroh, B., Lutzenberger, W., Kessler, M., Pietrowsky, R., and Birbaumer, N. (1988*b*). Baroreceptor stimulation alters pain sensation depending on tonic blood pressure. *Psychophysiology*, **25**, 25-29.

Eliakim, M., Bellet, S., Tawil, E., and Muller, O. (1961). Effect of vagal stimulation and acetylcholine on the ventricle. Studies in dogs with complete atrioventricular block. *Circulation Research*, **9**, 1372-1379.

Elisberg, E.I. (1963). Heart rate response to the Valsalva maneuver as a test of circulatory integrity. *Journal of the American Medical Association*, **186**, 200-205.

Elisberg, E.I., Goldberg, H., and Snider, G.L. (1951). Value of intraoral pressure as a measure of intrapleural pressure. *Journal of Applied Physiology*, **4**, 171-176.

Elisberg, E.I., Miller, G., Weinberg, S.L., and Katz, L.N. (1953). The effect of the Valsalva maneuver on the circulation. II. The role of the autonomic nervous system in the production of the overshoot. *American Heart Journal*, **45**, 227-236.

Ellenbogen, K.A., Mohanty, P.K., Szentpetery, S., and Thames, M.D. (1989). Arterial baroreflex abnormalities in heart failure. Reversal after orthotopic cardiac transplantation. *Circulation*, **79**, 51-58.

Ellenbogen, K.A., Smith, M.L., and Eckberg, D.L. (1990*a*). Increased vagal cardiac nerve traffic prolongs ventricular refractoriness in patients undergoing electrophysiology testing. *The American Journal of Cardiology*, **65**, 1345-1350.

Ellenbogen, K.A., Smith, M.L., Thames, M.D., and Mohanty, P.K. (1990*b*). Changes in regional adrenergic tone during sustained ventricular tachycardia associated with coronary artery disease or idiopathic dilated cardiomyopathy. *The American Journal of Cardiology*, **65**, 1334-1338.

Elliott, J.M., West, M.J,, and Chalmers, J. (1985). Effects of vasopressin on heart rate in conscious rabbits. *Journal of Cardiovascular Pharmacology*, **7**, 6-11.

Elliott, T.R. (1905). The action of adrenalin. *The Journal of Physiology*, **32**, 401-467.

El-Sherif, N. *et al.* (1976). Electrocardiographic antecedents of primary ventricular fibrillation. Value of the R-on-T phenomenon in myocardial infarction. *British Heart Journal*, **38**, 415-422.

Elsner, R., Millard, R.W., Kjekshus, J.K., White, F., Blix, A.S., and Kemper, W.S. (1985). Coronary blood flow and myocardial segment dimensions during simulated dives in seals. *American Journal of Physiology*, **249**, H1119-H1126.

Epstein, A.E., Hirschowitz, B.I., Kirklin, J.K., Kirk, K.A., Kay, G.N., and Plumb, V.J. (1990). Evidence for a central site of action to explain the negative chronotropic effect of atropine: studies on the human transplanted heart. *Journal of the American College of Cardiology*, **15**, 1610-1617.

Epstein, F.H., Post, R.S., and McDowell, M. (1953). The effect of an arteriovenous fistula on renal hemodynamics and electrolyte excretion. *The Journal of Clinical Investigation*, **32**, 233-241.

Epstein, M. (1976). Cardiovascular and renal effects of head-out water immersion in man. Application of the model in the assessment of volume homeostasis. *Circulation Research*, **39**, 619-628.

Epstein, M. (1978). Renal effects of head-out water immersion in man: implications for an understanding of volume homeostasis. *Physiological Reviews*, **58**, 529-581.

Epstein, S.E., *et al.* (1969). Circulatory effects of electrical stimulation of the carotid sinus nerves in man. *Circulation*, **40**, 269-275.

Eriksen, M., Waaler, B.A., Walløe, L., and Wesche, J. (1990). Dynamics and dimensions of cardiac output changes in humans at the onset and at the end of moderate rhythmic exercise. *The Journal of Physiology*, **426**, 423-437.

Ermirio, R., Ruggeri, P., Cogo, C.E., Molinari, C., and Calaresu, F.R. (1989). Neuronal and cardiovascular responses to ANF microinjected into the solitary nucleus. *American Journal of Physiology*, **256**, R577-R582.

Erne, P., Raine, A.E.G., Burgisser, E., Gradel, E., Burkart, F., and Buhler, F.R. (1987). Paradoxical inhibition of atrial natriuretic peptide release during pacing-induced hypotension. *Clinical Science*, **73**, 459-462.

Erne, P., Lipkin, D., and Maseri, A. (1988). Impaired beta-adrenergic receptor and normal postreceptor responsiveness in congestive heart failure. *The American Journal of Cardiology*, **61**, 1132-1134.

Ernsberger, P., Giuliano, R., Willette, R.N., Granata, A.R., and Reis, D.J. (1988). Hypotensive action of clonidine analogues correlates with binding affinity at imidazole and not alpha-2-adrenergic receptors in the rostral ventrolateral medulla. *Journal of Hypertension*, **6** (Suppl. 4), S554-S557.

Ernsting, J. (1966). *Some effects of raised intrapulmonary pressure in man*, pp.63-104. Technivision, Maidenhead, England.

Ernsting, J. and Parry, D.J. (1957). Some observations on the effects of stimulating the stretch receptors in the carotid artery of man (Abstract). *The Journal of Physiology*, **137**, 45P–46P.

Esente, P., Giambartolomei, A., Gensini, G.G., and Dator, C. (1983). Coronary reperfusion and Bezold–Jarisch reflex (bradycardia and hypotension). *The American Journal of Cardiology*, **52**, 221-224.

Esler, M. *et al.* (1984). Contribution of individual organs to total noradrenaline release in humans. *Acta Physiologica Scandinavica*, Suppl. **527**, 11-16.

Esler, M.D., Hasking, G.J., Willett, I.R., Leonard, P.W., and Jennings, G.L. (1985). Noradrenaline release and sympathetic nervous system activity. *Journal of Hypertension*, **3**, 117-129.

Esler, M., Jennings, G., Biviano, B., Lambert, G., and Hasking, G. (1986). Mechanism of elevated plasma noradrenaline in the course of essential hypertension. *Journal of Cardiovascular Pharmacology*, **8** (Suppl. 5), S39-S43.

Esler, M. *et al.* (1988). Assessment of human sympathetic nervous system activity from measurements of norepinephrine turnover. *Hypertension*, **11**, 3-20.

Esler, M., Jennings, G., Lambert, G., Meredith, I., Horne, M., and Eisenhofer, G. (1990). Overflow of catecholamine neurotransmitters to the circulation: source, fate, and functions. *Physiological Reviews*, **70**, 963-985.

Evans, J.M., Knapp, C.F., and Lowery, T.R. (1979). Pressor response buffering by β-adrenergic and cholinergic vasodilation in tranquilized dogs. *American Journal of Physiology*, **236**, H165-H173.

Evans, J.M., Randall, D.C., Funk, J.N., and Knapp, C.F. (1990). Influence of cardiac innervation on intrinsic heart rate in dogs. *American Journal of Physiology*, **258**, H1132-H1137.

Ewing, D.J., Kerr, F., Leggett, R., and Murray, A. (1976). Interaction between cardiovascular responses to sustained handgrip and Valsalva manoeuvre. *British Heart Journal*, **38**, 483-490.

Ewing, D.J., Hume, L., Campbell, I.W., Murray, A., Neilson, J.M.M., and Clarke, B.F. (1980). Autonomic mechanisms in the initial heart rate response to standing. *Journal of Applied Physiology*, **49**, 809-814.

Ewing, D.J., Borsey, D.Q., Bellavere, F., and Clarke, B.F. (1981). Cardiac autonomic neuropathy in diabetes: comparison of measures of R–R interval variation. *Diabetologia*, **21**, 18-24.

Eyzaguirre, C. and Lewin, J. (1961). The effect of sympathetic stimulation on carotid nerve activity. *The Journal of Physiology*, **159**, 251-267.

Eyzaguirre, C. and Uchizono, K. (1961). Observations on the fibre content of nerves reaching the carotid body of the cat. *The Journal of Physiology*, **159**, 268-281.

Faden, A.I. and Petras, J.M. (1978). An intraspinal sympathetic preganglionic pathway: anatomic evidence in the dog. *Brain Research*, **144**, 358-362.

Fagius, J. (1988). Muscle nerve sympathetic activity following ectopic heart beats — a note on the burst pattern of sympathetic impulses. *Journal of the Autonomic Nervous System*, **22**, 243-245.

Fagius, J. and Sundlöf, G. (1986). The diving response in man: effects on sympathetic activity in muscle and skin nerve fascicles. *The Journal of Physiology*, **377**, 429-443.

Fagius, J. and Wallin, B.G. (1980*a*). Sympathetic reflex latencies and conduction velocities in normal man. *Journal of the Neurological Sciences*, **47**, 433-448.

Fagius, J. and Wallin, B.G. (1980*b*). Sympathetic reflex latencies and conduction velocities in patients with polyneuropathy. *Journal of the Neurological Sciences*, **47**, 449-461.

Fagius, J., Wallin, B.G., Sundlöf, G., Nerhed, C., and Englesson, S. (1985). Sympathetic outflow in man after anaesthesia of the glossopharyngeal and vagus nerves. *Brain*, **108**, 423-438.

Fagius, J., Sundlöf, G., and Wallin, B.G. (1987). Variation of sympathetic reflex latency in man. *Journal of the Autonomic Nervous System*, **21**, 157-165.

Fagraeus, L. and Linnarsson, D. (1976). Autonomic origin of heart rate fluctuations at the onset of muscular exercise. *Journal of Applied Physiology*, **40**, 679-682.

Faraday, M. (1839). *Experimental Researches in Electricity*. Richard and John Edward Taylor, London.

Farhi, E.R., Cant, J.R., and Barger, A.C. (1982). Interactions between intrarenal epinephrine receptors and the renal baroreceptor in the control of PRA in conscious dogs. *Circulation Research*, **50**, 477-485.

Farhi, E.R., Cant, J.R., Paganelli, W.C., Dzau, V.J., and Barger, A.C. (1987). Stimulus–response curve of the renal baroreceptor: effect of converting enzyme inhibition and changes in salt intake. *Circulation Research*, **61**, 670-677.

Faris, I.B., Iannos, J., Jamieson, G.G., and Ludbrook, J. (1980*a*). Comparison of methods for eliciting the baroreceptor–heart rate reflex in conscious rabbits. *Clinical and Experimental Pharmacology and Physiology*, **7**, 281-291.

Faris, I.B., Iannos, J., Jamieson, G.G., and Ludbrook, J. (1980*b*). The carotid sinus baroreceptor reflex in conscious rabbits. *The Journal of Physiology*, **298**, 321-331.

Faris, I.B., Jamieson, G.G., and Ludbrook, J. (1982). Effect of exercise on gain of the carotid-sinus reflex in rabbits. *Clinical Science*, **63**, 115-119.

Farrell, T.G. *et al.* (1991). Baroreflex sensitivity and electrophysiological correlates in patients after acute myocardial infarction. *Circulation*, **83**, 945-952.

Farsang, C. *et al.* (1984). Reversal by naloxone of the antihypertensive action of clonidine: involvement of the sympathetic nervous system. *Circulation*, **69**, 461-467.

Fasola, A.F. and Martz, B.L. (1972). Peripheral venous renin activity during 70° tilt and lower body negative pressure. *Aerospace Medicine*, **43**, 713-715.

Faxon, D.P., Creager, M.A., Halperin, J.L., Gavras, H., Coffman, J.D., and Ryan, T.J. (1980*a*). Central and peripheral hemodynamic effects of angiotensin inhibition in patients with refractory congestive heart failure. *Circulation*, **61**, 925-930.

Faxon, D.P., Gross, S., Tilney, C., and Ryan, T.J. (1980*b*). The altered baroreceptor response to nitroglycerin in congestive heart failure. *Clinical Cardiology*, **3**, 7-13.

Faxon, D.P., Creager, M.A., Halperin, J.L., Bernard, D.B., and Ryan, T.J. (1984). Redistribution of regional blood flow following angiotensin-converting enzyme inhibition. Comparison of normal subjects and patients with heart failure. *The American Journal of Medicine*, **76**, 104-110.

Feigl, E. and Folkow, B. (1963). Cardiovascular responses in 'diving' and during brain stimulation in ducks. *Acta Physiologica Scandinavica*, **57**, 99-110.

Fein, J.M. and Frishman, W. (1980). Neurogenic hypertension related to vascular compression of the lateral medulla. *Neurosurgery*, **6**, 615-622.

Feldberg, W. and Gaddum, J.H. (1934). The chemical transmitter at synapses in a sympathetic ganglion. *The Journal of Physiology*, **81**, 305-319.

Feldberg, W. and Guertzenstein, P.G. (1976). Vasodepressor effects obtained by drugs acting on the ventral surface of the brain stem. *The Journal of Physiology*, **258**, 337-355.

Felder, R.B. (1986). Excitatory and inhibitory interactions among renal and cardiovascular afferent nerves in dorsomedial medulla. *American Journal of Physiology*, **250**, R580-R588.

Felder, R.B. and Heesch, C.M. (1987). Interactions in nucleus tractus solitarius between right and left carotid sinus nerves. *American Journal of Physiology*, **253**, H1127-H1135.

Felder, R.B. and Mifflin, S.W. (1988). Modulation of carotid sinus afferent input to nucleus tractus solitarius by parabrachial nucleus stimulation. *Circulation Research*, **63**, 35-49.

Felder, R.B. and Thames, M.D. (1979). Interaction between cardiac receptors and sinoaortic baroreceptors in the control of efferent cardiac sympathetic nerve activity during myocardial ischemia in dogs. *Circulation Research*, **45**, 728-736.

Felder, R.B., Heesch, C.M., and Thames, M.D. (1983). Reflex modulation of carotid sinus baroreceptor activity in the dog. *American Journal of Physiology*, **244**, H437-H443.

Feldman, J.L. and Ellenberger, H.H. (1988). Central coordination of respiratory and cardiovascular control in mammals. *Annual Review of Physiology*, **50**, 593-606.

Feldman, J.L. and Gautier, H. (1976). Interaction of pulmonary afferents and pneumotaxic center in control of respiratory pattern in cats. *Journal of Neurophysiology*, **39**, 31-44.

Feldman, R.L., Marx, J.D., Pepine, C.J., and Conti, C.R. (1982). Analysis of coronary responses to various doses of intracoronary nitroglycerin. *Circulation*, **66**, 321-327.

Feldman, T. *et al.* (1988). Hemodynamic recovery during simulated ventricular tachycardia: role of adrenergic receptor activation. *American Heart Journal*, **115**, 576-587.

Ferguson, D.W., Thames, M.D., and Mark, A.L. (1983). Effects of propranolol on reflex vascular responses to orthostatic stress in humans. Role of ventricular receptors. *Circulation*, **67**, 802-807.

Ferguson, D.W., Abboud, F.M., and Mark, A.L. (1984). Selective impairment of baroreflex-mediated vasoconstrictor responses in patients with ventricular dysfunction. *Circulation*, **69**, 451-460.

Ferguson, D.W., Abboud, F.M., and Mark, A.L. (1985). Relative contribution of aortic and carotid baroreflexes to heart rate control in man during steady state and dynamic increases in arterial pressure. *The Journal of Clinical Investigation*, **76**, 2265-2274.

Ferguson, D.W., Berg, W.J., Sanders, J.S., Roach, P.J., Kempf, J.S., and Kienzle, M.G. (1989). Sympathoinihbitory responses to digitalis glycosides in heart failure patients. Direct evidence from sympathetic neural recordings. *Circulation*, **80**, 65-77.

Fernandez, P.G., Snedden, W., Nath, C., Gault, H., and Vasdev, S. (1987). The responses of circulating catecholamines and the renin–angiotensin (R–A) axis to the Valsalva maneuver (Abstract). *Clinical Research*, **35**, 277A.

Fernandez, P.G., Snedden, W., Vasdev, S., Bolli, P., and Rabin, E.Z. (1988). Can plasma catecholamine levels be a useful index of sympathetic nervous system activity? *The Canadian Journal of Cardiology*, **4**, 169-176.

Fernandez de Molina, A., Kuno, M., and Perl, E.R. (1965). Antidromically evoked responses from sympathetic preganglionic neurones. *The Journal of Physiology*, **180**, 321-335.

Ferrante, F.L. and Opdyke, D.F. (1969). Mammalian ventricular function during submersion asphyxia. *Journal of Applied Physiology*, **26**, 561-570.

Ferrari, A., Bonazzi, O., Gardumi, M., Gregorini, L., Perondi, R., and Mancia, G. (1981*a*). Modulation of atrioventricular conduction by isometric exercise in human subjects. *Circulation Research*, **49**, 265-271.

Ferrari, A., Gregorini, L., Ferrari, M.C., Preti, L., and Mancia, G. (1981*b*). Digitalis and baroreceptor reflexes in man. *Circulation*, **63**, 279-285.

Ferrari, A., Bonazzi, O., Gregorini, L., Gardumi, M., Perondi, R., and Mancia, G. (1983). Modification of the baroreceptor control of atrio-ventricular conduction induced by digitalis in man. *Cardiovascular Research*, **17**, 633-641.

Ferrari, A.U., Cavallazzi, A., Gregorini, L., Perondi, R., Daffonchio, A., and Mancia, G. (1987). Effects of altering the interval between the stimulus and the reflex response in the analysis of the baroreceptor control of the sinus and atrioventricular nodes in man. *Cardiovascular Research*, **21**, 385-390.

Ferrario, C.M., Barnes, K.L., Szilagyi, J.E., and Brosnihan, K.B. (1979). Physiological and pharmacological characterization of the area postrema pressor pathways in the normal dog. *Hypertension*, **1**, 235-245.

Ferrario, C.M., Barnes, K.L., Diz, D.I., Block, C.H., and Averill, D.B. (1987). Role of area postrema pressor mechanisms in the regulation of arterial pressure. *Canadian Journal of Physiology and Pharmacology*, **65**, 1591-1597.

Ferrier, G.R. and Dresel, P.E. (1973). Role of the atrium in determining the functional and effective refractory periods and the conductivity of the atrioventricular transmission system. *Circulation Research*, **33**, 375-385.

Ferris, E.B., Jr., Capps, R.B., and Weiss, S. (1935). Carotid sinus syncope and its bearing on the mechanism of the unconscious state and convulsions. *Medicine*, **14**, 377-456.

Fidone, S.J. and Sato, A. (1969). A study of chemoreceptor and baroreceptor A and C-fibres in the cat carotid nerve. *The Journal of Physiology*, **205**, 527-548.

Fifer, M.A. *et al.* (1990). Hemodynamic and renal effects of atrial natriuretic peptide in congestive heart failure. *The American Journal of Cardiology*, **65**, 211-215.

Finke, R., Gross, R., Hackenthal, E., Huber, J., and Kirchheim, H.R. (1983). Threshold pressure for the pressure-dependent renin release in the autoregulating kidney of conscious dogs. *Pflügers Archiv für die Gesamte Physiologie des Menschen und der Tiere*, **399**, 102-110.

Finley, J.P., Bonet, J.F., and Waxman, M.B. (1979). Autonomic pathways responsible for bradycardia on facial immersion. *Journal of Applied Physiology*, **47**, 1218-1222.

Fiocchi, R. (1985). Carotid baroreceptor control of heart rate and blood pressure at rest and exercise in man (Ph.D. thesis), Katholieke Universiteit te Leuven, Leuven, Belgium.

Fiorica, V. and Kem, D.C. (1985). Plasma norepinephrine, blood pressure and heart rate response to graded change in body position. *Aviation, Space, and Environmental Medicine*, **56**, 1166-1171.

Fischer, G.M. and Llaurado, J.G. (1966). Collagen and elastin content in canine arteries selected from functionally different vascular beds. *Circulation Research*, **19**, 394-399.

Fitchett, D.H., Marin Neto, J.A., Oakley, C.M., and Goodwin, J.F. (1979). Hydralazine in the management of left ventricular failure. *The American Journal of Cardiology*, **44**, 303-309.

Fitzgerald, W. (1981). Labile hypertension and jogging: new diagnostic tool or spurious discovery? *British Medical Journal*, **282**, 542-544.

Fletcher, E.C., DeBehnke, R.D., Lovoi, M.S., and Gorin, A.B. (1985). Undiagnosed sleep apnea in patients with essential hypertension. *Annals of Internal Medicine*, **103**, 190-195.

Floras, J.S. (1990). Sympathoinhibitory effects of atrial natriuretic factor in normal humans. *Circulation*, **81**, 1860-1873.

Floras, J.S., Hassan, M.O., Jones, J.V., Osikowska, B.A., Sever, P.S., and Sleight, P. (1988*a*). Factors influencing blood pressure and heart rate variability in hypertensive humans. *Hypertension*, **11**, 273-281.

Floras, J.S., Hassan, M.O., Vann Jones, J., Osikowska, B.A., Sever, P.S., and Sleight, P. (1988*b*). Consequences of impaired arterial baroreflexes in essential hypertension: effects on pressor responses, plasma noradrenaline and blood pressure variability. *Journal of Hypertension*, **6**, 525-535.

Floras, J.S., Jones, J.V., Hassan, M.O., and Sleight, P. (1988*c*). Effects of acute and chronic beta-adrenoceptor blockade on baroreflex sensitivity in humans. *Journal of the Autonomic Nervous System*, **25**, 87-94.

Floras, J.S., Sinkey, C.A., Aylward, P.E., Seals, D.R., Thoren, P.N., and Mark, A.L. (1989). Postexercise hypotension and sympathoinhibition in borderline hypertensive men. *Hypertension*, **14**, 28-35.

Flynn, T.G. and Davies, P.L. (1985). The biochemistry and molecular biology of atrial natriuretic factor. *Biochemical Journal*, **232**, 313-321.

Folkow, B. (1978). Cardiovascular structural adaptation; its role in the initiation and maintenance of primary hypertension. *Clinical Science and Molecular Medicine*, **55**, 3s-22s.

Folkow, B. (1982). Physiological aspects of primary hypertension. *Physiological Reviews*, **62**, 347-504.

Folkow, B., Nilsson, N.J., and Yonce, L.R. (1967). Effects of 'diving' on cardiac output in ducks. *Acta Physiologica Scandinavica*, **70**, 347-361.

Folkow, B., Di Bona, G.F., Hjemdahl, P., Toren, P.H., and Wallin, B.G. (1983). Measurements of plasma norepinephrine concentrations in human primary hypertension. A word of caution on their applicability for assessing neurogenic contributions. *Hypertension*, **5**, 399-403.

Foreman, R.D. and Wurster, R.D. (1973). Localization and functional characteristics of descending sympathetic spinal pathways. *American Journal of Physiology*, **225**, 212-217.

Forsling, M.L. and Rees, M. (1975). Effects of hypoxia and hypercapnia on plasma vasopressin concentration. *Journal of Endocrinology*, **67**, 62P-63P.

Fouad, F.M., Tarazi, R.C., Ferrario, C.M., Fighaly, S., and Alicandri, C. (1984).

Assessment of parasympathetic control of heart rate by a noninvasive method. *American Journal of Physiology*, **246**, H838-H842.

Foux, A., Seliktar, R., and Valero, A. (1976). Effects of lower body negative pressure (LBNP) on the distribution of body fluids. *Journal of Applied Physiology*, **41**, 719-726.

Fowler, M.B., Laser, J.A., Hopkins, G.L., Minobe, W., and Bristow, M.R. (1986). Assessment of the β-adrenergic receptor pathway in the intact failing human heart: progressive receptor down-regulation and subsensitivity to agonist response. *Circulation*, **74**, 1290-1302.

Fox, G.P.P. and Matthews, T.G. (1989). Autonomic dysfunction at different ambient temperatures in infants at risk of sudden infant death syndrome. *The Lancet*, **2**, 1065-1067.

Francis, G.S., Goldsmith, S.R., and Cohn, J.N. (1982*a*). Relationship of exercise capacity to resting left ventricular performance and basal plasma norepinephrine levels in patients with congestive heart failure. *American Heart Journal*, **104**, 725-731.

Francis, G.S., Goldsmith, S.R., Ziesche, S.M., and Cohn, J.N. (1982*b*). Response of plasma norepinephrine and epinephrine to dynamic exercise in patients with congestive heart failure. *The American Journal of Cardiology*, **49**, 1152-1156.

Francis, G.S., Olivari, M.T., Goldsmith, S.R., Levine, T.B., Pierpont, G., and Cohn, J.N. (1983). The acute response of plasma norepinephrine, renin activity, and arginine vasopressin to short-term nitroprusside and nitroprusside withdrawal in patients with congestive heart failure. *American Heart Journal*, **106**, 1315-1320.

Francis, G.S., Goldsmith, S.R., Pierpont, G., and Cohn, J.N. (1984). Free and conjugated plasma catecholamines in patients with congestive heart failure. *The Journal of Laboratory and Clinical Medicine*, **103**, 393-398.

Francis, G.S., Goldsmith, S.R., Ziesche, S., Nakajima, H., and Cohn, J.N. (1985). Relative attenuation of sympathetic drive during exercise in patients with congestive heart failure. *Journal of the American College of Cardiology*, **5**, 832-839.

Francis, G.S., Rector, T.S., and Cohn, J.N. (1988). Sequential neurohumoral measurements in patients with congestive heart failure. *American Heart Journal*, **116**, 1464-1468.

Francis, G.S. *et al.* (1990). Comparison of neuroendocrine activation in patients with left ventricular dysfunction with and without congestive heart failure. A substudy of the studies of left ventricular dysfunction (SOLVD). *Circulation*, **82**, 1724-1729.

François-Franck (1876). Effets des excitations des nerfs sensibles sur le coeur, la respiration et la circulation artérielle. *Physiologie Experimentale Travaux du Laboratoire de M. Marey*, **2**, 221-228.

Franke, H. (1963). *Über das Karotissinus-Syndrom und den sogenannten hyperaktiven Karotissinus-Reflex*. Friedrich-Karl Schattauer-Verlag, Stuttgart.

Frankel, H.L., Michaelis, L.S., Golding, D.R., and Beral, V. (1972). The blood pressure in paraplegia I. *Paraplegia*, **10**, 193-198.

Frankel, H.L., Mathias, C.J., and Spalding, J.M.K. (1975). Mechanisms of reflex cardiac arrest in tetraplegic patients. *The Lancet*, **2**, 1183-1185.

Franz, G.N. (1969). Nonlinear rate sensitivity of the carotid sinus reflex as a consequence of static and dynamic nonlinearities in baroreceptor behavior. *Annals of the New York Academy of Sciences*, **156**, 811-824.

Fredericq, L. (1882). De l'influence de la respiration sur la circulation. Les oscillations respiratoires de la pression artérielle chez le chien. *Archives Biologie Paris*, **3**, 55-100.

Freedman, R.A., Swerdlow, C.D., Echt, D.S., Winkle, R.A., Soderholm-Difatte, V., and Mason, J.W. (1984). Facilitation of ventricular tachyarrhythmia induction by isoproterenol. *The American Journal of Cardiology*, **54**, 765-770.

Frenneaux, M.P., Counihan, P.J., Caforio, A.L.P., Chikamori, T., and McKenna, W.J. (1990). Abnormal blood pressure response during exercise in hypertrophic cardiomyopathy. *Circulation*, **82**, 1995-2002.

Freund, P.R., Rowell, L.B., Murphy, T.M., Hobbs, S.F., and Butler, S.H. (1979). Blockade of the pressor response to muscle ischemia by sensory nerve block in man. *American Journal of Physiology*, **237**, H433-H439.

Freyschuss, U. (1970). Cardiovascular adjustment to somatomotor activation. *Acta Physiologica Scandinavica*, Suppl. **342**, 1-63.

Freyschuss, U. and Melcher, A. (1975). Sinus arrhythmia in man: influence of tidal volume and oesophageal pressure. *Scandinavian Journal of Clinical and Laboratory Investigation* , **35**, 487-496.

Fries, W. and Zieglgänsberger, W. (1974). A method to discriminate axonal from cellbody activity and to analyse 'silent' cells. *Experimental Brain Research*, **21**, 441-445.

Frink, R.J., Merrick, B., and Lowe, H.M. (1975). Mechanism of the bradycardia during coronary angiography. *The American Journal of Cardiology*, **35**, 17-22.

Fritsch, J.M., Eckberg, D.L., Graves, L.D., and Wallin, B.G. (1986). Arterial pressure ramps provoke linear increases of heart period in humans. *American Journal of Physiology*, **251**, R1086-R1090.

Fritsch, J.M., Rea, R.F., and Eckberg, D.L. (1989). Carotid baroreflex resetting during drug-induced arterial pressure changes in humans. *American Journal of Physiology*, **256**, R549-R553.

Fritsch, J.M., Charles, J.B., Eckberg, D.L., Bennett, B.S., and Bungo, M.W. (1990). Effects of short-duration space flight on human carotid baroreceptor cardiac reflexes (Abstract). *The FASEB Journal*, **4**, A429.

Fritsch, J.M., Smith, M.L., Simmons, D.T.F., and Eckberg, D.L. (1991). Differential baroreflex modulation of human vagal and sympathetic activity. *American Journal of Physiology*, **260**, R635-R641.

Fujii, A.M. and Vatner, S.F. (1985). Direct versus indirect pressor and vasoconstrictor actions of angiotensin in conscious dogs. *Hypertension*, **7**, 253-261.

Fulton, M., Julian, D.C., and Oliver, M.F. (1969). Sudden death and myocardial infarction. *Circulation*, **39** (Suppl. IV), IV-182-IV-193.

Fumoux, F., Czernichow, P., Arnauld, E., Du Pont, J., and Vincent, J.D. (1978). Effect of hypotension induced by sodium nitrocyanoferrate (III) on the release of arginine-vasopressin in the unanaesthetized monkey. *Journal of Endocrinology*, **78**, 449-450.

Gabriel, M. and Seller, H. (1970). Interaction of baroreceptor afferents from carotid sinus and aorta at the nucleus tractus solitarii. *Pflügers Archiv für die Gesamte Physiologie des Menschen und der Tiere*, **318**, 7-20.

Galdston, M., Goldstein, R., and Steele, J.M. (1943). Studies of the variation in circulatory and respiratory responses to carotid sinus stimulation in man. *American Heart Journal*, **26**, 213-231.

Gale, K., Hamilton, B.L., Brown, S.C., Norman, W.P., Souza, J.D., and Gillis, R.A. (1980). GABA and specific GABA binding sites in brain nuclei associated with vagal outflow. *Brain Research Bulletin*, **5** (Suppl. 2), 325-328.

Gallo, L., Jr., Filho, J.M., Maciel, B.C., Marin Neto, J.A., Barreto Martins, L.E., and Lima Filho, E.C. (1987). Functional evaluation of sympathetic and parasympathetic system in Chagas' disease using dynamic exercise. *Cardiovascular Research*, **21**, 922-927.

Gaskell, W.H. (1883). On the innervation of the heart, with especial reference to the heart of the tortoise. *The Journal of Physiology*, **4**, 43-127.

Gasser, H.S. (1955). Properties of dorsal root unmedullated fibers on the two sides of the ganglion. *The Journal of General Physiology*, **38**, 709-728.

Gasser, H.S. and Meek, W.J. (1914). A study of the mechanisms by which muscular exercise produces acceleration of the heart. *American Journal of Physiology*, **34**, 48-71.

Gauer, O.H. and Henry, J.P. (1963). Circulatory basis of fluid volume control. *Physiological Reviews*, **43**, 423-481.

Gauer, O.H. and Henry, J.P. (1964). Negative (-Gz) acceleration in relation to arterial oxygen saturation, subendocardial hemorrhage and venous pressure in the forehead. *Aerospace Medicine*, **35**, 533-545.

Gauer, O.H. and Thron, H.L. (1965). Postural changes in the circulation. In *Handbook of physiology. Section 2: Circulation*, Volume III, ed. Hamilton, W.F., and Dow, P., pp. 2409-2439. American Physiological Society, Washington, D.C.

Gavras, H., Faxon, D.P., Berkoben, J., Brunner, H.R., and Ryan, T.J. (1978). Angiotensin converting enzyme inhibition in patients with congestive heart failure. *Circulation*, **58**, 770-776.

Gebber, G.L. (1990). Central determinants of sympathetic nerve discharge. In *Central regulation of autonomic functions*, ed. Loewy, A.D., and Spyer, K.M., pp. 126-144. Oxford University Press, New York.

Gebber, G.L. and Barman, S.M. (1980). Rhythmogenesis in the sympathetic nervous system. *Federation Proceedings*, **39**, 2526-2530.

Gebber, G.L., Taylor, D.G., and Weaver, L.C. (1973). Electrophysiological studies on organization of central vasopressor pathways. *American Journal of Physiology*, **224**, 470-481.

Geer, P.G., Wang, B.C., and Goetz, K.L. (1986). Blood pressure of sinoaortic-denervated dogs is not increased by cardiac denervation. *Proceedings of the Society for Experimental Biology and Medicine*, **181**, 33-40.

Gellhorn, E. (1940-1941). The internal environment and behavior. Part II. The influence of variations in the blood sugar on the functions of the brain. *The American Journal of Psychology*, **97**, 1204-1218.

Gellhorn, E., Yesinick, L., Kessler, M., and Hailman, H. (1942). Carotid sinus reflexes and convulsions. *American Journal of Physiology*, **137**, 396-403.

Geltman, E.M., Ehsani, A.A., Campbell, M.K., Schechtman, K., Roberts, R., and Sobel, B.E. (1979). The influence of location and extent of myocardial infarction on long-term ventricular dysrhythmia and mortality. *Circulation*, **60**, 805-814.

Gerard, M.W. and Billingsley, P.R. (1923). The innervation of the carotid body. *The Anatomical Record*, **25**, 391-400.

Gernandt, B., Liljestrand, G., and Zotterman, Y. (1945). Adrenaline apnoea. *Acta Physiologica Scandinavica*, **9**, 367-377.

Gero, J. and Gerová, M. (1962). The rôle of parameters of pulsating pressure in the stimulation of intracarotid receptors. *Archives Internationales de Pharmacodynamie et de Thérapie*, **140**, 35-44.

Geschwind, H.J., Lhoste, F., Scriven, A.J.I., Dhainaut, J.-F., Sabatier, C., and Laurent, D. (1984). Sympathetic nervous system activation in postextrasystolic potentiation: role of catecholamine release in enhancement of ventricular function. *Journal of the American College of Cardiology*, **4**, 216-225.

Giannattasio, C. *et al.* (1990). Cardiopulmonary receptor reflexes in normotensive athletes with cardiac hypertrophy. *Circulation*, **82**, 1222-1229.

Gilbey, M.P., Jordan, D., Richter, D.W., and Spyer, K.M. (1984). Synaptic mechanisms involved in the inspiratory modulation of vagal cardio-inhibitory neurones in the cat. *The Journal of Physiology*, **356**, 65-78.

Gildenberg, P.L., Ferrario, C.M., and McCubbin, J.W. (1973). Two sites of cardiovascular action of angiotensin II in the brain of the dog. *Clinical Science*, **44**, 417-420.

Gillis, R.A. (1971). Role of the nervous system in the arrhythmias produced by coronary occlusion in the cat. *American Heart Journal*, **81**, 677-684.

Gillis, R.A., DiMicco, J.A., Williford, D.J., Hamilton, B.L., and Gale, K.N. (1980). Importance of CNS GABAergic mechanisms in the regulation of cardiovascular function. *Brain Research Bulletin*, **5** (Suppl. 2), 303-315.

Gilman, A.G., Goodman, L.S., and Gilman, A. (1980). *Goodman and Gilman's The pharmacological basis of therapeutics*, 6th ed., pp. 124-125. Macmillan, New York.

Gilmore, J.P. and Zucker, I.H. (1974). Discharge of type B atrial receptors during changes in vascular volume and depression of atrial contractility. *The Journal of Physiology*, **239**, 207-223.

Gilmore, J.P., Peterson, T.V., Wesley, C.R., and Share, L. (1982). High versus low pressure receptors in modulating the volumetric control of antidiuretic hormone secretion in the monkey. *Basic Research in Cardiology*, **77**, 250-254.

Gilpin, E.A., Koziol, J.A., Madsen, E.B., Henning, H., and Ross, J., Jr. (1983). Periods of differing mortality distribution during the first year after acute myocardial infarction. *The American Journal of Cardiology*, **52**, 240-244.

Glasser, S.P., McCarty, R.J., and Iannone, L.A. (1974). Hyperactive carotid sinus reflex and His bundle electrography. *Journal of Electrocardiology*, **7**, 71-74.

Godin, D., Campeau, N., Nadeau, R., Cardinal, R., and de Champlain, J. (1985). Catecholamine release and ventricular arrhythmias during coronary occlusion and reperfusion in the dog. *Canadian Journal of Physiology and Pharmacology*, **63**, 1088-1095.

Goetz, K.L. (1988). Physiology and pathophysiology of atrial peptides. *American Journal of Physiology*, **254**, E1-E15.

Goetz, K.L., Bond, G.C., and Smith, W.E. (1974). Effect of moderate hemorrhage on plasma ADH and renin. *Proceedings of the Society for Experimental Biology and Medicine*, **145**, 277-280.

Goetz, K.L., Bond, G.C., and Bloxham, D.D. (1975). Atrial receptors and renal function. *Physiological Reviews*, **55**, 157-205.

Goldberg, H., Elisberg, E.I., and Katz, L.N. (1952). The effects of the Valsalva-like maneuver upon the circulation in normal individuals and patients with mitral stenosis. *Circulation*, **5**, 38-47.

Goldberg, J.M., Geesbreght, J.M., Randall, W.C., and Brynjolfsson, G. (1973). Sympathetically induced pacemaker shifts following sinus node excision. *American Journal of Physiology*, **224**, 1468-1474.

Goldberg, S. *et al.* (1983). Reperfusion arrhythmia: A marker of restoration of antegrade flow during intracoronary thrombolysis for acute myocardial infarction. *American Heart Journal*, **105**, 26-32.

Goldsmith, S.R., Francis, G.S., Cowley, A.W., and Cohn, J.N. (1982). Response of vasopressin and norepinephrine to lower body negative pressure in humans. *American Journal of Physiology*, **243**, H970-H973.

Goldsmith, S.R., Francis, G.S., Levine, T.B., and Cohn, J.N. (1983*a*). Regional blood flow response to orthostasis in patients with congestive heart failure. *The American Journal of Cardiology*, **6**, 1391-1395.

Goldsmith, S.R., Francis, G.S., Levine, T.B., Cowley, A.W. Jr., and Cohn, J.N. (1983*b*). Impaired response of plasma vasopressin to orthostatic stress in patients with congestive heart failure. *Journal of the American College of Cardiology*, **2**, 1080-1083.

Goldsmith, S.R., Francis, G.S., and Cohn, J.N. (1985). Norepinephrine infusions in congestive heart failure. *The American Journal of Cardiology*, **56**, 802-803.

Goldsmith, S.R., Dodge, D., and Cowley, A.W. (1987). Nonosmotic influences on osmotic stimulation of vasopressin in humans. *American Journal of Physiology*, **252**, H85-H88.

Goldstein, D.S. (1983*a*). Plasma catecholamines and essential hypertension. An analytical review. *Hypertension*, **5**, 86-99.

Goldstein, D.S. (1983*b*). Arterial baroreflex sensitivity, plasma catecholamines, and pressor responsiveness in essential hypertension. *Circulation*, **68**, 234-240.

Goldstein, D.S. and Keiser, H.R. (1984). Pressor and depressor responses after cholinergic blockade in humans. *American Heart Journal*, **107**, 974-979.

Goldstein, D.S., Horwitz, D., and Keiser, H.R. (1982). Comparison of techniques for measuring baroreflex sensitivity in man. *Circulation*, **66**, 432-439.

Goldstein, D.S., Nurnberger, J., Jr., Simmons, S., Gershon, E.S., Polinsky, R., and Keiser, H.R. (1983). Effects of injected sympathomimetic amines on plasma catecholamines and circulatory variables in man. *Life Sciences*, **32**, 1057-1063.

Goldstein, D.S., Levinson, P.D., Zimlichman, R., Pitterman, A., Stull, R., and Keiser, H.R. (1985). Clonidine suppression testing in essential hypertension. *Annals of Internal Medicine*, **102**, 42-48.

Goldstein, D.S., Cannon, R.O., Zimlichman, R., and Keiser, H.R. (1986). Clinical evaluation of impedance cardiography. *Clinical Physiology*, **6**, 235-251.

Goldstein, D.S., Eisenhofer, G., Sax, F.L., Keiser, H.R., and Kopin, I.J. (1987). Plasma norepinephrine pharmacokinetics during mental challenge. *Psychosomatic Medicine*, **49**, 591-605.

Goldstein, D.S., Brush, J.E., Jr., Eisenhofer, G., Stull, R., and Esler, M. (1988). In vivo measurement of neuronal uptake of norepinephrine in the human heart. *Circulation*, **78**, 41-48.

Goldstein, R.E., Beiser, G.D., Stampfer, M., and Epstein, S.E. (1975). Impairment of autonomically mediated heart rate control in patients with cardiac dysfunction. *Circulation Research*, **36**, 571-577.

Goltz, F. (1864). Ueber den Tonus der Gefässe und seine Bedeutung für die Blutbewegung. *Archiv für Pathologische Anatomie und Physiologie und für Klinische Medicin*, **29**, 394-432.

Gonzalez-Lima, F. (1988). Functional mapping of the brainstem during centrally evoked bradycardia: a 2-deoxyglucose study. *Behavioural Brain Research*, **28**, 325-336.

Goodchild, A.K., Dampney, R.A.L., and Bandler, R. (1982). A method for evoking physiological responses by stimulation of cell bodies, but not axons of passage, within localized regions of the central nervous system. *Journal of Neuroscience Methods*, **6**, 351-363.

Gooden, B.A., Stone, H.L., and Young, S. (1974). Cardiac responses to snout immersion in trained dogs. *The Journal of Physiology*, **242**, 405-414.

Gooden, B.A., Holdstock, G., and Hampton, J.R. (1978). The magnitude of the bradycardia induced by face immersion in patients convalescing from myocardial infarction. *Cardiovascular Research*, **12**, 239-242.

Goodwin, G.M., McCloskey, D.I., and Mitchell, J.H. (1972). Cardiovascular and respiratory responses to changes in central command during isometric exercise at constant muscle tension. *The Journal of Physiology*, **226**, 173-190.

Gordon, F.J. (1990). Opioids and central baroreflex control: a site of action in the nucleus tractus solitarius. *Peptides*, **11**, 305-309.

Gottlieb, S.H. and Engel, B.T. (1979). Autonomic interactions in the control of heart rate in the monkey. *Psychophysiology*, **16**, 528-536.

Gottlieb, S.S., Kukin, M.L., Ahern, D., and Packer, M. (1989). Prognostic importance of atrial natriuretic peptide in patients with chronic heart failure. *Journal of the American College of Cardiology*, **13**, 1534-1539.

Gould, L. and Lyon, A.F. (1967). Conversion of atrial flutter to sinus rhythm after stimulation of carotid sinus. *American Heart Journal*, **74**, 853-854.

Granata, A.R. and Kitai, S.T. (1989). Intracellular study of nucleus parabrachialis and nucleus tractus solitarii interconnections. *Brain Research*, **492**, 281-292.

Granata, A.R. and Reis, D.J. (1983). Release of [^{3}H]L-glutamine acid (L-Glu) and [^{3}H]D-aspartic acid (D-Asp) in the area of nucleus tractus solitarius in vivo produced by stimulation of the vagus nerve. *Brain Research*, **259**, 77-93.

Granata, A.R., Ruggiero, D.A., Park, D.H., Joh, T.H., and Reis, D.J. (1983). Lesions of epinephrine neurons in the rostral ventrolateral medulla abolish the vasodepressor components of baroreflex and cardiopulmonary reflex. *Hypertension*, **5** (Suppl. V), V80-V84.

Granata, A.R., Sved, A.F., and Reis, D.J. (1984). *In vivo* release by vagal stimulation of L-[^{3}H] glutamic acid in the nucleus tractus solitarius preloaded with L-[^{3}H] glutamine. *Brain Research Bulletin*, **12**, 5-9.

Granata, A.R., Kumada, M., and Reis, D.J. (1985*a*). Sympathoinhibition by A1-noradrenergic neurons is mediated by neurons in the C1 area of the rostral medulla. *Journal of the Autonomic Nervous System*, **14**, 387-395.

Granata, A.R., Ruggiero, D.A., Park, D.H., Joh, T.H., and Reis, D.J. (1985*b*). Brain stem area with C1 epinephrine neurons mediates baroreflex vasodepressor responses. *American Journal of Physiology*, **248**, H547-H567.

Granata, A.R., Numao, Y., Kumada, M., and Reis, D.J. (1986). A1 noradrenergic neurons tonically inhibit sympathoexcitatory neurons of C1 area in rat brainstem. *Brain Research*, **377**, 127-146.

Grandjean, B., Annat, G., Vincent, M., and Sassard, J. (1978). Influence of renal nerves on renin secretion in the conscious dog. *Pflügers Archiv für die Gesamte Physiologie des Menschen und der Tiere*, **373**, 161-165.

Grassi, G., Gavazzi, C., Cesura, A.M., Picotti, G.B., and Mancia, G. (1985). Changes in plasma catecholamines in response to reflex modulation of sympathetic vasoconstrictor tone by cardiopulmonary receptors. *Clinical Science*, **68**, 503-510.

Grassi, G. *et al.* (1988). Cardiopulmonary receptor modulation of plasma renin activity in normotensive and hypertensive subjects. *Hypertension*, **11**, 92-99.

Greeley, H.P., Smedal, M.I., and Most, W. (1955). The treatment of the carotid-sinus syndrome by irradiation. *The New England Journal of Medicine*, **252**, 91-94.

Green, J. (1980). A review of the Laceys' physiological hypothesis of heart rate change. *Biological Psychology*, **11**, 63-80.

Green, J.H. (1967). Physiology of baroreceptor function: mechanism of receptor stimulation. In *Barorecptors and hypertension*, ed. Kezdi, P., pp. 3-14. Pergamon Press, Oxford.

Green, J.H. and Heffron, P.F. (1968*a*). Studies upon patterns of activity in single post-ganglionic sympathetic fibres. *Archives Internationales de Pharmacodynamie et de Thérapie*, **173**, 232-243.

Green, J.H. and Heffron, P.F. (1968*b*). Studies upon the relationship between baroreceptor and sympathetic activity. *Quarterly Journal of Experimental Physiology and Cognate Medical Sciences*, **53**, 23-32.

Greenberg, T.T., Richmond, W.H., Stocking, R.A., Gupta, P.D., Meehan, J.P., and Henry, J.P. (1973). Impaired atrial receptor responses in dogs with heart failure due to tricuspid insufficiency and pulmonary artery stenosis. *Circulation Research*, **32**, 424-433.

Greene, A.S. and Brunner, M.J. (1988). Concerning reflex summation. *American Journal of Physiology*, **255**, H1244-H1246.

Greene, D.G. and Bunnell, I.L. (1953). The ciculatory response to the Valsalva maneuver of patients with mitral stenosis with and without autonomic blockade. *Circulation*, **8**, 261-268.

Greene, N.M. and Bachand, R.G. (1971). Vagal component of the chronotropic response to baroreceptor stimulation in man. *American Heart Journal*, **82**, 22-27.

Greenfield, A.D.M., Brown, E., Goei, J.S., and Plassaras, G.C. (1963). Circulatory responses to abrupt release of blood accumulated in the legs (Abstract). *The Physiologist*, **6**, 191.

Greenfield, J.C., Tindall, G.T., Dillon, M.L., and Mahaley, M.S. (1964). Mechanics of the human common carotid artery in vivo. *Circulation Research*, **15**, 240-246.

Greenfield, J.C., Cox, R.L., Hernandez, R.R., Thomas, C., and Schoonmaker, F.W. (1967). Pressure–flow studies in man during the Valsalva maneuver with observations on the mechanical properties of the ascending aorta. *Circulation*, **35**, 653-661.

Greenfield, J.C., Jr. and Patel, D.J. (1962). Relation between pressure and diameter in the ascending aorta of man. *Circulation Research*, **10**, 778-781.

Greenleaf, J.E. (1984). Physiological responses to prolonged bed rest and fluid immersion in humans. *Journal of Applied Physiology*, **57**, 619-633.

Greenway, C.V. and Innes, I.R. (1981). Effect of carotid sinus baroreceptor reflex on responses to phenylephrine and nitroprusside in anesthetized cats. *Journal of Cardiovascular Pharmacy*, **3**, 169-177.

Greenwood, R.J. and Dupler, D.A. (1962). Death following carotid sinus pressure. *The Journal of the American Medical Association*, **181**, 605-609.

Gregor, M. and Jänig, W. (1977). Effects of systemic hypoxia and hypercapnia on cutaneous and muscle vasoconstrictor neurones to the cat's hindlimb. *Pflügers Archiv für die Gesamte Physiologie des Menschen und der Tiere*, **368**, 71-81.

Gregor, M., Jänig, W., and Wiprich, L. (1977). Cardiac and respiratory rhythmicities in cutaneous and muscle vasoconstrictor neurones to the cat's hindlimb. *Pflügers Archiv für die Gesamte Physiologie des Menschen und der Tiere*, **370**, 299-302.

Gribbin, B., Pickering, T.G., Sleight, P., and Peto, R. (1971). Effect of age and high blood pressure on baroreflex sensitivity in man. *Circulation Research*, **29**, 424-431.

Grigor'eva, T.A. (1962). *The innervation of blood vessels*, pp. 20-50. Pergamon Press, New York.

Gross, P.M., Whipp, B.J., Davidson, J.T., Koyal, S.N., and Wasserman, K. (1976). Role of the carotid bodies in the heart rate response to breath holding in man. *Journal of Applied Physiology*, **41**, 336-340.

Gross, R., Hackenberg, H.-M., Hackenthal, E., and Kirchheim, H. (1981*a*). Interaction between perfusion pressure and sympathetic nerves in renin release by carotid baroreflex in conscious dogs. *The Journal of Physiology*, **313**, 237-250.

Gross, R., Kirchheim, H., and Ruffmann, K. (1981*b*). Effect of carotid occlusion and of perfusion pressure on renal function in conscious dogs. *Circulation Research*, **48**, 777-784.

Grossman, P., Stemmler, G., and Meinhardt, E. (1990). Paced respiratory sinus arrhythmia as an index of cardiac parasympathetic tone during varying behavioral tasks. *Psychophysiology*, **27**, 404-416.

Grossman, S.H., Davis, D., Gunnells, J.C., and Shand, D.G. (1982). Plasma norepinephrine in the evaluation of baroreceptor function in humans. *Hypertension*, **4**, 566-571.

Guazzi, M.D., Fiorentini, C., Olivari, M.T., Bartorelli, A., Magrini, F., and Biancardi, C. (1981). Circulatory and renin response in man to unilateral reduction of the renal perfusion pressure. *Cardiovascular Research*, **15**, 637-642.

Guertzenstein, P.G. and Silver, A. (1974). Fall in blood pressure produced from discrete regions of the ventral surface of the medulla by glycine and lesions. *The Journal of Physiology*, **242**, 489-503.

Guharay, F. and Sachs, F. (1984). Stretch-activated single ion channel currents in tissue-cultured embryonic chick skeletal muscle. *The Journal of Physiology*, **352**, 685-701.

Guidera, S.A., Bigger, J.T., Steinman, R.C., and Rolnitzky, L.M. (1990). Beta blockers increase vagal tone after myocardial infarction (MI) (Abstract). *Circulation*, **82** (Suppl. III), III-139.

Gundersen, T., Abrahamsen, A.M., Kjekshus, J., and Rønnevik, P.K. (1982). Timolol-related reduction in mortality and reinfarction in patients ages 65-75 years surviving acute myocardial infarction. *Circulation*, **66**, 1179-1184.

Gunn, C.G., Sevelius, G., Puiggari, M.J., and Myers, F.K. (1968). Vagal cardiomotor mechanisms in the hindbrain of the dog and cat. *American Journal of Physiology*, **214**, 258-262.

Guo, G.B., Thames, M.D., and Abboud, F.M. (1982). Differential baroreflex control of heart rate and vascular resistance in rabbits. Relative role of carotid, aortic, and cardiopulmonary baroreceptors. *Circulation Research*, **50**, 554-565.

Guo, G.B., Thames, M.D., and Abboud, F.M. (1983). Arterial baroreflexes in renal hypertensive rabbits. Selectivity and redundancy of baroreceptor influence on heart rate, vascular resistance, and lumbar sympathetic nerve activity. *Circulation Research*, **53**, 223-234.

Gupta, P.D. and Singh, M. (1987). Tachycardia originating from carotid chemoreceptors in apneic asphyxia in dogs. *American Journal of Physiology*, **253**, H591-H597.

Guthrie, G.P., Jr. and Kotchen, T.A. (1983). Effects of oral clonidine on baroreflex function in patients with essential hypertension. *Chest*, **83**, 327-328.

Guyenet, P.G., Filtz, T.M., and Donaldson, S.R. (1987). Role of excitatory amino acids in rat vagal and sympathetic baroreflexes. *Brain Research*, **407**, 272-284.

Guyenet, P.G., Darnall, R.A., and Riley, T.A. (1990). Rostral ventrolateral medulla and sympathorespiratory integration in rats. *American Journal of Physiology*, **259**, R1063-R1074.

Guyton, A.C., Granger, H.J., and Taylor, A.E. (1971). Interstitial fluid pressure. *Physiological Reviews*, **51**, 527-563.

Guz, A., Noble, M.I.M., Widdicombe, J.G., Trenchard, D., Mushin, W.W., and Makey, A.R. (1966). The role of vagal and glossopharyngeal afferent nerves in respiratory sensation, control of breathing and arterial pressure regulation in conscious man. *Clinical Science*, **30**, 161-170.

Habak, P.A., Mark, A.L., Kioschos, J.M., McRaven, D.R., and Abboud, F.M. (1974). Effectiveness of congesting cuffs ('rotating tourniquets') in patients with left heart failure. *Circulation*, **50**, 366-371.

Hade, J.S., Mifflin, S.W., Donta, T.S., and Felder, R.B. (1988). Stimulation of parabrachial neurons elicits a sympathetically mediated pressor response in cats. *American Journal of Physiology*, **255**, H1349-H1358.

Hagbarth, K.-E. and Vallbo, Å.B. (1968). Pulse and respiratory grouping of sympathetic impulses in human muscle nerves. *Acta Physiologica Scandinavica*, **74**, 96-108.

Hagbarth, K.-E., Hallin, R.G., Hongell, A., Torebjörk, H.E., and Wallin, B.G. (1972). General characteristics of sympathetic activity in human skin nerves. *Acta Physiologica Scandinavica*, **84**, 164-176.

Häggendal, J., Hartley, L.H., and Saltin, B. (1970). Arterial noradrenaline concentration during exercise in relation to the relative work levels. *Scandinavian Journal of Clinical and Laboratory Investigation*, **26**, 337-342.

Hainsworth, R. (1991). Reflexes from the heart. *Physiological Reviews*, **71**, 617–658.

Hajduczok, G., Chapleau, M.W., and Abboud, F.M. (1988). Rheoreceptors in the carotid sinus of dog. *Proceedings of the National Academy of Sciences USA*, **85**, 7399-7403.

Hakumäki, M.O.K., Wang, B.C., Sundet, W.D., and Goetz, K.L. (1985). Aortic baroreceptor discharge during nonhypotensive hemorrhage in anesthetized dogs. *American Journal of Physiology*, **249**, H393-H403.

Halberg, F., Johnson, E.A., Nelson, W., Runge, W., and Sothern, R. (1972). Autorhythmometry — procedures for physiologic self-measurements and their analysis. *The Physiology Teacher*, **1**, 1-11.

Hales, S. (1733). *Statical essays: containing haemastaticks; or, an account of some hydraulick and hydrostatical experiments made on the blood and blood-vessels of animals.* W. Innys, R. Manby, and T. Woodward, London.

Haller, A. (1778). *De partium corporis humani praecipuarum fabrica et functionibus.* Prelis Societatum Typographicarum, Berne.

Hallin, R.G. and Torebjörk, H.E. (1974). Single unit sympathetic activity in human skin nerves during rest and various manoeuvres. *Acta Physiologica Scandinavica*, **92**, 303-317.

Hamilton, R.B., Ellenberger, H., Liskowsky, D., and Schneiderman, N. (1981). Parabrachial area as mediator of bradycardia in rabbits. *Journal of the Autonomic Nervous System*, **4**, 261-281.

Hamilton, W.F., Woodbury, R.A., and Harper, H.T., Jr. (1936). Physiologic relationships between intrathoracic intraspinal and arterial pressures. *The Journal of the American Medical Association*, **107**, 853-856.

Hamilton, W.F., Woodbury, R.A., and Harper, H.T., Jr. (1944). Arterial, cerebrospinal and venous pressures in man during cough and strain. *American Journal of Physiology*, **141**, 42-50.

Hammer, M. and Engell, H.C. (1982). Episodic secretion of vasopressin in man. *Endocrinologica*, **101**, 517-523.

Hammill, S.C. *et al.* (1984). Electrophysiologic testing in the upright position: improved evaluation of patients with rhythm disturbances using a tilt table. *Journal of the American College of Cardiology*, **4**, 65-71.

Hansen, J.F., Hesse, B., and Christensen, N.J. (1978). Enhanced sympathetic nervous activity after intravenous propranolol in ischaemic heart disease: plasma noradrenaline splanchnic blood flow and mixed venous oxygen saturation at rest and during exercise. *European Journal of Clinical Investigation*, **8**, 31-36.

Hansen, J.T. (1987). Morphology of the carotid sinus wall in normotensive and spontaneously hypertensive rats. *The Anatomical Record*, **218**, 426-433.

Hargens, A.R., McClure, A.G., Skyhar, M.J., Lieber, R.L., Gershuni, D.H., and Akeson, W.H. (1987). Local compression patterns beneath pneumatic tourniquets applied to arms and thighs of human cadavera. *Journal of Orthopaedic Research* , **5**, 247-252.

Harris, P. (1987). Congestive cardiac failure: central role of the arterial blood pressure. *British Heart Journal*, **58**, 190-203.

Harrison, M.H., Rittenhouse, D., and Greenleaf, J.E. (1986). Effect of posture on arterial baroreflex control of heart rate in humans. *European Journal of Applied Physiology*, **55**, 367-373.

Haselton, J.R. and Guyenet, P.G. (1989*a*). Central respiratory modulation of medullary sympathoexcitatory neurons in rat. *American Journal of Physiology*, **256**, R739-R750.

Haselton, J.R. and Guyenet, P.G. (1989*b*). Electrophysiological characterization of putative C1 adrenergic neurons in the rat. *Neuroscience*, **30**, 199-214.

Hasking, G.J., Esler, M.D., Jennings, G.L., Burton, D., Johns, J.A., and Korner, P.I. (1986). Norepinephrine spillover to plasma in patients with congestive heart failure: evidence of increased overall and cardiorenal sympathetic nervous activity. *Circulation*, **73**, 615-621.

Hassen, A.H. and Feuerstein, G. (1987). μ-Opioid receptors in NTS elicit pressor responses via sympathetic pathways. *American Journal of Physiology*, **252**, H156-H162.

Hasser, E.M., Nelson, D.O., Haywood, J.R., and Bishop, V.S. (1987). Inhibition of renal sympathetic nervous activity by area postrema stimulation in rabbits. *American Journal of Physiology*, **253**, H91-H99.

Hauss, W.H. and Schmitt, G. (1961). Über vom Karotissinus des Menschen abgeleitete Elektroneurogramme mit Gedanken zur Pathogenese der arteriellen Blutdruckerkrankungen. *Zeitschrift für Kreislaufforschung*, **218**, 218-229.

Hauss, W.H. and Schmitt, G. (1962*a*). Die Bedeutung des Sinus caroticus für die Pathogenese der Blutdruckerkrankungen. In *Hypertonie*, ed. Hauss, W.H., and Losse, H., pp. 97-111. Georg Thieme Verlag, Stuttgart.

Hauss, W.H. and Schmitt, G. (1962*b*). Über Elektroneurogramme, die vom Carotissinus des Menschen abgeleitet wurden. *Archives Internationales de Pharmacodynamie et de Thérapie*, **139**, 487-493.

Hauss, W.H., Kreuziger, H., and Asteroth, H. (1949). Über die Reizung der Pressorezeptoren im Sinus caroticus beim Hund. *Zeitschrift für Kreislaufforschung*, **38**, 28-33.

Heath, D., Smith, P., Harris, P., and Winson, M. (1973). The atherosclerotic human carotid sinus. *Journal of Pathology*, **110**, 49-58.

Heath, R.G. and Mickle, W.A. (1960). Evaluation of seven years' experience with depth electrode studies in human patients. In *Electrical studies on the unanesthestized brain*, ed. Ramey, E.R., and O'Doherty, D.S., pp. 214-247. Paul B. Hoeber, New York.

Hedemark, L.L. and Kronenberg, R.S. (1982). Ventilatory and heart rate responses to hypoxia and hypercapnia during sleep in adults. *Journal of Applied Physiology*, **53**, 307-312.

Hedner, J., Ejnell, H., Sellgren, J., Hedner, T., and Wallin, G. (1988). Is high and fluctuating muscle nerve sympathetic activity in the sleep apnoea syndrome of pathogenetic importance for the development of hypertension? *Journal of Hypertension*, **6** (Suppl. 4), S529-S531.

Heistad, D.D. and Abboud, F.M. (1980). Circulatory adjustments to hypoxia. *Circulation*, **61**, 463-470.

Heistad, D.D. and Wheeler, R.C. (1970). Simulated diving during hypoxia in man. *Journal of Applied Physiology*, **28**, 652-656.

Heistad, D.D., Abboud, F.M., and Eckstein, J.W. (1968). Vasoconstrictor response to simulated diving in man. *Journal of Applied Physiology*, **25**, 542-549.

Heistad, D.D., Abboud, F.M., Mark, A.L., and Schmid, P.G. (1972). Impaired reflex vasoconstriction in chronically hypoxemic patients. *The Journal of Clinical Investigation*, **51**, 331-337.

Heistad, D.D., Abboud, F.M., Mark, A.L., and Schmid, P.G. (1974). Interaction of baroreceptor and chemoreceptor reflexes. Modulation of the chemoreceptor reflex by changes in baroreceptor activity. *The Journal of Clinical Investigation*, **53**, 1226-1236.

Hellner, K. and von Baumgarten, R. (1961). Über ein Endigungsgebiet afferenter, kardiovaskulärer Fasern des Nervus vagus im Rautenhirn der Katze. *Pflügers Archiv für die Gesamte Physiologie des Menschen und der Tiere*, **273**, 223-234.

Hendrickson, M. (1969). Carotid sinus reflex during radical neck surgery. *Southern Medical Journal*, **62**, 1195-1197.

Heng, M.B., Singh, B.N., Roche, A.H.G., Norris, R.M., and Mercer, C.J. (1975). Effects of intravenous verapamil on cardiac arrhythmias and on the electrocardiogram. *American Heart Journal*, **90**, 487-498.

Henriksen, O. and Rowell, L.B. (1986). Lack of effect of moderate hypoxemia on human postural reflexes to skeletal muscle. *Acta Physiologica Scandinavica*, **127**, 171-175.

Henry, J.L. and Calaresu, F.R. (1972). Distribution of cardioacceleratory sites in intermediolateral nucleus of the cat. *American Journal of Physiology*, **222**, 700-704.

Henry, J.L. and Calaresu, F.R. (1974). Excitatory and inhibitory inputs from medullary nuclei projecting to spinal cardioacceleratory neurons in the cat. *Experimental Brain Research*, **20**, 485-504.

Henry, J.P., Gauer, O.H., and Sieker, H.O. (1956). The effect of moderate changes in blood volume on left and right atrial pressures. *Circulation Research*, **4**, 91-94.

Henry, R.T. and Connor, J.D. (1989). Axons of passage may be reponsible for fastigial nucleus pressor response. *American Journal of Physiology*, **257**, R1436-R1440.

Hering, E. (1871). Über den Einfluss der Atmung auf den Kreislauf. Zweite Mittheilung. *Sitzungsberichte der Adademie der Wissenschaften, Mathematisch-Naturwissenschaftliche Classe, Abteilung II, Wien*, **64**, 333–353.

Hering, H.E. (1923). Der Karotisdruckversuch. *Münchener Medizinische Wochenschrift*, **70**, 1287-1290.

Hering, H.E. (1924*a*). Der Sinus caroticus an der Ursprungsstelle der Carotis interna als Ausgangsort eines hemmenden Herzreflexes und eines depressorischen Gefässreflexes. *Münchener Medizinische Wochenschrift*, **71**, 701-704.

Hering, H.E. (1924*b*). Die Sinusreflexe vom Sinus caroticus werden durch einen Nerven (Sinusnerv) vermittelt, der ein Ast des Nervus glossopharyngeus ist. *Münchener Medizinische Wochenschrift*, **71**, 1265-1266.

Hering, H.E. (1927). *Die Karotissinusreflexe auf Herz und Gefässe*, pp. 1-150. Steinkopff, Dresden.

Hering, H.E. (1930). Über die abhängigkeit der bei Lungenaufblasung und bei der respiratorischen Arrhythmie auftretenden Frequenzänderungen des Herzens von den Blutdruckzüglern. *Zeitschrift für Kreislaufforschung*, **22**, 593-598.

Herre, J.M. and Thames, M.D. (1987). Responses of sympathetic nerves to programmed ventricular stimulation. *Journal of the American College of Cardiology*, **9**, 147-153.

Hess, D.S., Hanlon, T., Scheinman, M., Budge, R., and Desai, J. (1982). Termination of ventricular tachycardia by carotid sinus massage. *Circulation*, **65**, 627-633.

Hess, W.R. and Brügger, M. (1943). Das subkortikale Zentrum der affektiven Abwehrreaktion. *Helvetica Physiologica et Pharmacologica Acta*, **1**, 33-52.

Hewlett, A.W. and van Zwaluwenburg, J.G. (1909). The rate of blood flow in the arm. *Heart*, **1**, 87-97.

Heymans, C. (1929). *Le sinus carotidien et les autres zones vasosensibles réflexogènes. Leur rôle en physiologie en pharmacologie et en pathologie.* Secrétariat de la Revue, Louvain.

Heymans, C. and Bouckaert, J.-J. (1934). Perfusion des sinus carotidiens isolés avec la pompe de Dale-Schuster. Réflexes vasomoteurs. *Comptes Rendus Heb-*

domadaires des Séances et Mémoires de la Société de Biologie et de ses Filiales et Associées, **86**, 31-33.

Heymans, C. and Bouckaert, J.J. (1930). Sinus caroticus and respiratory reflexes. I. Cerebral blood flow and respiration. Adrenaline apnoea. *The Journal of Physiology*, **69**, 254-266.

Heymans, C. and Bouckaert, J.-J. (1934). Modifications de la pression artérielle après section des quatre nerfs frénateurs chez le chien. *Comptes Rendus Hebdomadaires des séances et Mémoires de la Société de Biologie et de ses Filiales et Associées*, **86**, 252-255.

Heymans, C. and Ladon, A. (1925). Recherches physiologiques et pharmacologiques sur la tête isolée et le centre vague du chien. *Archives Internationales de Pharmacodynamie et de Thérapie*, **30**, 415-453.

Heymans, C. and Neil, E. (1958). *Reflexogenic areas of the cardiovascular system*, pp. 1-271. Little, Brown and Company, Boston, Massachusetts.

Heymans, C. and van den Heuvel-Heymans, G. (1950). Action of drugs on arterial wall of carotid sinus and blood pressure. *Archives Internationales de Pharmacodynamie et de Thérapie*, **83**, 520-528.

Heymans, C., Bouckaert, J.J., and Dautrebande, L. (1931). Sur la régulation réflexe de la circulation par les nerfs vasosensibles du sinus carotidien. *Archives Internationales de Pharmacodynamie et de Thérapie*, **40**, 292-343.

Heymans, C., Bouckaert, J.J., and Regniers, P. (1933). *Le sinus carotidien et la zone homologue cardio-aortique*. G. Doin, Paris.

Heymans, C., Bouckaert, J.-J., and Pannier, R. (1944). Sur la localisation des éléments chimio- et pressosensibles de la zone réflexogène sino-carotidienne. *Bulletin de l'Académie Royale de Médecine de Belgique*, **9**, 42-56.

Heyndrickx, G.R., Boettcher, D.H., and Vatner, S.F. (1976). Effects of angiotensin, vasopressin, and methoxamine on cardiac function and blood flow distribution in conscious dogs. *American Journal of Physiology*, **231**, 1579-1587.

Hickler, R.B., Wells, R.E., Jr., Tyler, R., and Hamlin, J.T. (1959). Plasma catecholamine and electroencephalographic responses to acute postural change. *The American Journal of Medicine*, **26**, 410-423.

Higgins, C.B., Vatner, S.F., Eckberg, D.L., and Braunwald, E. (1972*a*). Alterations in the baroreceptor reflex in conscious dogs with heart failure. *The Journal of Clinical Investigation*, **51**, 715-724.

Higgins, C.B., Vatner, S.F., Franklin, D., and Braunwald, E. (1972*b*). Effects of experimentally produced heart failure on the peripheral vascular response to severe exercise in conscious dogs. *Circulation Research*, **31**, 186-194.

Hildebrandt, G. (1960-1961). Die Durchblutung der menschlichen Wademuskulatur bein orthostatischer Belastung. *Pflügers Archiv für die Gesamte Physiologie des Menschen und der Tiere*, **272**, 6.

Hildebrandt, J.R. (1974). Central connections of aortic depressor and carotid sinus nerves. *Experimental Neurology*, **45**, 590-605.

Hilgenberg, F. (1958). Neurohistologische Studien über die Pressorezeptoren des Carotissinus bei Hypertonikern. *Acta Neurovegetativa*, **19**, 1-14.

Hilton, F., Giordano, J., and Fortney, S. (1989). Vasodepressor syncope induced by lower body negative pressure: possible relevance to +Gz-stress training — a case report. *Aviation, Space, and Environmental Medicine*, **60**, 61-63.

Hilton, S.M. (1963). Inhibition of baroceptor reflexes on hypothalamic stimulation. (Abstract) *The Journal of Physiology*, **165**, 56P-57P.

Hilton, S.M. and Redfern, W.S. (1986). A search for brain stem cell groups integrating the defence reaction in the rat. *The Journal of Physiology*, **378**, 213-228.

Hilton, S.M. and Spyer, K.M. (1971). Participation of the anterior hypothalamus in the baroreceptor reflex. *The Journal of Physiology*, **218**, 271-293.

Hilton, S.M. and Zbrożyna, A.W. (1963). Amygdaloid region for defence reactions and its efferent pathway to the brain stem. *The Journal of Physiology*, **165**, 160-173.

Hines, S., Houston, M., and Robertson, D. (1981). The clinical spectrum of autonomic dysfunction. *The American Journal of Medicine*, **70**, 1091-1096.

Hintze, T.H. (1987). Reflex regulation of the circulation after stimulation of cardiac receptors by prostaglandins. *Federation Proceedings*, **46**, 73-80.

Hirooka, Y. *et al.* (1990). Attenuated forearm vasodilative response to intra-arterial atrial natriuretic peptide in patients with heart failure. *Circulation*, **82**, 147-153.

Hirsch, J.A. and Bishop, B. (1981). Respiratory sinus arrhythmia in humans: how breathing pattern modulates heart rate. *American Journal of Physiology*, **241**, H620-H629.

Hjemdahl, P., Daleskog, M., and Kahan, T. (1979). Determination of plasma catecholamines by high performance liquid chromatography with electrochemical detection: comparison with a radioenzymatic method. *Life Sciences*, **25**, 131-138.

Hjemdahl, P., Freyschuss, U., Juhlin-Dannfelt, A., and Linde, B. (1984). Differentiated sympathetic activation during mental stress evoked by the Stroop test. *Acta Physiologica Scandinavica*, Suppl. **527**, 25-29.

Hodsman, G.P., Phillips, P.A., Ogawa, K., and Johnston, C.I. (1986). Atrial natriuretic factor in normal man: effects of tilt, posture, exercise and haemorrhage. *Journal of Hypertension*, **4** (Suppl. 6), S503-S505.

Hökfelt, T., Fuxe, K., Goldstein, M., and Johansson, O. (1974). Immunohistochemical evidence for the existence of adrenaline neurons in the rat brain. *Brain Research*, **66**, 235-251.

Hökfelt, T., Elde, R., Johansson, O., Terenius, L., and Stein, L. (1977). The distribution of enkephalin-immunoreactive cell bodies in the rat central nervous system. *Neuroscience Letters*, **5**, 25-31.

Hökfelt, T., Johansson, O., Ljungdahl, Å., Lundberg, J.M., and Sultzberg, M. (1980). Peptidergic neurones. *Nature*, **284**, 515-521.

Holaday, J.W. (1983). Cardiovascular effects of endogenous opiate systems. *Annual Review of Pharmacology and Toxicology*, **23**, 541-594.

Hollander, A.P. and Bouman, L.N. (1975). Cardiac acceleration in man elicited by a muscle–heart reflex. *Journal of Applied Physiology*, **38**, 272-278.

Hollenberg, N.K. and Uvnäs, B. (1963). The role of the cardiovascular response in the resistance to asphyxia of avian divers. *Acta Physiologica Scandinavica*, **58**, 150-161.

Holmberg, M.J., Gorman, A.J., Cornish, K.G., and Zucker, I.H. (1983). Attentuation of arterial baroreflex control of heart rate by left ventricular receptor stimulation in the conscious dog. *Circulation Research*, **52**, 597-607.

Holmgren, A. (1956). Circulatory changes during muscular work in man with special reference to arterial and central venous pressures in the systemic circulation. *Scandinavian Journal of Clinical and Laboratory Investigation*, Suppl. **24**, 1-97.

Holton, P. and Wood, J.B. (1965). The effects of bilateral removal of the carotid bodies and denervation of the carotid sinuses in two human subjects. *The Journal of Physiology*, **181**, 365-378.

Holtz, J., Sommer, O., and Bassenge, E. (1987). Inhibition of sympathoadrenal activity by atrial natriuretic factor in dogs. *Hypertension*, **9**, 350-354.

Hong, S.K., Moore, T.O., Seto, G., Park, H.K., Hiatt, W.R., and Bernauer, E.M. (1970). Lung volumes and apneic bradycardia in divers. *Journal of Applied Physiology*, **29**, 172-176.

Hopkins, D.A. (1987). The dorsal motor nucleus of the vagus nerve and the nucleus ambiguus: structure and connections. In *Cardiogenic reflexes*, ed. Hainsworth, R., McWilliam, P.N., and Mary, D.A.S.G., pp. 185-203. Oxford University Press, Oxford.

Hopkins, D.A. and Holstege, G. (1978). Amygdaloid projections to the mesencephalon, pons and medulla oblongata in the cat. *Experimental Brain Research*, **32**, 529-547.

Hordof, A.J., Spotnitz, A., Mary-Rabine, L., Edie, R.N., and Rosen, M.R. (1978). The cellular electrophysiologic effects of digitalis on human atrial fibers. *Circulation*, **57**, 223-229.

Horn, E.M. *et al.* (1988). Reduced lymphocyte stimulatory guanine nucleotide regulatory protein and β-adrenergic receptors in congestive heart failure and reversal with angiotensin converting enzyme inhibitor therapy. *Circulation*, **78**, 1373-1379.

Horvath, J.S. *et al.* (1976). Plasma renin activity, plasma angiotensin II and extracellular fluid volume in patients after renal transplantation. *Clinical Science and Molecular Medicine*, **51**, 227s-230s.

Horwitz, I.S. and Kaufman, A.I. (1979). Effects of baroreceptor activation on spontaneous activity in the sweat glands and nictitating membrane of the cat. *Brain Research*, **178**, 411-424.

Hoshi, T., Kojima, T., Kameyama, S., and Matsuda, K. (1955). On the influence of postural change upon the cardiac rate in man. *The Tohoku Journal of Experimental Medicine*, **62**, 221-234.

Hoshino, P.K., Blaustein, A.S., and Gaasch, W.H. (1988). Effect of propranolol on the left ventricular response to the Valsalva maneuver in normal subjects. *The American Journal of Cardiology*, **61**, 400-404.

Housley, G.D., Martin-Body, R.L., Dawson, N.J., and Sinclair, J.D. (1987). Brain stem projections of the glossopharyngeal nerve and its carotid sinus branch in the rat. *Neuroscience*, **22**, 237-250.

Howard, P. and Leathart, G.L. (1951). Changes of pulse pressure and heart rate induced by changes of posture in subjects with normal and failing hearts. *Clinical Science*, **10**, 521-526.

Howe, P.R.C. (1985). Blood pressure control by neurotransmitters in the medulla oblongata and spinal cord. *Journal of the Autonomic Nervous System*, **12**, 95-115.

Hubbard, S.J. (1958). A study of rapid mechanical events in a mechanoreceptor. *The Journal of Physiology*, **141**, 198-218.

Hudson, W.M., Morley, C.A., and Perrins, E.J. (1985). Is a hypersensitive carotid sinus reflex relevant? *Clinical Progress in Electrophysiology and Pacing*, **3**, 155-159.

Huikuri, H.V. *et al.* (1989). Changes in spontaneous sinus node rate as an estimate of cardiac autonomic tone during stable and unstable ventricular tachycardia. *Journal of the American College of Cardiology*, **13**, 646-652.

Hulet, W.H. and Smith, H.W. (1961). Postural natriuresis and urine osmotic concentration in hydropenic subjects. *The American Journal of Medicine*, **30**, 8-25.

Hume, L., Ewing, D.J., Campbell, I.W., Reuben, S.R., and Clarke, B.F. (1979). Heart-rate response to sustained hand grip: comparison of the effects of cardiac autonomic blockade and diabetic autonomic neuropathy. *Clinical Science*, **56**, 287-291.

Humphrey, D.R. (1967). Neuronal activity in the medulla oblongata of cat evoked by stimulation of the carotid sinus nerve. In *Baroreceptors and hypertension*, ed. Kezdi, P., pp. 131-167. Pergamon Press, Oxford.

Humphrey, S.J. and McCall, R.B. (1984). Evidence that L-glutamic acid mediates baroreceptor function in the cat. *Clinical and Experimental Hypertension. Part A — Theory and Practice*, **A6**, 1311-1329.

Humphreys, P.W., and Lind, A.R. (1963). The blood flow through active and inactive muscles of the forearm during sustained hand-grip contractions. *The Journal of Physiology*, **166**, 120-135.

Huxley, F.M. (1913). On the reflex nature of apnoea in the duck in diving: I. The reflex nature of submersion apnoea. *Quarterly Journal of Experimental Physiology*, **6**, 147-157.

Huxley, V.H., Tucker, V.L., Verburg, K.M., and Freeman, R.H. (1987). Increased capillary hydraulic conductivity induced by atrial natriuretic peptide. *Circulation Research*, **60**, 304-307.

Hyde, T.M., Gibbs, M., and Peroutka, S.J. (1988). Distribution of muscarinic cholinergic receptors in the dorsal vagal complex and other selected nuclei in the human medulla. *Brain Research*, **447**, 287-292.

Iano, T.L., Levy, M.N., and Lee, M.H. (1973). An acceleratory component of the parasympathetic control of heart rate. *American Journal of Physiology*, **224**, 997-1005.

Ibrahim, M.M. (1975). Localization of lesion in patients with idiopathic orthostatic hypotension. *British Heart Journal*, **37**, 868-872.

Ibrahim, M.M., Tarazi, R.C., Dustan, H.P., and Bravo, E.L. (1974). Idiopathic orthostatic hypotension: circulatory dynamics in chronic autonomic insufficiency. *The American Journal of Cardiology*, **34**, 288-294.

Igler, F.O., Donegan, J.H., Hoo, K.-C., Korns, M.E., and Kampine, J.P. (1981). Chronic localized hypotension and resetting of carotid sinus baroreceptors. Electrophysiological and histological studies in the dog. *Circulation Research*, **49**, 649-654.

Imai, Y., Nolan, P.L., and Johnston, C.I. (1983). Restoration of suppressed baroreflex sensitivity in rats with hereditary diabetes insipidus (Brattleboro rats) by arginine-vasopressin and DDAVP. *Circulation Research*, **53**, 140-149.

Imaizumi, T. and Thames, M.D. (1986). Influence of intravenous and intracerebroventricular vasopressin on baroreflex control of renal nerve traffic. *Circulation Research*, **58**, 17-25.

Imaizumi, T., Brunk, S.D., Gupta, B.N., and Thames, M.D. (1984*a*). Central effect of intravenous phenylephrine on baroreflex control of renal nerves. *Hypertension*, **6**, 906-914.

Imaizumi, T., Takeshita, A., Makino, N., Ashihara, T., Yamamoto, K., and Nakamura, M. (1984*b*). Impaired baroreflex control of vascular resistance and heart rate in acute myocardial infarction. *British Heart Journal*, **52**, 418-421.

Imholz, B.P.M., van Montfrans, G.A., Settels, J.J., van der Hoeven, G.M.A., Karemaker, J.M., and Wieling, W. (1988). Continuous non-invasive blood pressure monitoring: reliability of Finapres device during the Valsalva manoeuvre. *Cardiovascular Research*, **22**, 390-397.

Inoue, H. and Zipes, D.P. (1987*a*). Changes in atrial and ventricular refractoriness and in atrioventricular nodal conduction produced by combinations of vagal and sympathetic stimulation that result in a constant spontaneous sinus cycle length. *Circulation Research*, **60**, 942-951.

Inoue, H. and Zipes, D.P. (1987*b*). Results of sympathetic denervation in the canine heart: supersensitivity that may be arrhythmogenic. *Circulation*, **75**, 877-887.

Inoue, H. and Zipes, D.P. (1988). Time course of denervation of efferent sympathetic and vagal nerves after occlusion of the coronary artery in the canine heart. *Circulation Research*, **62**, 1111-1120.

Inoue, H., Skale, B.T., and Zipes, D.P. (1988). Effects of ischemia on cardiac afferent sympathetic and vagal reflexes in dog. *American Journal of Physiology*, **255**, H26-H35.

Inoue, K., Miyake, S., Kumashiro, M., Ogata, H., and Yoshimura, O. (1990). Power spectral analysis of heart rate variability in traumatic quadriplegic humans. *American Journal of Physiology*, **258**, H1722-H1726.

Irisawa, H. and Ninomiya, I. (1967). Comparison of the averaged nervous activities of aortic and carotid sinus nerves. *American Journal of Physiology*, **213**, 504-510.

Iriuchijima, J. and Kumada, M. (1963). Efferent cardiac vagal discharge of the dog in response to electrical stimulation of sensory nerves. *The Japanese Journal of Physiology*, **13**, 599-605.

Iriuchijima, J. and Kumada, M. (1964). Activity of single vagal fibers efferent to the heart. *The Japanese Journal of Physiology*, **14**, 479-487.

Irving, L. (1934). On the ability of warm-blooded animals to survive without breathing. *The Scientific Monthly*, **38**, 422-428.

Irving, L., Scholander, P.F., and Grinnell, S.W. (1942). The regulation of arterial blood pressure in the seal during diving. *American Journal of Physiology*, **135**, 557-566.

Isaacson, J.S. and Reid, I.A. (1990). Importance of endogenous angiotensin II in the cardiovascular responses to sympathetic stimulation in conscious rabbits. *Circulation Research*, **66**, 662-671.

Ishizuka, Y. *et al.* (1987). The baroreflex-mediated changes in plasma norepinephrine and heart rate in patients with essential hypertension. *Japanese Circulation Journal*, **51**, 1165-1173.

Ismay, M.J.A., Lumbers, E.R., and Stevens, A.D. (1979). The action of angiotensin II on the baroreflex response of the conscious ewe and the conscious fetus. *The Journal of Physiology*, **288**, 467-479.

Ito, C.S. and Scher, A.M. (1979). Hypertension following denervation of aortic baroreceptors in unanesthetized dogs. *Circulation Research*, **45**, 26-34.

Iversen, L.L. (1983). Nonopioid neuropeptides in mammalian CNS. *Annual Review of Pharmacology and Toxicology*, **23**, 1-27.

Iwase, S., Mano, T., and Saito, M. (1987). Effects of graded head-up tilting on muscle sympathetic activities in man. *The Physiologist*, **30** (Suppl.), S-62–S-65.

Izzo, J., Jr. (1983). Cardiovascular hormonal effects of circulating norepinephrine. *Hypertension*, **5**, 787-789.

Jain, S.K., Trenchard, D., Reynolds, F., Noble, M.I.M., and Guz, A. (1973). The effect of local anaesthesia of the airway on respiratory reflexes in the rabbit. *Clinical Science*, **44**, 519-538.

Jalife, J. and Moe, G.K. (1979). Phasic effects of vagal stimulation on pacemaker activity of the isolated sinus node of the young cat. *Circulation Research*, **45**, 595-608.

Jalife, J. and Moe, G.K. (1983). Role of the autonomic nervous system in the genesis of arrhythmias. In *Frontiers of cardiac electrophysiology*, ed. Rosenbaum, M.B., and Elizari, M.V., pp. 501-521. Martinus Nijhoff Publishers, Boston, Massachusetts.

James, T.N. (1961). *Anatomy of the coronary arteries*. Paul B. Hoeber, New York.

James, T.N. and Nadeau, R.A. (1963). Sinus bradycardia during injections directly into the sinus node artery. *American Journal of Physiology*, **204**, 9-15.

James, T.N. and Spence, C.A. (1966). Distribution of cholinesterase within the sinus node and AV node of the human heart. *The Anatomical Record*, **155**, 151-162.

Jänig, W. (1985). Organization of the lumbar sympathetic outflow to skeletal muscle and skin of the cat hindlimb and tail. *Reviews of Physiology, Biochemistry and Pharmacology*, **102**, 119-213.

Jänig, W. and Kümmel, H. (1977). Functional discrimination of postganglionic neurones to the cat's hindpaw with respect to the skin potentials recorded from the hairless skin. *Pflügers Archiv für die Gesamte Physiologie des Menschen und der Tiere*, **371**, 217-225.

Jänig, W. and Szulczyk, P. (1980). Functional properties of lumbar preganglionic neurones. *Brain Research*, **186**, 115-131.

Jänig, W., Sundlöf, G., and Wallin, B.G. (1983). Discharge patterns of sympathetic neurons supplying skeletal muscle and skin in man and cat. *Journal of Autonomic Nervous System*, **7**, 239-256.

Jarecki, M., Thorén, P.N., and Donald, D.E. (1978). Release of renin by the carotid baroreflex in anesthetized dogs. Role of cardiopulmonary vagal afferents and renal arterial pressure. *Circulation Research*, **42**, 614-619.

Jeffrey, F.E., Fareeduddin, K., and Abelmann, W.H. (1970). Increased tolerance of patients with circulatory congestion to orthostatic stress. *American Journal of the Medical Sciences*, **259**, 323-332.

Jenkins, J.S., Ang, V.T.Y., Hawthorn, J., Rossor, M.N., and Iversen, L.L. (1984). Vasopressin, oxytocin and neurophysins in the human brain and spinal cord. *Brain Research*, **291**, 111-117.

Jennett, S. (1969). Changes in ventilation and heart rate during and after brief moderate hypoxia in resting man (Abstract). *The Journal of Physiology*, **200**, 106P-107P.

Jennett, S., Lamb, J.F., and Travis, P. (1982). Sudden large and periodic changes in heart rate in healthy young men after short periods of exercise. *British Medical Journal*, **285**, 1154-1156.

Jewett, D.L. (1964). Activity of single efferent fibres in the cervical vagus nerve of the dog, with special reference to possible cardio-inhibitory fibres. *The Journal of Physiology*, **175**, 321-357.

Jie, K., van Brummelen, P., Vermey, P., Timmermans, P.B.M.W.M., and van

Zwieten, P.A. (1986). Alpha$_1$- and alpha$_2$-adrenoceptor mediated vasoconstriction in the forearm of normotensive and hypertensive subjects. *Journal of Cardiovascular Pharmacology*, **8**, 190-196.

Jie, K., van Brummelen, P., Vermey, P., Timmermans, P.B.M.W.M., and van Zwieten, P.A. (1987). Modulation of noradrenaline release by peripheral presynaptic α_2-adrenoceptors in humans. *Journal of Cardiovascular Pharmacology*, **9**, 407-413.

Joels, N. (1975). Reflex respiratory effects of circulating catecholamines. In *Handbook of physiology. Section 7: Endocrinology. Volume VI: Adrenal gland*, ed. Blaschko, H., Sayers, G., and Smith, A.D., pp. 491-505. American Physiological Society, Washington, D.C.

Johansen, K. and Aakhus, T. (1963). Central cardiovascular responses to submersion asphyxia in the duck. *American Journal of Physiology*, **205**, 1167-1171.

Johnson, A.K. and Loewy, A.D. (1990). Circumventricular organs and their role in visceral functions. In *Central regulation of autonomic functions*, ed. Loewy, A.D., and Spyer, K.M., pp. 247-267. Oxford University Press, New York.

Johnson, J.M., Rowell, L.B., Niederberger, M., and Eisman, M.M. (1974). Human splanchnic and forearm vasoconstrictor responses to reductions of right atrial and aortic pressures. *Circulation Research*, **34**, 515-524.

Johnson, M.D., Smith, P.G., Mills, E., and Schanberg, S.M. (1983). Paradoxical elevation of sympathetic activity during catecholamine infusion in rats. *Journal of Pharmacology and Experimental Theraputics*, **227**, 254-259.

Johnston, B.D. (1968). Nerve endings in the human endocardium. *The American Journal of Anatomy*, **122**, 621-629.

Jones, D.R. (1973). Systemic arterial baroreceptors in ducks and the consequences of their denervation on some cardiovascular responses to diving. *The Journal of Physiology*, **234**, 499-518.

Jones, D.R. and Purves, M.J. (1970). The carotid body in the duck and the consequences of its denervation upon the cardiac responses to immersion. *The Journal of Physiology*, **211**, 279-294.

Jordan, D. and Spyer, K.M. (1977). Studies on the termination of sinus nerve afferents. *Pflügers Archiv für die Gesamte Physiologie des Menschen und der Tiere*, **369**, 65-73.

Jordan, D. and Spyer, K.M. (1979). Studies on the excitability of sinus nerve afferent terminals. *The Journal of Physiology*, **297**, 123-134.

Jordan, D. and Spyer, K.M. (1980). Observations on the terminations of sinus and aortic nerve afferents. In *Arterial baroreceptors and hypertension*, ed. Sleight, P., pp. 229-237. Oxford University Press, Oxford.

Jordan, D., Mifflin, S.W., and Spyer, K.M. (1988). Hypothalamic inhibition of neurones in the nucleus tractus solitarius of the cat is GABA mediated. *The Journal of Physiology*, **399**, 389-404.

Jose, A.D. and Taylor, R.R. (1969). Autonomic blockade by propranolol and atropine to study intrinsic myocardial function in man. *The Journal of Clinical Investigation*, **48**, 2019-2031.

Josephson, M.E., Spielman, S.R., Greenspan, A.M., and Horowitz, L.N. (1979). Mechanism of ventricular fibrillation in man. Observations based on electrode catheter recordings. *The American Journal of Cardiology*, **44**, 623-631.

Joy, M.D. and Lowe, R.D. (1970). Evidence that the area postrema mediates the central cardiovascular response to angiotensin II. *Nature*, **228**, 1303-1304.

Joyner, M.J., Shepherd, J.T., and Seals, D.R. (1990*a*). Sustained increases in sympathetic outflow during prolonged lower body negative pressure in humans. *American Journal of Physiology*, **68**, 1004-1009.

Joyner, M.J., Suwarno, N.O., Seals, D.R., and Shepherd, J.T. (1990*b*). Effects of graded lower body negative pressure on forearm blood flow in patients after heart–lung transplantation (Abstract). *The FASEB Journal*, **4**, A428.

Judson, W.E., Hatcher, J.D., and Wilkins, R.W. (1955). Blood pressure responses to the Valsalva maneuver in cardiac patients with and without congestive failure. *Circulation*, **11**, 889-899.

Julien, C., Kandza, P., Barres, C., Lo, M., Cerutti, C., and Sassard, J. (1990). Effects of sympathectomy on blood pressure and its variability in conscious rats. *American Journal of Physiology*, **259**, H1337-H1342.

Julius, S., Pascual, A.V., and London, R. (1971). Role of parasympathetic inhibition in the hyperkinetic type of borderline hypertension. *Circulation*, **44**, 413-418.

Jung, R. and Katona, P.G. (1990). Cardiovascular and respiratory responses to slow ramp carotid sinus pressures in the dog. *Journal of Applied Physiology*, **68**, 1465-1474.

Kahn, R.H. (1929). Die Blutdruckregler. *Zeitschrift für die Gesamte Experimentelle Medizin*, **68**, 201-215.

Kakembo, G.H.K. (1982). The factors affecting inter-individual differences in response to bethanidine (Ph.D. thesis), Medizinischen Universitatsklinik, Bonn.

Kakurin, L.I., Lobachik, V.I., Mikhailov, V.M., and Senkevich, Y.A. (1976). Antiorthostatic hypokinesia as a method of weightlessness simulation. *Aviation, Space, and Environmental Medicine*, **47**, 1083-1086.

Kalia, M. (1981). Brain stem localization of vagal preganglionic neurons. *Journal of the Autonomic Nervous System*, **3**, 451-481.

Kalia, M. and Mesulam, M.-M. (1980). Brain stem projections of sensory and motor components of the vagus complex in the cat: II. Laryngeal, tracheobronchial, pulmonary, cardiac, and gastrointestinal branches. *The Journal of Comparative Neurology*, **193**, 467-508.

Kalia, M. *et al.* (1984). Distribution of neuropeptide immunoreactive nerve terminals within the subnuclei of the nucleus of the tractus solitarius of the rat. *The Journal of Comparative Neurology*, **222**, 409-444.

Kalkoff, W. (1957). Pressorezeptorische Aktionspotentiale und Blutdruckregulation. *Verhandlungen der Deutschen Gesellschaft für Kreislaufforschung*, **23**, 397-401.

Kalkoff, W. (1959). Die Reflexbeeinflussung des Blutdruckes während Adrenalineinwirkung am Karotissinus — Äusserung einer spezifischen, von der Pressorezeption unterscheidbaren Funktion. *Verhandlungen der Deutschen Gesellschraft für Kreislaufforschung*, **25**, 173-178.

Kalkoff, W., Plischke, K.H., and Rath, F.W. (1966). Chemoreception und lokale Adrenalinwirkung am Sinus caroticus. *Pflügers Archiv für die Gesamte Physiologie des Menschen und der Tiere*, **289**, 122-140.

Kalkoff, W., Kalkoff-Ganse, B., and Plumbohm, P. (1971). Dichteverteilung der Pressorezeptorenimpulse bei adäquater Reizung des Karotissinus. *Acta Biologica et Medica Germanica*, **26**, 711-732.

Kamiya, K., Shimizu, T., Toyama, J., and Yamada, K. (1983). Reflex control of ventricular refractoriness in the nonischemic myocardium during coronary occlusion. *Japanese Journal of Physiology*, **33**, 801-810.

Kammerling, J.J., Green, F.J., Watanabe, A.M., Inoue, H., Barber, M.J., Henry, D.P., and Zipes, D.P. (1987). Denervation supersensitivity of refractoriness in noninfarcted areas apical to transmural myocardial infarction. *Circulation*, **76**, 383-393.

Kaneko, Y. *et al.* (1970). Effect of ganglion-blocking agents on renin release in hypertensive patients. *Circulation Research*, **27**, 97-103.

Kannon, H. and Yagi, K. (1978). Supraoptic neurosecretory neurons: evidence for the existence of converging inputs both from carotid baroreceptors and osmoreceptors. *Brain Research*, **145**, 385-390.

Kapoor, W.N., Karpf, M., Wieand, S., Peterson, J.R., and Levey, G.S. (1983). A prospective evalauation and follow-up of patients with syncope. *The New England Journal of Medicine*, **309**, 197-204.

Kappagoda, C.T. and Ravi, K. (1989). Plasmapheresis affects responses of slowly and rapidly adapting airway receptors to pulmonary venous congestion in dogs. *The Journal of Physiology*, **416**, 79-91.

Kappagoda, C.T., Linden, R.J., and Snow, H.M. (1972). The effect of stretching the superior vena caval–right atrial junction on right atrial receptors in the dog. *The Journal of Physiology*, **227**, 875-887.

Kappagoda, C.T., Man, G.C.W., and Teo, K.K. (1987). Behaviour of canine pulmonary vagal afferent receptors during sustained acute pulmonary venous pressure elevation. *The Journal of Physiology*, **394**, 249-265.

Karam, M., Wise, R.A., Natarajan, T.K., Permutt, S., and Wagner, H.N. (1984). Mechanism of decreased left ventricular stroke volume during inspiration in man. *Circulation*, **69**, 866-873.

Kardon, M.B., Peterson, D.F., and Bishop, V.S. (1975). Reflex heart rate control via specific aortic nerve afferents in the rabbit. *Circulation Research*, **37**, 41-47.

Karemaker, J.M. and Borst, C. (1980). Measurement of baroreflex sensitivity in hypertension research. In *Arterial baroreceptors and hypertension*, ed. Sleight, P., pp. 455-461. Oxford University Press, Oxford.

Karlsberg, R.P., Penkoske, P.A., Cryer, P.E., Corr, P.B., and Roberts, R. (1979). Rapid activation of the sympathetic nervous system following coronary artery occlusion: relationship to infarct size, site, and haemodynamic impact. *Cardiovascular Research*, **13**, 523-531.

Kassis, E. (1987). Cardiovascular response to orthostatic tilt in patients with severe congestive heart failure. *Cardiovascular Research*, **21**, 362-368.

Kassis, E. (1989). Baroreflex control of the circulation in patients with congestive heart failure. *Danish Medical Bulletin*, **36**, 195-211.

Kassis, E. and Amtorp, O. (1987). Cardiovascular and neurohumoral postural responses and baroreceptor abnormalities during a course of adjunctive vasodilator therapy with felodipine for congestive heart failure. *Circulation*, **75**, 1204-1213.

Kassis, E., Jacobsen, T.M., Mogensen, F., and Amtorp, O. (1986). Sympathetic reflex control of skeletal muscle blood flow in patients with congestive heart failure: evidence for β-adrenergic circulatory control. *Circulation*, **74**, 929-938.

Kasting, G.A., Eckberg, D.L., Fritsch, J.M., and Birkett, C.L. (1987). Continuous resetting of the human carotid baroreceptor–cardiac reflex. *American Journal of Physiology*, **252**, R732-R736.

Katholi, R.E., Carey, R.M., Ayers, C.R., Vaughan, E.D., Jr., Yancey, M.R., and Morton, C.L. (1977). Production of sustained hypertension by chronic intrarenal norepinephrine infusion in conscious dogs. *Circulation Research*, **40** (Suppl. I), I-118–I-126.

Katona, P.G. and Barnett, G.O. (1969). Central origin of asymmetry in the carotid sinus reflex. *Annals of the New York Academy of Sciences*, **165**, 779-786.

Katona, P.G. and Jih, F. (1975). Respiratory sinus arrhythmia: noninvasive measure of parasympathetic cardiac control. *Journal of Applied Physiology*, **39**, 801-805.

Katona, P.G., Poitras, J.W., Pantelakis, N., Jensen, E.W., and Barnett, G.O. (1968). Deterministic nature of baroreceptor firing. *American Journal of Physiology*, **215**, 1-7.

Katona, P.G., Poitras, J.W., Barnett, G.O., and Terry, B.S. (1970). Cardiac vagal efferent activity and heart period in the carotid sinus reflex. *American Journal of Physiology*, **218**, 1030-1037.

Katona, P.G., Lipson, D., and Dauchot, P.J. (1977). Opposing central and peripheral effects of atropine on parasympathetic cardiac control. *American Journal of Physiology*, **232**, H146-H151.

Katona, P.G., McLean, M., Dighton, D.H., and Guz, A. (1982). Sympathetic and parasympathetic cardiac control in athletes and nonathletes at rest. *Journal of Applied Physiology*, **52**, 1652-1657.

Kauf, E. (1923). Untersuchungen über das Verhalten des Herzens nach Muskelarbeit. *Wiener Archiv für Innere Medizin*, **5**, 567-580.

Kaufman, M.P., Longhurst, J.C., Rybicki, K.J., Wallach, J.H., and Mitchell, J.H. (1983). Effects of static muscular contraction on impulse activity of groups III and IV afferents in cats. *American Journal of Physiology*, **55**, 105-112.

Kaufman, M.P., Rybicki, K.J., Waldrop, T.G., and Mitchell, J.H. (1984). Effect on arterial pressure of rhythmically contracting the hindlimb muscles of cats. *Journal of Applied Physiology*, **56**, 1265-1271.

Kawakami, Y., Natelson, B.H., and DuBois, A.B. (1967). Cardiovascular effects of face immersion and factors affecting diving reflex in man. *Journal of Applied Physiology*, **23**, 964-970.

Kee, D.B., Jr. and Wood, J.H. (1985). Physiological neuroendocrinology of peptides, steroids and other hormones in cerebrospinal fluid. *Progress in Neuro-Psychopharmacology Biological Psychiatry*, **9**, 55-72.

Keith, I.C., Kidd, C., Linden, R.J., and Snow, H.M. (1975). Modification of neuronal activity in the dog medulla oblongata by stimulation of the left atrial receptors (Abstract). *The Journal of Physiology*, **245**, 80P-81P.

Kellogg, D.L., Jr., Johnson, J.M., and Kosiba, W.A. (1990). Baroreflex control of the cutaneous active vasodilator system in humans. *Circulation Research*, **66**, 1420-1426.

Kelly, D. (1972). Physiological changes during operations on the limbic system in man. *Conditional Reflex*, **7**, 127-138.

Kemp, E. (1957). Arterial hypertension in poliomyelitis. *Acta Medica Scandinavica*, **157**, 109-118.

Kempf, F.C., Jr. and Josephson, M.E. (1984). Cardiac arrest recorded on ambulatory electrocardiograms. *The American Journal of Cardiology*, **53**, 1577-1582.

Kendrick, J.E., Matson, G.L., and Lalley, P.M. (1979). Central interaction between the baroreceptor reflexes from the carotid sinus and aortic arch. *American Journal of Physiology*, **236**, H127-H133.

Kenney, W.L. (1985). Parasympathetic control of resting heart rate: relationship to aerobic power. *Medicine and Science in Sports and Exercise*, **17**, 451-455.

Kent, B.B., Drane, J.W., and Manning, J.W. (1971). Suprapontine contributions to the carotid sinus reflex in the cat. *Circulation Research*, **29**, 534-541.

Kent, K.M., Smith, E.R., Redwood, D.R., and Epstein, S.E. (1973). Electrical stability of acutely ischemic myocardium. Influences of heart rate and vagal stimulation. *Circulation*, **47**, 291-298.

Kent, K.M., Epstein, S.E., Cooper, T., and Jacobowitz, D.M. (1974). Cholinergic innervation of the canine and human ventricular conducting system. Anatomic and electrophysiologic correlations. *Circulation*, **50**, 948-955.

Kerr, F.W.L. (1967). Function of the dorsal motor nucleus of the vagus. *Science*, **157**, 451-452.

Keys, A. and Violante, A. (1942). The cardio-circulatory effects in man of neo-synephrin. *The Journal of Clinical Investigation*, **21**, 1-12.

Kezdi, P. (ed.) (1967*a*). *Baroreceptors and hypertension*. Pergamon, Oxford.

Kezdi, P. (1967*b*). Resetting of the carotid sinus in experimental renal hypertension. In *Barorecptors and hypertension*, ed. Kezdi, P., pp. 301-308. Pergamon, Oxford.

Kezdi, P. and Geller, E. (1968). Baroreceptor control of postganglionic sympathetic nerve discharge. *American Journal of Physiology*, **214**, 427-435.

Kezdi, P., Spickler, J.W., and Kordenat, R.K. (1973). Neurogenic factors in renal hypertension. In *Hypertension: mechanisms and management*, ed. Onesti, G., Kim, K.E., and Moyer, J.H., pp. 681-692. Grune and Stratton, New York.

Kezdi, P., Kordenat, R.K., and Arnold, W.L. (1980). Role of vagal afferents in the tonic regulation of blood-pressure in sino-aortic deafferented dogs. In *Arterial baroreceptors and hypertension*, ed. Sleight, P., pp. 432-435. Oxford, New York.

Kilbinger, H. and Löffelholz, K. (1976). The isolated perfused chicken heart as a tool for studying acetylcholine output in the absence of cholinesterase inhibition. *Journal of Neural Transmission*, **38**, 9-14.

Killip, T. and Gault, J.H. (1965). Mode of onset of atrial fibrillation in man. *American Heart Journal*, **70**, 172-179.

Kimura, S., Bassett, A.L., Kohya, T., Kozlovskis, P.L., and Myerburg, R.J. (1987). Automaticity, triggered activity, and responses to adrenergic stimulation in cat subendocardial Purkinje fibers after healing of myocardial infarction. *Circulation*, **75**, 651-660.

King, B.D., Sack, D., Kichuk, M.R., and Hintze, T.H. (1987). Absence of hypertension despite chronic marked elevations in plasma norepinephrine in conscious dogs. *Hypertension*, **9**, 582-590.

Kiowski, W. and Julius, S. (1978). Renin response to stimulation of cardiopulmonary mechanoreceptors in man. *The Journal of Clinical Investigation*, **62**, 656-663.

Kiowski, W., van Brummelen, P., and Bühler, F.R. (1979). Plasma noradrenaline correlates with α-adrenoreceptor-mediated vasoconstriction and blood pressure in patients with essential hypertension. *Clinical Science*, **57**, 177B-180B.

Kiowski, W., Erne, P., Bertel, O., Bolli, P., and Bühler, F. (1986). Acute and chronic sympathetic reflex activation and antihypertensive response to nifedipine. *Journal of the American College of Cardiology*, **7**, 344-348.

Kirchheim, H., Ehmke, H., and Persson, P. (1988). Physiology of the renal baroreceptor mechanism of renin release and its role in congestive heart failure. *The American Journal of Cardiology*, **62**, 68E-71E.

Kirchheim, H.R. (1976). Systemic arterial baroreceptor reflexes. *Physiological Reviews*, **56**, 100-176.

Kirchheim, H.R., Finke, R., Hackenthal, E., Lowe, W., and Persson, P. (1985). Baroreflex sympathetic activation increases threshold pressure for the pressure-dependent renin release in conscious dogs. *Pflügers Archiv für die Gesamte Physiologie des Menschen und der Tiere*, **405**, 127-135.

Kitahama, K., Pearson, J., Denoroy, L., Kopp, N., Ulrich, J., Maeda, T., and Jouvet, M. (1985). Adrenergic neurons in human brain demonstrated by immunohistochemistry with antibodies to phenylethanolamine-n-methyltransferase (PNMT): discovery of a new group in the nucleus tractus solitarius. *Neuroscience Letters*, **53**, 303-308.

Kitney, R., Linkens, D., Selman, A., and McDonald, A. (1982). The interaction between heart rate and respiration: Part II - Nonlinear analysis based on computer modelling. *Automedica*, **4**, 141-153.

Kitney, R.I. (1975). Entrainment of the human RR interval by thermal stimuli. *The Journal of Physiology*, **252**, 37P-38P.

Kitney, R.I. (1984). New findings in the analysis of heart rate variability in infants. *Automedica*, **5**, 289-310.

Kitney, R.I. and Rompelman, O. (1987). *The beat-by-beat investigation of cardiovascular function*. Clarendon Press, Oxford.

Kleiger, R.E., Miller, J.P., Bigger, J.T., Jr., Moss, A.J., and the Multicenter Post-Infarction Research Group (1987). Decreased heart rate variability and its association with increased mortality after acute myocardial infarction. *The American Journal of Cardiology*, **59**, 256-262.

Klevans, L.R. and Gebber, G.L. (1970). Facilitatory forebrain influence on cardiac component of baroreceptor reflexes. *American Journal of Physiology*, **219**, 1235-1241.

Kluge, K.A., Harper, R.M., Schechtman, V.L., Wilson, A.J., Hoffman, H.J., and Southall, D.P. (1988). Spectral analysis assessment of respiratory sinus arrhythmia in normal infants and infants who subsequently died of sudden infant death syndrome. *Pediatric Research*, **24**, 677-682.

Kluger, J., Cody, R.J., and Laragh, J.H. (1982). The contributions of sympathetic tone and the renin–angiotensin system to severe chronic congestive heart failure: response to specific inhibitors (prazosin and captopril). *The American Journal of Cardiology*, **49**, 1667-1674.

Knoche, H. and Schmitt, G. (1964). Beitrag zur Kenntnis des Nervengewebes in der Wand des Sinus caroticus. I. Mitteilung. *Zeitschrift für Zellforschung und Mikroskopische Anatomie*, **63**, 22-36.

Knoche, H., Wiesner-Menzel, L., and Addicks, K. (1980). Ultrastructure of baroreceptors in the carotid sinus of the rabbit. *Acta Anatomica*, **106**, 63-83.

Knoll, P. (1885). Über periodische Athmungs- und Blutdruckschwankungen. *Sitzungsberichte der Mathematisch–Naturwissenschaftlichen Classe*, **92**, 439-460.

Kober, G. and Arndt, J.O. (1970). Die Druck-Durchmesser-Beziehung der A. carotis communis des wachen Menschen. *Pflügers Archiv für die Gesamte Physiologie des Menschen und der Tiere*, **314**, 27-39.

Kobinger, W. (1978). Central α-adrenergic systems as targets for hypotensive drugs. *Reviews of Physiology, Biochemistry and Pharmacology*, **81**, 39-100.

Kobinger, W. and Oda, M. (1969). Effects of sympathetic blocking substances on the diving reflex of ducks. *European Journal of Pharmacology*, **7**, 289-295.

Koch, E. (1923). Klinische Beobachtungen zum Karotisdruckversuch. *Münchener Medizinische Wochenschrift*, **43**, 1316-1318.

Koch, E. (1931). Die reflektorische Selbststeuerung des Kreislaufes. In *Ergebnnisse der Kreislaufforschung*, ed. Kisch, B. Steinkopff, Dresden.

Koch, E. (1932). Die Irradiation der pressoreceptorischen Kreislaufreflexe. *Klinische Wochenschrift*, **11**, 225-227.

Koch, E. and Mattonet, K. (1934). Versuche zur Frage der arteriellen Hypertonie nach Dauerausschaltung von pressoreceptorischen Kreislaufnerven. *Zeitschrift für Experimentelle Medizin*, **94**, 105-113.

Koch, E. and Mies, H. (1928). Chronischer arterieller Hochdruck durch experimentelle Dauerausschaltung der Blutdruckzügler. *Krankheitsforschung*, **7**, 241-256.

Kochar, M.S., Ebert, T.J., and Kotrly, K.J. (1984). Primary dysfunction of the afferent limb of the arterial baroreceptor reflex system in a patient with severe supine hypertension and orthostatic hypotension. *Journal of the American College of Cardiology*, **4**, 802-805.

Koch-Weser, J. (1976). Hydralazine. *The New England Journal of Medicine*, **295**, 320-323.

Kocsis, B., Fedina, L., and Pasztor, E. (1989). Two-phase change of sympathetic rhythms in brain ischemia, Cushing reaction, and asphyxia. *American Journal of Physiology*, **256**, R120-R132.

Koepchen, H.P. (1984). History of studies and concepts of blood pressure waves. In *Mechanisms of blood pressure waves*, ed. Miyakawa, K., Koepchen, H.P., and Polosa, C., pp. 3-23. Springer-Verlag, Berlin.

Koepchen, H.P., Wagner, P.-H., and Lux, H.D. (1961). Über die Zusammenhänge zwischen zentraler Erregbarkeit, reflektorischem Tonus und Atemrhythmus bei der nervösen Steuerung der Herzfrequenz. *Pflügers Archiv für die Gesamte Physiologie des Menschen und der Tiere*, **273**, 443-465.

Koepchen, H.P., Langhorst, P., Seller, H., Polster, J., and Wagner, P.H. (1967). Neuronale Aktivität im unteren Hirnstamm mit Beziehung zum Kreislauf. *Pflügers Archiv für die Gesamte Physiologie des Menschen und der Tiere*, **294**, 40-64.

Koepchen, H.P., Langhorst, P., and Seller, H. (1975). The problem of indentification of autonomic neurons in the lower brain stem. *Brain Research*, **87**, 375-393.

Koepchen, H.-P., Klüssendorf, D., and Sommer, D. (1981). Neurophysiological background of central neural cardiovascular–respiratory coordination: basic remarks and experimental approach. *Journal of the Autonomic Nervous System*, **3**, 335-368.

Koizumi, K. and Kollai, M. (1981). Control of reciprocal and non-reciprocal action of vagal and sympathetic efferents: study of centrally induced reactions. *Journal of the Autonomic Nervous System*, **3**, 483-501.

Koizumi, K. and Sato, A. (1969). Influence of sympathetic innervation on carotid sinus baroreceptor activity. *American Journal of Physiology*, **216**, 321-329.

Koizumi, K. and Sato, A. (1972). Reflex activity of single sympathetic fibres to skeletal muscle produced by electrical stimulation of somatic and vago-depressor afferent nerves in the cat. *Pflügers Archiv für die Gesamte Physiologie des Menschen und der Tiere*, **332**, 283-301.

Koizumi, K., Terui, N., and Kollai, M. (1985). Effect of cardiac vagal and sympathetic nerve activity on heart rate in rhythmic fluctuations. *Journal of the Autonomic Nervous System*, **12**, 251-259.

Kollai, M. and Koizumi, K. (1979). Reciprocal and non-reciprocal action of the vagal and sympathetic nerves innervating the heart. *Journal of the Autonomic Nervous System*, **1**, 33-52.

Kollai, M. and Koizumi, K. (1980). Patterns of single unit activity in sympathetic postganglionic nerves. *Journal of the Autonomic Nervous System*, **1**, 305-312.

Kollai, M. and Koizumi, K. (1989). Cardiac vagal and sympathetic nerve responses to baroreceptor stimulation in the dog. *Pflügers Archiv für die Gesamte Physiologie des Menschen und der Tiere*, **413**, 365-371.

Kollai, M. and Mizsei, G. (1990). Respiratory sinus arrhythmia is a limited measure of cardiac parasympathetic control in man. *The Journal of Physiology*, **424**, 329-342.

Kolman, B.S., Verrier, R.L., and Lown, B. (1975). The effect of vagus nerve stimulation upon vulnerability of the canine ventricle. Role of sympathetic–parasympathetic interactions. *Circulation*, **52**, 578-585.

Kontos, H.A., Levasseur, J.E., Richardson, D.W., Mauck, H.P., Jr., and Patterson, J.L., Jr. (1967). Comparative circulatory responses to systemic hypoxia in man and in unanesthetized dog. *Journal of Applied Physiology*, **23**, 381-386.

Koren, G., Weiss, A.T., Ben-David, Y., Hasin, Y., Luria, M.H., and Gotsman, M.S. (1986). Bradycardia and hypotension following reperfusion with streptokinase (Bezold–Jarisch reflex): A sign of coronary thrombolysis and myocardial salvage. *American Heart Journal*, **112**, 468-471.

Korner, P.I. (1971). Integrative neural cardiovascular control. *Physiological Reviews*, **51**, 312-367.

Korner, P.I. (1982). Causal and homeostatic factors in hypertension. *Clinical Science*, **63**, 5s-26s.

Korner, P.I., Shaw, J., West, M.J., and Oliver, J.R. (1972). Central nervous system control of baroreceptor reflexes in the rabbit. *Circulation Research*, **31**, 637-652.

Korner, P.I., Shaw, J., Uther, J.B., West, M.J., McRitchie, R.J., and Richards, J.G. (1973). Autonomic and non-autonomic circulatory components in essential hypertension in man. *Circulation*, **48**, 107-117.

Korner, P.I., West, M.J., Shaw, J., and Uther, J.B. (1974). 'Steady-state' properties of the baroreceptor–heart rate reflex in essential hypertension in man. *Clinical and Experimental Pharmacology and Physiology*, **1**, 65-76.

Korner, P.I., Tonkin, A.M., and Uther, J.B. (1976). Reflex and mechanical circulatory effects of graded Valsalva maneuvers in normal man. *Journal of Applied Physiology*, **40**, 434-440.

Korner, P.I., Tonkin, A.M., and Uther, J.B. (1979). Valsalva constrictor and heart rate reflexes in subjects with essential hypertension and with normal blood pressure. *Clinical and Experimental Pharmacology and Physiology*, **6**, 97-110.

Korsukewitz, J., Wagner, J., Dittberner, K.-H., and Zerbst, E. (1974). Open and closed loop carotid sinus nerve stimulation in man. In *Central rhythmic and regulation*, ed. Umbach, W., and Koepchen, H.P., pp. 159-164. Hippokrates Verlag, Stuttgart.

Kortas, C., Bichet, D.G., Rouleau, J.L., and Schrier, R.W. (1986). Vasopressin in congestive heart failure. *Journal of Cardiovascular Pharmacology*, **8** (Suppl. 7), S107-S110.

Kōsaka, K. (1909). Über die Vaguskerne des Hundes. *Neurologisches Centralblatt*, **28**, 406-410.

Köster, G. and Tschermak, A. (1902). Ueber den Nervus depressor als Reflexnerv der Aorta. *Archiv für Anatomie und Entwickelungsgeschichte Anatomische Abtheilung des Archives für Anatomie und Physiologie*, 255-294.

Kostis, J.B. and Rosen, R.C. (1987). Central nervous system effects of β-adrenergic-blocking drugs: the role of ancillary properties. *Circulation*, **75**, 204-212.

Kostreva, D.R. (1983). Functional mapping of cardiovascular reflexes and the heart using 2-[^{14}C]deoxyglucose. *The Physiologist*, **26** (Suppl.), 333-350.

Kotchen, T.A., Guthrie, G.P., Jr., McKean, H., and Kotchen, J.M. (1982). Adrenergic responsiveness in prehypertensive subjects. *Circulation*, **65**, 285-290.

Koushanpour, E. and Behnia, R. (1987). Partition of carotid baroreceptor response in two-kidney renal hypertensive dogs. *American Journal of Physiology*, **253**, R568-R575.

Koushanpour, E. and Kelso, D.M. (1972). Partition of the carotid sinus baroreceptor response in dogs between the mechanical properties of the wall and the receptor elements. *Circulation Research*, **31**, 831-845.

Koushanpour, E. and McGee, J.P. (1969). Effect of mean pressure on carotid sinus baroceptor response to pulsatile pressure. *American Journal of Physiology*, **216**, 599-603.

Kowey, P.R., Verrier, R.L., and Lown, B. (1983*a*). Effect of alpha-adrenergic receptor stimulation on ventricular electrical properties in the normal canine heart. *American Heart Journal*, **105**, 366-371.

Kowey, P.R., Verrier, R.L., and Lown, B. (1983*b*). The repetitive extrasystole as an index of vulnerability to ventricular fibrillation during myocardial ischemia in the canine heart. *American Heart Journal*, **106**, 1321-1325.

Kozelka, J.W., Chung, J.M., and Wurster, R.D. (1981). Ascending spinal pathways mediating somato-cardiovascular reflexes. *Journal of the Autonomic Nervous System*, **3**, 171-175.

Kramer, R.S., Mason, D.T., and Braunwald, E. (1968). Augmented sympathetic neurotransmitter activity in the peripheral vascular bed of patients with congestive heart failure and cardiac norepinephrine depletion. *Circulation*, **38**, 629-634.

Krauhs, J.M. (1979). Structure of rat aortic baroreceptors and their relationship to connective tissue. *Journal of Neurocytology*, **8**, 401-414.

Krayer, O. (1961). The history of the Bezold–Jarisch effect. *Naunyn-Schmiedebergs Archiv für Pharmakologie und Experimentelle Pathologie*, **240**, 361-368.

Krediet, R.T. and Dunning, A.J. (1979). Baroreflex sensitivity in hypertension during beta-adrenergic blockade. *British Heart Journal*, **41**, 106-110.

Krieger, E.M. (1970). Time course of baroreceptor resetting in acute hypertension. *American Journal of Physiology*, **218**, 486-490.

Krogh, A. and Lindhard, J. (1913-1914). The regulation of respiration and circulation during the initial stages of muscular work. *The Journal of Physiology*, **47**, 112-136.

Kronenberg, R.S. and Drage, C.W. (1973). Attenuation of the ventilatory and heart rate responses to hypoxia and hypercapnia with aging in normal men. *The Journal of Clinical Investigation*, **52**, 1812-1819.

Kronik, G., Slany, J., and Moesslacher, H. (1979). Contrast M-mode echocardiography in diagnosis of atrial septal defect in acyanotic patients. *Circulation*, **59**, 372-378.

Kubo, S.H. and Cody, R.J. (1983). Circulatory autoregulation in chronic congestive heart failure: responses to head-up tilt in 41 patients. *The American Journal of Cardiology*, **52**, 512-518.

Kubo, T. and Kihara, M. (1987). Evidence for the presence of GABAergic and glycine-like systems responsible for cardiovascular control in the nucleus tractus solitarii of the rat. *Neuroscience Letters*, **74**, 331-336.

Kubo, S.H., Rector, T.S., Heifetz, S.M., and Cohn, J.N. (1989). α_2-receptor-mediated vasoconstriction in patients with congestive heart failure. *Circulation*, **80**, 1660-1667.

Kuller, L., Lilienfeld, A., and Fisher, R. (1966). Epidemiological study of sudden and unexpected deaths due to arteriosclerotic heart disease. *Circulation*, **34**, 1056-1068.

Kumada, M., Dampney, R.A.L., and Reis, D.J. (1977). The trigeminal depressor response: a novel vasodepressor response originating from the trigeminal system. *Brain Research*, **119**, 305-326.

Kummer, W. (1990). Three types of neurochemically defined autonomic fibres innervate the carotid baroreceptor and chemoreceptor regions in the guinea-pig. *Anatomy and Embryology*, **181**, 477-489.

Kunze, D.L. (1972). Reflex discharge patterns of cardiac vagal efferent fibres. *The Journal of Physiology*, **222**, 1-15.

Kunze, D.L. (1980). Regulation of activity of cardiac vagal motoneurons. *Federation Proceedings*, **39**, 2513-2518.

Kunze, D.L. (1986). Acute resetting of baroreceptor reflex in rabbits: a central component. *American Journal of Physiology*, **250**, H866-H870.

Kunze, D.L. and Brown, A.M. (1978). Sodium sensitivity of baroreceptors. Reflex effects on blood pressure and fluid volume in the cat. *Circulation Research*, **42**, 714-720.

Kunze, D.L., Krauhs, J.M., and Orlea, C.J. (1984). Direct action of norepinephrine on aortic baroreceptors of rat adventitia. *American Journal of Physiology*, **247**, H811-H816.

Lacey, J.I. and Lacey, B.C. (1962). The law of initial value in the longitudinal study of autonomic constitution: reproducibility of autonomic responses and response patterns over a four-year interval. *Annals of the New York Academy of Sciences*, **98**, 1257-1290.

Lacey, J.I., Kagan, J., Lacey, B.C., and Moss, H.A. (1963). The visceral level: situational determinants and behavioral correlates of autonomic response patterns. In *Expression of the emotions in man*, ed. Knapp, P.H., pp. 161-196. International Universities Press, New York.

Lagerlöf, H., Eliasch, H., Werkö, L., and Berglund, E. (1951). Orthostatic changes of the pulmonary and peripheral circulation in man. A preliminary report. *Scandinavian Journal of Clinical Laboratory Investigation*, **3**, 85-91.

La Grange, R.G., Sloop, C.H., and Schmid, H.E. (1973). Selective stimulation of renal nerves in the anesthetized dog. Effect on renin release during controlled changes in renal hemodynamics. *Circulation Research*, **33**, 704-712.

Lake, C.R., Ziegler, M.G., and Kopin, I.J. (1976). Use of plasma norepinephrine for evaluation of sympathetic neuronal function in man. *Life Sciences*, **18**, 1315-1326.

Lake, C.R., Ziegler, M.G., Coleman, M.D., and Kopin, I.J. (1977). Age-adjusted plasma norepinephrine levels are similar in normotensive and hypertensive subjects. *The New England Journal of Medicine*, **296**, 208-209.

Lamb, J.F. (1963). Oscillation of heart rate after exercise in man (Abstract). *The Journal of Physiology*, **168**, 55P-56P.

Lamberti, J.J., Jr., Urquhart, J., and Siewers, R.D. (1968). Observations on the regulation of arterial blood pressure in unanesthetized dogs. *Circulation Research*, **23**, 415-428.

Lampen, H. (1949). Über Entzügelungshochdruck bei Polyneuritis. *Deutsche Medizinische Wochenschrift*, **74**, 536-540.

Lampen, H., Kezdi, P., Koppermann, E., and Kaufmann, L. (1949). Experimenteller Entzügelungshochdruck bei arterieller Hypertonie. *Zeitschrift für Kreislaufforschung*, **38**, 577-592.

Lancet (1981). Clinical implications of the diving response. *The Lancet*, **1**, 1403-1404.

Landgren, S. (1952*a*). On the excitation mechanism of the carotid baroceptors. *Acta Physiologica Scandinavica*, **26**, 1-34.

Landgren, S. (1952*b*). The baroceptor activity in the carotid sinus nerve and the distensibility of the sinus wall. *Acta Physiologica Scandinavica*, **26**, 35-56.

Landgren, S., Neil, E., and Zotterman, Y. (1951). The response of the carotid baroceptors to the local administration of drugs. *Acta Physiologica Scandinavica*, **25**, 24-37.

Lang, R.E., Thölken, H., Ganten, D., Luft, F.C., Ruskoaho, H., and Unger, Th. (1985). Atrial natriuretic factor — a circulating hormone stimulated by volume loading. *Nature*, **314**, 264-266.

Lange, R.L., Reid, M.S., Tresch, D.D., Keelan, M.H., Bernhard, V.M., and Coolidge, G. (1972). Nonatheromatous ischemic heart disease following withdrawal from chronic industrial nitroglycerin exposure. *Circulation*, **46**, 666-678.

Langford, H.G. *et al.* (1987). Neurogenic hypertension in man: Reis, Barman-Gebber or Carey syndrome. *Journal of Hypertension*, **5** (Suppl. 5), S467-S469.

Langhorst, P., Schulz, G., Lambertz, M., and Krienke, B. (1980). Processing of the information from baroreceptors by neurons in the dorso-medial part of the nucleus of the solitarii tract. In *Arterial baroreceptors and hypertension*, ed. Sleight, P., pp. 268-275. Oxford University Press, Oxford.

Langrehr, D. (1964). Receptor-Afferenzen im Halsvagus des Menschen. *Klinische Wochenschrift*, **42**, 239-244.

Lappe, R.W., Todt, J.A., and Wendt, R.L. (1987). Effects of atrial natriuretic factor on the vasoconstrictor actions of the renin–angiotensin system in conscious rats. *Circulation Research*, **61**, 134-140.

LaRovere, M.T., Specchia, G., Mazzoleni, C., Mortara, A., and Schwartz, P.J. (1986). Baroreflex sensitivity in post-myocardial infarction patients. Correlation with physical training and prognosis (Abstract). *Circulation*, **74** (Suppl. II), II-514.

LaRovere, M.T., Specchia, G., Mortara, A., and Schwartz, P.J. (1988). Baroreflex sensitivity, clinical correlates, and cardiovascular mortality among patients with a first myocardial infarction. A prospective study. *Circulation*, **78**, 816-824.

Lavallée, M., de Champlain, J., Nadeau, R.A., and Yamaguchi, N. (1978). Muscarinic inhibition of endogenous myocardial catecholamine liberation in the dog. *Canadian Journal of Physiology and Pharmacology*, **56**, 642-649.

Leatham, A. (1982). Carotid sinus syncope. *British Heart Journal*, **47**, 409-410.

LeBlanc, J., Blais, B., Barabé, B., and Côté, J. (1976). Effects of temperature and wind on facial temperature, heart rate, and sensation. *Journal of Applied Physiology*, **40**, 127-131.

Leclercq, J.F., Maisonblanche, P., Cauchemez, B., and Coumel, P. (1988). Respective role of sympathetic tone and of cardiac pauses in the genesis of 62 cases of ventricular fibrillation recorded during Holter monitoring. *European Heart Journal*, **9**, 1276-1283.

Ledsome, J.R., Ngsee, J., and Wilson, N. (1983). Plasma vasopressin concentration in the anaesthetized dog before, during and after atrial distension. *The Journal of Physiology*, **338**, 413-421.

Lee, D.D.-P., Kimura, S., and DeQuattro, V. (1989). Noradrenertic activity and silent ischaemia in hypertensive patients with stable angina: effect of metoprolol. *The Lancet*, **1**, 403-406.

Lee, G.deJ. and Gimlette, T.M.D. (1957). A simple test for interatrial communication. *British Medical Journal*, **1**, 1278-1281.

Lee, G.deJ., Matthews, M.B., and Sharpey-Schafer, E.P. (1954). The effect of the Valsalva Manoeuvre on the systemic and pulmonary arterial pressure in man. *British Heart Journal*, **16**, 311-316.

Lee, M.-E., Thrasher, T.N., Keil, L.C., and Ramsay, D.J. (1986). Cardiac receptors, vasopressin, and corticosteriod release during arterial hypotension in dogs. *American Journal of Physiology*, **251**, R614-R620.

Lee, T.M., Kuo, J.S., and Chai, C.Y. (1972). Central integrating mechanism of the Bezold–Jarisch and baroceptor reflexes. *American Journal of Physiology*, **222**, 713-720.

Lee, W.H. and Packer, M. (1986). Prognostic importance of serum sodium concentration and its modification by converting-enzyme inhibition in patients with severe chronic heart failure. *Circulation*, **73**, 257-267.

Lehmann, K.G., Lane, J.G., Piepmeier, J.M., and Batsford, W.P. (1987). Cardiovascular abnormalities accompanying acute spinal cord injury in humans: incidence, time course and severity. *Journal of the American College of Cardiology*, **10**, 46-52.

Leimbach, W.N., Jr., Schmid, P.G., and Mark, A.L. (1984). Baroreflex control of plasma arginine vasopressin in humans. *American Journal of Physiology*, **247**, H638-H644.

Leimbach, W.N., Jr., Wallin, B.G., Victor, R.G., Aylward, P.E., Sundlöf, G., and Mark, A.L. (1986). Direct evidence from intraneural recordings for increased central sympathetic outflow in patients with heart failure. *Circulation*, **73**, 913-919.

Leithe, M.E., Margorien, R.D., Hermiller, J.B., Unverferth, D.V., and Leier, C.V. (1984). Relationship between central hemodynamics and regional blood flow in normal subjects and in patients with congestive heart failure. *Circulation*, **69**, 57-64.

Lendrum, B., Feinberg, H., Boyd, E., and Katz, L.N. (1960). Rhythm effects on contractility of the beating isovolumic left ventricle. *American Journal of Physiology*, **199**, 1115-1120.

Leon, D.F., Shaver, J.A., and Leonard, J.J. (1970). Reflex heart rate control in man. *American Heart Journal*, **80**, 729-739.

Leone, C. and Gordon, F.J. (1989). Is L-glutamate a neurotransmitter of baroreceptor information in the nucleus of the tractus solitarius? *The Journal of Pharmacology and Experimental Therapeutics*, **250**, 953-962.

Leriche, R., Meyer, P., and Fontaine, R. (1932). Tachycardie paroxystique du type auriculaire. Ablation bilatérale du ganglion étoilé. Resultat au bout de sept ans. *La Presse Medicale*, **69**, 1317-1318.

Leuenberger, U., Kenney, G., Davis, D., Deiling, S., Strzlecka, D., and Zelis, R. (1989). Nonneuronal norepinephrine clearance is low in congestive heart failure (Abstract). *Clinical Research*, **37**, 274A.

Leveston, S.A., Shah, S.D., and Cryer, P.E. (1979). Cholinergic stimulation of norepinephrine release in man. Evidence of a sympathetic postganglionic axonal lesion in diabetic adrenergic neuropathy. *The Journal of Clinical Investigation*, **64**, 374-380.

Levin, A.B. (1966). A simple test of cardiac function based upon the heart rate changes induced by the Valsalva maneuver. *The American Journal of Cardiology*, **18**, 90-99.

Levine, B.D. *et al.* (1991). Physical fitness and cardiovascular regulation: mechanisms of orthostatic intolerance. *Journal of Applied Physiology*, **70**, 112-122.

Levine, S.A. (1962). Carotid sinus massage. A new diagnostic test for angina pectoris. *The Journal of the American Medical Association*, **182**, 1332-1334.

Levine, T.B., Francis, G.S., Goldsmith, S.R., and Cohn, J.N. (1983). The neurohumoral and hemodynamic response to orthostatic tilt in patients with congestive heart failure. *Circulation*, **67**, 1070-1075.

Levine, T.B., Olivari, M.T., Garberg, V., Sharkey, S.W., and Cohn, J.N. (1984). Hemodynamic and clinical response to enalapril, a long-acting converting-enzyme inhibitor, in patients with congestive heart failure. *Circulation*, **69**, 548-533.

Levine, T.B., Olivari, M.T., and Cohn, J.N. (1986). Effects of orthotopic heart transplantation on sympathetic control mechanisms in congestive heart failure. *The American Journal of Cardiology*, **58**, 1035-1040.

Levites, R., Baker, T., and Anderson, G.J. (1978). The significance of hypotension developing during treadmill exercise testing. *American Heart Journal*, **95**, 747-753.

Levy, M.N. (1971). Sympathetic–parasympathetic interactions in the heart. *Circulation Research*, **29**, 437-445.

Levy, M.N. and Blattberg, B. (1976). Effect of vagal stimulation on the overflow of norepinephrine into the coronary sinus during cardiac sympathetic nerve stimulation in the dog. *Circulation Research*, **38**, 81-85.

Levy, M.N. and Zieske, H. (1969). Effect of enhanced contractility on the left ventricular response to vagus nerve stimulation in dogs. *Circulation Research*, **24**, 303-311.

Levy, M.N., Ng, M., Martin, P., and Zieske, H. (1966). Sympathetic and parasympathetic interactions upon the left ventricle of the dog. *Circulation Research*, **19**, 5-10.

Levy, M.N., Martin, P.J., Iano, T., and Zieske, H. (1970). Effects of single vagal stimuli on heart rate and atrioventricular conduction. *American Journal of Physiology*, **218**, 1256-1262.

Lew, J.Y., Matsumoto, Y., Pearson, J., Goldstein, M., Hökfelt, T., and Fuxe, K. (1977). Localization and characterization of phenylethanolamine N-methyl transferase in the brain of various mammalian species. *Brain Research*, **119**, 199-210.

Lewandowsky, M. (1898). Ueber Schwankungen des Volumänderungen der Lunge. *Pflügers Archiv für die Gesamte Physiologie des Menschen und der Tiere*, **73**, 288-296.

LeWinter, M.M., Karliner, J.S., and Covell, J.W. (1978). Alteration in heart rate response to hemorrhage in conscious dogs with volume overload. *American Journal of Physiology*, **235**, H422-H428.

Liang, B., Verrier, R.L., Melman, J., and Lown, B. (1979). Correlation between circulating catecholamine levels and ventricular vulnerability during psychological stress in conscious dogs. *Proceedings of the Society for Experimental Biology and Medicine*, **161**, 266-269.

Liang, C.-s., Imai, N., Stone, C.K., Woolf, P.D., Kawashima, S., and Tuttle, R.R. (1987). The role of endogenous opioids in congestive heart failure: effects of nalmefene on systemic and regional hemodynamics in dogs. *Circulation*, **75**, 443-451.

Liard, J.F. (1984). Atropine potentiates the pressor effect of arginine-vasopressin in conscious dogs. *Journal of Cardiovascular Pharmacology*, **6**, 867-871.

Liard, J.F. and Spadone, J.C. (1984). Hemodynamic effects of antagonists of the vasoconstrictor action of vasopressin in conscious dogs. *Journal of Cardiovascular Pharmacology*, **6**, 713-719.

Liard, J.F., Dériaz, O., Tschopp, M., and Schoun, J. (1981). Cardiovascular effects of vasopressin infused into the vertebral circulation of conscious dogs. *Clinical Science*, **61**, 345-347.

Lilly, L.S., Dzau, V.J., Williams, G.H., Rydstedt, L., and Hollenberg, N.K. (1984). Hyponatremia in congestive heart failure: Implications for neurohumoral activation and responses to orthostasis. *Journal of Clinical Endocrinology and Metabolism*, **59**, 924-930.

Lind, A.R., Taylor, S.H., Humphreys, P.W., Kennelly, B.M., and Donald, K.W. (1964). The circulatory effects of sustained voluntary muscle contraction. *Clinical Science*, **27**, 229-244.

Lind, R.W. (1987). Neural connections. In *Circumventricular organs and body fluids*. Volume **1**, ed. Gross, P.M., pp. 27-42. CRC Press, Boca Raton, Florida.

Lindblad, L.E. (1980). Baroreceptor reflexes in man. A study on cardiovascular and neural effector mechanisms (Ph.D. thesis), Repro Print, pp. 1-62. Stockholm.

Lindblad, L.E., Bevegård, S., Castenfors, J., and Tranesjo, J. (1981). Circulatory effects of carotid sinus stimulation and changes in blood volume distribution in hypertensive man. *Acta Physiologica Scandinavica*, **111**, 299-306.

Lindblad, L.E., Wallin, B.G., and Bevegård, S. (1982). Transient vasodilatation in forearm on stimulation of carotid baroreceptors in man. *Journal of the Autonomic Nervous System*, **5**, 373-379.

Linden, R.J. (1979). Atrial receptors and heart rate. In *Cardiac receptors*, ed. Hainsworth, R., Kidd, C., and Linden, R.J., pp. 165-191. Cambridge University Press, Cambridge.

Linden, R.J. and Kappagoda, C.T. (1982). *Atrial receptors*, Cambridge University Press, Cambridge. pp. 1-363.

Linder, L., Kiowski, W., Bühler, F.R., and Lüscher, T.F. (1990). Indirect evidence for release of endothelium-derived relaxing factor in human forearm circulation in vivo. Blunted response in essential hypertension. *Circulation*, **81**, 1762-1767.

Lindqvist, A. (1990). Noninvasive methods to study autonomic nervous control of circulation. *Acta Physiologica Scandinavica*, **138** (Suppl. 588), 1-272.

Lipski, J. and Trzebski, A. (1975). Bulbo-spinal neurons activated by baroreceptor afferents and their possible role in inhibition of preganglionic sympathetic neurons. *Pflügers Archiv für die Gesamte Physiologie des Menschen und der Tiere*, **356**, 181-192.

Lipski, J., McAllen, R.M., and Spyer, K.M. (1975). The sinus nerve and baroreceptor input to the medulla of the cat. *The Journal of Physiology*, **251**, 61-78.

Lipski, J., McAllen, R.M., and Trzebski, A. (1976). Carotid baroreceptor and chemoreceptor inputs onto single medullary neurones. *Brain Research*, **107**, 132-136.

Lisander, B. and Martner, J. (1971). Interaction between the fastigial pressor response and the baroreceptor reflex. *Acta Physiologica Scandinavica*, **83**, 505-514.

Litchfield, R.L., Kerber, R.E., Benge, J.W., Mark, A.L., Sopko, J., Bhatnager, R.K., and Marcus, M.L. (1982). Normal exercise capacity in patients with severe left ventricular dysfunction: compensatory mechanisms. *Circulation*, **66**, 129-134.

Little, W.C., Barr, W.K., and Crawford, M.H. (1985). Altered effect of the Valsalva maneuver on left ventricular volume in patients with cardiomyopathy. *Circulation*, **71**, 227-233.

Ljung, B. (1969). Local transmitter concentrations in vascular smooth muscle during vasoconstrictor nerve activity. *Acta Physiologica Scandinavica*, **77**, 212-223.

Lloyd-Mostyn, R.H. and Watkins, P.J. (1976). Total cardiac denervation in diabetic autonomic neuropathy. *Diabetes*, **25**, 748-751.

Loeb, J.M., deTarnowsky, J.M., Warner, M.R., and Whitson, C.C. (1985). Dynamic interactions between heart rate and atrioventricular conduction. *American Journal of Physiology*, **249**, H505-H511.

Loeschcke, H.H. and Koepchen, H.P. (1958). Versuch zur Lokalisation des Angriffsortes des Atmungs- und Kreislaufwirkung von Novocain im Liquor cerebrospinalis. *Pflügers Archiv für die Gesamte Physiologie des Menschen und der Tiere*, **266**, 628-641.

Loewi, O. (1921). Über humorale Übertragbarkeit der Herznervenwirkung. *Pflügers Archiv für die Gesamte Physiologie des Menschen und der Tiere*, **189**, 239-242.

Loewy, A.D. (1982). Descending pathways to the sympathetic preganglionic neurons. In *Progress in brain research, Vol. 57. Anatomy of descending pathways to the spinal cord*, ed. Kuypers, H.G.J.M., and Martin, G.F., pp. 267-277. Elsevier Biomedical Press, Amsterdam.

Loewy, A.D. (1990). Central autonomic pathways. In *Central regulation of autonomic functions*, ed. Loewy, A.D., and Spyer, K.M., pp. 88-103. Oxford University Press, New York.

Loewy, A.D. and Burton, H. (1978). Nuclei of the solitary tract: Efferent projections to the lower brain stem and spinal cord of the cat. *The Journal of Comparative Neurology*, **181**, 421-450.

Loewy, A.D. and McKellar, S. (1980). The neuroanatomical basis of central cardiovascular control. *Federation Proceedings*, **39**, 2495-2503.

Loewy, A.D. and Spyer, K.M. (1990). *Central regulation of autonomic functions.* Oxford University Press, New York.

Löffelholz, K. and Muscholl, E. (1970). Inhibition by parasympathetic nerve stimulation of the release of the adrenergic transmitter. *Naunyn-Schmiedeberg's Archives of Pharmacology*, **267**, 181-184.

Lombardi, F., Verrier, R.L., and Lown, B. (1983). Relationship between sympathetic neural activity, coronary dynamics, and vulnerability to ventricular fibrillation during myocardial ischemia and reperfusion. *American Heart Journal*, **105**, 958-965.

Lombardi, F. *et al.* (1987). Heart rate variability as an index of sympathovagal interaction after acute myocardial infarction. *The American Journal of Cardiology*, **60**, 1239-1245.

Lombardi, F., Ruscone, T.G., and Malliani, A. (1989). Premature ventricular contractions and reflex sympathetic activation in cats. *Cardiovascular Research*, **23**, 205-212.

Lombroso, U. (1913). Über die Reflexhemmung des Herzens während der reflektorischen Atmungshemmung bei verschiedenen Tieren. *Zeitschrift für Biologie*, **61**, 517-538.

London, G.M., Weiss, Y.A., Pannier, B.P., Laurent, S.L., and Safar, M.E. (1987). Tilt test in essential hypertension. Differential responses in heart rate and vascular resistance. *Hypertension*, **10**, 29-34.

Longhurst, J., Gifford, W., and Zelis, R. (1976). Impaired forearm oxygen consumption during static exercise in patients with congestive heart failure. *Circulation*, **54**, 477-480.

Longhurst, J.C. (1984). Cardiac receptors: their function in health and disease. *Progress in Cardiovascular Diseases*, **27**, 201-222.

Lopes, O.U. and Palmer, J.F. (1976). Proposed respiratory 'gating' mechanism for cardiac slowing. *Nature*, **264**, 454-456.

Loup, F., Tribollet, E., Dubois-Dauphin, M., Pizzolato, G., and Dreifuss, J.J. (1989). Localization of oxytocin binding sites in the human brainstem and upper spinal cord: an autoradiographic study. *Brain Research*, **500**, 223-230.

Lovelace, D.E. and Knoebel, S.B. (1986). Power spectrum analysis of heart rate variations; relevance to arrhythmia prevalence (Abstract). *Circulation*, **74** (Suppl. II), II-189.

Lovick, T.A. (1987). Differential control of cardiac and vasomotor activity by neurones in nucleus paragigantocellularis lateralis in the cat. *The Journal of Physiology*, **389**, 23-35.

Lowensohn, H.S. (1986). *Atropine's effect upon the heart and its systemic output.* Walter Reed Army Institute of Research, Washington, D.C.

Lower, R. (1669). *Tractatus de corde. Item de motu colore sanguinis et chyli in eum transitu.* Daniel Elzevir, Amsterdam.

Lown, B. and DeSilva, R.A. (1978). Roles of psychologic stress and autonomic nervous system changes in provocation of ventricular premature complexes. *The American Journal of Cardiology*, **41**, 979-985.

Lown, B. and Levine, S.A. (1961). The carotid sinus. Clinical value of its stimulation. *Circulation*, **23**, 766-789.

Lown, B. and Verrier, R.L. (1976). Neural activity and ventricular fibrillation. *The New England Journal of Medicine*, **294**, 1165-1170.

Ludbrook, J. (1983). Reflex control of blood pressure during exercise. *Annual Review of Physiology*, **45**, 155-168.

Ludbrook, J. (1984). Comparison of the reflex effects of arterial baroreceptors and cardiac receptors on the heart rate of conscious rabbits. *Clinical and Experimental Pharmacology and Physiology*, **11**, 245-260.

Ludbrook, J., Mancia, G., Ferrari, A., and Zanchetti, A. (1977). The variable-pressure neck-chamber method for studying the carotid baroreflex in man. *Clinical Science and Molecular Medicine*, **53**, 165-171.

Ludbrook, J., Faris, I.B., Iannos, J., Jamieson, G.G., and Russell, W.J. (1978). Lack of effect of isometric handgrip exercise on the responses of the carotid sinus baroreceptor reflex in man. *Clinical Science and Molecular Medicine*, **55**, 189-194.

Ludbrook, J., Mancia, G., and Zanchetti, A. (1980). Does the baroreceptor–heart rate reflex indicate the capacity of the arterial baroreceptors to control blood pressure? *Clinical and Experimental Pharmacology and Physiology*, **7**, 499-503.

Ludwig, C. (1847). Beiträge zur Kenntniss des Einflusses der Respirationsbewegungen auf den Blutlauf im Aortensysteme. *Mullers Archive fuer Anatomie Physiologie und Medizin*, 242-302.

Ludwig, D.A. and Convertino, V.A. (1991). A statistical note on the redundancy of nine standard baroreflex parameters. *Aviation, Space, and Environmental Medicine*, **62**, 172-175.

Lugliani, R., Whipp, B.J., Seard, C., and Wasserman, K. (1971). Effect of bilateral carotid-body resection on ventilatory control at rest and during exercise in man. *The New England Journal of Medicine*, **285**, 1105-1111.

Lugliani, R., Whipp, B.J., and Wasserman, K. (1973). A role for the carotid body in cardiovascular control in man. *Chest*, **63**, 744-750.

Lumbers, E.R. and Potter, E.K. (1982). The effects of vasoactive peptides on the carotid cardiac baroreflex. *Clinical and Experimental Pharmacology and Physiology*, Suppl. **7**, 45-49.

Lumbers, E.R., McCloskey, D.I., and Potter, E.K. (1979). Inhibition by angiotensin II of baroreceptor-evoked activity in cardiac vagal efferent nerves in the dog. *The Journal of Physiology*, **294**, 69-80.

Lundberg, J.M. *et al.* (1982). Neuropeptide Y (NPY)-like immunoreactivity in peripheral noradrenergic neurons and effects of NPY on sympathetic function. *Acta Physiologica Scandinavica*, **116**, 477-480.

Lundberg, J.M., Terenius, L., Hökfelt, T., and Tatemoto, K. (1984). Comparative immunohistochemical and biochemical analysis of pancreatic polypeptide-like peptides with special reference to presence of neuropeptide Y in central and peripheral neurons. *The Journal of Neuroscience*, **4**, 2376-2386.

Luschka, H. (1862). *Die Anatomie des Menschen in Rücksicht auf die Bedürfnisse der praktischen Heilkunde*. H. Laupp, Tübingen.

Lynch, J.J., Schuchard, G.H., Gross, C.M., and Wann, L.S. (1984). Prevalence of right-to-left atrial shunting in a healthy population: detection by Valsalva maneuver contrast echocardiography. *The American Journal of Cardiology*, **53**, 1478-1480.

Maack, T. *et al.* (1984). Effects of auriculin (atrial natriuretic factor) on blood pressure, renal function, and the renin–aldosterone system in dogs. *The American Journal of Medicine*, **77**, 1069-1075.

MacDougall, J.D., Tuxen, D., Sale, D.G., Moroz, J.R., and Sutton, J.R. (1985). Arterial blood pressure response to heavy resistance exercise. *Journal of Applied Physiology*, **58**, 785-790.

Mackenzie, L.F., Wells, P.V., Dewis, E.G., and Ylvisaker, L.S. (1931). Interpretation of flarimeter tests. *American Journal of the Medical Sciences*, **182**, 497-513.

Magendie, M. (1854). *Phénomènes physiques de la vie.* J.-B. Baillière, Paris.

Magid, N.M. *et al.* (1985). Diminished heart rate variability in sudden cardiac death (Abstract). *Circulation*, **72** (Suppl. III), III-241.

Magnus, O., Koster, M., and van der Drift, J.H.A. (1977). Cerebral mechanisms and neurogenic hypertension in man, with special reference to baroreceptor control. *Progress in Brain Research*, **47**, 199-218.

Magrini, F. and Niarchos, A.P. (1983). Hemodynamic effects of massive peripheral edema. *American Heart Journal*, **105**, 90-97.

Maisel, A.S. *et al.* (1989). Elevation of plasma neuropeptide Y levels in congestive heart failure. *The American Journal of Medicine*, **86**, 43-48.

Malayan, S.A., Ramsay, D.J., Keil, L.C., and Reid, I.A. (1980). Effects of increases in plasma vasopressin concentration on plasma renin activity, blood pressure, heart rate, and plasma corticosteroid concentration in conscious dogs. *Endocrinology*, **107**, 1899-1904.

Maley, B. and Newton, B.W. (1985). Immunohistochemistry of γ-aminobutyric acid in the cat nucleus tractus solitarius. *Brain Research*, **330**, 364-368.

Malliani, A. (1982). Cardiovascular sympathetic afferent fibers. *Reviews of Physiology, Biochemistry, and Experimental Pharmacology*, **94**, 11-74.

Malliani, A. and Pagani, M. (1983). The role of the sympathetic nervous system in congestive heart failure. *European Heart Journal*, **4** (Suppl. A), 49-54.

Mancia, G. and Donald, D.E. (1975). Demonstration that the atria, ventricles, and lungs each are responsible for a tonic inhibition of the vasomotor center in the dog. *Circulation Research*, **36**, 310-318.

Mancia, G., Ferrari, A., Gregorini, L., Valentini, R., Ludbrook, J., and Zanchetti, A. (1977). Circulatory reflexes from carotid and extracarotid baroreceptor areas in man. *Circulation Research*, **41**, 309-315.

Mancia, G., Iannos, J., Jamieson, G.G., Lawrence, R.H., Sharman, P.R., and Ludbrook, J. (1978*a*). Effect of isometric hand-grip exercise on the carotid sinus baroreceptor reflex in man. *Clinical Science and Molecular Medicine*, **54**, 33-37.

Mancia, G., Ludbrook, J., Ferrari, A., Gregorini, L., and Zanchetti, A. (1978*b*). Baroreceptor reflexes in human hypertension. *Circulation Research*, **43**, 170-177.

Mancia, G. *et al.* (1979*a*). Baroreceptor control of atrioventricular conduction in man. *Circulation Research*, **44**, 752-758.

Mancia, G. *et al.* (1979*b*). Control of blood pressure by carotid sinus baroreceptors in human beings. *The American Journal of Cardiology*, **44**, 895-902.

Mancia, G. *et al.* (1979*c*). Plasma catecholamines and blood pressure responses to the carotid baroreceptor reflex in essential hypertension. *Clinical Science*, **57**, 165s-167s.

Mancia, G. *et al.* (1981). Carotid sinus reflex control of renin release in hypertensive subjects with high renin secretion. *Clinical Science*, **61**, 505-509.

Mancia, G., Grassi, G., Ferrari, A., and Zanchetti, A. (1985). Reflex cardiovascular regulation in humans. *Journal of Cardiovascular Pharmacology*, **7** (Suppl. 3), S152-S159.

Mancia, G., Parati, G., Pomidossi, G., Casadei, R., di Rienzo, M., and Zanchetti, A. (1986). Arterial baroreflexes and blood pressure and heart rate variabilities in humans. *Hypertension*, **8**, 147-153.

Mancia, G., Grassi, G., and Giannattasio, C. (1988). Cardiopulmonary receptor reflex in hypertension. *American Journal of Hypertension*, **1**, 249-255.

Mandel, W.J., Hayakawa, H., Allen, H.N., Danzig, R., and Kermaier, A.I. (1972). Assessment of sinus node function in patients with the sick sinus syndrome. *Circulation*, **46**, 761-769.

Mandelstamm, M. and Lifschitz, S. (1932). Die Wirkung der Karotissinusreflexe auf den Blutdruck beim Menschen. *Wiener Archiv für Innere Medizin*, **22**, 397-444.

Mangiapane, M.L. and Simpson, J.B. (1980*a*). Subfornical organ: forebrain site of pressor and dipsogenic action of angiotensin II. *American Journal of Physiology*, **239**, R382-R389.

Mangiapane, M.L. and Simpson, J.B. (1980*b*). Subfornical organ lesions reduce the pressor effect of systemic angiotensin II. *Neuroendocrinology*, **31**, 380-384.

Mannard, A. and Polosa, C. (1973). Analysis of background firing of single sympathetic preganglionic neurons of cat cervical nerve. *Journal of Neurophysiology*, **36**, 398-408.

Manning, P.T., Schwartz, D., Katsube, N.C., Holmberg, S.W., and Needleman, P. (1985). Vasopressin-stimulated release of atriopeptin: endocrine antagonists in fluid homeostasis. *Science*, **229**, 395-397.

Mano, T. (1990). Sympathetic nerve mechanisms of human adaptation to environment — findings obtained by recent microneurographic studies. *Environmental Medicine*, **34**, 1-35.

Manson, P. (1866). Peculiar affection of the internal carotid artery in connexion with disease of the brain. *Medical Times Gazette*, **53**, 336-337.

Mäntysaari, M., Antila, K., and Peltonen, T. (1984). Relationship between the changes in heart rate and cardiac output during the Valsalva maneuver. *Acta Physiologica Scandinavica*, Suppl. **537**, 45-49.

Marcel, M.-P. (1956). *Le syndrome des nerfs vasosensibles sino-carotidiens et cardio-aortiques. L'anarchie tensionelle*, pp. 1-136. J.-B. Baillière, Paris.

Marchlinski, F.E. (1987). Characterization of oscillations in ventricular refractoriness in man after an abrupt increment in heart rate. *Circulation*, **75**, 550-556.

Marey, J. (1859). Recherches sur le pouls au moyen d'un nouvel appareil enregistreur: le sphygmographe. *Memoires lus a la Société de Biologie*, 281-309.

Marin Neto, J.A., Gallo, L. Jr., Manço, J.C., Rassi, A., and Amorim, D.S. (1975). Postural reflexes in chronic Chagas's heart disease. *Cardiology*, **60**, 343-357.

Marin Neto, J.A., Gallo, L., Jr., Manço, J.C., Rassi, A., and Amorim, D.S. (1980). Mechanisms of tachycardia on standing: studies in normal individuals and in chronic Chagas' heart patients. *Cardiovascular Research*, **14**, 541-550.

Marin-Neto, J.A., Pintya, A.O., Gallo, L. Jr., and Maciel, B.C. (1991). Abnormal baroreflex control of heart rate in decompensated congestive heart failure and reversal after compensation. *The American Journal of Cardiology*, **67**, 604-610.

Mark, A.L. (1983). The Bezold–Jarisch reflex revisited: clinical implications of inhibitory reflexes originating in the heart. *Journal of the American College of Cardiology*, **1**, 90-102.

Mark, A.L. and Kerber, R.E. (1982). Augmentation of cardiopulmonary baroreflex control of forearm vascular resistance in borderline hypertension. *Hypertension*, **4**, 39-46.

Mark, A.L. and Mancia, G. (1983). Cardiopulmonary baroreflexes in humans. In *Handbook of physiology. Section 2: The cardiovascular system. Volume III. Peripheral circulation and organ blood flow, Part 2, ed. Shepherd, J.T., and Abboud, F.M., pp. 795-813. American Physiological Society, Bethesda.*

Mark, A.L., Kioschos, J.M., Abboud, F.M., Heistad, D.D., and Schmid, P.G. (1973). Abnormal vascular responses to exercise in patients with aortic stenosis. The Journal of Clinical Investigation, **52**, 1138-1146.

Mark, A.L., Abboud, F.M., and Fitz, A.E. (1978). Influence of low- and high-pressure baroreceptors on plasma renin activity in humans. *American Journal of Physiology*, **235**, H29-H33.

Mark, A.L., Victor, R.G., Nerhed, C., and Wallin, B.G. (1985). Microneurographic studies of the mechanisms of sympathetic nerve responses to static exercise in humans. *Circulation Research*, **57**, 461-469.

Markel, M.L., Miles, W.M., Zipes, D.P., and Prystowsky, E.N. (1988). Parasympathetic and sympathetic alterations of Mobitz type II heart block. *Journal of the American College of Cardiology*, **11**, 271-275.

Marlot, D., Mortola, J.P., and Duron, B. (1982). Functional localization of pulmonary stretch receptors in the tracheobronchial tree of the kitten. *Canadian Journal of Physiology and Pharmacology*, **60**, 1073-1077.

Marshall, D.W., Westmoreland, B.F., and Sharbrough, F.W. (1983). Ictal tachycardia during temporal lobe seizures. *Mayo Clinic Proceedings*, **58**, 443-446.

Martens, U., Nikl, Ch., Beck, O.A., and Hochrein, H. (1981). Über das Vorkommen eines hyperaktiven Karotissinus bei akutem Myokardinfarkt. *Zeitschrift für Kardiologie*, **70**, 28-31.

Martin, C.E., Shaver, J.A., Leon, D.F., Thompson, M.E., Reddy, P.S., and Leonard, J.J. (1974). Autonomic mechanisms in hemodynamic responses to isometric exercise. *The Journal of Clinical Investigation*, **54**, 104-115.

Martin, G.J. *et al.* (1987). Heart rate variability and sudden death secondary to coronary artery disease during ambulatory electrocardiographic monitoring. *The American Journal of Cardiology*, **60**, 86-89.

Martin, P. (1977). Paradoxical dynamic interaction of heart period and vagal activity on atrioventricular conduction in the dog. *Circulation Research*, **40**, 81-89.

Martin, P., Levy, M.N., and Matsuda, Y. (1982). Fade of cardiac responses during tonic vagal stimulation. *American Journal of Physiology*, **243**, H219-H225.

Martins, J.B. (1984). Carotid sinus baroreflex influence on electrophysiologic properties of the canine atrioventricular node and ventricle. *The American Journal of Cardiology*, **54**, 431-436.

Martins, J.B. and Zipes, D.P. (1980). Effects of sympathetic and vagal nerves on recovery properties of the endocardium and epicardium of the canine left ventricle. *Circulation Research*, **46**, 100-110.

Martins, J.B., Kienzle, M.G., McCue, M.L., and Constantin, L. (1988). Propranolol blocks ventricular refractory period changes with orthostatic stress in humans. *Journal of the American College of Cardiology*, **12**, 1488-1493.

Martins, J.B., Kim, W., and Marcus, M.L. (1989*a*). Chronic hypertension and left ventricular hypertrophy facilitate induction of sustained ventricular tachycardia in dogs 3 hours after left circumflex coronary artery occlusion. *Journal of the American College of Cardiology*, **14**, 1365-1373.

Martins, J.B., Lewis, R., Wendt, D., Lund, D.D., and Schmid, P.G. (1989*b*). Subendocardial infarction produces epicardial parasympathetic denervation in canine left ventricle. *American Journal of Physiology*, **256**, H859-H866.

Mason, D.T. and Braunwald, E. (1964). Studies on digitalis. X. Effects of ouabain on forearm vascular resistance and venous tone in normal subjects and in patients in heart failure. *The Journal of Clinical Investigation*, **43**, 532-543.

Mason, D.T., Braunwald, E., Ross, J., Jr., and Morrow, A.G. (1964). Diagnostic value of the first and second derivatives of the arterial pressure pulse in aortic valve disease and in hypertrophic subaortic stenosis. *Circulation*, **30**, 90-100.

Mason, J.W., Stinson, E.B., and Harrison, D.C. (1976). Autonomic nervous system and arrhythmias: studies in the transplanted denervated human heart. *Cardiology*, **61**, 75-87.

Massie, B.M. (1988). Is neurohumoral activation deleterious to the long-term outcome of patients with congestive heart failure? *Journal of the American College of Cardiology*, **12**, 547-558.

Massie, B.M. *et al.* (1988). Skeletal muscle metabolism during exercise under ischemic conditions in congestive heart failure. Evidence for abnormalities unrelated to blood flow. *Circulation*, **78**, 320-326.

Masuda, Y., Pace, D.G., and Levy, M.N. (1984). The rate of arterial blood pressure change as a factor in the hysteresis of the baroreceptor reflex. *Journal of Hypertension*, **2**, 189-194.

Mathias, C.J., Christensen, N.J., Corbett, J.L., Frankel, H.L., Goodwin, T.J., and Peart, W.S. (1975). Plasma catecholamines, plasma renin activity and plasma aldosterone in tetraplegic man, horizontal and tilted. *Clinical Science and Molecular Medicine*, **49**, 291-299.

Mathias, C.J., Reid, J.L., Wing, L.M.H., Frankel, H.L., and Christensen, N.J. (1979). Antihypertensive effects of clonidine in tetraplegic subjects devoid of central sympathetic control. *Clinical Science*, **57**, 425s-428s.

Matsuguchi, H. and Schmid, P.G. (1982). Pressor response to vasopressin and impaired baroreflex function in DOC-salt hypertension. *American Journal of Physiology*, **242**, H44-H49.

Matsukawa, S. and Reid, I.A. (1990). Role of the area postrema in the modulation of the baroreflex control of heart rate by angiotensin II. *Circulation Research*, **67**, 1462-1473.

Matsukawa, T. *et al.* (1988). Angiotensin II inhibits baroreflex control of muscle sympathetic nerve activity and the heart rate in patients with essential hypertension. *Journal of Hypertension*, **6** (Suppl. 4), S501-S504.

Matta, R.J., Verrier, R.L., and Lown, B. (1976). Repetitive extrasystole as an index of vulnerability to ventricular fibrillation. *American Journal of Physiology*, **230**, 1469-1473.

Matthes, K. and Ebeling, J. (1948). Untersuchungen über die Atemschwankungen des Blutdrucks und der Pulsfrequenz beim Menschen. *Pflügers Archiv für die Gesamte Physiologie des Menschen und der Tiere*, **250**, 747-768.

Mattila, J. and Buñag, R.D. (1986). Sympathomimetic pressor responses to thyrotropin-releasing hormone in rats. *American Journal of Physiology*, **251**, H86-H92.

May, G.S., Eberlein, K.A., Furberg, C.D., Passamani, E.R., and DeMets, D.L. (1982). Secondary prevention after myocardial infarction: a review of long-term trials. *Progress in Cardiovascular Diseases*, **24**, 331-352.

Mazgalev, T., Dreifus, L.S., Michelson, E.L., Pelleg, A., and Price, R. (1986). Phasic effects of postganglionic vagal stimulation on atrioventricular nodal conduction. *American Journal of Physiology*, **251**, H619-H630.

McAllen, R.M. and Dampney, R.A.L. (1990). Vasomotor neurons in the rostral ventrolateral medulla are organized topographically with respect to type of vascular bed but not body region. *Neuroscience Letters*, **110**, 91-96.

McAllen, R.M. and May, C.N. (1990). The conduction velocity of the descending spinal pathway to the renal sympathetic nerve in the cat. *Journal of the Autonomic Nervous System*, **30**, 139-142.

McAllen, R.M. and Spyer, K.M. (1976). The location of cardiac vagal preganglionic motoneurones in the medulla of the cat. *The Journal of Physiology*, **258**, 187-204.

McAllen, R.M. and Spyer, K.M. (1978*a*). The baroreceptor input to cardiac vagal motoneurones. *The Journal of Physiology*, **282**, 365-374.

McAllen, R.M. and Spyer, K.M. (1978*b*). Two types of vagal preganglionic motoneurones projecting to the heart and lungs. *The Journal of Physiology*, **282**, 353-364.

McAllen, R.M., Neil, J.J., and Loewy, A.D. (1982). Effects of kainic acid applied to the ventral surface of the medulla oblongata on vasomotor tone, the baroreceptor reflex and hypothalamic autonomic responses. *Brain Research*, **238**, 65-76.

McAreavey, D., Neilson, J.M.M., Ewing, D.J., and Russell, D.C. (1989). Cardiac parasympathetic activity during the early hours of acute myocardial infarction. *British Heart Journal*, **62**, 165-170.

McCaa, R.E., Hall, J.E., and McCaa, C.S. (1978). The effects of angiotensin I-converting enzyme inhibitors on arterial blood pressure and urinary sodium excretion. Role of the renal renin–angiotensin and kallikrein–kinin systems. *Circulation Research*, **43** (Suppl. I), I-32–I-39.

McCall, R.B. (1986). Lack of involvement of GABA in baroreceptor-mediated sympathoinhibition. *American Journal of Physiology*, **250**, R1065-R1073.

McCall, R.B., Gebber, G.L., and Barman, S.M. (1977). Spinal interneurons in the baroreceptor reflex arc. *American Journal of Physiology*, **232**, H657-H665.

McCance, A.J. and Forfar, J.C. (1989). Selective enhancement of the cardiac sympathetic response to exercise by anginal chest pain in humans. *Circulation*, **80**, 1642-1651.

McCloskey, D.I. and Potter, E.K. (1981). Excitation and inhibition of cardiac vagal motoneurones by electrical stimulation of the carotid sinus nerve. *The Journal of Physiology*, **316**, 163-175.

McCubbin, J.W., Green, J.H., and Page, I.H. (1956). Baroceptor function in chronic renal hypertension. *Circulation Research*, **4**, 205-210.

McDonald, D.M. (1983*a*). Morphology of the rat carotid sinus nerve. I. Course, connections, dimensions and ultrastructure. *Journal of Neurocytology*, **12**, 345-372.

McDonald, D.M. (1983*b*). Morphology of the rat carotid sinus nerve. II. Number and size of axons. *Journal of Neurocytology*, **12**, 373-392.

McDonald, D.M. and Larue, D.T. (1983). The ultrastructure and connections of blood vessels supplying the rat carotid body and carotid sinus. *Journal of Neurocytology*, **12**, 117-153.

McDonald, D.M. and Mitchell, R.A. (1975). The innervation of glomus cells, ganglion cells and blood vessels in the rat carotid body: a quantitative ultrastructural analysis. *Journal of Neurocytology*, **4**, 177-230.

McDonald, L., Baker, C., Bray, C., McDonald, A., and Restieaux, N. (1969). Plasma-catecholamines after cardiac infarction. *The Lancet*, **2**, 1021-1023.

McDowall, R.J.S. (1924). A vago-pressor reflex. *The Journal of Physiology*, **59**, 41-47.

McElroy, P.A., Janicki, J.S., and Weber, K.T. (1988). Physiologic correlates of the heart rate response to upright isotonic exercise: relevance to rate-responsive pacemakers. *Journal of the American College of Cardiology*, **11**, 94-99.

McGrattan, P.A., Brown, J.H., and Brown, O.M. (1987). Parasympathetic effects on in vivo rat heart can be regulated through an αV1-adrenergic receptor. *Circulation Research*, **60**, 465-471.

McGregor, M., Adam, W., and Sekelj, P. (1961). Influence of posture on cardiac output and minute ventilation during exercise. *Circulation Research*, **9**, 1089-1092.

McIntosh, H.D., Burnum, J.F., Hickam, J.B., and Warren, J.V. (1954). Circulatory changes produced by the Valsalva maneuver in normal subjects, patients with mitral stenosis, and autonomic nervous system alterations. *Circulation*, **9**, 511-520.

McKay, R.G. *et al.* (1984). Left ventricular pressure–volume diagrams and end-systolic pressure–volume relations in human beings. *Journal of the American College of Cardiology*, **3**, 301-312.

McLachlan, E.M. and Hirst, G.D.S. (1980). Some properties of preganglionic neurons in upper thoracic spinal cord of the cat. *Journal of Neurophysiology*, **43**, 1251-1265.

McLaran, C.J. *et al.* (1986). Increased vagal tone as an isolated finding in patients undergoing electrophysiological testing for recurrent syncope: response to long term anticholinergic agents. *British Heart Journal*, **55**, 53-57.

McMichael, J. and Sharpey-Schafer, E.P. (1944). Cardiac output in man by a direct Fick method. Effects of posture, venous pressure change, atropine, and adrenaline. *British Heart Journal*, **6**, 33-40.

McNamara, H.I., Sikorski, J.M., and Clavin, H. (1969). The effects of lower body negative pressure on hand blood flow. *Cardiovascular Research*, **3**, 284-291.

McRitchie, R.J., Vatner, S.F., Boettcher, D., Heyndrickx, G.R., Patrick, T.A., and Braunwald, E. (1976). Role of arterial baroreceptors in mediating cardiovascular response to exercise. *American Journal of Physiology*, **230**, 85-89.

Mechelke, K. (1953). Über die Atemschwankungen des Blutdrucks und der Pulsfrequenz beim Menschen. *Archiv für Kreislaufforschung*, **19**, 204-220.

Meek, W.J. and Eyster, J.A.E. (1913-1914). Experiments on the origin and propagation of the impulse in the heart. *Heart*, **5**, 227-243.

Meeley, M.P., Ruggiero, D.A., Ishitsuka, T., and Reis, D.J. (1985). Intrinsic γ-aminobutyric acid neurons in the nucleus of the solitary tract and the rostral ventrolateral medulla of the rat: an immunocytochemical and biochemical study. *Neuroscience Letters*, **58**, 83-89.

Meguro, Y. *et al.* (1989). A sympathetic component of baroreflex function in patients with essential hypertension. *Clinical and Experimental Pharmacology and Physiology*, **15** (Suppl.), 93-95.

Mehrmann, K. (1925). Der Heringsche Karotisdruckversuch am Menschen (Ph.D. thesis), Bonn.

Mehta, D., Wafa, S., Ward, D.E., and Camm, A.J. (1988). Relative efficacy of various physical manoeuvres in the termination of junctional tachycardia. *The Lancet*, **1**, 1181-1185.

Meier, A. *et al.* (1981). Pressor factors and cardiovascular pressor responsiveness in borderline hypertension. *Hypertension*, **3**, 367-372.

Meijling, H.A. (1938). Bau und innervation von Glomus caroticum und Sinus caroticus. *Acta Neerlandica Morphologiae Normalis et Pathologicae*, **1**, 193-287.

Melcher, A. (1976). Respiratory sinus arrhythmia in man. A study in heart rate regulating mechanism. *Acta Physiologica Scandinavica*, Suppl. **435**, 1-35.

Melcher, A. (1980). Carotid baroreflex heart rate control during the active and the assisted breathing cycle in man. *Acta Physiologica Scandinavica*, **108**, 165-171.

Melcher, A. and Donald, D.E. (1981). Maintained ability of carotid baroreflex to regulate arterial pressure during exercise. *American Journal of Physiology*, **241**, H838-H849.

Mellander, S. and Johansson, B. (1968). Control of resistance, exchange, and capacitance functions in the peripheral circulation. *Pharmacological Reviews*, **20**, 117-196.

Mengesha, Y.A. and Bell, G.H. (1979). Forearm and finger blood flow responses to passive body tilts. *Journal of Applied Physiology*, **46**, 288-292.

Mense, S. and Stahnke, M. (1983). Responses in muscle afferent fibres of slow conduction velocity to contractions and ischaemia in the cat. *The Journal of Physiology*, **342**, 383-397.

Merrill, A.J. (1946). Edema and decreased renal blood flow in patients with chronic congestive heart failure: evidence of 'forward failure' as the primary cause of edema. *The Journal of Clinical Investigation*, **25**, 389-400.

Merrill, A.J., Morrison, J.L., and Brannon, E.S. (1946). Concentration of renin in renal venous blood in patients with chronic heart failure. *The American Journal of Medicine*, **1**, 468-472.

Meyer, L. (1876). Ueber aneurymatische Veränderungen der Carotis interna Geisteskranker. *Archiv für Psychiatrie und Nervenkrankheiten*, **6**, 84-109.

Mifflin, S.W. and Felder, R.B. (1988). An intracellular study of time-dependent cardiovascular afferent interactions in nucleus tractus solitarius. *Journal of Neurophysiology*, **59**, 1798-1813.

Mifflin, S.W. and Felder, R.B. (1990). Synaptic mechanisms regulating cardiovascular afferent inputs to solitary tract nucleus. *American Journal of Physiology*, **259**, H653-H661.

Mifflin, S.W., Spyer, K.M., and Withington-Wray, D.J. (1988*a*). Baroreceptor inputs to the nucleus tractus solitarius in the cat: postsynaptic actions and the influence of respiration. *The Journal of Physiology*, **399**, 349-367.

Mifflin, S.W., Spyer, K.M., and Withington-Wray, D.J. (1988*b*). Baroreceptor inputs to the nucleus tractus solitarius in the cat: modulation by the hypothalamus. *The Journal of Physiology*, **399**, 369-387.

Milgram, N.W., MacLeod, C.M., and Petit, T.L. (1987). Neuroplasticity, learning and memory. In *Neuroplasticity, learning, and memory*, ed. Milgram, N.W., MacLeod, C.M., and Petit, T.L., pp. 1-16. Alan R. Liss, New York.

Millhorn, D.E. and Eldridge, F.L. (1986). Role of ventrolateral medulla in regulation of respiratory and cardiovascular systems. *Journal of Applied Physiology*, **61**, 1249-1263.

Minardo, J.D. *et al.* (1988). Scintigraphic and electrophysiological evidence of canine myocardial sympathetic denervation and reinnervation produced by myocardial infarction or phenol application. *Circulation*, **78**, 1008-1019.

Minson, R.B., McRitchie, R.J., and Chalmers, J.P. (1990). Effects of neuropeptide Y on baroreflex control of heart rate and myocardial contractility in conscious rabbits. *Clinical and Experimental Pharmacology and Physiology*, **17**, 39-49.

Miselis, R., Shapiro, R.E., and Hand, P.J. (1979). Subfornical organ efferents to neural systems for control of body water. *Science*, **205**, 1022-1025.

Miselis, R.R., Shapiro, R.E., and Hyde, T.M. (1987). The area postrema. In *Circumventricular organs and body fluids*, Volume II, ed. Gross, P.M., pp. 185-207. CRC Press, Boca Raton, Florida.

Mitchell, G.A.G. and Warwick, R. (1955). The dorsal vagal nucleus. *Acta Anatomica*, **25**, 371-395.

Mitchell, J.H., Payne, F.C., Saltin, B., and Schibye, B. (1980). The role of muscle mass in the cardiovascular response to static contractions. *The Journal of Physiology*, **309**, 45-54.

Mitchell, J.H., Reeves, D.R., Jr., Rogers, H.B., Secher, N.H., and Victor, R.G. (1989). Autonomic blockade and cardiovascular responses to static exercise in partially curarized man. *The Journal of Physiology*, **413**, 433-445.

Mitsutake, A., Takeshita, A., Kuroiwa, A., and Nakamura, M. (1981). Usefulness of the Valsalva maneuver in management of the long QT syndrome. *Circulation*, **63**, 1029-1035.

Miura, M. and Reis, D.J. (1969). Termination and secondary projections of carotid sinus nerve in the cat brain stem. *American Journal of Physiology*, **217**, 142-153.

Miura, M. and Reis, D.J. (1970). A blood pressure response from fastigial nucleus and its relay pathway in brainstem. *American Journal of Physiology*, **219**, 1330-1336.

Miura, M. and Reis, D.J. (1972). The role of the solitary and paramedian reticular nuclei in mediating cardiovascular reflex responses from carotid baro- and chemoreceptors. *The Journal of Physiology*, **223**, 525-548.

Miyajima, E., Yamada, Y., Matsukawa, T., Tochikubo, O., Ishii, M., and Kaneko, Y. (1988). Neurogenic abnormalities in young borderline hypertensives. *Clinical and Experimental Hypertension. Part A — Theory and Practice*, **A10** (Suppl. 1), 209-223.

Miyajima, E. *et al.* (1989). Inverse relationship between baroreflex sensitivity and basal muscle sympathetic nerve activity in hypertensives (Abstract). *Acta Physiologica Scandinavica*, **136** (Suppl. 584), 30.

Moe, G.K. (1975). Evidence for reentry as a mechanism of cardiac arrhythmias. *Reviews of Physiology, Biochemistry and Pharmacology*, **72**, 55-81.

Mohanty, P.K., Thames, M.D., Arrowood, J.A., Sowers, J.R., McNamara, C., and

Szentpetery, S. (1987). Impairment of cardiopulmonary baroreflex after cardiac transplantation in humans. *Circulation*, **75**, 914-921.

Mohanty, P.K., Sowers, J.R., McNamara, C., and Thames, M.D. (1988). Reflex effects of prolonged cardiopulmonary baroreceptor unloading in humans. *American Journal of Physiology*, **254**, R320-R324.

Mohanty, P.K., Arrowood, J.A., Ellenbogen, K.A., and Thames, M.D. (1989). Neurohumoral and hemodynamic effects of lower body negative pressure in patients with congestive heart failure. *American Heart Journal*, **118**, 78-85.

Mohiaddin, R.H. *et al.* (1989). Regional aortic compliance studied by magnetic resonance imaging: the effects of age, training, and coronary artery disease. *British Heart Journal*, **62**, 90-96.

Möhring, J. *et al.* (1980). Greatly enhanced pressor response to antidiuretic hormone in patients with impaired cardiovascular reflexes due to idiopathic orthostatic hypotension. *Journal of Cardiovascular Pharmacology*, **2**, 367-376.

Moissejeff, E. (1926). Zur Kenntnis des Carotissinusreflexes. *Zeitschrift für die gesamte Experimentelle Medizin*, **53**, 696-704.

Molina, C.R., Fowler, M.B., McCrory, S., Peterson, C., Myers, B.D., Schroeder, J.S., and Murad, F. (1988). Hemodynamic, renal and endocrine effects of atrial natriuretic peptide infusion in severe heart failure. *Journal of the American College of Cardiology*, **12**, 175-186.

Monaghan, D.T. and Cotman, C.W. (1985). Distribution of *N*-methyl-D-aspartate-sensitive L-[^{3}H]-glutamate-binding sites in rat brain. *The Journal of Neuroscience*, **5**, 2909-2919.

Montani, J.-P., Liard, J.-F., Schoun, J., and Möhring, J. (1980). Hemodynamic effects of exogenous and endogenous vasopressin at low plasma concentrations in conscious dogs. *Circulation Research*, **47**, 346-355.

Moore, T.O., Lin, Y.C., Lally, D.A., and Hong, S.K. (1972). Effects of temperature, immersion, and ambient pressure on human apneic bradycardia. *Journal of Applied Physiology*, **33**, 36-41.

Morady, F. *et al.* (1987*a*). Role of myocardial ischemia during programmed stimulation in survivors of cardiac arrest with coronary artery disease. *Journal of the American College of Cardiology*, **9**, 1004-1012.

Morady, F., Halter, J.B., DiCarlo, L.A., Jr., Baerman, J.M., and de Buitleir, M. (1987*b*). The interplay between endogenous catecholamines and induced ventricular tachycardia during electrophysiologic testing. *American Heart Journal*, **113**, 227-233.

Morady, F. *et al.* (1988). Accentuated antagonism between β-adrenergic and vagal effects on ventricular refractoriness in humans. *Circulation*, **77**, 289-297.

Morand, P. (1972). La manoeuvre de Valsalva: test global de la fonction myocardique. *Archives des Maladies du Coeur et des Vaisseaux*, **65**, 1483-1495.

Morest, D.K. (1967). Experimental study of the projections of the nucleus of the tractus solitarius and the area postrema in the cat. *The Journal of Comparative Neurology*, **130**, 277-300.

Morita, H. and Vatner, S.F. (1985). Effects of hemorrhage on renal nerve activity in conscious dogs. *Circulation Research*, **57**, 788-793.

Morley, C.A. and Sutton, R. (1984). Carotid sinus syncope. *International Journal of Cardiology*, **6**, 287-293.

Morley, C.A., Perrins, E.J., Grant, P., Chan, S.L., McBrien, D.J., and Sutton, R.

(1982). Carotid sinus syncope treated by pacing. Analysis of persistent symptoms and role of atrioventricular sequential pacing. *British Heart Journal*, **47**, 411-418.

Morley, C.A., Hudson, W.M., Kowk, H.T., Perrins, E.J., and Sutton, R. (1983). Is there a difference between sick sinus syndrome and carotid sinus syndrome? *British Heart Journal*, **49**, 620-621.

Morley, C.A., Dehn, T.C.B., Perrins, E.J., Chan, S.L., and Sutton, R. (1984). Baroreflex sensitivity measured by the phenylephrine pressor test in patients with carotid sinus and sick sinus syndromes. *Cardiovascular Research*, **18**, 752-761.

Mörlin, C., Wallin, B.G., and Eriksson, B.M. (1983). Muscle sympathetic activity and plasma noradrenaline in normotensive and hypertensive man. *Acta Physiologica Scandinavica*, **119**, 117-121.

Morris, M. and Alexander, N. (1989). Baroreceptor influences on oxytocin and vasopressin secretion. *Hypertension*, **13**, 110-114.

Morris, S.A., Tanowitz, H.B., Wittner, M., and Bilezikian, J.P. (1990). Pathophysiological insights into the cardiomyopathy of Chagas' disease. *Circulation*, **82**, 1900-1909.

Morrison, S.F., Barman, S.M., and Gebber, G.L. (1984). Baroreceptor influences on cardiac-related sympathetic nerve activity. *Brain Research*, **301**, 175-178.

Morrison, S.F., Milner, T.A., and Reis, D.J. (1988). Reticulospinal vasomotor neurons of the rat rostral ventrolateral medulla: Relationship to sympathetic nerve activity and the C1 adrenergic cell group. *The Journal of Neuroscience*, **8**, 1286-1301.

Moruzzi, G. (1940). Paleocerebellar inhibition of vasomotor and respiratory carotid sinus reflexes. *Journal of Neurophysiology*, **3**, 20-32.

Moskowitz, R.M., Kinney, E.L., and Zelis, R.F. (1979). Hemodynamic and metabolic responses to upright exercise in patients with congestive heart failure. Treatment with nitroglycerin ointment. *Chest*, **76**, 640-646.

Mosqueda-Garcia, R., Beck, C., Appalsamy, M., and Robertson, D. (1990). Cardiovascular effects of microinjection of endothelin in the brainstem of the rat (Abstract). *Clinical Research*, **38**, 414A.

Mraovitch, S., Kumada, M., and Reis, D.J. (1982). Role of the nucleus parabrachialis in cardiovascular regulation in cat. *Brain Research*, **232**, 57-75.

Mueller, H.S. and Ayres, S.M. (1980). Propranolol decreases sympathetic nervous activity reflected by plasma catecholamines during evolution of myocardial infarction in man. *The Journal of Clinical Investigation*, **65**, 338-346.

Muers, M.F. and Sleight, P. (1972). Action potentials from ventricular mechanoreceptors stimulated by occlusion of the coronary sinus in the dog. *The Journal of Physiology*, **221**, 283-309.

Mukharji, J. *et al.* (1987). Contrast injection bradycardia during coronary angiography: effects in the denervated human heart. *The Journal of Heart Transplantation*, **6**, 44-48.

Mukherjee, A., Wong, T.M., Buja, L.M., Lefkowitz, R.J., and Willerson, J.T. (1979). Beta adrenergic and muscarinic cholinergic receptors in canine myocardium. Effects of ischemia. *The Journal of Clinical Investigation*, **64**, 1423-1428.

Müller, E. (1891). *Ueber die Wirkung des Atropin auf das gesunde und kranke menschliche Herz*. Schnakenburg's Buchdruckerei, Dorpat.

Munch, P.A., Thoren, P.N., and Brown, A.M. (1987). Dual effects of norepinephrine and mechanisms of baroreceptor stimulation. *Circulation Research*, **61**, 409-419.

Münzel, T., Drexler, H., Holtz, J., Kurtz, S., and Just, H. (1991). Mechanisms involved in the response to prolonged infusion of atrial natriuretic factor in patients with chronic heart failure. *Circulation*, **83**, 191-201.

Muratori, G. (1967). Histological observations on the structure of the carotid sinus in man and mammals. In *Baroreceptors and hypertension*, ed. Kezdi, P., pp. 253-265. Pergamon, Oxford.

Murdaugh, H.V., Jr., Cross, C.E., Millen, J.E., Gee, J.B.L., and Robin, E.D. (1968). Dissociation of bradycardia and arterial constriction during diving in the seal phoca vitulina. *Science*, **162**, 364-365.

Murgo, J.P., Westerhof, N., Giolma, J.P., and Altobelli, S.A. (1981). Manipulation of ascending aortic pressure and flow wave reflections with the Valsalva maneuver: relationship to input impedance. *Circulation*, **63**, 122-132.

Murray, A., Ewing, D.J., Campbell, I.W., Neilson, J.M.M., and Clarke, B.F. (1975). RR interval variations in young male diabetics. *British Heart Journal*, **37**, 882-885.

Murray, R.H. and Shropshire, S. (1970). Effect of atropine on circulatory responses to lower body negative pressure and vasodepressor syncope. *Aerospace Medicine*, **41**, 717-722.

Murray, R.H., Krog, J., Carlson, L.D., and Bowers, J.A. (1967). Cumulative effects of venesection and lower body negative pressure. *Aerospace Medicine*, **38**, 243-247.

Murray, R.H., Thompson, L.J., Bowers, J.A., and Albright, C.D. (1968). Hemodynamic effects of graded hypovolemia and vasodepressor syncope induced by lower body negative pressure. *American Heart Journal*, **76**, 799-811.

Murray, R.H., Thompson, L.J., Bowers, J.A., Steinmetz, E.F., and Albright, C.D. (1969). Hemodynamic effects of hypovolemia in normal subjects and patients with congestive heart failure. *Circulation*, **39**, 55-63.

Musch, T.I., Moore, R.L., Smaldone, P.G., Riedy, M., and Zelis, R. (1989). Cardiac adaptations to endurance training in rats with a chronic myocardial infarction. *Journal of Applied Physiology*, **66**, 712-719.

Musgrave, F.S., Zechman, F.W., and Mains, R.C. (1969). Changes in total leg volume during lower body negative pressure. *Aerospace Medicine*, **40**, 602-606.

Musgrave, F.S., Zechman, F.W., and Mains, R.C. (1971). Comparison of the effects of 70° tilt and several levels of lower body negative pressure on heart rate and blood pressure in man. *Aerospace Medicine*, **42**, 1065-1069.

Myers, M.G. and de Champlain, J. (1983). Effects of atenolol and hydrochlorothiazide on blood pressure and plasma catecholamines in essential hypertension. *Hypertension*, **5**, 591-596.

Nabel, E.G. *et al.* (1987). Relationship of cardiac chamber volume to baroreflex activity in normal humans. *Journal of Clinical Endocrinology and Metabolism*, **65**, 475-481.

Nadel, J.A. and Widdicombe, J.G. (1962). Effect of changes in blood gas tensions and carotid sinus pressure on tracheal volume and total lung resistance to airflow. *The Journal of Physiology*, **163**, 13-33.

Nakano, J. (1964). Effects of atrial and ventricular tachycardias on the cardiovascular dynamics. *American Journal of Physiology*, **206**, 547-552.

Nakano, J., Zekert, H., Griege, C.W., Wang, K.-M., Schaefer, H.S., and Wégria, R. (1961). Effect of ventricular tachycardia and arteriovenous fistula on catecholamines blood level. *American Journal of Physiology*, **200**, 413-416.

Nakao, H., Ballim, H.M., and Gellhorn, E. (1956). The role of the sino-aortic receptors in the action of adrenaline, nor-adrenaline and acetylcholine on the cerebral cortex. *Electroencephalography and Clinical Neurophysiology*, **8**, 413-420.

Napier, M.A. *et al.* (1984). Specific membrane receptors for atrial natriuretic factor in renal and vascular tissues. *Proceedings of the National Academy of Sciences*, **81**, 5946-5950.

Needleman, P., Kauffman, A.H., Douglas, J.R., Jr., Johnson, E.M., Jr., and Marshall, G.R. (1973). Specific stimulation and inhibition of renal prostaglandin release by angiotensin analogs. *American Journal of Physiology*, **224**, 1415-1419.

Needleman, P., Bronson, S.D., Wyche, A., Sivakoff, M., and Nicolaou, K.C. (1978). Cardiac and renal prostaglandin I_2. Biosynthesis and biological effects in isolated perfused rabbit tissues. *The Journal of Clinical Investigation*, **61**, 839-849.

Neely, B.H., Gantenberg, N.S., and Hageman, G.R. (1990). Enhanced cardiac efferent sympathetic activity during programmed stimulation (Abstract). *The FASEB Journal*, **4**, A832.

Negoescu, R.M. and Csiki, I.-E. (1989). Model of respiratory sinus arrhythmia in man. *Medical Biological Engineering Computing*, **27**, 260-268.

Neil, E. (1979). Cardiac vagal efferent activity. In *Cardiac receptors*, ed. Hainsworth, R., Kidd, C., and Linden, R.J., pp. 361-374. Cambridge University Press, Cambridge.

Neil-Dwyer, G., Bartlett, J., McAinsh, J., and Cruickshank, J.M. (1981). β-Adrenoceptor blockers and the blood-brain barrier. *British Journal of Clinical Pharmacology*, **11**, 549-553.

Nicod, P. *et al.* (1985). Acute hemodynamic effect of a vascular antagonist of vasopressin in patients with congestive heart failure. *The American Journal of Cardiology*, **55**, 1043-1047.

Niebauer, M. and Zucker, I.H. (1985). Static and dynamic responses of carotid sinus baroreceptors in dogs with chronic volume overload. *The Journal of Physiology*, **369**, 295-310.

Niebauer, M.J., Holmberg, M.J., and Zucker, I.H. (1986). Aortic baroreceptor characteristics in dogs with chronic high output failure. *Basic Research in Cardiology*, **81**, 111-122.

Nikolic, G., Bishop, R.L., and Singh, J.B. (1982). Sudden death recorded during Holter monitoring. *Circulation*, **66**, 218-224.

Nilsson, G., Pettersson, A., Hedner, J., and Hedner, T. (1987). Increased plasma levels of atrial natriuretic peptide (ANP) in patients with paroxysmal supraventricular tachyarrhythmias. *Acta Medica Scandinavica*, **221**, 15-21.

Ninomiya, I. and Fujita, S. (1976). Reflex effects of thermal stimulation on sympathetic nerve activity to skin and kidney. *American Journal of Physiology*, **230**, 271-278.

Ninomiya, I., Nisimaru, N., and Irisawa, H. (1971). Sympathetic nerve activity to the spleen, kidney, and heart in response to baroceptor input. *American Journal of Physiology*, **221**, 1346-1351.

Nixon, J.V., Murray, R.G., Leonard, P.D., Mitchell, J.H., and Blomqvist, C.G.

(1982). Effect of large variations in preload on left ventricular performance characteristics in normal subjects. *Circulation*, **65**, 698-703.

Nomura, H., Shiosaka, S., and Tohyama, M. (1987). Distribution of substance P-like immunoreactive structures in the brainstem of the adult human brain: an immunocytochemical study. *Brain Research*, **404**, 365-370.

Nonidez, J.F. (1937). Distribution of the aortic nerve fibers and the epithelioid bodies (supracardial 'paraganglia') in the dog. *The Anatomical Record*, **69**, 299-317.

Nonidez, J.F. (1941). Studies on the innervation of the heart. II. Afferent nerve endings in the large arteries and veins. *The American Journal of Anatomy*, **68**, 151-189.

Nordmann, J.J., Dayanithi, G., and Cazalis, M. (1985). Coupling between the bioelectrical activity of a neurosecretory cell and the release at its terminals of neuropeptides. In *Vasopressin*, ed. Schrier, R.W., pp. 375-383. Raven Press, New York.

Norgren, R. (1978). Projections from the nucleus of the solitary tract in the rat. *Neuroscience*, **3**, 207-218.

Norgren, R. and Leonard, C.M. (1973). Ascending central gustatory pathways. *The Journal of Comparative Neurology*, **150**, 217-237.

Norsk, P., Bonde-Petersen, F., and Warberg, J. (1987). Plasma arginine vasopressin during neck suction in upright sitting man. *Acta Endocrinologica (Copenhagen)*, **114**, 243-248.

Norwegian Multicenter Study Group. (1981). Timolol-induced reduction in mortality and reinfarction in patients surviving acute myocardial infarction. *The New England Journal of Medicine*, **304**, 801-807.

Öberg, B. and Thorén, P. (1972). Increased activity in left ventricular receptors during hemorrhage or occlusion of caval veins in the cat. — A possible cause of the vaso-vagal reaction. *Acta Physiologica Scandinavica*, **85**, 164-173.

Öberg, B. and White, S. (1970). The role of vagal cardiac nerves and arterial baroreceptors in the circulatory adjustments to hemorrhage in the cat. *Acta Physiologica Scandinavica*, **80**, 395-403.

Oberholzer, J.H. (1960). Circulatory centers in medulla and midbrain. *Physiological Reviews*, **40**, 179-195.

Ogawa, J. *et al.* (1981). Daily profile of baroreflex sensitivity and the variability of blood pressure in essential hypertensive patients. *Clinical Science*, **61**, 157s-159s.

O'Leary, D.S. and Scher, A.M. (1990). Time course of recovery of arterial pressure control after carotid denervation. *American Journal of Physiology*, **258**, H73-H79.

O'Leary, D.S., Rowell, L.B., and Scher, A.M. (1991). Baroreflex-induced vasoconstriction in active skeletal muscle of conscious dogs. *American Journal of Physiology*, **260**, H37-H41.

Olivari, M.T., Levine, T.B., and Cohn, J.N. (1983). Abnormal neurohumoral response to nitroprusside infusion in congestive heart failure. *Journal of the American College of Cardiology*, **2**, 411-417.

Olivari, M.T., Levine, T.B., Ring, W.S., Simon, A., and Cohn, J.N. (1987). Normalization of sympathetic nervous system function after orthotopic cardiac transplant in man. *Circulation*, **76** (Suppl. V), V-62–V-64.

Oliver, J.A., Sciacca, R.R., Pinto, J., and Cannon, P.J. (1981). Participation of the prostaglandins in the control of renal blood flow during acute reduction of cardiac output in the dog. *The Journal of Clinical Investigation*, **67**, 229-237.

Oliver, J.R., Korner, P.I., Woods, R.L., and Zhu, J.L. (1990). Reflex release of vasopressin and renin in hemorrhage is enhanced by autonomic blockade. *American Journal of Physiology*, **258**, H221-H228.

Olney, J.W. (1978). Neurotoxicity of excitatory amino acids. In *Kainic acid as a tool in neurobiology*, ed. McGeer, E.G., Olney, J.W., and McGeer, P.L., pp. 95-121. Raven Press, New York.

Olsen, C.R., Fanestil, D.D., and Scholander, P.F. (1962). Some effects of breath holding and apneic underwater diving on cardiac rhythm in man. *Journal of Applied Physiology*, **17**, 461-466.

Olshansky, B. and Martins, J.B. (1987). Usefulness of isoproterenol facilitation of ventricular tachycardia induction during extrastimulus testing in predicting effective chronic therapy with beta-adrenergic blockade. *The American Journal of Cardiology*, **59**, 573-577.

Olszewski, J. and Baxter, D. (1954). *Cytoarchitecture of the human brain stem*, pp. 1-159, J.B. Lippincott Company, Philadelphia.

Onrot, J., Wiley, R.G., Fogo, A., Biaggioni, I., Robertson, D., and Hollister, A.S. (1987). Neck tumour with syncope due to paroxysmal sympathetic withdrawal. *Journal of Neurology, Neurosurgery, and Psychiatry*, **50**, 1063-1066.

Ookuwa, H., Takata, S., Ogawa, J., Iwase, N., Ikeda, T., and Hattori, N. (1987). Abnormal cardiopulmonary baroreflexes in normotensive young subjects with a family history of essential hypertension. *Journal of Clinical Hypertension*, **3**, 596-604.

Oparil, S., Vassaux, C., Sanders, C.A., and Haber, E. (1970). Role of renin in acute postural homeostasis. *Circulation*, **41**, 89-95.

Osterziel, K.J., Dietz, R., Schmid, W., and Kübler, W. (1988). Reflektorische Regulation der Herzfrequenz bei herzinsuffizienten Patienten. *Zeitschrift für Kardiologie*, **77**, 576-581.

Osterziel, K.J., Dietz, R., Manthey, J., Schmid, W., and Kübler, W. (1990*a*). Haemodynamic changes caused by alteration of autonomic activity in patients with heart failure. *British Heart Journal*, **63**, 221-224.

Osterziel, K.J., Dietz, R., Schmid, W., Mikulaschek, K., Manthey, J., and Kübler, W. (1990*b*). ACE inhibition improves vagal reactivity in patients with heart failure. *American Heart Journal*, **120**, 1120-1129.

O'Toole, M.F., Wurster, R.D., Phillips, J.G., and Randall, W.C. (1984). Parallel baroreceptor control of sinoatrial rate and atrioventricular conduction. *American Journal of Physiology*, **246**, H149-H153.

Ott, N.T., Tarhan, S., and McGoon, D.C. (1975). Circulatory effects of vagal inflation reflex in man. *Zeitschrift für Kardiologie*, **64**, 1066-1070.

Owsjannikow, P. (1871). Die tonischen und reflektorischen Centren der Gefässnerven. *Berichte der Sächsischen Gesellschaft der Wissenschaften*, **23**, 135-143.

Pace, D.G. and Martin, P. (1985). Effect of brief bursts of carotid sinus stimuli on heart rate and atrioventricular conduction. *Journal of Electrocardiology*, **18**, 63-70.

Packer, M., Meller, J., Medina, N., Gorlin, R., and Herman, M.V. (1979). Rebound hemodynamic events after the abrupt withdrawal of nitroprusside in patients with severe chronic heart failure. *The New England Journal of Medicine*, **301**, 1193-1197.

Paes de Carvalho, A., de Mello, W.C., and Hoffman, B.F. (1959). Electrophysiological evidence for specialized fiber types in rabbit atrium. *American Journal of Physiology*, **196**, 483-488.

Pagani, M. *et al.* (1986). Power spectral analysis of heart rate and arterial pressure variabilities as a marker of sympatho-vagal interaction in man and conscious dog. *Circulation Research*, **59**, 178-193.

Pagani, M. *et al.* (1988). Changes in autonomic regulation induced by physical training in mild hypertension. *Hypertension*, **12**, 600-610.

Pagano, G. (1900). Sur la sensibilité du coeur et des vaisseaux sanguins. *Archives Italiennes de Biologie*, **33**, 1-36.

Page, I.H. and Helmer, O.M. (1940). A crystalline pressor substance (angiotonin) resulting from the reaction between renin and renin-activator. *The Journal of Experimental Medicine*, **71**, 29-42.

Paintal, A.S. (1963). Vagal afferent fibres. *Ergebnisse der Physiologie Biologischen Chemie und Experimentellen Pharmakologie*, **52**, 74-156.

Paintal, A.S. (1972). Cardiovascular receptors. In *Handbook of sensory physiology*, Volume III/1, *Enteroceptors*, ed. Neil, E., pp. 1-45. Springer-Verlag, Berlin.

Palacios, J.M., Probst, A., and Cortes, R. (1983). The distribution of serotonin receptors in the human brain: high density of [^{3}H]LSD binding sites in the raphe nuclei of the brainstem. *Brain Research*, **274**, 150-155.

Paley, H.W., Weissler, A.M., and Schoenfeld, C.D. (1964). The effect of upright posture on left ventricular volume in man (Abstract). *Clinical Research*, **12**, 105.

Palkovits, M. (1981). Neuropeptides and biogenic amines in central cardiovascular control mechanisms. In *Central nervous mechanisms in hypertension*, ed. Buckley, J.P., Ferrario, C.M., and Lokhandwala, M.F., pp. 73-87. Raven Press, New York.

Palkovits, M. and Jacobowitz, D.M. (1974). Topographic atlas of catecholamine and acetylcholinesterase-containing neurons in the rat brain. II. Hindbrain (mesencephalon, rhombencephalon). *The Journal of Comparative Neurology*, **157**, 29-42.

Palkovits, M. and Záborszky, L. (1977). Neuroanatomy of central cardiovascular control. Nucleus tractus solitarii: afferent and efferent neuronal connections in relation to the baroreceptor reflex arc. *Progress in Brain Research*, **47**, 9-34.

Palme, F. (1936). Über die Wirkung des Adrenalins auf den Sinus caroticus. *Zeitschrift für Kreislaufforschung*, **28**, 173-174.

Palmero, H.A. and Caeiro, T.E. (1976). Statistical analysis of the different factors which could affect postextrasystolic potentiation in the human heart. *American Heart Journal*, **92**, 427-434.

Palmero, H.A., Caeiro, T.F., Iosa, D.J., and Bas, J. (1981). Baroreceptor reflex sensitivity index derived from phase 4 of the Valsalva maneuver. *Hypertension*, **3** (Suppl. II), II-134–II-137.

Pamphlett, R. and Harper, C. (1985). Leigh's disease: a cause of arterial hypertension. *The Medical Journal of Australia*, **143**, 306-308.

Panneton, W.M. and Loewy, A.D. (1980). Projections of the carotid sinus nerve to the nucleus of the solitary tract in the cat. *Brain Research*, **191**, 239-244.

Pantridge, J.F., Webb, S.W., Adgey, A.A.J., and Geddes, J.S. (1973). The first hour after the onset of acute myocardial infarction. *Progress in Cardiology*, **3**, 173-188.

Pantridge, J.F., Adgey, A.A.J., Geddes, J.S., and Webb, S.W. (1975). *The acute coronary attack*, pp. 26-42. Pitman Medical, Bath.

Parati, G., Pomidossi, G., Ramirez, A., Gavazz, C., Bertinieri, G., and Mancia, G. (1983). Reproducibility of laboratory tests evaluating neural cardiovascular regulation in man. *Journal of Hypertension*, **1** (Suppl. 2), 88-90.

Parati, G., Grassi, G., Coruzzi, P., Musiari, L., Ravogli, A., Novarini, A., *et al.* (1987). Influence of cardiopulmonary receptors on the bradycardic responses to carotid baroreceptor stimulation in man. *Clinical Science*, **72**, 639-645.

Parati, G. *et al.* (1988). Evaluation of the baroreceptor-heart rate reflex by 24-hour intra-arterial blood pressure monitoring in humans. *Hypertension*, **12**, 214-222.

Parati, G., Casadei, R., Groppelli, A., di Rienzo, M., and Mancia, G. (1989). Comparison of finger and intra-arterial blood pressure monitoring at rest and during laboratory testing. *Hypertension*, **13**, 647-655.

Parisi, A.F., Harrington, J.J., Askenazi, J., Pratt, R.C., and McIntyre, K.M. (1976). Echocardiographic evaluation of the Valsalva maneuver in healthy subjects and patients with and without heart failure. *Circulation*, **54**, 921-927.

Parker, P., Celler, B.G., Potter, E.K., and McCloskey, D.I. (1984). Vagal stimulation and cardiac slowing. *Journal of the Autonomic Nervous System*, **11**, 226-231.

Parry, C.H. (1799). *An inquiry into the symptoms and causes of the syncope anginosa, commonly called angina pectoris*. R. Cruttwell, Bath.

Patakas, D., Louridas, G., and Kakavelas, E. (1982). Reduced baroreceptor sensitivity in patients with chronic obstructive pulmonary disease. *Thorax*, **37**, 292-295.

Patel, L. and Turner, P. (1981). Central actions of β-adrenoceptor blocking drugs in man. *Medicinal Research Reviews*, **1**, 387-410.

Paton, J.F.R. and Spyer, K.M. (1990). Brain stem regions for mediating the cardiovascular responses elicited from the posterior cerebellar cortex in the rabbit. *The Journal of Physiology*, **427**, 533-552.

Paton, J.F.R., Silva-Carvalho, L., Goldsmith, G.E., and Spyer, K.M. (1990). Inhibition of barosensitive neurones evoked by lobule IXb of the posterior cerebellar cortex in the decerebrate rabbit. *The Journal of Physiology*, **427**, 553-565.

Patrick, J.M., Gibson, Z., and Hanley, S.P. (1984). Within-breath modulation of the triggering of extrasystoles in man. *Clinical Science*, **67**, 225-227.

Paulev, P.-E. (1969). Respiratory and cardiovascular effects of breath-holding. *Acta Physiologica Scandinavica*, Suppl. **324**, 1-116.

Paulev, P.-E., Honda, Y., Sakakibara, Y., Morikawa, T., Tanaka, Y., and Nakamura, W. (1988). Brady- and tachycardia in light of the Valsalva and the Mueller maneuver (apnea). *Japanese Journal of Physiology*, **38**, 507-517.

Pawelczyk, J.A. and Raven, P.B. (1989). Reductions in central venous pressure improve carotid baroreflex responses in conscious men. *American Journal of Physiology*, **257**, H1389-H1395.

Pearson, J., Goldstein, M., Markey, K., and Brandeis, L. (1983). Human brainstem catecholamine neuronal anatomy as indicated by immunocytochemistry with antibodies to tyrosine hydroxylase. *Neuroscience*, **8**, 3-32.

Peňáz, J. (1962). Frequency response of the cardiac chronotropic action of the vagus in the rabbit. *Archives de Internationales de Physiologie et de Biochimie*, **70**, 636-650.

Peňáz, J. (1973). Photoelectric measurement of blood pressure, volume and flow in the finger. In *Digest of the 10th international conference on medical and biologic engineering*, p. 104. Dresden.

Peňáz, J., Honzíková, N., and Fišer, B. (1978). Spectral analysis of resting variability of some circulatory parameters in man. *Physiologia Bohemoslovaca*, **27**, 349-357.

Perez-Gomez, F. and Garcia-Aguado, A. (1977). Origin of ventricular reflexes caused by coronary arteriography. *British Heart Journal*, **39**, 967-973.

Perez-Gomez, F., De Dios, R.M., Rey, J., and Garcia Aguado, A. (1979). Prinzmetal's angina: reflex cardiovascular response during episode of pain. *British Heart Journal*, **42**, 81-87.

Pernow, J. (1988). Co-release and functional interactions of neuropeptide Y and noradrenaline in peripheral sympathetic vascular control. *Acta Physiologica Scandinavica*, Suppl. **568**, 1-56.

Persson, P.B. and Kirchheim, H.R. (1991). *Baroreceptor reflexes, integrative functions and clinical aspects.* Springer-Verlag, Heidelberg.

Persson, P., Ehmke, H., Kirchheim, H., and Seller, H. (1987). The influence of cardiopulmonary receptors on long-term blood pressure control and plasma renin activity in conscious dogs. *Acta Physiologica Scandinavica*, **130**, 553-561.

Persson, P., Ehmke, H., Kirchheim, H., and Seller, H. (1988). Effect of sino-aortic denervation in comparison to cardiopulmonary deafferentiation on long-term blood pressure in conscious dogs. *Pflügers Archiv für die Gesamte Physiologie des Menschen und der Tiere*, **411**, 160-166.

Peterson, D.F. and Bishop, V.S. (1974). Reflex blood pressure control during acute myocardial ischemia in the conscious dog. *Circulation Research*, **34**, 226-232.

Peterson, D.F., Kaspar, R.L., and Bishop, V.S. (1973). Reflex tachycardia due to temporary coronary occlusion in the conscious dog. *Circulation Research*, **32**, 652-659.

Petit,–. (1727). Dans lequel il est démontré que les nerfs intercostaux fornissent des rameaux qui portent des esprits dans les yeux. *Mémoires de Mathématique et de Physique, Tirés des Registres de l'Académie Royale des Sciences*, 1-19.

Petro, J.K., Hollander, A.P., and Bouman, L.N. (1970). Instantaneous cardiac acceleration in man induced by a voluntary muscle contraction. *Journal of Applied Physiology*, **29**, 794-798.

Petrov, L., Karachiviev, D., and Bredy-Dobreva, G. (1979). Participation of the cardiopulmonary baroreceptors in the buffering of fast changes in the arterial pressure. *Acta Physiologica et Pharmacologica Bulgarica*, **5**, 19-26.

Peveler, R.C., Bergel, D.H., Robinson, J.L., and Sleight, P. (1983). The effect of phenylephrine upon arterial pressure, carotid sinus radius and baroreflex sensitivity in the conscious greyhound. *Clinical Science*, **64**, 455-461.

Pfeifer, M.A., Weinberg, C.R., Cook, D., Best, J.D., Reenan, A., and Halter, J.B. (1983). Differential changes of autonomic nervous system function with age in man. *The American Journal of Medicine*, **75**, 249-258.

Philip, A.P.W. (1817). *An experimental inquiry into the laws of the vital functions*, pp. 1-347. Thomas and George Underwood, London.

Philipp, T., Distler, A., and Cordes, U. (1978). Sympathetic nervous system and blood-pressure control in essential hypertension. *The Lancet*, **2**, 959-963.

Pickering, G.W., Kissin, M., and Rothschild, P. (1936). The relationship of the carotid sinus mechanism to persistent high blood pressure in man. *Clinical Science*, **2**, 193-200.

Pickering, T.G. (1970). Baroreceptor reflexes in man in health and disease (Ph.D. thesis), Oxford, pp. 19-21.

Pickering, T.G. and Davies, J. (1973). Estimation of the conduction time of the baroreceptor–cardiac reflex in man. *Cardiovascular Research*, **7**, 213-219.

Pickering, T.G. and Sleight, P. (1969). Quantitative index of baroreflex activity in normal and hypertensive subjects using Valsalva's manoeuvre (Abstract). *British Heart Journal*, **31**, 392.

Pickering, T.G., Gribbin, B., Petersen, E.S., Cunningham, D.J.C., and Sleight, P. (1971). Comparison of the effects of exercise and posture on the baroreflex in man. *Cardiovascular Research*, **5**, 582-586.

Pickering, T.G., Gribbin, B., and Oliver, D.O. (1972*a*). Baroreflex sensitivity in patients on long-term haemodialysis. *Clinical Science*, **43**, 645-657.

Pickering, T.G., Gribbin, B., Petersen, E.S., Cunningham, D.J.C., and Sleight, P. (1972*b*). Effects of autonomic blockade on the baroreflex in man at rest and during exercise. *Circulation Research*, **30**, 177-185.

Pickering, T.G., Gribbin, B., and Sleight, P. (1972*c*). Comparison of the reflex heart rate response to rising and falling arterial pressure in man. *Cardiovascular Research*, **6**, 277-283.

Podrid, P.J., Fuchs, T., and Candinas, R. (1990). Role of the sympathetic nervous system in the genesis of ventricular arrhythmia. *Circulation*, **82** (Suppl. I), I-103–I-113.

Poliner, L.R., Dehmer, G.J., Lewis, S.E., Parkey, R.W., Blomqvist, C.G., and Willerson, J.T. (1980). Left ventricular performance in normal subjects: a comparison of the responses to exercise in the upright and supine positions. *Circulation*, **62**, 528-534.

Polosa, C. (1967). The silent period of sympathetic preganglionic neurons. *Canadian Journal of Physiology and Pharmacology*, **45**, 1033-1045.

Polosa, C., Mannard, A., and Laskey, W. (1979). Tonic activity of the autonomic nervous system: functions, properties, origins. In *Integrative functions of the autonomic nervous system*, ed. Brooks, C.McC., Koizumi, K., and Sato, A., pp. 342-354. University of Tokyo Press, Tokyo.

Pomeranz, B. *et al.* (1985). Assessment of autonomic function in humans by heart rate spectral analysis. *American Journal of Physiology*, **248**, H151-H153.

Pomidossi, G. *et al.* (1989). Baroreceptor control of heart rate during myocardial ischemia in man (Abstract). *Circulation*, **80** (Suppl. II), II-551.

Pool, J.L. and Ransohoff, J. (1949). Autonomic effects on stimulating rostral portion of cingulate gyri in man. *Journal of Neurophysiology*, **11**, 385-392.

Porges, S.W. (1986). Respiratory sinus arrhythmia: physiological basis, quantitative methods, and clinical implications. In *Cardiorespiratory and cardiosomatic psychophysiology*, ed. Grossman, P., Janssen, K.H.L., and Vaitl, D., pp. 101-115. Plenum Press, New York.

Pórszász, J. and Pórszász-Gibiszer, K. (1968). Effect of blood pressure changes on single unit activity in the bulbar reticular formation. *Acta Physiologica Academiae Scientiarum Hungaricae*, **34**, 249-258.

Pórszász, J., Barankay, T., Szolcsányi, J., Gibiszer-Pórszász, K., and Madarász, K.

(1962). Studies of the neural connexion between the vasodilatator and vasoconstrictor centres in the cat. *Acta Physiologica Hungarica*, **22**, 29-41.

Porter, T.R. *et al.* (1990). Autonomic pathophysiology in heart failure patients. Sympathetic–cholinergic interrelations. *The Journal of Clinical Investigation*, **850**, 1362-1371.

Porus, R.L. and Marcus, F.I. (1963). Ventricular fibrillation during carotid-sinus stimulation. *The New England Journal of Medicine*, **268**, 1338-1342.

Potter, E.K. (1982). Adaptation of cardiac vagal responses to baroreceptor stimuli in the dog. *The Journal of Physiology*, **329**, 411-423.

Pratt, C.M., Francis, M.J., Luck, J.C., Wyndham, C.R., Miller, R.R., and Quinones, M.A. (1983). Analysis of ambulatory electrocardiograms in 15 patients during spontaneous ventricular fibrillation with special reference to preceding arrhythmic events. *Journal of the American College of Cardiology*, **2**, 789-797.

Price, H.L. (1957). Circulating adrenaline and noradrenaline during diethyl ether anaesthesia in man. *Clincial Science*, **16**, 377-387.

Prinzmetal, M., Ekmekci, A., Kennamer, R., Kwoczynski, J.K., Shubin, H., and Toyoshima, H. (1960). Variant form of angina pectoris. Previously undelineated syndrome. *The Journal of the American Medical Association*, **174**, 1794-1800.

Priori, S.G. and Corr, P.B. (1990). Mechanisms underlying early and delayed afterdepolarizations induced by catecholamines. *American Journal of Physiology*, **258**, H1796-H1805.

Priori, S.G., Yamada, K.A., and Corr, P.B. (1991). Influence of hypoxia on adrenergic modulation of triggered activity in isolated adult canine myocytes. *Circulation*, **83**, 248-259.

Probst, A., Cortes, R., and Palacios, J.M. (1985). Distribution of α_2-adrenergic receptors in the human brainstem: an autoradiographic study using [^{3}H]*p*-aminoclonidine. *European Journal of Pharmacology*, **106**, 477-488.

Pryor, S.L., Strange, S., Richter, E.A., Christensen, N.J., Saltin, B., and Victor, R.G. (1990). Sympathetic discharge and norepinephrine spillover in humans (Abstract). *Clinical Research*, **38**, 413A.

Prystowsky, E.N. and Zipes, D.P. (1985). Postvagal tachycardia. *The American Journal of Cardiology*, **55**, 995-999.

Prystowsky, E.N., Jackman, W.M., Rinkenberger, R.L., Heger, J.J., and Zipes, D.P. (1981). Effect of autonomic blockade on ventricular refractoriness and atrioventricular nodal conduction in humans. Evidence supporting a direct cholinergic action on ventricular muscle refractoriness. *Circulation Research*, **49**, 511-518.

Prystowsky, E.N., Naccarelli, G.V., Jackman, W.M., Rinkenberger, R.L., Heger, J.J., and Zipes, D.P. (1983). Enhanced parasympathetic tone shortens atrial refractoriness in man. *The American Journal of Cardiology*, **51**, 96-100.

Przybylski, J. (1981). Do arterial chemoreceptors play a role in the pathogenesis of hypertension? *Medical Hypotheses*, **7**, 127-131.

Pullan, P.T., Johnston, C.I., Anderson, W.P., and Korner, P.I. (1980). Plasma vasopressin in blood pressure homeostasis and in experimental renal hypertension. *American Journal of Physiology*, **239**, H81-H87.

Quail, A.W., Woods, R.L., and Korner, P.I. (1987). Cardiac and arterial baroreceptor influences in release of vasopressin and renin during hemorrhage. *American Journal of Physiology*, **252**, H1120-H1126.

Quincke, H. (1875). Über Vagusreizung beim Menschen. *Berliner Klinische Wochenschrift*, **12**, 189-191.

Raczkowska, M., Eckberg, D.L., and Ebert, T.J. (1983). Muscarinic cholinergic receptors modulate vagal cardiac responses in man. *Journal of the Autonomic Nervous System*, **7**, 271-278.

Randall, O.S., van den Bos, G.C., and Westerhof, N. (1984). Systemic compliance: does it play a role in the genesis of essential hypertension? *Cardiovascular Research*, **18**, 455-462.

Ranson, S.W. and Billlingsley, P.R. (1916). Vasomotor reactions from stimulation of the floor of the fourth ventricle. *American Journal of Physiology*, **41**, 85-90.

Rapaport, E., Wong, M., Escobar, E.E., and Martinez, G. (1966). The effect of upright posture on right ventricular volumes in patients with and without heart failure. *American Heart Journal*, **71**, 146-152.

Raper, A.J., Richardson, D.W., Kontos, H.A., and Patterson, J.L., Jr. (1967). Circulatory responses to breath holding in man. *Journal of Applied Physiology*, **22**, 201-206.

Rawles, J.M., Pai, G.R., and Reid, S.R. (1989). A method of quantifying sinus arrhythmia: parallel effect of respiration on P–P and P–R intervals. *Clinical Science*, **76**, 103-108.

Rea, R.F. and Eckberg, D.L. (1987). Carotid baroreceptor–muscle sympathetic relation in humans. *American Journal of Physiology*, **253**, R929-R934.

Rea, R.F. and Hamdan, M. (1990). Baroreflex control of muscle sympathetic nerve activity in borderline hypertension. *Circulation*, **82**, 856-862.

Rea, R.F. and Wallin, B.G. (1989). Sympathetic nerve activity in arm and leg muscles during lower body negative pressure in humans. *Journal of Applied Physiology*, **66**, 2778-2781.

Rea, R.F., Eckberg, D.L., Fritsch, J.M., and Goldstein, D.S. (1990). Relation of plasma norepinephrine and sympathetic traffic during hypotension in humans. *American Journal of Physiology*, **258**, R982-R986.

Rector, T.S., Olivari, M.T., Levine, T.B., Francis, G.S., and Cohn, J.N. (1987). Predicting survival for an individual with congestive heart failure using the plasma norepinephrine concentration. *American Heart Journal*, **114**, 148-152.

Redfield, M.M., Edwards, B.S., McGoon, M.D., Heublein, D.M., Aarhus, L.L., and Burnett, J.C., Jr. (1989). Failure of atrial natriuretic factor to increase with volume expansion in acute and chronic heart failure in the dog. *Circulation*, **80**, 651-657.

Rees, P.M. (1967*a*). The distribution of biogenic amines in the carotid bifurcation region. *The Journal of Physiology*, **193**, 245-253.

Rees, P.M. (1967*b*). Observations on the fine structure and distribution of presumptive baroreceptor nerves at the carotid sinus. *The Journal of Comparative Neurology*, **131**, 517-548.

Rees, P.M. (1968). Electron microscopical observations on the architecture of the carotid arterial walls, with special reference to the sinus portion. *The Journal of Anatomy*, **103**, 35-47.

Rees, P.M., Sleight, P., Robinson, J.L., Bonchek, L.I., and Doctor, A. (1978). Histology and ultrastructure of the carotid sinus in experimental hypertension. *The Journal of Comparative Neurology*, **181**, 245-252.

Reid, J.L. and Rubin, P.C. (1987). Peptides and central neural regulation of the circulation. *Physiological Reviews*, **67**, 725-749.

Reid, J.L., Wing, L.M.H., Mathias, C.J., Frankel, H.L., and Neill, E. (1977). The central hypotensive effect of clonidine. Studies in tetraplegic subjects. *Clinical Pharmacology and Therapeutics*, **21**, 375-381.

Reid, J.V.O. (1969). The cardiac pacemaker: effects of regularly spaced nervous input. *American Heart Journal*, **78**, 58-64.

Reiffel, J.A. and Bigger, J.T., Jr. (1983). Current status of direct recordings of the sinus node electrogram in man. *Pace*, **6**, 1143-1150.

Rein, H. and Schneider, M. (1956). *Einführung in die Physiologie des Menschen*, pp. 178-179. Springer-Verlag, Berlin.

Reis, D.J. and Cuenod, M. (1965). Central neural regulation of carotid baroreceptor reflexes in the cat. *American Journal of Physiology*, **209**, 1267-1277.

Reis, D.J. and Fuxe, K. (1968). Adrenergic innervation of the carotid sinus. *American Journal of Physiology*, **215**, 1054-1057.

Reis, D.J., Doba, N., Snyder, D.W., and Nathan, M.A. (1977). Brain lesions and hypertension: Chronic lability and elevation of arterial pressure produced by electrolytic lesions and 6-hydroxydopamine treatment of nucleus tractus solitarii (NTS) in rat and cat. *Progress in Brain Research*, **47**, 169-188.

Reis, D.J., Granata, A.R., Joh, T.H., Ross, C.A., Ruggiero, D.A., and Park, D.H. (1984*a*). Brain stem catecholamine mechanisms in tonic and reflex control of blood pressure. *Hypertension*, **6** (Suppl. II), II-7–II-15.

Reis, D.J., Ross, C.A., Ruggiero, D.A., Granata, A.R., and Joh, T.H. (1984*b*). Role of adrenaline neurons of ventrolateral medulla (the C1 group) in the tonic and phasic control of arterial pressure. *Clinical and Experimental Hypertension. Part A — Theory and Practice*, **A6**, 221-241.

Remensnyder, J.P., Mitchell, J.H., and Sarnoff, S.J. (1962). Functional sympatholysis during muscular activity. Observations on influence of carotid sinus on oxygen uptake. *Circulation Research*, **11**, 370-380.

Remington, J.W. (1955). Hysteresis loop behavior of the aorta and other extensible tissues. *American Journal of Physiology*, **180**, 83-95.

Rey, N. (1971). Influence of age and training on the diving reflex response. *Acta Physiologica et Pharmacologica Latino-Americana*, **21**, 244-251.

Ricardo, J. and Koh, E.T. (1978). Anatomical evidence of direct projections from the nucleus of the solitary tract to the hypothalamus, amygdala, and other forebrain structures in the rat. *Brain Research*, **153**, 1-26.

Ricci, D.R., Orlick, A.E., Cipriano, P.R., Guthaner, D.F., and Harrison, D.C. (1979). Altered adrenergic activity in coronary arterial spasm: insight into mechanism based on study of coronary hemodynamics and the electrocardiogram. *The American Journal of Cardiology*, **43**, 1073-1079.

Richards, A.M., Nicholls, M.G., Ikram, H., Webster, M.W.I., Yandle, T.G., and Espiner, E.A. (1985). Renal, haemodynamic, and hormonal effects of human alpha atrial natriuretic peptide in healthy volunteers. *The Lancet*, **1**, 545-549.

Richardson, D.W., Kontos, H.A., Raper, A.J., and Patterson, J.L., Jr. (1967). Modification by beta-adrenergic blockade of the circulatory responses to acute hypoxia in man. *The Journal of Clinical Investigation*, **46**, 77-85.

Richet, C. (1894). Influence de l'atropine sur la durée de l'asphyxie chez le canard.

Comptes Rendus Hebdomadaires des Séances et Mémoires de la Société de Biologie, **1**, 789-790.

Richet, C. (1898). De la résistance des canards à l'asphyxie. *Comptes Rendus Hebdomadaires des Séances et Mémoires de la Société de Biologie*, **5**, 685-686.

Richter, D.W., Keck, W., and Seller, H. (1970). The course of inhibition of sympathetic activity during various patterns of carotid sinus nerve stimulation. *Pflügers Archiv für die Gesamte Physiologie des Menschen und der Tiere*, **317**, 110-123.

Richter, D.W., Jordan, D., Ballantyne, D., Meesmann, M., and Spyer, K.M. (1986). Presynaptic depolarization in myelinated vagal afferent fibres terminating in the nucleus of the tractus solitarius in the cat. *Pflügers Archiv für die Gesamte Physiologie des Menschen und der Tiere*, **406**, 12-19.

Riegger, G.A.J., Liebau, G., and Kochsiek, K. (1982). Antidiuretic hormone in congestive heart failure. *The American Journal of Medicine*, **72**, 49-52.

Rigel, D.F. (1988). Effects of neuropeptides on heart rate in dogs: comparison of VIP, PHI, NPY, CGRP, and NT. *American Journal of Physiology*, **255**, H311-H317.

Rigel, D.F., Lipson, D., and Katona, P.G. (1984). Excess tachycardia: heart rate after antimuscarinic agents in conscious dogs. *American Journal of Physiology*, **246**, H168-H173.

Riggs, C.E., Jr., Johnson, D.J., Konopka, B.J., and Kilgour, R.D. (1981). Exercise heart rate response to facial cooling. *European Journal of Applied Physiology*, **47**, 323-330.

Rikard-Bell, G.C., Törk, I., Sullivan, C., and Scheibner, T. (1990). Distribution of substance P-like immunoreactive fibres and terminals in the medulla oblongata of the human infant. *Neuroscience*, **34**, 133-148.

Riley, H.A., German, W.J., Wortis, H., Herbert, C., Zahn, D., and Eichna, L. (1942). Glossopharyngeal neuralgia initiating or associated with cardiac arrest. *Transactions of the American Neurological Association*, **68**, 28-29.

Rimoldi, O., Pierini, S., Ferrari, A., Cerutti, S., Pagani, M., and Malliani, A. (1990). Analysis of short-term oscillations of R–R and arterial pressure in conscious dogs. *American Journal of Physiology*, **258**, H967-H976.

Robbe, H.W.J., Mulder, L.J.M., Rüddel, H., Langewitz, W.A., Veldman, J.B.P., and Mulder, G. (1987). Assessment of baroreceptor reflex sensitivity by means of spectral analysis. *Hypertension*, **10**, 538-543.

Robertson, D., Stevens, R.M., Friesinger, G.C., and Oates, J.A. (1977). The effect of the Valsalva maneuver on echocardiographic dimensions in man. *Circulation*, **55**, 596-602.

Robertson, G.L. (1977). The regulation of vasopressin function in health and disease. In *Recent progress in hormone research*, ed. Greep, R.O., pp. 333-374. Academic Press, New York.

Robertson, G.L. (1983). Thirst and vasopressin function in normal and disordered states of water balance. *The Journal of Laboratory and Clinical Medicine*, **101**, 351-371.

Robertson, G.L. (1984). Abnormalities of thirst regulation. *Kidney International*, **25**, 460-469.

Robertson, G.L. and Athar, S. (1976). The interaction of blood osmolality and blood volume in regulating plasma vasopressin in man. *Journal of Clinical Endocrinology and Metabolism*, **42**, 613-620.

Robertson, G.L., Mahr, E.A., Athar, S., and Sinha, T. (1973). Development and clinical application of a new method for the radioimmunoassay of arginine vasopressin in human plasma. *The Journal of Clinical Investigation*, **52**, 2340-2352.

Robertson, G.L., Shelton, R.L., and Athar, S. (1976). The osmoregulation of vasopressin. *Kidney International*, **10**, 25-37.

Robertson, R.M., Susawa, T., Sugiura, M., Haile, V., and Inagami, T. (1990). Circulating endothelin levels are modulated by heart failure in man (Abstract). *Clinical Research*, **38**, 26A.

Robin, E.D., Whaley, R.D., Crump, C.H., and Travis, D.M. (1958). Alveolar gas tensions, pulmonary ventilation and blood pH during physiologic sleep in normal subjects. *The Journal of Clinical Investigation*, **37**, 981-989.

Robinson, B.F., Epstein, S.E., Beiser, G.D., and Braunwald, E. (1966). Control of heart rate by the autonomic nervous system. Studies in man on the interrelation between baroreceptor mechanisms and exercise. *Circulation Research*, **19**, 400-411.

Rocchini, A.P. and Barger, A.C. (1979). Renin release with carotid occlusion in the conscious dog: role of renal artery pressure. *American Journal of Physiology*, **236**, H108-H111.

Roddie, I.C. and Shepherd, J.T. (1956). The reflex nervous control of human skeletal muscle blood vessels. *Clinical Science*, **15**, 433-440.

Roddie, I.C. and Shepherd, J.T. (1957). The effects of carotid artery compression in man with special reference to changes in vascular resistance in the limbs. *The Journal of Physiology*, **139**, 377-384.

Roddie, I.C. and Shepherd, J.T. (1958). Receptors in the high-pressure and low-pressure vascular systems. Their role in the reflex control of the human circulation. *The Lancet*, **1**, 493-496.

Roddie, I.C., Shepherd, J.T., and Whelan, R.F. (1957). Reflex changes in vasoconstrictor tone in human skeletal muscle in response to stimulation of receptors in a low-pressure area of the intrathoracic vascular bed. *The Journal of Physiology*, **139**, 369-376.

Roelandt, J., Klootwijk, P., Lubsen, J., and Janse, M.J. (1984). Sudden death during longterm ambulatory monitoring. *European Heart Journal*, **5**, 7-20.

Roever, G. (1869). *Kritische und exppcrimentelle Untersuchung des Nerveinflusses auf die Erweiterung und Verengerung der Blutgefässe*, pp. 1-71. Koch, Rostock.

Rogge, J.D. and Moore, W.W. (1968). Influence of lower body negative pressure on peripheral venous ADH levels in man. *Journal of Applied Physiology*, **25**, 134-138.

Rosenbaum, M.B. and Lepeschkin, E. (1955). The effect of ventricular systole on auricular rhythm in auriculoventricular block. *Circulation*, **11**, 240-261.

Rosenblueth, A. and Simeone, F.A. (1934). The interrelations of vagal and accelerator effects on the cardiac rate. *American Journal of Physiology*, **110**, 42-55.

Roskam, J. (1930). Un syndrome nouveau: syncopes cardiaques graves et syncopes répétées par hyperréflectivité sino-carotidienne. *La Presse Medicale*, **39**, 590-593.

Ross, A. and Steptoe, A. (1980). Attenuation of the diving reflex in man by mental stimulation. *The Journal of Physiology*, **302**, 387-393.

Ross, C.A., Armstrong, D.M., Ruggiero, D.A., Pickel, V.M., Joh, T.H., and Reis, D.J. (1981). Adrenaline neurons in the rostral ventrolateral medulla innervate thoracic spinal cord: a combined immunocytochemical and retrograde transport demonstration. *Neuroscience Letters*, **25**, 257-262.

Ross, C.A., Ruggiero, D.A., Joh, T.H., Park, D.H., and Reis, D.J. (1983). Adrenaline synthesizing neurons in the rostral ventrolateral medulla: a possible role in tonic vasomotor control. *Brain Research*, **273**, 356-361.

Ross, C.A., Ruggiero, D.A., Joh, T.H., Park, D.H., and Reis, D.J. (1984*a*). Rostral ventrolateral medulla: selective projections to the thoracic autonomic cell column from the region containing C1 adrenaline neurons. *The Journal of Comparative Neurology*, **228**, 168-185.

Ross, C.A. *et al.* (1984*b*). Tonic vasomotor control by the rostral ventrolateral medulla: effect of electrical or chemical stimulation of the area containing C1 adrenaline neurons on arterial pressure, heart rate, and plasma catecholamines and vasopressin. *The Journal of Neuroscience*, **4**, 474-494.

Ross, C.A., Ruggiero, D.A., and Reis, D.J. (1985). Projections from the nucleus tractus solitarii to the rostral ventrolateral medulla. *The Journal of Comparative Neurology*, **242**, 511-534.

Rossberg, F. and Peňáz, J. (1988). Initial cardiovascular response on change of posture from squatting to standing. *European Journal of Applied Physiology*, **57**, 93-97.

Rossi, L. (1954-1955). Atrioventricular conduction system and nerves in the human heart. *Scientia Medica Italica*, **3**, 514-554.

Rothe, C.F. (1979). Reflex control of the veins in cardiovascular function. *The Physiologist*, **22**, 28-35.

Rowell, L.B. (1973). Regulation of splanchnic blood flow in man. *The Physiologist*, **16**, 127-142.

Rowell, L.B. (1974). Human cardiovascular adjustments to exercise and thermal stress. *Physiological Reviews*, **54**, 75-159.

Rowell, L.B. (1986). *Human circulation regulation during physical stress*. Oxford University Press, Oxford.

Rowell, L.B. and Blackmon, J.R. (1987). Human cardiovascular adjustments to acute hypoxaemia. *Clinical Physiology*, **7**, 349-376.

Rowell, L.B. and Sheriff, D.D. (1988). Are muscle 'chemoreflexes' functionally important? *News in Physiological Sciences*, **3**, 250-253.

Rowell, L.B., Wyss, C.R., and Brengelmann, G.L. (1973). Sustained human skin and muscle vasoconstriction with reduced baroreceptor activity. *Journal of Applied Physiology*, **34**, 639-643.

Rowell, L.B., Hermansen, L., and Blackmon, J.R. (1976). Human cardiovascular and respiratory responses to graded muscle ischemia. *Journal of Applied Physiology*, **41**, 693-701.

Rowell, L.B., Freund, P.R., and Hobbs, S.F. (1981). Cardiovascular responses to muscle ischemia in humans. *Circulation Research*, **48** (Suppl. I), I-37–I-47.

Rowell, L.B., Saltin, B., Kiens, B., and Christensen, N.J. (1986). Is peak quadriceps blood flow in humans even higher during exercise with hypoxemia. *American Journal of Physiology*, **251**, H1038-H1044.

Rowell, L.B., Johnson, D.G., Chase, P.B., Comess, K.A., and Seals, D.R. (1989). Hypoxemia raises muscle sympathetic activity but not norepinephrine in resting humans. *Journal of Applied Physiology*, **66**, 1736-1743.

Rudomin, P. (1967). Presynaptic inhibition induced by vagal afferent volleys. *Journal of Neurophysiology*, **30**, 964-981.

Ruggiero, D.A., Meeley, M.P., Anwar, M., and Reis, D.J. (1985*a*). Newly identified GABAergic neurons in regions of the ventrolateral medulla which regulate blood pressure. *Brain Research*, **339**, 171-177.

Ruggiero, D.A., Ross, C.A., Anwar, M., Park, D.H., Joh, T.H., and Reis, D.J. (1985*b*). Distribution of neurons containing phenylethalnolamine *N*-methyltransferase in medulla and hypothalamus of rat. *The Journal of Comparative Neurology*, **239**, 127-154.

Ruggiero, D.A., Cravo, S.L., Arango, V., and Reis, D.J. (1989). Central control of the circulation by the rostral ventrolateral reticular nucleus: anatomical substrates. *Progress in Brain Research*, **81**, 49-79.

Rusch, N.J., Shepherd, J.T., Webb, R.C., and Vanhoutte, P.M. (1981). Different behavior of the resistance vessels of the human calf and forearm during contralateral isometric exercise, mental stress, and abnormal respiratory movements. *Circulation Research*, **46** (Suppl. I), I-118–I-130.

Ruscone, T.G., Lombardi, F., Malfatto, G., and Malliani, A. (1987). Attenuation of baroreceptive mechanisms by cardiovascular sympathetic afferent fibers. *American Journal of Physiology*, **253**, H787-H791.

Ruskin, J., Harley, A., and Greenfield, Jr., J.C. (1968). Pressure flow studies in patients having a pressor response to the Valsalva maneuver. *Circulation*, **38**, 277-281.

Ryan, C., Hollenberg, M., Harvey, D.B., and Gwynn, R. (1976). Impaired parasympathetic responses in patients after myocardial infarction. *The American Journal of Cardiology*, **37**, 1013-1018.

Saavedra, J.M., Palkovits, M., Brownstein, M.J., and Axelrod, J. (1974). Localisation of phenylethanolamine *N*-methyl transferase in the rat brain nuclei. *Nature*, **248**, 695-696.

Sachs, F. (1987). Baroreceptor mechanisms at the cellular level. *Federation Proceedings*, **46**, 12-16.

Sagawa, K. (1978). Concerning 'gain'. *American Journal of Physiology*, **235**, H117.

Sagawa, K. (1983). Baroreflex control of systemic arterial pressure and vascular bed. In *Handbook of physiology. The cardiovascular system*. Volume III. *Peripheral circulation and organ blood flow*, Part **2**, ed. Shepherd, J.T., and Abboud, F.M., pp. 453-496. American Physiological Society, Bethesda.

Saito, M., Mano, T., Abe, H., and Iwase, S. (1986). Responses in muscle sympathetic nerve activity to sustained hand-grips of different tensions in humans. *European Journal of Applied Physiology*, **55**, 493-498.

Saito, M., Mano, T., Iwase, S., Koga, K., Abe, H., and Yamazaki, Y. (1988). Responses in muscle sympathetic activity to acute hypoxia in humans. *Journal of Applied Physiology*, **65**, 1548-1552.

Saito, M., Mano, T., and Iwase, S. (1989). Sympathetic nerve activity related to local fatigue sensation during static contraction. *Journal of Applied Physiology*, **67**, 980-984.

Saito, Y. *et al.* (1987). Clinical application of atrial natriuretic polypeptide in patients with congestive heart failure: beneficial effects on left ventricular function. *Circulation*, **76**, 115-124.

Sakamoto, S. and Liang, C.-s. (1989). Opiate receptor inhibition improves the blunted baroreflex function in conscious dogs with right-sided congestive heart failure. *Circulation*, **80**, 1010-1015.

Salata, J.J. and Jalife, J. (1985). 'Fade' of hyperpolarizing responses to vagal stimulation at the sinoatrial and atrioventricular nodes of the rabbit heart. *Circulation Research*, **56**, 718-727.

Salata, J.J., Gill, R.M., Gilmour, R.F., Jr., and Zipes, D.P. (1986). Effects of sympathetic tone on vagally induced phasic changes in heart rate and atrioventricular node conduction in the anesthetized dog. *Circulation Research*, **58**, 584-594.

Salmoiraghi, G.C. (1962). 'Cardiovascular' neurones in brain stem of cat. *Journal of Neurophysiology*, **25**, 182-197.

Samaan, A. (1935). The antagonistic cardiac nerves and heart rate. *The Journal of Physiology*, **83**, 332-340.

Samet, P., Fritts, H.W., Jr., Fishman, A.P., and Cournand, A. (1957). The blood volume in heart disease. *Medicine*, **36**, 211-235.

Samodelov, L.F., Godehard, E., and Arndt, J.O. (1979). A comparison of the stimulus–response curves of aortic and carotid sinus baroreceptors in decerebrated cats. *Pflügers Archiv für die Gesamte Physiologie des Menschen und der Tiere*, **383**, 47-53.

Sampson, S.R. and Mills, E. (1970). Effects of sympathetic stimulation on discharges of carotid sinus baroreceptors. *American Journal of Physiology*, **218**, 1650-1653.

Samueloff, S.L., Browse, N.L., and Shepherd, J.T. (1966). Response of capacity vessels in human limbs to head-up tilt and suction on lower body. *Journal of Applied Physiology*, **21**, 47-54.

Sander-Jensen, K. (1986*a*). Hypotension induced by passive head-up tilt: endocrine and circulatory mechanisms. *American Journal of Physiology*, **251**, R742-R748.

Sander-Jensen, K., Secher, N.H., Bie, P., Warberg, J., and Schwartz, T.W. (1986*b*). Vagal slowing of the heart during haemorrhage: observations from 20 consecutive hypotensive patients. *British Medical Journal*, **292**, 364-366.

Sanders, J.S. and Ferguson, D.W. (1989). Profound sympathoinhibition complicating hypovolemia in humans. *Annals of Internal Medicine*, **111**, 439-441.

Sanders, J.S., Ferguson, D.W., and Mark, A.L. (1988). Arterial baroreflex control of sympathetic nerve activity during elevation of blood pressure in normal man: dominance of aortic baroreflexes. *Circulation*, **77**, 279-288.

Sanders, J.S., Mark, A.L., and Ferguson, D.W. (1989*a*). Importance of aortic baroreflex in regulation of sympathetic responses during hypotension. Evidence from direct sympathetic nerve recordings in humans. *Circulation*, **79**, 83-92.

Sanders, J.S., Mark, A.L., and Ferguson, D.W. (1989*b*). Evidence for cholinergically mediated vasodilation at the beginning of isometric exercise in humans. *Circulation*, **79**, 815-824.

Sandler, H. (1980). Effects of bedrest and weightlessness on the heart. In *Hearts and heart-like organs*, Vol. **2**, ed. Bourne, G.H., pp. 435-524. Academic Press, New York.

Sanmarco, M.E., Pontius, S., and Selvester, R.H. (1980). Abnormal blood pressure response and marked ischemic ST-segment depression as predictors of severe coronary artery disease. *Circulation*, **61**, 572-578.

Saper, C.B., Loewy, A.D., Swanson, L.W., and Cowan, W.M. (1976). Direct hypothalamo-autonomic connections. *Brain Research*, **117**, 305-312.

Sapru, H.N. and Krieger, A.J. (1979). Role of receptor elements in baroreceptor resetting. *American Journal of Physiology*, **236**, H174-H182.

Sarnoff, S.J. and Yamada, S.I. (1959). Evidence for reflex control of arterial pressure from abdominal receptors with special reference to the pancreas. *Circulation Research*, **7**, 325-335.

Sarnoff, S.J., Hardenbergh, E., and Whittenberger, J.L. (1948). Mechanism of the arterial pressure response to the Valsalva test: the basis for its use as an indicator of the intactness of the sympathetic outflow. *American Journal of Physiology*, **154**, 316-327.

Sato, A. and Schmidt, R.F. (1973). Somatosympathetic reflexes: afferent fibers, central pathways, discharge characteristics. *Physiological Reviews*, **53**, 916-947.

Saul, J.P., Albrecht, P., Berger, R.D., and Cohen, R.J. (1988*a*). Analysis of long term heart rate variability: methods, 1/f scaling and implications. *Computers in Cardiology*, 419-422.

Saul, J.P., Arai, Y., Berger, R.D., Lilly, L.S., Colucci, W.S., and Cohen, R.J. (1988*b*). Assessment of autonomic regulation in chronic congestive heart failure by heart rate spectral analysis. *The American Journal of Cardiology*, **61**, 1292-1299.

Saul, J.P., Berger, R.D., Chen, M.H., and Cohen, R.J. (1989). Transfer function analysis of autonomic regulation. II. Respiratory sinus arrhythmia. *American Journal of Physiology*, **256**, H153-H161.

Saum, W.R., Brown, A.M., and Tuley, F.H. (1976). An electrogenic sodium pump and baroreceptor function in normotensive and spontaneously hypertensive rats. *Circulation Research*, **39**, 497-505.

Saunders, D.E., Jr. and Ord, J.W. (1962). The hemodynamic effects of paroxysmal supraventricular tachycardia in patients with the Wolff–Parkinson–White syndrome. *The American Journal of Cardiology*, **9**, 223-236.

Savard, G.K., Richter, E.A., Strange, S., Kiens, B., Christensen, N.J., and Saltin, B. (1989). Norepinephrine spillover from skeletal muscle during exercise in humans: role of muscle mass. *American Journal of Physiology*, **257**, H1812-H1818.

Sawchenko, P.E. and Swanson, L.W. (1982). The organization of noradrenergic pathways from the brainstem to the paraventricular and supraoptic nuclei in the rat. *Brain Research Reviews*, **4**, 275-325.

Sayers, B.McA., Cicchiello, L.R., Raftery, E.B., Mann, S., and Green, H. (1982). The assessment of continuous ambulatory blood pressure records. *Medical Informatics*, **7**, 93-108.

Scarpa, A. (1794). *Tabulae nevrologicae ad illustrandam historiam anatomicam cardiacorum nervorum, noni nervorum cerebri, glossopharyngaei et pharyngaei ex octavo cerebri.* Balthassarem Comini, Ticini.

Schaal, S.F., Wallace, A.G., and Sealy, W.C. (1969). Protective influence of cardiac denervation against arrhythmias of myocardial infarction. *Cardiovascular Research*, **3**, 241-244.

Schäfer,–. (1877). Ueber die aneurysmatische Erweiterung der Carotis interna an ihrem Ursprung. *Allgemeine Zeitschrift für Psychiatrie und Psychisch-gerichtliche Medicin*, **34**, 438-451.

Scher, A.M. and Ito, C.S. (1980). Hypertension is produced by arterial baroreceptor denervation in the dog. In *Arterial baroreceptors and hypertension*, ed. Sleight, P., pp. 405-410. Oxford University Press, Oxford.

Scherrer, U., Pryor, S.L., Bertocci, L.A., and Victor, R.G. (1990*a*). Arterial baroreceptor buffering of sympathetic activation during exercise-induced elevations of arterial pressure. *The Journal of Clinical Investigation*, **86**, 1855-1861.

Scherrer, U., Vissing, S., Morgan, B.J., Hanson, P., and Victor, R.G. (1990*b*). Vasovagal syncope after infusion of a vasodilator in a heart-transplant recipient. *The New England Journal of Medicine*, **322**, 602-604.

Schiebinger, R.J. and Linden, J. (1986). The influence of resting tension on immunoreactive atrial natriuretic peptide secretion by rat atria superfused *in vitro*. *Circulation Research*, **59**, 105-109.

Schiff, M. (1853). Ueber den Einfluss der Nerven auf die Gefässe der Zunge. *Archiv für Physiologische Heilkunde*, **12**, 377-392.

Schiffrin, E.L. (1988). Decreased density of binding sites for atrial natriuretic peptide on platelets of patients with severe congestive heart failure. *Clinical Science*, **74**, 213-218.

Schiffrin, E.L., Gutkowska, J., Kuchel, O., Cantin, M., and Genest, J. (1985). Plasma concentration of atrial natriuretic factor in a patient with paroxysmal atrial tachycardia. *The New England Journal of Medicine*, **312**, 1196-1197.

Schlager, E. and Meier, T. (1947). A strange Balinese method of inducing sleep (with some notes about balyans). *Acta Tropica*, **4**, 127-134.

Schlepper, M., Weppner, H.G., and Merle, H. (1978). Haemodynamic effects of supraventricular tachycardias and their alterations by electrically and verapamil induced termination. *Cardiovascular Research*, **12**, 28-33.

Schlomka, G. (1937). Untersuchungen über die physiologische Unregelmässigkeit des Herzschlages. III. Mitteilung. Über die Abhängigkeit der respiratorischen (Ruhe-)Arrhythmie von der Schlagfrequenz und vom Lebensalter. *Zeitschrift für Kreislaufforschung*, **29**, 510-524.

Schmedtje, J.F., Jr. and Eckberg, D.L. (1991). Preservation of oscillations in postocclusive reactive hyperemia. *Journal of Applied Physiology*, **70**, 363-367.

Schmidt, C.F. (1932). Carotid sinus reflexes to the respiratory center. I. Identification. *American Journal of Physiology*, **102**, 94-118.

Schmitt, F.O. (1984). Molecular regulators of brain function: a new view. *Neuroscience*, **13**, 991-1001.

Schneider, R.A. and Costiloe, J.P. (1965). Relationship of sinus arrythmia to age and its prognostic significance in ischemic heart disease (Abstract). *Clinical Research*, **13**, 219.

Schneider, R.A. and Costiloe, J.P. (1966). Level and variability of serial resting blood pressures in subjects with and without ischemic heart disease. *Angiology*, **17**, 201-211.

Schneyer, K. (1935*a*). Der pressorezeptorische Herzreflex. *Zeitschrift für Kreislaufforschung*, **7**, 217-226.

Schneyer, K. (1935*b*). Zur Frage der Blutdruckcharakteristik. *Pflügers Archiv für die Gesamte Physiologie des Menschen und der Tiere*, **235**, 176-183.

Scholander, P.F. (1940). Experimental investigations on the respiratory function in diving mammals and birds. *Hvalrådets Skrifter*, **22**, 1-131.

Scholander, P.F., Hammel, H.T., LeMessurier, H., Hemmingsen, E., and Garey, W. (1962). Circulatory adjustment in pearl divers. *Journal of Applied Physiology*, **17**, 184-190.

Scholander, P.F., Hargens, A.R., and Miller, S.L. (1968). Negative pressure in the interstitial fluid of animals. *Science*, **161**, 321-328.

Schömig, A. (1990). Catecholamines in myocardial ischemia. Systemic and cardiac release. *Circulation*, **82** (Suppl II), II-13–II-22.

Schömig, A., Ness, G., Mayer, E., Katus, H., and Dietz, R. (1984). Sympathetic activity in patients with acute myocardial infarction before and after intracoronary thrombolytic therapy (Abstract). *European Heart Journal*, **5** (Suppl. 1), 39.

Schraufek, S.R. *et al.* (1990). Enhancement of respiratory sinus arrhythmia by moderate exercise. In *Chronobiology: its role in clinical medicine, general biology, and agriculture*, Part A, ed. Hayes, D.K., Pauly, J.E., and Reiter, R.J., pp. 283-296. Wiley-Liss, New York.

Schuessler, R.B., Canavan, T.E., Boineau, J.P., and Cox, J.L. (1988). Baroreflex modulation of heart rate and initiation of atrial activation in dogs. *American Journal of Physiology*, **255**, H503-H513.

Schultz, H.D., Pisarri, T.E., Coleridge, H.M., and Coleridge, J.C.G. (1987). Carotid sinus baroreceptors modulate tracheal smooth muscle tension in dogs. *Circulation Research*, **60**, 337-345.

Schultzberg, M. and Lindh, B. (1988). Transmitters and peptides in autonomic ganglia. In *Handbook of chemical neuroanatomy*, Volume 6: *The peripheral nervous system*, ed. Björklund, A., Hökfelt, T., and Owman, C., pp. 297-326. Elsevier, Amsterdam.

Schwaber, J.S., Kapp, B.S., and Higgins, G. (1980). The origin and extent of direct amygdala projections to the region of the dorsal motor nucleus of the vagus and the nucleus of the solitary tract. *Neuroscience Letters*, **20**, 15-20.

Schwartz, J. and Reid, I.A. (1981). Effect of vasopressin blockade on blood pressure regulation during hemorrhage in conscious dogs. *Endocrinology*, **109**, 1778-1780.

Schwartz, L.S., Goldfischer, J., Sprague, G.J., and Schwartz, S.P. (1969). Syncope and sudden death in aortic stenosis. *The American Journal of Cardiology*, **23**, 647-658.

Schwartz, P.J. (1985). Idiopathic long QT syndrome: Progress and questions. *American Heart Journal*, **109**, 399-411.

Schwartz, P.J. and Stone, H.L. (1985). The analysis and modulation of autonomic reflexes in the prediction and prevention of sudden death. In *Cardiac electrophysiology and arrhythmias*, ed. Zipes, D.P., and Jalife, J., pp. 165-176. Grune and Stratton, New York.

Schwartz, P.J. and Zaza, A. (1986). The rational basis and the clinical value of selective cardiac sympathetic denervation in the prevention of malignant arrhythmias. *European Heart Journal*, **7** (Suppl. A), 107-118.

Schwartz, P.J., Pagani, M., Lombardi, F., Malliani, A., and Brown, A.M. (1973). A cardiocardiac sympathovagal reflex in the cat. *Circulation*, **32**, 215-220.

Schwartz, P.J., Vanoli, E., Stramba-Badiale, M., De Ferrari, G.M., Billman, G.E., and Foreman, R.D. (1988*a*). Autonomic mechanisms and sudden death. New insights from analysis of baroreceptor reflexes in conscious dogs with and without a myocardial infarction. *Circulation*, **78**, 969-979.

Schwartz, P.J., Zaza, A., Pala, M., Locati, E., Beria, G., and Zanchetti, A. (1988*b*). Baroreflex sensitivity and its evolution during the first year after myocardial infarction. *Journal of the American College of Cardiology*, **12**, 629-636.

Schwartz, P.J., Zaza, A., Locati, E., and Moss, A.J. (1991). Stress and sudden death. The case of the long QT syndrome. *Circulation*, **83** (Suppl II), II-71–II-80.

Schwartz, S.I., Griffith, L.S.C., Neistadt, A., and Hagfors, N. (1967). Chronic carotid sinus nerve stimulation in the treatment of essential hypertension. *American Journal of Surgery*, **114**, 5-15.

Schweitzer, A. (1935). Zur Frage der respiratorischen Arrhythmie. *Verhandlungen der Deutschen Gesellschaft für Kreislaufforschung*, **8**, 148-154.

Schweitzer, A. (1937). *Die Irradiation autonomer Reflexe. Untersuchungen zur Funktion des autonomen Nervensystems*, pp. 286-288. Verlag von S. Karger, Basel.

Scott, A.R., Bennett, T., and MacDonald, I.A. (1988). Effects of hyperinsulinaemia on the cardiovascular responses to graded hypovolaemia in normal and diabetic subjects. *Clinical Science*, **75**, 85-92.

Scroop, G.C., Katic, F., Joy, M.D., and Lowe, R.D. (1971). Importance of central vasomotor effects in angiotensin-induced hypertension. *British Medical Journal*, **1**, 324-326.

Seagard, J.L., Van Brederode, J.F.M., Dean, C., Hopp, F.A., Gallenberg, L.A., and Kampine, J.P. (1990). Firing characteristics of single-fiber carotid sinus baroreceptors. *Circulation Research*, **66**, 1499–1509.

Seals, D.R., Chase, P.B., and Taylor, J.A. (1988). Autonomic mediation of the pressor responses to isometric exercise in humans. *Journal of Applied Physiology*, **64**, 2190-2196.

Secher, N.J., Thomsen, A., and Arnsbo, P. (1977). Measurement of rapid changes in cardiac stroke volume. An evaluation of the impedance cardiography method. *Acta Anaesthesiologica Scandinavica*, **21**, 353-358.

Sedgwick, A.W., Craig, R.J., Crouch, R., and Dowling, B. (1974). The effects of physical training on the day and night long-term heart-rates of middle-aged men. *European Journal of Applied Physiology*, **33**, 307-314.

Seller, H. (1973). The discharge pattern of single units in thoracic and lumbar white rami in relation to cardiovascular events. *Pflügers Archiv für die Gesamte Physiologie des Menschen und der Tiere*, **343**, 317-330.

Seller, H. and Illert, M. (1969). The localization of the first synapse in the carotid sinus baroreceptor reflex pathway and its alteration of the afferent input. *Pflügers Archiv für die Gesamte Physiologie des Menschen und der Tiere*, **306**, 1-19.

Seller, H., Langhorst, P., Richter, D., and Koepchen, H.P. (1968). Über die Abhängigkeit der pressoreceptorischen Hemmung des Sympathicus von der Atemphase und ihre Auswirkung in der Vasomotorik. *Pflügers Archiv für die Gesamte Physiologie des Menschen und der Tiere*, **302**, 300-314.

Selman, A., McDonald, A., Kitney, R., and Linkens, D. (1982). The interaction between heart rate and respiration. Part I — Experimental studies in man. *Automedica*, **4**, 131-139.

Sever, P.S., Birch, M., Osikowska, B., and Tunbridge, R.D.G. (1977). Plasma-noradrenaline in essential hypertension. *The Lancet*, **1**, 1078-1081.

Severs, W.B. and Daniels-Severs, A.E. (1973). Effects of angiotensin on the central nervous system. *Pharmacological Reviews*, **25**, 415-449.

Shade, R.E., Bishop, V.S., Haywood, J.R., and Hamm, C.K. (1990). Cardiovascular and neuroendocrine responses to baroreceptor denervation in baboons. *American Journal of Physiology*, **258**, R930-R938.

Shannon, D.C., Carley, D.W., and Benson, H. (1987). Aging of modulation of heart rate. *American Journal of Physiology*, **253**, H874-H877.

Share, L. (1967). Role of peripheral receptors in the increased release of vasopressin in response to hemorrhage. *Endocrinology*, **81**, 1140-1146.

Sharpey-Schafer, E.P. (1955). Effects of Valsalva's manoeuvre on the normal and failing circulation. *British Medical Journal*, **1**, 693-695.

Sharpey-Schafer, E.P. (1956). Circulatory reflexes in chronic disease of the afferent nervous system. *The Journal of Physiology*, **134**, 1-10.

Sharpey-Schafer, E.P. (1965). Effect of respiratory acts on the circulation. In *Handbook of physiology*. Section 2: *Circulation*. Volume III, ed. Hamilton, W.F. and Dow, P., pp. 1875-1886. American Physiological Society, Washington, D.C.

Shenker, Y., Sider, R.S., Ostafin, E.A., and Grekin, R.J. (1985). Plasma levels of immunoreactive atrial natriuretic factor in healthy subjects and patients with edema. *The Journal of Clinical Investigation*, **76**, 1684-1687.

Shepherd, A.M.M., Lin, M.-S., Keeton, T.K., and McNay, J.L. (1981). Plasma noradrenaline as a measure of baroreflex sensitivity in hypertensive man. *Clinical Science*, **61**, 165s-168s.

Sheriff, D.D., O'Leary, D.S., Scher, A.M., and Rowell, L.B. (1990). Baroreflex attenuates pressor response to graded muscle ischemia in exercising dogs. *American Journal of Physiology*, **258**, H305-H310.

Shimada, K., Kitazumi, T., Ogura, H., Sadakane, N., and Ozawa, T. (1986). Effects of age and blood pressure on the cardiovascular responses to the Valsalva maneuver. *Journal of the American Geriatric Society*, **34**, 431-434.

Shin, H.S., Hulbert, W.C., and Biggs, D.F. (1987). Observations on the fine structure of the baroreceptors and adrenergic innervation of the guinea-pig carotid sinus. *Journal of Morphology*, **194**, 65-74.

Shubrooks, S.J., Jr. (1972). Carotid sinus counterpressure as a baroreceptor stimulus in the intact dog. *Journal of Applied Physiology*, **32**, 12-19.

Siciliano,–. (1900). Les effets de la compression des carotides sur la pression, sur le coeur et sur la respiration. *Archives Italiennes de Biologie*, **33**, 338-344.

Sieber, G., Kissling, G., Miller, D.T., and Jacob, R. (1973). Indirekter Nachweis parasympathischer Nervenfasern im Ventrikelmyokard des Menschen. *Naunyn-Schmiedeberg's Archives of Pharmacology*, **276**, 211-221.

Siemers, E.R., Rea, M.A., Felten, D.L., and Aprison, M.H. (1982). Distribution and uptake of glycine, glutamate and γ-aminobutyric acid in the vagal nuclei and eight other regions of the rat medulla oblongata. *Neurochemical Research*, **7**, 455-468.

Sigler, L.H. (1935-1936). Further observations on the carotid sinus reflex. *Annals of Internal Medicine*, **9**, 1380-1392.

Sigler, L.H. (1941). Hyperactive cardioinhibitory carotid sinus reflex. *Archives of Internal Medicine*, **67**, 177-193.

Sigler, L.H. (1942). The hyperactive cardioinhibitory carotid-sinus reflex as an aid in the diagnosis of coronary disease. *The New England Journal of Medicine*, **226**, 46-51.

Silverberg, A.B., Shah, S.D., Haymond, M.W., and Cryer, P.E. (1978). Norepinephrine: hormone and neurotransmitter in man. *American Journal of Physiology*, **234**, E252-E256.

Simon, A.C., Safar, M.E., Weiss, Y.A., London, G.M., and Milliez, P.I. (1977).

Baroreflex sensitivity and cardiopulmonary blood volume in normotensive and hypertensive patients. *British Heart Journal*, **39**, 799-805.

Simon, E. and Riedel, W. (1975). Diversity of regional sympathetic outflow in integrative cardiovascular control: patterns and mechanisms. *Brain Research*, **87**, 323-333.

Simon, G., Kiowski, W., and Julius, S. (1977). Effect of beta adrenoceptor antagonists on baroreceptor reflex sensitivity in hypertension. *Clincal Pharmacology and Therapeutics*, **22**, 293-298.

Singer, D.R.J., Shore, A.C., Markandu, N.D., Buckley, M.G., Sagnella, G.A., and MacGregor, G.A. (1987). Dissociation between plasma atrial natriuretic peptide levels and urinary sodium excretion after intravenous saline infusion in normal man. *Clinical Science*, **73**, 285-289.

Sinoway, L. and Prophet, S. (1990). Skeletal muscle metaboreceptor stimulation opposes peak metabolic vasodilation in humans. *Circulation Research*, **66**, 1576-1584.

Sinoway, L.I. *et al.* (1988). Sympathetic tone affects human limb vascular resistance during a maximal metabolic stimulus. *American Journal of Physiology*, **255**, H937-H946.

Skagen, K., Henriksen, O., and Bonde-Petersen, F. (1981). Effect of lower body negative pressure upon local regulation of blood flow in human subcutaneous tissue. *Acta Physiologica Scandinavica*, **111**, 113-120.

Skinner, S.L., McCubbin, J.W., and Page, I.H. (1964). Control of renin secretion. *Circulation Research*, **15**, 64-76.

Skinner, J.E. and Reed, J.C. (1981). Blockade of frontocortical-brain stem pathway prevents ventricular fibrillation of ischemic hearts. *American Journal of Physiology*, **240**, H156-H163.

Skoog, K.M. and Mangiapane, M.L. (1988). Area postrema and cardiovascular regulation in rats. *American Journal of Physiology*, **254**, H963-H969.

Skowsky, W.R., Rosenbloom, A.A., and Fisher, D.A. (1974). Radioimmunoassay measurement of arginine vasopressin in serum: development and application. *Journal of Clinical Endocrinology and Metabolism*, **38**, 278-287.

Sleight, P. (1964). A cardiovascular depressor reflex from the epicardium of the left ventricle in the dog. *The Journal of Physiology*, **173**, 321-343.

Sleight, P. (1976). Reflex cardiac arrest in tetraplegia. *The Lancet*, **1**, 253.

Sleight, P. (1979). Reflex control of the heart. *The American Journal of Cardiology*, **44**, 889-894.

Sleight, P. (1980). *Arterial baroreceptors and hypertension*. Oxford University Press, Oxford.

Sleight, P. (1981). Cardiac vomiting. *British Heart Journal*, **46**, 5-7.

Sleight, P. and Widdicombe, J.G. (1965). Action potentials in fibres from receptors in the epicardium and myocardium of the dog's left ventricle. *The Journal of Physiology*, **181**, 235-258.

Sleight, P., West, M.J., Korner, P.I., Oliver, J.R., Chalmers, J.P., and Robinson, J.L. (1975). The action of clonidine on the baroreflex control of heart rate in conscious animals and man, and on single aortic baroreceptor discharge in the rabbit. *Archives Internationales de Pharmacodynamie et de Thérapie*, **214**, 4-11.

Sleight, P., Robinson, J.L., Brooks, D.E., and Rees, P.M. (1977). Characteristics of single carotid sinus baroreceptor fibers and whole nerve activity in the normotensive and the renal hypertensive dog. *Circulation Research*, **41**, 750-758.

Sleight, P. *et al.* (1979). Baroreflex control of blood pressure and plasma noradrenaline during exericse in essential hypertension. *Clinical Science*, **57**, 169s-171s.

Slutsky, A.S. and Rebuck, A.S. (1978). Heart rate response to isocapnic hypoxia in conscious man. *American Journal of Physiology*, **234**, H129-H132.

Smirk, F.H. and Palmer, D.G. (1960). A myocardial syndrome with particular reference to the occurrence of sudden death and of premature systoles interrupting antecedent T waves. *The American Journal of Cardiology*, **6**, 620-629.

Smith, F.M. and Jones, D.R. (1990). Effects of acute and chronic baroreceptor denervation on diving responses in ducks. *American Journal of Physiology*, **258**, R895-R902.

Smith, M.L., Ellenbogen, K.A., Eckberg, D.L., Szentpetery, S., and Thames, M.D. (1989). Subnormal heart period variability in heart failure: Effect of cardiac transplantation. *Journal of the American College of Cardiology*, **14**, 106-111.

Smith, M.L., Ellenbogen, K.A., Eckberg, D.L., Sheehan, H.M., and Thames, M.D. (1990). Subnormal parasympathetic activity after cardiac transplantation. *The American Journal of Cardiology*, **66**, 1243-1246.

Smith, M.L., Ellenbogen, K.A., Eckberg, D.L., and Beightol, L.A. (1991). Human sympathetic neural responses to induced ventricular tachycardia. *Journal of the American College of Cardiology*, 18, 1015–1024

Smith, O.A., Astley, C.A., DeVito, J.L., Stein, J.M., and Walsh, K.E. (1980). Functional analysis of hypothalamic control of the cardiovascular responses accompanying emotional behavior. *Federation Proceedings*, **39**, 2487-2494.

Smith, S.A., Stallard, T.J., and Littler, W.A. (1986). Estimation of sinoaortic baroreceptor heart rate reflex sensitivity and latency in man: a new microcomputer assisted method of analysis. *Cardiovascular Research*, **20**, 877-882.

Smith, S.A., Salih, M.M., and Littler, W.A. (1987*a*). Assessment of beat to beat changes in cardiac output during the Valsalva manoeuvre using electrical bioimpedance cardiography. *Clinical Science*, **72**, 423-428.

Smith, S.A., Stallard, T.J., Salih, M.M., and Littler, W.A. (1987*b*). Can sinoaortic baroreceptor heart rate reflex sensitivity be determined from phase IV of the Valsalva manoeuvre? *Cardiovascular Research*, **21**, 422-427.

Smith, W.C., Burwell, C.S., and DeVito, M.J. (1929). The effect of atropine upon the output of the hearts of normal men. *The Journal of Clinical Investigation*, **6**, 237-245.

Smyth, H.S., Sleight, P., and Pickering, G.W. (1969). Reflex regulation of arterial pressure during sleep in man. A quantitative method of assessing baroreflex sensitivity. *Circulation Research*, **24**, 109-121.

Snyder, M.D., DeBoard, J.W., Beger, T.H., and Gibbons, J.J. (1989). Anatomy of the common peroneal nerve at the knee: a cadaver study (Abstract). *Regional Anesthesia*, **14** (Suppl.), 38.

Sofroniew, M.V., Weindl, A., Schrell, U., and Wetzstein, R. (1981). Immunohistochemistry of vasopressin, oxytocin and neurophysin in the hypothalamus and extrahypothalamic regions of the human and primate brain. *Acta Histochemica*, **24** (Suppl.), 79-95.

Sollman, T. and Brown, E.D. (1912). The blood pressure fall produced by traction on the carotid artery. *American Journal of Physiology*, **30**, 88-104.

SOLVD Investigators. (1991). Effect of enalapril on survival in patients with reduced left ventricular ejection fractions and congestive heart failure. *The New England Journal of Medicine*, **325**, 293–302.

Somana, R. and Walberg, F. (1979). The cerebellar projection from the parabrachial nucleus in the cat. *Brain Research*, **172**, 144-149.

Somers, V.K., Conway, J., LeWinter, M., and Sleight, P. (1985). The role of baroreflex sensitivity in post-exercise hypotension. *Journal of Hypertension*, **3** (Suppl. 3), S129-S130.

Somers, V.K., Conway, J., and Sleight, P. (1986). The effect of physical training on home blood pressure measurements in normal subjects. *Journal of Hypertension*, **4** (Suppl. 6), S657-S658.

Somers, V.K., Mark, A.L., and Abboud, F.M. (1988*a*). Potentiation of sympathetic nerve responses to hypoxia in borderline hypertensive subjects. *Hypertension*, **11**, 608-612.

Somers, V.K., Mark, A.L., and Abboud, F.M. (1988*b*). Sympathetic activation by hypoxia and hypercapnia — implications for sleep apnea. *Clinical and Experimental Hypertension. Part A — Theory and Practice*, **A10** (Suppl. 1), 413-422.

Somers, V.K., Mark, A.L., Zavala, D.C., and Abboud, F.M. (1989*a*). Contrasting effects of hypoxia and hypercapnia on ventilation and sympathetic activity in humans. *Journal of Applied Physiology*, **67**, 2101-2106.

Somers, V.K., Mark, A.L., Zavala, D.C., and Abboud, F.M. (1989*b*). Influence of ventilation and hypocapnia on sympathetic nerve responses to hypoxia in normal humans. *Journal of Applied Physiology*, **67**, 2095-2100.

Somers, V.K., Conway, J., Johnston, J., and Sleight, P. (1991). Effects of endurance training on baroreflex sensitivity and blood pressure in borderline hypertension. *The Lancet*, **337**, 1363-1368.

Sonnenberg, H. and Veress, A.T. (1984). Cellular mechanism of release of atrial natriuretic factor. *Biochemical and Biophysical Research Communications*, **124**, 443-449.

Sopher, S.M., Smith, M.L., Eckberg, D.L., Fritsch, J.M., and Dibner-Dunlap, M.E. (1990). Autonomic pathophysiology in heart failure: carotid baroreceptor–cardiac reflexes. *American Journal of Physiology*, **259**, H689-H696.

Spear, J.F. and Moore, E.N. (1973). Influence of brief vagal and stellate nerve stimulation on pacemaker activity and conduction within the atrioventricular conduction system of the dog. *Circulation Research*, **32**, 27-41.

Spear, J.F., Kronhaus, K.D., Moore, E.N., and Kline, R.P. (1979). The effect of brief vagal stimulation on the isolated rabbit sinus node. *Circulation Research*, **44**, 75-88.

Sprangers, R.L.H. (1990). On the role of cardiopulmonary receptors at the onset of muscular exercise (Ph.D. thesis), Rodopi, Amsterdam.

Sprangers, R.L.H., Wesseling, K.H., Imholz, A.L.T., Imholz, B.P.M., and Wieling, W. (1991). Initial blood pressure fall on stand up and exercise explained by changes in total peripheral resistance. *Journal of Applied Physiology*, **70**, 523-530.

Sprenkle, J.M., Eckberg, D.L., Goble, R.L., Schelhorn, J.J., and Halliday, H.C. (1986). Device for rapid quantification of human carotid baroreceptor–cardiac reflex responses. *Journal of Applied Physiology*, **60**, 727-732.

Spyer, K.M. (1972). Baroreceptor sensitive neurones in the anterior hypothalamus of the cat. *The Journal of Physiology*, **224**, 245-257.

Spyer, K.M. (1981). Neural organisation and control of the baroreceptor reflex. *Reviews of Physiology, Biochemistry, and Experimental Pharmacology*, **88**, 23-124.

Spyer, K.M. and Jordan, D. (1987). Electrophysiology of the nucleus ambiguus. In *Cardiogenic reflexes*, ed. Hainsworth, R., McWilliam, P.N., and Mary, D.A.S.G., pp. 237-249. Oxford University Press, Oxford.

Spyer, K.M., Donoghue, S., Felder, R.B., and Jordan, D. (1984). Processing of afferent inputs in cardiovascular control. *Clinical and Experimental Hypertension. Part A — Theory and Practice*, **A6**, 173-184.

Stanton, M.S. *et al.* (1989). Regional sympathetic denervation after myocardial infarction in humans detected noninvasively using I-123-metaiodobenzylguanidine. *Journal of the American College of Cardiology*, **14**, 1519-1526.

Stanton, P.D. and Hinrichsen, C.F.L. (1980). Monoaminergic innervation of the carotid sinus in mammals: a histochemical study. *Acta Anatomica*, **108**, 34-44.

Starling, E.H. (1897). Points on pathology of heart disease. *The Lancet*, **1**, 669-655.

Staszewska-Barczak, J. (1971). The reflex stimulation of catecholamine secretion during the acute stage of myocardial infarction in the conscious dog. *Clinical Science*, **41**, 419-439.

Stefadouros, M.A., Shahawy, M.E., Stefadouros, F., and Witham, A.C. (1975). The effect of upright tilt on the volume of the failing human left ventricle. *American Heart Journal*, **90**, 735-743.

Stegemann, J. and Müller-Bütow, H. (1966). Zur regeltheoretischen Analyse des Blutkreislaufes. I. Die zentrale Verrechnung der Signale aus den einzelnen Pressoreceptorenfeldern. *Pflügers Archiv für die Gesamte Physiologie des Menschen und der Tiere*, **287**, 247-256.

Stegemann, J. and Tibes, U. (1969). Sinusoidal stimulation of carotid sinus baroreceptors and peripheral blood pressure in dogs. *Annals of the New York Academy of Sciences*, **156**, 787-795.

Stegemann, J., Baer, F.M., and Hoffman, U. (1988). The Valsalva maneuver as an indirect, non-invasive indicator of central blood volume shift. *Aviation, Space, and Environmental Medicine*, **59**, 422-427.

Stein, E., Damato, A.N., Kosowsky, B.D., Lau, S.H., and Lister, J.W. (1966). The relation of heart rate to cardiovascular dynamics. Pacing by atrial electrodes. *Circulation*, **33**, 925-932.

Stein, M., Schwartz, R., and Mirsky, I.A. (1954). The antidiuretic activity of plasma of patients with hepatic cirrhosis, congestive heart failure, hypertension and other clinical disorders. *The Journal of Clinical Investigation*, **33**, 77-81.

Steinhaus, A.H. (1933). Chronic effects of exercise. *Physiological Reviews*, **13**, 103-147.

Stella, A., Dampney, R.A.L., Golin, R., and Zanchetti, A. (1978). Afferent vagal control of renin release in the anesthetized cat. *Circulation Research*, **43** (Suppl. I), I-107–I-111.

Stella, G. (1931). Some observations on the effect of pressure in the carotid sinus upon the arterial pressure and upon the coronary circulation. *The Journal of Physiology*, **73**, 45-53.

Stephenson, R.B. and Tagett, M.G. (1984). Sympathetic and parasympathetic contributions to carotid baroreflexes in angiotensin-induced hypertension (Abstract). *Federation Proceedings*, **43**, 443.

Steptoe, A. and Sawada, Y. (1989). Assessment of baroreceptor reflex function during mental stress and relaxation. *Psychophysiology*, **26**, 140-147.

Stevens, P.M. and Lamb, L.E. (1965). Effects of lower body negative pressure on the cardiovascular system. *The American Journal of Cardiology*, **16**, 506-515.

Stiles, G.L., Strasser, R.H., Lavin, T.N., Jones, L.R., Caron, M.G., and Lefkowitz, R.J. (1983). The cardiac β-adrenergic receptor. Structural similarities of β_1 and β_2 receptor subtypes demonstrated by photoaffinity labeling. *The Journal of Biological Chemistry*, **258**, 8443-8449.

Stinnett, H.O., Bishop, V.S., and Peterson, D.F. (1976). Reduction in baroreflex cardiovascular responses due to venous infusion in the rabbit. *Circulation Research*, **39**, 766-772.

Stinson, E.B., Griepp, R.B., Schroeder, J.S., Dong, E., Jr., and Shumway, N.E. (1972). Hemodynamic observations one and two years after cardiac transplantation in man. *Circulation*, **45**, 1183-1194.

Stjernberg, L. and Wallin, B.G. (1983). Sympathetic neural outflow in spinal man. A preliminary report. *Journal of the Autonomic Nervous System*, **7**, 313-318.

Stjernberg, L., Blumberg, H., and Wallin, B.G. (1986). Sympathetic activity in man after spinal cord injury. Outflow to muscle below the lesion. *Brain*, **109**, 695-715.

Stone, D.J., Lyon, A.F., and Teirstein, A.S. (1965). A reappraisal of the circulatory effects of the Valsalva maneuver. *The American Journal of Medicine*, **39**, 923-933.

Stone, G.W., Kubo, S.H., and Cody, R.J. (1986). Adverse influence of baroreceptor dysfunction on upright exercise in congestive heart failure. *The American Journal of Medicine*, **80**, 799-802.

Stone, H.L., Dormer, K.J., Foreman, R.D., Thies, R., and Blair, R.W. (1985). Neural regulation of the cardiovascular system during exercise. *Federation Proceedings*, **44**, 2271-2278.

Stornetta, R.L., Morrison, S.F., Ruggiero, D.A., and Reis, D.J. (1989). Neurons of rostral ventrolateral medulla mediate somatic pressor reflex. *American Journal of Physiology*, **256**, R448-R462.

Strandell, T. and Shepherd, J.T. (1967). The effect in humans of increased sympathetic activity on the blood flow to active muscles. *Acta Medica Scandinavica*, Suppl. **472**, 146-167.

Strasberg, B., Sagie, A., Erdman, S., Kusniec, J., Sclarovsky, S., and Agmon, J. (1989). Carotid sinus hypersensitivity and the carotid sinus syndrome. *Progress in Cardiovascular Diseases*, **31**, 379-391.

Stratton, J.R., Pfeifer, M.A., and Halter, J.B. (1987). The hemodynamic effects of sympathetic stimulation combined with parasympathetic blockade in man. *Circulation*, **75**, 922-929.

Stroh-Werz, M., Langhorst, P., and Camerer, H. (1976). Neuronal activity with relation to cardiac rhythm in the lower brain stem of the dog. *Brain Research*, **106**, 293-305.

Stroh-Werz, M., Langhorst, P., and Camerer, H. (1977). Neuronal activity with cardiac rhythm in the nucleus of the solitary tract in cats and dogs. II. Activity modulation in relation to the respiratory cycle. *Brain Research*, **133**, 81-93.

Stucki, P., Hatcher, J.D., Judson, W.E., and Wilkins, R.W. (1955). Studies of circulation time during the Valsalva test in normal subjects and in patients with congestive heart failure. *Circulation*, **11**, 900-908.

Stuesse, S.L. and Fish, S.E. (1984). Projections to the cardioinhibitory region of the nucleus ambiguus of rat. *The Journal of Comparative Neurology*, **229**, 271-278.

Stuesse, S.L., Wallick, D.W., Zieske, H., and Levy, M.N. (1981). Changes in vagal phasic chronotropic responses with sympathetic stimulation in the dog. *American Journal of Physiology*, **241**, H850-H856.

Sugrue, D.D., Gersh, B.J., Holmes, D.R., Jr., Wood, D.L., Osborn, M.J., and Hammill, S.C. (1986). Sympatomatic 'isolated' carotid sinus hypersensitivity: natural history and results of treatment with anticholinergic drugs or pacemaker. *Journal of the American College of Cardiology*, **7**, 158-162.

Sullebarger, J.T., Liang, C.-s., Woolf, P.D., Willick, A.E., and Richeson, J.F. (1990). Comparison of phenylephrine bolus and infusion methods in baroreflex measurements. *Journal of Applied Physiology*, **69**, 962-967.

Sullivan, M.J., Knight, J.D., Higginbotham, M.B., and Cobb, F.R. (1989). Relation between central and peripheral hemodynamics during exercise in patients with heart failure. Muscle blood flow is reduced with maintenance of arterial perfusion pressure. *Circulation*, **80**, 769-781.

Sun, M.-K. and Guyenet, P.G. (1985). GABA-mediated baroreceptor inhibition of reticulospinal neurons. *American Journal of Physiology*, **249**, R672-R680.

Sun, M.-K. and Guyenet, P.G. (1987). Arterial baroreceptor and vagal inputs to sympathoexcitatory neurons in rat medulla. *American Journal of Physiology*, **252**, R699-R709.

Sunder-Plassmann, P. (1930). Untersuchungen über den Bulbus carotidis bei Mensch und Tier im Hinblick auf die 'Sinusreflexe' nach H.E. Hering; ein Vergleich mit anderen Gefässstreken; die Histopathologie des Bulbus carotidis; das Glomus caroticum. *Zeitscrift für Anatomie und Entwicklungsgeschicht*, **93**, 567-622.

Sundlöf, G. and Wallin, B.G. (1977). The variability of muscle nerve sympathetic activity in resting recumbent man. *The Journal of Physiology*, **272**, 383-397.

Sundlöf, G. and Wallin, B.G. (1978*a*). Effect of lower body negative pressure on human muscle nerve sympathetic activity. *The Journal of Physiology*, **278**, 525-532.

Sundlöf, G. and Wallin, B.G. (1978*b*). Human muscle nerve sympathetic activity at rest. Relationship to blood pressure and age. *The Journal of Physiology*, **274**, 621-637.

Sundlöf, G. and Wallin, B.G. (1978*c*). Muscle–nerve sympathetic activity in man. Relationship to blood pressure in resting normo- and hyper-tensive subjects. *Clinical Science and Molecular Medicine*, **55**, 387s-389s.

Sundlöf, G., Wallin, B.G., Strömgren, E., and Nerhed, C. (1983). Acute effects of metoprolol on muscle sympathetic activity in hypertensive humans. *Hypertension*, **5**, 749-756.

Sung, C.-S., Mathur, V.S., Garcia, E., De Castro, C.M., and Hall, R.J. (1980). Is postextrasystolic potentiation dependent on Starling's law? *Circulation*, **62**, 1032-1035.

Swanson, L.W. and Hartman, B.K. (1975). The central adrenergic system. An immunofluorescence study of the location of cell bodies and their efferent connections in the rat utilizing dopamine-β-hydroxylase as a marker. *The Journal of Comparative Neurology*, **163**, 467-506.

Swedberg, K. *et al.* (1984). Comparison of myocardial catecholamine balance in chronic congestive heart failure and in angina pectoris without failure. *The American Journal of Cardiology*, **54**, 783-786.

Swedberg, K., Eneroth, P., Kjekshus, J., and Wilhelmsen, L. (1990). Hormones regulating cardiovascular function in patients with severe congestive heart failure and their relation to mortality. *Circulation*, **82**, 1730-1736.

Syrota, A. *et al.* (1985). Muscarinic cholinergic receptor in the human heart evidenced under physiological conditions by positron emission tomography. *Proceedings of the National Academy of Sciences*, **82**, 584-588.

Szatalowicz, V.L., Arnold, P.E., Chaimovitz, C., Bichet, D., Berl, T., and Schrier, R.W. (1981). Radioimmunoassay of plasma arginine vasopressin in hyponatremic patients with congestive heart failure. *The New England Journal of Medicine*, **305**, 263-266.

Szczepańska-Sadowska, E. (1973). Hemodynamic effects of a moderate increase of the plasma vasopressin level in conscious dogs. *Pflügers Archiv für die Gesamte Physiologie des Menschen und der Tiere*, **338**, 313-322.

Szentágothai, J. (1952). The general visceral efferent column of the brain stem. *Acta Morphologica*, **2**, 313-328.

Tafil-Klawe, M., Raschke, F., and Hildebrandt, G. (1990). Functional asymmetry in carotid sinus cardiac reflexes in humans. *European Journal of Applied Physiology*, **60**, 402-405.

Takata, S., Iwase, N., Okuwa, H., Ogawa, J., Ikeda, T., and Hattori, N. (1985). Baroreflex function and pressor responsiveness in normotensive young subjects with a family history of hypertension. *Japanese Circulation Journal*, **49**, 990-996.

Takata, S. *et al.* (1986). Effect of propranolol on baroreflex control of heart rate in young patients with essential hypertension. *Angiology*, **37**, 106-111.

Takemura, G. *et al.* (1989). Ventricular expression of atrial natriuretic polypeptide and its relations with hemodynamics and histology in dilated human hearts. Immunohistochemical study of the endomyocardial biopsy specimens. *Circulation*, **80**, 1137-1147.

Takeshita, A., Tanaka, S., Kuroiwa, A., and Nakamura, M. (1975). Reduced baroreceptor sensitivity in borderline hypertension. *Circulation*, **51**, 738-742.

Takeshita, A., Mark, A.L., Eckberg, D.L., and Abboud, F.M. (1979). Effect of central venous pressure on arterial baroreflex control of heart rate. *American Journal of Physiology*, **236**, H42-H47.

Takeshita, A., Matsuguchi, H., and Nakamura, M. (1980). Effect of coronary occlusion on arterial baroreflex control of heart rate. *Cardiovascular Research*, **14**, 303-306.

Takeshita, A., Jingu, S., Imaizumi, T., Kunihiko, Y., Koyanagi, S., and Nakamura, M. (1986). Augmented cardiopulmonary baroreflex control of forearm vascular resistance in young athletes. *Circulation Research*, **59**, 43-48.

Takeshita, A. *et al.* (1987). Attenuation of reflex forearm vasoconstriction by α-human atrial natriuretic peptide in men. *Circulation*, **61**, 555-559.

Talman, W.T. (1985). Cardiovascular regulation and lesions of the central nervous system. *Annals of Neurology*, **18**, 1-12.

Talman, W.T. (1989). Kynurenic acid microinjected into the nucleus tractus solitarius of rat blocks the arterial baroreflex but not responses to glutamate. *Neuroscience Letters*, **102**, 247-252.

Talman, W.T., Perrone, M.H., and Reis, D.J. (1980). Evidence for L-glutamate as the neurotransmitter of baroreceptor afferent nerve fibers. *Science*, **209**, 813-815.

Talman, W.T., Perrone, M.H., Scher, P., Kwo, S., and Reis, D.J. (1981). Antagonism of the baroreceptor reflex by glutamate diethyl ester, an antagonist to L-glutamate. *Brain Research*, **217**, 186-191.

Talman, W.T., Granata, A.R., and Reis, D.J. (1984). Glutamatergic mechanisms in the nucleus tractus solitarius in blood pressure control. *Federation Proceedings*, **43**, 39-44.

Tanaka, N., Tanaka, H., Toyama, Y., Kashima, T., Niimura, T., and Kanehisa, T. (1975). Paroxysmal hypertension in aortitis syndrome. *American Heart Journal*, **90**, 369-375.

Tang, S.-S., Stevenson, L., and Dzau, V.J. (1990). Endothelial renin–angiotensin pathway. Adrenergic regulation of angiotensin secretion. *Circulation Research*, **66**, 103-108.

Tansey, M.J.B., Opie, L.H., Vinik, A., and Kennelly, B.M. (1981). Plasma pancreatic polypeptide and gastrin in the assessment of autonomic activity in acute myocardial infarction. *European Journal of Cardiology*, **12**, 243-259.

Taquini, A.C., Fermoso, J.D., and Aramendia, P. (1961). The effects of bleeding and tilting in congestive heart failure. *Malattie Cardiovascolari*, **2**, 461-473.

Taugner, R., Bührle, C.P., Hackenthal, E., Mannek, E., and Nobiling, R. (1984). Morphology of the juxtaglomerular apparatus and secretory mechanisms. *Contributions to Nephrology*, **43**, 76-101.

Taylor, D.G. and Gebber, G.L. (1975). Baroreceptor mechanisms controlling sympathetic nervous rhythms of central origin. *American Journal of Physiology*, **228**, 1002-1013.

Taylor, R.R. and Hopkins, B.E. (1972). Influence of respiration and respiratory sinus arrhythmia on aortic regurgitation. *American Journal of Physiology*, **223**, 668-672.

Tello, J.F. (1924). Développement et terminaison du nerf dépresseur. *Travaux du Laboratoire de Recherches Biologiques de l'Université de Madrid*, **22**, 295-308.

Ten Harkel, A.D.J., van Lieshout, J.J., van Lieshout, E.J., and Wieling, W. (1990). Assessment of cardiovascular reflexes: influence of posture and period of preceding rest. *Journal of Applied Physiology*, **68**, 147-153.

Ter Horst, G.J., Luiten, P.G.M., and Kuipers, F. (1984). Descending pathways from hypothalamus to dorsal motor vagus and ambiguus nuclei in the rat. *Journal of the Autonomic Nervous System*, **11**, 59-75.

Thames, M.D. and Abboud, F.M. (1979). Reflex inhibition of renal sympathetic nerve activity during myocardial ischemia mediated by left ventricular receptors with vagal afferents in dogs. *The Journal of Clinical Investigation*, **63**, 395-402.

Thames, M.D. and Schmid, P.G. (1979). Cardiopulmonary receptors with vagal afferents tonically inhibit ADH release in the dog. *American Journal of Physiology*, **237**, H299-H304.

Thames, M.D., Jarecki, M., and Donald, D.E. (1978*a*). Neural control of renin secretion in anesthetized dogs. Interaction of cardiopulmonary and carotid baroreceptors. *Circulation Research*, **42**, 237-245.

Thames, M.D., Klopfenstein, H.S., Abboud, F.M., Mark, A.L., and Walker, J.L. (1978*b*). Preferential distribution of inhibitory cardiac receptors with vagal afferents to the inferoposterior wall of the left ventricle activated during coronary occlusion in the dog. *Circulation Research*, **43**, 512-519.

Theile, W. (1852). Ueber das Arteriensystem von *Simia Inuus*. *Archiv für Anatomie, Physiologie und Wissenschlaftliche Medicin Leipzig*, 419-449.

Thomas, J.A. and Marks, B.H. (1978). Plasma norepinephrine in congestive heart failure. *The American Journal of Cardiology*, **41**, 233-243.

Thomas, J.E. (1972). Disease of the carotid sinus — syncope. In *Vascular diseases of the nervous system*. Part 1, ed. Vinken, P.J., and Bruyn, G.W., pp. 532-551. North-Holland Publishing Company, Amsterdam.

Thomas, M.R. and Calaresu, F.R. (1972). Responses of single units in the medial hypothalamus to electrical stimulation of the carotid sinus nerve in the cat. *Brain Research*, **44**, 49-62.

Thompson, C.A., Tatro, D.L., Ludwig, D.A., and Convertino, V.A. (1990). Baroreflex responses to acute changes in blood volume in humans. *American Journal of Physiology*, **259**, R792-R798.

Thompson, M.E., Felsten, G., Yavorsky, J., and Natelson, B.H. (1987). Differential effect of stimulation of nucleus ambiguus on atrial and ventricular rates. *American Journal of Physiology*, **253**, R150-R157.

Thorén, P. (1973). Evidence for a depressor reflex elicited from left ventricular receptors during occlusion of one coronary artery in the cat. *Acta Physiologica Scandinavica*, **88**, 23-34.

Thorén, P.N. (1976). Activation of left ventricular receptors with nonmedullated vagal afferent fibers during occlusion of a coronary artery in the cat. *The American Journal of Cardiology*, **37**, 1046-1051.

Thorén, P. (1979). Role of cardiac vagal C-fibers in cardiovascular control. *Reviews of Physiology, Biochemistry, and Pharmacology*, **86**, 1-94.

Thoren, P. and Jones, J.V. (1977). Characteristics of aortic baroreceptor C-fibres in the rabbit. *Acta Physiologica Scandinavica*, **99**, 448-456.

Thorén, P., Saum, W.R., and Brown, A.M. (1977). Characteristics of rat aortic baroreceptors with nonmedullated afferent nerve fibers. *Circulation Research*, **40**, 231-237.

Thron, H.L., Brechmann, W., Wagner, J., and Keller, K. (1967). Quantitative Untersuchungen über die Bedeutung der Gefässdehnungsreceptoren im Rahmen der Kreislaufhomoiostase beim wachen Menschen. I. Das Verhalten von arteriellem Blutdruck und Herzfrequenz bei abgestufter Veränderung des transmuralen Blutdrucks im Bereich des Carotissinus. *Pflügers Archiv für die Gesamte Physiologie des Menschen und der Tiere*, **293**, 68-99.

Tikkanen, I., Fyhrquist, F., Metsärinne, K., and Leidenius, R. (1985). Plasma atrial natriuretic peptide in cardiac disease and during infusion in healthy volunteers. *The Lancet*, **2**, 66-69.

Timmis, A.D., Kenny, J.F., Smyth, P., Campbell, S., Kerkez, S.A., and Jewitt, D.E. (1984). Restoration of normal reflex responses to orthostatic stress during felodipine therapy in congestive heart failure. *Cardiovascular Research*, **18**, 613-619.

Timms, R.J. (1981). A study of the amygdaloid defence reaction showing the value of althesin anaesthesia in studies of the functions of the fore-brain in cats. *Pflügers Archiv für die Gesamte Physiologie des Menschen und der Tiere*, **391**, 49-56.

Tobian, L., Janecek, J., Tomboulian, A., and Ferreira, D. (1961). Sodium and potassium in the walls of arterioles in experimental renal hypertension. *The Journal of Clinical Investigation*, **40**, 1922-1925.

Tominaga, S., Curnish, R.R., Belardinelli, L., Rubio, R., and Berne, R.M. (1980). Adenosine release during early and sustained exercise of canine skeletal muscle. *American Journal of Physiology*, **238**, H156-H163.

Torvik, A. (1957). The spinal projection from the nucleus of the solitary tract. An experimental study in the cat. *Journal of Anatomy*, **91**, 314-322.

Towne, J.B. and Bernhard, V.M. (1980). The relationship of postoperative hypertension to complications following carotid endarterectomy. *Surgery*, **88**, 575-579.

Trank, J.W. and Visscher, M.B. (1962). Carotid sinus baroceptor modifications associated with endotoxin shock. *American Journal of Physiology*, **202**, 971-977.

Trimarco, B. *et al.* (1981). Baroreflex responsiveness in borderline hypertensives: a study with neostigmine. *Cardiovascular Research*, **15**, 436-442.

Trimarco, B. *et al.* (1983). Valsalva maneuver in the assessment of baroreflex responsiveness in borderline hypertensives. *Cardiology*, **70**, 6-14.

Trimarco, B. *et al.* (1986). Impaired responsiveness of the ventricular sensory receptor in hypertensive patients with left ventricular hypertrophy. *Circulation*, **74**, 980-990.

Trimarco, B. *et al.* (1987). Effects of coronary occlusion on arterial baroreflex control of heart rate and vascular resistance. *American Journal of Physiology*, **252**, H749-H759.

Trimarco, B. *et al.* (1988). Reflex control of coronary vascular tone by cardiopulmonary receptors in humans. *Journal of the American College of Cardiology*, **11**, 944-952.

Trimarco, B. *et al.* (1990). Prolonged duration of myocardial ischemia in patients with coronary heart disease and impaired cardiopulmonary baroreceptor sensitivity. *Circulation*, **81**, 1792-1802.

Tripathi, A. and Nadel, E.R. (1986). Forearm skin and muscle vasoconstriction during lower body negative pressure. *Journal of Applied Physiology*, **60**, 1535-1541.

Tripathi, A., Mack, G., and Nadel, E.R. (1989). Peripheral vascular reflexes elicited during lower body negative pressure. *Aviation, Space, and Environmental Medicine*, **60**, 1187-1193.

Trzebski, A. (1987). Feliks Nawrocki. *Acta Physiologica Polonica*, **38**, 66-73.

Trzebski, A., Lipski, J., Majcherczyk, S., Szulczyk, P., and Chruścielewski, L. (1975). Central organization and interaction of the carotid baroreceptor and chemoreceptor sympathetic reflex. *Brain Research*, **87**, 227-237.

Trzebski, A., Raczkowska, M., and Kubin, L. (1980). Carotid baroreceptor reflex in man, its modulation over the respiratory cycle. *Acta Neurobiologiae Experimentalis*, **40**, 807-820.

Trzebski, A., Tafil, M., Zoltowski, M., and Przybylski, J. (1982). Increased sensitivity of the arterial chemoreceptor drive in young men with mild hypertension. *Cardiovascular Research*, **16**, 163-172.

Tsunoda, K., Mendelsohn, F.A.O., Sexton, P.M., Chai, S.Y., Hodsman, G.P., and Johnston, C.I. (1988). Decreased atrial natriuretic peptide binding in renal medulla in rats with chronic heart failure. *Circulation Research*, **62**, 155-161.

Tsunoda, S.-i., Shiozawa, Z., and Mano, T. (1988). Microneurographic analysis of thyrotropin-releasing hormone effects on sympathetic outflow in human muscle nerves. *Journal of the Autonomic Nervous System*, **23**, 207-211.

Turner, A.H., Newton, M.I., and Haynes, F.W. (1930). The circulatory reaction to gravity in healthy young women. Evidence regarding its precision and its instability. *American Journal of Physiology*, **94**, 507-519.

Turner, L.A., Zuperku, E.J., Bosnjak, Z.J., and Kampine, J.P. (1985). Autonomic modulation of refractoriness in canine specialized His–Purkinje system. *American Journal of Physiology*, **248**, R515-R523.

Tydén, G., Samnegård, H., and Thulin, L. (1980*a*). Rational treatment of hypotension after carotid endarterectomy by carotid sinus nerve blockade. *Acta Chirurgica Scandinavica*, Suppl. **500**, 61-64.

Tydén, G., Samnegård, H., Thulin, L., and Muhrbeck, O. (1980*b*). Effect of carotid endarterectomy on baroreflex sensitivity in man. Intraoperative studies. *Acta Chirurgica Scandinavica*, Suppl. **500**, 67-69.

Tydén, G., Samnegård, H., Melcher, A., and Thulin, L. (1981). Effect of carotid endarterectomy on the antihypertensive properties of the carotid sinus reflex. *Acta Chirurgica Scandinavica*, **147**, 15-17.

Ueda, H., Uchida, Y., Ueda, K., Gondaira, T., and Katayama, S. (1969). Centrally mediated vasopressor effect of angiotensin II in man. *Japanese Heart Journal*, **10**, 243-247.

Ulrych, M., Frohlich, E.D., Tarazi, R.C., Dustan, H.P., and Page, I.H. (1969). Cardiac output and distribution of blood volume in central and peripheral circulations in hypertensive and normotensive man. *British Heart Journal*, **31**, 570-574.

Undesser, K.P., Hasser, E.M., Haywood, J.R., Johnson, A.K., and Bishop, V.S. (1985). Interactions of vasopressin with the area postrema in arterial baroreflex function in conscious rabbits. *Circulation Research*, **56**, 410-417.

Unnerstall, J.R., Kopajtic, T.A., and Kuhar, M.J. (1984). Distribution of α_2 agonist binding sites in the rat and human central nervous system: analysis of some functional, anatomic correlates of the pharmacologic effects of clonidine and related adrenergic agents. *Brain Research Reviews*, **7**, 69-101.

Uretsky, B.F., Verbalis, J.G., Generalovich, T., Valdes, A., and Reddy, P.S. (1985). Plasma vasopressin response to osmotic and hemodynamic stimuli in heart failure. *American Journal of Physiology*, **248**, H396-H402.

Vallbo, A.B. and Hagbarth, K.-E. (1967). Impulses recorded with micro-electrodes in human muscle nerves during stimulation of mechanoreceptors and voluntary contractions. *Electroencephalography and Clinical Neurophysiology*, **23**, 389-396.

Vallbo, Å.B., Hagbarth, K.-E., Torebjörk, H.E., and Wallin, B.G. (1979). Somatosensory, proprioceptive, and sympathetic activity in human peripheral nerves. *Physiological Reviews*, **59**, 919-957.

Van Citters, R.L., Franklin, D.L., Smith, O.A., Jr., Watson, N.W., and Elsner, R.W. (1965). Cardiovascular adaptations to diving in the northern elephant seal *Mirounga Angustirostris*. Comparative Biochemistry and Physiology, **16**, 267-276.

Van Lieshout, J.J. (1989). Cardiovascular reflexes in orthostatic disorders (Ph.D. thesis), Rodopi, Amsterdam.

Van Lieshout, J.J., Wieling, W., Karemaker, J.M., and Eckberg, D.L. (1991). The vasovagal responses. *Clinical Science*, 1991.

Vanhoutte, P.M. and Levy, M.N. (1980). Prejunctional cholinergic modulation of adrenergic neurotransmission in the cardiovascular system. *American Journal of Physiology*, **238**, H275-H281.

Varma, S., Johnsen, S.D., Sherman, D.E., and Youmans, W.B. (1960). Mechanisms of inhibition of heart rate by phenylephrine. *Circulation Research*, **8**, 1182-1186.

Varma, S., Jaju, B.P., and Bhargava, K.P. (1969). Mechanism of vasopressin-induced bradycardia in dogs. *Circulation Research*, **24**, 787-792.

Vatner, D.E., Vatner, S.F., Fujii, A.M., and Homcy, C.J. (1985). Loss of high affinity cardiac beta adrenergic receptors in dogs with heart failure. *The Journal of Clinical Investigation*, **76**, 2259-2264.

Vatner, D.E. *et al.* (1988). Impaired cardiac muscarinic receptor function in dogs with heart failure. *The Journal of Clinical Investigation*, **81**, 1836-1842.

Vatner, S.F., Franklin, D., and Braunwald, E. (1971). Effects of anesthesia and sleep on circulatory response to carotid sinus nerve stimulation. *American Journal of Physiology*, **220**, 1249-1255.

Vatner, S.F., Higgins, C.B., Franklin, D., and Braunwald, E. (1972). Extent of carotid sinus regulation of the myocardial contractile state in conscious dogs. *The Journal of Clinical Investigation*, **51**, 995-1008.

Vatner, S.F., Higgins, C.B., and Braunwald, E. (1974). Sympathetic and parasympathetic components of reflex tachycardia induced by hypotension in conscious dogs with and without heart failure. *Cardiovascular Research*, **8**, 153-161.

Vecht, R.J., Graham, G.W.S., and Sever, P.S. (1978). Plasma noradrenaline concentrations during isometric exercise. *British Heart Journal*, **40**, 1216-1220.

Vendsalu, A. (1960). Studies on adrenaline and noradrenaline in human plasma. *Acta Physiologica Scandinavica*, **49** (Suppl. 173), 70-77.

Verney, E.B. (1947-1948). The antidiuretic hormone and the factors which determine its release. *Proceedings of the Royal Society of London. Series B — Biological Sciences*, **135**, 25-105.

Verrier, R.L. and Lown, B. (1984). Behavioral stress and cardiac arrhythmias. *Annual Review of Physiology*, **46**, 155-176.

Verrier, R., Calvert, A., Lown, B., and Axelrod, P. (1974). Effect of acute blood pressure elevation on the ventricular fibrillation threshold. *American Journal of Physiology*, **226**, 893-897.

Victor, R.G. and Leimbach, W.N., Jr. (1987). Effects of lower body negative pressure on sympathetic discharge to leg muscles in humans. *Journal of Applied Physiology*, **63**, 2558-2562.

Victor, R.G. and Mark, A.L. (1985). Interaction of cardiopulmonary and carotid baroreflex control of vascular resistance in humans. *The Journal of Clinical Investigation*, **76**, 1592-1598.

Victor, R.G. and Seals, D.R. (1989). Reflex stimulation of sympathetic outflow during rhythmic exercise in humans. *American Journal of Physiology*, **257**, H2017-H2024.

Victor, R.G., Leimbach, W.N., Jr., Seals, D.R., Wallin, B.G., and Mark, A.L. (1987*a*). Effects of the cold pressor test on muscle sympathetic nerve activity in humans. *Hypertension*, **9**, 429-436.

Victor, R.G., Seals, D.R., and Mark, A.L. (1987*b*). Differential control of heart rate and sympathetic nerve activity during dynamic exercise. *The Journal of Clinical Investigation*, **79**, 508-516.

Victor, R.G., Bertocci, L.A., Pryor, S.L., and Nunnally, R.L. (1988). Sympathetic nerve discharge is coupled to muscle cell pH during exercise in humans. *The Journal of Clinical Investigation*, **82**, 1301-1305.

Victor, R.G., Pryor, S.L., Secher, N.H., and Mitchell, J.H. (1989). Effects of partial neuromuscular blockade on sympathetic responses to static exercise in humans. *Circulation Research*, **65**, 468-476.

Vierordt, K. (1854). Die bildliche Darstellung des menschlichen Arterienpulses. *Archiv für Physiologische Heilkunde*, **13**, 284-287.

Vigier, D. and Rouvière, A. (1979). Afferent and efferent connections of the area postrema demonstrated by the horseradish peroxidase method. *Archives Italiennes de Biologie*, **117**, 325-339.

Vogt, A., Unterberg, C., and Kreuzer, H. (1987). Acute effects of the new angiotensin converting enzyme inhibitor ramipril on hemodynamics and carotid sinus baroreflex activity in congestive heart failure. *The American Journal of Cardiology*, **59**, 149D-153D.

Volkmann, A.W. (1853). Ueber den Ursprung des Sympathicus vom Rückenmarke. *Archiv für Physiologische Heilkunde*, **12**, 92-98.

Volpe, M. *et al.* (1982). The autonomic nervous tone abnormalities in the genesis of the impaired baroreflex responsiveness in borderline hypertensive subjects. *Clinical Science*, **62**, 581-588.

Volpe, M. *et al.* (1985). Carotid sinus reflex control of coronary blood flow in human subjects. *Journal of the American College of Cardiology*, **5**, 1312-1318.

Volpe, M. *et al.* (1987). Influence of atrial natriuretic factor on forearm reflex vasoconstriction induced by cardiopulmonary or arterial receptor unloading. *Journal of Hypertension*, **5** (Suppl. 5), S63-S65.

Volpe, M. *et al.* (1988*a*). Reduction of atrial natriuretic factor circulating levels by endogenous sympathetic activation in hypertensive patients. *Circulation*, **77**, 997-1002.

Volpe, M. *et al.* (1988*b*). Atrial natriuretic factor potentiates forearm reflex vasoconstriction induced by cardiopulmonary receptor deactivation in man. *Circulation*, **77**, 849-855.

Volpe, M. (1990). Converting enzyme inhibition prevents the effects of atrial natriuretic factor on baroreflex responses in humans. *Circulation*, **82**, 1214-1221.

Vølund, A. (1978). Application of the four-parameter logistic model to bioassay: comparison with slope ratio and parallel line models. *Biometrics*, **34**, 357-365.

Von Euler, U.S. and Liljestrand, G. (1943). The rôle of the chemoreceptors of the sinus region for the occlusion test in the cat. *Acta Physiologica Scandinavica*, **5**, 319-323.

Von Euler, U.S., Luft, R., and Sundin, T. (1955). The urinary excretion of noradrenaline and adrenaline in healthy subjects during recumbency and standing. *Acta Physiologica Scandinavica*, **34**, 169-174.

Von Ungern-Sternberg, A., Schwarz, W., and Schuster, C.J. (1975). Die Gefässmechanik der Arteria carotis communis bei arterieller Hypertonie. *Zeitschrift für Kardiologie*, **64**, 1053-1065.

Vybiral, T. *et al.* (1990). Parasympathetic stimulation with transdermal scopolamine increases heart rate variability in acute myocardial infarction — a preliminary report (Abstract). *Clinical Research*, **38**, 337A.

Wagner, H.N., Jr. and Braunwald, E. (1956). The pressor effect of the antidiuretic principle of the posterior pituitary in orthostatic hypotension. *The Journal of Clinical Investigation*, **35**, 1412-1418.

Wagner, J., Wackerbauer, J., and Hilger, H.H. (1968). Arterielles Blutdruck- und Herzfrequenzverhalten bei Hypertonikern unter Änderung des transmuralen Druckes im Karotissinusbereich. *Zeitschrift für Kreislaufforschung*, **57**, 701-712.

Wagner, R. and Kapal, E. (1951). Über Eigenschaften des Aortenwindkessels. *Zeitschrift für Biologie*, **104**, 169-202.

Walgenbach, S.C. (1984). Sustained interruption of the aortic baroreflex by left cervical vagotomy in dogs. *American Journal of Physiology*, **246**, H319-H323.

Walgenbach, S.C. and Donald, D.E. (1983). Inhibition by carotid baroreflex of exercise-induced increases in arterial pressure. *Circulation Research*, **52**, 253-262.

Walker, J.L., Abboud, F.M., Mark, A.L., and Thames, M.D. (1980). Interaction of cardiopulmonary and somatic reflexes in humans. *The Journal of Clinical Investigation*, **65**, 1491-1497.

Waller, A. (1862). Experimental researches on the functions of the vagus and the cervical sympathetic nerves in man. *Proceedings of the Royal Society*, **11**, 302-303.

Wallin, B.G. and Eckberg, D.L. (1982). Sympathetic transients caused by abrupt alterations of carotid baroreceptor activity in humans. *American Journal of Physiology*, **242**, H185-H190.

Wallin, B.G. and Frisk-Holmberg, M. (1981). The antihypertensive mechanism of clonidine in man. Evidence against a generalized reduction of sympathetic activity. *Hypertension*, **3**, 340-346.

Wallin, B.G. and Nerhed, C. (1982). Relationship between spontaneous variations of muscle sympathetic activity and succeeding changes of blood pressure in man. *Journal of the Autonomic Nervous System*, **6**, 293-302.

Wallin, B.G. and Rea, R. (1988). Spinal sympathetic conduction velocity in humans. *Journal of the Autonomic Nervous System*, **24**, 221-225.

Wallin, B.G. and Sundlöf, G. (1979). A quantitative study of muscle nerve sympathetic activity in resting normotensive and hypertensive subjects. *Hypertension*, **1**, 67-77.

Wallin, B.G. and Sundlöf, G. (1982). Sympathetic outflow to muscles during vasovagal syncope. *Journal of the Autonomic Nervous System*, **6**, 287-291.

Wallin, B.G., Delius, W., and Hagbarth, K.-E. (1973*a*). Comparison of sympathetic nerve activity in normotensive and hypertensive subjects. *Circulation Research*, **33**, 9-21.

Wallin, B.G., Delius, W., and Hagbarth, K.-E. (1973*b*). Sympathetic activity in peripheral nerves of normo- and hypertensive subjects. *Clinical Science and Molecular Medicine*, **45**, 127s-130s.

Wallin, G., Delius, W., and Hagbarth, K.E. (1974*a*). Regional control of sympathetic outflow in human skin and muscle nerves. In *Central rhythmic and regulation*, ed. Umbach, W., and Koepchen, H.P., pp. 190-195. Hippokrates Verlag, Stuttgart.

Wallin, B.G., Delius, W., and Sundlöf, G. (1974*b*). Human muscle nerve sympathetic activity in cardiac arrhythmias. *Scandinavian Journal of Clinical and Laboratory Investigation*, **34**, 293-300.

Wallin, B.G., Sundlöf, G., and Delius, W. (1975). The effect of carotid sinus nerve stimulation on muscle and skin nerve sympathetic activity in man. *Pflügers Archiv für die Gesamte Physiologie des Menschen und der Tiere*, **358**, 101-110.

Wallin, B.G., Sundlöf, G., Eriksson, B.-M., Dominiak, P., Grobecker, H., and Lindblad, L.E. (1981). Plasma noradrenaline correlates to sympathetic muscle nerve activity in normotensive man. *Acta Physiologica Scandinavica*, **111**, 69-73.

Wallin, B.G., Sundlöf, G., Strömgren, E., and Åberg, H. (1984*a*). Sympathetic outflow to muscles during treatment of hypertension with metoprolol. *Hypertension*, **6**, 557-562.

Wallin, B.G., Westerberg, C.-E., and Sundlöf, G. (1984*b*). Syncope induced by glossopharyngeal neuralgia: sympathetic outflow to muscle. *Neurology*, **34**, 522-524.

Wallin, B.G., Mörlin, C., and Hjemdahl, P. (1987). Muscle sympathetic activity and venous plasma noradrenaline concentrations during static exercise in normotensive and hypertensive subjects. *Acta Physiologica Scandinavica*, **129**, 489-497.

Walter, P.F., Crawley, I.S., and Dorney, E.R. (1978). Carotid sinus hypersensitivity and syncope. *The American Journal of Cardiology*, **42**, 396-403.

Wang, B.C., Sundet, W.D., Hakumäki, M.O.K., and Goetz, K.L. (1983). Vasopressin and renin responses to hemorrhage in conscious, cardiac-denervated dogs. *American Journal of Physiology*, **245**, H399-H405.

Wang, B.C., Share, L., and Goetz, K.L. (1985). Factors influencing the secretion of vasopressin into cerebrospinal fluid. *Federation Proceedings*, **44**, 72-77.

Wang, Q. and Li, P. (1988). Inhibition of baroreflex following microinjection of GABA or morphine into the nucleus tractus solitarii in rabbits. *Journal of the Autonomic Nervous System*, **25**, 165-172.

Wang, S.C. and Borison, H.L. (1947). An analysis of the carotid sinus cardiovascular reflex mechanism. *American Journal of Physiology*, **150**, 712-721.

Wang, W., Chen, J.-S., and Zucker, I.H. (1990). Carotid sinus baroreceptor sensitivity in experimental heart failure. *Circulation*, **81**, 1959-1966.

Warner, H.R. and Russell, R.O., Jr. (1969). Effect of combined sympathetic and vagal stimulation on heart rate in the dog. *Circulation Research*, **24**, 567-573.

Warner, M.R. and Levy, M.N. (1989). Neuropeptide Y as a putative modulator of the vagal effects on heart rate. *Circulation Research*, **64**, 882-889.

Warner, M.R. and Loeb, J.M. (1986). Beat-by-beat modulation of AV conduction. I. Heart rate and respiratory influences. *American Journal of Physiology*, **251**, H1126-H1133.

Warner, M.R., deTarnowsky, J.M., Whitson, C.C., and Loeb, J.M. (1986). Beat-by-beat modulation of AV conduction. II. Autonomic neural mechanisms. *American Journal of Physiology*, **251**, H1134-H1142.

Waterfield, R.L. (1931). The effect of posture on the volume of the leg. *The Journal of Physiology*, **72**, 121-131.

Watkins, L., Jr., Burton, J.A., Haber, E., Cant, J.R., Smith, F.W., and Barger, A.C. (1976). The renin–angiotensin–aldosterone system in congestive heart failure in conscious dogs. *The Journal of Clinical Investigation*, **57**, 1606-1617.

Watson, R.D.S., Stallard, T.J., and Littler, W.A. (1979). Effects of β-adrenoreceptor antagonists on sino-aortic baroreflex sensitivity and blood pressure in hypertensive man. *Clinical Science*, **57**, 241-247.

Watson, R.D.S., Stallard, T.J., Flinn, R.M., and Littler, W.A. (1980). Factors determining direct arterial pressure and its variability in hypertensive man. *Hypertension*, **2**, 333-341.

Watson, W.E. (1962). Some circulatory responses to Valsalva's manoeuvre in patients with polyneuritis and spinal cord disease. *Journal of Neurosurgery and Psychiatry*, **25**, 19-23.

Waxman, M.B. and Wald, R.W. (1977*a*). Recurrent paroxysmal supraventricular tachycardia: a complication of ventricular pacing in a patient with occult Wolff–Parkinson–White syndrome. *The Journal of Electrocardiology*, **10**, 291-298.

Waxman, M.B. and Wald, R.W. (1977*b*). Termination of ventricular tachycardia by an increase in cardiac vagal drive. *Circulation*, **56**, 385-391.

Waxman, M.B., Bonet, J.F., Finley, J.P., and Wald, R.W. (1980*a*). Effects of respiration and posture on paroxysmal supraventricular tachycardia. *Circulation*, **62**, 1011-1020.

Waxman, M.B., Wald, R.W., Finley, J.P., Bonet, J.F., Downar, E., and Sharma, A.D. (1980*b*). Valsalva termination of ventricular tachycardia. *Circulation*, **62**, 843-851.

Waxman, M.B., Wald, R.W., Sharma, A.D., Huerta, F., and Cameron, D.A. (1980*c*). Vagal techniques for termination of paroxysmal supraventricular tachycardia. *The American Journal of Cardiology*, **46**, 655-664.

Waxman, M.B., Staniloff, H., and Wald, R.W. (1981). Respiratory and vagal modulation of ventricular tachycardia. *Journal of Electrocardiology*, **14**, 83-89.

Waxman, M.B., Sharma, A.D., Cameron, D.A., Huerta, F., and Wald, R.W. (1982). Reflex mechanisms responsible for early spontaneous termination of paroxysmal supraventricular tachycardia. *The American Journal of Cardiology*, **49**, 259-272.

Waxman, M.B., Wald, R.W., and Cameron, D. (1983). Interactions between the autonomic nervous system and tachycardias in man. *Cardiology Clinics*, **1**, 143-185.

Waxman, M.B., Sharma, A.D., Asta, J., Cameron, D.A., and Wald, R.W. (1989). The protective effect of vagus nerve stimulation on catecholamine–halothane-induced ventricular fibrillation in dogs. *Canadian Journal of Physiology and Pharmacology*, **67**, 801-809.

Weaver, W.D., Cobb, L.A., and Hallstrom, A.P. (1982). Ambulatory arrhythmias in resuscitated victims of cardiac arrest. *Circulation*, **66**, 212-218.

Webb, S.W., Adgey, A.A.J., and Pantridge, J.F. (1972). Autonomic disturbance at onset of acute myocardial infarction. *British Medical Journal*, **3**, 89-92.

Weber, E. and Weber, E.H. (1845). Experimenta, quibus probatur nervos vagos rotatione machinae galvano-magneticae irritatos, motum cordis retardare et adeo intercipere. *Omedeis Annali: Universali di Medicina, Milan*, **116**, 227-230.

Weber, K.T., Kinasewitz, G.T., Janicki, J.S., and Fishman, A.P. (1982). Oxygen utilization and ventilation during exercise in patients with chronic cardiac failure. *Circulation*, **65**, 1213-1223.

Wei, J.Y., Markis, J.E., Malagold, M., and Braunwald, E. (1983). Cardiovascular reflexes stimulated by reperfusion of ischemic myocardium in acute myocardial infarction. *Circulation*, **67**, 796-801.

Weidmann, P. *et al.* (1980). Cardiovascular pressor reactivity as related to plasma catecholamines: role in the pathogenesis of essential hypertension and in the antihypertensive mechanism of diuretic treatment. In *Hypertension: mechanisms and management*, ed. Philipp, T., and Distler, A., pp. 23-36. Springer-Verlag, Berlin.

Weiner, D.A., McCabe, C.H., Cutler, S.S., and Ryan, T.J. (1982). Decrease in systolic blood pressure during exercise testing: reproducibility, response to coronary bypass surgery and prognostic significance. *The American Journal of Cardiology*, **49**, 1627-1631.

Weiss, S. and Baker, J.P. (1933). The carotid sinus reflex in health and disease. *Medicine*, **12**, 297-354.

Weissler, A.M., Leonard, J.J., and Warren, J.V. (1957). Effects of posture and atropine on the cardiac output. *The Journal of Clinical Investigation*, **36**, 1656-1662.

Welch, W.J., Bauernfeind, R.A., Rea, R.F., Appel, R.A., and Eckberg, D.L. (1986). Effects of carotid sinus stimulation on the atrioventricular node in man (Abstract). *Clinical Research*, **34**, 210A.

Welch, W.J., Smith, M.L., Rea, R.F., Bauernfeind, R.A., and Eckberg, D.L. (1989). Enhancement of sympathetic nerve activity by single premature ventricular beats in humans. *Journal of the American College of Cardiology*, **13**, 69-75.

Wellens, H.J.J. and Durrer, D. (1975). The role of an accessory atrioventricular pathway in reciprocal tachycardia. Observations in patients with and without the Wolff–Parkinson–White syndrome. *Circulation*, **52**, 58-72.

Wellens, H.J.J., Vermeulen, A., and Durrer, D. (1972). Ventricular fibrillation occuring on arousal from sleep by auditory stimuli. *Circulation*, **46**, 661-665.

Welply, N.C., Mathias, C.J., and Frankel, H.L. (1975). Circulatory reflexes in tetraplegics during artificial ventilation and general anaesthesia. *Paraplegia*, **13**, 172-182.

Wendt, D.J. and Martins, J.B. (1990). Autonomic neural regulation of intact Purkinje system of dogs. *American Journal of Physiology*, **258**, H1420-H1426.

Werkö, L., Bucht, H., and Josephson, B. (1949). The renal extraction of para-aminohippuric acid and oxygen in man during postural changes of the circulation. *Scandinavian Journal of Clinical and Laboratory Investigation*, **1**, 321-327.

West, T.C. (1955). Ultramicroelectrode recording from the cardiac pacemaker. *The Journal of Pharmacology and Experimental Therapeutics*, **115**, 283-290.

Wetzel, G.T. and Brown, J.H. (1985). Presynaptic modulation of acetylcholine release from cardiac parasympathetic neurons. *American Journal Physiology*, **248**, H33-H39.

Wetzel, G.T., Goldstein, D., and Brown, J.H. (1985). Acetylcholine release from rat atria can be regulated through an α_1-adrenergic receptor. *Circulation Research*, **56**, 763-766.

Wexler, L., Bergel, D.H., Gabe, I.T., Makin, G.S., and Mills, C.J. (1968). Velocity of blood flow in normal human venae cavae. *Circulation Research*, **23**, 349-359.

Whayne, T.F., Jr. and Killip, T. (1967). Simulated diving in man: comparison of facial stimuli and response in arrhythmia. *Journal of Applied Physiology*, **22**, 800-807.

Whelan, R.F. and Young, I.M. (1953). The effect of adrenaline and noradrenaline infusions on respiration in man. *British Journal of Pharmacology and Chemotherapy*, **8**, 98-102.

White, C.W. (1981). Reversibility of abnormal arterial baroreflex control of heart rate in heart failure. *American Journal of Physiology*, **241**, H778-H782.

White, C.W., Eckberg, D.L., and Inasaka, T. (1973*a*). Direct effects of methoxamine and phenylephrine on sinus mode function (Abstract). *The American Journal of Cardiology*, **31**, 164.

White, C.W., Eckberg, D.L., Kioschos, J.M., and Abboud, F.M. (1973*b*). A study of coronary artery reflexes in man (Abstract). *Circulation*, **48** (Suppl.IV), IV-65.

White, C.W., Eckberg, D.L., Inasaka, T., and Abboud, F.M. (1976). Effects of angiographic contrast media on sino-atrial nodal function. *Cardiovascular Research*, **10**, 214-223.

Whytt, R. (1757). *An essay on the vital and other involuntary motions of animals*. Hamilton, Balfour, and Neill, Edinburgh.

Widimiský, J., Fejfarová, M.H., and Fejfar, Z. (1957). Changes of cardiac output in hypertensive disease. *Cardiologia*, **31**, 381-389.

Wilcox, R.G., Bennett, T., Brown, A.M., and MacDonald, I.A. (1982). Is exercise good for high blood pressure? *British Medical Journal*, **285**, 767-769.

Wilcox, R.G., Bennett, T., MacDonald, I.A., Pipkin, F.B., and Baylis, P.H. (1987). Post-exercise hypotension: the effects of epanolol or atenolol on some hormonal and cardiovascular variables in hypertensive men. *British Journal of Clinical Pharmacology*, **24**, 151-162.

Wilkins, R.W. and Culbertson, J.W. (1947). The effects of surgical sympathectomy upon certain vasopressor responses in hypertensive patients. *Transactions of the Association of American Physicians*, **60**, 195-207.

Willis, T. (1664). *Cerebri anatome: cui accessit nervorum descriptio et usus*. J. Flesher, London.

Wilson, F.N. (1924). Report of a case of auricular flutter in which vagus stimulation was followed by an increase in the rate of the circus rhythm. *Heart*, **11**, 61-66.

Wilson, J.R. and Ferraro, N. (1983). Exercise intolerance in patients with chronic left heart failure: relation to oxygen transport and ventilatory abnormalities. *The American Journal of Cardiology*, **51**, 1358-1363.

Wilson, J.R., Martin, J.L., Schwartz, D., and Ferraro, N. (1984). Exercise intolerance in patients with chronic heart failure: role of impaired nutritive flow to skeletal muscle. *Circulation*, **69**, 1079-1087.

Wilson, J.R., Ferraro, N., and Wiener, D.H. (1985). Effect of the sympathetic nervous system on limb circulation and metabolism during exercise in patients with heart failure. *Circulation*, **72**, 72-81.

Wilson, J.R., Frey, M.J., Mancini, D.M., Ferraro, N., and Jones, R. (1989*a*). Sympathetic vasoconstriction during exercise in ambulatory patients with left ventricular failure. *Circulation*, **79**, 1021-1027.

Wilson, J.R., Mancini, D.M., McCully, K., Ferraro, N., Lanoce, V., and Chance, B. (1989*b*). Noninvasive detection of skeletal muscle underperfusion with near-infrared spectroscopy in patients with heart failure. *Circulation*, **80**, 1668-1674.

Winder, C.V. (1937). Pressoreceptor reflexes from the carotid sinus. *American Journal of Physiology*, **118**, 379-388.

Winder, C.V. (1938). Isolation of the carotid sinus pressoreceptive respiratory reflex. *American Journal of Physiology*, **122**, 306-318.

Winder, W.W., Hickson, R.C., Hagberg, J.M., Ehsani, A.A., and McLane, J.A. (1979). Training-induced changes in hormonal and metabolic responses to sub-maximal exercise. *Journal of Applied Physiology*, **46**, 766-771.

Winer, N., Chokshi, D.S., Yoon, M.S., and Freedman, A.D. (1969). Adrenergic receptor mediation of renin secretion. *Journal of Clinical Endocrinology*, **29**, 1168-1175.

Winkle, R.A., Gradman, A.H., and Fitzgerald, J.W. (1978). Antiarrhythmic drug effect assessed from ventricular arrhythmia reduction in the ambulatory electrocardiogram and treadmill test: comparison of propranolol, procainamide and quinidine. *The American Journal of Cardiology*, **42**, 473-480.

Winson, M., Heath, D., and Smith, P. (1974). Extensibility of the human carotid sinus. *Cardiovascular Research*, **8**, 58-64.

Winter, B. (1972). Bilateral carotid body resection for asthma and emphysema. *International Surgery*, **57**, 458-466.

Winter, R., Wilson, C., and Bennett, T. (1982). Cardiovascular responses to lateral semicircular canal stimulation in man. *Cardiovascular Research*, **16**, 610-612.

Wolf, M.M., Varigos, G.A., Hunt, D., and Sloman, J.G. (1978). Sinus arrhythmia in acute myocardial infarction. *The Medical Journal of Australia*, **2**, 52-53.

Wolf, S., Schneider, R.A., and Groover, M.E. (1965). Further studies on the circulatory and metabolic alterations of the oxygen-conserving (diving) reflex in man. *Transactions of the American Association of Physicians*, **78**, 242-254.

Wolf, W.A., Kuhn, D.M., and Lovenberg, W. (1981). Blood pressure responses to local application of serotonergic agents in the nucleus tractus solitarii. *European Journal of Pharmacology*, **69**, 291-299.

Wolinsky, H. (1971). Effects of hypertension and its reversal on the thoracic aorta of male and female rats. Morphological and chemical studies. *Circulation Research*, **28**, 622-637.

Wolthuis, R.A., Bergman, S.A., and Nicogossian, A.E. (1974). Physiological effects of locally applied reduced pressure in man. *Physiological Reviews*, **54**, 566-595.

Woo, M.A., Stevenson, W.G., Moser, D.K., Harper, R.M., Garfinkel, A., and Trelease, R. (1990). Patterns of heart rate variability in heart failure (Abstract). *Circulation*, **82** (Suppl.III), III-709.

Wood, J.E., Litter, J., and Wilkins, R.W. (1956). Peripheral venoconstriction in human congestive heart failure. *Circulation*, **13**, 524-527.

Wood, P. (1963). Polyuria in paroxysmal tachycardia and paroxysmal atrial flutter and fibrillation. *British Heart Journal*, **25**, 273-282.

Wurster, R.D. (1977). Spinal sympathetic control of the heart. In *Neural regulation of the heart*, ed. Randall, W.C., pp. 210-246. Oxford University Press, New York.

Yamada, K.A., McAllen, R.M., and Loewy, A.D. (1984). GABA antagonists applied to the ventral surface of the medulla oblongata block the baroreceptor reflex. *Brain Research*, **297**, 175-180.

Yamada, Y. *et al.* (1988). Impaired baroreflex changes in muscle sympathetic nerve activity in adolescents who have a family history of essential hypertension. *Journal of Hypertension*, **6** (Suppl. 4), S525-S528.

Yamada, Y., Miyajima, E., Tochikubo, O., Matsukawa, T., and Ishii, M. (1989). Age-related changes in muscle sympathetic nerve activity in essential hypertension. *Hypertension*, **13**, 870-877.

Yamaji, T., Ishibashi, M., Nakaoka, H., Imataka, K., Amano, M., and Fujii, J. (1985). Possible role for atrial natriuretic peptide in polyuria associated with paroxysmal atrial arrhythmias (Letter to the editor). *The Lancet*, **1**, 1211.

Yamane, Y. (1968). Plasma ADH level in patients with chronic congestive heart failure. *Japanese Circulation Journal*, **32**, 745-759.

Yamazaki, T., Brunner, M.J., and Sagawa, K. (1988). Pulsatile vs. mean component of baroreflex compensation for posthemorrhage hypotension. *American Journal of Physiology*, **254**, H1074-H1080.

Yamour, B.J., Sridharan, M.R., Rice, J.F., and Flowers, N.C. (1980). Electrocardiographic changes in cerebrovascular hemorrhage. *American Heart Journal*, **99**, 294-300.

Yew, D.T., Luo, C.B., Zheng, D.R., Guan, Y.L., Lin, Y.Q., and Chen, W.Z. (1990). Development and localization of enkephalin and substance P in the nucleus of tractus solitarius in the medulla oblongata of human fetuses. *Neuroscience*, **34**, 491-498.

Yoran, C., Higginson, L., Romero, M.A., Covell, J.W., and Ross, J., Jr. (1981). Reflex sympathetic augmentation of left-ventricular inotropic state in the conscious dog. *American Journal of Physiology*, **241**, H857-H863.

Yusuf, S., Lopez, R., and Sleight, P. (1980). Heart rate and ectopic prematurity in relation to sustained ventricular arrhythmias. *British Heart Journal*, **44**, 233-239.

Zacest, R., Gilmore, E., and Koch-Weser, J. (1972). Treatment of essential hypertension with combined vasodilation and beta-adrenergic blockade. *The New England Journal of Medicine*, **286**, 617-622.

Zandberg, P., De Jong, W., and De Wied, D. (1979). Effect of catecholamine-receptor stimulating agents on blood pressure after local application in the nucleus tractus solitarii of the medulla oblongata. *European Journal of Pharmacology*, **55**, 43-56.

Zbilut, J.P., Murdock, D.K., Lawless, C.E., Lawson, L., Von Dreele, M.M., and Porges, S.W. (1986). Spectral analysis of heart rate periodicities in heart transplant recipients (Abstract). *Clinical Research*, **34**, 906A.

Zelis, R. (1974). The contribution of local factors to the elevated venous tone of congestive heart failure. *The Journal of Clinical Investigation*, **54**, 219-224.

Zelis, R. and Mason, D.T. (1970). Diminished forearm arteriolar dilator capacity produced by mineralocorticoid-induced salt retention in man. Implications concerning congestive heart failure and vascular stiffness. *Circulation*, **61**, 589-92.

Zelis, R., Mason, D.T., and Braunwald, E. (1968). A comparison of the effects of vasodilator stimuli on peripheral resistance vessels in normal subjects and in patients with congestive heart failure. *The Journal of Clinical Investigation*, **47**, 960-970.

Zelis, R., Delea, C.S., Coleman, H.N., and Mason, D.T. (1970). Arterial sodium content in experimental congestive heart failure. *Circulation*, **41**, 213-216.

Zelis, R., Longhurst, J., Capone, R.J., and Lee, G. (1973). Peripheral circulatory control mechanisms in congestive heart failure. *The American Journal of Cardiology*, **32**, 481-490.

Zelis, R., Lee, G., and Mason, D.T. (1974*a*). Influence of experimental edema on metabolically determined blood flow. *Circulation Research*, **34**, 482-490.

Zelis, R., Longhurst, J., Capone, R.J., and Mason, D.T. (1974*b*). A comparison of regional blood flow and oxygen utilization during dynamic forearm exercise in normal subjects and patients with congestive heart failure. *Circulation*, **50**, 137-143.

Zelis, R., Caudill, C.C., Baggette, K., and Mason, D.T. (1976). Reflex vasodilation induced by coronary angiography in human subjects. *Circulation*, **53**, 490-493.

Zema, M.J., Restivo, B., Sos, T., Sniderman, K.W., and Kline, S. (1980). Left ventricular dysfunction — bedside Valsalva manoeuvre. *British Heart Journal*, **44**, 560-569.

Žemaityté, D., Varoneckas, G., and Sokolov, E. (1984). Heart rhythm control during sleep. *Psychophysiology*, **21**, 279-289.

Zhang, X., Abdel-Rahman, A.A., and Wooles, W.R. (1990). Selective sensitization by L-glutamate of baroreflex-mediated bradycardia following microinjection into the rostral ventrolateral medulla. *Brain Research*, **520**, 141-150.

Zimmermann, M., Maisonblanche, P., Cauchemez, B., Leclercq, J.-F., and Coumel, P. (1986). Determinants of the spontaneous ectopic activity in repetitive monomorphic idiopathic ventricular tachycardia. *Journal of the American College of Cardiology*, **7**, 1219-1227.

Zipes, D.P., Mihalick, M.J., and Robbins, G.T. (1974). Effects of selective vagal and stellate ganglion stimulation on atrial refractoriness. *Cardiovascular Research*, **8**, 647-655.

Zoller, R.P., Mark, A.L., Abboud, F.M., Schmid, P.G., and Heistad, D.D. (1972). The role of low pressure baroreceptors in reflex vasoconstrictor responses in man. *The Journal of Clinical Investigation*, **51**, 2967-2972.

Zucker, I.H. and Cornish, K.G. (1981). The Bezold–Jarisch reflex in the conscious dog. *Circulation Research*, **49**, 940-948.

Zucker, I.H. and Gilmore, J.P. (1981). Atrial receptor modulation of renal function in heart failure. In *Disturbances in neurogenic control of the circulation*, ed. Abboud, F.M., Fozzard, H.A., Gilmore, J.P., and Reis, D.J., pp. 1-16. American Physiological Society, Bethesda.

Zucker, I.H., Earle, A.M., and Gilmore, J.P. (1977). The mechanism of adaptation of left atrial stretch receptors in dogs with chronic congestive heart failure. *The Journal of Clinical Investigation*, **60**, 323-331.

Index